# STEM CELLS
## Biology and Application

# STEM CELLS
## Biology and Application

MARY L. CLARKE

JONATHAN FRAMPTON

**CRC Press**
Taylor & Francis Group
Boca Raton  London  New York

CRC Press is an imprint of the
Taylor & Francis Group, an **informa** business

A GARLAND SCIENCE BOOK

CRC Press
Taylor & Francis Group
6000 Broken Sound Parkway NW, Suite 300
Boca Raton, FL 33487-2742

© 2020 by Taylor & Francis Group, LLC
CRC Press is an imprint of Taylor & Francis Group, an Informa business

No claim to original U.S. Government works

International Standard Book Number-13: 978-0-367-48172-8 (Hardback)
978-0-8153-4511-4 (Paperback)

**Visit the Taylor & Francis Web site at**
**http://www.taylorandfrancis.com**

**and the CRC Press Web site at**
**http://www.crcpress.com**

**eResources:** All figures from the book are available for instructors to download. Access can be requested at www.routledge.com/9780815345114.

# Contents

# Preface

Life-long maintenance of multicellular organisms in the face of tissue and cell loss through injury, infection, and normal processes is a biological challenge. Organisms in both the plant and animal kingdoms have largely dealt with this challenge by acquiring cells that are able to persist throughout life, while at the same time acting as the reservoir feeding pathways of cell expansion, differentiation, and maturation necessary for tissue homeostasis and repair. These cells, which we refer to as stem cells, have been the subject of research over more than two centuries, although not always knowingly focused on their special characteristics, and have become the subject of an ever-growing field of investigation and application.

The field of stem cell science is vast, both in terms of the number of people involved in research and the number of disciplines that it encompasses, from molecular and cell biology to tissue engineering, animal husbandry, and politics and ethics. The level of interest in stem cells, especially over the last two decades, and the multidisciplinary nature of the subject have led to the publication of a multitude of books. Most of these books are quite focused and often delve very deeply into detail. Surprisingly, there have been few texts aimed at undergraduate and postgraduate courses that introduce stem cells, and even fewer that have tried to cover the breadth of stem cell science, incorporating plants and animals, invertebrates and vertebrates, fundamental biology and methodology, applications, and social science considerations. This is the task we set ourselves—to create a one-stop shop to enable the student to gain a holistic understanding of stem cells. Our aim is to excite students with the wonders of stem cells and the crucial part that they play in multicellular biology. We have also strived to give a sense of the research that has led to the understanding we have, addressing such questions as: how is stem cell biology explored, what sorts of people have been the thought leaders and what motivated them, what mistakes have been made along the way, and what is controversial?

The book is made up of 20 chapters, which are extensively cross-referenced but can be read in isolation or serve as a backdrop for a lecture. We have divided the chapters into five broad sections, the first two of which form a solid foundation of basic principles of stem cell biology and the methodology used to investigate them. The behavior of stem cells illustrates well how multiple molecular processes come together to dictate very specific cell characteristics, including the ability of the cells to communicate with and respond to their environment and changing circumstances. The biology of stem cells and the way that they have evolved to perform a myriad of roles in different species and different tissues is in itself a fascinating subject, and in the third section, "Stem Cells across the Animal and Plant Kingdoms," we look at the ways in which stem cell characteristics have evolved and become part of multicellular life. This section also challenges the boundaries of what are and are not stem cells. The fourth section, "Stem Cells in Specialized Adult Animal Tissues," represents an extensive survey and comparison of the involvement of stem cells in animals, largely deriving from studies on mammals but incorporating other model organisms, including invertebrates. An even greater driving force toward understanding the biology of stem cells has come about because of their potential applications, including improved therapies for many diseases, but also affecting economically important plants and animals. Therefore, in the final section, we give a perspective on the biomedical and economic applications and societal implications of stem cell science.

As mentioned, each chapter can be taken in isolation, and all adopt the same structure, involving an introduction followed by a series of major topics, which are extensively illustrated and accompanied by a number of "boxes" that provide additional material. This material includes "Background" describing concepts and underpinning information; "Case Studies" to elaborate through specific examples; "Pioneers," by which we mean some of the people who have had a major impact on progress in the respective area; and "Controversies," which speaks for itself. At the end of each chapter are a list of "Key points," which could also be called the learning outcomes, and "Further reading," which is not meant to be comprehensive or too detailed on any one topic but instead serves as an entry into wider reflection on the subject and illustration of some of the major discoveries in the field.

Although our principal goal is to provide a comprehensive introduction to stem cell biology to those new to the discipline, we also hope that our book can be an opportunity to a wider audience who have already engaged in some aspects of the field but who may benefit by taking a broader view. All too often in academic pursuit, the trend is to focus and specialize in a way that, although necessary to achieve a sufficient level of scholarship and discovery, can limit thinking outside the box. We are sure that we are not alone in having encountered, when giving a general stem cell lecture, somewhat surprised comments about the evolution of stem cell properties or the very fact that plants have stem cells! We hope there is something in our book for all, both newcomers and those who can already claim to be stem cell biologists.

# Acknowledgments

This book came about during the Spring of 2012 following a meeting with Gina Almond, then a Senior Editor at Taylor & Francis Group. Over the course of a coffee in a restaurant down one of those pleasant side streets in Oxford, a plan was hatched to produce a comprehensive textbook on stem cell biology. Gina came to that meeting armed with a market analysis showing that such a book was much in demand, especially for final year undergraduates and postgraduate students, most books on stem cells at that time being highly focussed and specialized. Within a couple of months Gina moved to other things at Taylor & Francis Group, but she passed us into the immensely capable hands of Summers Scholl, who was at that time an Editor for Garland Science based in New York. We never looked back, as Summers expertly guided us through what it takes to put together such a book. Without Summers's unflinching enthusiasm and confidence in our ability to complete the task we surely would have failed. We lost count of the number of monthly "progress" calls we had between New York and Birmingham, which adopted a recurrent pattern: we would apologize for missing our target for the month, Summers would be very understanding, and we would feel relieved and then make unachievable promises for the productivity over the coming month! We learnt a huge amount from Summers, both technical and stylistic, in particular that not every native or non-native English speaker in the world appreciates multi-clause sentences that run to over 50 words! Reorganization of the Taylor & Francis Group meant that we lost Summers just as we had completed the first drafts of all of the chapters, but her legacy remains and we like to think of her as a life-long friend.

The next at Taylor & Francis whose task it was to keep us on the right path and stick to something approximating to a tight schedule was Jordan Wearing, the Development Editor for CRC Press/Garland Science. His determination to get the job done was just what was needed, as he helped us through completion of the initial drafting and the external reviewing processes towards the completion of a manuscript to be handed to the Production Team. We knew that he meant business when he gave us a very tight timeline for delivery of the completed manuscript, which quite honestly filled us with trepidation but was just what was needed.

As we neared the final stages of the writing, Jordan introduced us to two colleagues, regrettably neither of whom we have met in person, who took our ideas about the illustrations and made them into the much more professional versions that you will see on the pages that follow. Patrick Lane, who works for his own company, ScEYEnce Studios, took our initial thoughts and crafted the figures you see, often having to tolerate several changes in our thinking. We are also indebted to Becky Hainz-Baxter, our Permissions Coordinator. She worked all hours imaginable, including over Thanksgiving, to track down the sources of our figures, often relying on minimal information and frequently having to identify suitable alternatives—we had absolutely no idea what a huge task the permissions process would be!

We should also say that the anonymous reviewers, solicited by Summers and Jordan, deserve our considerable thanks for helping to shape the revision of the drafts, ranging from a total rewrite in one or two instances to illustrating to us that many of our terribly English idioms might result in blank faces amongst the majority of the students who might be reading our words. Many of these reviewers also gave us huge encouragement when the going was tough through their often very positive comments about the way that we were tackling particular aspects of the subject.

As we reached the very final stages of production, Kyle Meyer, the Medical Specialist Project Editor at CRC Press, and Arun Kumar at Novatechset oversaw the details of the final published form, ironing out all those things that sneaked past what we thought was our intense gaze, including expunging a host of more archaic English spelling.

Lastly, we would both like to thank our families and friends who have been supportive throughout and always full of encouraging comments, although at times we are sure that they really did wonder if the dream would ever become a reality. Their belief, and the tolerance they showed, as yet again we had to spend weekends trying to achieve our own deadlines, played a large part in making sure that we did see this project through to its conclusion.

# AN INTRODUCTION TO STEM CELLS

# PART ONE

# The essential principles of stem cell science

1

In this chapter, we will define what constitutes a stem cell, while making clear where there can be misunderstanding or confusion that leads to some non-stem cells being misrepresented. We will distinguish the basic stem cell types and how they are characterized, focusing on the sorts of experimental approaches employed, many of which have been specifically developed for the purpose of investigating these cells. At this point, it will be pertinent to consider the reasons stem cells have become an important part of plant and animal biology. Then we will discuss the broad principles of the ways in which stem cell function is tightly controlled, in particular the need for a specific environment and interaction with external factors. Finally, we will introduce how stem cells can or might be utilized for practical purposes, including improved disease understanding and treatment.

## 1.1 THE DEFINING PROPERTIES OF A STEM CELL

Not surprisingly, it is important at the beginning to have a very clear idea of what constitutes a stem cell. This requires not only unambiguous criteria by which to define the properties of a cell but also needs an understanding of the principles that underlie stem cell properties and knowledge of the methods employed to identify, isolate, and characterize stem cells.

### Two biological properties can be used to define the essence of a stem cell

Inherent within the broad requirements of different stem cells are the two defining properties that distinguish them from other cells. Whatever its specific role, a stem cell must be able to give rise to (**differentiate** into) one or more cell types, but at the same time, it must replace itself so as not to become depleted, through what is usually referred to as **self-renewal**. The first of these characteristics involves the **commitment** of a stem cell to a specialized lineage leading toward one or more functional cell types.

Understanding the mechanisms of self-renewal and the process by which a cell can become committed is at the heart of stem cell science and occupies the efforts of a very large proportion of the stem cell research community. Self-renewal is simply a variant of the normal cellular process by which one cell becomes two. Termed cell division, or "proliferation," this process involves copying of the nuclear chromosomal DNA, separation of the duplicate copies into two halves of the nucleus, and finally splitting of the nucleus and the rest of the cell to produce two so-called "daughter" cells. Cell division occurs throughout development and in all tissues of the adult, to a greater or lesser extent, both to increase cell number and as part of the **differentiation** process that leads to a cell with a defined function. Both daughters can be equally different from the parental cell, but it is also possible that they are at the same time distinct from one another. A stem cell can be very easily defined with these ideas of change in mind. When a stem cell divides, at least one of the daughter cells has to retain an identical pattern of genes that are turned on as in the parental cell. The other daughter can be either subtly

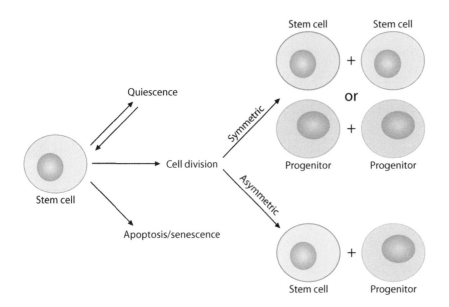

**Figure 1.1 Defining characteristics of a stem cell.** Stem cells are defined by the two fundamental characteristics of self-renewal and ability to differentiate along specific lineages. The diagram illustrates the possible fates of a stem cell. The stem cell can be in a state of quiescence, during which it does not enter the cell cycle, but this process is reversible. When a stem cell undergoes cell division, this can be either symmetrically or asymmetrically. Symmetric division results in the formation of two identical daughter stem cells (stem cell expansion) or two progenitors committed to differentiation (stem cell depletion). Asymmetric division results in the formation of one identical daughter stem cell and one more committed daughter cell. In some circumstances, a stem cell is irreversibly removed from any capacity to divide through entry into senescence or programmed cell death (apoptosis).

different (and committed to a pathway of differentiation) or it can also be identical to the parental cell. These two possibilities are known, respectively, as **asymmetrical** and **symmetrical** self-renewal (Figure 1.1). Some stem cells, such as those associated with early development, largely display symmetric self-renewal, while others, including most of those supporting adult tissues, predominantly divide asymmetrically, although these may occasionally turn to symmetric division if circumstances demand, for example, in an acute crisis such as is required of skin stem cells after wounding.

## Commitment of stem cells gives rise to progenitors that serve to expand the numbers of differentiating cells

The property of incremental change also applies to many cell types referred to as **progenitors** that are involved in lineage-specific cell differentiation, although these cells are often erroneously called stem cells, often because of lack of formal proof. Progenitors provide a crucial link between stem cells and their fully differentiated derivatives because they allow for massive expansion of cell number, which is vital given that stem cells are relatively rare components of particular tissues and mature functional cells are generally required in huge numbers, often very rapidly. This role in expansion of numbers has led to progenitors often being referred to as **transit amplifying cells**, and both terms can be used interchangeably, although the potential for expansion of progenitors diminishes as they become progressively more differentiated (Figure 1.2).

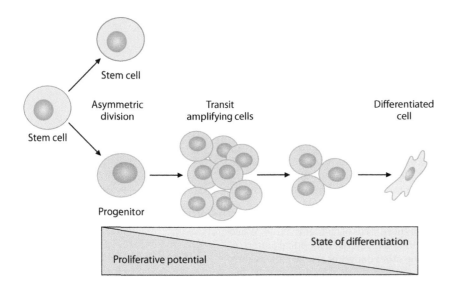

**Figure 1.2 Transit amplifying cells.** Under normal circumstances of asymmetric division of an adult stem cell, the immediate differentiated progenitor progeny cell is able to undergo extensive proliferation to expand the available cells that will go on to terminally differentiate. The diagram shows how the proliferative potential of these so-called transit amplifying cells progressively declines as differentiation proceeds towards the mature state.

## Although stem cells are defined by specific criteria, they are very varied in their origins and differentiation potential

Stem cells are frequently grouped into the two broad categories of **embryonic stem cells** (ES cells), that is, those obtained from pre-implantation embryos, and **adult stem cells**, which are present in most (probably all) adult tissues.

In mammals, after fertilization and up to the 8-cell stage embryo, all cells can be regarded as stem cells, even if their existence is only transient, since each is capable of giving rise to a complete embryo of the **blastocyst** stage and beyond. These cells can not only give rise to the embryo but also contribute to the extra-embryonic tissues (the **trophectoderm**). ES cells are derived from the blastocyst stage embryo and have much the same differentiation potential (Chapter 3). ES cells are adapted to perpetual self-renewal in laboratory culture, dividing rapidly and symmetrically, whereas the cells from which they originate (the **inner cell mass**) have only a transient existence. ES cells are therefore not really a naturally occurring stem cell, but nevertheless they have raised the most interest in the field of stem cell science in terms of their possible uses (Chapter 17).

In parallel with the establishment of the fetus, a wide variety of stem cells become associated with the newly forming tissues. These cells are usually referred to as adult stem cells, although the terms "tissue" or "somatic" stem cells are sometimes used. The requirements and behavior of adult stem cells are quite distinct from ES cells. Adult stem cells mainly divide asymmetrically, although infrequently in many circumstances. Generally, the role of adult stem cells is to provide a perpetual source of replacement cells for the tissue of which they are a part (Chapters 9 to 16).

The embryonic versus adult categorization of stem cells is an oversimplification and disguises their true diversity. Moreover, these definitions are essentially of relevance to vertebrates, around which most research into stem cells has been performed. In many of the examples of stem cells that we will be considering, comparing plants and animals, both invertebrates and vertebrates, the terms embryonic and adult may be irrelevant or inaccurate, and descriptions focusing on the differentiation capacity might be more meaningful.

## Adult stem cells may have a mechanism for life-long protection of their genome integrity

To exist throughout life, yet continuously be capable of generating cells to undergo lineage-specific differentiation, stem cells should ideally not accumulate potentially damaging mutations in their genomic DNA, which could, for instance, lead to cancer. Many adult stem cells actually spend much of their time in a non-dividing state of **quiescence** because the level of differentiated cell turnover is low, and this in itself may limit the opportunity for DNA damage. More specifically, though, it has been suggested that the long lifespan of adult stem cells has led to a mechanism of DNA replication that protects the genome against the replication-associated accumulation of mutations. This so-called immortal strand mechanism was hypothesized by John Cairns in 1975 and envisages that instead of segregating their DNA during mitosis in a random manner, adult stem cells divide their DNA asymmetrically and retain a distinct template set of DNA strands (parental strands) in each division (Figure 1.3). It is envisaged that adult stem cells pass mutations arising from errors in DNA replication on to non-stem cell daughters that soon differentiate to a functional cell. The hypothesis remains highly controversial, although limited supporting evidence has been provided in some stem cell systems.

## Individual stem cell types have distinct capacities in terms of the number of differentiated lineages to which they can contribute

Different stem cells, whether embryonic or adult (or not necessarily falling into either category), can be further categorized according to the number of distinct cell types to which they can give rise. A given stem cell's **potency**, as this capability is known, can range from **totipotency**, as seen for the cells up to the 8-cell stage that are each capable

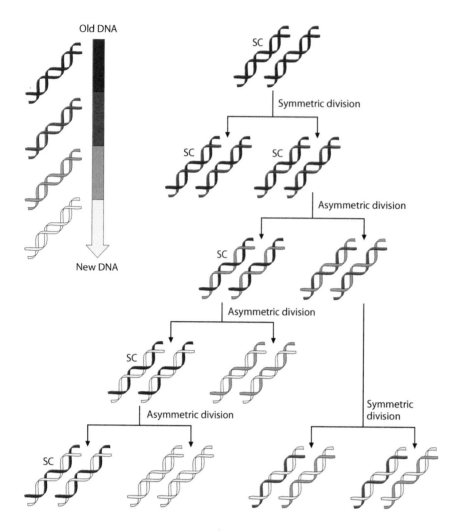

Figure 1.3 **The immortal strand hypothesis.** The immortal strand hypothesis proposes that the genome can be protected from replication associated mutational damage by continual sequestration of the non-replicated template strand. The illustration shows how during asymmetric stem cell (SC) divisions, chromosomes containing the oldest template DNA (dark red) are segregated to the stem cell. DNA is replicated semi-conservatively; each chromosome contains one older template strand. Complements of old DNA-containing chromosomes are co-segregated through many rounds of asymmetric cell division, although symmetric stem cell divisions segregate chromosomes randomly. Over time, stem cells contain proportionally more template-containing chromosomes than any other cells in the population, which contain mostly newer synthesized DNA. (From Karpowicz P, Morshead C, Kam C et al. (2005) *J Cell Biol* 170: 721–732. [doi:10.1083/jcb.200502073]. Permission conveyed through Copyright Clearance Center, Inc.)

of developing into an embryo and its extra-embryonic structures, to just a single cell type (**monopotency**) that characterizes certain tissue-specific stem cells like those found in skeletal muscle. In between these two extremes, there are stem cells that can differentiate into all fetal and adult cell types (**pluripotency** as seen in ES cells) or into only a small number of distinct cell types, for example, **hematopoietic stem cells** (HSC) that are **multipotent** (Table 1.1). Some care is needed when these terms are encountered, as there is sometimes confusion about their meaning. Most often, this concerns the misuse of "totipotency" to describe ES cells when what is meant is "pluripotency" since these cells are not able to give rise to extra-embryonic tissue. Although the vast majority of adult stem cells are limited to differentiation toward no more than a few specialized cell types, there are claims, some of which are still

**TABLE 1.1  STEM CELL POTENCY**

| Potency | Differentiation capability | Examples |
|---|---|---|
| Totipotent | All cell types of both the embryo and the extra-embryonic tissues | Cells of the developing embryo up to the 8-cell stage |
| Pluripotent | All cell types of the embryo proper | Embryonic stem cells |
| Multipotent | Two or more cell types, usually restricted to a single tissue | Hematopoietic stem cells |
| Monopotent | A single cell type | Skeletal muscle stem cells |

*Note:* The table indicates the four types of potency seen in stem cells with representative examples.

disputed, that cells with as extensive a potential as ES cells exist in various locations in the adult body.

## Pluripotent stem cells can be generated experimentally from somatic cells

An additional category of stem cell, which like ES cells represents cells that are **pluripotent**, includes cells generated from normal differentiated cells. Through a process generally referred to as **reprogramming,** stem cell properties can be reactivated by artificially altering the pattern of expressed genes so that the earliest stages of development are recapitulated (Chapter 5). Reprogramming of adult cells, such as skin fibroblasts, can be achieved through essentially two types of approach (Figure 1.4). First, in a procedure that is now generally referred to as **therapeutic cloning**, the nucleus of an adult cell is injected into a fertilized egg (oocyte) that has had its own nucleus removed. Proteins, and possibly other molecules, present in the cytoplasm of the oocyte act to revert the gene expression profile of the adult cell nucleus to that of an early pre-implantation embryo, effectively resetting the developmental clock. Since this includes passage through the inner cell mass stage, from which ES cells can be generated, this means that stem cell behavior and pluripotency are rekindled. Such cells are generally referred to as **somatic cell nuclear transfer (SCNT) ES cells**. The second approach involves introduction into a somatic cell of three or four regulatory proteins (transcription factors) that act as switches to control the transcription of RNA from a gene. The particular transcription factors employed are chosen based on their known association with the maintenance of pluripotency in ES cells. As a result of the enforced alteration in gene expression, some of the somatic cells adopt a phenotype that is highly similar to that of ES cells, and as such these cells have been termed **induced pluripotent stem cells** (iPS cells).

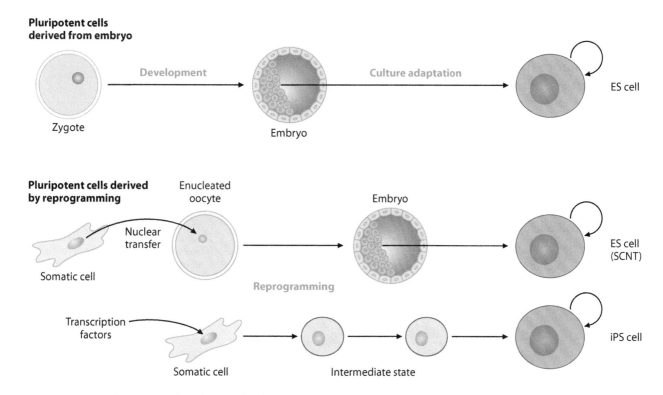

**Figure 1.4 Reprogramming of a somatic cell nucleus to a pluripotent state.** Pluripotent stem cells can be generated either directly from early embryos or from somatic cells through a process of reprogramming. The upper part of the diagram summarizes the process by which cells from a developing embryo can be adapted in culture to generate pluripotent embryonic stem (ES) cells. The lower half of the illustration shows how a somatic cell can be reprogrammed to a pluripotent state that is very similar to that of an ES cell. This can be achieved by transfer of the cell nucleus to an enucleated oocyte, which is allowed to develop into an embryo that can then be used to derive a so-called somatic cell nuclear transfer (SCNT) ES cell. Alternatively, transcription factors associated with pluripotency can be introduced into the somatic cell nucleus, where they bring about a reprogramming involving progressive selection for pluripotency, leading ultimately to the expansion of induced pluripotent stem (iPS) cells. Self-renewal capacity is indicated by the circular arrows.

## Somatic cell characteristics and stem cell potency may not be fixed quantities

The ability to reprogram somatic cells back to a pluripotent state has stimulated a wider interest in whether differentiated cells and even stem cells in the adult are more generally flexible in terms of their phenotype. In addition to the routes for reprogramming of an adult somatic cell to a pluripotent state, there are circumstances, both natural and experimental, under which a somatic cell can convert to an alternative phenotype, including through the intermediary of an adult stem cell type (Figure 1.5). In some specialized scenarios in biology, fully differentiated cells can either **de-differentiate** to a more immature state of the lineage of which they are a part or can **trans-differentiate** to a mature cell type on a completely different cell lineage. Such processes will be considered further in Chapter 7, in particular in relation to naturally occurring regeneration, seen, for example, in some vertebrate limbs. Trans-differentiation can be achieved experimentally, adopting much the same approach as used to achieve reprogramming to iPS cells, either via an adult stem cell state or directly to a differentiated cell phenotype, for example, changing a fibroblast into a neural cell, a cardiomyocyte, or a hepatocyte (Chapter 5). In this case, the selection of transcription factors to bring about trans-differentiation is guided by knowledge of the proteins associated with the particular differentiation pathway of the desired cell type.

The final instance of cell phenotype flexibility concerns the possibility that individual adult stem cell types might be able to switch their ability to differentiate into cells of the tissue of which they are a part into completely different specialized cell types, even those that normally derive from a different germ layer. This idea of adult stem cell **plasticity**, as it is usually referred to, is highly contentious, having been very topical over a decade ago following the publication of a few prominent studies purporting, for example, to show the conversion of HSC into neural or liver cells or of **neural stem cells** (NSC) into blood. However, the conversions of cell phenotype observed tended to be very rare events, different laboratories often having difficulty reproducing the results, and most investigators eventually concluded that the very occasional events were too infrequent to be of practical use and often could be explained by mechanisms other than a direct change in cell fate.

## Stem cell science has been driven by and drives experimental methodology, but a degree of caution should always be exercised in the interpretation of the results

As we will describe in Chapter 2, stem cell science, like most areas of modern biology research, is being driven by technological advancements. In essence, the study of stem cells is usually an iterative process, requiring the development and

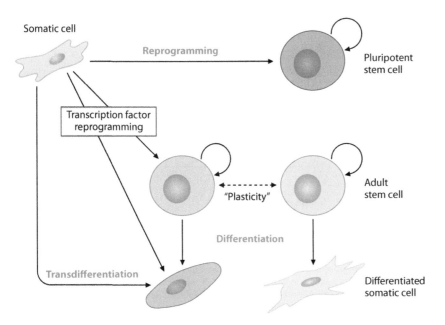

**Figure 1.5 Somatic cells and adult stem cells may exhibit phenotypic flexibility.** The ability to reprogram somatic cells back to a pluripotent state has raised interest in the natural or induced capacity of differentiated adult cells to undergo a number of phenotypic conversions. The process of trans-differentiation of one type of differentiated cell (green) into another (brown) has been observed in some situations. Experimentally, and paralleling the transcription factor-driven generation of iPS cells, it is possible to convert some somatic cells into either another differentiated cell type or into a specific adult stem cell (purple). The most contentious example of cell phenotype conversion is that of "plasticity," in which it has been proposed to be a natural occurrence between different adult stem cell types, although this is now generally regarded to be an artifact or at best a very rare event. Self-renewal capacity is indicated by the circular arrows.

application of a means to identify, isolate, and assay stem cells. Much is performed *ex vivo* on isolated cells, such as *in vitro* assays of proliferative and differentiation potential, while the "gold standard" determination of a stem cell has to be an assay that shows that it can do what it is supposed to be capable of in the context of the organism *in vivo*. Such ultimate tests of **stem cell potential** have often involved a transplantation assay, as exemplified by the restoration of hematopoiesis by transplantation of HSC or the generation of all tissue types in a chimeric mouse produced by injection of ES cells into a mouse pre-implantation embryo. These are powerful assays, but they are performed in an unnatural environment and therefore always need to be viewed with caution. Hence, the HSC transplant requires "conditioning" of the recipient bone marrow, that is, creation of **niche** space through removal of the host's resident stem cells, and when engraftment is achieved, the niche occupancy does not reach anywhere near the normal level. If the HSC being assayed are human, then the necessity to utilize immune compromised mouse recipients adds a further unnatural dimension. Likewise, the ability of an ES cell to produce all adult mouse tissues, including the germ cells, is a powerful demonstration of pluripotency, but in the case of human ES cells, we have to assume that a surrogate assay, **teratoma** formation in an immune compromised mouse, reflects the same sort of potential. As the ability to trace stem cells *in vivo* has improved, both as the result of ways to use reporters of stem cells and microscopic technologies allowing observation of live cells *in situ*, so investigators have often been able to obtain an even better picture.

Although researchers continually strive to develop equivalent types of assays suitable for the investigation of different types of stem cells, it is not always possible to reach the level of sophistication afforded by the various *in vivo* assays, forcing reliance on surrogates of stem cell potential ranging from the *in vitro* growth and differentiation characteristics of the cells to the profiles of molecules that they express. The latter approach can be valuable, as there are undoubtedly some elements of a stem cell "signature" that can be found in the range of genes and proteins that are expressed, but these can only be pointers and are often very misleading, especially when the interpretation is stretched beyond what the data is capable of showing. For example, surface proteins such as the stem cell factor (SCF) receptor Kit (CD117), or the property of small molecule exclusion assayed as a "**side population**" using a flow cytometer, are often associated with *bona fide* stem cells, but they can equally well be found on progenitors and more mature cells. Similarly, expression of RNAs, sometimes only at very low levels, corresponding to transcription factors linked to stemness is no proof that those factors are expressed as proteins or that they are part of a regulatory network controlling stem cell potential.

## 1.2  THE NECESSITY FOR STEM CELLS

As we have defined previously, stem cells can be divided into two broad classes, that is, embryonic and adult. The nature of ES cells means that in a sense, these cells do not have a natural "purpose"; rather they are an artificial representation of a transient state of pluripotency during early development. Nevertheless, their pluripotent character does reflect on what might be regarded as more natural circumstances such as the maintenance of germ cell capacity (which, strictly speaking, lies within our category of adult stem cells) and possibly some scenarios involving injury-associated reprogramming. The necessity for adult stem cells, whether in plants or animals, is plain to be seen since they provide a potential for the life-long functioning of the various tissues that constitute the organism.

### Stems cells became necessary as multicellular life emerged during evolution

The transition during evolution from solely single cell eukaryotes to multicellularity may have heralded the arrival of stem cells. Multicellularity is seen in several branches of the eukaryotes and can vary from a coalescence of cells, often transiently, as seen, for instance, in slime molds, to the complex interaction of distinct tissues that characterize land plants and invertebrate and vertebrate animals (Figure 1.6). Distinct forms of multicellular life have clearly evolved on several occasions in the eukaryotes, suggesting that there must have been two or more routes for the evolution

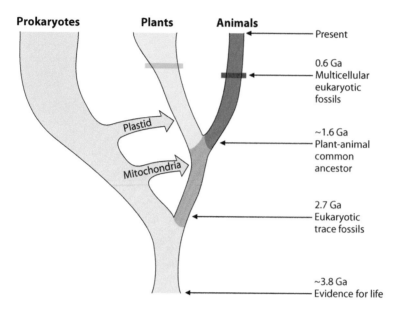

**Figure 1.6 The evolution of multicellularity in land plants and animals.** The schematic phylogenetic tree illustrates the evolution of prokaryotes, plants, and animals from 3.8 billion (Ga) years ago to the present. The approximate points at which multicellularity arose in the plant and animal branches leading to present day species are marked by the green and red lines, respectively.

of stem cell characteristics. Although not definitively settled, it seems likely that stem cells came into play in animals around the time of the evolution of the sponges, over 600 million years ago. These relatively simple organisms have no more than ten cell types but show clear evidence of cells with stem cell characteristics that depend on the expression of genes whose homologs are often encountered in the context of the regulation of pluripotency and germ cells in current invertebrates and vertebrates (Chapter 8). The first multicellular land plants with forms akin to species that we see today arose around 450 million years ago, so we should assume perhaps that the stem cell requirements that are seen in the non-vascular, non-flowering species (Chapter 7) must have arisen around this time.

Just as a single cell organism replicates itself to perpetuate its species, why is it that multicellularity has been accompanied by the evolution of stem cells? At least part of the answer has to be longevity of the organism. A single cell organism is only as old as the time between each replication. In contrast, the longevity of multicellular organisms is dictated by the cooperation of many different cell types that are organized into tissues or organs, each of which will have a characteristic cell turnover and need for replacement. It is the combined need for development of the complexity of the multicellular organism and its ability continually to replace cells in specific component tissues that has been made possible by the evolution of stem cells. Long-lived, often relatively inactive stem cells persist in most, if not all, tissues of the body, where they serve as a source of replacement cells when required or remain active in key growth points of plants to permit continual growth of roots, shoots, leaves, and flowers. Stem cells that were laid down during development need to persist throughout life, which can range from years to centuries for animals, and in some cases even millennia in plant species.

The fundamental molecular cell processes, which evolved over billions of years, are common between plants and animals (e.g., glucose metabolism and DNA replication). Given the vast separation in evolutionary time between the beginnings of multicellularity in plants and animals (Figure 1.6), it is perhaps no surprise that the molecular basis of stem cell behavior seems to have arisen through different routes. This is apparent from comparisons between the molecular processes that dictate stem cell function through the control of gene expression.

That there is more than one way to achieve a stem cell phenotype extends beyond the profound differences between the molecular mechanisms that dictate animal and plant stem cell characteristics. Hence, comparison of the gene expression patterns and underlying mechanisms determining these between individual adult stem cells shows some similarities but also a lot of differences. For instance, in vertebrates, HSC are regulated by distinct mechanisms compared to NSC (Chapter 14), and similarly stem cells in a plant shoot apical meristem utilize a different set of transcription factors than are seen in the equivalent cells in the root apical meristem (Chapter 7).

## Tissues vary in the demands that they place on their resident stem cells

It has become clear over the last few years that all tissues exhibit cell loss and replacement and consequently require an underpinning reserve capacity. This capacity is more often than not achieved through stem cells, although some tissues can maintain homeostasis through mechanisms that do not require stem cells but may instead depend on them in "emergency" circumstances, such as traumatic damage.

The demand for cell replacement very much depends on the individual tissue, both in terms of complexity and the rate of turnover of cells. Some tissues, most notably the hematopoietic system and epithelial tissues such as the skin and the lining of the intestine, exhibit truly extraordinary rates of cell loss and replacement. Hence, it has been estimated that up to $10^{12}$ mature blood cells are produced every day in humans, while the lining of the small intestine is replaced every five days. The turnover of many organs is much less obvious, often to the point where there has been long-standing debate about whether any cell replacement occurs. For example, both the brain and the heart are regarded as tissues that contain very long-lived cells. This does not appear so surprising for the brain, in which neural pathways are laid down in our early development, but it seems too incredible to imagine that an individual cardiomyocyte in the heart can keep functioning by contracting 70 times a minute throughout life (this would be about $3 \times 10^9$ times in an average human lifespan!) and not need to be replaced.

## 1.3    THE CONTROL OF STEM CELL FUNCTION IS COMPLEX AND VARIED

Stem cells are like any other cell in a eukaryote in that they have a nucleus, cytoplasm, and all of the organelles that are necessary to sustain their existence and perform their specific function. In a sense, the specialization that makes a stem cell what it is can be viewed as no different than what makes a muscle cell what it is and distinguishes it from a neuron or a white blood cell. The nature of a given cell type and what makes it different from any other is the profile of genes that are turned on to produce RNA and ultimately proteins out of the many tens of thousands present in the genome. For example, in a skeletal muscle cell, these proteins would include components of the contractile apparatus such as myosin, actin, and troponin, but what might the equivalent be for a stem cell in that its function is to be maintained as a stem cell while remaining able to differentiate? It is probably fair to assume that the critical determinants of a stem cell being what it is are regulatory factors that maintain the genome in a state of readiness to express genes characteristic of specific lineage differentiation and at the same time proliferate without changing. The crucial point about all cells that are not stem cells, that is, any cell downstream of the stem cell in the differentiation hierarchy, is that at each cellular division, both of the daughter cells are to some degree distinct from the parental cell in terms of their profile of expressed genes.

### Stem cell characteristics are determined by molecular networks

The nature of stem cells and the responses that are required of them necessitate complicated mechanisms of control. Failure correctly to regulate stem cells can have dire consequences, ranging from their untimely depletion, which could lead to premature tissue aging and failure, to uncontrolled self-renewal, which can be a key step in the progression of normal cells towards a malignant state (Chapter 18). The regulation of stem cells, which will be considered in detail in Chapters 4 and 6, involves components that are extrinsic, that is, soluble systemic factors (growth factors, cytokines, chemokines) and a variety of other cells and cell-associated materials, and intrinsic mechanisms, including the regulation of gene transcription (synthesis of mRNA), the processing and stability of the transcribed mRNAs, and protein translation and modification.

Ultimately, what gives a stem cell its special characteristics or determines its commitment to differentiation is its profile of gene expression and the way in which this leads to a specific pattern of functional proteins. As researchers have delved into the finer details of the molecular mechanisms that control stem cell behavior, there has been considerable interest in trying to identify common features between different stem cell types, whether at the level of signaling pathways from cell surface receptors or in terms of the regulation of gene transcription, RNA processing, translation, or post-translational processes. As so often in biology, there is no simple unifying conclusion. For sure, there are mechanisms that have been conserved through animal evolution that dictate stem cell pluripotency. These mechanisms are complex, involving the interaction of **epigenetic** modifications, specific transcription factors, non-coding RNAs, protein translation, modification and stability, and probably many other components. However, it is easy to go away with the impression that only a few factors are crucial when considering the dramatic consequences of enforced expression of individual transcription factors in the context of reprogramming of adult somatic cells (Chapter 5). However, this would be a false conclusion, since such artificial expression of key factors likely disturbs the regulatory status quo, thereby setting in motion a whole series of altered interactions within the regulatory network.

There are some unifying principles in stem cell regulation, especially in the use of signal transduction pathways, even though the downstream targets of these and the precise combination of their use varies between different stem cell types. The most widely utilized signaling pathways that are frequently encountered in the context of stem cell regulation are Wnt/β-Catenin, Notch, Hedgehog, and Transforming growth factor-β/Bone morphogenetic protein (TGF-β /BMP) (Chapter 4). In addition, a number of lineage-restricted growth factor receptors and surface adhesion molecules are often seen on stem cells, although their presence on particular populations of cells has often been overinterpreted as an indication of stem cell identity. Those that have elicited the most interest include Kit, the receptor tyrosine kinase that binds SCF, and Prominin (CD133), which is a pentaspan transmembrane glycoprotein that specifically localizes to cellular protrusions and has been proposed to act as an organizer of cell membrane topology. Both Kit and Prominin are present on several adult stem cell types, but they are neither exclusive to stem cells nor are they present on all adult stem cells. As for the components of the machinery within the cell, the overall picture of stem cell regulation through signaling pathways and growth factors is one of many possible combinations achieving the same end result, so that each type of stem cell has to be functionally defined and then characterized in its own right.

## The external environment of a stem cell is critical for its correct functioning

Crucial to the correct regulation of a stem cell is the environment, or "niche," within which it exists. The niche in a general sense can be made up of a number of components, including cells and **extracellular matrix** with which the stem cell makes direct contacts since this provides many of the direct interactions that it has with other cells and extra-cellular substances and includes molecules that may have ultimately derived from distant sites in the body, including signals from neural inputs (Figure 1.7, Chapter 6). The physical orientation of a stem cell in its niche can also play an important part in achieving the balance between self-renewal and commitment to differentiation. Hence, there are documented examples (see the discussion of

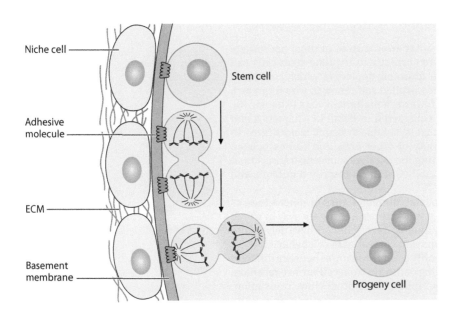

**Figure 1.7 The stem cell niche.** Diagram of a hypothetical stem cell niche highlighting some key components thought to maintain a stem cell and regulate the commitment toward differentiated cells. The stem cell is often physically attached, and even orientated specifically, through adhesion receptors connecting to niche cells or the extracellular matrix (ECM). In the example illustrated, one stem cell is shown undergoing symmetric division, both daughter stem cells remaining attached to the niche. Another stem cell is shown following asymmetric division, in which case the stem cell daughter remains attached to the niche while the progenitor daughter moves away from the niche. (From Spradling A, Drummond-Barbosa D & Kai T (2001) *Nature* 414: 98–104. [doi: 10.1038/35102160]. Adapted by permission from Springer Nature.)

germ line stem cells in Chapter 12) where the stem cell connects to the exterior via transmembrane proteins that link to the cytoskeleton in such a way that the plane of cell division is directed causing one daughter cell to move away from contact with the niche, a feature that in itself can be an instruction to that daughter cell to differentiate.

## 1.4 THE THERAPEUTIC AND ECONOMIC POTENTIAL OF STEM CELLS

The understanding of stem cell biology promises a whole range of new approaches to the understanding and treatment of disease, both in humans and other animals. Stem cell science is also set to deliver a variety of economically important benefits, including improvements in crops and livestock. Both the healthcare and economic possibilities will be elaborated upon in Chapters 17 to 19. The societal impact of such applications of stem cells is fairly obvious in respect to the positive benefits, but the latter do not come without problems, especially in considering various ethical aspects of stem cell derivation and use, a topic that we will explore in Chapter 20.

### The potential for the application of stem cell knowhow in healthcare is vast

The wealth of knowledge about stem cells and our capacity to manipulate them, built up over more than two decades of intensive investigation and experimentation, have opened up many new avenues for the understanding and treatment of human disease.

Stem cells and stem cell properties are at the heart of many human conditions, the inappropriate loss or gain of function, which is often very much linked to aging processes, leading to diseases ranging from tissue failure to cancer (Chapter 18). Hence, loss of adult stem cell capacity could lead to the decline in the functioning of a particular tissue and is thought to be at the root of some aspects of the aging process, such as muscle wastage or bone marrow failure. Similarly, the self-renewal properties of stem cells are very similar to the way in which a cancer cell behaves, so that mutational changes in stem cells or the gain of stem cell-like properties in an adult somatic cell might readily lead to malignancy. Understanding such losses and gains of stem cell characteristics in itself has the potential to provide new treatment options, and the ability to experiment on stem cells provides the additional benefits for modeling and the testing of potential new drug or cell-based therapies.

The most anticipated, and potentially most significant, application of stem cell science in a clinical context is the prospect of repairing or replacing tissues and organs that have been damaged through disease or traumatic injury. Collectively, such approaches are usually grouped in the broad category that is usually referred to as **regenerative medicine**, but as we will describe in Chapter 17, this term covers a multitude of possibilities.

### Tissue repair and replacement can be potentially achieved using either embryonic or adult stem cells

As often in science and its translation to practical application, the theory is straightforward, but the actual realization of stem cell science benefits in regenerative medicine is a long process with many hurdles to be overcome. The sorts of challenges faced include being able to generate sufficient, fully functional cells from an appropriate stem cell source, correctly incorporating the differentiated cells into a fully functioning tissue, and overcoming or avoiding immune system incompatibility of the replacement cells if the stem cell source is derived from an individual other than the patient. Nevertheless, the potential benefits generally outweigh the problems that need to be solved, and some options may in fact be easier to implement than others, especially when these involve manipulation of the patient's own cells.

Both embryonic and adult stem cells offer routes to regenerative therapy (Figure 1.8), which we will summarize briefly here. The potential of ES cells to give rise to all cell types of the body, coupled with their essentially limitless ability to proliferate in culture, has led to considerable excitement and hope that they could be a source of cells and tissues as an alternative to the current availability of organs for transplantation. ES cells without doubt have great therapeutic potential, and the high level of interest this has raised has been a driving force for the rapid expansion of research in this area. The biggest issue, as it is in transplantation

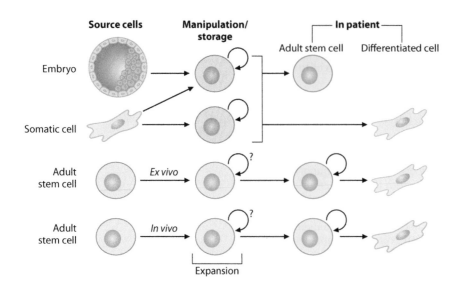

Figure 1.8 **Strategies for the generation of differentiated cells for regenerative medicine purposes.** Various options are being explored for the use of stem cells in therapies that aim to repair or replace damaged or diseased tissues. Pluripotent stem cells (pink) derived either from embryos or reprogramming of somatic cells can be used to produce the desired differentiated cell (green) either directly or via an intermediate adult stem cell or progenitor (blue), the latter option providing the possibility for long-term cell replacement. Alternatively, as illustrated in the lower part of the figure, adult stem cells can be employed for regeneration, either following purification and possible *ex vivo* expansion followed by transplantation or through treatments of the adult stem cells in the patient so as to stimulate their expansion or differentiation in the desired manner. Self-renewal capacity is indicated by the circular arrows.

using donated organs, is avoiding rejection resulting from the immune response elicited by histocompatibility mismatch between the donor cells and the recipient. There are a variety of possible solutions to this latter problem, but without doubt the most interest, and a consequent explosion of research activity, has focused on the discovery in 2006 that adult cells can be reprogrammed into iPS cells, which have essentially the same properties and potential as ES cells but the advantage that they can be derived from the patient for whom replacement cells need to be generated. Such a patient-specific approach to therapy involving pluripotent stem cells would also avoid many of the limitations and ethical problems that are inevitably associated with the use of ES cells.

Stem cell biologists are always being asked when they will actually deliver new therapies. The question arises largely from thinking around ES cells but fails to appreciate that there are already several therapeutic applications of adult stem cells, and it is a fairly safe bet that there will be a rapid increase over the next decade in the treatment of both stem cell related diseases and in the use of adult stem cells in a regenerative context. As far back as 1957, E. Donnell Thomas pioneered human bone marrow transplantation to treat malignant hematological conditions, a procedure that relied upon the presence of HSC. The use of HSC continues and is ever more successful and certainly paves the way for other adult stem cell therapies. Moreover, as we learn more about adult stem cells in their natural environment, it is quite probable that in some scenarios, it will be possible to manipulate their function *in situ*, without the need for replacement. An example of such a situation might be to use knowledge about the ways in which a given stem cell and its niche interact and are affected by aging, both processes that might feasibly be amenable to treatment through a pharmacological route.

## The benefits that can be had through stem cell science go far beyond healthcare

It is not just human health prospects that can expect to gain from stem cell science (Chapter 19). Many aspects of agriculture are already seeing stem cell science-based approaches, some of which mirror areas of human application, such as the use of mesenchymal stem cells to repair tendon injuries in horses. Pluripotent stem cell technology in particular is finding many uses, ranging from better ways to genetically modify livestock to last ditch efforts to save critically endangered species of wild animals. Horticulture is likewise seeing increasing prospects for the use of stem cell technology to which it is anyway no stranger, having for centuries utilized the power of plant regeneration for propagation, but is now incorporating direct manipulation of stem cells for crop improvement or selective generation of medically and economically important drugs.

## 1.5   SUMMARY

In this introductory chapter, we have aimed to set the scene for the other parts of the book. We have provided a set of definitions and terminology that will be important throughout as well as some basic principles of the experimental methods that are commonly used. We have tried to give some initial sense of the extent of and reasons for stem cell involvement in plant and animal biology and some thoughts about how this may have evolved and been controlled. Lastly, we have touched on the wealth of practical applications of stem cell science, which, although quite obviously dominated by healthcare considerations, actually range widely across horticulture and agriculture.

## KEY POINTS

- Stem cells are characterized by their ability to make exact copies of themselves (self-renewal) and to differentiate into one or more functional cell types

- Stem cells are a necessary part of complex multicellular life in plants and animals

- Different stem cells are able to differentiate into a range of cell types, which defines their potency and can vary from a single lineage to all cell types of the organism

- Stem cells may have a transient existence during development or exist throughout life in tissues

- Embryonic stem (ES) cells, which are pluripotent in that they can give rise to all cells of the fetus and adult, only exist in the laboratory, having been derived from the early embryo

- Adult stem cells, which can range from having single to multiple potencies, are located in specific tissues and serve to replenish normal cellular loss or repair damage, although they often spend most of their time in a state of quiescence

- Adult differentiated cells can be reprogrammed to a pluripotent state either through cloning (SCNT-ES cells) or over expressing specific transcription factors (iPS cells)

- Isolated stem cells can be assayed in culture for some aspects of their self-renewal and differentiation properties, but the gold-standard test involves their life-long perpetuation and contribution to a tissue following transplantation

- Stem cells need to be tightly controlled in order to retain their properties, and this requires their interaction both with soluble systemic factors and a variety of other cells in their "niche"

- External input into stem cells is interpreted as a stem cell-specific pattern of gene expression

- Loss or gain of stem cell properties underlies both malignant and non-malignant disease, including age-related tissue decline

- Stem cell biology offers possibilities for regenerative medicine either through provision of cells for transplantation or by enhancement of endogenous stem cell capability

- Knowledge of stem cell biology is increasingly being translated into not only medical practice but also many aspects of horticulture and agriculture

## FURTHER READING

Alberts, B., Johnson, A., Lewis, J., Raff, M., Roberts, K. and Walter, P. (2007) *Molecular Biology of the Cell*, 5th ed. Garland Science, New York & London.

California Institute for Regenerative Medicine Stem Cell Basics Primer. http://www.cirm.ca.gov/our-progress/stem-cell-basics

EuroStemCell Fact Sheets. http://www.eurostemcell.org/stem-cell-factsheets

Harvard Stem Cell Institute Stem Book. http://www.stembook.org

Knoepfler, P. (2013) *Stem Cells: An Insider's Guide*. World Scientific Co. Pte. Ltd., Singapore.

# 2

# Stem cell identification, isolation, and characterization

Research on stem cells is dependent on standard approaches and technologies common to the study of cell biology. Many of the strategies that we will describe in this chapter for stem cell identification, isolation, and characterization rely on general cell characteristics. In addition, as we have seen in Chapter 1, a stem cell has a number of defining properties that can also be utilized, most importantly the ability to both **self-renew** and **differentiate** such that it can reconstitute the cells of its tissue for the life of the individual. It is important to remember that stem cells, especially those residing within adult tissues, are usually rare components of very mixed populations of cells and that removal from their specialized environment may well have an effect on their normal functioning. It is also worth bearing in mind that the precise definition of what constitutes a stem cell can very much depend upon the assays and criteria that are applied and that the progress of stem cell research can mean that what may once have been thought of as a "pure" stem cell can often later be regarded as a mixed population of cells with various degrees of "stemness." Identification, isolation, and functional testing often require an iterative approach, allowing increasing refinement of the definition of the most potent stem cell component.

## 2.1   IDENTIFICATION OF STEM CELLS

There is no one strategy for the identification and isolation of stem cells, each circumstance building upon a body of evidence from often very different directions, but, that said, there are nevertheless some commonly adopted approaches. Without doubt, stem cell identification and isolation often rely heavily upon combinations of surface biomarkers, together with some other features such as particular metabolic activities or cell cycle activity, or the lack of it, since stem cells frequently reside in a state of **quiescence**. Whatever combination of techniques are applied, these are always used in conjunction with assays that demonstrate the key properties that are characteristic of stem cells, namely self-renewal and differentiation.

### Stem cells express surface molecules that can be exploited in their isolation

Stem cells, like other cell types, have a surface that is covered with a plethora of different proteins. Most of these proteins traverse the cell membrane, but some do not and are either covalently linked to other proteins or are more loosely associated. Many of the surface proteins are also "decorated" with a variety of non-protein modifications, by far the most common of which is carbohydrate. The cell surface proteins that are used to identify cells are often known by two names, and this can sometimes cause confusion, not least because there can be a lack of exact correspondence between them. Most marker proteins have a gene name that describes them based on their origin, function, structure, and so on, as for any other cellular protein; for example, Glycoprotein IIb or GPIIb, an integrin made up from two protein chains, $\alpha_{IIb}$ and $\beta3$. Cell surface proteins also often have a designation if recognized by a specific antibody. In humans, about 400 different proteins to date are designated by a **cluster of differentiation** (CD) number. Taking again the example of the GPIIb protein, this has two CD numbers: CD41, which corresponds to the $\alpha_{IIb}$ protein chain, and CD61, which indicates recognition of the $\beta3$ component.

Some proteins are present on a wide range of cells, whereas others are more restricted in their distribution, sometimes only being present on one cell lineage of differentiation or, in the extreme case, only on a particular cell type. Such highly specific expression of a cell surface protein would obviously serve as a unique marker of that cell; however, this is not generally the case, but a given cell can often be discriminated because of the unique combination of proteins that are present or absent at the surface (Figure 2.1).

Many of the concepts of stem cell identification and isolation have been elaborated through research into **hematopoietic stem cells** (HSC, Chapter 9) and this has relied heavily upon the use of combinations of antibodies against CD antigens. HSC are undoubtedly the best characterized **adult stem cells** in terms of the level of refinement of surface markers that can be used in their identification and purification. The majority of studies on HSC have been performed in the mouse, but have of necessity been translated to investigation of the equivalent human cells. In doing so, it became obvious that there are numerous discrepancies in surface protein expression between mouse and human (Table 2.1). The first major difference described was in the expression of CD34, which is present on human long-term repopulating HSC but is absent from the equivalent cells in the mouse. In 2005, the signaling lymphocytic activation molecule (SLAM) 150 was much hailed as a way of refining the isolation of mouse HSC; however, this proved to be of no value with respect to human HSC, as they do not express it.

Following the lead set by research on HSC, biologists have sought to identify equivalent patterns of marker expression that uniquely define other types of stem cells. Some markers that are present on HSC are found on other stem cell types (e.g., CD117 [c-Kit] and CD133 [Prominin]); however, different stem cells are also characterized by proteins that are not associated with HSC, and more will be said about this in subsequent chapters on specific adult stem cell types.

Key to making use of surface proteins are monoclonal antibodies, which bind very specifically to each of them and can be conjugated to one of the wide range of available fluorochromes that emit a discrete spectrum when stimulated with light of a shorter wavelength produced from a laser source (Box 2.1). It is then an easy matter to detect cells to which the antibody has bound either through a fluorescent microscope or by analyzing the cells in a flow cytometer, a machine that passes a stream of single cells through a narrow beam of laser light and then is able to measure the emitted fluorescence from each cell.

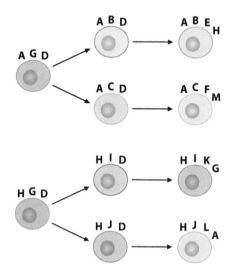

**Figure 2.1 Antigen combinations define specific cells.** Stem cells, like any cell, express a combination of protein and carbohydrate-based antigens on their surface, many of which can be identified using specific antibodies. Seldom, if at all, is a particular antigen restricted to a given cell type; however, the combination of several antigens can often be used to define one cell versus another. The diagram illustrates two hypothetical cell differentiation hierarchies, each starting from a stem cell on the left. Individual antigens are indicated by letters. The two hierarchies are defined by antigens A or H. G and D in combination represent the immature stem cell stage, expression of A or H distinguishing one stem cell from another. Other antigens define more differentiated states.

## TABLE 2.1  CD ANTIGENS THAT CHARACTERIZE HEMATOPOIETIC STEM CELLS

| CD antigen | Mouse | Human | Other adult stem cell types |
|---|---|---|---|
| CD34 | − | + | Subset of mesenchymal stem cells; muscle satellite cells |
| CD38 | + | − | |
| CD48 | − | − | |
| CD117 | + | + | Prostate stem cells; cardiac stem cells |
| CD135 | − | + | |
| CD133 | − | + | Prostate stem cells; neuronal and glial stem cells |
| CD150 | + | − | |

*Note:* Individual antigens expressed at the cell surface can be used in combination to define specific stem cell types. The table provides examples commonly used to identify hematopoietic stem cells, although there is often not a strict correspondence between mouse and human cells. Also, several of these antigens are expressed on other adult stem cell types.

# Cutting edge microscopy techniques enable precise location of stem cells

Observation of stem cells in their tissue location is crucial in understanding how they interact with their environment, and various microscopy platforms in combination with diverse labeling strategies can be applied to this end. By far the most used approach involves observation of fluorescently labeled cells in combination with confocal microscopy (Figure 2.2). The technology of laser scanning confocal microscopy (LSCM) has revolutionized fluorescent imaging in cell biology and has

## BOX 2.1 BACKGROUND: FLUOROCHROMES

Many compounds can be excited by light and emit light at a different wavelength, causing them to fluoresce (Figure 1),

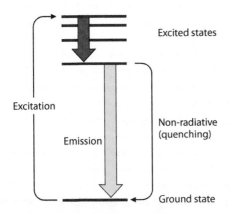

Box 2.1 Figure 1 **Excitation and emission spectra of fluorochromes.** Fluorochromes absorb light at specific wavelengths defined by the compound's properties. During excitation, electrons are raised from the ground state to an excited state and then return to the ground state, emitting light at a specific wavelength for the fluorochrome. (From Ormerod M & Novo D. (2008) *Flow Cytometry – A Basic Introduction.* (http://flowbook.denovo-software.com/chapter-3-fluorescence-fluorochromes). © 2008 Michael G. Ormerod. Used with permission.)

and for this reason are generally called fluorochromes. Individual fluorochromes vary in the difference between excitation and emission spectra, a quantity known as the Stokes shift (Figure 2). More than one fluorochrome can be excited by the same wavelength of light provided that their Stokes shifts are different. Fluorochrome-containing compounds can be used to stain structures such as the plasma membrane or DNA. For example, carboxyfluorescein succinimidyl ester (CFSE) is taken up by cells and upon cleavage by cellular esterases releases a fluorescent moiety that is retained in the cytoplasm, while diamidinophenylindole (DAPI) binds between the base pairs of DNA in the nucleus and fluoresces blue when excited with UV light. More frequently, fluorochromes are covalently bound to antibodies for use in flow cytometry and microscopy. The latter uses have led to the development of a wide range of molecules that emit light across the visible spectrum, and those commonly employed can be grouped into five classes, each with particular advantages and limitations (Table 1). With the advent of more sophisticated laser technology, there is an increasing demand to develop an even broader spectrum of fluorochromes that can be used successfully in combination to identify the rarest population of stem cells within a tissue.

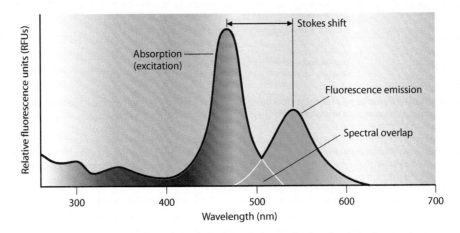

Box 2.1 Figure 2 **Excitation and emission spectra of fluorochromes.** The spectral profile of a given fluorochrome shows the excitation curves at a specific wavelength and the Stokes shifted emission curve. (From https://www.olympus-lifescience.com/en/microscoperesource/primer/lightandcolor/fluoroexcitation/. Used with permission from Olympus Corporation.)

### BOX 2.1 TABLE 1  CLASSES OF FLUOROCHROMES

| Fluorochrome class | Example | Source | Advantages | Limitations |
|---|---|---|---|---|
| Fluorescent proteins | Green fluorescent protein (GFP) Phycobiliproteins – allophycocyanin (APC), phycoerythrin (PE) | Jellyfish *Aequorea Victoria* Cyanobacteria and algae | Brightest fluorochromes | Not great for microscopy and rapidly photobleached |
| Synthetic small molecules | Fluorescein isothiocyanin (FITC), Cyanine 5 (Cy5) | Manufactured organic compound | | |
| Quantum dots | Qdot 565, Qdot 705 | Inorganic nanoparticles | Photostable and bright | Large size, toxicity |
| Polymer dyes | Brilliant series dyes | Manufactured polymer dye | Extremely bright | Require specific buffers to prevent interactions between multiple polymer dyes |
| Tandem dyes or Förster resonance energy transfer (FRET) dyes | Phycoerythrin-cyanine5 (PE-Cy5), allophycocyanine-cyanine7 (APC-Cy7) | | Useful for extending usable spectrum off one laser | FRET efficiency not 100%, tandem dyes can degrade or uncouple |

*Note:* The table lists different types of fluorescent molecules that are used in biological research. Examples are given for each category, together with some of their respective advantages and limitations.

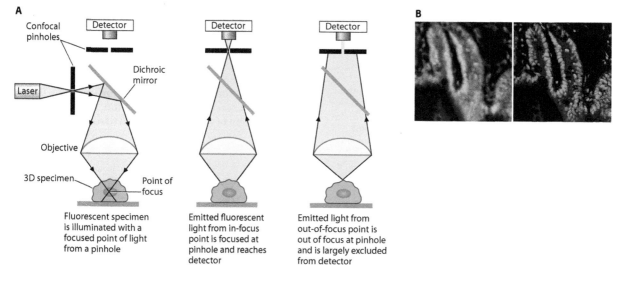

**Figure 2.2 Confocal microscopy.** Confocal microscopy is an essential tool in cell biology and importantly provides far greater resolution than conventional fluorescence microscopy. **(A)** Diagram to show the basic concept of confocal microscopy. Left: An excitation laser suitable for the fluorochromes being used is focused through a pinhole to illuminate a specimen at the desired point of focus. Middle: Fluorescent light emitted from the in-focus point passes through a dichroic mirror and then through a second pinhole, where it is registered by the detector. Right: Any out-of-focus light emitted from the specimen is largely excluded from the detector by failure to pass through the second pinhole. **(B)** Comparison of wide field (left) with confocal microscopy (right) of a section of mouse intestine to demonstrate the increased resolution gained through confocal microscopy. (A: From Alberts B, Johnson A, Lewis J et al. (2015) *Molecular Biology of the Cell*, 6th ed. Garland Science. With permission from WW Norton Company. B: Used with permission from Carl Zeiss.)

proven crucial in many advances in the understanding of stem cells, especially their location within and relationship to their **niche**. Traditional wide-field microscopy was plagued by limitations in the degree of detail that could be seen, largely due to the thickness of the specimen, the illumination of all tissue layers resulting in a predominantly out-of-focus image. Use of a confocal microscope increases optical resolution and contrast through point illumination and elimination of out-of-focus light, making it possible to image multiple planes of the specimen. Fluorochrome-conjugated antibodies or fluorescent proteins are excited within a section of tissue or cell preparation on a slide by illumination with lasers of specific wavelengths. The point of excitation light is focused by an objective lens at the desired focal plane within the specimen. The emitted fluorescence from the sample passes through a mirror that selectively allows certain wavelengths of light to pass through while reflecting others (a "dichroic" mirror); the emitted light is then focused at a second pinhole, where it is measured by a photomultiplier that generates a signal related to the brightness of light from the specimen. Any light emission from the sample that is outside of the desired plane of focus fails to hit the second pinhole and is therefore omitted as out-of-focus light, creating a much sharper image. The resulting "optical slices" can be used to make an effective three-dimensional reconstruction. Although confocal imaging produces more detailed and sharper images than traditional wide-field microscopy, this is at the expense of signal intensity, as a lot of the light is filtered out when passing through the pinhole, but in stem cell biology, the detail and sharpness are of greater importance than the intensity of the signal. There are some significant limitations of LSCM, one of which is photobleaching and phototoxicity due to the exposure of a restricted area to intense laser wavelengths for several minutes while an image is captured.

Two-photon microscopy (multiphoton microscopy) is a variant of confocal microscopy that allows imaging of living tissue with a better depth of resolution, enabling examination of cells in their normal environment, for instance, the stem cell niche, overcoming the problem of phototoxicity encountered using regular LSCM. Two-photon microscopy uses red-shifted light, each excitation of the fluorochrome requiring absorption of two photons, the emitted single photon being of shorter wavelength. The use of infrared light in this way minimizes scattering in the tissue, while the multiphoton absorption and shorter wavelength emission strongly reduce the background signal and allow increased depths to be visualized.

## Flow cytometry allows multi-parameter analysis of stem cell populations

Flow cytometry enables analysis of astronomical numbers of cells (>50,000 cells per second), providing information on the amount of a protein present on the surface of each cell (or internally if cells are fixed and permeabilized). Modern day flow cytometers work around three key features: fluidics, optics, and electronics. The fluidics create hydrodynamic focusing, ensuring that a narrow stream of labeled cells passes through one or more laser sources in single file, allowing excitation of each individual cell as it passes the light beams (Figure 2.3). The optics allow for the cells to be excited by multiple wavelengths of light and for the fluorescent light emitted from the excited fluorochromes to pass through specific filters, called band-pass filters, which permit only the desired wavelengths to reach the respective photoreceptors, where the signal is converted through the electronics into a digital image. Several antibodies that recognize

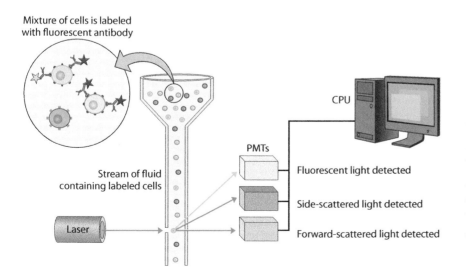

Mixture of cells is labeled with fluorescent antibody

Stream of fluid containing labeled cells

Laser

PMTs

CPU

Fluorescent light detected

Side-scattered light detected

Forward-scattered light detected

Figure 2.3 **Flow cytometry.** Flow cytometry combines fluorescent labeling of cells with fluid dynamics to enable single-cell interrogation at high speed, providing multiple parameter readouts that give information on not only antigen expression but also cell size, shape, and cytoplasmic composition. The diagram illustrates the basic principles of a flow cytometer. A mixture of fluorescently labeled cells (red, blue, and green circles) in the sample being analyzed enters a flow of sheath fluid in the broad part of the nozzle. The nozzle progressively constricts in the direction of flow, leading to a pressure difference between the outer layers and the center, which has the effect of focusing the sample flow (hydrodynamic focusing). Cells are forced into a stream that is one cell wide so that they can then pass through the laser light beam one at a time. Some of the light passing through the cell is scattered in a way dependent on the size and shape of the cell (forward and side scatter) and is measured by a specific photomultiplier tube (PMT). If the cell carries a fluorochrome (stars) that is excited by the laser light, then it emits light that can be measured by a separate detector.

different proteins can be used together if each is labeled with a distinct fluorochrome that emits a unique spectrum of light, making it possible to define patterns of protein expression on cells and to discriminate a particular cell type from a complex mix.

## Cytoplasmic stem cell features can be detected using fluorescence-based approaches

In addition to surface marker expression, there are a number of other characteristics of stem cells, many of them shared by a wide range of stem cell types, which can be used for their identification and isolation. It must, however, be emphasized that these methods are generally only useful as an adjunct to other techniques and are not suitable as a sole means of isolating stem cells.

One method, pioneered by Margaret Goodell in 1996 for the isolation of HSC, uses flow cytometry to measure the ability of many stem cells to actively expel small molecules from their cytoplasm. A family of transmembrane proteins known as ABC transporters are involved in a wide variety of both normal and stem cells with the purpose of removing diverse chemicals. One family member, ABC-G2, is often expressed by stem cells and has the ability to export certain chemical dyes that have entered the cytoplasm. Empirically, it was found that one such dye, Hoechst 33342, is removed by ABC-G2 and that this can be visualized by measuring red and blue fluorescent light emissions upon stimulating with a UV laser. In the complex pattern of light emitted by a mixture of cells treated with Hoechst 33342, many stem cells appear as a population, usually called the **side population** (SP), which exhibits low red and blue fluorescence because the dye has been largely removed by the transporter (Figure 2.4). Since SP characteristics are not restricted to stem cells, the property is best utilized as a means to refine stem cell identification and isolation in combination with other parameters, especially surface marker staining.

Certain metabolic features of stem and **progenitor** cells can be used in their identification and isolation. For instance, high levels of aldehyde dehydrogenase (ALDH) activity have been reported for normal and cancer precursor cells of various lineages, including hematopoietic, mammary, endothelial, mesenchymal, and neural. The level of ALDH can be determined using the substrate BODIPY aminoacetaldehyde (Aldefluor) that is able to diffuse into cells and, following enzymatic cleavage, releases the carboxylate form, which becomes trapped because of its negative charge and can be quantified by flow cytometry.

The DNA replication and proliferation characteristics of stem cells have also been employed both to identify them, especially *in situ*, and infrequently as a means of isolation. In particular, the replicative quiescence of many adult stem cells compared to normal cells can be made use of through either the labeling of DNA as it is synthesized or by measuring the extent of cell division as a membrane-bound or cytoplasm-confined label becomes diluted. Hence, if thymine, which is one of the basic building blocks of DNA, is substituted by the related molecule bromodeoxyuridine (BrdU), then it is possible to mark those cells that have undergone replication (usually visualized by attaching a fluorochrome to a monoclonal antibody that recognizes BrdU). When cells are exposed to the BrdU for just a short period, then only cells that are dividing during that time will become labeled. Subsequent cell divisions in the absence of BrdU then lead to a progressive dilution of the amount of the label present in the cell. Since in most circumstances adult stem cells spend a majority of their time in a quiescent state, this means that if a stem cell can be labeled as it replicates its DNA, then the BrdU incorporated will not be diluted out as rapidly as in normal cells. Stem cells can therefore be identified as the ones that retain the label after a protracted period following removal of the BrdU (hence the term **label retaining cells**). This technique can be performed in live animals following injection of BrdU, providing a means to label adult stem cells *in situ* with minimal disturbance to their environment. Alternatively, as described previously, certain enzyme substrates such as CFSE, which are taken up by cells and render the cytoplasm fluorescent upon enzymatic cleavage, can be used to measure cell division, since the amount of fluorescence in each of the daughters is half of what was present in the parental cell, and at each subsequent division, the amount of fluorescence drops by half (Figure 2.5).

## Fluorescent proteins can be used to label stem cell populations

A commonly used approach to label cells in a lineage-restricted manner, including stem cells, involves the expression of a gene whose encoded protein can be either directly or indirectly detected. Such "reporter" proteins are generally either enzymes, which can be detected using appropriate substrates, or fluorescent proteins

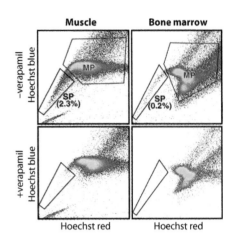

**Figure 2.4 Side population cells identified in muscle and bone marrow.** Flow cytometry profiles of muscle (left) and bone marrow (right) cells stained with the DNA binding dye Hoechst 33342. Upon excitation with an ultraviolet laser, the Hoechst dye emits light in both the blue and red wavelengths. Those cells that are able to pump out the dye (it enters the cells passively) exhibit a low intensity of red and blue light emission and are described as being within the side population (SP), while other cells that are less efficient at active dye exclusion fluoresce more intensely and fall within the main population (MP). The SP phenotype can be ablated in the presence of the calcium channel blocker Verapamil (lower panels). Empirically, the SP is often significantly enriched in cells with stem cell properties. (From Asakura A, Seale P, Girgis-Gabardo A et al. (2002) *J Cell Biol* 159:123–134 [doi: 10.1083/jcb.200202092]. Permission conveyed through Copyright Clearance Center, Inc.)

Figure 2.5 **Determination of the frequency of cell division in a population.** Within a population of cells, even if this is defined by flow cytometry as being homogenous, there will be variations in the rate of cell division. Such differences can be an indicator of the relative stem cell potency of a given cell, since stem cells are often relatively quiescent compared to more differentiated progeny. One way to measure the relative rate of division involves loading cells with the substrate carboxyfluorescein succinimidyl ester (CFSE), which fluorescently labels the cytoplasm. At each cell division, the fluorescence emitted from the daughter cells is halved, which is indicated in the diagram by the reduction in green shading from right to left. If, after a period of growth in culture, the population is passed through a flow cytometer, the proportion of cells that have undergone different numbers of divisions can be quantified, shown schematically here as 0 to 4 divisions. Cells that have undergone no divisions in such an experiment might represent stem cells. (From Romar G, Kupper T & Divito S. (2016) *J Invest Dermatol* 136: e1–e7 [doi:10.1016/j.jid.2015.11.020]. Copyright 2016 with permission from Elsevier.)

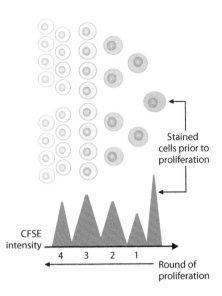

(Box 2.2). An example of an enzyme frequently used as a reporter is β-galactosidase from the bacterium *E. coli*, which can be detected using a substrate that, upon enzymatic cleavage, releases either a blue compound (5-bromo-4-chloro-3-indolyl-β-D-galactopyranoside, abbreviated to BCIG and commonly referred to as X-gal) that is visualized by light microscopy or a fluorescent moiety (fluorescein Di-β-D-galactopyranoside, usually known as FDG) that can be detected by flow cytometry. The firefly-derived enzyme luciferase is also favored as a reporter, its substrate being ATP that in the presence of the co-factor luciferin results in the release of visible light that can be quantified in a luminometer or visualized by microscopy.

The reporter gene can be introduced into a cell either ectopically using a DNA vector (usually plasmid or virus-based, most often now as a lentivirus) or through being incorporated into the genome, either at a random location as a **transgene** or at a specific site through homologous recombination in an **embryonic stem (ES) cell** (Chapter 3). Whatever means is used to introduce a reporter, its coding sequence has to be linked to suitable DNA elements that are able to control its expression, including a gene "promoter" at which transcription can be initiated and regulatory sequences that control the initiation of transcription. The exact nature of the attached promoter and regulatory sequences determines in what cells the reporter is expressed, that is, widely or restricted to specific cells such as stem cells.

## Lineage tracing is an essential tool in stem cell research

The idea of **lineage tracing** considerably predates the advent of molecular biology and reporter genes. Back in the nineteenth century, Charles Otis Whitman used light microscopy to observe that in the development of the leech, cells arise from pre-existing cells rather than by spontaneous generation. Since then, lineage tracing has become a valuable means of studying cell ancestry and fate in developmental and differentiation processes. The basic idea is that a single cell is marked in such a way that its progeny can be traced. The important properties of the labeling tracer are that it: (1) should not change the properties of the labeled cell, its progeny or neighboring cells; (2) must be retained over time; (3) must be passed onto all progeny of the founder cell; and (4) must never transfer to unrelated neighboring cells. Applied to stem cells, lineage tracing enables researchers to identify the progeny derived from a single founder stem cell, as well as their location and differentiation status, including *in vivo* in intact tissues and whole organisms.

During the 1920s, Walter Vogt developed "vital dyes" that revolutionized lineage tracing. These dyes label living cells without any toxicity, making it possible to track them over time. Forty years later, the technology was further developed when radioactively labeled compounds such as tritiated thymidine (which is incorporated into the DNA of dividing cells) became available. Vital dyes are prone to the problem of transfer between cells because of their water solubility, which can be alleviated by conjugation to high-molecular weight dextrans; however, the use of vital dyes comes with the additional problem that the compounds become diluted during cellular division and so cannot be used for long-term tracing.

As discussed previously, DNA-encoded markers can now be introduced into cells, either ectopically through transfection and viral transduction or by genetic engineering to incorporate DNA into the genome. Both approaches provide sophisticated possibilities for the labeling and tracing of cells without the problem of transference between cells. Most importantly, stable integration of a genetic marker in the genome means that it will be inherited by all of the progeny of the founder stem cell, making long-term lineage tracing studies more straightforward. The most commonly used reporters are either lacZ or fluorescent proteins such as GFP, the expression of

**BOX 2.2 BACKGROUND: FLUORESCENT PROTEINS**

The detection of cells, including stem cells, has been revolutionized through the identification of naturally occurring proteins that have inherent fluorescent properties. Cloned versions of these fluorescent proteins can be used either in isolation as reporters of gene expression and cell location or fused to proteins of interest as a way of tracking their location, stability, and so on.

Fluorescent proteins have an inherent capacity to form a visible wavelength "chromophore" through post-translational modifications of a sequence of three amino acids within their own polypeptide sequence. The first example described of such a fluorescent protein was green fluorescent protein (GFP), derived from the jellyfish *Aequorea victoria*. GFP exhibits bright green fluorescence upon excitation by light in the blue wavelengths (excitation peak: 395 nm, emission peak: 509 nm). The utility of GFP realized in the 1990s prompted scientists to hunt for other proteins that fluoresce at different wavelengths. Fluorescent proteins with emissions ranging from cyan to red were identified in marine organisms such as the reef coral Anthozoa (e.g., the red fluorescent protein from *Discosoma*, commonly called DsRed). At least five natural chromophores have now been discovered that are associated with a range of fluorescence hues, although the precise excitation and emission wavelengths for each depend strongly on the microenvironment created by the overall protein structure surrounding the chromophore.

Striving for ideal properties and multiple spectral variants has led researchers to generate multiple engineered mutations of the naturally occurring proteins. Properties manipulated this way include the ability to efficiently fold at 37°C, prevention of protein multimerization, and efficient translation using amino acid codons that are commonly used in vertebrates. Most significantly, the fact that excitation and emission of a particular chromophore are influenced by the three-dimensional structure of the protein has stimulated the engineering of a wide range of spectral variants. The best example of these mutational changes is perhaps the DsRed protein, which has been altered so that it exists as a monomer rather than the natural tetrameric form and can be obtained as a series of variants, known as the mFruits (m for monomeric), with colors that span from yellow to far-red (**Figure 1**).

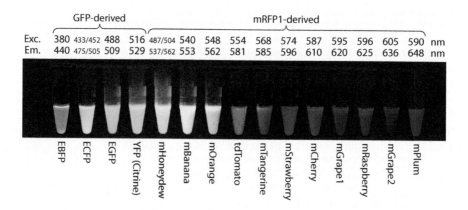

**Box 2.2 Figure 1** **Genetically engineered fluorescent proteins.** A wide range of fluorescent proteins has been generated by genetic modification of naturally occurring proteins. The image shows samples of several of these modified proteins, ranked according to their fluorescent wavelength. The four on the left are mutated versions of green fluorescent protein (GFP) derived from the jellyfish *A. victoria*. The other proteins are variants of the coral *Discosoma* red fluorescent protein (RFP). The excitation and emission maxima for each protein are shown. In the protein names, the prefix "E" indicates enhanced fluorescence and "m" that the protein exists as a monomer. (From Tsien, RY. (2009) *Les Prix Nobel 2008*. Copyright © The Nobel Foundation 2008. Used with permission.)

which is under the control of a tissue- or cell-specific gene promoter. The power of this form of lineage tracing is further increased by using a regulated activation strategy, thereby giving precise temporal control over when the reporter is expressed. A typical combined cell type-specific and temporally controlled reporter system utilizes **Cre recombinase** protein fused to the estrogen receptor (ER) that is itself driven from a cell type specific promoter (**Figure 2.6**). When the ER agonist 4-hydroxy tamoxifen (4-OHT) is introduced into the experimental animal, this leads to activation of the Cre enzyme, which then is able to act upon target loxP recombination sites that are positioned around a transcriptional STOP sequence upstream of a second transgenic reporter (lacZ in the example illustrated). The expression of the second transgenic reporter is constitutive once activated by Cre so that all downstream cells remain lacZ positive, whereas as differentiation proceeds, the cell type-specific transgene expressing Cre may become silent, which itself can be visualized if Cre expression is

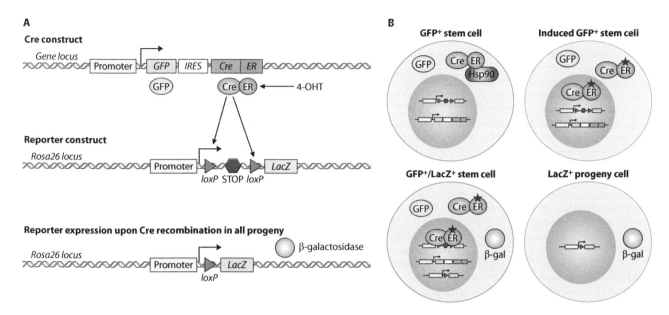

**Figure 2.6 Lineage tracing using Cre-activated reporters.** The expression of reporter genes from transgenes has become a very popular means to label stem cells and their differentiated progeny in several experimental animal species, especially in the mouse. Combining cell specific promoter elements with hormone-dependent Cre recombinase to control genes expressing either fluorescent proteins or enzymes that can yield a detectable product provides a very powerful means to achieve precise tracking of stem cell behavior. The schematic illustrates one example of the way in which such a combination of transgenes can be used. **(A)** Two transgenes are shown in this example. In the transgene at the top, sequences encoding green fluorescent protein (GFP) and Cre recombinase fused to an estrogen receptor (CreER) are driven from a cell specific promoter. An internal ribosome entry site (IRES) allows expression of CreER independent of GFP. A second transgene contains the lacZ gene encoding β-galactosidase (β-Gal) driven from a ubiquitously expressed promoter. Transcription across the LacZ gene is blocked by a STOP sequence, which is flanked by Cre recombination targets (loxP, arrow heads). **(B)** In this hypothetical example, GFP and CreER are both expressed in the stem cell (upper left cell), the activity of the latter being blocked by the binding of heat shock protein (Hsp90). Upon binding of the estrogen agonist 4-hydroxy-tamoxifen (4-OHT; asterisk) to the ER (upper right cell), the CreER becomes active and is able to act on the loxP sites in the second transgene, bringing about expression of LacZ/β-Gal (lower left cell). LacZ is then expressed constitutively in both the stem cell and its downstream progeny, whereas the expression of GFP is turned off once the stem cell differentiates (lower right cell) and the cell specific promoter becomes inactive. (From Kretzschmar K & Watt F. (2012) *Cell* 148:33–45 [doi:10.1016/j.cell.2012.01.002]. Copyright 2012 with permission from Elsevier.)

combined with a distinct reporter (GFP in the example illustrated). This technique has been used to track the fate of many defined stem cell populations in adult tissue, such as the Lgr5+ stem cells of the intestinal crypt (Box 2.3).

An alternative and increasingly used temporally activated strategy for lineage tracing downstream of adult stem cells allows observation of the way in which a single cell gives rise to a clonal expansion of progenitors and their more differentiated progeny. This technique, which makes use of the so-called "Confetti" mouse line (Figure 2.7), also relies on Cre-mediated removal of a STOP sequence in a second ubiquitously expressed transgene. Cre-ER expressed from one transgene, either constitutively or from a cell type-specific promoter, is transiently activated by 4-OHT and can then act upon loxP recombination sites in a second transgene. In the Confetti mouse, the second transgene contains four distinct fluorescent protein encoding genes arranged in two pairs. The genes in each pair are in opposite transcriptional orientations and are flanked by loxP sites. On a random basis in some cells, the transient Cre activity leads to loss of the STOP sequence and one of the pairs of genes, while the retained gene pair can adopt one of two orientations by being "flipped" as a result of the Cre activity. The outcome is the expression of one of the four fluorescent proteins, which is then retained in all progeny, enabling assessment of the way in which a single stem cell can give rise to the component cells of the tissue, a feature that is again nicely illustrated by the behavior of stem cells in the small intestine (Figure 2.7B).

## 2.2  ISOLATION OF STEM CELLS

In order to test the functional properties of a stem cell, it generally has first to be isolated from a highly heterogeneous population derived from the tissue of interest. Various techniques can be used, but ultimately a high purity of stem cells must be achieved (95%–100% enriched). Such a high degree of purity is essential for determining the

## BOX 2.3 CASE STUDY: LINEAGE TRACING OF INTESTINAL STEM CELLS

The intestinal epithelium is ordered into crypts and villi, with a turnover rate in the mouse of every 3–5 days. The stem cell population responsible for intestinal replenishment has been difficult to functionally identify due to the lack of unique markers and absence of a definitive stem cell assay. The development of lineage tracing techniques has enabled the identification of **intestinal stem cells** through the generation of knock in transgenic mice expressing the green fluorescent protein (GFP) and LacZ reporter under the control of the *Lgr5* gene promoter (Figure 1). *Lgr5* is a reported target gene of Wnt signaling, which is known to play an important regulatory role within the intestinal crypt and therefore is a key candidate marker of intestinal stem cells (Chapter 11). Confocal imaging of intestinal sections for GFP expression, and therefore Lgr5+ cells, highlights that *Lgr5* expression is restricted to the 14 slender cells sandwiched between the **Paneth** cells at the crypt base of the small intestine (Figure 2). Since GFP is expressed only in Lgr5+ cells, it cannot be used for lineage tracing since the progeny of cells gradually lose Lgr5 expression. The presence of the tamoxifen inducible CreERT2 cassette and Rosa26-LacZ reporter in the cells means that in the presence of tamoxifen, Cre-recombinase is activated and excises the genetic roadblock in front of the Rosa26-LacZ reporter, which irreversibly marks (in the presence of X-gal) all of the progeny derived from the population of Lgr5+ stem cells, enabling lineage tracing over several months (Figure 3). This elaborate labeling technique highlighted that not only was *Lgr5* exclusively expressed in cycling columnar cells at the crypt base but that these Lgr5+ crypt base columnar cells generated all of the epithelial lineages of the intestine over a period of at least 60 days.

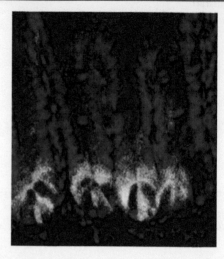

**Box 2.3 Figure 2 Lineage tracing of intestinal stem cells.** Confocal fluorescent image of the small intestine from a mouse containing the Lgr5-EGFP-Cre transgene and Rosa26-lacZ reporter transgene. The image shows intestinal villi expressing GFP in Lgr5-positive cells, which is restricted to the six to eight slender cells (green) sandwiched between the Paneth cells at the crypt base. All cells are counterstained red with a DNA-binding dye. (From Barker N, van Es J, Kuipers J et al. (2007) *Nature* 449:1003–1007 [doi: 10.1038/nature06196]. Reprinted by permission from Springer Nature.)

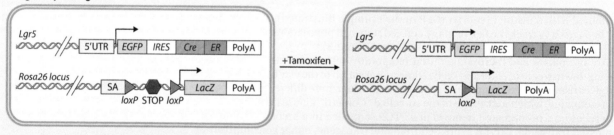

**Box 2.3 Figure 1 Lineage tracing of intestinal stem cells.** Intestinal stem cells can be identified and their descendants tracked through selective activation of the expression of reporter genes. In this example, a mouse has been generated containing a knock-in transgene that is regulated by the gene promoter from the intestinal stem cell-specific Lgr5 gene and encodes both enhanced green fluorescent protein (EGFP) and a Cre recombinase that can be controlled through a linked hormone binding domain (CreERT2). The transgene contains sequences that ensure efficient transcription and translation, including a 5′ untranslated sequence (5′UTR), an internal ribosome entry site (IRES), and a polyadenylation signal (PolyA). The mouse contains a second knock-in transgene at the ubiquitously expressed *Rosa26* locus. This transgene contains the lacZ reporter gene, the expression of which is prevented by transcriptional stop sequences (STOP). The STOP sequence is flanked by the recognition elements for Cre recombinase (loxP, brown arrows). The Lgr5-EGFP-CreER transgene is expressed in intestinal stem cells so that cells can be visualized from their green fluorescence. In the presence of Tamoxifen, the Cre recombinase becomes active and excises the STOP sequence from the lacZ transgene, thereby allowing transcription of the lacZ and expression of β-galactosidase, which can be detected by its action on a substrate such as X-gal. The expression of the lacZ gene is permanently activated so that all cells derived from the stem cell can be labeled. (From Barker N, van Es J, Kuipers J et al. (2007) *Nature* 449:1003–1007 [doi: 10.1038/nature06196]. Reprinted by permission from Springer Nature.)

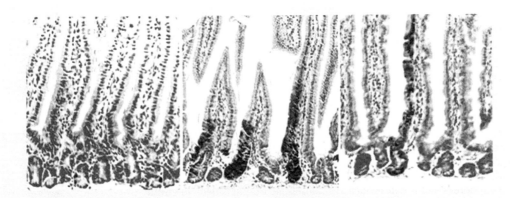

Box 2.3 Figure 3 **Lineage tracing of intestinal stem cells.** Images of the small intestine from a mouse containing the Lgr5-EGFP-Cre transgene and Rosa26-lacZ reporter transgene. The image shows a histological analysis of β-galactosidase activity (blue) after tamoxifen induction for 1 day (left), 5 days (middle), and 60 days (right). (From Barker N, van Es J, Kuipers J et al. (2007) *Nature* 449:1003–1007 [doi: 10.1038/ nature06196]. Reprinted by permission from Springer Nature.)

function and molecular properties of the stem cell since contaminating cells of other types can severely mask the outcome of transplantation assays or gene expression analysis. The issue of heterogeneity has been tackled more recently through the isolation and analysis of individual cells. Here we will describe the principal methods used to isolate stem cells by cell sorting, making use of either specific marker expression or distinct cell functional properties of stem cells such as quiescence and metabolism.

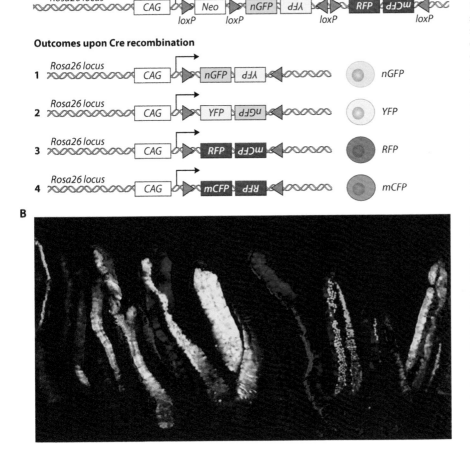

Figure 2.7 **Lineage tracing of intestinal stem cells using Confetti Mouse.** The so-called Confetti Mouse provides a way randomly to label individual stem cells and their downstream progeny, which can be detected as a clone of expanding progenitors and differentiated cells. **(A)** The diagram illustrates the structure of the reporter encoding four fluorescent proteins (FPs) driven by a strong ubiquitous promoter (CAG engineered in the Rosa26 locus). Upon Cre action at loxP sites (arrow heads) flanking a stop sequence (Neo) and the FP genes, the construct recombines randomly to remove the Neo sequence and simultaneously remove or flip the FP genes. This recombination results in four possible outcomes with different FPs being expressed (GFP—green; YFP—yellow; RFP—red; or CFP—blue). **(B)** The fluorescent image shows a section of small intestine villi and crypts from a Confetti Mouse in which a CreER recombinase has been expressed in the stem cells from a Lgr5 promoter and activated by administration of 4-OHT. The recombination process is independent in each cell and also transient and random. After recombination, each stem cell continuously produces one FP, which its eventual daughter cells will continue to produce, creating clonal patches with the same color. (A: From Kretzschmar K & Watt F. (2012) *Cell* 148:33–45 [doi: 10.1016/j.cell.2012.01.002]. Copyright 2012 with permission from Elsevier. B: From http://jackson.jax.org/rs/444-BUH-304/ images/013731_images.pdf. Copyright © The Jackson Laboratory. Used with permission.)

## Fluorescence-activated cell sorting enables the separation of rare stem cells from a heterogeneous population

Beyond the analytical application of immunofluorescent marker detection, the flow cytometer has been adapted to enable separation of cells in the technique known as **fluorescence-activated cell sorting (FACS)**. A FACS machine discriminates cells in a population based on their specific pattern of binding to a combination of fluorescently labeled antibodies. As in an analytical flow cytometer, cells are constrained to the center of a very narrow, rapidly flowing stream of liquid that is adjusted so that single cells are well separated (Figure 2.8). The difference from an analytical machine is that in FACS, the stream is broken by vibration into droplets, each of which contains no more than one cell (many contain no cells). Just before the stream is about to break into droplets, it passes through a beam of laser light that excites the fluorescent moieties attached to the antibodies on the cell surface. If a fluorescent output is detected from that cell, indicating that this cell should be collected, then an electrical charge is imparted onto the droplet containing the cell as the stream breaks up. The charged droplet is then deflected away from the main fluid stream by electrostatically charged plates and can be collected in a tube. The extremely fast rate of flow of cells through the machine (the most modern machines can exceed 60,000 events per second) means that it is feasible to collect sufficient cells for further investigation in a reasonable space of time, even if these cells are very rare (the desired cells can represent less than 0.1% of the total population and be isolated to a level of purity in excess of 95%).

The use of FACS as the preferred sorting method for certain populations is dictated by the requirement for >95% pure stem cell populations, which other methods cannot achieve. As described earlier, many stem cell populations are isolated based on cell surface staining with multiple antibodies or by intracellular staining, which some of the simpler methods cannot accommodate. However, in order to enhance the purity of the desired stem cell fraction, FACS can be used in combination with methods such as magnetic bead separation (see subsequently) in essence to deplete ("debulk") the cell preparation of cells that have the potential to contaminate the sort and hence reduce purity.

## Single-cell index sorting facilitates retrospective correlation between a surface phenotype and function

Another advantage of FACS is the possibility of depositing single cells into small wells in a plastic dish, enabling the analysis of individual stem cells, an ever

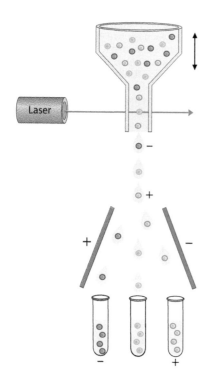

Hydrodynamic focusing in a nozzle vibrated by a transducer produces a stream breaking into droplets

Laser interrogation and signal processing followed by sort decision; blue sort right, red sort left, green no sort

Electronic delay until cell reaches break-off point. Then the stream is charged

Charged droplets deflect by electrostatic field from plates held at high voltage (+/– 3000 volts)

Various collection devices can be attached: tubes, slides, multiwell plates

**Figure 2.8 Cell sorting.** The technique of flow cytometry has been adapted to enable separation of cells on the basis of any of the parameters that can be measured, including combinations of surface antigen expression as well as cell size and shape. Having passed through the laser beam, individual cells in a heterogeneous population can be discriminated on the basis of the fluorescently labeled antibodies attached to their surface and possibly also their size and shape. The laser interrogation and subsequent analysis of the light output signals allows a sort decision to be made for the desired cell type. After passing through the laser beam, the cell stream reaches a vibrating nozzle that causes it to break into droplets that contain single cells. A positive or negative charge is applied to the droplet containing the desired cell, and when the droplets then pass through an electrostatic field, they are deflected according to their charge into collection tubes.

expanding and necessary technique given the complexity and heterogeneity of stem cell populations. However, despite multiparameter cell sorting and the resolution of very rare subpopulations of cells, there is often a great degree of heterogeneity that is not simply reflected in the expression of cell surface markers. Recent developments in FACS technology enable scientists to track back the precise surface phenotype of each sorted cell through a process known as index sorting. This technique enables a specific cell attribute, such as stem cell **potency** or gene expression profile determined by single-cell RNA sequencing, to be associated to a precise pattern of surface antigen expression, including variations in the absolute levels of individual markers.

## Magnetic cell separation is an alternative means to isolate antibody-labeled stem cells

Although FACS is a popular method for stem cell isolation, it can be both costly and time consuming. Another commonly employed strategy for the isolation of stem cells both in the research laboratory and within the clinical setting is the use of magnetic cell separation. Magnetic cell separation has the advantage of isolating the desired cell population in a short period of time and can be used on both small and large scales, for abundant or rare cells, although the degree of purity does not reach the levels achievable by FACS. As mentioned already, magnetic sorting is often used as a prelude to FACS in order to debulk a population. One of two variant methods can be used, that is, Magnetic-Activated Cell Sorting (MACS) or Dynabeads. Each method utilizes superparamagnetic particles, which only become magnetized in the presence of a magnetic field (Figure 2.9). Dynabeads are approximately 1–5 μm in diameter, whereas MACS beads are much smaller, at only 50 nm. Both methods can be used for either negative selection (unwanted cells are magnetically bound) or positive selection (cells of interest are magnetically bound).

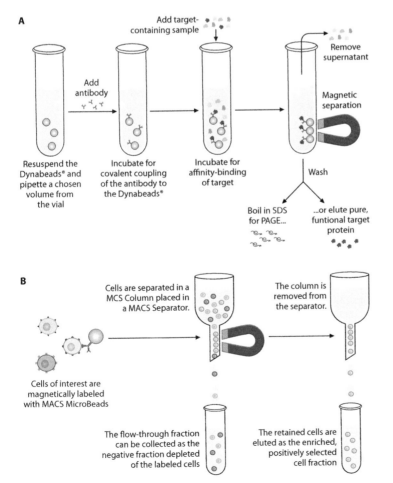

**Figure 2.9 Magnetic cell separation.** Two methods are commonly used to isolate cells using specific antibodies attached to magnetic beads. The main difference between these methods is the size of the beads. (**A**) In the isolation of cells using the Dynabeads magnetic cell sorting system, the magnetic beads are incubated with an antibody against the desired cell antigen. Once covalently bound to the antibody, the beads are then incubated with the mixed population of cells before a magnetic field is applied to select the bead-bound cells, leaving any unbound cells to be removed in the supernatant. (**B**) For separation using the MACS microbead system, cells are labeled with an antibody bound to MACS microbeads. Cells are separated as they pass through a magnetic field, with the flow-through containing the depleted cell population. In the final step, cells are removed from the magnetic field and are eluted into a tube as a positively selected fraction.

An important consideration is whether the presence of the magnetic particles interferes with the function of the cells being isolated; unlike the beads used in the MACS system, the much larger size of Dynabeads requires that they be removed before functional studies can be conducted. The ability of the MACS nanoparticles to be internalized has made them the preferred option for the isolation of cells to be studied with respect to functions such as cell–cell interaction, although the development of a polyclonal anti-Fab antibody that enables release of the Dynabead antibody from its antigen has overcome this particular issue. A disadvantage of MACS is that it requires a more complicated separation apparatus consisting of columns placed in a magnetic field.

Immunomagnetic sorting has advantages over FACS in terms of simplicity and cost; however, its use is far more limited than FACS, as cells can only be separated on the basis of the expression of one antigen at a time and cannot be discriminated on the basis of the level of antigen expression. Other disadvantages of magnetic sorting compared to FACS are the degree of purity achievable and the fact that it is limited in its use of cell surface molecules.

## 2.3   CHARACTERIZATION OF STEM CELLS

The characteristics of cells isolated by the methods described in the previous section, and in particular assessment of features that define their stemness, can be explored through a myriad of approaches, some of which are applicable in a wide range of contexts and some of which will be quite specific to the stem cell type of interest. In this section, we will give examples of the sorts of methods that are commonly employed, dividing them into three broad categories, that is, *in vivo* assays, *in vitro* cell culture assays, and *in vitro* molecular analyses.

Many of the techniques used to characterize specific **tissue stem cells** often need to be applied iteratively, especially when existing knowledge of their characteristics is limited or when trying to refine an already existing isolation strategy. Even having defined a population of cells in a tissue that have stem cell properties, further subdivision often allows definition of a minor subpopulation that is actually the most biologically significant and reveals any features of hierarchical organization. Analysis of a population of stem cells, however pure it may appear, inevitably leads to averaged measurements of phenotype and functional characteristics so that the drive toward greater refinement has led to a trend of ever-increasing adaptation of techniques to enable them to be applied to single cells, although this remains challenging in many circumstances.

### The true definition of a stem cell requires *in vivo* assays

The "gold standard" test of **stem cell potential** generally requires an *in vivo* assay that is capable of assessing properties such as the ability to locate into and be maintained by the appropriate stem cell niche as well as the long-term potential to provide a self-renewing population that differentiates appropriately into the cell types of the relevant tissue.

Such *in vivo* assays require an animal model into which the test cells can be transplanted with the expectation that they reach the tissue of interest and lodge in a suitable niche where they can proliferate and differentiate. Necessary prerequisites are that the transplanted ("donor") cells can be distinguished from those of the recipient ("host"); that there is available space in the appropriate niche, which might require some ablation of resident host stem cells prior to transplantation; and that the donor and host are immunologically compatible. These types of assays are most frequently performed in mice, but in some circumstances, other species such as sheep have been employed. Immune rejection in mice can be avoided because of the use of inbred strains, while in more complex circumstances such as sheep, the solution has been found in performing transplantation into the fetal stage before the time in development when tolerance is established.

For most purposes, not surprisingly, it is not feasible to test human stem cells in people, but *in vivo* testing in non-human hosts is still possible if the inevitable rejection by the immune system can be avoided. Most commonly, this is achieved through the use of hosts, almost always mice, which are immunologically deficient. Human into mouse **xenografts**, as they are called, have been facilitated by the development of several immunocompromised strains, in particular relying upon the severe combined immunodeficiency (SCID) mutant. SCID mice are deficient in both

B- and T-cell mediated immunity, and their usefulness has been enhanced in various ways through combination with other spontaneous or engineered mutations. The strain that has been most commonly used for xenografts is a combination of the non-obese diabetic (NOD) mutation with SCID, usually referred to as NOD/SCID, which lacks not only functional B and T lymphocytes but also has low levels of natural killer cell activity.

The precise way in which a transplantation assay of stem cells is performed depends on the type of cell being characterized and whether it is a same species assay or a xenograft. The examples of HSC and ES cells illustrate some of the basic principles and differences. The assay of HSC (Box 2.4) involves irradiation of the host mouse to completely ablate its resident bone marrow stem cells and hematopoietic system. Test cells are then introduced either via the bloodstream or directly into the bone marrow cavity. Using such an assay, stem cell potential can be graded depending on how long the introduced cells are able to provide a functioning hematopoietic system, life-long reconstitution being the reflection of ultimate potency. Often, these assays also involve the co-injection of a reference wild type population, especially when the aim is to determine how some genetic manipulation may have influenced stem cell behavior. The most rigorous test involves assessment of the ability of cells to be serially transplanted from the primary reconstituted recipient to a secondary irradiated host, thereby demonstrating that engrafting cells are undergoing persistent self-renewal. Essentially the same approaches can be taken when a xenograft is being generated, except in this case, it is important that not all of the host's bone marrow be ablated by the irradiation so that it can maintain mouse-specific hematopoiesis while at the same time creating space in the niche for the integration of the foreign stem cells in the bone marrow niche.

The strategies developed for the assay of HSC have been adapted for many other adult stem cell types, derived from both mouse and human sources, including those from tissues such as brain, liver, and breast, examples of which will be described in the relevant chapters later in this book.

## BOX 2.4  CASE STUDY: TRANSPLANTATION ASSAY OF HEMATOPOIETIC STEM CELL ACTIVITY

Hematopoietic stem cells are assayed by bone marrow transplantation assay (Figure 1). This assay involves the partial or complete ablation of the hematopoietic system of the recipient mouse by ionizing radiation. Following irradiation, test stem cells from a congenic strain of mouse are injected into the recipient and engraftment allowed to become established over the following weeks. Traditionally,

two different congenic strains of mouse are used in these assays that differ in the allelic expression of the pan hematopoietic marker CD45. The two types of CD45 alleles, namely CD45.1 and CD45.2, are functionally identical and thus serve as useful markers to track the engraftment potential of the donor cells by antibody detection and flow cytometric analysis.

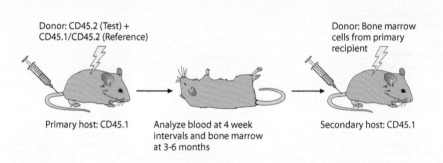

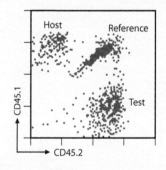

Box 2.4 Figure 1 **Transplantation assay of hematopoietic stem cell activity.** The "gold standard" test of any presumed stem cell is its ability to regenerate a given adult tissue. In the case of hematopoietic stem cells (HSC), this can be demonstrated by transplantation of cells into lethally irradiated mice. Primary host mice expressing the hematopoietic antigen CD45.1 are lethally irradiated (lightning bolt) to eradicate host bone marrow before being injected with test hematopoietic cells expressing the CD45.2 antigen together with reference bone marrow from mice that are heterozygous for CD45.1 and CD45.2. Engraftment is assayed by blood sampling over several months and sacrificing the primary hosts to analyze their bone marrow at a defined time. Bone marrow cells can be transplanted into secondary hosts to assay the long-term stem cell potential. The right-hand panel shows a flow cytometry dot plot from a typical blood analysis of a recipient mouse. Using antibodies against CD45.1 and CD45.2 conjugated to fluorochromes, it is possible to discriminate the three populations of bone marrow cells.

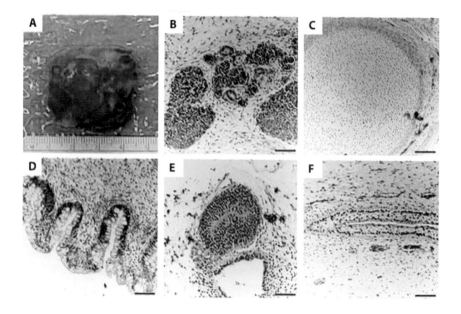

Figure 2.10 **Teratoma formation as an *in vivo* assay of human embryonic stem cell differentiation capacity.** Proliferating cultures of human embryonic stem cells can be assayed for their pluripotency by teratoma formation following transplantation into immunocompromised mice. (**A**) Photograph of an explanted teratoma. Images in panels (**B–F**) are sections of a teratoma stained with hematoxylin and eosin to identify embryonic tissues that represent all three embryonic germ layers. Mesoderm is depicted in (**B**) and (**C**) as nascent renal tubules and glomeruli within a bed of primitive renal epithelium and as cartilage surrounded by condensing mesenchyme, respectively. Endoderm is shown in (**D**) as glandular intestinal tissue. Ectoderm is shown in (**E**) and (**F**) as nascent neural tube and primitive squamous epithelium, respectively. Bars = 100 μm. (From Yabut O & Bernstein H. (2011) *Aging* 3:494–508 [doi: 10.18632/aging.100328]. Published under the terms of the Creative Commons Attribution License (http://creativecommons.org/licenses/).)

The unique properties of ES cells are most dramatically demonstrated for those derived from mouse embryos because of their ability to become incorporated into the inner cell mass of an embryo and then go on to contribute to the tissues of the resulting adult **chimera**. Although human ES cells can clearly not be assayed in this way, it is nevertheless important to have some indicator of their *in vivo* **pluripotency**. Fortunately, this can be achieved to some extent by injection of ES cells into an immunocompromised mouse (usually into a muscle). In these circumstances, the ES cells expand into a tumor-like mass (a **teratoma**) that spontaneously differentiates into a broad range of cell types in a way that essentially reflects embryonic development (**Figure 2.10**).

Xenotransplantation models have also been crucial in **cancer stem cell** (CSC) research (see Chapter 18), initially in the study of acute myeloid leukemia, but extending now to many examples of tumors for which it is possible to demonstrate the presence of an underpinning CSC (**Figure 2.11**). Although xenotransplantation is the gold standard method of determining stem cell activity in human cells, its use in cancer stem cell biology and treatment is met with some problems:

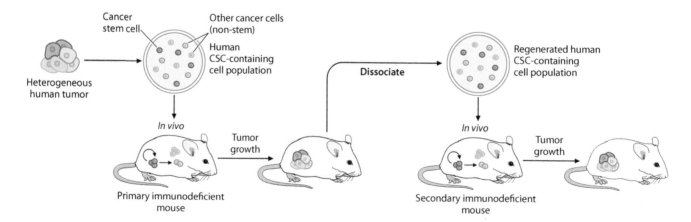

Figure 2.11 **Assay of cancer stem cells.** Stem cells that underpin the growth of tumors can be assayed *in vivo* in much the same way as adult stem cells using a transplantation strategy. The diagram illustrates how transplantation can be used to determine cancer stem cell (CSC) activity within a heterogeneous population of human tumor cells. The tumor cells are dissociated into a single cell suspension and transplanted into a primary immunodeficient mouse. Following the formation of tumors in the recipient mouse, these can be further dissociated and transplanted into a secondary immunodeficient mouse to assay for further CSC activity. (From Nguyen L, Vanner R, Dirks P et al. (2012) *Nat Rev Cancer* 12:133–143 [doi: 10.38/nrc3184]. Reprinted by permission from Springer Nature.)

- Primary tumor cells depend on growth factors that may not be cross-species active
- Immunodeficiency facilitates efficient engraftment without cell rejection, but the resulting system lacks many elements thought to influence tumor growth
- The mutated gene responsible for the SCID phenotype (*Prkdc^scid*) is involved in DNA damage repair so that the mice are hypersensitive to DNA damaging treatments, ameliorating the usefulness of this model in drug testing
- The limited lifespan of the mouse may not allow detection of slow growing human tumors

## Stem cells can be visualized *in vivo* with advanced imaging systems

There is an ever-growing requirement to be able to observe the behavior of stem cells within living tissue and organisms since the information gained from other techniques provides only a static insight and therefore does not reflect the natural dynamics of biological processes. Techniques such as multiphoton confocal microscopy offer some possibility of observing living stem cells in their natural environment, but in order to gain a full understanding of processes such as stem cell mobilization, homing, and engraftment, it has proved necessary to adopt and develop novel technologies of microscopy to incorporate the ability to image stem cells in living organisms. A range of imaging methods are now available, each having particular benefits and shortcomings, with the result that investigators often take a "multimodal" approach, integrating the outputs from several technologies. As illustrated in Figure 2.12, imaging modalities can be compared based on their sensitivity and resolution, both of which need to be optimal if small numbers of stem cells are to be tracked and observed in specialized locations including their niche.

Intravital microscopy (IVM) has proven particularly useful in studies of stem cell recruitment to injured organs such as the liver and enables imaging of the fluorescent cells over periods of time on the order of hours. IVM is essentially a modification of confocal microscopy, with adaptations to the microscope setup that allow tissues in anesthetized animals to be observed from their surface (e.g., in the bone marrow – Box 2.5). IVM is based on the detection of fluorescence from individual cells and is primarily used to study location, motility, adhesion, and interactions of individual cells in three dimensions over time. IVM is limited in its application to a time window of a few hours and is also restricted by the types of vascular structures within tissues

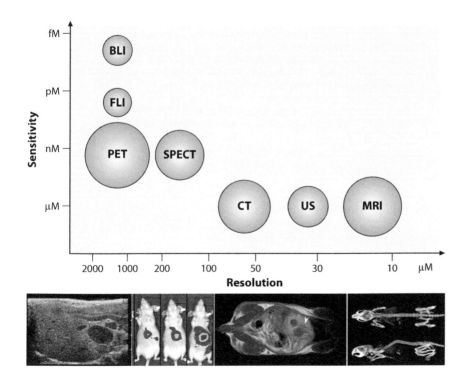

**Figure 2.12 Relative sensitivity and resolution provided by different *in vivo* imaging modalities.** Different imaging technologies have distinct capabilities with respect to the resolution and sensitivity that they can achieve. The choice of a particular imaging method will depend on what is the most desirable balance of resolution versus sensitivity. The two-dimensional representation of *in vivo* imaging methods in the top panel compares relative sensitivity and resolution by position in the diagram as well as the cost as indicated by the size of the relevant circle. BLI—bioluminescence; FLI—fluorescence; PET—positron emission tomography; SPECT—single photon emission computed tomography; CT—computed tomography; US—ultrasound; MRI—magnetic resonance imaging. The bottom panels are examples of images obtained using some of the methods. From left to right these are US, optical (BLI or FLI), MRI, and SPECT-CT. (From Wu T, Heuillard E, Lindner V et al. (2016) *Sci Rep* 6:35230 [doi:10.1038/srep35230]. Published under the terms of the Creative Commons Attribution License (https://creativecommons.org/licenses/by/4.0). And from van der Have F, Ivashchenko O, Goorden M et al. (2016) *Nucl Med Biol* 43:506–511 [doi:10.1016/j.nucmedbio.2016.05.015]. Copyright 2016 with permission from Elsevier.)

**BOX 2.5  CASE STUDY: INTRAVITAL MICROSCOPY INVESTIGATION OF STEM CELL INTERACTIONS WITH BONE MARROW IN THE SKULL**

It is difficult to visualize stem cell interactions within the bone marrow environment of the long bones, the conventional site for investigation of HSC, whereas the skull has become a preferred site for studying the dynamics of their homing and mobilization characteristics because of the relative transparency of the frontoparietal bone, which is less than 100 microns thick. By exposing the surface of the skull, it has proven possible directly to observe dynamic cell interactions by fluorescence confocal microscopy in live rodents, for example, to assess how hematopoietic stem cells introduced into the circulation are taken up through the endothelium of the blood vessels that pass through the bone marrow (Figure 1).

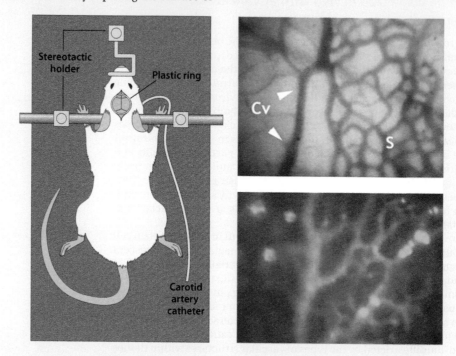

Box 2.5 Figure 1 **IVM investigation of stem cell interactions with bone marrow in the skull.** Due to the relatively thin nature of the bone of the skull, it is possible to make some optically based observations of the behavior of hematopoietic stem cells (HSC) in this bone marrow environment. The cartoon on the left shows how a small rodent can be immobilized to make possible confocal microscopic observation through the surface of the skull once the covering tissue has been removed. The top panel on the left shows a phase-contrast image of the vasculature permeating the bone marrow in the skull of a mouse. The fluorescent image in the panel below depicts an experiment in which CFSE-labeled HSC have been introduced into the carotid artery of the mouse, and these can be seen as the bright green dots at several points along individual vessels (Images: Used with permission from Dr. Neena Kalia, University of Birmingham. Cartoon: From Hidalgo A, Weiss L & Frenette P. (2002) *J Clin Invest* 110:559–569 [doi: 10.1172/JCI200214047]. Permission conveyed through Copyright Clearance Center, Inc.).

that can be observed because of issues of depth within the structure and the optical transparency of the tissue.

Non-invasive tracking is an evolving technology being driven by the need to understand the long-term fate and behavior of stem cells *in vivo*. Such modalities are required to provide data on stem cell homing, survival, proliferation, and differentiation and should have the ability to detect very low numbers of cells. Labeling strategies and imaging technologies are closely linked, each having advantages and disadvantages in terms of either sensitivity or resolution. These include bioluminescence and fluorescence, detected using sensitive cameras in the *in vivo* imaging system (IVIS), radionuclide labeling as applied in positron emission tomography (PET), single-photon emission computed tomography (SPECT), and magnetic resonance imaging (MRI).

In IVIS, a cooled couple-charged camera device is used to collect photons emitted from cells within the body of an experimental animal. The source of the photons is ultimately a genetic reporter that produces either a bioluminescent signal from the action of luciferase in the presence of ATP substrate and cofactor or fluorescence emitted from a fluorescent protein following excitation with a suitable wavelength. By labeling stem cells with fluorescent membrane dyes, IVIS can be used to track the migration and engraftment of stem cells following transplantation (Figure 2.13). Fluorescent imaging in whole animals has many limitations compared to bioluminescent imaging, such as higher autofluorescence and limited depth penetration, particularly in dark-skinned mice, where the signal gets quenched by the skin pigmentation. However, in recent years, the development of more advanced fluorophores that emit in the near infrared wavelengths, and therefore have better tissue penetration, is making this type of analysis possible.

Radionuclide imaging includes PET and SPECT, in which cells are directly labeled with a radioisotope prior to transplantation and subsequent imaging. These methods have the advantage of being highly sensitive but are not suitable for longitudinal studies due to the short half-life of the radionuclides and the adverse effects on stem cell viability and genetics.

One of the most readily accessible techniques for *in vivo* imaging is MRI, which works by the detection of the nuclear spin of molecules within the tissue. This is a powerful technique, as it provides high resolution, but suffers from being of relatively low sensitivity.

## *In vitro* culture of stem cells can reveal their differentiation capacity

The situation with respect to culture conditions is somewhat less difficult when the objective is to look at the differentiation potential of a stem cell. This is again best illustrated with respect to HSC, for which very sophisticated differentiation assays have been developed. The culture conditions rely on the presence of growth factors and nutrients that will permit complete differentiation along one or more of the pathways of differentiation that the stem cell is expected to be capable of. In the case of HSC, this includes cytokines that act on one or more of the eight possible lineages that give rise to mature blood cells. The basic principle of such assays is to determine what the stem cells are capable of giving rise to, as can be recognized after a number of days by the specific features of the differentiated cells (e.g., surface marker expression, the presence of characteristic cytoplasmic enzyme activities, etc.). In the right conditions, a HSC can give rise to multiple cell lineages, whereas a more mature hematopoietic progenitor cell will have a more restricted capability. Since it would not be possible to discriminate from where the individual differentiated cells originated if such assays were performed in a liquid culture of the whole sorted population, these assays are generally carried out in one of two ways so that the potential of individual cells can be observed. Most commonly, the purified stem cell population is "seeded" into the media with the appropriate growth conditions but which also contains a substance that is like a soft gel (usually methylcellulose). This prevents the cells from moving around extensively, and if seeded at the correct density means that the differentiated derivatives from each cell are clearly separated as colonies that can be scored based on their morphology as **colony forming units (CFUs)** or blast forming units (BFUs) and can eventually be collected for phenotypic analysis (Figure 2.14). Alternatively, using FACS, the sorted stem cells can be deposited as single cells into tiny individual wells in a plastic dish where they then can be allowed to grow and differentiate in liquid conditions.

Equivalent strategies exist for most other adult stem cell types. In principle, ES cells, like adult stem cells, can be assessed for many aspects of their differentiation potential *in vitro*. However, as will be explained in more detail in Chapter 3, the fact that they are pluripotent and have to recapitulate several stages of embryonic development before reaching any one of the possible 200 or so differentiated cell lineages requires special conditions that mimic the environment of the early embryo.

## *In vitro* assays of stem cell function are inherently flawed due to the inability to replicate stem cell niche conditions

A crucial aspect of any investigation of stem cells that requires *in vitro* culture after isolation, whether to study the cell cycle, self-renewal, or differentiation, is to mimic as closely as possible their normal environment and the input of extrinsic signals that

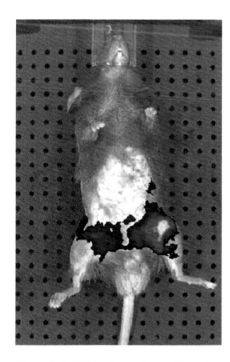

**Figure 2.13 IVM tracing cell engraftment.** Following the isolation of a stem cell population (HSC in this instance), cells are labeled with a near-infrared fluorescent dye and injected directly into one femur of lethally irradiated recipient mice. Stem cell migration and engraftment can be tracked in the short term by anesthetizing the mice and performing *in vivo* fluorescent imaging. The signal shown here comes from direct HSC injection into the left femur of the mouse, with some migration to the opposite leg after 24 hours.

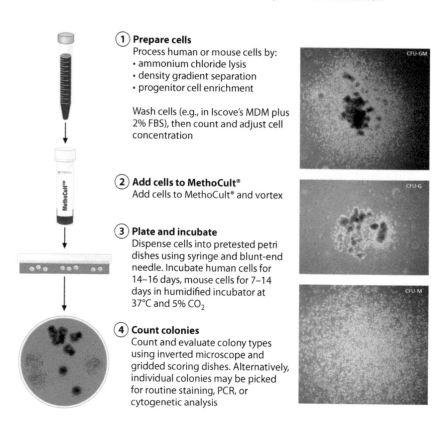

**(1) Prepare cells**

Process human or mouse cells by:
- ammonium chloride lysis
- density gradient separation
- progenitor cell enrichment

Wash cells (e.g., in Iscove's MDM plus 2% FBS), then count and adjust cell concentration

**(2) Add cells to MethoCult®**

Add cells to MethoCult® and vortex

**(3) Plate and incubate**

Dispense cells into pretested petri dishes using syringe and blunt-end needle. Incubate human cells for 14–16 days, mouse cells for 7–14 days in humidified incubator at 37°C and 5% $CO_2$

**(4) Count colonies**

Count and evaluate colony types using inverted microscope and gridded scoring dishes. Alternatively, individual colonies may be picked for routine staining, PCR, or cytogenetic analysis

**Figure 2.14** *In vitro* **colony assay in semi-solid medium.** On the left is a schematic to show the procedure of colony assays of human or mouse hematopoietic cells: (1) Cells are prepared prior to the assay and then washed and counted. (2) Cells are added to semisolid Methocult medium and mixed to distribute the cells evenly; this medium contains all of the nutrients and growth factors required for growth and differentiation of hematopoietic progenitors. (3) Medium containing the cells is plated into Petri dishes and incubated for 14–16 days or 7–14 days for human and mouse cells, respectively. (4) Colonies are counted and scored based on their morphology to determine colony formation and differentiation potential. On the right are photos of three typical colonies obtained in this assay. CFU-GM – colony-forming unit granulocyte/macrophage; BFU-E – blast-forming unit erythroid; and CFU pre-B – colony forming unit pre-B cell. (From *Mouse Colony-Forming Unit (CFU) Assays Using MethoCult™*. Version 3.5.0. Technical Manual. STEMCELL Technologies Inc. Cat No. 28405. Copyright © 2017 STEMCELL Technologies Inc.)

they usually encounter. The concept that all stem cells require a specific niche to ensure their correct regulation has significant implications for their culture and investigation in the laboratory. How can isolated, purified stem cells be maintained in an artificial environment if this does not completely recapitulate the appropriate niche? Culture media can be defined that will maintain the properties of some but not all stem cells. ES cells are relatively easily maintained, partly perhaps because they are not the exact counter part of a normal stem cell and their generation involves adaptation to culture. Some adult stem cells, most notably **neural stem cells**, can also be kept in culture for long periods, while for many others, it may be possible to expand a cell that retains some stem cell characteristics yet loses a crucial property, most usually self-renewal potential. So, even though HSC have been studied for several decades, conditions remain to be identified that prevent their rapid loss of potential for self-renewal.

## The formation of organoids in culture can mimic normal tissue development from stem cells

A promising and rapidly expanding *in vitro* approach to the study of stem cell potential and the downstream differentiation to form the cellular components of a specific tissue involves so-called "**organoid**" culture. Organoids are three-dimensional structures that arise in culture from a single cell, which both mimic the formation of a tissue from stem cells and also recreate the microenvironment that ensures persistence of the stem cells, thereby representing a much improved situation compared to more conventional cell culture conditions (**Figure 2.15**). Organoid cultures are now often providing the key to deciphering the interactions of stem cells within their niche and mechanisms of disease formation and progression. Although organoid cultures are likely to be more representative of *in vivo* conditions, they do lack essential components that are required within a stem cell niche, such as the vasculature that we will describe later (Chapter 6).

## Molecular analysis can reveal both common and unique features that dictate both the fundamental characteristics of stem cells as well as type-specific properties

A major objective in research on stem cells is to define molecular mechanisms that underlie their specific characteristics and their responses to their environment, and

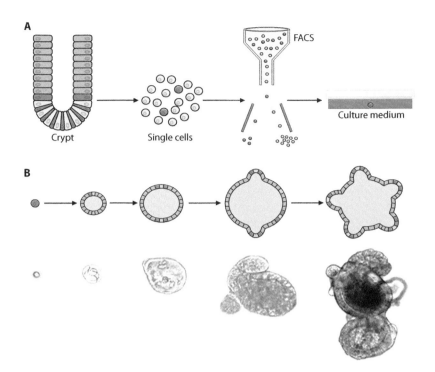

Figure 2.15 **Generation of organoids.** The generation of three-dimensional organoids from a single stem cell demonstrates their immense regenerative capacity. (**A**) Single Lgr5+ intestinal stem cells genetically labeled with GFP are isolated from intestinal crypts by fluorescent activated cell sorting (FACS) and cultured in a supportive medium of growth factors. (**B**) The single Lgr5+ stem cell (purple) begins to form a symmetric cyst structure that later begins to bud into structures resembling the crypts of the intestine consisting of Lgr5+ stem cells and Paneth cells (blue). (From Sato T & Clevers H. (2013) *Science*, 340:1190–1194 [doi: 10.1126/science.1234852]. Reprinted with permission from AAAS.)

there has certainly been considerable interest in trying to define a common signature for stem cell properties. Although, there are some shared features among different stem cells (Chapter 4), these should not be relied upon as predictive indicators. As we have already highlighted when discussing stem cell isolation, the issue of heterogeneity pervades the study of stem cells so that results are often an averaged quantity that may hide details of mechanisms that can only be deduced through single-cell analyses. Another cautionary point is that the dependence of stem cells on their microenvironment (Chapter 6) means that as soon as they are isolated in a pure form, they may well undergo molecular changes as specific contacts and signals are removed.

Major questions that molecular analyses are usually aimed at addressing include: What defines the self-renewal and differentiation properties of a stem cell and how does this differ between different stem cell types? How does a stem cell react in molecular terms to its microenvironment and systemic factors that it encounters? What processes are affected as stem cells (or their environment) age? In addition, much research on stem cells involves manipulation of presumed critical regulatory or functional molecules (e.g., by genetic manipulation), and it is important to be able to assess how such alterations affect molecular processes. Answers to these sorts of questions can be facilitated through detailed knowledge of gene structure and expression as RNA and proteins, the nature of modifications that occur to RNA post-transcription or to proteins post-translation, and the overall molecular and macromolecular profile of the cell.

As we have seen previously, the surface antigens used in the purification of specific stem cells are obviously part of the defining molecular signature of stem cells; however, a more detailed insight into the distinguishing features of stem cells can be achieved using techniques that encompass as broad a range as possible of the molecular make-up of the cell. Such techniques are known as "omics" in the current jargon and most notably include genomics and **epigenomics**, which respectively relate to genomic DNA sequence and modifications to both the DNA and the histones to which it is bound; transcriptomics and proteomics, which aim to describe the profile of expressed RNAs and the proteins encoded by the messenger RNA component; and metabolomics, that is, the measurement of the levels of small molecules (metabolites).

## Genomics involves deciphering the DNA sequence and polymorphisms that distinguish individuals

Genomics can encompass many aspects of the study of the structure and function of the genome, although we will restrict ourselves to consideration of the primary sequence of the DNA in the sense that this might impact stem cell behavior. Knowledge of the primary sequence of DNA obviously has a multitude of implications in terms of designing experiments to manipulate and study gene expression in relation to stem cell function, but more specifically inherited or acquired differences in the genome can be of importance if they affect genes that are essential for self-renewal, differentiation, and maintenance of normal stem cells throughout life or the acquisition of stem cell properties in cancer, as will be discussed in Chapter 18. DNA sequence variations between genomes have been assessed by a variety of means, including micro satellite and "SNP" (single nucleotide polymorphism) analysis, but these have now been largely superseded due to the phenomenal acceleration over the last two decades in the speed and accuracy with which DNA can be sequenced, making it feasible now to compare literally hundreds of genomes in the space of a few weeks. The first rough draft of the human genome was completed in 2000 and was declared complete on 14 April 2003. Since then, the advent of high-throughput or "next generation" sequencing (NGS) has shifted the rate of new sequencing by several orders of magnitude. The dominant technology for NGS, DNA fragments and primers are first attached on a slide and amplified with polymerase so that local clonal DNA "colonies" are generated. These then act as the templates for sequencing reactions using four fluorescently labeled reversible terminator bases, the individual fluorescent signal of which is imaged after every nucleotide extension. Analysis of the images across the slide through the progression of the extension reactions enables a direct readout of multiple millions of bases per second. For the purposes of work relating to stem cells, sequence reads will be assembled by alignment with a reference genome, thereby identifying any natural or mutational differences that might be responsible for altered stem cell properties.

## Epigenomics determines crucial interactions between proteins and DNA that affect the pattern of gene expression

The interaction between proteins and DNA is absolutely essential for many cellular functions such as DNA replication, DNA repair, and maintenance of genomic stability and for determining regulatory networks in both normal and diseased cell types. Epigenomics addresses the crucial question: Where do proteins bind within the genome? A method termed chromatin immunoprecipitation (ChIP) has been used to resolve such questions (Figure 2.16). As originally developed, the ChIP protocol requires quite a large number of cells, which raises problems when researching rare populations of stem cells. Modifications to the ChIP procedure have been made to assist in the analysis of protein binding within small populations of cells, down to as low as 100 cells.

ChIP is utilized for determining the location of post-translationally modified histones within the genome and for mapping target sites for transcription factors and other chromatin-associated proteins. This pioneering technique was developed in 1984 and has undergone modifications to refine the number of cells required and to reduce the time it takes to perform the procedure. Initially, there were two types of ChIP that differed in the preparation of the starting chromatin: Native ChIP (NChIP) and the more widely used cross-linked ChIP (XChIP). Both methods rely on the same principles, XChIP varying only in that it utilizes reversible cross-linking of protein and DNA with formaldehyde. Chromatin fragmentation (after cross-linking in the case of XChIP) is achieved either by enzymatic digestion (micrococcal nuclease) or by sonication of the whole cells/nuclei into fragments of 20–1000 base pairs. The sheared lysate is cleared by sedimentation, and the protein-DNA complexes are immunoprecipitated from the supernatant using specific antibodies raised against the protein of interest. Proteins are removed by enzymatic digestion (after reversal of cross-links in the case of XChIP) and the ChIP-enriched DNA is purified and then identified and quantified by means of either polymerase chain reaction (PCR), microarray (ChIP-on-chip), molecular **cloning** and sequencing, or direct sequencing using NGS (ChIP-seq).

## Transcriptomics is a rapidly evolving field that is essential for characterizing stem cells

To date, the best molecular definition of stem cells has been provided by determination of the profile of expressed genes represented in the RNA population. Transcriptomics provides a catalogue of all the species of gene transcripts expressed within a cell,

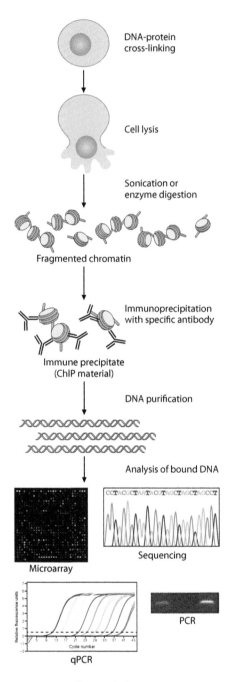

Figure 2.16 **Chromatin immunoprecipitation.** Schematic diagram illustrating the process of chromatin immunoprecipitation (ChIP). First, DNA and proteins are reversibly cross-linked before cells are lysed, and the chromatin is fragmented by sonication or enzymatic digestion into fragments of 20–1000 base pairs. Specific antibodies are then used to immunoprecipitate protein-DNA complexes. Proteins are removed by enzymatic digestion, and the enriched DNA is purified and subsequently used for analysis by quantitative polymerase chain reaction (qPCR), on microarrays or by sequencing.

including mRNAs, non-coding RNAs, and small RNAs. Furthermore, transcriptomics enables the determination of transcriptional structure of genes in terms of start sites, 5′ and 3′ ends, splicing patterns and other post-transcriptional modifications. There are several technologies that have been developed over the years in order to analyze the transcriptome of cells, and these are based on either hybridization between complimentary sequences or direct sequence determination.

## Hybridization-based analysis can provide valuable information about the expression genes expressed in stem cells

The technique adopted for a given population of stem cells is dependent on the degree of information that is required. The more in depth analyses involve a great deal more time, effort, and expense than standard methods such as the reverse transcription polymerase chain reaction and its quantitative variant (RT-qPCR).

During the early 1990s, the technique of reverse-transcription polymerase chain reaction (RT-PCR) became prominent in many laboratories. In these early years, this technique was only semi-quantitative, giving an indication whether the gene of interest was expressed and differentially regulated in the cell of interest. This technique was later made quantitative, using a combination of thermal cycling and fluorescence detection to measure the accumulation of the PCR product (Figure 2.17). The PCR cycle number at which the fluorescence crosses that of the background level relates to the amount of starting RNA. Cells expressing higher levels of RNA from a given gene reach the threshold more rapidly. The RT-qPCR method is very sensitive, to the extent that it can be performed on single cells, but has a drawback in that it is gene specific and does not assess the full gene expression signature of a cell.

Microarray screening (Figure 2.18) provides a rapid, reliable, reproducible quantification of the expression of thousands of genes simultaneously. This hybridization-based technique uses a single array or "chip" that can contain DNA probes suitable to determine the transcript levels for every known gene. The gene expression level is reflected by the intensity of fluorescence emitted from a labeled cDNA that is hybridized to a specific DNA probe spot on the array. Although microarray technologies provide relatively inexpensive high-throughput data, there are limitations to the transcript information obtained, since it relies heavily upon existing knowledge of genome sequence, is not ideally suited to discriminate post-transcriptional variants, and results in high background signal levels as a result of cross-hybridization.

## Sequence-based analysis through RNA sequencing provides a greater depth of information compared to other transcriptomic methods

More recently, NGS applied to RNA sequencing (RNA-seq) has become the method of choice for characterizing the RNA transcribed from a particular genome. RNA-seq provides direct access to the differential expression of genes, including the expression of varying gene alleles and differently spliced transcripts, non-coding RNAs, post-transcriptional mutations or editing, and gene fusions. These attributes of RNA-seq provide a much higher degree of sensitivity and quantitative range than microarray analysis.

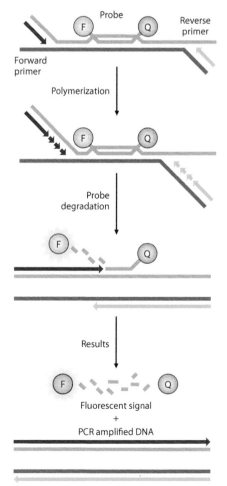

Figure 2.17 **Reverse-transcription quantitative real time polymerase chain reaction.** Real time reverse-transcription polymerase chain reaction (qRT-PCR) utilizes fluorescence as an indirect measure of transcript abundance. A gene-specific probe (blue line) that hybridizes to a region of the sequence of interest carries modifications at either end that give a fluorescent readout in proportion to the amount of complimentary DNA (cDNA) and cDNA present in the sample. When hybridized, the fluorescent moiety (F) on the probe fails to fluoresce due to the presence of a quencher (Q) at the other end. As the probe becomes displaced by the Taq polymerase as it copies the RNA sequence (progressing purple arrow), it is degraded by the exonuclease activity associated with the polymerase, activating the fluorophore, which can then be quantified by detectors within a sensitive PCR machine. (From Strachan T & Read A. (2011) *Human Molecular Genetics*, 4th ed. Garland Science, Taylor & Francis Group, LLC. Used with permission.)

Figure 2.18 **Microarray analysis.** A simplified diagram to show the process of microarray analysis. Complementary DNA (cDNA) is prepared from isolated RNA obtained from a control and experimental population of cells. The cDNAs from both populations are labeled with different fluorescent probes before being hybridized to a microarray "chip" containing DNA probes for a set of genes. The chip is exposed to laser excitation at the dye-specific wavelengths, and the emitted light is detected and analyzed by computer technology. The intensities of fluorescence are measured and compared for the two fluorescences, enabling differences in the gene expression profile between the control and experimental samples to be deduced.

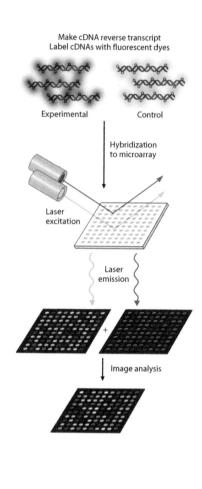

The technology of RNA-seq (Figure 2.19) begins with a population of RNA that is converted to a library of cDNA fragments with adaptors attached to one or both ends. Each molecule, with or without amplification, is sequenced to obtain short sequences from one end (single-end sequence) or both ends (pair-end sequence). These readouts are typically between 30 and 400 base pairs, and the resulting reads are aligned to a reference genome or reference transcript. Other advantages of RNA-seq over other transcriptomic methods in addition to sensitivity are that it is not limited to detecting transcripts that correspond to an annotated transcribed region and that it reveals the precise location of transcription boundaries up to single-base resolution.

## Single-cell transcriptomics is presently the best way for the definition of an expressed gene signature of stemness

As we have already discussed, most isolated stem cell populations are heterogeneous to some extent, and therefore it can be difficult precisely to determine the gene expression profile(s) that relates to specific stem cell function. For this reason, single-cell RNA-seq (scRNA-seq) has become increasingly popular as a means to determine the gene expression signature of a particular stem cell type and the extent to which this varies within a population of cells with similar stem cell characteristics. There are two main challenges associated with scRNA-seq, that is, how to capture single stem cells and how to amplify and sequence the minuscule amounts of mRNA that can be recovered (averaging 10–30 pg per cell). A number of methods have been developed for capturing a single cell, each having advantages and limitations (Figure 2.20). The most recent method consists of a droplet microfluidic–based approach (Drop-seq) that encapsulates single cells into an oil microdroplet (1 pL to 10 nL) together with a microbead covered with barcoded primers. The single cell is automatically subjected to cell lysis within the droplet, and the mRNA hybridizes to the bead. Beads from multiple single cells are pooled together and undergo reverse transcription, amplification, and sequencing. Following sequencing, the data can be deconvoluted using the unique barcode assigned to each bead, allowing the sequence reads to be assigned to the cell of origin.

## Proteomics

Although a complete catalogue of the RNAs present within a cell and their level of expression can provide important information about the nature of a stem cell, most of the functional characteristics of any cell are actually determined by its complement of proteins. The profile of RNA expressed by a cell to a large extent dictates the complexity of proteins, but of course this is greatly increased by the precise way in which RNA is translated, the proteolytic cleavage and stability of individual proteins, and the extent to which a protein acquires post-translational modifications (PTMs) that will have an impact on its function. PTMs include small chemical modifications such as phosphorylation, acetylation, methylation, and oxidation, as well as the attachment of larger moieties like sugar residues and small proteins that dictate protein stability (ubiquitin and SUMO). Collectively, the mechanisms governing

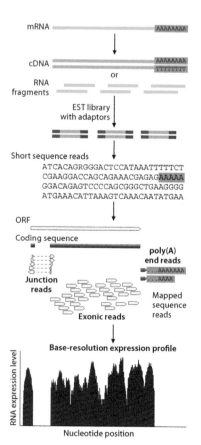

Figure 2.19 **RNA-seq.** Illustration of a typical RNA-seq experiment. Messenger RNA (mRNA) is first converted into a complementary DNA (cDNA) library and fragmented. Sequencing adaptors (blue and red) are added to each cDNA fragment, and a short sequence is obtained from each cDNA using high-throughput sequencing technology. The sequence reads are aligned with a reference transcriptome and classified as either junction, exonic, or poly(A) end reads. The expression level is quantified from the cumulative number of reads that align to the reference sequence. (From Wang Z, Gerstein M & Snyder M. (2009) *Nat Rev Genetics* 10:57–63 [doi: 10.1038/nrg2484]. Reprinted by permission from Springer Nature.)

| Capture method | | Advantage | Limitations |
|---|---|---|---|
| Serial dilution | 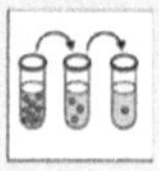 | • Simple<br>• Low cost | • Time consuming<br>• Requires expertise |
| Micromanipulation |  | • Good for samples with low cell numbers | • Relies on visualization of cells under a microscope<br>• Low throughput |
| Laser capture microdissection |  | • Preserves spatial relationship within tissue | • Tissue must be sectioned causing loss of chromsomal material<br>• Relies on visualization of cells under a microscope |
| FACS | 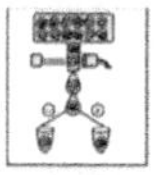 | • High throughput<br>• High specificity | • Large amount of dissociated cells needed |
| Microfluidics | 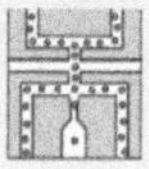 | • High throughput<br>• Highly selective | • Cells need to be homogenous in size<br>• Requires at least the capture of a thousand cells |

Figure 2.20 **Methods of single cell capture.** Single-cell analysis of stem cell populations is becoming increasingly important better to define the stem cell capacity of heterogeneous populations of cells. Single-cell RNA-seq is one technique that requires the capture and isolation of one cell. There are multiple approaches to capturing single cells, depicted here in this table, each method having particular advantages and limitations.

translation, PTM addition (and removal), and stability generate what has become known as the cell proteome, the analysis of which is encompassed by the field of proteomics. Several laboratory techniques can be said to contribute information about the presence and levels of particular proteins, but as for other omics strategies, researchers are striving to capture as much information as possible about a cell's proteome in different scenarios. The use of mass spectrometry (MS) is the gold standard for identification and quantification of proteins in complex mixtures and has been modified to enable detection of specific categories of PTM.

## Metabolomics

The portfolio of omics technologies is completed by metabolomics, that is, the systematic analysis of the collection of small molecule metabolites (often defined as less than 1 kDa in size) that are the end products of the collected processes in a cell. Like the transcriptome and the proteome, the metabolome is dynamic, potentially changing on a timescale of seconds, and has recently begun to be studied in various stem cell-related contexts (Figure 2.21). It is not presently possible to identify the entire range of metabolites by a single method, but generally a separation method is combined with a detection method, often in a single technology platform. Gas chromatography, high performance liquid chromatography, or capillary electrophoresis are the preferred methods of separation, each having specific advantages and limitations. Mass spectrometry is the most widely used means to detect and identify the fractionated metabolites. The other commonly used detection method is nuclear magnetic resonance (NMR) spectroscopy, which has the advantage that it does not rely on separation of the components of the metabolome but is relatively insensitive compared to MS-based approaches.

## 2.4   SUMMARY

In this chapter, we have attempted to provide an overview of the methods and strategies that are often employed in the identification, isolation, and characterization of stem cells from adult tissues. The historical influence of investigation of hematopoietic stem cells is reflected in the way that many other stem cells have been investigated, but each stem cell type can have unique features and may raise specific challenges. Without doubt, patterns of surface antigen expression are crucial to the identification and isolation of many adult stem cells, and lineage tracing has utility in many contexts when trying to understand how a given stem cell contributes to tissue maintenance. Recent advances such as single-cell RNA sequencing and high-resolution *in vivo* observation are without a doubt going to give us an even greater understanding of stem cell complexity and function in the near future.

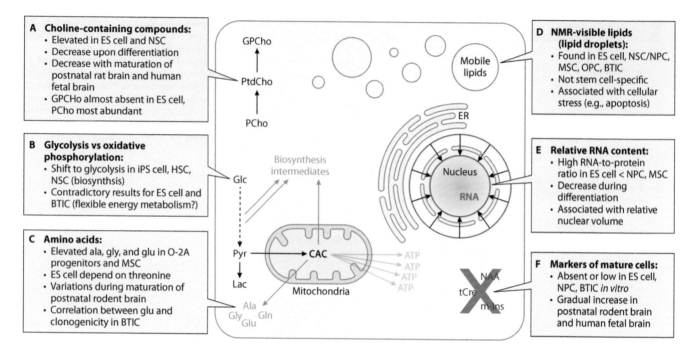

**Figure 2.21 Major stem cell-specific findings of metabolic profiling approaches. (A)** Choline-containing compounds elevated in embryonic stem cells (ESC) and neural stem cells (NSC). PCho – phosphocholine; PtdCho – phosphatidylcholine; GPCho – glycerophosphocholine. **(B)** Glycolysis versus oxidative phosphorylation in induced pluripotent stem cells (iPSC), hematopoietic stem cells (HSC), and NSC. BTIC – brain tumor-initiating cell; Glc – glycine; Pyr – Pyruvate; Lac – Lactate; CAC – citric acid cycle; ATP – adenosine triphosphate. **(C)** Amino acids associated with stem cell metabolism. MSC – mesenchymal stem cell; Ala – alanine; Glu – glutamate; Gln – glutamine. **(D)** NMR-visible lipids found under stress conditions in stem cells. NPC – Neural progenitor cell; OPC – oligodendrocyte. **(E)** Relative RNA content during differentiation. **(F)** Markers of mature cells. NAA – N-acetyl aspartate; tCre – creatine-containing compounds, m-Ins – myoinositol. (From Ramm Sander P, Hau P, Koch S et al (2013) *Trends Biotechnol* 31:204–213 [doi: 10.1016/j.tibtech.2013.01.008]. Copyright 2013 with permission from Elsevier.)

## KEY POINTS

- The stem cell surface is covered with a mixture of proteins that can be used to identify and isolate them from a heterogeneous population; however, no known unique stem cell marker has been identified to date.

- Monoclonal antibodies conjugated to fluorochromes are raised against specific surface antigens and can be used in technologies such as flow cytometry and microscopy to identify and isolate rare populations of stem cells.

- Other stem cell features can be exploited, such as their ability to efflux DNA binding dyes through membrane protein pumps, resulting in a distinctive side population profile. Their replicative quiescence can also be utilized by staining techniques (BrdU and CFSE) to identify the true stem cell population.

- Ultimately, stem cells are defined functionally by their ability to differentiate toward all lineages of a particular tissue type that can be demonstrated in *in vitro* assays such as colony forming assays. The true test of stem cell function is their ability to transplant into recipient animal tissues and reconstitute the full range of differentiated cell types while maintaining their own integrity to self-renew.

- Transgene reporters can be used to lineage trace stem cells and their progeny in a particular tissue.

- "Omics" approaches can be used to further define stem cell features and provide a molecular signature of a particular stem cell population. Such approaches include genomics, transcriptomics, and proteomics.

## FURTHER READING

Hoppe, P.S., Coutu, D.L. and Schroeder, T. (2014) Single-cell technologies sharpen up mammalian stem cell research. *Nat. Cell Biol.*, 16: 919–927.

Hsu, Y.-C. (2015) The theory and practice of lineage tracing. *Stem Cells*, 33: 3197–3204.

Hu, Y., An, Q., Sheu, K., Trejo, B., Fan, S. and Guo, Y. (2018) Single cell multi-omics technology: Methodology and application. *Front. Cell Dev. Biol.*, 6: 28.

Krishan, A., Krishnamurthy, H. and Totey, S. (Eds) (2012) *Applications of Flow Cytometry in Stem Cell Research and Tissue Regeneration*, 1st ed. Wiley-Blackwell.

Marx, V. (2018) Stem cells: Lineage tracing lets single cells talk about their past. *Nat. Meth.*, 15: 411–414.

Nguyen, P.D. and Currie, P.D. (2018) *In vivo* imaging: Shining a light on stem cells in the living animal. *Development*, 145, dev150441.

de Souza, N. (2018) Organoids. *Nat Meth*, 15: 23

# STEM CELL POTENCY AND

# CONTROL

# PART TWO

# Embryonic stem cells, cloning, and pluripotency

# 3

Without a doubt, the greatest interest, debate, and controversy in the world of stem cell biology have been around the topic of human **embryonic stem (ES) cells** and human cell **cloning**. All levels in society seem to have had something to say, from presidents and religious leaders to politicians and the man on the street. This high level of interest has not solely been due to the absolutely amazing scientific discoveries made in stem cell biology but also because of the ethical implications of potentially interfering with or trying to replicate the earliest stages of human life. The magnitude and significance of these scientific findings are without precedent in the biological sciences. Critical discoveries and advances in the understanding of ES cells and their defining property of pluripotency (Chapter 1) have already been recognized with two Nobel Prizes, and more may well follow.

Although the excitement and public interest have focused on human embryo-derived cells and human cloning, these are just a part of a much broader field of discovery and application involving cells with pluripotent characteristics, spanning many vertebrate species and applications ranging from experimental genetic engineering through to veterinary medicine and rare species conservation. Apart from some transiently existing cells during specific stages of vertebrate development and pluripotent cells that have been characterized in some invertebrates (Chapter 7), **pluripotent stem cells** are mainly artificially derived through the culture of embryo-derived cells or through cloning or **reprogramming** of adult somatic cells (Table 3.1).

In this chapter, we will focus on pluripotent cells derived directly from embryos or through cloning. After having revealed some of the mechanistic principles that underlie pluripotency in Chapter 4, we will discuss reprogrammed pluripotent stem cells in Chapter 5. We start by looking at stem cells derived from vertebrate

**TABLE 3.1  CELL LINES WITH PLURIPOTENT CHARACTERISTICS HAVE BEEN ISOLATED FROM A VARIETY OF DIFFERENT SOURCES IN VERTEBRATES**

| Pluripotent stem cell type | Origin |
| --- | --- |
| Embryonal carcinoma cells | Tumors (m, h) |
| Embryonic stem cells | Pre-implantation blastocyst (many mammals), blastodisc (c), blastula (f) |
| Epiblast stem cells | Post-implantation epiblast (m) |
| Embryonic germ stem cells | Primordial germ cells (m, h, c) |
| SCNT embryonic stem cells | Adult somatic cells (many mammals) |
| Induced pluripotent stem cells | Adult somatic cells (many mammals) |

*Note:* The table lists the main types of pluripotent stem cells that have been derived from vertebrates together with the normal embryonic or adult source tissues (m—mouse; h—human; c—chicken; f—fish).

embryos, paying attention to the historical development of the field culminating in the generation of the first mouse ES cells in 1981 and the subsequent extension to humans. Then we consider how cloning can be used as a way to produce ES cells from an individual, raising the possibility of personalized stem cell production. Last, we touch on the growing interest in making ES cells from non-mammalian vertebrates, which at present has seen most progress for a variety of species of fish and birds.

## 3.1    PLURIPOTENT CELLS DERIVED FROM VERTEBRATE EMBRYOS

The research that led to the first isolation of pluripotent stem cells, that is, cells capable of giving rise to all cell types of the developing embryo, has a long history, evolving from many decades of investigation of the process of embryogenesis and the realization that tumors arising from embryonic tissues represent a stem cell-like state. As is so often the case, studies on the laboratory mouse were paramount both in the definition of embryonic processes and the subsequent work on pluripotent stem cells. This has been and remains the most active area of stem cell research, and our present level of understanding of the cell and molecular characteristics of pluripotent stem cells is awe inspiring. Not surprisingly given the medical implications, pluripotent stem cell research has become focused on humans, and more recent advances in the technology are opening up ever more avenues for the manipulation of the genome both for fundamental and translational studies.

### The early stages of vertebrate embryo development are characterized by pluripotency

Vertebrate pluripotent stem cell lines are a laboratory creation, although they reflect inherent pluripotency in cells that are seen at certain stages of embryonic development. In other words, there have to be cells at some point in the embryo that are capable of giving rise to all of the tissues, and by definition are pluripotent, which will develop into the fetus and eventually become the adult. However, these development-related pluripotent cells need to be "captured" through specialized culture techniques and conditions in order for them to become stem cell lines.

Although the precise timings and the end results of embryogenesis in various vertebrate species are different, the early stages (Figure 3.1) show remarkable similarity across species. From the one-cell stage following fertilization (the zygote) through to the eight-cell embryo, all cells are considered **totipotent**, that is, able to produce both embryonic and extra-embryonic tissues. By the next defined stage, the morula, the individual cell potential becomes more restricted. Soon after, the cells form either a hollow sphere in mammals (the **blastocyst**) consisting of an outer extra-embryonic layer (the **trophectoderm**) and an inner cluster of cells (the **inner cell mass** or ICM for short) or a germinal disc on the surface of the yolk in birds and reptiles (the **blastodisc**). It is the cells of the mammalian ICM that, in mice at least, can be cultured in the laboratory as ES cells. As the blastodisc or ICM develops further, it differentiates into the hypoblast and **epiblast**, the latter giving rise to the three embryonic germ layers (**ectoderm, mesoderm,** and **endoderm**) in the process called gastrulation. The mouse epiblast can also be cultured to derive apparently pluripotent stem cells, and the human epiblast is thought to be the source of ES cells obtained from human embryos. Specification of cells that go on to form the germ cells of the developing testes or ovaries occurs after gastrulation. These **primordial germ cells (PGC)** and the cells they mature into are pluripotent and can likewise be used to derive cell lines with pluripotent characteristics.

Figure 3.1 **Early stages in the development of the vertebrate embryo from which pluripotent stem cells can be isolated.** The figure shows the early embryonic development of the mouse from the zygote through to the pre-streak embryo. The progression of development is shown as phase-contrast microscope images of embryos (**A**). The schematic representation (**B**) shows the gradual change in the differentiation potential of individual cells within the embryo and their commitment to specific lineages, which is indicated by color shading (white – totipotent blastomeres; purple – pluripotent inner cell mass; green – trophectoderm; red – pluripotent epiblast; blue – hypoblast/extraembryonic endoderm). E: embryonic day. Scale bars – 50 μm. (From Niakan K, Schrode N, Cho L et al. *Nat Protoc* 8:1028–1041 [doi:10.1038/nprot.2013.049]. Reprinted by permission from Springer Nature.)

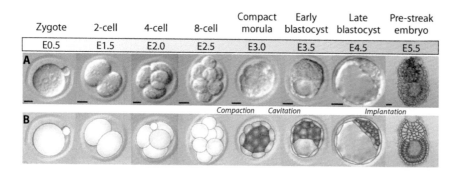

| Zygote | 2-cell | 4-cell | 8-cell | Compact morula | Early blastocyst | Late blastocyst | Pre-streak embryo |
|---|---|---|---|---|---|---|---|
| E0.5 | E1.5 | E2.0 | E2.5 | E3.0 | E3.5 | E4.5 | E5.5 |

## Tumors of embryonic origin point the way to embryonic stem cells

The history of the derivation and culture of pluripotent stem cells actually dates back to research performed on tumors of germ cell origin in the 1950s and 1960s (Figure 3.2). Germ cell tumors can be benign or malignant, usually occurring in the gonads, but are also seen at other sites throughout the body, possibly as a consequence of abnormal migration of germ cells during embryogenesis. A little-mentioned researcher, Leroy Stevens (Box 3.1), was the first to recognize the link between these tumors and pluripotency and led the way forward for the isolation of ES cells some 25 years or so later. As part of Stevens' early work at the Jackson Laboratory in Bar Harbor, Maine, he used inbred strains of mice that exhibited a high incidence of spontaneous testicular **teratocarcinomas**, tumors containing a mix of immature cells and differentiated cells of multiple tissue types. Stevens went on to show that these tumors were transplantable and that similar tumors could be induced by injection of PGC from the developing fetal testis. In his publication of 1954, he concluded that his experiments with germ cell tumors demonstrated that "Pluripotential embryonic cells appear to give rise to both rapidly differentiating cells and others which like themselves, remain undifferentiated."

**Figure 3.2 The timescale of murine embryonic stem cell discovery.** Observations on the pluripotent nature of mouse embryo-derived cells and the subsequent isolation of pluripotent stem cell lines occurred over a period of more than two decades. The starting point was the seminal findings of Leroy Stevens, which led eventually to isolation of ES cells through the work of Gail Martin and Martin Evans and their subsequent application as a vehicle for genome modifications. This ability to genetically engineering the genome continues to this day as the central means of understanding gene function.

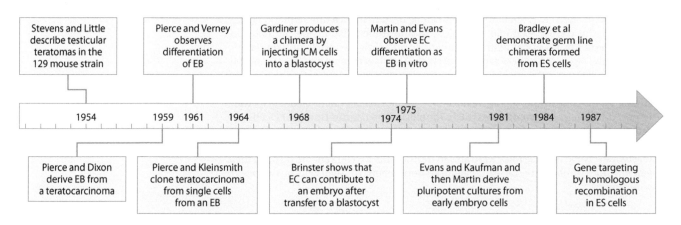

| Stevens and Little describe testicular teratomas in the 129 mouse strain | Pierce and Verney observes differentiation of EB | Gardiner produces a chimera by injecting ICM cells into a blastocyst | Martin and Evans observe EC differentiation as EB in vitro | Bradley et al demonstrate germ line chimeras formed from ES cells |

1954 — 1959 1961 1964 — 1968 — 1975 1974 — 1981 1984 1987

| Pierce and Dixon derive EB from a teratocarcinoma | Pierce and Kleinsmith clone teratocarcinoma from single cells from an EB | Brinster shows that EC can contribute to an embryo after transfer to a blastocyst | Evans and Kaufman and then Martin derive pluripotent cultures from early embryo cells | Gene targeting by homologous recombination in ES cells |

---

### BOX 3.1 PIONEER: LEROY STEVENS

Leroy Stevens (Figure 1) is one of those figures in stem cell research that have perhaps not received all the credit they deserve for having made ground-breaking discoveries that significantly shaped future research. In his case, Stevens can quite reasonably be said to have been the father of embryonic stem cell research, even though those who built on his findings in the subsequent quarter of a century received most of the limelight. Stevens started in the Jackson Laboratory in 1953, where his first task was not exactly what you might have expected for somebody who was about to define the nature of pluripotent stem cells. The laboratory had received funding from the tobacco industry in the hope that it could be shown that the paper in cigarettes, rather than the tobacco itself, was responsible for smoking-related lung cancer. Stevens did as was requested of him and exposed mice to the separated components of cigarettes. Of course, as we now know well enough, he did not get the result the tobacco company had wanted, but one day he noticed a mouse with a huge scrotum. When he looked at the testes, he found a mass of different tissues including a peculiar mix of hair and teeth, cartilage, and tiny tubules, that is, a teratoma. Stevens was excited by his observations, being quoted "This stuff was extremely interesting, and it sure beat studying cigarette papers!" and he soon got his own funding further to explore teratomas. The mouse strain Stevens had used to select for the teratoma tendency, known by the rather dull name of 129, is familiar to anybody who has been involved in ES cell research. This mouse was, and continues to be, used for ES cell derivation and the massive field of murine genetic engineering that

Box 3.1 Figure 1 **Leroy Stevens.** (Reproduced with permission from The Jackson Laboratory.)

rapidly emerged in the 1980s. Another key element to Stevens' experiments, which again helped pave the way to ES cell discovery, was that he developed a serial transfer technique in order to be able to study the rare cells in some of the teratomas that rendered them cancerous. This technical advance allowed Stevens to describe the presence of teratoma-forming pluripotent stem cells both in the tumors and in developing germ cell tissues of the embryo. He also looked at earlier stages of embryo development, including the inner cell mass of the pre-implantation embryo, again seeing the formation of teratomas – and the rest is history! Profiling Stevens in 2000 in *The Scientist*, Dr. Ricki Lewis noted, "Leroy Stevens is truly the unsung hero of stem cell research."

Subsequent to Stevens' work on the transplantation of teratomas, pluripotent stem cell lines were established. These so-called **embryonal carcinoma (EC)** lines were initially derived from the undifferentiated component of transplantable teratocarcinomas of mice and later from analogous human tumors. EC cell lines, which have near but not absolutely normal karyotype (the chromosomal content as defined in cells undergoing mitosis), retain the capacity to differentiate into a variety of tissue types, both in tumors formed following the injection of cells into animals and *in vitro* under appropriate culture conditions. One such line, known simply as P19, was created by Michael McBurney in Ottawa and was used for well over a decade as a standard model for the **differentiation** of pluripotent stem cells. Small molecules can be used to induce differentiation along specific pathways; for example, retinoic acid and dimethylsulfoxide promote the **commitment** toward either neuronal or cardiac and skeletal muscle differentiation, respectively (**Figure 3.3**). These experiments with EC cells generated important concepts around pluripotent cells, including the realization that their differentiation followed the normal pathways of early embryonic development. As part of some differentiation strategies, EC cells would aggregate once their supportive media or feeder cells were removed, forming **embryoid bodies (EBs)**. The pluripotent nature of murine EC cells was further demonstrated through their ability to contribute to mouse tissues in **chimeras** produced by incorporation of EC cells into the ICM following their injection into normal blastocysts. None of these mice were able to transmit the EC cell genome through their gametes, most probably because the cells used were aneuploid. Such work on the *in vitro*

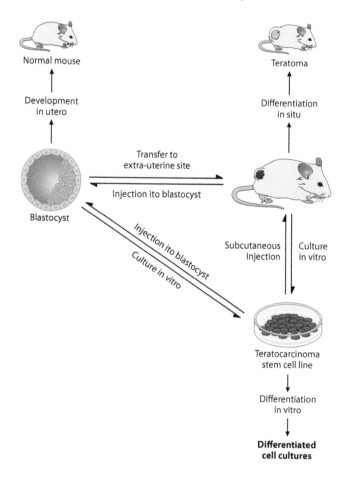

Figure 3.3 **Relationship between pluripotent cells in mouse teratocarcinomas, embryos, and cells derived from them.** The diagram summarizes the differentiation potentials and relationships between pluripotent cells from the blastocyst, teratocarcinomas, and cell lines derived from them as deduced by researchers prior to the first description of ES cell lines in 1981 (Adapted from Martin G. (1981) *Proc Natl Acad Sci USA*, 78: 7634–7638 [doi:10.1073/pnas.78.12.7634].)

differentiation and chimera contribution of EC cells paved the way for the beginnings of work on ES cells at the beginning of the 1980s.

## The first embryonic stem cells were derived from early mouse embryos

Following the way being paved by Stevens, McBurney and others based on their work with teratocarcinomas and EC cells, the group of Martin Evans in Cambridge found that certain antigens expressed on EC cells highlighted pluripotent cells in the early pre- and post-implantation mouse embryo. In 1981, two groups using slightly different approaches independently derived pluripotent stem cell lines from the mouse embryo ICM (Figure 3.4). Using cell culture techniques based on experience gained from studies on EC cells, Martin Evans (Box 3.2) and Matthew Kaufman in Cambridge were able to isolate pluripotent cells by the outgrowth of recognizable EC-like cells from mouse blastocysts. Expansion of the desired cells was facilitated using a monolayer of supportive cells ("feeder cells") that were presumed to stimulate the proliferation

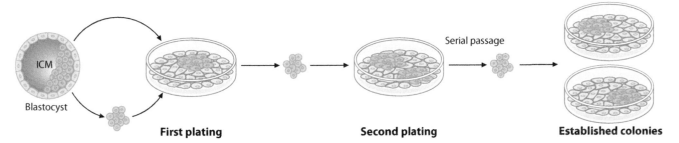

**First plating**    **Second plating**    **Established colonies**

Figure 3.4 **The process of obtaining ES cells from a mouse embryo.** ES cells can be isolated by culturing of the inner cell mass (ICM) of the pre-implantation blastocyst. The blastocyst stage embryo consists of an outer layer of trophectodermal cells (grey) surrounding the ICM, which contains pluripotent cells (light brown). The ICM is grown on a layer of supportive fibroblasts (blue) either (i) within the intact blastocyst, from which the expanding ES cells emerge, or "hatch," to form a colony on the feeder cells (as performed by Matthew Kaufman and Martin Evans) or (ii) after isolation from the trophectoderm, for example, using immunosurgery, as was carried out by Gail Martin. Stable ES cell lines eventually develop after several weeks of culture and serial passage onto fresh layers of fibroblasts.

### BOX 3.2 PIONEERS: MARTIN EVANS AND GAIL MARTIN

Martin Evans (Figure 1A) graduated from Cambridge University in 1963 with a degree in natural sciences, and then studied in University College London, where he worked on developmentally controlled mRNA and was awarded a PhD in 1969. He continued there as a lecturer in anatomy and embryology and then in 1978 returned to Cambridge in the Department of Genetics, where he began to work with Matthew Kaufman in 1980. In July of the following year, Evans and Kaufman published their experimental results on the isolation of pluripotent stem cells from mouse blastocysts in the journal *Nature*, just ahead of Gail Martin (Figure 1B), who published similar findings in December of the same year. Gail had obtained her PhD from the University of California at Berkeley in 1971. Following postdoctoral research at University College London, where she worked in the laboratory of Martin Evans, she joined the faculty at the University of California San Francisco in 1976 and is currently a professor in the Department of Anatomy. She received many awards of recognition for her work on ES cells, including being appointed as a member of the National Academy of Sciences and receiving the EG Conklin Medal from the Society for Developmental Biology in 2002, but it was Evans who was elevated to "Sir Martin" by the Queen in 2004 and who gained the ultimate accolade from the science community by being made a Nobel Laureate in 2007. His Nobel Prize was in Physiology or Medicine, which was actually awarded jointly to Mario Capecchi and Oliver Smithies for their

combined efforts in developing the technology for introducing homologous recombination in mice employing ES cells. In 1999, Evans left Cambridge to become professor of mammalian genetics and director of the School of Biosciences at Cardiff University, where he worked until he retired at the end of 2007. Evans was appointed president of Cardiff University in 2009 and subsequently became chancellor in 2012.

Box 3.2 Figure 1 **Martin Evans and Gail Martin.** (**A**: Photo credit: Cardiff University. Licensed under the Creative Commons Attribution 3.0 Unported license. **B**: Used with permission from UCSF. Photo credit: Susan Merrell, UCSF, 2012.)

or inhibit the differentiation of normal pluripotent embryonic cells. These cells could be continually grown in the laboratory and maintained a normal karyotype. Working in San Francisco, Gail Martin (see Box 3.2), who had previously performed work on teratocarcinoma cells as a postdoctoral researcher with Evans in London, also isolated pluripotent cell lines from blastocysts, for which she coined the term embryonic stem cells. In deriving these ES cells, Gail Martin had used a slightly different approach entailing removal of the outer trophectoderm by complement-mediated lysis following binding of specific antibodies ("immunosurgery") and culture on feeder cells in the presence of medium "conditioned" by EC cells. Although they differentiated *in vitro* and formed teratomas like those that Evans and Kaufman had derived earlier in the year, the cells derived by Gail Martin did not retain a normal karyotype.

## The properties of mouse embryonic stem cells reflect their embryonic origin

The original ES cell lines isolated by Evans and Kaufman were derived from Leroy Stevens' 129 strain of mouse and were male. Chimera formation with these ES cells extended to the germ cells and could even shift the sex of the offspring derived from injected female blastocysts. Murine ES cells are adherent and grow as tight, slightly domed colonies (Figure 3.5), either on the surface of a monolayer of feeder cells or directly on a plastic surface that has been treated with **extracellular matrix (ECM)** proteins such as collagen (in the form of gelatin, which is perhaps more familiar in the kitchen!). Apart from their colony morphology, murine ES cells, like other pluripotent cells, are alkaline phosphatase positive and express key regulatory molecules characteristic of the early blastocyst, such as the transcription factors Nanog and Oct4 (Chapter 4).

When using feeder cells, which are either primary mouse embryo-derived fibroblasts (MEFs) or a fibroblastic cell line, they are in general initially treated to prevent their proliferation (by gamma irradiation or using cell cycle blockers such as Mitomycin C), and there is no absolute requirement to provide additional growth factors in order to maintain ES cells. However, a few years after the initial isolation of ES cells, it was found that a polypeptide factor could substitute for the feeder cell requirement. There were two different routes to the discovery of this factor that specifically suppresses the spontaneous differentiation of ES cells, one through analysis of medium conditioned by Buffalo Rat liver cells, the identified factor being named differentiation inhibitory activity (DIA), and the second involving characterization of a molecule that induced differentiation in myeloid leukemia cells, which became known as leukemia inhibitory factor (LIF). Rather strangely, it is the less relevant name of LIF that has stuck. As understanding of the nature of ES cells grew, particularly in terms of the signaling pathways controlling differentiation versus maintenance of the pluripotent state (Chapter 4), the role of LIF was enhanced or even substituted for by small molecule inhibitors. Most effective has been the two-inhibitor or "2i" combination of chemicals acting on mitogen-activated protein (MAP) kinase and glycogen synthase kinase 3 (GSK3).

## The availability of mouse embryonic stem cells was crucial for the emergence of the genetic engineering revolution

The ability to immortalize pluripotent cells from mouse embryos as ES cells opened the door on several avenues of experimental research. Over the last three decades, mouse ES cells have revolutionized both our understanding of developmental processes and individual gene function as well as fueling great expectations for similar cells in humans, especially as a means to reach a medical Utopia of unrestricted **regenerative medicine**.

Without a doubt, the biggest revolution has been in the engineering of the mouse genome using ES cells, which became, if not routine, then fairly straightforward once a few basic methods had been mastered. Building on the initial work of Evans and his colleagues in Cambridge, pioneering developments by Oliver Smithies and Mario Capecchi enabled targeted modifications to the genome, leading to their joint award with Evans of the Nobel Prize in Physiology or Medicine in 2007. The starting point was the observation by Liz Robertson and Allan Bradley, working with Evans, that viral DNA introduced in tissue culture into the ES cell genome could be transmitted to germ cells in chimeric embryos formed following injection of the cells into a blastocyst, thereby allowing transmission to another generation of mice. Around the same time as these studies were being performed in Cambridge, Oliver Smithies and Mario Capecchi in America had come up with a way specifically to alter endogenous gene loci by homologous recombination between the genome and an introduced DNA construct containing sequences homologous to the target gene. When performed in ES cells,

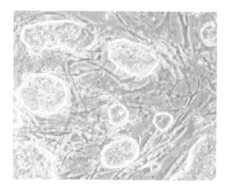

**Figure 3.5 ES cells can be cultured in a number of ways to maintain their pluripotent state.** The most common culture conditions utilize a one-cell-thick layer of supportive cells, usually mouse embryo fibroblasts (MEF), upon which the ES cells grow as tight clusters. The bright field image shows mouse ES cell colonies growing on a Mitomycin C treated MEF cell feeder layer. They characteristically have a phase-bright rim with the individual ES cells being very tightly packed together, distinguishing them from the irregular, often filamentous fibroblasts. (With permission from Rik Derynck's laboratory.)

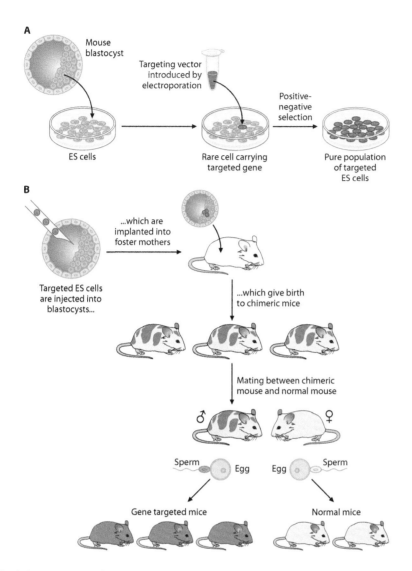

**Figure 3.6 General strategy for the utilization of ES cells in gene targeting in mice.** (**A**) Gene targeting of ES cells (light brown) in culture is followed by isolation of an ES cell clone containing the desired mutation (dark brown). Selection using cytotoxic drugs for insertion of the gene sequences into the genome is used to enrich for ES cells containing the correctly modified gene. Selected ES cell clones are checked, usually by Southern blotting of restriction enzyme digested genomic DNA, to ensure that homologous recombination has led to the correct exchange of sequences. (**B**) ES cells with the desired genetic modification are injected into blastocysts, which are then transplanted into the uterus of a foster mother. Depending on the number of ES cells incorporated into the embryo, pups are born that are chimeric, being a mix of cells derived from the host embryo ICM and the injected ES cells carrying the genetic modification. To facilitate isolation of the desired pups, the ES cells and recipient blastocysts are derived from mice with different coat color genes, represented here as cream for the host blastocyst strain and brown for the genetically modified ES cells. If ES cells become incorporated into the germ cells of the chimera, then subsequent breeding to a wild type mouse (beige) will allow establishment of a mouse line (brown progeny). Since most ES cell lines are derived from male mice, it is usual to select a male chimera for this stage of breeding. Since only one allele is affected by the gene targeting event, the first-generation progeny need to be cross-bred in order to achieve homozygosity (From https://www.nobelprize.org/nobel_prizes/medicine/laureates/2007/advanced.html. (Used with permission. Copyright © The Nobel Committee for Physiology or Medicine.)

which have a particularly high propensity to undergo homologous recombination, cells in which recombination has occurred between the DNA construct sequences and the corresponding region of the genome can be selected for. Gene targeting is generally accomplished by insertion of a selectable marker (most often a neomycin resistance gene) into the gene of interest, causing disruption of splicing, promoter function, or the reading frame, with or without deletion of some of the gene. Successful targeting of one allele of genomic DNA depends upon replacement of the endogenous gene by homologous recombination through sequences on either side of the modification carried in the targeting construct. Once fully characterized in culture, the modified ES cells can be injected into mouse blastocysts so that they become incorporated into a chimeric animal that, if the chimerism extends to the germ cells, can be bred so as to establish a stable, genetically modified mouse line (Figure 3.6). This method is used to manipulate a single gene, in most cases "**knock out**" of the target gene (creating a "null" allele), although more subtle genetic manipulation can occur (e.g., only changing single nucleotides). The diploid nature of the ES cell genome means that both sets of a gene must be inactivated for successful phenotype-driven forward genetic screens, an end point that in most cases is achieved by crossing parents that are each carrying the modified allele on one chromosome.

## Creation of human embryonic stem cells followed a strategy modeled on that used to generate the mouse equivalents

Once it became clear that ES cells could be made from mouse embryos and that these could be valuable tools in research into development and gene function, it was the natural extension to try to replicate the process in other species, most importantly for humans. Following the historical progression of discovery in mice, cells with stem cell-like properties capable of giving rise to the three germ cell lineages were obtained from human teratocarcinomas during the 1980s, and then in the mid 1990s, James Thomson at the University of Wisconsin (Box 3.3) succeeded in deriving primate ES cells first from the rhesus monkey and then the marmoset. These lines had the desired karyotype and differentiation potential and resembled the human EC cells previously described, suggesting that human ES cells should be obtainable but that they would be of a somewhat different phenotype compared to murine ES cells.

In 1998, James Thomson became the first person to generate human cells that satisfied three of the four essential and testable criteria (Table 3.2). The method he and his colleagues employed used surplus blastocysts donated by couples undergoing

### BOX 3.3 PIONEER: JAMES THOMSON

James Thomson (Figure 1) graduated with a degree in biophysics from the University of Illinois in 1981. He then trained as a veterinary medical scientist at the University of Pennsylvania, receiving his doctorate in veterinary medicine in 1985, followed by graduate research on genetic imprinting in early mammalian development, for which he was awarded a PhD in 1988. Thomson then spent two years as a postdoctoral research fellow in the Primate *In Vitro* Fertilization and Experimental Embryology Laboratory at the Oregon National Primate Research Center and completed a residency in veterinary pathology at the University of Wisconsin–Madison in 1994. Thomson serves as director of regenerative biology at the Morgridge Institute for Research in Madison, Wisconsin; is a professor in the Department of Cell and Regenerative Biology at the University of Wisconsin School of Medicine and Public Health; and is also a professor in the Molecular, Cellular, and Developmental Biology Department at the University of California, Santa Barbara. He is a founder and chief scientific officer for Cellular Dynamics International, a Madison-based company producing derivatives of human induced pluripotent stem cells for drug discovery and toxicity testing. Thomson is a member of the National Academy of Sciences and the recipient of numerous awards and prizes. He was on the cover of TIME magazine's "America's Best in Science & Medicine" feature in 2001 for his work with human embryonic stem cells and again in 2008 when the magazine named him one of the world's 100 most influential people for his derivation of human induced pluripotent stem cells.

Box 3.3 Figure 1 **James Thomson.** (Used with permission. Photo credit: David Nevala for Morgridge Institute for Research.)

### TABLE 3.2  CRITERIA USED TO DEFINE ES CELL STATUS

| Criteria for ES cells | Mouse ES cells | Human ES cells |
|---|---|---|
| Self-renew indefinitely *in vitro* and maintain a normal karyotype | + | +[a] |
| Differentiate *in vitro* into many cell types representative of the three embryonic germ layers | + | + |
| Contribute *in vivo* to the formation of every cell in the embryo, including the germ line, when reinjected into an early embryo | + | − |
| Form tumors (teratomas) when injected into an adult organism | + | + |

[a] Human ES cells can exhibit frequent chromosomal changes as they adapt to culture.
*Notes:* The table highlights the four main criteria that need to be satisfied in order to define an ES cell. The extent to which either mouse or human ES cell lines satisfy these criteria is indicated.

**A**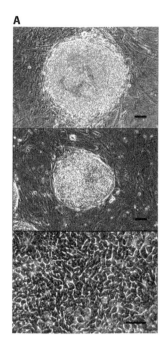

**B**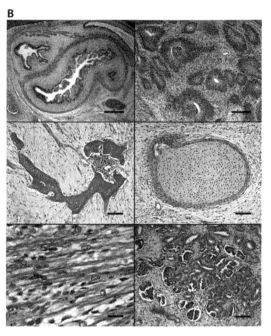

Figure 3.7 **The first isolation of human ES cells.** In 1998, James Thomson was the first researcher to demonstrate that ES cells could be produced from human embryos. The light microscope images represented in this figure illustrate the human ES cells created and the testing of their capability to differentiate into a wide range of tissues. (**A**) Inner cell mass attached to a mouse embryonic fibroblast feeder layer after 8 days of culture (top), when they form into a colony (middle), and after dispersion of the initial colony growing on fresh feeders (bottom). Scale bar: 100 μm. (**B**) Micrographs of histological sections of teratomas formed by human ES cells following their injection into the rear leg muscles of immunocompromised mice. The images show gut-like structures (top left), neural epithelium (top right), bone (mid left), cartilage (mid right), striated muscle (bottom left), and fetal glomeruli (bottom right). (From Thomson J, Itskovitz-Eldor J, Shapiro S et al. (1998) *Science* 282: 1145–1147 [doi:10.1126/science.282.5391.1145]. Reprinted with permission from AAAS.)

infertility treatment and was essentially the same as that originally used to produce the first murine ES cells. Their method included removal of the trophectoderm by immunosurgery, as had been used by Gail Martin in her derivation of mouse ES cells, followed by plating on MEF.

Let us step backward a moment to the defining features of an ES cell that arose out of the work on the mouse. A strict definition of a truly pluripotent ES cell requires four criteria to be satisfied (see Table 3.2). All are met by murine ES cells, but human equivalents could only ever be expected fully to satisfy three of the criteria and partially adhere to the fourth, that is, demonstration of differentiation potential *in vitro* and *in vivo*, the latter including contribution to the germ line in a chimera. Any test of pluripotency of the human ES cell-like cells had to rely on surrogate assays since there is no possibility of conducting the sorts of chimera generating experiments that so clearly demonstrate the potential of murine ES cells. The gold standard in the case of human ES cells involves injection of cells into immune-compromised mice to test for the formation of teratomas and evidence that these tumors contain differentiated cells representative of the three embryonic germ layers, as was indeed demonstrated by James Thomson in the first lines that he generated (Figure 3.7).

## Embryonal germ cells, derived from developing gonads, are also pluripotent

The early studies on teratocarcinomas in mice showed that they could arise from primordial germ cells, either spontaneously or following transplantation, and this led in 1992 to Brigid Hogan and her colleagues deriving what they termed **embryonal germ (EG)** cells using conditions very similar to those used to produce ES cells. These EG cells satisfy the criteria for pluripotent cell lines, being very similar to their mouse ES cell counterparts, with the exception of some subtle differences in gene imprinting correlating with the stage of development from which the two cell types originate. Following the mouse work, and close on the heels of the isolation of human ES cell lines, human equivalents of the mouse EG cells were first cultured in 1998, using pre-meiotic fetal gonads dissected from first trimester fetuses obtained following pregnancy terminations.

The derivation of EG lines from PGC requires a poorly understood process involving **epigenetic** reprogramming. The efficiency of this process was found to be very much enhanced using the 2i technique that we already mentioned in connection with improvements to ES cell generation. The mouse EG cells obtained this way demonstrate every aspect of a fully pluripotent stem cell line, including germ line contribution in chimeras, and point the way to an easier route to obtain EG cells from other species. Rat EG cells with all of the same properties have already been produced, and future research will determine whether the same is possible in other mammals.

## Although pluripotent, there are differences between mouse and human embryonic stem cells and between embryonic stem cells and embryonal germ cells

Notwithstanding the fact that it is not possible to make an exact comparison between human pluripotent stem cell lines and the presumed equivalents from mice and other mammalian species, it was obvious to those who isolated such cells subsequent to mouse ES cells that everything was perhaps not the same. For a start, the basic appearance of cells as they grew in culture was different (see Figure 3.7), forming flatter, less tightly packed colonies compared to the mouse equivalents. More significantly, there were obvious differences in the requirements for growth factors to aid maintenance of **self-renewal** in culture. Hence, although LIF very effectively supports murine ES and EG cells, the human counterparts require fibroblast growth factor (FGF) and activin (Table 3.3).

Although at the gross karyotype level, the majority of human ES cell lines are normal upon derivation, they show a greater propensity than mouse ES cells to acquire chromosome abnormalities as they adapt to culture (Figure 3.8). The predominant genetic changes found involve alterations in chromosome number and structure (particularly of chromosomes 1, 12, 17, and 20). Global gene expression analyses reveal that the frequent gain of chromosome 12 renders the ES cells more similar to germ cell tumors, something that is obviously not desirable if human ES cells are to have clinical application.

## The isolation of mouse epiblast stem cells helped to resolve why embryonic stem cells from mice and humans appear to differ

A likely explanation for the distinctive characteristics of murine and human ES cells was found in 2007 when Ron McKay and his team showed that they could isolate the pluripotent tissue represented by the epiblast in the early post-implantation mouse embryo and from this grow cells on MEF with features of pluripotent stem cells. These cells expressed the transcription factors Oct4 and Nanog that are characteristic of pluripotency and were termed **epiblast-derived stem cells (EpiSC).** Compared to the small, domed colonies seen when murine ES cells are cultured, the mouse EpiSC colonies were larger and grew as a monolayer similar to human ES cells (Figure 3.9). Analysis of their global gene expression profile distinguishes EpiSC from ES cells and the ICM, and, tellingly, the murine EpiSC have a requirement for the same growth factors that support human ES cells, namely FGF and activin, rather than LIF. EpiSC can be differentiated *in vitro* to multiple lineages and will form teratomas *in vivo*, but are unable to incorporate into the ICM to produce a chimera. The current view is that human ES cells are derived from the epiblast rather than the ICM, presumably because the embryos used mature *in vitro* to the later stage of development, while murine ES cells represent a more immature stage. These two states of pluripotency are referred to as "primed" and "naïve," respectively.

**TABLE 3.3  DIFFERENCES IN THE PROPERTIES OF PLURIPOTENT STEM CELL LINES DERIVED FROM DISTINCT SOURCES**

| Pluripotent stem cell line | Colony morphology | Growth factor requirement |
|---|---|---|
| mES | Tight, rounded, multi-layer clumps | LIF, BMP |
| mEG | Tight, rounded, multi-layer clumps | FGS/SCF, then LIF |
| mEpiSC | Flat, loose aggregates | FGF, activin |
| hES | Flat, loose aggregates | FGF, activin |
| hEG | Tight, rounded, multi-layer clumps | FGF |

*Notes:* Each pluripotent stem cell type has characteristic features when growing as colonies, including their morphology and their dependence on specific growth factor combinations. The table highlights how different pluripotent cell types vary in their colony morphology and growth factor requirements (mES—murine embryonic stem cells; mEG—murine embryonal germ cells; mEpiSC—murine epiblast stem cells; hES—human embryonic stem cells; and hEG—human embryonal germ cells).
*Abbreviations*: LIF—leukemia inhibitory factor; BMP—bone morphogenetic protein; FGF—fibroblast growth factor; SCF—stem cell factor.

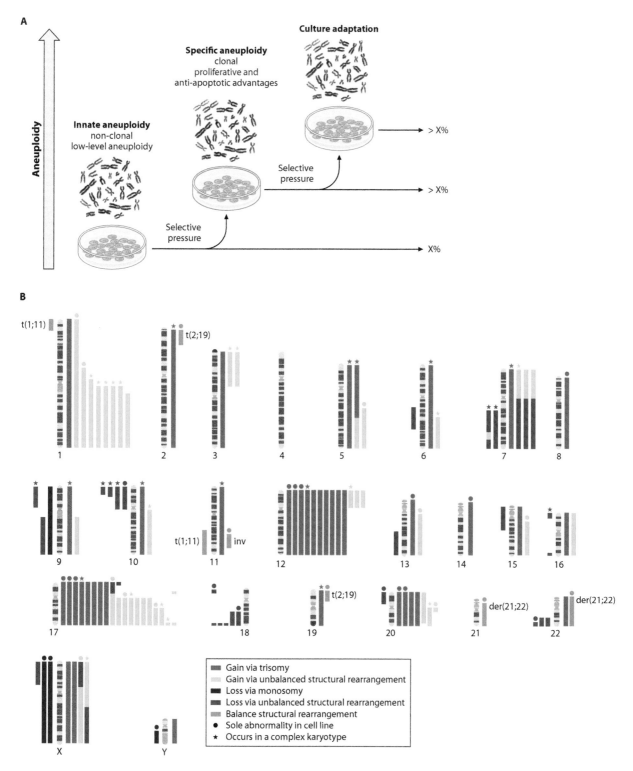

**Figure 3.8 Human ES cells exhibit genetic instability.** The culturing of any cell for prolonged periods, which inevitably imposes a selective pressure for improved growth and consequent genetic change, is a particular problem when growing human ES cells. (**A**) Changes to the genetic content of cells through partial or complete losses or gains of chromosomes leads to a state of aneuploidy. This will occur at a low frequency under normal growth but can become enriched for if the culture conditions are not optimal and a selective pressure is applied for proliferative and anti-apoptotic advantage. (**B**) Diagrammatic representation summarizing a large number of analyses of the chromosomal content of human ES cell lines. Against each of the 22 autosomal and 2 sex chromosomes, bars represent losses (red) or gains (green) of chromosome segments. Each colored bar represents one chromosome change occurrence in one cell line, dark colors being whole chromosomal changes and lighter colors representing a partial chromosomal change. Blue bars depict balanced rearrangements, that is, where chromosomal segments have been swapped between different chromosomes. (A: From Devalle S, Sartore R, Paulsen B et al. (2012) *Front Cell Neurosci.* 6:36 [doi:10.3389/fncel.2012.00036]. Published under the terms of the CC BY license. B: From The International Stem Cell Initiative, Amps K, Andrews P et al. (2011) *Nat Biotech*, 29: 1132–1144 [doi:10.1038/nbt.2051]. Adapted by permission from Springer Nature.)

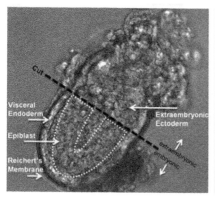

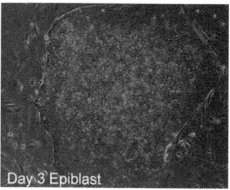

Figure 3.9 **Mouse epiblast stem cells.** Pluripotent cells exist in the developing embryo beyond the blastocyst stage and, like the ICM, can be isolated and cultured as stable cell lines. The epiblast, which derives directly from the ICM and gives rise to the embryo proper through its differentiation into the three primary germ layers, contains pluripotent cells that can yield cell lines resembling ES cells yet with a number of distinct properties. Mouse EpiSC lines are derived from isolated E5.5 murine epiblasts (image on left, indicated by the dashed white line and dissected at the dashed black line), which, when grown on feeders, generate large flat colonies after 3 days in culture (right). (From Chenoweth J & Tesar P. (2010) Isolation and maintenance of mouse epiblast stem cells. In: Ding S (ed) *Cellular Programming and Reprogramming.* Methods in Molecular Biology (Methods and Protocols), vol 636. Humana Press. Reprinted by permission from Springer Nature.)

## Embryonic stem cells can be differentiated into most tissue cell types

Aside from their capacity for indefinite self-renewal under the right conditions, the most important feature of ES cells is their ability to differentiate into multiple mature cell types. Although a few cells will spontaneously differentiate within a culture of murine or human ES cells, the key issue is how this can be controlled, especially to achieve commitment of cells to just one of the myriad of possible pathways of differentiation. Simple removal of those aspects of the culture conditions favoring self-renewal, that is, the specific growth factors and adhesion to the culture vessel, is enough to initiate differentiation, although some of the possible differentiation protocols involve culture of ES cells either as one-cell-thick layers on extracellular matrix proteins or directly on supportive **stromal cells** (Figure 3.10).

The most widely used technique for ES cell differentiation involves aggregation of the cells in suspension culture, leading to the formation of hollow spheres, or embryoid bodies. A variant of the basic method places the ES cells in individual droplets of culture medium, in the so-called hanging drop method. Embryoid bodies formed either way contain cells of the three embryonic germ layers, the differentiated cells appearing with a timing that roughly mirrors post-implantation development of an **embryo**. Although not completely random, making this process of EB differentiation quantitative and directed (that is, each ES cell becoming the desired cell type) has required the design of detailed culture strategies. Further outgrowth and differentiation of specific cell lineages entails dissociation of the EB and culture in defined media containing combinations of growth factors and other selective components. Knowledge of the cell and molecular mechanisms underpinning embryogenesis has helped define the ideal conditions for specific lineage differentiation, and it is now possible to obtain most mature adult cells from both mouse and human ES cells. The development of "reporter" ES cell lines (Chapter 2), both for murine and human cells, has very much helped the development of differentiation strategies, enabling observation and purification of cells as they commit to the desired lineage. A good example is the production of fully functional

Figure 3.10 **Differentiation of ES cells.** ES cells can be differentiated into cells representing the three germ layers under a variety of conditions. The three basic protocols are illustrated. Left: Aggregation of ES cells in conditions that do not allow them to adhere to the culture dish. A variant uses suspension of the ES cells in a drop of medium that is suspended from the surface of an inverted culture dish (the "hanging drop" method). Aggregates of ES cells develop into an embryo-like architecture in the form of an embryoid body with primitive endoderm outside and other lineages inside. Middle: A monolayer of feeder cells, such as OP9 stromal cells, is used to seed the dispersed ES cells, providing support and growth factors conducive to differentiation. Right: Simple culture in which ES cells are placed on a defined matrix, such as collagen IV. (From Nishikawa S, Jakt L & Era T. (2007) *Nat Rev Mol Cell Biol* 8; 502–507 [doi: 10.1038/nrm2189]. Adapted by permission from Springer Nature. Images From Lei X, Deng Z & Duan E. (2016) Uniform embryoid body production and enhanced mesendoderm differentiation with murine embryonic stem cells in a rotary suspension bioreactor. In: Turksen K (ed) *Bioreactors in Stem Cell Biology.* Methods in Molecular Biology, vol 1502. Humana Press. Reprinted by permission from Springer (embryoid body) and https://www.sigmaaldrich.com/life-science/stem-cell-biology/mouse-embryonic-fibroblasts.html (stromal cells) and https://www.sigmaaldrich.com/life-science/stem-cell-biology/stem-cell-protocols.html (matrix). Reproduced with permission from Sigma-Aldrich Co. LLC.)

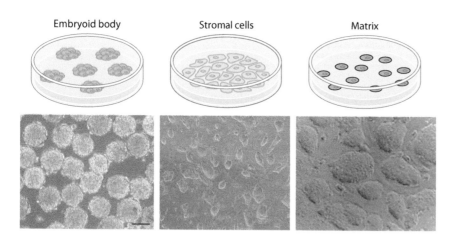

Embryoid body          Stromal cells          Matrix

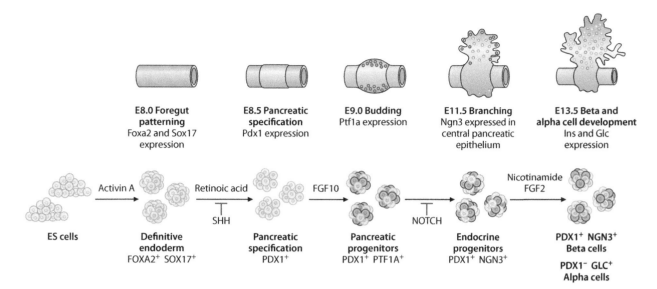

Figure 3.11 **ES cell differentiation parallels normal embryo development.** Reflecting the fact that ES cells are cultured equivalents of the ICM, they theoretically have the capacity to differentiate into any of the cell types that characterize the developing embryo. However, since ES cells are not in the context of the embryo, culture conditions need to be identified that can replicate the critical environmental influences. Such conditions have been defined for many differentiation pathways, one of the most highly investigated being that which gives rise to the endocrine and exocrine cells of the pancreas. The upper schematic represents normal development of the pancreas in the mouse embryo from embryonic foregut endoderm. Endocrine progenitors and mature Beta cells are indicated by red and blue dots, respectively. The lower part of the figure shows a differentiation strategy used to generate insulin-producing Beta cells from human ES cells. The different stages of the process are aligned with the approximately equivalent stages of development in the mouse embryo. The key transcription factors involved in the specification of the pancreas, FOXA2, SOX17, PDX1, PTF1a, and NGN3, are indicated where they are important. The first indication of the emerging pancreas is marked by expression of PDX1. PTF1A marks the population (red) that will give rise to the pancreas. Subsequent expression of NGN3 within the pancreatic epithelium defines a progenitor population for all endocrine lineages, including the Beta cells. (From Murry CE & Keller G (2008) *Cell* 132: 661–680 [doi:10.1016/j.cell.2008.02.008]. Copyright 2008 with permission from Elsevier.)

pancreatic β cells (Figure 3.11), a process that as recently refined with human ES cells is making regenerative therapy for type I diabetes a very real prospect.

## The generation of parthenogenetic and haploid embryonic stem cells increases the potential avenues for genetic manipulation

All of the pluripotent cells that we have considered until this point are derived from diploid cells, with the possible exception of EC cells that are variably aneuploid. As we have seen, mouse ES cells have been especially useful for "reverse" genetic screening, that is, manipulating a gene and determining the consequent phenotype *in vivo* (as opposed to classical "forward" genetics in which mutagenesis causing a phenotype leads toward the identification of the responsible genes). If the genome being manipulated were haploid instead of diploid, then forward genetics, including on genes with potentially recessive effects, would be feasible in ES cells. The first attempts to produce haploid ES cells were made by Martin Evans and his colleagues not long after their first ES cells had been created from murine blastocysts. Their approach was to make use of **parthenogenesis** (Figure 3.12A,B), the process in which an oocyte develops into an embryo without fertilization. This can be achieved by chemical exposure of mouse eggs and selection of eggs that developed a single pronucleus following polar body extrusion. Such eggs were transferred to suitable "foster" recipients and blastocysts subsequently recovered and used to generate haploid-derived pluripotent cell lines that were capable of forming well-differentiated teratocarcinomas. However, chromosome analysis of early passage cell lines revealed that all were diploid and therefore of no advantage for genetic engineering. More recently, stable haploid ES cell lines have been produced, again following chemical activation of an egg but now incorporating LIF and the 2i inhibitor cocktail combined with cell sorting to select haploid cells (Figures 3.12C and 3.13). Once stabilized, these haploid ES cells can become part of a chimera when injected into blastocysts but rapidly become diploid as they commit to differentiation. In 2007, a California-based stem cell research company announced that they had made human ES cells from unfertilized human eggs using parthenogenesis.

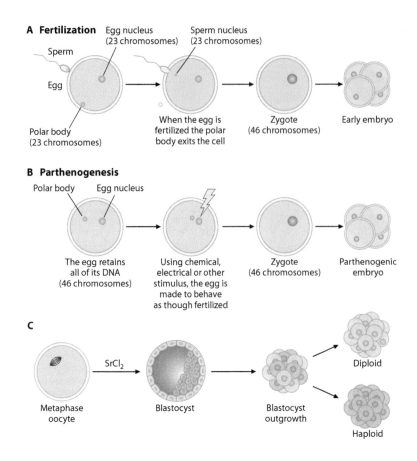

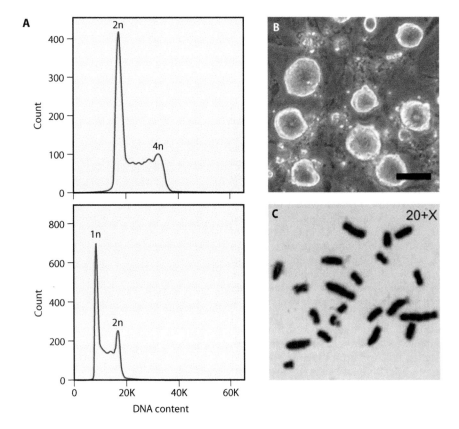

**Figure 3.12 Parthenogenesis.**
Parthenogenesis is a natural form of asexual reproduction in which development of an embryo occurs from an unfertilized egg cell. Since the egg only has one copy of each chromosome, and haploid individuals are usually non-viable, this means that diploidy has to be restored in order to achieve parthenogenetic offspring, although this may only apply with respect to some of the chromosomes. Comparing normal fertilization (**A**) with parthenogenesis (**B**) shows how the normal requirement for biparental acquisition of a full set of chromosomes can be circumvented through stimulation of the oocyte such that the polar body is retained, although if the polar body is not recruited, then the alternative is haploid cell growth. (**C**) Parthenogenesis can be utilized to generate haploid embryos and hence ES cells by a variety of means. In the example given, the simplest strategy involves treatment of an oocyte with strontium chloride, which can give rise to haploid embryo development, although some cells will convert to diploidy within the ICM. By cloning cells from the blastocyst outgrowth, it is possible to isolate ES cell clones that are either uniparental diploid or haploid (A, B: From President's Council on Bioethics 2002, p. 61. C: From Wu B., Li W., Wang L. et al. (2013) *Acta Pharmacologica Sinica*, 34: 725–731 [doi:10.1038/aps.2013.26]. Adapted by permission from Springer Nature.)

**Figure 3.13 Haploid ES cells.** Haploid ES cells can be generated from mouse and rat embryos and serve as a powerful tool for genetic engineering. Haploid embryos can be produced by activation of unfertilized oocytes, for example, using strontium chloride. Once developed *in vitro* to the blastocyst stage, the ICM can be isolated and cultured in defined medium in the presence of LIF and 2i medium for ES cell derivation. (**A**) Flow cytometric analysis of DNA after propidium iodide staining of diploid control ES cells (top) or a haploid ES cell line (bottom). The overall DNA content is indicated (2n is the normal diploid DNA content in non-dividing cells). (**B**) Micrograph showing the morphology of haploid rat ES cells colonies. (**C**) Chromosome spreads of haploid rat ES cells revealing the normal complement of 21 chromosomes (A: From Leeb M & Wutz A (2011) *Nature*, 479:131–134 [doi:10.1038/nature10448]. Reprinted by permission from Springer Nature. B and C: From Li W, Li X, Li T et al. (2014) *Cell Stem Cell* 14: 404–414 [doi: 10.1016/j.stem.2013.11.016]. Copyright 2014 with permission from Elsevier.)

# 3.2    PLURIPOTENT CELLS DERIVED BY CLONING

The generation and application of ES cells derived from human embryos surplus to *in vitro* fertilization programs raises not only potential ethical issues but yields cells whose differentiated derivatives can only be used in a limited number of potential recipients because of tissue type mismatch. Spurred on largely by these two major problems, considerable research effort has been focused on ways to revert adult somatic cells to a pluripotent state. Such reversion of differentiated cells requires a complete reprogramming with respect to their epigenetic status and gene transcription profile and can be achieved by one of two broad approaches. The first utilizes an oocyte to elicit reprogramming, presumably through regulatory molecules (proteins, RNAs, etc.) present in the egg cytoplasm, while the second takes advantage of knowledge of the molecular basis of pluripotency, using over expression of key transcription factors in adult somatic cells to duplicate the conditions present in pluripotent stem cells. The latter approach produces **induced pluripotent stem (iPS) cells**, which we will discuss in much greater detail in Chapter 5.

## Cloning of an adult cell nucleus involves resetting of development and recapitulation of embryogenesis

The way in which oocyte cytoplasm is employed to reprogram an adult differentiated cell is through the process generally known as **nuclear cloning**, or just **cloning**. The essential features of cloning are that the nucleus of an oocyte is removed and replaced by a nucleus obtained from the cell to be cloned. The reconstructed zygote can then develop as an embryo. At this point, two paths are possible, namely one that leads to the development of a live animal if the cloned embryo is implanted into the uterus of a foster mother (so-called **reproductive cloning**) and one that takes the embryo only as far as the blastocyst stage, which can serve as a source of pluripotent ICM cells to be used for the generation of ES cells in culture (usually termed **therapeutic cloning**) (Figure 3.14).

   The origins of reproductive cloning from adult cells go back to seminal studies in the early 1960s by John Gurdon (Box 3.4) in Cambridge, England. Gurdon demonstrated that a nucleus from the intestinal epithelium of a tadpole (from a species known as the African Clawed Toad, even though it is a frog) could be reprogrammed following introduction into an enucleated oocyte, leading to the formation of a viable embryo that developed into an exact copy of the donor animal (Figure 3.15). This approach to cloning is usually termed **somatic cell nuclear transfer (SCNT)**. The significance of Gurdon's work was eventually recognized through the award of a Nobel Prize, which he shared with Shinya Yamanaka who, as we will see in Chapter 5, led the field of iPS cell research.

## The cloning of a sheep from an adult cell changed the course of animal cloning

Of course, any discussion of reproductive cloning cannot neglect Dolly the Sheep (Box 3.5). Dolly was not the first cloned mammal, but she was the first mammal to be cloned from an adult cell. Since 1996, when Dolly was born, other mammals have been cloned from adult cells, including mice, rabbits, horses and donkeys, pigs, goats and cattle. In most cases, the donor nucleus has been derived from a cell that was still capable of division, but in 2004, a mouse was cloned using an olfactory neuron, showing that the donor nucleus can come from a tissue of the body that does not normally divide.

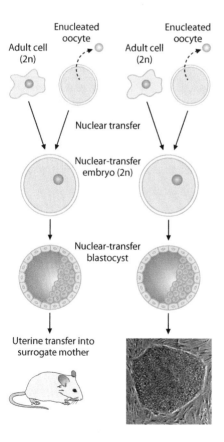

**Figure 3.14 Reproductive and therapeutic cloning.** Organisms, including mammals, can be cloned through reprogramming an adult cell nucleus by introducing it into an oocyte from which its own nucleus has been removed. Once the nucleus has been transferred, usually by direct injection, the embryo can develop *in vitro* to the blastocyst stage. At this point, it can either be transferred into the uterus of a foster mother, where it can develop and be born (Left: "reproductive cloning"), or the ICM can be isolated and cultured to enable derivation of ES cells (Right: "therapeutic cloning"). (Adapted from Hochedlinger K & Jaenisch R. (2003) *N Engl J Med* 349: 275–286 [doi: 10.1056/NEJMra035397].)

**BOX 3.4 PIONEER: JOHN GURDON**

John Bertrand Gurdon (Figure 1), born in 1933 into a long lineage of Gurdons, one ancestor even having shot Richard the Lionheart, can be thought of as the father of animal cloning. His path toward this accolade was not straightforward. His father encouraged him to join the army or to study business, while his mother wanted him to study science. The army was ruled out by a bout of bronchitis. Famously, his teacher at Eton reported that he was the worst student he had ever taught in biology, and he was prevented from studying science for the last three years of his school education (Figure 2), focusing instead on the Classics. In spite of this discouragement, Gurdon ended up studying zoology at Oxford University. His independent thinking spirit shone through even in his early days as an undergraduate when he discovered a new species of sawfly in a wood near Oxford. In spite of this success, the professor of entomology in Oxford refused to accept Gurdon as a PhD student, which turned out to be fortuitous since he instead worked toward a PhD in embryology under the direction of Michael Fischberg, an eminent developmental biologist. Gurdon's supervisor suggested that he should work on somatic cell nuclear transfer using a procedure that had been published in 1952 by Briggs and King. The rest is history, as they say! The cloning of a nucleus derived from an epithelial cell of the frog Xenopus was the basis for all reproductive and therapeutic cloning, leading ultimately to Gurdon's award of the Nobel Prize for Physiology or Medicine in 2007, which he shared with Shinya Yamanaka, the originator of reprogramming of adult cells using pluripotency transcription factors (Chapter 5).

Box 3.4 Figure 1 **John Gurdon.** (Used with permission from Dr. John Gurdon.)

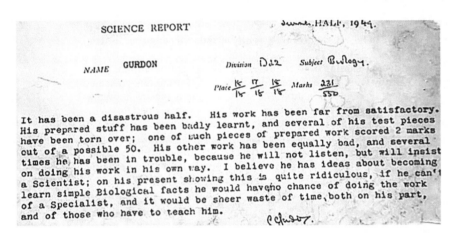

Box 3.4 Figure 2 **John Gurdon's school report.** (Used with permission from Dr. John Gurdon.)

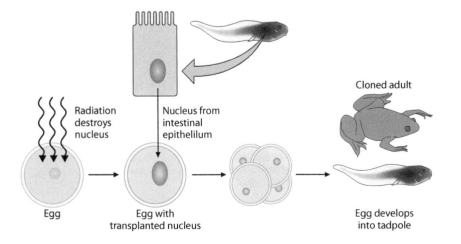

Figure 3.15 **Cloning of a frog from an intestinal epithelium cell.** The first successful cloning of a vertebrate from an adult cell was performed by the seminal experiment performed by John Gurdon. In this experiment, an intestinal epithelium cell taken from a tadpole provided a nucleus that was transferred into an egg that had had its own nucleus destroyed by radiation. The reconstituted egg was then allowed to develop and duly gave rise to a tadpole, which was a perfect copy, or clone, of the donor tadpole. (From illustration by Mattias Karlen for the 2012 Nobel Committee for Physiology or Medicine. Used with permission. Copyright © The Nobel Committee for Physiology or Medicine.)

BOX 3.5 BACKGROUND: IAN WILMUT AND THE CLONING OF DOLLY THE SHEEP

Ian Wilmut (Figure 1), born in the English Midlands, had an early interest in farming, which led him to study for a degree in agriculture at the University of Nottingham. He soon began to concentrate on embryology and received his doctorate at the University of Cambridge in 1971 for work on the freezing of boar semen. In 1973, he was part of the team that produced the first calf from a frozen embryo, an animal the team named Frosty (preluding a series of amusing names for the animals that he generated through

his experiments). In 1974, he moved to the Animal Breeding Research Station in Edinburgh, which later became the Roslin Institute. Wilmut's work on cloning with his colleague Keith Campbell covered over a decade, culminating first in 1996 with the production of a pair of lambs, Megan and Morag, from embryonic cells. This only really impacted the scientific community, but a very different reaction resulted February 1997, when Wilmut and Campbell announced the birth of a very special lamb (Figure 2). This lamb had been created from the fusion of an enucleated egg from a Scottish Blackface ewe with a mammary epithelial cell from a six-year-old Finn Dorset white sheep, which was then cultured for about a week before being implanted into the uterus of a surrogate mother, another Scottish Blackface ewe. Again, Wilmut indulged his whimsical side in naming the lamb Dolly, combining the tissue origin of the cloned cell (the udder) with thoughts of the country and western singer Dolly Parton! The announcement created a sensation and stirred fears in the general public that the cloning of a human being would be the next step. Dolly died of a respiratory ailment in 2003, a possible consequence of the imperfect nature of the reprogramming process during the cloning. Wilmut and Campbell's achievement opened the door on the field of both reproductive cloning of livestock and other commercially important species and therapeutic cloning for research purposes and ultimately the generation of personalized ES cells.

Box 3.5 Figure 1 **Ian Wilmut with Dolly.** (Photo courtesy of The Roslin Institute, The University of Edinburgh.)

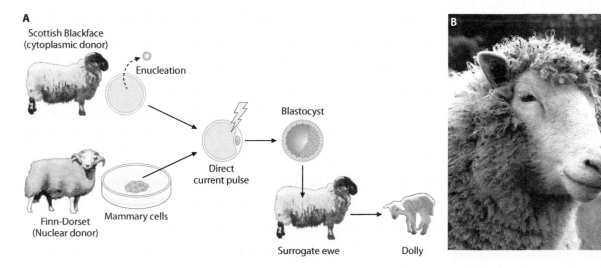

Box 3.5 Figure 2 **Dolly the sheep.** In much the same way as John Gurdon cloned an adult frog skin cell, so Ian Wilmut cloned a sheep, and therefore the first adult mammal, although technologically this proved far more challenging. The now infamous Dolly the Sheep was cloned from a mammary epithelial cell and survived as a fully developed adult for almost seven years. (**A**) Diagram showing the reproductive cloning strategy using a mammary epithelial cell from a donor Finn-Dorset ewe, a recipient egg from a Scottish Blackface ewe, and a Scottish Blackface ewe foster mother. (**B**) Dolly the Sheep. (Photo courtesy of The Roslin Institute, The University of Edinburgh.)

## Creating somatic cell nuclear transfer embryonic stem cells has proved to be technically challenging and controversial

As already mentioned, cloning of adult cells through the route of somatic cell nuclear transfer offers the possibility of generating pluripotent stem cells with the genetic characteristics of individual animals, which have obvious benefits if the ultimate aim is to produce replacement cells and tissues for transplantation purposes. In humans,

### BOX 3.6 CONTROVERSY: HWANG WOO-SUK'S CLAIMS OF SCNT-ES CELL GENERATION

Hwang Woo-suk (Figure 1) is a South Korean veterinarian and researcher who, working as a professor at Seoul National University, became infamous for fabricating a series of experiments purporting to demonstrate the generation of personalized human ES cells by SCNT. Hwang was a recognized expert in animal cloning who had secured celebrity status in South Korea in the late 1990s. In February 2004, quite to the surprise of the human ES cell research community, Hwang announced that he had created an ES cell using the SCNT method and subsequently published his results in the journal *Science*. Hwang's team announced an even greater achievement a year later, which was also published in *Science*, claiming to have created 11 human ES cell lines using 185 eggs. This was seen as an incredible breakthrough because the ES cells had supposedly been created with somatic cells from patients of different ages and genders, while the first ES cell line derived from eggs and somatic cells from a single female donor. He was called the "Pride of Korea" in South Korea. However, a few months later, in November 2005, Gerald Schatten, a University of Pittsburgh researcher who had collaborated with Hwang, announced that he was cutting off his links with Hwang, commenting "my decision is grounded solely on concerns regarding oocyte (egg) donations in Hwang's research reported in 2004." Following an intense media probe, Roh Sung-il, one of Hwang's close collaborators, admitted that he had paid women co-workers US$1,400 each for donating their eggs to be used in Hwang's research. In a press conference shown by all major South Korean television stations, Hwang denied coercing his researchers into donating eggs, and there was an immediate groundswell of national support for him, resulting in an increase in the number of women who wanted to donate their eggs for his research. However, investigations continued in the background, and his university concluded that all 11 of Hwang's ES cell lines had been fabricated, declaring in January 2006 that Hwang's two *Science* papers were fraudulent. The journal retracted both papers on unconditional terms. In May of that year, Hwang was charged with embezzlement and bioethics law violations, and he was

Box 3.6 Figure 1 **Hwang Woo-suk.** Front cover of *Time* magazine from January 2006 featuring Hwang Woo-suk and the story behind the fraudulent claims of human somatic cell nuclear transfer (SCNT) ES cells. (From KIM KYUNG-HOON/REUTERS/Newscom.)

subsequently fired by Seoul National University, and the South Korean government canceled his funding and banned him from conducting further stem cell research.

Ironically, the examination of the chromosomes of the fraudulent ES cell lines led to the discovery in 2007 that Hwang had unknowingly produced embryos resulting from parthenogenesis. This made Hwang the first to successfully perform the process of parthenogenesis to create a human embryo and, ultimately, a human parthenogenetic stem cell line.

---

SCNT has been regarded as a possible means to generate personalized ES cells from a patient's own somatic cells, which could then be used potentially to study disease mechanism or for cell-based regenerative therapies. However, in spite of well over a decade of research, attempts to produce SCNT-ES cells were unsuccessful, and in some cases blatantly fraudulent (Box 3.6), and it took until 2013 to devise a method that would enable what had been already been achieved in other mammals.

The stumbling block to the successful generation of SCNT ES cells was usually arrest of embryos at the eight-cell stage. The solution, discovered by Shoukhrat Mitalipov and colleagues in Oregon, was to modify oocyte enucleation and donor nucleus introduction so as to retain meiosis factors during the reprogramming process. This resulted in embryos capable of progression to the point from which ES cells could readily be derived.

### Mitochondrial contamination means that somatic cell nuclear transfer clones and embryonic stem cells derived from them are not exact copies of the donor individual

The mitochondria in the cytoplasm also contain DNA, and during SCNT, this mitochondrial DNA is largely derived from the enucleated oocyte, with very little carry over of mitochondria together with the donor nucleus (Figure 3.16). This distinct genetic origin of the mitochondrial and nuclear genomes in SCNT clones and ES cells

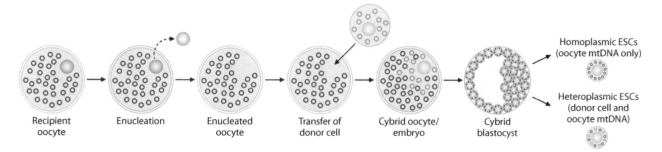

**Figure 3.16 Mitochondrial heteroplasmy.** The introduction of a cell nucleus into an enucleated oocyte also results in the transfer of mitochondria and their genetic material from the donor cell. As the reconstituted zygote develops into an embryo, mitochondria from the recipient oocyte and those introduced from the donor both replicate. The diagram depicts an oocyte reconstructed from a donor somatic cell nucleus (blue) and an enucleated host egg (red) showing the two populations of mitochondrial (mt) DNA (heteroplasmy) due to carry over of some donor mitochondria with the transplanted nucleus. The population of mtDNA present in ES cells derived from such embryos will initially be representative of the mtDNA populations present at oocyte reconstruction, but one population may be preferentially selected for, which could in turn result in homoplasmy. (From St John J & Lovell-Badge R (2007) *Nat Cell Biol* 9: 988–992 [doi: 10.1038/ncb436]. Adapted by permission from Springer Nature.)

derived from them has a number of important implications. First, this means that any cells derived through this route for personalized therapy are not exactly equivalent to the patient, although the genetic component deriving from the reprogramming oocyte mitochondria is miniscule in comparison to the donor's nuclear DNA and will not impact function or immunological acceptance. More importantly, when the reprogramming oocyte is derived from a mammalian species distinct from the donor nucleus, nuclear-mitochondrial incompatibilities may lead to death.

## 3.3 PLURIPOTENT CELLS FROM NON-MAMMALIAN SPECIES

Pluripotent cells are not restricted to mammals, and, as we will see in Chapter 7, there is good evidence for stem cells capable of giving rise to the whole organism as far back in the evolutionary tree as the dawn of multicellular animals. For example, some species of invertebrate flatworms contain stem cells known as neoblasts that are able to repopulate all differentiated tissues. The greatest concentration of research into non-mammalian pluripotent stem cells, especially attempts to derive the equivalents of ES cells, has focused on birds and fish, largely because of the potential agricultural uses that such cells could be put to.

### Embryonic stem cells can be produced from pluripotent cells that arise during embryogenesis in birds

Embryonic development in birds, which has been predominantly studied in chickens, is in many respects the same as in mammals, including the importance of pluripotent cells. After fertilization, embryo growth occurs rapidly in the oviduct before the egg is laid about 20 hours later. This phase of development is divided into fourteen stages (I–XIV), the crucial one from our perspective being Stage X (Figure 3.17). At this point, the embryo consists of 20,000 to 50,000 blastodermal cells that, like the ICM in the mouse pre-implantation embryo, are pluripotent and if injected into a recipient embryo will contribute to a chimera including the germline. Again paralleling work with mice, it was found that Stage X blastodermal cells could be cultured on irradiated embryonic feeder cells and, with the supportive action of a cocktail of growth factors, would yield ES cells. These chicken ES cells exhibit similar features to their murine and human counterparts, including intense alkaline phosphatase and telomerase activities, both of which are characteristic of the pluripotent state, and expression of homologs of many of the genes associated with pluripotency in mammals, such as Nanog, Sox2, and Klf4. Chicken ES cells also readily differentiate upon removal of

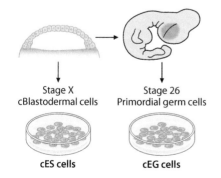

**Figure 3.17 Chicken ES cells.** The derivation of pluripotent stem cells from birds such as the chicken follows very much the same route as used for mammals. Chicken embryonic stem (cES) cells can be obtained from Stage X blastodermal cells that are cultured over feeder cells. As for mammals, it is also possible to derive pluripotent stem cell lines from primordial germ cells. In the chicken, PGC are isolated from Stage 26 of development and cultured on feeders with growth factor supplements to give rise to embryonal germ (cEG) cells.

growth factors and culture in suspension, forming embryoid bodies including cells of the three germ layers. Testing the potential of chicken ES cells for chimera formation is actually an easier undertaking in birds compared to mammals since they can simply be injected into a Stage X embryo, which will then continue to develop *in situ* in the egg without the need for a foster mother. Chimeras involving somatic cells are readily produced, although germline chimerism is harder to achieve; nevertheless, this does demonstrate the true pluripotent credentials of these cells.

Just as primordial germ cells can be obtained from the genital ridge of the mammalian embryo and cultured to produce embryonal germ cell lines, so it is possible to isolate chicken PGC from the developing gonads of a chicken embryo using much the same culture techniques. These EG cells can produce somatic chimeras when injected into Stage X embryos, but germline capability, which is rapidly lost in culture, can be realized if the cells are introduced into the forming gonads.

## Fish are the most primitive vertebrates so far to yield embryonic stem cells

As for birds, considerable effort has gone into the derivation of ES cells from fish, partly driven by an agricultural imperative. Some species of fish are also excellent models for the study of development and gene function, and ES cell technology would greatly enhance the range of possible investigations. For this latter reason, the majority of research on fish ES cells over the last two decades has focused on zebrafish and medaka (Figure 3.18). Their derivation is the same story all over again: isolated early embryo cells (from the blastula), feeder cells, and combinations of conditioned media and growth factors. In fact, it is relatively easy to obtain ES cells from fish, which adhere to all of those criteria established through the study of mouse ES cells.

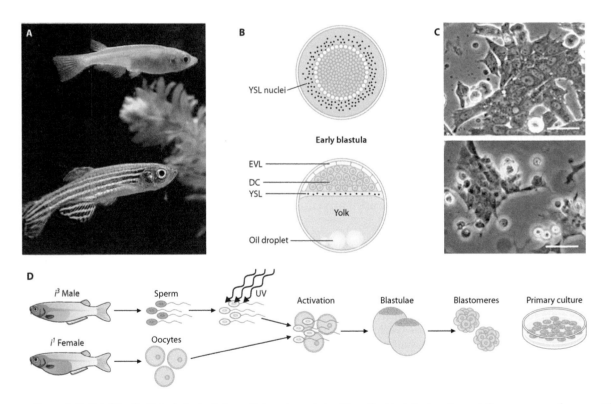

**Figure 3.18 Fish ES cells.** Fish pluripotent stem cells have been derived from the blastula stage in much the same way as for mammals, and likewise it has proved possible to manipulate embryonic development so as to create cells with a haploid chromosomal content. (**A**) The experimental fish medaka (top) and zebrafish (bottom) have been used in the derivation of fish equivalents of mammalian ES cells. (**B**) Fish ES cells can be obtained through the culture on feeders of cells from the early blastula, which is illustrated viewed from above (top) and the side (bottom). (**C**) An example of a culture of diploid (top) and haploid (bottom) medaka ES cells. The haploid fish ES cells have very similar properties to their diploid counterpart, although they are smaller. (**D**) It is relatively straightforward to derive haploid ES cells from medaka embryos. The diagram summarizes the process, crucially showing how sperm are inactivated by UV irradiation. (A: From http://www.bs.s.u-tokyo.ac.jp/~hassei/English/index.html. Reproduced with permission from Professor Hiroyuki Takeda. B: From Li Z, Bhat N, Manali D et al. (2011) *Int J Biol Sci* 7: 418–425 [doi: 10.7150/ijbs.7.418]. With permission from Ivyspring International Publisher. C, D: From Yi M, Hong N & Hong Y. (2010) *Nat Protoc* 5: 1418–1430 [doi:10.1038/nprot.2010.104. Reprinted by permission from Springer Nature.)

Haploid ES cells have also been derived from medaka, specifically a haploid version of the blastula stage embryo being produced using sperm that had had their nuclei destroyed by UV irradiation but still retained the ability to trigger egg activation (see Figure 3.18).

## 3.4 SUMMARY

In this chapter, we have described the origin and nature of pluripotent stem cells that can be derived in a laboratory using early embryos, a process which has predominantly been carried out in the mouse and humans but is now being extended to other species such as fish and birds. Based on knowledge of vertebrate development, embryos can be used as a source of naturally pluripotent cells, which exist transiently but can be "captured" through cell culture to produce essentially immortal cell lines. In addition, we have illustrated how the very earliest stage, that is, the egg before fertilization, can be employed following enucleation as a means to reprogram an adult cell nucleus, enabling it to once more progress through development. Since this reiterated embryogenesis includes the stage from which pluripotent stem cells can be derived, this provides a means to personalize the production of pluripotent stem cells. The themes of reprogramming and pluripotency will be covered further in Chapter 5 when we consider how pluripotent stem cells can actually be created without the need for an embryo, instead making use of our knowledge about the molecular mechanisms that dictate the pluripotent state.

---

## KEY POINTS

- Vertebrate embryo development includes cells with pluripotent characteristics in the pre-implantation and early post-implantation stages as well as the primordial germ cells (PGC) that emerge soon after.

- Pluripotent cells from the early mouse embryos give rise to tumors (teratomas) if transplanted into another mouse.

- Tumors known as teratocarcinomas can arise spontaneously from pluripotent cells, usually derived from PGC. Stem cell lines can be derived from these tumors and exhibit many features of pluripotent cells, although they are usually aneuploid.

- Pluripotent cells from the pre-implantation embryo inner cell mass can be used to derive embryonic stem (ES) cells by culture on feeder cells. These cells retain a normal karyotype and are capable of indefinite self-renewal, differentiation to all lineages, and contribution to both somatic and germ lines.

- Human ES cells are more similar to a distinct pluripotent stem cell isolated from the epiblast of the post-implantation mouse embryo.

- Embryonal germ (EG) cells can be derived from PGC in much the same way as ES cells and possess many of the same properties.

- ES cells will differentiate spontaneously *in vitro* when removed from stromal cell and growth factor support, forming embryoid bodies (EB) that mimic the early stages of development.

- ES cell differentiation can be directed through a combination of selective use of growth factors and other media components as well as incorporated reporters that can be used to highlight and select differentiation along specific pathways.

- Murine ES cells have been invaluable for genetic engineering, that is, the targeted manipulation of genes as a means to investigate their function.

- Cloning by nuclear transfer raises the possibility of being able to generate personalized ES cells, which has recently been successful in humans.

---

## FURTHER READING

### Developmental biology

Gilbert, S.F. and Barresi, M.J.F. (2016) *Developmental Biology*, 11th ed. Oxford University Press, Oxford.

### Cloning

Campbell, K.H., McWhir, J., Ritchie, W.A. and Wilmut, I. (1996) Sheep cloned by nuclear transfer from a cultured cell line. *Nature*, 380: 64–66.
Gurdon, J.B. (1962) The developmental capacity of nuclei taken from intestinal epithelium cells of feeding tadpoles. *J Embryol Exp Morphol.*, 10: 622–640.

Narbonne, P., Miyamoto, K. and Gurdon, J.B. (2012) Reprogramming and development in nuclear transfer embryos and in interspecific systems. *Curr. Opin. Genet. Dev.*, 22: 450–458.

### Creation of pluripotent stem cells

25 years of embryonic stem cells. https://www.nature.com/collections/lfxwyyhgrn#current
Bustamante-Marín, X., Garness, J.A. and Capel, B. (2013) Testicular teratomas: An intersection of pluripotency, differentiation and cancer biology. *Int J Dev Biol.*, 57: 201–210.

Evans, M.J. and Kaufman, M.H. (1981) Establishment in culture of pluripotential cells from mouse embryos. *Nature*, 292: 154–156.

Hong, N., Zhendong, L. and Hong, Y. (2011) Fish stem cell cultures. *Int J Biol Sci.*, 7: 392–402.

Lavial, F. and Pain, B. (2010) Chicken embryonic stem cells as a non-mammalian embryonic stem cell model. *Develop. Growth Differ.*, 52: 101–114.

Leib, M. and Wutz, A. (2011) Derivation of haploid embryonic stem cells from mouse embryos. *Nature*, 479: 131–134.

Martin, G.R. (1981) Isolation of a pluripotent cell line from early mouse embryos cultured in medium conditioned by teratocarcinoma stem cells. *Proc. Natl. Acad. Sci. USA*, 78: 7634–7638.

Tachibana, M., Amato, P., Sparman, M., Gutierrez, N.M., Tippner-Hedges, R., Ma, H., Kang, E. et al. (2013) Human embryonic stem cells derived by somatic cell nuclear transfer. *Cell*, 153: 1228–1238.

Tesar, P.J., Chenoweth, J.G., Brook, F.A., Davies, T.J., Evans, E.P., Mack, D.L., Gardner, R.L. and McKay, R.D.G. (2007) New cell lines from mouse epiblast share defining features with human embryonic stem cells. *Nature*, 448: 196–199.

Thomson, J.A., Itskovitz-Eldor, J., Shapiro, S.S., Waknitz, M.A., Swiergiel, J.J., Marshall, V.S. and Jones, J.M. (1998) Embryonic stem cell lines derived from human blastocysts. *Science*, 282: 1145–1147.

## Differentiation of embryonic stem cells

Murry, C.E. and Keller, G. (2008) Differentiation of embryonic stem cells to clinically relevant populations: Lessons from embryonic development. *Cell*, 132: 661–680.

# Controlling stem cells: Receipt and interpretation of external cues

**4**

The basic principles that define a cell as a stem cell are quite precise, yet they are only a description of cell behavior. How does the molecular machinery of a stem cell determine **self-renewal** and balance this against determination to differentiate? Moreover, given that the same apparently simple features of self-renewal and **differentiation** potential characterize all stem cells, are the underpinning mechanisms the same for all stem cells or do they vary between different categories of stem cell?

The clearest division of stem cells is between those that are related to a specific point in embryonic development and are pluripotent (Chapter 3) and those that are associated with the maintenance of adult tissues that range in their **potency** from **mono-** to **multipotent** (Chapters 9 to 16). Fundamental differences in not just the extent of potency but also the level of cell cycle activity and the ability or not to enter a **quiescent** state suggest that there must be some considerable differences in the molecular processes that control these two broad classes of stem cell. The diversity of requirements of **adult stem cells**, both in terms of day-to-day production of downstream progeny as well as the number of cell types that they can give rise to, also implies that all is not equal in terms of their control mechanisms. Then, there is the fact that stem cells must have arisen at least twice during evolution, once in the animal kingdom roughly at the time when the sponges emerged, and once in the plant kingdom (Chapter 8). Although animal and plant stem cells share the defining properties, does this necessarily mean that the molecular details are the same or even at all related?

Ultimately, the properties of a given cell, whether it is a stem cell or not, are determined by the activity of proteins, the precise combinations of protein functions cumulatively leading to the specific cell phenotypic characteristics. Each feature that is unique to stem cells, including quiescence, self-renewal, **asymmetry** or **symmetry** of division, potency, and **commitment** to differentiation, is the consequence of a combination of protein activities. The way in which a given level of activity of a particular protein is reached can be achieved as the result of many different steps that can go right back to the gene encoding the protein. Although oversimplified, these steps include gene transcription, post-transcriptional processing of the resultant RNA, protein translation, and protein stability and modification (Figure 4.1), and these processes are in turn influenced by inputs that originate outside of the cell, which cumulatively constitute its microenvironment or "**niche**." A more detailed description of the nature of the niche and the controlling inputs that it provides to a stem cell will be provided in Chapter 6.

In this chapter, we will focus on the types of inputs received by stem cells and how these influence cellular processes that ultimately determine their specific characteristics. The concepts outlined and illustrated will be picked up throughout the rest of the book, especially as individual stem cell types are described in greater detail. Also, what we describe here are processes that relate to the regulation of stem cells under normal conditions of homeostasis or that may be required for repair and regeneration, but they also have great relevance in the context of abnormal stem

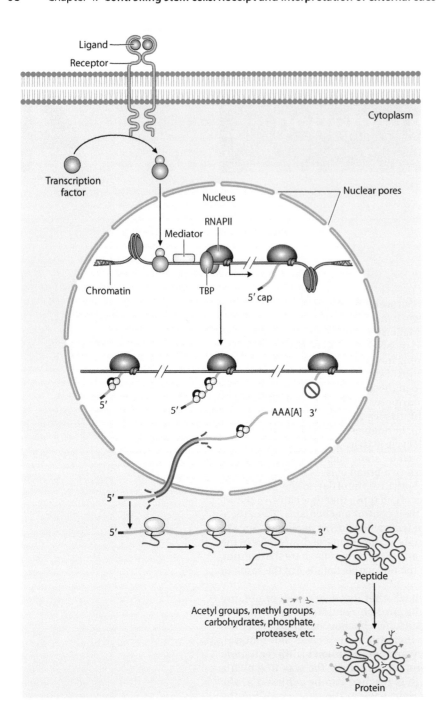

Figure 4.1 **Molecular processes involved in signal transduction.** The diagram illustrates the key components involved in signal transduction, starting at the cell surface with receptor recognition of a small molecule signal, through effects on gene transcription, RNA processing, translation and folding, and post-translational modifications. (From Gilbert SF. *Developmental Biology*, 8th ed. Sunderland, MA, USA: Sinauer Associates, Inc. With permission from Oxford University Press.)

cell behavior, often being the root cause of cancer or age-related stem cell deficits, so that we will return to some of them when we specifically address these issues in Chapter 18.

We will start by covering the major signal inputs that the stem cell can be exposed to, whether these be soluble factors or molecules presented by the niche, including signaling with a mechanical component. Illustrative examples will highlight how a relatively small number of alternatives, employed in many different cell contexts, can actually combine in complementary or antagonistic ways to produce a stem cell specific outcome. We will also consider how cell metabolism plays a critical part in stem cell-specific function and can represent another response to the cell's environment. In the second part of this chapter, we will take a closer look at some of the cellular processes that interpret the signaling inputs, including **epigenetic** modifications of histones, transcriptional control of gene expression, and micro RNAs as transcriptional and post-transcriptional regulators.

# 4.1 SIGNAL TRANSDUCTION PATHWAYS UTILIZED BY STEM CELLS

Only a small handful of signaling pathways appear to underpin how stem cells receive and respond to information from their external environment, and most of these pathways are also utilized in a wide range of cell types that do not have stem cell properties. The explanation behind this apparent paradox has to lie either in the way in which the activities of individual pathways are combined specifically or in the nature of the modulating factors that impact the signal transduction process in stem cells. We will focus on seven principal signaling pathways that are highly conserved in animal species, at least back to the early members of the bilaterans (animals with bilateral symmetry), and have each been shown to be crucial in several animal stem cell types; namely Wnt/$\beta$-catenin, Hedgehog (Hh), Notch, transforming growth factor $\beta$ (TGF$\beta$)/bone morphogenetic protein (BMP), receptor tyrosine kinase (RTK), cytokines, and adhesion receptors.

## Wnt/$\beta$-catenin signaling is very prevalent among different stem cell types

Wnt signaling (Box 4.1) affects adult stem cells present in a variety of tissues, including the gut, skin, mammary gland, nervous system, and hematopoiesis. Depending on the specific context, Wnt signaling can either help to maintain the undifferentiated self-renewing state or be a part of the commitment and differentiation process. The best-studied example of the essential role that can be played by Wnt signaling in stem cells is in the crypts at the base of villi in the small intestine (Chapter 11). Both the stem cells and the niche components of the crypt are engaged in canonical Wnt signaling, and experimental manipulation of several of the component molecules can lead to loss of stem cells, aberrant differentiation, or uncontrolled cell expansion depending on whether a particular protein is reduced or over expressed. Two distinct **intestinal stem cell** (ISC) types are present in the crypt, namely the +4 **label-retaining** cells, which are activated only during injury, and the crypt basis columnar cells (CBC), which are responsible for sustained tissue homeostasis. CBC are characterized by the expression of Lgr5, an orphan G-protein coupled receptor, which is a Wnt/$\beta$-catenin target gene that can associate with Wnt/$\beta$-catenin/Fzd/Lrp, thereby amplifying Wnt signaling (Figure 4.2). **Paneth cells**, the major niche component crucial for the maintenance of the stem cells, are an important source for Wnt ligands, secreting various Wnt factors (e.g., Wnt3, 6, 9b).

In ISC and other adult stem cells, such as **hematopoietic stem cells** (HSC), Wnt signaling promotes self-renewal, whereas in pluripotent **embryonic stem (ES) cells**, it has been shown to be essential for cellular differentiation. Whether canonical Wnt signaling leads to self-renewal or differentiation is our first illustration of how apparently simple pathways used in different contexts can lead to distinct outcomes. Recent research suggests that this could be the result of a choice between different co-activator proteins that cooperate with $\beta$-catenin. Hence, the transcriptional coactivators CREB-binding protein (CBP) and p300, which are highly homologous protein acetyltransferases that acetylate histones and other proteins, have opposing influences on the $\beta$-catenin transactivator complex in stem cells, CBP facilitating transcription of genes that support self-renewal and p300 being important in the control of commitment to differentiation (Figure 4.3).

## The Hedgehog family of developmental morphogens affects a variety of stem cells

Hedgehog family signaling (Box 4.2) is important in many aspects of the development of both invertebrates and vertebrates but also has a significant role in a variety of adult stem cells and **progenitors**, including in tissues such as the brain, skin, prostate, and bladder. The best understood involvement of a Hedgehog family protein in adult stem cells is in the brain. Sonic Hedgehog (Shh) serves to control aspects of proliferation and commitment to differentiation in **neural stem cells** (NSC) located in both the subventricular zone (SVZ) and the subgranular zone (SGZ) (Chapter 14). Conditional loss- and gain-of-function experiments in mice revealed that the number of neurons populating the olfactory bulb, which are derived from the SVZ, are very sensitive to the

## BOX 4.1 BACKGROUND: Wnt/β-CATENIN SIGNALING

First defined in *Drosophila* through the gene *Wingless*, the influence of which is in the name, what was eventually termed Wnt signaling is evolutionarily conserved and crucial for many aspects of development and fundamental cellular processes. In mammals, the Wnt family is composed of 19 secreted glycoproteins, the cognate receptors for which are seven-pass transmembrane proteins called Frizzled (Fz), which also associate with a variety of co-receptors. There are three branches of Wnt signaling downstream of the Wnt-receptor interaction, of which so-called canonical Wnt signaling that controls the transactivator protein β-catenin is

central in stem cell biology. Binding of Wnt to Fz recruits the scaffold protein Dishevelled, which then relays the signal along the canonical or two alternative non-canonical routes. Along the canonical path, the result of Dishevelled binding is disruption of the β-catenin complex (GSK-3β, adenomatous polyposis coli [APC], and Axin), which normally prevents degradation of β-catenin (Figure 1). The ensuing accumulation of β-catenin and its translocation to the nucleus allows formation of a transactivator complex together with T-cell factor (TCF)/lymphoid enhancer factor (LEF) and consequent effects on gene expression.

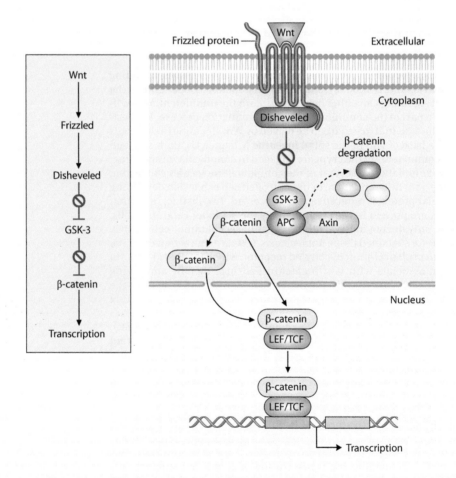

Box 4.1 Figure 1 **Wnt/β-catenin signaling.** The diagram illustrates the key components involved in signal transduction initiated by the binding of a Wnt ligand to the Frizzled protein, leading eventually to binding of β-catenin to the LEF/TCF transcription regulator. Abbreviations: Wnt, Wingless; GSK-3, glycogen synthase kinase 3; APC, adenomatous polyposis coli; LEF/TCF, lymphoid enhancer binding factor/T-cell-specific factor. (From Gilbert SF. *Developmental Biology*, 8th ed. Sunderland, MA, USA: Sinauer Associates, Inc. With permission from Oxford University Press.)

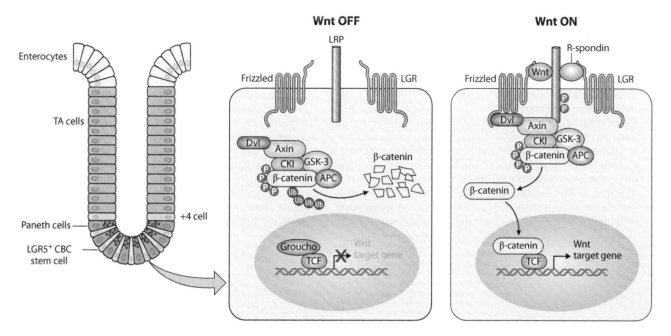

**Figure 4.2 Wnt signaling in intestinal crypt stem cells.** The schematic diagram on the left represents the cell composition of an intestinal villus basal crypt. The stem cells reside at the bottom of the crypt and are of two types: (i) those at the +4 position (blue) and (ii) the LGR5+ cells (gray). These cells are interleaved with Paneth cells (purple), which constitute the niche elements. The diagrams on the right illustrate how Wnt signaling in the LGR5+ cells, which follows binding of Wnt to Frizzled and interaction with ligand-bound LGR5, leads to stabilization of β-catenin, enabling it to de-repress TCF and to activate Wnt target genes. Abbreviations: TA, transit amplifying; β-cat, β-catenin; GSK-3, glycogen synthase kinase 3; APC, adenomatous polyposis coli complex. (Left: From Abo A & Clevers H. (2012) *Nat Biotech* 30: 835–836 [doi: 10.1038/nbt.2361]. Reprinted by permission from Springer Nature. Right: From Schuijers J & Clevers H. (2012) *EMBO J* 31: 2685 -2696 [doi: 10.1038/emboj.2012.149]. With permission from Wiley. © 2012 European Molecular Biology Organization.)

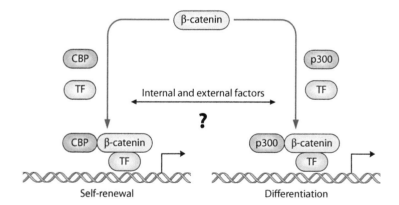

**Figure 4.3 Alternative co-activator binding may determine the specificity of Wnt signaling in stem cells.** Upon nuclear localization, the ability of β-catenin to drive gene expression and stem cell fate decision is dependent on its association with either the CBP or p300 co-activators. Binding of CBP to β-catenin maintains cell proliferation and inhibits differentiation, while binding to p300 triggers the cell to exit the cell cycle and initiate differentiation. (From Ring A, Kim Y & Kahn M. (2014) *Stem Cell Rev* 10:512–525 [doi:10.1007/s12015-014-9515-2]. Reprinted by permission from Springer Nature.)

level of Shh signaling (Figure 4.4). Genetic ablation of the gene encoding Shh resulting in NSC depletion, whereas enhanced self-renewal at the expense of commitment to neural precursors and neuron differentiation is seen if Shh is over expressed. In the NSC niche, the main source of Shh appears to be neurons. Interestingly, such a neural input through Shh seems to influence stem cell fate decisions in some other tissue niches.

Apart from a role in the brain, Hh signaling regulates stem cell maintenance in several other tissues, including the skin, bone, and muscle. In the skin, as in the brain, Shh is the main Hh ligand present, being detected in the bulge region of the hair follicle, which represents the niche for stem cells that not only allow regeneration of hairs but also contribute to the sebaceous glands and the interfollicular epithelium (Chapter 15). Ablation of Shh expression has the expected effect of reducing proliferation of stem cells and progenitors. There are several sources of Shh in the hair follicle; under normal circumstances, epithelial cells produce Shh that promotes a new cycle of hair renewal, a positive feedback loop resulting because the progenitors themselves are able to release Shh. Shh is also

## BOX 4.2 BACKGROUND: HEDGEHOG SIGNALING

Originally identified as a developmental **morphogen** in *Drosophila*, the Hedgehog (Hh) family of secreted proteins play diverse roles in development and the adult. There are three homologs in mammals, namely Sonic Hedgehog (Shh), Indian Hedgehog (Ihh), and Desert Hedgehog (Dhh). The Hh proteins bind to the 12-pass transmembrane receptor Patched (Ptch), thereby relieving its inhibitory effect on the G-coupled receptor Smoothened (Smo) (**Figure 1**). Release of Smo leads to activation of the Gli (glioma-associated oncogene family members) transcription factors. In the absence of Hh ligand, Gli2 and Gli3 undergo limited proteasomal degradation, resulting in the cleavage and removal of the Gli C-terminal activator domain, which leads to the conversion of Gli3, and to a lesser extent Gli2, into transcriptional repressors (GliR and Gli2R). Gli transcriptional activators (GliA) are formed only in response to Hh stimulation. Thus, Hh signaling functions through modulating the balance between GliA and GliR. GliA then triggers expression of Hh target genes such as *Gli1*, the protein product of which functions only as a transcriptional activator and thus amplifies Hh signaling. Recent analysis of the cis-regulatory modules of Hh-regulated genes has revealed that cells interpret the levels of Hh signaling through differential affinity Gli-binding sites in target genes, whereas tissue specificity is achieved through the participation of co-activators.

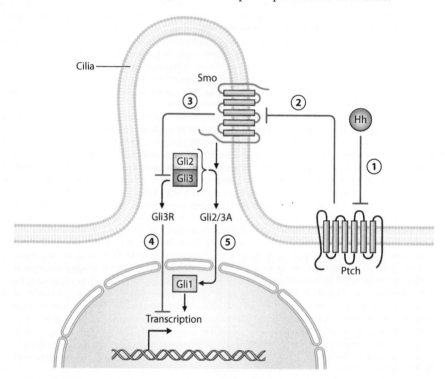

Box 4.2 Figure 1 **Hedgehog signaling.** The diagram illustrates the key components involved in signal transduction initiated by the binding (1) of Hedgehog (Hh) to Patched (Ptch), releasing Smoothened (Smo) from constitutive inhibition (2). Smo then shuttles through the cilia and activates the Gli2/3 complex (3), leading to gene expression via Gli1 (4,5). (From Pan A, Chang L, Nguyen A et al. (2013). *Front Physiol* 4: 61 [doi: 10.3389/fphys.2013.00061]. Published under the terms of the Creative Commons Attribution License http://creativecommons.org/licenses/.)

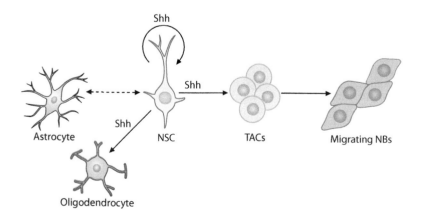

Figure 4.4 **Sonic Hedgehog is a key component regulating neural stem cells.** Neural stem cells (NSC) in the mouse forebrain subventricular zone produce lineage-restricted progenitors and respond to Sonic Hedgehog (Shh). Shh is important for maintaining the undifferentiated and proliferative state of the NSC (circular arrow). Mature astrocytes and the NSC from which they derive both respond to Shh signaling. NSC also give rise to oligodendrocytes and transit amplifying cells (TAC), processes that are augmented by Shh, which go onto produce migrating neuroblasts (NB). (Adapted from Petrova R & Joyner A. (2014) *Development* 141: 3445–3457 [doi: 10.1242/dev.083691].)

delivered to the bulge region from the sensory nerves that wrap around the hair follicle (Figure 4.5). Since denervation abolishes the ability of the follicular stem cells in this region to contribute to interfollicular stem cells, it may be that Shh helps to maintain the **plasticity** of the stem cell population. The role of peripheral nerves in delivering Shh to regulate normal and regenerating non-neuronal tissues is only just beginning to emerge. Thus, nerve-derived Shh stands out as a putative crucial mediator of organ homeostasis and regeneration with the potential to target stem cell populations located in different organs.

## Notch signaling can affect either proliferation or fate decisions of adult stem cells, often in collaboration with other signaling pathways

Notch signaling (Box 4.3) plays an important part in the control of stem cells in some tissues, regulating either proliferation or binary differentiation alternatives, often in concert with other signals, most notably Wnt/β-catenin. The best understood systems in which Notch has a major determining role are the intestine, skin, and muscle, but there is evidence for involvement in several other tissues.

Conditional gene **knockout** or over expression studies in the small intestinal basal crypts of mice have shown that Notch works at two main points in conjunction with Wnt, which we have already seen has a critical signaling influence (Figure 4.6). First, Notch promotes the proliferation of the ISC or immediate downstream **transit amplifying cells**, and second, its level of activity dictates the balance of the binary fate decision between goblet cells and enterocytes, respectively inhibiting or promoting these alternative differentiation routes as cells migrate away from the transit amplifying zone of the crypt. In another epithelial tissue, the skin, Notch also collaborates with Wnt to control proliferation versus differentiation, working at the level of both the epidermal stem cells in the hair follicle and the interfollicular epithelium as well as on the quiescent population of stem cells present in the bulge region (Chapter 15).

The other tissue that has a clearly defined critical dependence on Notch for tissue maintenance in the adult is skeletal muscle. Upon injury, Notch is activated in **satellite stem cells** (Chapter 10), stimulating their proliferation, and then needs to be attenuated in the daughter cell destined for myoblast differentiation through the asymmetric accumulation of the Notch antagonist Numb. Interestingly, compromise of Notch signaling is related to the age-related deficit in skeletal muscle regeneration (known as sarcopena in a clinical context) (Chapter 18) in that satellite cells have reduced capability correlating with decreased expression of Delta-like 1 ligand.

Other tissues have, or are expected to have, a requirement for Notch signaling within their stem cell component. Not surprisingly, given its importance during neural development, Notch signaling plays a part in adult neurogenesis in the SVZ and SGZ. However, a similar parallel expected for hematopoiesis seems not to exist. Even though Notch is a key determinant of fate decisions during aspects of hematopoietic differentiation, most notably favoring T-cell versus B-cell differentiation, there has been some controversy about its influence on HSC. At present, the consensus view is that Notch signaling is not required for the maintenance or functioning of HSC, although over expression experiments suggest that there could be a practical benefit for using imposed Notch signaling as a means to expand the stem cell population *ex vivo*.

## Bone morphogenetic proteins have important roles in both embryonic and adult stem cells

The family of proteins related to TGFβ is large (Box 4.4), with an equally enormous range of functions throughout embryonic development and adult life. The breadth of involvement of TGFβ-related factors in animal biology is reflected in their roles in the regulation of both embryonic and adult stem cells, in particular involving members of the biggest subfamily of BMPs, which across invertebrate and vertebrate species is constituted of more than 20 proteins.

The part played by BMP proteins in stem cells follows the pattern that you are by now becoming very familiar with; that is, their roles vary in different scenarios. It is difficult to define wherein lies the specificity of BMP action in terms of eliciting a stem cell response, and they work in close concert with other signaling pathways, including Wnt and Notch. Another familiar trend is that studies on stem cells in *Drosophila* have

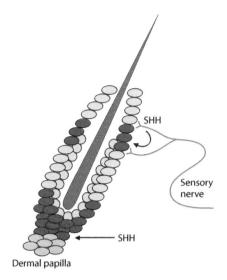

**Figure 4.5 A neural source of Sonic Hedgehog affects stem cells in the hair follicle bulge region.** The diagram shows a section through a hair follicle. Stem cell domains (red) in the upper and the lower bulge of are exposed to Sonic Hedgehog (Shh) ligands during the telogen phase of the hair cycle. An unidentified source of Shh acts on the lower bulge, whereas the stem cell domain at the top of the bulge receives Shh from cutaneous nerves (orange line) originating from dorsal root ganglia adjacent to the spinal cord.

The Notch gene was named for the phenotype of a mutant *Drosophila* with an indentation in the wings and is highly conserved from nematodes to humans. Notch signaling influences cell fate decisions, such as differentiation, apoptosis, and cell cycle. In mammals, there are four Notch receptors (Notch 1 to 4) and five structurally similar Notch ligands (Delta-like 1, 3, and 4, and Jagged 1 and 2). Both Notch and Notch ligands are single-pass transmembrane proteins. Notch protein undergoes an intramolecular cleavage (S1) while still in the Golgi body to form a heterodimer, one portion being extracellular and the other remaining embedded in the cell membrane. Following binding of Notch ligand presented by one cell to Notch on the receiving cell (Figure 1), a second cleavage of Notch (S2) involving ADAM10 metalloproteinase occurs in the extracellular domain, facilitating a third cleavage (S3) by γ-secretase that allows translocation of the intracellular domain to the nucleus. Once in the nucleus, the Notch intracellular domain (NICD) associates with the DNA-binding protein CSL, thereby converting a transcriptional repressor to an activator, which acts upon key target genes such as those encoding the HES and HRT families of basic helix-loop-helix transcription factors.

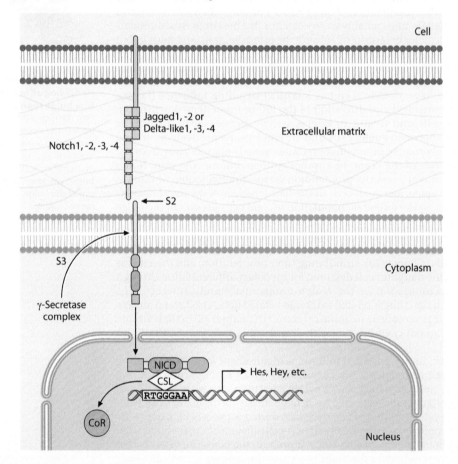

Box 4.3 Figure 1 **Notch signaling.** Notch ligands Jagged1, or 2, and Delta-like 1, 3, or 4 can bind to Notch receptors 1, 2, 3, or 4 on adjacent cells. Ligand binding leads to two cleavage events: S2 immediately outside the transmembrane domain, which cleaves the extracellular domain from the intracellular domain, followed by an S3 cleavage event within the transmembrane domain by the γ-secretase complex that liberates the Notch intracellular domain (NICD). NICD moves to the nucleus, where it binds to the CSL transcriptional repressor protein, causing release of transcriptional co-repressor (CoR) proteins and subsequent transcription of target genes, including those encoding the basic helix-loop-helix protein families HES and HEY. (From Rehman A & Wang C-Y. (2006) *Trends Cell Biol* 16: 293–300 [doi:10.1016/j.tcb.2006.04.003]. Copyright 2006 with permission from Elsevier.)

led the way in terms of our understanding of the molecular details of how BMPs can act on stem cells. The maintenance of **germ line stem cells** (GSC) in both female and male *Drosophila* (Chapter 12) requires the BMP2/4 homolog Decapentaplegic (Dpp), which is produced by niche cells and acts upon the stem cells to prevent expression of the *Bam* (bag of marbles) gene that is necessary for differentiation (Figure 4.7).

There are many examples of a dependence upon BMP signaling among adult stem cells in vertebrates. Again, studies on some systems have led the way in our

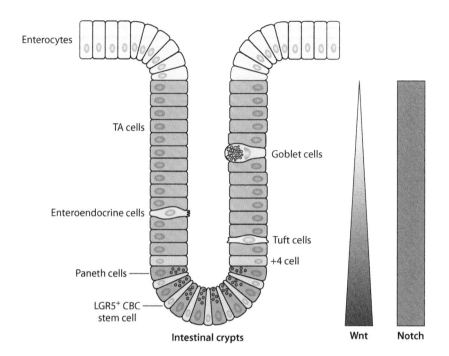

Enterocytes

TA cells

Goblet cells

Enteroendocrine cells

Tuft cells

+4 cell

Paneth cells

LGR5⁺ CBC
stem cell

**Intestinal crypts**

Wnt    Notch

**Figure 4.6 Notch and Wnt signaling in the intestinal stem cell niche.** Self-renewal and differentiation of intestinal stem cells in the crypts of intestinal villi requires a balance of Wnt and Notch signaling. The schematic diagram shows the crypt, consisting of the +4 position (blue) and LGR5+ (gray) stem cells, interleaved with Paneth cells (purple) and giving rise to transit amplifying (TA) progenitor cells (orange). Notch signaling operates throughout the stem cell and progenitor region, overlapping with the zone of Wnt signaling, which is strongest at the base of the crypt. (From Spit M, Koo B-K & Maurice M. (2018) *Open Biol* 8:180120 [doi: 10.1098/rsob.180120]. Published under the terms of the Creative Commons Attribution License (http://creativecommons.org/licenses/by/4.0/.)

understanding of this pathway of regulation, most notably epithelial stem cells, with the popular paradigm of the HSC lagging somewhat in this case. Like *Drosophila* GSC, stem cells of the hair follicle exhibit a requirement for BMP signaling to inhibit differentiation (Chapter 15, Figure 15.17), although downstream at the transit amplifying cell stage, BMPs serve the opposite function by helping to promote differentiation along different lineages as the hair cycle progresses. Also, contrasting with *Drosophila* GSC, the cells of the intestinal crypt exhibit a requirement for BMP signaling to promote differentiation, its action being actively inhibited in the basal crypt where the stem cells reside. A role for BMP signaling in HSC was initially dismissed when gene knockouts of some pathway components failed to give a clear phenotypic consequence. However, there is good evidence for an indirect effect of BMP signaling on HSC, while more recent studies have now refuted the initial conclusions about a lack of a cell autonomous influence. The indirect effect relates to the differentiation of **mesenchymal stem cells** (MSC) to osteoblasts, which represent one of the principal cell types constituting the HSC niche (Chapters 6 and 9). This, however, is not straightforward, the effect of BMP being either pro-differentiation if binding to the Type I BMP receptor BMPR1b or inhibitory if the BMPR1a homolog is engaged. Evidence for a direct influence of BMP on HSC came first from gene inactivation studies on the BMP Type II receptor, which when ablated severely curtails the self-renewal capacity of HSC. More recently,

---

## BOX 4.4 BACKGROUND: TGFβ SUPERFAMILY SIGNALING

TGFβ superfamily (Figure 1) is a large group of structurally related cell regulatory proteins that was named after its first member, TGF-β1. TGF-β superfamily proteins are found in both invertebrates and vertebrates and can be divided into four main subfamilies: (i) the TGFβ subfamily; (ii) the BMP family, which has the largest number of members; (iii) the activin subfamily; and (iv) a group of divergent members. TGFβ superfamily members are small proteins of about 110–120 residues, which are derived by proteolytic cleavage from the C-terminal of a precursor protein. As ligands, these proteins interact with a conserved family of cell surface serine/threonine-specific protein kinase receptors to generate an intracellular signal. Dimers of a Type II receptor bind the protein ligand and are then able to recruit and phosphorylate a Type I receptor dimer, which

in turn phosphorylates a protein in the SMAD family (this is not a simple acronym by any means; rather, it derives from the *Drosophila* protein, mothers against decapentaplegic [MAD], and the *Caenorhabditis elegans* protein from the gene sma, for small body size!). The first SMAD affected, also known as the receptor- or R-SMAD, then binds SMAD4, also known as the co-SMAD, forming a complex that accumulates in the nucleus and directly controls gene regulation (Figure 2). Adding to the complexity, BMPs can also signal through a SMAD-independent route mediated by the mitogen-activated protein kinase (MAPK) pathway downstream of TAK1 (TGF-β activated kinase-1), a tyrosine kinase that has multiple substrates. TAB1/2 (TAK1 binding protein 1/2) is a key effector in the activation of the NF-κB and JNK (Jun N-terminal kinase) pathways.

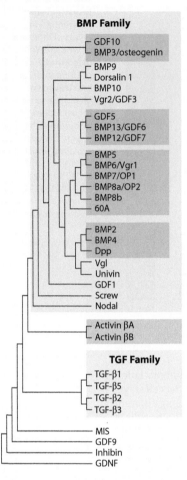

**BMP Family**

GDF10
BMP3/osteogenin
BMP9
Dorsalin 1
BMP10
Vgr2/GDF3
GDF5
BMP13/GDF6
BMP12/GDF7
BMP5
BMP6/Vgr1
BMP7/OP1
BMP8a/OP2
BMP8b
60A
BMP2
BMP4
Dpp
Vgl
Univin
GDF1
Screw
Nodal
Activin βA
Activin βB

**TGF Family**

TGF-β1
TGF-β5
TGF-β2
TGF-β3
MIS
GDF9
Inhibin
GDNF

**Box 4.4 Figure 1 TGFβ-related super-family of protein factors.** The diagram represents the entire TGFβ superfamily of signal peptides, showing their evolutionary relationship to one another and grouping into four categories. (From Gilbert SF. *Developmental Biology*, 8th ed. Sunderland, MA, USA: Sinauer Associates, Inc. With permission from Oxford University Press.)

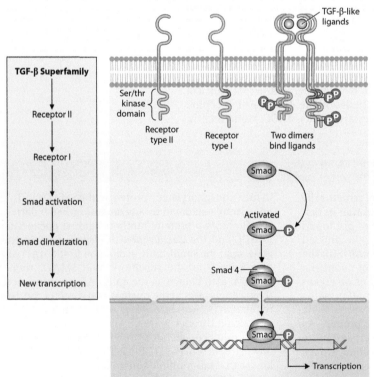

**TGF-β Superfamily**

Receptor II
↓
Receptor I
↓
Smad activation
↓
Smad dimerization
↓
New transcription

TGF-β-like ligands

Ser/thr kinase domain

Receptor type II

Receptor type I

Two dimers bind ligands

Smad

Activated

Smad

Smad 4

Smad

Smad

Transcription

**Box 4.4 Figure 2 Signaling involving TGFβ-related proteins.** The diagram illustrates the key components involved in signal transduction initiated by the binding of TGFβ-like ligands to combined dimers of Type I and Type II receptors. Ligand binding leads to autophosphorylation of the intracellular domains of the two receptors, initiating the downstream phosphorylation, activation, and dimerization of Smad proteins, enabling them to act upon their target genes. (Adapted from Gilbert SF. *Developmental Biology*, 8th ed. Sunderland, MA, USA: Sinauer Associates, Inc.)

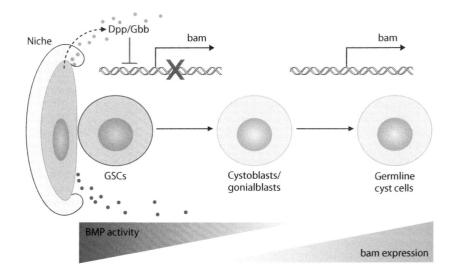

Figure 4.7 **BMP-related factors maintain stem cells in the *Drosophila* ovary.** BMP signaling maintains germ stem cell (GSC) stemness by blocking differentiation in the *Drosophila* ovary and testis. GSC niche cell secreted BMP family proteins Decapentaplegic and Glass bottomed boat (Dpp and Gbb) form a gradient, which mediates repression of transcription of the Bag of marbles (bam) gene in GSC and their cystoblasts/gonialblasts. In germline cyst cells, there is no detectable BMP activity and bam expression is released from BMP inhibition, allowing bam to play its part in differentiation. (From Zhang J & Li L. (2005) *Dev Biol* 284: 1–11 [doi: 10.1016/j.ydbio.2005.05.009]. Copyright 2005 with permission from Elsevier.)

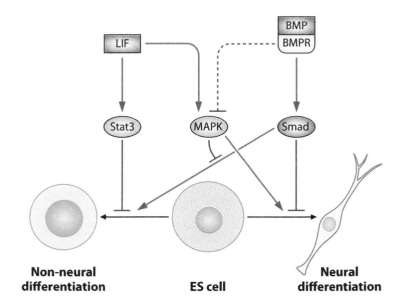

Figure 4.8 **BMP blocks ES cell neural differentiation and maintains pluripotency.** BMP binding to its receptor (BMPR) in ES cells blocks neural differentiation and maintains ES cell pluripotency by creating a balance with growth factor signaling from factors including leukemia inhibitory factor (LIF) and fibroblast growth factors (FGF). The BMP signal inhibits growth factor/MAPK signal-mediated neural differentiation and induces non-neural differentiation. LIF and FGF block BMP-stimulated non-neural differentiation and cooperate with BMP in the regulation of self-renewal through the MAPK pathway or LIF/Stat3 signaling. (From Zhang J & Li L. (2005) *Dev Biol* 284: 1–11 [doi:10.1016/j.ydbio.2005.05.009]. Copyright 2005 with permission from Elsevier.)

and highly interesting because of the light shed on HSC heterogeneity (Chapter 9), it has been discovered that a small subpopulation of adult HSC can respond to BMP and have distinct functional properties. Hence, the laboratory of Elaine Dzierzak showed that 5% of HSC respond to BMP and have a differentiation preference for so-called "balanced" or lymphoid differentiation, as opposed to myeloid or lymphoid differentiation for those HSC that are unable to react to BMP. The take-home message from this then is, keep looking! Certainly, there is much more to be discovered about how adult stem cells respond to BMPs and of course all of the other signal molecules that we are describing in this chapter.

BMP signaling also has an important role in **pluripotent stem cells**, although there are species differences that add to the complexity. The majority of studies have been performed on mouse ES cells in which BMP4 supports self-renewal, although this is a consequence of its action on blocking neural differentiation. This effect works in opposition to signaling initiated by the cytokine leukemia inhibitory factor (LIF), which blocks the non-neural differentiation potential of the cells and counteracts the contrasting action of BMP4, which is to promote **mesodermal** and **trophoblast** differentiation (Figure 4.8).

## Receptor tyrosine kinases allow stem cells to respond to multiple systemic signals

Receptor tyrosine kinases are the cell surface receptors for many polypeptide growth factors, cytokines, and hormones (Box 4.5). The RTKs have been highly conserved

BOX 4.5 BACKGROUND: RECEPTOR TYROSINE KINASE SIGNALING

The receptor tyrosine kinase (RTK) signal transduction pathway is found throughout the plant and animal kingdoms, responding to a vast array of polypeptide growth factors and affecting most cellular processes in one way or another. Ligands that bind to RTKs include the fibroblast growth factors (FGF), epidermal growth factors (EGF), platelet-derived growth factors (PDGF), vascular endothelial growth factors (VEGF), and stem cell factor (SCF), which can originate locally or systemically. Each RTK can bind only one or a small set of these ligands.

Binding of the ligand to the extracellular domain of the RTK monomer causes it to dimerize and to activate the inherent kinase activity of the receptor, leading to phosphorylation of each other on specific tyrosine residues in the cytoplasmic portion of the protein. These phosphorylated tyrosine residues are then able to bind to adaptor proteins that link to a downstream kinase cascade. Once bound to the RTK, the adaptor protein activates the guanine nucleotide releasing protein to facilitate conversion of the so-called G protein (Ras) to the GTP bound form, which is then able to activate the next kinase in the cascade (Raf). After delivering the signal to Raf, complexing of Ras with the GTPase-activating protein (GAP) leads to

hydrolysis of the GTP, returning the G protein to its inactive state. A series of subsequent activating phosphorylations of kinases ultimately leads to an active kinase (ERK) that enters the nucleus and is able to phosphorylate one or more transcription factors and thereby alter their activity and the expression of the genes that they control (Figure 1). In addition to the Ras pathway, some RTKs can also engage alternative routes to transduce ligand binding into a signal, including the PI3 K and JAK-STAT pathways. The PI3 K (phosphoinositol-3-kinase) pathway is important in regulating the cell cycle. PI3 K activation phosphorylates and activates AKT, localizing it in the plasma membrane from where it can activate a variety of proteins, including mTOR. The JAK-STAT pathway is initiated by binding of a Janus kinase (JAK) to each of the receptor proteins in the dimer. JAKs are cytoplasmic tyrosine kinases that, once bound, phosphorylate the receptor and also autophosphorylate. In this way, the activated receptor can then phosphorylate transcription factors of the signal transducers and activators of transcription (STAT) family. Phosphorylation of the STAT proteins causes them to dimerize, which is the active form that is then able to translocate into the nucleus and bind to its target genes (Figure 2).

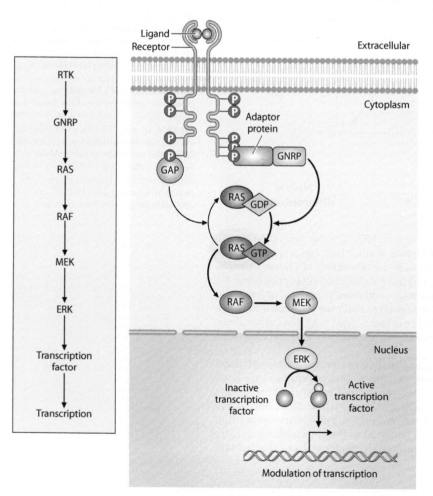

Box 4.5 Figure 1 **Receptor tyrosine kinase signaling.** The diagram illustrates the key components involved in signal transduction initiated by the binding of growth factors to a homodimer of their cognate receptor. Autophosphorylation (P) of tyrosine residues allows recruitment of adaptor proteins and sets in train the downstream cascade of activating phosphorylations leading eventually to transcription factor activation and changes in gene expression. (Adapted from Gilbert SF. *Developmental Biology*, 8th ed. Sunderland, MA, USA: Sinauer Associates, Inc.)

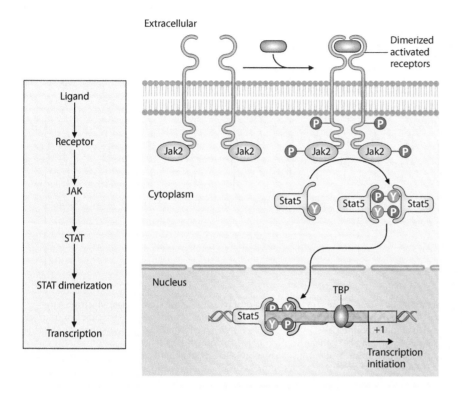

Box 4.5 Figure 2 **JAK-STAT signaling downstream of a receptor tyrosine kinase.** The diagram illustrates how receptor tyrosine kinase homodimerization following ligand binding and autophosphorylation of intracellular tyrosine residues can lead to recruitment and activation of Janus kinases (Jak), which in turn activates Stat transcription factor proteins by tyrosine phosphorylation that results in dimerization and their translocation to the nucleus. (Adapted from Gilbert SF. *Developmental Biology*, 8th ed. Sunderland, MA, USA: Sinauer Associates, Inc.)

throughout evolution and in the context of stem cell biology have a role in both the animal and plant kingdoms. The variety of receptors is very large, 59 receptor tyrosine kinase encoding genes having been identified in the human genome. Among all of the signaling pathways that we describe in this chapter, the RTKs probably represent the best example where the specificity of response that affects stem cell properties lies at least to some extent in the expression of particular receptors and their ligands. The fact that stem cell factor (SCF) binds as the sole ligand to the RTK known as KIT, which is expressed on several stem cell types, highlights this assertion.

Human RTK genes can be divided into more than 20 classes on the basis of their modular makeup (Figure 4.9), only some of which have as yet been found to have a role in the context of stem cells. We will focus on just three of these classes, as they illustrate some of the particular features of this group of signal transducing receptors and also represent some of the best characterized stem cell relevant RTKs.

Class I RTKs are receptors for EGFs, and, for example, in *Drosophila* are involved in regulating the proliferation of ISC in the midgut, a role that is quite simple compared to many signal transduction pathways. Following damage or stress, the intestinal epithelium produces ligands (Vein) that activate the EGF receptor (EGFR) in ISC. This leads to Ras/MAPK signaling and phosphorylation of the HMG-box transcriptional repressor Capicua (Cic), which negates its inhibitory effects and causes an increase in expression of cell cycle-related genes that promote ISC growth and division (Figure 4.10).

The Class III of RTK proteins, which includes proteins containing immunoglobulin homology domains in the extracellular portion, incorporates PDGF receptors (PDGFR) α and β, the FLT3 receptor, and the aforementioned KIT receptor. As surface markers of stem cells, these RTK proteins have proven to be useful in the prospective identification and purification of stem cells, especially the HSC. Although quite widely expressed, PDGFRs are of particular significance in stem cells of mesenchymal origin. In MSC, PDGFRα is a selective marker for the most immature, proliferative forms, while PDGFRβ appears to play a role in their migration. Probably reflecting relatedness to MSC, one recently described population of cardiac progenitors expresses PDGFRα, which in combination with expression of Sca1 is highly useful in facilitating their isolation (Chapter 13). Both KIT and the FLT3 protein have been pivotal in the characterization of HSC and are indispensable for their full stem cell capability. Signaling by SCF through KIT supports HSC survival and self-renewal, the receptor being expressed on all

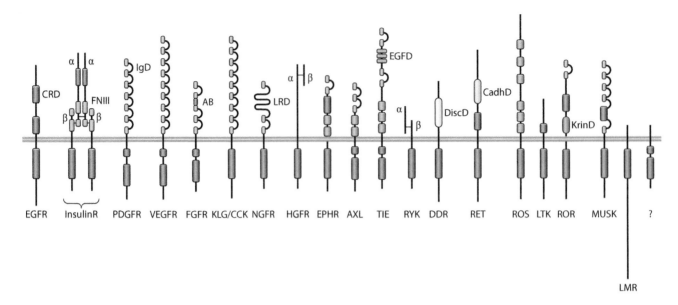

Figure 4.9 **Human receptor tyrosine kinases.** Diagrammatic representation of human receptor tyrosine kinase subfamilies, which are largely defined by their characteristic extracellular domains. Domains represented by boxes are: intracellular kinase domain (red); AB, acidic box (purple); CadhD, cadherin-like domain (green); CRD, cysteine-rich domain (purple); DiscD, discoidin-like domain (yellow); EGFD, epidermal growth factor-like domain; FNIII, fibronectin type III-like domain; IgD, immunoglobulin-like domain; KrinD, kringle-like domain; LRD, leucine-rich domain. The abbreviations of the RTK groups are: EGFR, epidermal growth factor receptor; InsulinR, insulin receptor; PDGFR, platelet-derived growth factor receptor; VEGFR; vascular endothelial growth factor receptor; FGFR, fibroblast growth factor receptor; KLG/CCK, colon carcinoma kinase; NGFR, nerve growth factor receptor; HGFR, hepatocyte growth factor receptor, EphR, ephrin receptor; Axl, a phosphotyrosine kinase; TIE, tyrosine kinase receptor; RYK, receptor related to tyrosine kinases; DDR, discoidin domain receptor; Ret, rearranged during transfection; ROS, RTK expressed in some epithelial cell types; LTK, leukocyte tyrosine kinase; ROR, receptor orphan; MuSK, muscle-specific kinase; and LMR, Lemur. (From Blume-Jensen P & Hunter T. (2001) *Nature* 411: 355–365 [doi: 10.1038/35077225]. Reprinted by permission from Springer Nature.)

stages of the hematopoietic hierarchy from most immature stem cell through to the committed progenitors. FLT3, on the other hand, promotes survival following binding to its ligand FL (FLT3 ligand) and has a more limited expression pattern, being restricted to multipotent progenitors in the mouse, although in humans, the cells preceding this stage are also positive. Both SCF and FL are produced by cells within the HSC niche and exist in both a soluble form or membrane bound, perhaps as a means of restricting their action to the immediate vicinity of the niche. KIT is expressed on a number of non-stem cell types, and interestingly in the male testis, where it is an essential factor; it is found on spermatogonial cells and their immediate precursor stem cells.

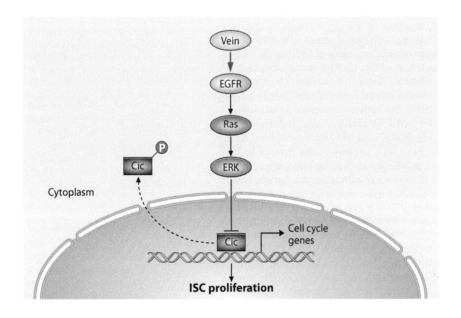

Figure 4.10 **Control of *Drosophila* intestinal stem cells through EGFR signaling.** EGF receptor (EGFR) signaling in *Drosophila* intestinal stem cells is initiated by binding of ligands such as Vein and mediates activation of ERK, leading to phosphorylation of the HMG-box transcriptional repressor Capicua (Cic). Phosphorylation of Cic causes it to relocate to the cytoplasm, allowing activation of genes encoding proteins crucial for proliferation. (From Jin Y, Ha N, Fores M et al. (2015) *PLoS Genet* 11:e1005634 [doi: 10.1371/journal. pgen.1005634]. Published under the terms of the Creative Commons Attribution License (http://creativecommons.org/licenses/.)

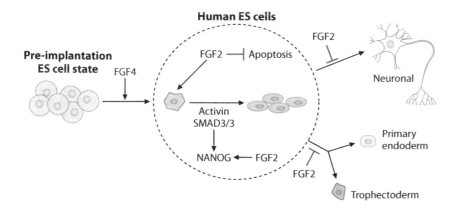

Figure 4.11 **FGF signaling stabilizes the metastable state of epiblast stem cells and human ES cells.** FGF2 signaling in human ES cells (analogous to murine epiblast stem cells) maintains the pluripotent state and blocks neuronal, trophectoderm, and primitive endoderm differentiation. FGF2 regulates NANOG expression both directly and indirectly via activin/NODAL induction. The pre-implantation ES cell state, which is analogous to murine ES cells, is directed toward the ES cell state by FGF4 signaling. Arrows and T-bars indicate activation and inhibition, respectively. (Adapted from Lanner F & Rossant J. (2010) *Development* 137: 3351–3360 [doi: 10.1242/dev.050146].)

The third group of RTKs with well-defined involvement in stem cells belong to Class V and also contain immunoglobulin homology domains. This group includes the FGF receptors (FGFRs), which have a huge number of roles, especially in development but also continuing in the adult. As a group, FGFRs illustrate how an enormous diversity of receptor signaling possibilities can be generated from a relatively small number of protein coding genes. Hence, the four human FGFR genes can actually yield 48 different receptor isoforms through alternative splicing. Adding to the combinatorial complexity, there are 23 recognized FGFs. There is certainly much scope for cell signaling specificity, but it is probably fair to say that this has yet to be fully explored in the context of stem cell biology. One stem cell type that has a clear requirement for FGF signaling is the ES cell, particularly as represented by human cells (although, as we describe in Chapter 3, mouse/human differences in the response of ES cells to growth factors are largely the result of the different stage in embryogenesis from which they are derived). In human ES cells, FGFs transduce signals to MAPK to maintain the pluripotent state (Figure 4.11). FGF-2 blocks neuronal, **trophectoderm**, and primitive **endoderm** differentiation, while FGF-4 promotes progression of cells from the pre-implantation ES cell state to human ES cells (equivalent to **epiblast** stem cells in mouse). FGF-2 also regulates NANOG expression both directly and indirectly via activin/NODAL induction in ES cell-derived feeder cells.

## Cytokine signaling overlaps mechanistically with receptor tyrosine kinases and similarly has a more restricted tissue relevance

Like many before us, we are making a distinction between RTK signaling and cytokine signaling, but in many ways, this is something of a semantic separation. The terms "cytokine" and "growth factor" are often interchanged, and many aspects of the downstream components of growth factor/RTK and cytokine/cytokine receptor signaling are the same. So, why make a distinction? We have done so because cytokine receptors are different in that dimerization as the initial event following ligand binding involves two types of receptor peptides, rather the homodimerization that characterizes RTK signaling, and the intracellular domains lack inherent tyrosine kinase activity. Downstream events occurring from activated cytokine receptors can involve any of those pathways that are seen for RTKs, but the JAK-STAT route predominates, association with JAK being key to tyrosine phosphorylations in JAK itself, on the cytokine receptor, and of STAT leading directly to a transcriptional response (Figure 4.12).

Studies in *Drosophila* first implicated JAK-STAT signaling in the control of stem cell maintenance in the male GSC niche (Chapter 12); subsequently, it has been shown to play a role in other niches in both *Drosophila* and mammals. In the adult *Drosophila* midgut, the surrounding visceral muscle maintains ISC through a combination of EGFR/Ras/ERK signaling, stimulated by the EGF ligand Vein and through stimulating Wingless and JAK-STAT pathway activities. In mammals, though, cytokine signaling has been defined through its frequent association with the hematopoietic system, including both in HSC and committed cells. NSC self-renewal and progenitor cell division and differentiation are affected by the neuropoietic cytokine family, which includes interleukin-6 (IL-6) and LIF. Among other stem cell targets, IL-6 inhibits the differentiation of MSC to chondrocytes, while LIF also has an important effect on ES

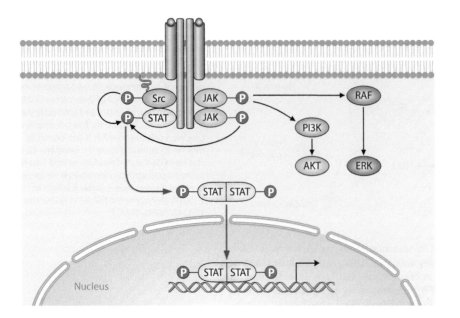

Figure 4.12 **JAK-STAT signaling from cytokine receptors.** The diagram illustrates how cytokine receptors once engaged with their respective ligand activate JAK and can then signal by one of several routes, although the most predominant pathway involves phosphorylation (P) and dimerization of STAT. Following activation of STAT, the dimer moves to the nucleus where it is able to affect gene expression. (From Baker S, Rane SG & Reddy EP. (2007) *Oncogene* 26: 6724–6737 [doi:10.1038/sj.onc.1210757]. Reprinted by permission from Springer Nature.)

cells in that the JAK-STAT downstream signal that it stimulates inhibits non-neuronal differentiation, thereby playing a part in the maintenance of self-renewal.

## Adhesion molecules perform a dual function, which includes delivery of signals to the cell interior

Some of the surface proteins on cells fall into the category of adhesion receptors; that is, their perceived principal function is to help link the cell to other cells or the **extracellular matrix** (ECM) through interaction with a partner protein or motif. We will limit our description here because these types of receptor engagements on stem cells are considered in some detail when we discuss the stem cell microenvironment (Chapter 6). Nevertheless, it is important here to highlight the fact that engagement of adhesion receptors can lead to induction of an intracellular signal just like any of the other receptors that we have covered so far. The two most important groups of adhesion receptor proteins present on stem cells are the integrins and the cadherins, which crop up on many stem cell types, often operating together to elicit a specific response. Integrins can signal through complexes incorporating focal adhesion kinase (FAK) or integrin-linked kinase (ILK), but also modulate signaling through RTKs, while cadherins signal via β-catenin (Figure 4.13). As an illustration of the roles that can be played by integrin- and cadherin-mediated signaling in stem cells, we can take the example of the differentiation of osteoblasts from MSC in the context of the endosteal niche that is key to the functioning of HSC (Chapter 9). Integrins and cadherins control cell adherence and cell fate in the endosteal niche, linking with Wnt/β-catenin signaling to regulate osteogenic differentiation and mechanotransduction (Figure 4.14). The way in which the transmission of mechanical stress can dictate stem cell fate is further considered in Chapter 10 in relation to the influence of ECM stiffness on MSC differentiation.

## Energy metabolism and energy-sensing pathways have a direct influence on stem cell behavior

The availability of energy substrates and the way in which these dictate how the cell responds appear to have a significant influence in discriminating pluripotent stem cells and certain adult stem cells from their differentiated counterparts and might even be considered to be promoting stemness.

Unlike differentiated cells, many stem cells rely to a greater extent on glycolysis rather than oxidative phosphorylation to generate adenosine triphosphate (ATP), a feature that is most certainly an adaptation to low oxygen levels present *in vivo* either during development or in the adult stem cell microenvironment (Figure 4.15). For example, adult HSC and NSC are thought to reside in niches in which the oxygen levels are only 1%–6% compared with 20% in the atmosphere. These dependencies are reflected by the fact that low oxygen levels favor the *in vitro* maintenance of stem cells

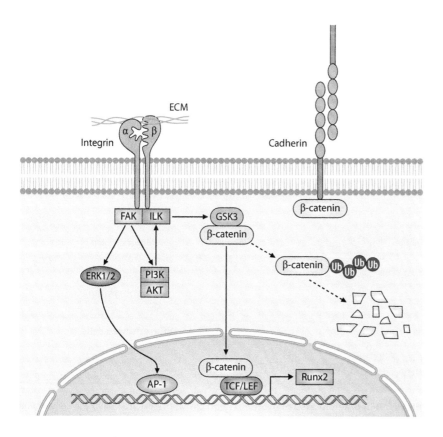

Figure 4.13 **Integrins and cadherins cooperate to elicit osteogenic differentiation from mesenchymal stem cells.** The diagram illustrates how stem cell surface adhesion receptor engagement with the extracellular matrix (ECM) through integrins or with another cell through cadherin homodimerization not only affect the localization of the cell but also pass signals to the nucleus to elicit changes in gene expression. The specific example shown applies during the differentiation of mesenchymal stem cells to osteogenic cells. Abbreviations: AKT, protein kinase B; AP-1, activator protein 1; ILK, integrin-linked kinase; ERK, extracellular signal–regulated kinase; FAK, focal adhesion kinase; GSK3, glycogen synthase kinase 3; PI3 K, phosphatidylinositol 3-kinase; TCF/LEF, T-cell factor/lymphoid enhancer-binding factor; Ub, ubiquitinylation. (From Marie P, Hay E & Saidak Z. (2014) *Trends Endocrin Metabol* 25: 567–575 [doi:10.1016/j.tem.2014.06.009]. Copyright 2014 with permission from Elsevier.)

such as ES cells and NSC, avoiding their premature differentiation, and facilitate the process of **reprogramming** fibroblasts to the pluripotent state. The hypoxia-inducible transcription factors (HIFs), which are stabilized and activated under low oxygen conditions, are crucial in relaying the effect that oxygen has on stem cell fate.

In addition to oxygen levels, cells are able to sense the availability of nutrients in their environment utilizing a number of interconnected signaling pathways

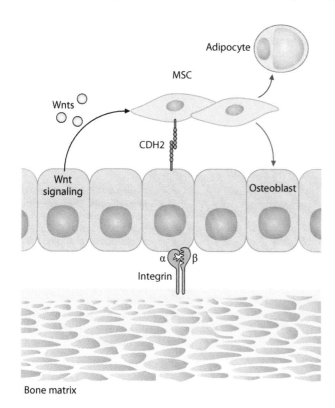

Figure 4.14 **Integrins and cadherins cooperate in the formation of the endosteal niche in bones.** The diagram shows how osteoblasts sitting at the interface between the bone and the bone marrow space in the endosteum communicate with mesenchymal stem cells (MSC) either through direct adhesion via cadherin2 (CDH2) or by paracrine signaling from Wnt. The consequences of these interactions include MSC differentiation either to osteoblasts (green arrow) or adipocytes (brown arrow). (From Marie P, Hay E & Saidak Z. (2014) *Trends Endocrin Metabol* 25: 567–575 [doi:10.1016/j.tem.2014.06.009]. Copyright 2014 with permission from Elsevier.)

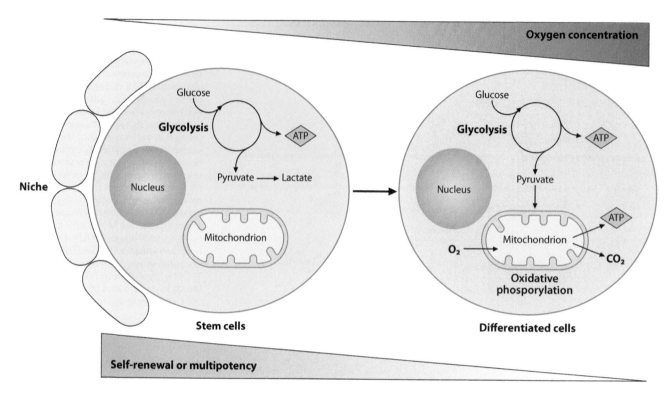

**Figure 4.15 Energy sources in stem cells and their differentiated derivatives.** Stem cells often exist in a low-oxygen environment and generate adenosine triphosphate (ATP) mainly through glycolysis, which is independent of oxygen. In contrast, differentiated cells tend to exist in environments that experience a higher level of oxygen and generate ATP largely through oxidative phosphorylation.

(Figure 4.16). For example, reflecting glucose levels, insulin signals via a Class II RTK to inhibit the FOXO transcription factors, which are important for the maintenance of adult stem cells and the pluripotency of ES cells. In HSC and NSC, FOXOs maintain quiescence through their ability to up regulate proteins involved in cell cycle arrest, whereas in ES cells, one FOXO isoform regulates the expression of the key pluripotency transcription factors OCT4 and SOX2. The level of amino acids that a stem cell is exposed to impinge on the serine/threonine kinase mTOR (target of rapamycin), which in ES cells is necessary for self-renewal, while in adult stem cells, lower activity of the kinase favors stem cell maintenance. The mechanism involves a family of GTPases related to Ras, which exist as dimers that are converted to the active form able to influence mTOR following the binding of amino acids, in particular leucine. Last, AMP-activated protein kinase (AMPK) serves as the central

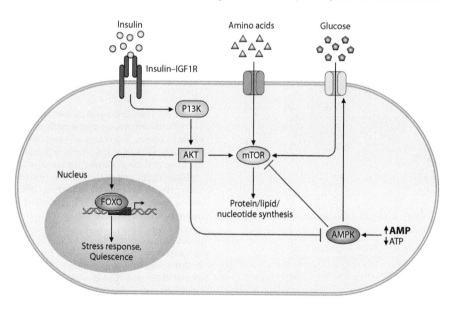

**Figure 4.16 Nutrient-sensing pathways in stem cells.** The schematic illustration summarizes many of the cellular components that enable stem cells to respond to energy availability by changing their metabolism and potentially their fate. Blue shading highlights molecules active in a high-energy state. Red shading highlights molecules that are inactive in a high-energy state or become active in response to cellular stresses, such as low oxygen and low energy. Abbreviations: AKT, protein kinase B; AMPK, AMP-activated protein kinase; FOXO, Forkhead Box O transcription factor; Insulin-IGF1R, insulin-insulin-like growth factor 1 receptor; mTOR, mammalian target of rapamycin; PI3 K, phosphatidylinositol 3-kinase.

"fuel gauge," responding to low energy. When AMP/ADP predominate over ATP, AMPK switches pathways from energy consumption to production. Although not yet fully understood, AMPK activity appears to be essential in both ES cells and adult stem cells in order to maintain their proliferation.

## 4.2    HOW THE STEM CELL MACHINERY RESPONDS TO SIGNAL INPUTS

Signal transduction pathways, whether from growth factors or metabolites, are clearly critical in relaying information about the stem cell's environment, enabling it to respond in an appropriate way in terms of self-renewal or differentiation in order to maintain tissue and organismal homeostasis. In the second part of this chapter, we will argue that these inputs are by themselves capable of dictating stemness, but before reaching that point, we need to consider how the signal input is converted into a change in protein activity and whether the processes involved in any way have an inherent contribution to stem cell-specific properties. As depicted in Figure 4.1, the regulation of gene expression can be influenced in many ways, including through chromatin structure and epigenetic modifications; transcription factors; post-transcriptional and translational regulation, especially via miRNAs; and a host of post-translational modifications involving additions of small chemical groups, peptides, and complex molecules, or cleavage and degradation (Figure 4.17). For each of these processes, researchers have tried to identify if there are any signature patterns that correlate with stem cells, discriminate pluripotent characteristics from the more restricted potency of adult stem cells, or define particular states of a stem cell such as quiescence versus self-renewal versus commitment to differentiation. At the risk of being oversimplistic, it is probably fair to say that transitions between different stem cell states can be linked to changes in one or more of the processes that lead to alterations in gene expression and protein activity.

The amount of detail available about the processes that are influenced in stem cells as a consequence of the various signal inputs they receive is truly vast and well beyond the scope of this brief introduction to the subject. Instead of trying to cover everything in depth, we will focus on a few examples of molecular processes in the cell where there is good evidence for stem cell specificity or distinctiveness. We will start with epigenetic modifications of histone proteins, especially how these help us to define the pluripotent state, then consider a few cases in which the action of transcription factors determines stem cell characteristics, and finally, we will summarize the current understanding of micro RNAs (miRNAs) and their role in fine-tuning specific stem cell properties.

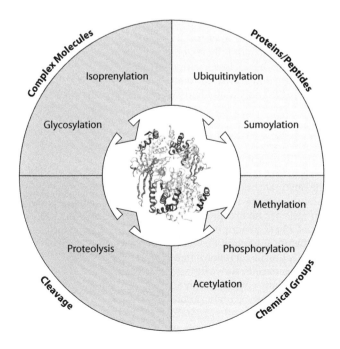

Figure 4.17 **Types of post-translational modification.** Summary of the modifications that can apply to a protein. (From Wang Y-C, Peterson SE & Loring J. (2014) *Cell Res* 24:143–160 [doi: 10.1038/cr.2013.151]. Reprinted by permission from Springer Nature.)

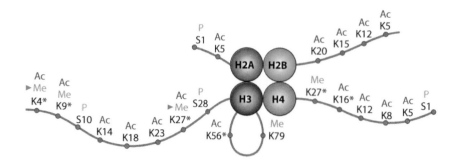

Figure 4.18 **Histone modifications associated with stem cell function.** The diagram represents the four core histones (H2A, H2B, H3, and H4) that constitute the nucleosome. Each histone has a "tail" that contains the amino acids lysine (K) and serine (S) that are potential targets for chemical modification by phosphorylation (P), acetylation (Ac), or methylation (Me). Specific modifications at several of these residues are linked either to fate determination in embryonic or adult stem cells (red star) or are involved in "bivalent domains" in pluripotent stem cells that appear to mark genes poised for activation once the cell commits to a particular pathway of differentiation.

## Histone modifications play an important part in determining stem cell fate

Both the maintenance of stemness and lineage commitment are tightly controlled by reversible and highly dynamic epigenetic mechanisms, including DNA methylation and histone modifications. These modifications alter the structure of chromatin by facilitating or preventing the binding of transcription factors as well as a number of co-factors. Focusing on the histone modifications, it has been found that a subset of these on certain amino acid residues in the tails of the core proteins (H2A, H2B, H3, and H4) are associated with cell fate determination in either ES cells or adult stem cells (Figure 4.18), and genome-wide studies on these modifications have shed light on how chromatin contributes to the maintenance of stem cells and their eventual commitment toward lineage progression.

The most clear-cut association between specific histone modifications and stem cell characteristics is seen in pluripotent stem cells in relation to the trimethylation of histone H3 at lysine residues 4 (H3K4me3) and 27 (H3K27me3). In non-stem cells, these histone modifications are generally recognized respectively as indicators of active or repressed transcription, but in ES cells and iPS cells, they paradoxically occur together on a large number of genes in close proximity to transcription start sites. Such **"bivalent domains,"** as they have become known, occur around genes encoding master regulators of cell lineages and serve to maintain pluripotent cells in a poised state to allow flexibility for eventual lineage choice.

Exploring whether a specific pattern of histone modifications characterize adult stem cells, either in the quiescent state or linked to commitment to differentiation, has been technically more challenging because of the difficulty in obtaining enough cells. However, sufficient cells have been isolated from tissues such as muscle and skin, making it possible to use genome-wide chromatin immunoprecipitation followed by sequencing (ChIP-seq) to obtain epigenetic profiles of the resident stem cells. In contrast to observations in pluripotent cells, few genes appear to carry the bivalent domain in these adult stem cells, although in both quiescent hair follicle stem cells and muscle satellite cells, thousands of genes are marked with H3K4me3, which is associated with active gene transcription. However, given the low transcriptional output in quiescent stem cells, it is likely that in this case, H3K4me3 may identify genes that will be transcribed upon activation.

## Transcription factors define stem cell types

Whatever cell type is concerned, it is always the case that the main arbiter of its phenotype is the combination of genes that are expressed, which is in turn largely dictated by the activity of the transcriptional apparatus, although the upstream and downstream influences of epigenetics and miRNAs respectively modulate how transcription factors act and the fate of the transcribed RNA sequences. Stem cells are no exception, and research effort has been intense toward trying to understand what transcription factor machinery operates and how similar or distinct this is in different situations. By far the clearest picture, and probably the most signature-like transcription factor involvement in defining stem cell characteristics, is the case of pluripotent stem cells. As we discuss in Chapter 5, the pluripotent state can be induced in adult somatic cells by enforced expression of several alternative combinations of transcription factors, involving a core contribution from Oct4, Klf4, and Sox2. These factors, together with another protein critical for pluripotency in ES cells, namely Nanog, form an interconnected network of cross-regulation that interacts with other transcriptional regulators that fine-tune the

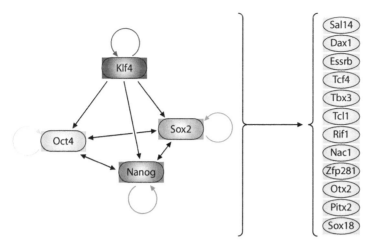

**Figure 4.19 Core pluripotency transcription factor network in ES cells.** Many regulatory proteins dictate the pluripotent state of embryonic stem (ES) cells, but four transcription factors are central to the overall control. The diagram illustrates these four pluripotency factors, namely Oct4, Sox2, Nanog, and Klf4. In addition to the core factors, several other transcription factors (shown on the right) closely cooperate with them to maintain pluripotency and self-renewal. The linear and circular arrows indicate cross-regulation and self-regulation of the expression of the transcription factor network. (Adapted from Greenow K and Clarke AR (2012) *Physiol Rev*, 92: 75–99.)

pluripotent state (Figure 4.19). This mechanism for establishing pluripotency is highly conserved, certainly operating in all vertebrates.

Adult stem cells can be shown to be highly dependent on specific combinations of transcription factors, but unlike the way in which a defined set of transcription factors determine the pluripotent state, each stem cell type seems to have evolved its own molecular strategy to enable conversion of the signal inputs into gene expression that supports properties such as self-renewal and potency. Hence, NSC rely on a combination of the activities of the Sox, Gli, and Hes families of transcription factors, among others, for self-renewal, whereas other adult stem cells can utilize different selections from the transcriptional regulatory protein armamentarium, and many of these will be highlighted in the subsequent chapters that deal with individual adult stem cell types.

## miRNAs fine-tune gene expression in a stem cell-specific manner

Mature miRNAs are single-stranded non-protein coding RNAs of 20–23 nucleotides in length that regulate protein translation, and less often RNA degradation, through complementarity binding to mRNA transcripts (Box 4.6). Such regulation of gene expression is implicated in one way or another with all cellular processes, and therefore it is no surprise that miRNAs play a critical role in the differentiation and maintenance of stem cells.

Of the cellular processes that receive inputs from signal transduction, the expression of miRNAs offers plenty of opportunities for fine-tuning of stem cell properties in a way that evolution has tailored in a specific number of ways. Even though there is not a one-to-one correspondence between the miRNA and the target, there being families of miRNAs, each member of which might act upon groups of mRNAs, there can be a degree of stem cell specificity in the overall profile. For example, ES cells express a profile of miRNAs that act in concert to regulate their self-renewal and differentiation. This miRNA signature, which differs between mouse and human cells, includes miRNAs expressed in undifferentiated ES cells and those that are activated during differentiation. Interestingly, the promoters of most of the miRNAs expressed in undifferentiated ES cells can be activated by the pluripotency transcription factors Oct4, Sox2, and Nanog (Figure 4.20). miRNAs whose expression increases during differentiation are also of importance since their low expression might serve to maintain ES cells in an undifferentiated state. For example, miR-145 controls ES cell differentiation by directly targeting the stem cell factors, thereby silencing the self-renewal program. The miRNA let-7 is also an important pro-differentiation factor that tightly controls the level of stem cell factors. Interestingly, let-7 processing is negatively regulated by lin28, which is a reprogramming factor (Chapter 5), and whose expression is under the control of the pluripotency factors. miRNAs such as these are therefore intertwined with the molecular network of pluripotency. Yet other miRNAs are involved in direct differentiation of ES cells toward specialized lineages, such as miR-133 and miR-1 that are required for cardiomyocyte differentiation.

miRNAs are also important for the normal functioning of some adult stem cells. For instance, GSC in the *Drosophila* ovary make use of miRNAs in the regulation of their self-renewal and differentiation, a fact that is clearly revealed in mutant flies made defective for the Dicer-1 protein, in which case the stem cells are rapidly lost from the niche.

micro RNAs (miRNAs) are single-stranded RNAs of 20–23 nucleotides in length that are complementary to sequences present in mRNA transcripts of protein coding genes. miRNAs are evolutionarily conserved and are scattered throughout the genome, where they are either coded as isolated transcript units or clustered and co-transcribed as polycistronic primary transcripts. Transcription of miRNAs is regulated in a similar manner to that of protein coding genes. RNA polymerase II transcribes large primary transcripts from the miRNA genomic sequences, which are then processed by sequential endonucleolytic cleavages (Figure 1). The first cleavage occurring in the nucleus involves the RNAse III enzyme Drosha, which results in

60–70 nucleotide hairpins that are then transported into the cytoplasm, where they are further cleaved by another RNAse III enzyme, Dicer, to produce a short double-stranded RNA duplex of 20–23 bp. This duplex is embedded with an Argonaute (Ago) protein to form the miRNA-induced silencing complex (RISC), in which the mRNA complementary strand is protected and the other strand is released and degraded. The mature miRNA in the RISC binds to the 3′ UTR, or in few cases the coding region, of the target mRNA transcript based on complementarity, which if perfect induces degradation of the target mRNA through Ago endonuclease activity. Partial pairing results in repression of target mRNA translation.

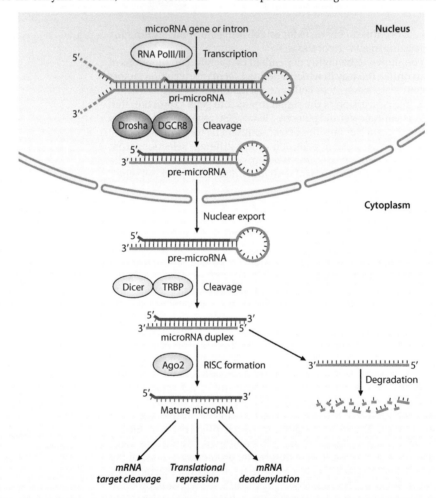

**Box 4.6 Figure 1 miRNA processing pathway.** The pathway of miRNA maturation includes the production of the primary transcript (pri-miRNA) by RNA polymerase II or III and its cleavage in the nucleus by the Drosha-DGCR8 complex. The resulting pre-miRNA precursor hairpin is exported from the nucleus and in the cytoplasm is further cleaved by the RNase Dicer in complex with the double-stranded RNA-binding protein TRBP. The functional strand of the mature miRNA is loaded together with Argonaute (Ago2) proteins into the RNA-induced silencing complex (RISC), in which it is then able to silence target mRNAs through mRNA cleavage, bring about translational repression, or catalyse deadenylation. (From Winter J, Jung S, Keller S et al. (2009) *Nat Cell Biol* 11: 228–234 [doi: 10.1038/ncb0309-228]. Adapted by permission from Springer Nature.)

There is abundant evidence for the involvement of miRNAs in various **tissue stem cells**. Hence, HSC population size is regulated by miR-125a, which brings about translational repression of the pro-apoptotic protein Bak1. In the brain, several miRNAs are linked to the maintenance of NSC. Hence, miR-124 is expressed at low levels in NSC and is up regulated in adult neurons, in which it induces neural differentiation by directly targeting two transcription factors (Sox9 and REST). Another brain-associated miRNA, let-7b,

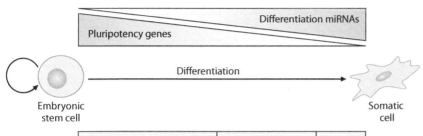

| miRNAs | Targets | Species |
|---|---|---|
| miR-134, miR-296, miR-470 | Oct4, Sox2, Nanog | M |
| miR-200c, miR-183, miR-203 | Sox2, Klf4 | M |
| miR-145 | Oct4, Sox2, Klf4 | H |
| Let-7 | cMyc, Lin28, Sall4 | M |
| miR-125, miR-181 | Cbx7 | M, H |

**Figure 4.20 miRNAs modulate pluripotency determining transcription factors in embryonic stem cells to determine self-renewal versus differentiation.** The activity of pluripotency genes, including core transcription factors, must be shut down as embryonic stem cells differentiate. The diagram illustrates some examples of miRNAs that are up regulated to support the exit from pluripotency by directly targeting the core and associated pluripotency transcription factors. M: mouse; H: human. (From Rosa A & Brivanlou A. (2013) *Int J Mol Sci* 14: 14346–14373 [doi:10.3390/ijms140714346]. Published under the terms of the Creative Commons Attribution License http://creativecommons.org/licenses/by/3.0/.)

also accelerates neural differentiation, but in this case by inhibiting NSC proliferation including through the targeting of cyclinD1 involved in cell cycle progression. miRNAs have been shown to be important regulators of muscle cell fate decision. For example, miR-1, miR-206, and miR-486 down regulate Pax7, a protein required to maintain the **muscle stem cell** population (Chapter 10). Last, miR-203, the most abundant miRNA in mammalian skin, promotes epidermal differentiation by repressing the transcription factor p63, resulting in restrictive proliferative potential and induction of cell cycle exit, while miR-125b is preferentially expressed in skin stem cells and is implicated in the balance between self-renewal and early lineage commitment. Interestingly, the same miRNA is also employed in HSC and muscle stem cells, suggesting that some common principles of stem cell behavior might be underpinned by miRNAs.

## Signal inputs and their intracellular interpretation create stem cell specific signatures

A question that has been asked frequently over the last two decades or so of research into the fundamental basis of stem cells is whether the unifying cell biological principles of what defines stemness are reflected at the level of molecular regulation. Our brief survey in this chapter of the key features of signal transduction pathways from the cell exterior to changes in the functional activity of proteins suggests that we probably have some answers, even if not what was originally anticipated by the research community.

Not unsurprisingly, much attention has focused on what distinguishes pluripotent stem cells from adult stem cells as a group. The molecular signature that defines vertebrate pluripotent stem cell characteristics is without doubt becoming clear, including what signaling pathways promote proliferation versus differentiation and the detailed epigenetic and transcriptional mechanisms that control gene expression. Much less certain, though, is whether there are common features among the broad range of adult stem cells that ensure tissue homeostasis throughout life. It is probably reasonable to conclude that there is nothing underpinning adult stem cell regulation that is as clear cut as the pluripotency factors and bivalent epigenetic modifications that determine ES cell and iPS cell properties. In fact, in considering how stem cells interact with their environment (Chapter 6), it can be argued that stemness in adult stem cells is more an environmentally defined principle rather than molecular hard wiring in the cell. In this view, all of the inputs that are received by a stem cell collectively determine properties such as self-renewal, quiescence, asymmetric division, and differentiation commitment, but the combination of inputs achieving this result can be different depending on the particular tissue circumstance. For sure, as we have described, there are only a relatively small number of potential signaling pathways, but these can be combined in many and varied ways that essentially mean that each stem cell type has a unique profile of inputs.

## The independent evolution of stem cells in plants and animals points to the essential nature of stemness

The study of how stem cells have evolved might give some additional insight into the question of whether there is fundamental molecular regulation of stemness.

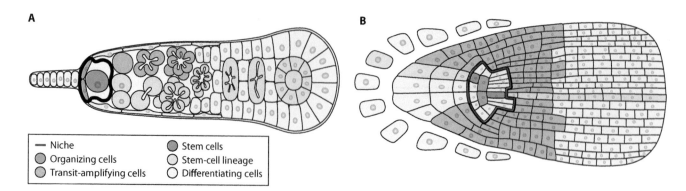

Figure 4.21 **Comparison of stem cell niche organization in animals and plants.** Although stem cells evolved independently in plants and animals, there are a number of similarities in the way in which the respective stem cell niches are organized. The schematic compares stem cell niches in an ovary of *Drosophila* (**A**) and in the plant root apical meristem (RAM) (**B**). The stem cell domains are outlined in red. Individual stem cells are highlighted (dark blue), and their downstream progeny are indicated in pale blue. In the case of the RAM, two stem cells are shown, one on either side of the organizing center (purple), to illustrate how their progeny are directed either toward the root tip or the vascular bundle. (From Scheres B. (2007) *Nat Rev Mol Cell Biol* 8: 345–354 [doi: 10.1038/nrm2164]. Reprinted by permission from Springer Nature.)

We have seen in this chapter that many of the signal transduction pathways and downstream molecular processes involved in the regulation of stem cell features in animals are highly conserved from some of the earliest multicellular organisms through to humans. Most importantly in this context, though, is the fact that the evolution of multicellularity, and hand-in-hand with this the need for stem cells, has occurred independently in the plant and animal kingdoms. The origin of stemness can therefore be considered an example of convergent evolution. Even with a fairly cursory comparison of the signal transduction pathways and molecular processes involved, it is fairly obvious that the parallels between plant and animal stem cells are functional, underpinned by common environmental features, and that conserved molecular features play little or no part in defining stemness. Comparison of the *Drosophila* ovary GSC niche (Chapter 12) with the niche supporting stem cells in the root apical **meristem** (RAM, Chapter 8) illustrates this commonality of structure-function relationship (Figure 4.21). In both cases, a small number of stem cells occupy a well-defined domain with adjacent supporting niche and "organizer" cells. In committing to differentiation, first transit amplifying cells and then cells of defined differentiated lineages move away from the stem cell niche. The structure-function homology between plant and animal stem cell niches is reflected in the types of signal inputs and the ultimate consequences for molecular control of gene expression. Hence, just as animal stem cells utilize the set of receptor interactions and signaling, leading to changes in epigenetics and transcription factor activity, so plants employ analogous components, but with little overlap in terms of the involvement of homologous proteins.

## 4.3   SUMMARY

Our aim in this chapter has been to give a flavor of the ways in which stem cells receive information from their surroundings and how the signals generated can be interpreted in terms of defining the properties of the cells, including self-renewal potential and commitment to differentiation. There is no overall common theme to the signal transduction or resulting alterations in gene expression and protein activity, but there is a take-home message that the stem cell type-specific regulation is achieved through combinations of signals, some of which are cooperative and some of which are antagonistic. The achievement of pluripotency in ES cells and the like does seem to involve a fairly invariant set of transcription factors and epigenetic modifiers, but adult stem cells have found many ways to achieve the same endpoint of the level of control that they require in order to persist throughout life. This diversity of solutions is most obvious when comparing plant and animal stem cells, which share essentially no common molecular mechanisms to regulate their stemness but nevertheless do reveal some analogous features in terms of their microenvironments and downstream behavior following commitment to differentiate. The potential predominance of the influence of the niche will be explored further in Chapter 6.

## KEY POINTS

- The characteristic features of stem cells are dictated by a combination of the activities of specific proteins, which, as in any cell, are the result of molecular processes that control gene expression, post-transcriptional processing and translation, and post-translational modifications.

- Molecular processes that control the steps leading to a given level of protein activity are themselves regulated by influences external to the cell, which are relayed through a variety of signal transduction pathways.

- Input from a relatively small number of signal transduction pathways impacts stem cell behavior, raising the apparent paradox of how specificity is achieved to distinguish stem cells from one another and from differentiated cells.

- Specificity in signal transduction appears to be largely a consequence of the combination of individual pathways employed in a given stem cell.

- In addition to external inputs in the form of peptide factors that interact with specific receptors, stem cells are also responsive to signals generated through cell-cell contact mediated through adhesion receptors and to the sensing of the metabolic substrate status in the cell's environment.

- Downstream of the signal transduction inputs, additional specificity in the interpretation of the signal can be ascribed to several of the molecular processes that lead to an eventual change in the overall activity of proteins that influence stem cell properties. Prominent components include epigenetic effects of modifications to histones, the profile of transcription factors present, and the modulating influence of micro RNAs.

- Stem cell properties have evolved twice at least in multicellular eukaryotes, once in plants and once in animals. Comparison between plants and animals of the factors eliciting and controlling stem cell behavior reveals that shared characteristics such as self-renewal and the control of differentiation are underpinned by distinct molecules. However, the structures of the stem cell environments are very similar, suggesting that this is the ultimate source of stemness in most tissue scenarios.

## FURTHER READING

### Signal transduction pathways in stem cells

Bigas, A. and Porcheri, C. (2018) Notch and stem cells. *Adv Exp Med Biol.*, 1066: 235–263.

Petrova, R. and Joyner, A.L. (2014) Roles for Hedgehog signaling in adult organ homeostasis and repair. *Development*, 141: 3445–3457.

Ring, A., Kim, Y-M. and Kahn, M. (2014) Wnt/catenin signaling in adult stem cell physiology and disease. *Stem Cell Rev.*, 10: 512–525.

Wagner, T.U. (2007) Bone morphogenetic protein signaling in stem cells – one signal, many consequences. *FEBS J.*, 274: 2968–2976.

Zhanga, J. and Li, L. (2005) BMP signaling and stem cell regulation. *Dev Biol.*, 284: 1–11.

### Transcription factors and miRNAs in stem cells

Mathieu, J. and Ruohola-Baker, H. (2013) Regulation of stem cell populations by microRNAs. *Adv Exp Med Biol.*, 786: 329–351.

Niwa, H. (2018) The principles that govern transcription factor network functions in stem cells. *Development*, 145. pii: dev157420.

### The effects of metabolism on stem cells

Clevers, H., Loh, K.M. and Nusse, R. (2014) Stem cell signaling. An integral program for tissue renewal and regeneration: Wnt signaling and stem cell control. *Science*, 346:1248012.

Ito, K. and Suda, T. (2014) Metabolic requirements for the maintenance of self-renewing stem cells. *Nat Rev Mol Cell Biol.*, 15: 243–256.

Ochocki, J.D. and Simon, M.C. (2013) Nutrient-sensing pathways and metabolic regulation in stem cells. *Cell Biol.*, 203: 23–33.

Ryall, J.G., Cliff, T., Dalton, S. and Sartorelli, V. (2015) Metabolic reprogramming of stem cell epigenetics. *Cell Stem Cell.*, 17: 651–662.

# Cell plasticity, reprogramming, and induced pluripotent stem cells

**5**

The combination of the promise of a medical panacea and ethical controversies made **embryonic stem (ES) cells** the most publicized and discussed topic in stem cell science at the beginning of the twenty-first century (Chapter 3). Stimulated to a degree by the ethical debate around the use of human embryos, but also the practical limitations inherent in the use of human ES cells, there was a push to find a way by which it might be possible to generate **pluripotent stem cells** from any individual. As we describe in Chapter 3, **therapeutic cloning** through the technique of **somatic cell nuclear transfer** and subsequent generation of ES cells can achieve the personalization of pluripotent stem cells; however, this is not a straightforward technique and still requires embryos.

The main subject of this chapter is how our understanding of the molecular processes that determine the ES cell phenotype (Chapter 4) has provided a solution to the generation of personalized stem cells, enabling reversion of an adult differentiated cell to an early stage of embryogenesis in which a cell is pluripotent. This reversal from a differentiated to a pluripotent state is termed **"reprogramming."** In a sense, reprogramming could also be termed **transdifferentiation**, that is, one defined cell type converting to another (Figure 5.1). Various studies over the last three decades have shown that the forced expression of one or more genes encoding transcription factors can not only restore pluripotency in a fully differentiated cell state but can also induce a post-mitotic cell to proliferate (**"dedifferentiation"**) or elicit transdifferentiation. In fact, it has been suggested that transdifferentiation should be termed "lineage reprogramming," thereby both distinguishing it from and drawing parallels with "pluripotent reprogramming." So before describing how the **induced pluripotent stem cell** revolution took off, we will take an historical perspective on how various attempts to manipulate transdifferentiation paved the way for reprogramming. The rapid advances in pluripotent reprogramming technology actually reinvigorated interest in the potential of lineage reprogramming, and we will conclude this chapter by looking at recent successes in experimentally switching one differentiated adult cell type to another. Although strictly speaking not a stem cell issue, avoiding the need to pass through a stem cell stage, lineage reprogramming could become an alternative strategy for the production of specific cell types for therapeutic application.

## 5.1 REVERSIBILITY OF THE DIFFERENTIATED STATE OF ADULT CELLS

It is easy to think of the differentiated, usually non-dividing, state of adult somatic cells as being fixed following their passage through **progenitor** stages and securing by **epigenetic** factors such as DNA methylation and histone modifications. However, a number of quite different observations of natural biological processes or the consequences of experimental manipulation have shown in fact that the epigenetic lock-down on the genome is surprisingly flexible, allowing all manner of cell conversions. We will look at how the status of a differentiated cell can be altered in a natural context of dedifferentiation and transdifferentiation, which involve changes in cell phenotype, which can be either within a given lineage or between distinct lineages. Such phenotype

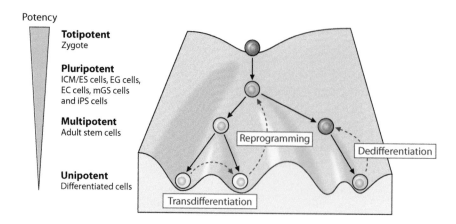

Potency

**Totipotent**
Zygote

**Pluripotent**
ICM/ES cells, EG cells,
EC cells, mGS cells
and iPS cells

**Multipotent**
Adult stem cells

**Unipotent**
Differentiated cells

Reprogramming

Dedifferentiation

Transdifferentiation

**Figure 5.1 Theoretical energy barriers to the reversal or switching of the differentiation state of a cell.** Conrad Waddington, who was a leading embryologist and geneticist from the 1930s to the 1950s, developed the now famous concept of the "epigenetic landscape" controlling cell fate. This concept is usually depicted in a diagram, as shown here, of a hillside with crests and valleys representing the likelihood and effort involved in a cell making a transition from one state to another. These transitions of cell status need to overcome energetic barriers, which take the form of changes in regulatory networks including a variety of epigenetic modifications. Dedifferentiation, transdifferentiation, and reprogramming are highlighted together with the way in which these can relate to the potency and stem cell nature of cells. (From Eguizabal C, Montserrat N, Veiga A et al. (2013) *Semin Reprod Med* 31: 82–94 [doi: 10.1055/s-0032-1331802]. With permission © Georg Thieme Verlag KG.)

changes can also be elicited by experimental intervention either through **nuclear cloning** or by manipulation of factors able to act on lineage-specific gene regulation in the nucleus.

## Dedifferentiation and transdifferentiation of adult somatic cells occur naturally

Dedifferentiation can be regarded as the reversion of a cell to an earlier progenitor and hence proliferative stage at which cells can expand and then go on to differentiate back to the original cell type. Several examples of non-mammalian vertebrate regeneration, which we discuss in Chapter 7, involve dedifferentiation, including in the fish heart following amputation of a portion of the ventricle and regrowth of severed amphibian limbs. Limb regrowth involves formation of the **blastema**, which then gives rise to the multiple cell types required to regenerate a precise replacement of all components of the limb, but, notably, cells retain a "memory" of their differentiated origin and only produce that same cell in the new tissue. This memory is most certainly epigenetic in nature and emphasizes that dedifferentiation can only be regarded as a relatively minor reversal of the differentiated state. There are also examples of dedifferentiation in mammalian tissues, although nothing quite so obviously dramatic as seen in the lower vertebrates. For example, when a nerve is damaged, the surrounding Schwann cells, which provide the insulating myelin sheath, dedifferentiate to a progenitor that is able to proliferate to expand cell numbers before once again differentiating to the mature cell.

Transdifferentiation is distinguished from dedifferentiation purely by the fact that the outcome of reversion to a proliferative state with subsequent differentiation leads to one or more cell types that are different from the original cell. This clearly implies a significant change in the epigenetic status, although limitations on the range of possible new differentiation fates suggest that some elements of the original epigenetic lockdown are retained. Transdifferentiation is undoubtedly used quite widely in animals, especially in circumstances of tissue damage and loss of specific functional cell types. A good example, once again seen in amphibians, is their ability to regenerate a new lens in the eye if it is experimentally removed. In these circumstances, pigmented epithelial cells in the eye are able to transdifferentiate via a proliferative stage ultimately becoming new lens tissue. Further, as we describe in Chapter 14, both the retina of the eye and the inner ear provide examples of transdifferentiation in mammals. The Müller glial cells of the retina appear to have evolved to provide two essential functions, that is, to support all of the neural cell types to which they have contacts spreading throughout the retina, and to act as an emergency progenitor when there is significant loss of these cells through damage, a function that requires reversion to a proliferative progenitor stage that can then go on to produce the various neural cell types, including the rod and cone photoreceptor cells. Likewise, if the sensory hair cells in the vestibular system of the inner ear are lost through damage, then the cells that normally support them can either directly transdifferentiate into new hair cells, or alternatively, as for Müller glial cells, they first reach a proliferative stage and then differentiate into sensory cells.

## Animal cloning has informed us about the plasticity of the nucleus and reprogramming

The most significant indication that the differentiated state is reversible, and that the interpretation of the information in the genome can in fact be reset to the earliest point in development, comes from **cloning** through the transfer of a somatic cell nucleus into an enucleated egg (Chapter 3). The seminal experiments were performed in the 1950s and 1960s, with the first successful attempts being by Briggs and King, who generated frog tadpoles from enucleated oocytes into which they had transplanted nuclei derived from frog blastulas. Although these cloning experiments did not produce adult frogs, they nevertheless paved the way for the Nobel Prize-winning work of John Gurdon, who obtained adult frogs first from tadpole somatic cell nuclei and subsequently in 1962 from nuclei taken from intestinal epithelium cells of an adult feeding stage. It took more than three decades before it was possible to achieve the same with mammals, when in 1996, Ian Wilmut and colleagues famously cloned a nucleus extracted from a mammary epithelial cell to obtain Dolly the Sheep.

## Cell fusion experiments pointed the way toward lineage reprogramming

The fusion of two cells to one another is a natural process in certain tissues but is also both useful in the laboratory for the generation of hybrid cells. Experimentally, fusion of cells can be brought about in the laboratory by applying an electrical pulse or through use of fusogenic agents, most often either polyethylene glycol or Sendai virus. Cell-cell fusion can be categorized on the basis of the cell types involved and the distribution of the genetic material that is retained following the merging of the two cells (Figure 5.2). Fusion between cells of the same type is termed homotypic and is seen naturally between osteoclasts in bone and myocytes in the formation of skeletal muscle myofibers. If the two cells are of different types, this is called a heterotypic fusion, which is more relevant in an experimental context, as we will see in more detail subsequently. Following fusion, the two nuclei either merge, forming what is known as a synkaryon, or remain as separate entities, in which case this is termed a **heterokaryon**.

Induced cell fusion experiments have, like nuclear cloning, provided evidence for the flexibility rather than irreversibility of epigenetically controlled gene expression in fully differentiated cells. Especially revealing have been heterotypic fusions resulting in heterokaryons in which it has been possible to monitor the genes expressed by the individual donor nuclei. In this way, it was determined that some components present in the cytoplasm of one fusion partner can switch the pattern of genes expressed by the nucleus of the other cell type. The seminal findings were made in 1985 by Helen Blau in Stanford University, who showed that when muscle cells are fused with non-muscle cells, muscle gene expression is activated in the non-muscle cell type (Figures 5.3 and 5.4). Blau tested a variety of non-muscle cell types, representing the three germ layers and different stages of development, and in all cases, muscle gene expression was induced, although there were differences in the kinetics and extent of muscle gene activation (Figure 5.5). The conclusion Blau reached was that gene expression from a nucleus could be reprogrammed in the presence of appropriate combinations of trans-acting regulatory molecules. In her paper describing this work, Blau stated

Figure 5.2 **Types of cell fusion.** Fusion between animal cells can occur naturally or experimentally using a number of different techniques. The cells involved in the fusion event can be either the same (homotypic fusion) or of different types (heterotypic fusion). Both types of cell fusion can result in a cell containing either a fused nucleus (synkaryon) containing the combined chromosomal information (4N) or one that is bi-nucleated (heterokaryon), each nucleus containing a normal chromosome number (2N). (From Kemp K, Wilkins A & Scolding N. (2014) *Acta Neuropathol* 128: 629–638 [doi: 10.1007/s00401-014-1303-1]. Published under the terms of the Creative Commons Attribution License http://creativecommons.org/licenses/.)

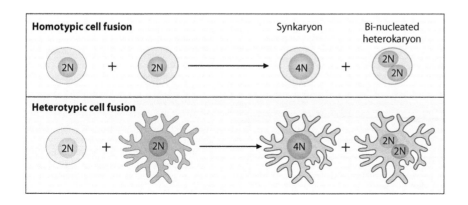

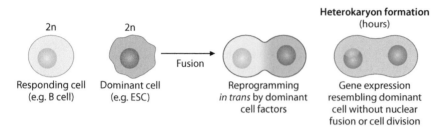

Responding cell
(e.g. B cell)

Dominant cell
(e.g. ESC)

Fusion

Heterokaryon formation
(hours)

Reprogramming
*in trans* by dominant
cell factors

Gene expression
resembling dominant
cell without nuclear
fusion or cell division

Figure 5.3 **Nuclear reprogramming in a heterotypic heterokaryon.** In a heterotypic cell fusion involving cells that each have a normal (2N) genetic component, the nuclei remain separate but interact by influencing each other's pattern of gene expression. In the diagram, the red cell nucleus is shown dominating the mechanisms of gene expression in the nucleus of the blue cell, indicated as a shift from blue to red.

that "the isolation of the genes encoding the tissue-specific trans-acting regulators responsible for muscle gene activation should now be possible," a prediction, as we will see in the next section, which was to come true within just over a year.

## Transcription factor proteins can act as master regulators to determine cell phenotype

The breakthrough in defining what factors are responsible for the nuclear reprogramming induced by muscle cells that Blau had described, and by extension how it is possible for an adult nucleus to be cloned to produce a perfect adult animal, as seen by Gurdon, among others, came about in 1987 in the laboratory of Andrew Lassar at the Hutchinson Cancer Research Center in Seattle. Lassar and his co-workers Robert Davis and Harold Weintraub were building on observations that treatment of mouse embryonic fibroblasts with the anti-cancer compound 5-azacytidine causes them to adopt a muscle cell phenotype. Reasoning that this shift in phenotype was brought about by the activation of one or a few key genes, they looked for muscle cell-specific RNAs by screening cDNA libraries and identified three candidates that were specifically expressed in myocytes and appeared in fibroblasts following 5-azacytidine-induced conversion. When these cDNAs, designated MyoA, MyoD, and MyoH, were introduced into fibroblasts, it was found that one of them, MyoD, was sufficient to convert them to stable myoblasts (Figure 5.6). The protein encoded by the MyoD cDNA belongs to a group of transcription factors known as helix-loop-helix proteins.

The finding that MyoD could single-handedly induce transdifferentiation led to the concept of transcription factor "master regulators," the idea being that each and every differentiated cell type might be generated following the expression of a key regulator protein. Although the devil is in the details, so to speak, and it is now known that one factor alone is often not sufficient for lineage reprogramming, it is in essence true that enforced expression of transcription factors associated with a particular cell type can lead to lineage reprogramming, including the ultimate achievement of pluripotent reprogramming.

Many cell types were subsequently shown to be susceptible to reprogramming through expression of lineage-associated transcription factors (Figure 5.7). Among these was the conversion of macrophage-like blood cells into a number of distinct blood cells, including erythrocytes, eosinophils, and megakaryocytes, by introduction of DNA sequences encoding the transcription factor GATA-1. This study, from the laboratory of Thomas Graf in Heidelberg, Germany, was notable not only because it demonstrated once again that a single factor could elicit transdifferentiation but also

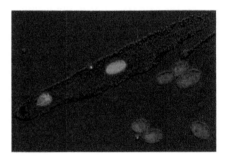

Figure 5.4 **Reprogramming of a fibroblast nucleus following fusion with a muscle cell.** Heterotypic cell fusions can be visualized by following morphological and antigenic features of the two contributing cell types. In the example illustrated, a mouse muscle cell and a human hepatocyte were fused using polyethylene glycol. The fluorescent microscope image shows staining of the heterokaryon with an antibody against a human muscle protein (red). In addition, it is possible to distinguish the two nuclei of the heterokaryon, which are stained blue, since that derived from the mouse cell is punctate, whereas the human nucleus has a more even staining pattern. Other nuclei in the image are part of a tri-nucleated heterokaryon that has not activated human muscle gene expression. (From Blau H, Pavlath G, Hardeman E et al. (1985) *Science* 230: 758–766 [doi:10.1126/science.2414846]. Reprinted with permission from AAAS.)

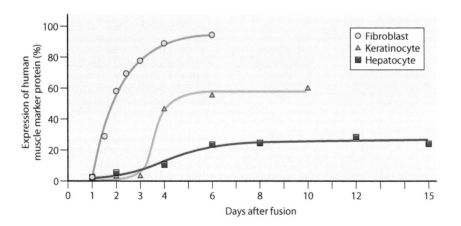

Figure 5.5 **Different cell types are reprogrammed with variable efficiency following fusion with a muscle cell.** Heterotypic fusion of cells can be performed with cells derived from any of the three major germ layers, but the success of fusion is dependent on the specific cell type. The graphs represent the efficiency of fusion between mouse muscle cells and different human cell types as indicated by the expression of a human muscle protein. The three human cell types represent cells of mesodermal (fibroblast), ectodermal (keratinocyte), and endodermal (hepatocyte) origins. Reprogramming to a muscle (mesodermal) cell fate is most efficient from the mesodermal human cell derivative. (From Blau H, Pavlath G, Hardeman E et al. (1985) *Science* 230: 758–766 [doi: 10.1126/science.2414846]. Reprinted with permission from AAAS.)

for the implication that in being reprogrammed, the cells were potentially passing through an immature, **multipotent** progenitor stage.

## 5.2 REPROGRAMMING ADULT CELLS TO BECOME PLURIPOTENT STEM CELLS

Around the turn of the millennium, and running in parallel with the research that was building a picture of the flexibility of the nucleus in adult differentiated cells, the derivation and characterization of pluripotent ES cells from mice, humans, and many other mammalian species was progressing apace. As we have seen, these independent research paths did not remain completely apart, since it became possible to generate ES cells from cloned embryos derived through nuclear transfer (Chapter 3). The second convergence, which it is fair to say rocked the world of stem cell biology, came about by combining the growing understanding of the molecular mechanisms that regulate pluripotency with the by then established fact that enforced expression of transcription factors can bring about lineage reprogramming, even crossing presumed boundaries between different germ layer divisions.

### Pluripotent stem cells were first created artificially by introducing specific transcription factors into fibroblasts

In 2006, Shinya Yamanaka (Box 5.1) from Kyoto University in Japan announced to a stunned audience of several thousand people at the International Society for Stem Cell Research annual meeting in Toronto that he had succeeded in generating ES cell-like pluripotent stem cells from fibroblasts. Yamanaka's approach was to screen available databases of expressed mRNAs in order to identify genes that correlate with the pluripotent state. He defined a set of 24 ES cell-associated transcripts (ECAT) that were all highly expressed in mouse ES cells and early embryos, and then introduced cDNAs for these candidate genes into mouse fibroblasts. An extra trick that Yamanaka used was to employ sequences from one of the ECAT genes, Fbx15, to drive expression of neomycin resistance as a means to select for cells that had been converted to an ES cell-like state (Figure 5.8). In this way, Yamanaka and his co-worker Kazutoshi Takahashi identified a combination of four transcription factors, Oct3/4, Sox2, Klf4, and Myc ("OSKM" for short), now generally called the "Yamanaka factors," which led to the formation of colonies of cells that looked remarkably similar to ES cells (Figure 5.9). Although initially not able to incorporate into **blastocyst chimeras** that could develop into adult mice, the reprogrammed cells did form **teratomas** when injected into mice, thereby satisfying a major criterion of pluripotency and leading Yamanaka to coin the term **induced pluripotent stem (iPS) cells**.

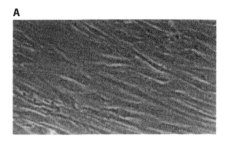
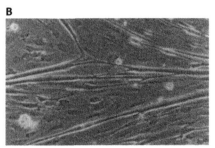

Figure 5.6 **MyoD converts fibroblasts to muscle.** Transcription factor proteins that characterize cells of a particular phenotype are in some cases able to convert a cell of another type to that phenotype. MyoD is one such factor associated with differentiating muscle cells. The phase contrast microscopic images show how rat fibroblasts can be converted to muscle cells by introduction of MyoD. The cells in (**A**) were infected with a control virus, whereas those in (**B**) were treated with a virus expressing the MyoD protein, inducing myogenesis causing the cells to elongate and fuse to form multinucleated myotubes. (From Weintraub H, Tapscott S, Davis R et al. (1989) *PNAS* 86:5434–5438 [doi: 10.1073/pnas.86.14.5434]. With permission.)

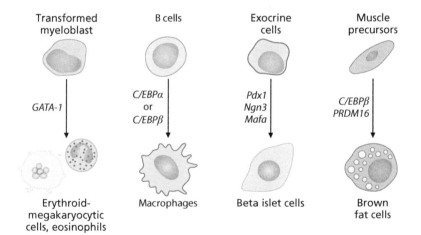

Figure 5.7 **Transcription factor-driven reprogramming between developmentally related cell types.** Enforced expression of one or more transcription factors associated with a specific differentiated cell type can drive the transdifferentiation of a different cell toward that phenotype. The diagram shows some examples of differentiated adult cell types along the top that can be transformed using the indicated transcription factors to enable reprogramming to a distinct cell type. (From Graf T. (2011) *Cell Stem Cell* 9:504–516 [doi: 10.1016/j.stem.2011.11.012]. Copyright 2011 with permission from Elsevier.)

**BOX 5.1 PIONEER: SHINYA YAMANAKA**

Shinya Yamanaka (Figure 1) is one of those figures who have truly revolutionized stem cell science. He was born in 1962, the year that John Gurdon, with whom he would eventually share the Nobel Prize in Physiology or Medicine in 2012, generated the first live animal cloned from an adult somatic cell. Yamanaka went to Kobe University's School of Medicine in 1981, where he decided to become an orthopedic surgeon, he says because he suffered so many broken bones while doing sports! But his surgical skills did not come up to his expectations, and he felt that he would serve the patients he had seen with incurable diseases much better if he sought solutions through basic science. So in 1989, he became a pharmacology PhD student at Osaka City University Graduate School of Medicine, which he followed with postdoctoral research on mouse molecular genetics at the Gladstone Institute of Cardiovascular Diseases in San Francisco. Yamanaka flourished at the Gladstone, and he says that it was the best decision he ever made, setting him on the path that would eventually lead to his Nobel Prize-winning research. At the Gladstone, he learned how to culture mouse ES cells and make chimeras, but he also gained the mentality that the most important attributes in research are to have clear vision and to work very hard toward realizing that vision.

**Box 5.1 Figure 1 Shinya Yamanaka.** (Adapted from http://pibphoto.nic.in/photo//2014/Aug/I2014083156207.jpg. This file is a copyrighted work of the Government of India, licensed under the Government Open Data License – India (GODL).)

In 1996, Yamanaka returned to Japan as assistant professor in the department of pharmacology at Osaka City University Medical School. He became interested in trying to understand how ES cells maintain their differentiation ability while rapidly proliferating. In 1999, he got his first laboratory at the Nara Institute of Science and Technology, continuing the same line of investigation. At this time, Yamanaka set his laboratory's goal as being to generate ES cell-like pluripotent cells from somatic cells, without using embryos, but he was unsure that he could achieve this in his research lifetime. Working with the initial hypothesis that factors that maintain the pluripotency of ES cells might induce pluripotency in somatic cells, Yamanaka and his team collected 24 candidate genes and developed an assay system based on a gene that they had studied earlier, Fbx15, which, although expressed in ES cells, failed to produce a phenotype when knocked out. This meant that they could tag a reporter to the gene (resistance to a cytotoxic drug) that would be expressed in pluripotent cells but would not be necessary for their maintenance. In 2004, while still engaged in this research, Yamanaka moved to the Institute of Frontier Medical Sciences at Kyoto University, at least partly because at the time it was the only institute in Japan that had succeeded in culturing human ES cells. When Yamanaka's colleague Kazutoshi Takahashi introduced a mixture of all 24 genes into fibroblasts using retroviral vectors, he observed drug-resistant colonies in a culture dish but was then able to whittle down to an essential group of only four factors, Oct3/4, Sox2, Klf4, and c-Myc, which ever since have been known as the "Yamanaka factors." In the 2006 paper he published in *Cell*, Yamanaka named the cells they had produced "induced pluripotent stem cells or iPS cells."

In November 2007, very quickly following the description of mouse iPS cells, Yamanaka's group reported the generation of the human equivalents, and within two months, Kyoto University founded a Center for iPS Cell Research and Applications (CiRA), the world's first such organization focusing on iPS cell technology, which they placed under Yamanaka's direction. This was inevitably a huge change in Yamanaka's research life, the staff complement of 250 soaking up a lot of his time in more administrative duties. The ever-present need for sufficient funding remains a major pressure for him, Yamanaka showing how dedicated he has been to finding solutions when he ran the full Kyoto Marathon in 2012 to raise more than 10 million yen in online donations from the public. In spite of all these pressures and diversions, Yamanaka manages to continue with his own research, including a small laboratory at the Gladstone Institute.

What followed the first description of iPS cells was nothing short of an explosion in the number of laboratories focusing their research on improving and exploiting this new technique. Yamanaka's first iPS cells turned out to be only partially reprogrammed, differing from ES cells with regard to gene expression and DNA methylation patterns, but he was soon able to extend the capability of iPS cells so that, like conventional ES cells, they could yield adult chimeras capable of germ line transmission. Yamanaka achieved this by screening the transduced fibroblasts for expression of a gene, Nanog, which serves as a more stringent indicator of the pluripotent state. In order simply to visualize Nanog expression, Yamanka employed a Nanog-GFP reporter, although, as shown by others, he could equally well have used a similar reporter driven by Oct3/4 gene elements.

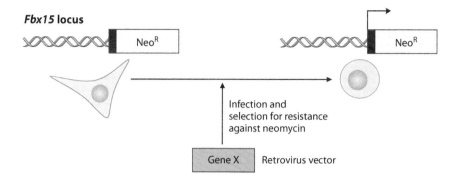

**Fbx15 locus**

Infection and selection for resistance against neomycin

Gene X    Retrovirus vector

Figure 5.8 **Strategy to test the pluripotent reprogramming potential of ECAT genes.** Schematic representation of the assay developed by Yamanaka to test the reprogramming potential of ES cell-associated genes (ECAT). A cassette expressing a bacterial neomycin resistance (Neo^R) gene was inserted into the mouse *Fbx15* gene locus. This locus is only expressed in pluripotent cells and therefore provides a way to select for acquisition of an ES cell phenotype because the cells become resistant to high concentrations of the compound G418. Candidate ECAT genes were introduced into fibroblasts from Fbx15-Neo^R mice by retroviral infection. (From Takahashi K & Yamanaka S. (2006) *Cell* 126: 663–676 [doi:10.1016/j.cell.2006.07.024]. Copyright 2006 with permission from Elsevier.)

Although live adults could be generated from the improved, fully reprogrammed mouse iPS cells, it was found that about a fifth of them went on to develop tumors because of the re-expression of the Myc **transgene**. It was therefore important that alternative combinations of reprogramming factors soon be discovered that avoided

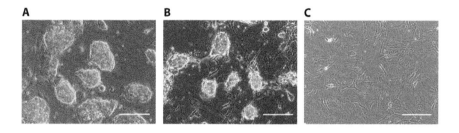

A    B    C

Figure 5.9 **iPS cell colonies appear very similar to ES cells.** Phase-contrast microscopy images showing the morphology of ES cells (**A**), iPS cells derived from mouse embryo fibroblasts (MEF) (**B**), and MEF (**C**). Scale bars = 200 μm. (From Takahashi K & Yamanaka S. (2006) *Cell* 126: 663–676 [doi:10.1016/j.cell.2006.07.024]. Copyright 2006 with permission from Elsevier.)

the requirement for Myc (Figure 5.10). There was found to be a consistent need for Oct4 and Sox2, but factors other than Klf4 and Myc could also enhance their ability to reprogram, some of which were in Yamanaka's original list of 24 ECAT genes. These include the pluripotency gene Nanog, the micro RNA binding protein Lin28, estrogen-related receptor beta (Esrrb), and the transcriptional repressor T-box protein 3 (Tbx3).

Within a year of the first description of mouse iPS cells, Yamanaka and his team successfully used the OSKM factors to produce human iPS cells from skin fibroblasts (Figure 5.11). Independently, Jamie Thomson in Madison, Wisconsin, derived similar iPS cells from both fetal and foreskin fibroblasts. Thomson's choice of reprogramming factors differed in that although they included Oct3/4 and Sox2, Klf4 and Myc were replaced by the combination of Nanog and Lin28. Both groups showed that the human iPS cells they were able to generate were similar to those obtained from mouse cells and had the requisite pluripotent characteristics of multi-lineage differentiation *in vitro* and teratoma formation *in vivo* (Figure 5.12).

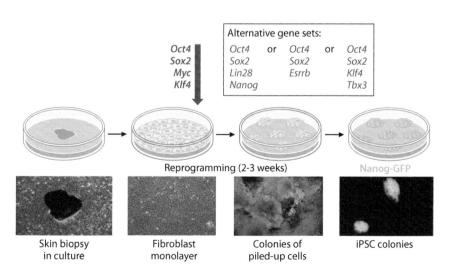

| Oct4 | Alternative gene sets: | | |
| Sox2 | Oct4 or | Oct4 or | Oct4 |
| Myc | Sox2 | Sox2 | Sox2 |
| Klf4 | Lin28 | Esrrb | Klf4 |
|  | Nanog |  | Tbx3 |

Reprogramming (2–3 weeks)    Nanog-GFP

Skin biopsy in culture    Fibroblast monolayer    Colonies of piled-up cells    iPSC colonies

Figure 5.10 **Identification and selection of fully reprogrammed iPS cells.** As a way to facilitate screening for cells that are undergoing conversion to a stem cell capable of populating the mouse germ line, researchers have employed transgenic reporters that reflect the pluripotent state. Such a strategy has usually employed fluorescent protein reporter expression driven by either Oct3/4 or Nanog gene regulatory elements. The diagram illustrates the use of a Nanog-GFP transgenic mouse line, showing how GFP-positive colonies emerge in culture following introduction of reprogramming factors into fibroblasts. Phase-contrast and fluorescent images illustrate the appearance of such cultures. Classically, reprogramming was achieved with the Yamanaka factors, Oct3/4, Sox2, Klf4, and Myc, but other combinations can give the same result, including the alternative gene sets illustrated. (From Graf T. (2011) *Cell Stem Cell* 9:504–516 [doi:10.1016/j.stem.2011.11.012]. Copyright 2011 with permission from Elsevier.)

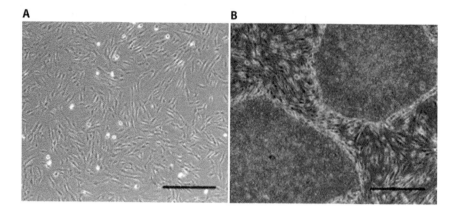

**Figure 5.11 Human iPS cells.** The reprogramming of human somatic cells leads to formation of iPS cell colonies that have a distinctive morphology, similar to that seen for human ES cells. The phase-contrast microscopy images show human adult skin fibroblasts (**A**) and the resultant iPS cells following reprogramming (**B**), which in this image appear in two tightly clustered cell colonies. Scale bars = 200 μm. (From Takahashi K, Tanabe K, Ohnuki M et al. (2007) *Cell* 131: 861–872 [doi:10.1016/j.cell.2007.11.019]. Copyright 2007 with permission from Elsevier.)

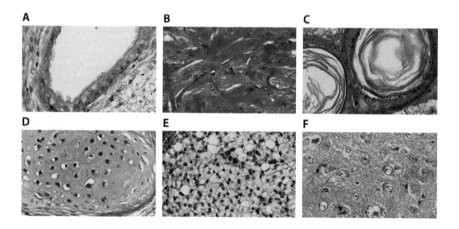

**Figure 5.12 Differentiated tissue types found in teratomas formed from human iPS cells.** The pluripotency of human iPS cells can be assessed by transplantation into immune-compromised mice with the consequent development of teratomas containing tissue types representing the three major germ cell lineages. The light microscopy images shown are of histologically stained sections of distinct tissue types detected in such iPS cell-derived teratomas. (**A**) Gut epithelium; (**B**) muscle; (**C**) epidermis; (**D**) cartilage; (**E**) adipose tissue; (**F**) neural tissue. (From Takahashi K, Tanabe K, Ohnuki M et al. (2007) *Cell* 131: 861–872 [doi:10.1016/j.cell.2007.11.019]. Copyright 2007 with permission from Elsevier.)

Reprogramming of both mouse and human cells has not been restricted to fibroblasts, and there have been two principal objectives behind trying to generate iPS cells from different adult differentiated cells. First, if iPS cell technology is to have therapeutic application, then it is highly desirable to have a readily available source of cells that is minimally invasive. Isolation of fibroblasts requires a painful biopsy, so it would be preferable to obtain cells for example from a blood sample. This was duly achieved, and in fact almost every cell that has been tried has yielded iPS cells, including mature lymphoid cells, which adds the interesting possibility of the "capture" and expansion of highly specific immune effector cells, such as cytotoxic T-cells, following differentiation of the iPS cells back to the T-lymphoid lineage (Chapter 17). Some cells, especially of epithelial or endothelial origin, proved to be more readily reprogrammed, and perhaps the best example of non-invasive cell derivation is from bladder epithelium, from which cells are shed that can be collected simply from a sample of urine. Comparison of the potential of iPS cells derived from different cell types showed that some degree of memory exists, the consequence of which is that they have a greater tendency to differentiate toward the same cell type.

## Improved methods for the generation of induced pluripotent stem cells were soon forthcoming

An early objective toward improvement of the reprogramming process was to find alternatives to retroviruses as a means of introducing the reprogramming factors, the reason for this being that retroviral vectors integrate into the genome and therefore represent a potential mutagenic risk. Non-integrating virus vectors, including adenoviruses and Sendai virus, have been used, but a host of non-virally based methods have also been developed, including episomal vectors, transposons, and direct introduction of reprogramming factors in the form of mRNA encoding them or as recombinant proteins.

The availability of reporter transgenes, such as those driven by Oct3/4 or Nanog gene sequences, facilitated the screening for chemicals that can enhance the reprogramming process or actually substitute for individual reprogramming factors.

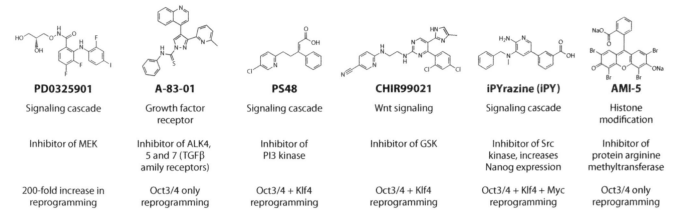

| PD0325901 | A-83-01 | PS48 | CHIR99021 | iPYrazine (iPY) | AMI-5 |
|---|---|---|---|---|---|
| Signaling cascade | Growth factor receptor | Signaling cascade | Wnt signaling | Signaling cascade | Histone modification |
| Inhibitor of MEK | Inhibitor of ALK4, 5 and 7 (TGFβ amily receptors) | Inhibitor of PI3 kinase | Inhibitor of GSK | Inhibitor of Src kinase, increases Nanog expression | Inhibitor of protein arginine methyltransferase |
| 200-fold increase in reprogramming | Oct3/4 only reprogramming | Oct3/4 + Klf4 reprogramming | Oct3/4 + Klf4 reprogramming | Oct3/4 + Klf4 + Myc reprogramming | Oct3/4 only reprogramming |

Figure 5.13 **Chemicals that enhance pluripotent reprogramming.** Screening of libraries of small molecules has identified a large number of compounds that are able to enhance the process of pluripotent reprogramming. These compounds impinge on specific signaling pathways that influence self-renewal or differentiation. Six examples of such compounds are illustrated, together with the overall effect they have on reprogramming and what is known about their specific site of action. (From Jung D-W, Kim W-H, Williams D. (2014) *ACS Chem Biol* 9: 80–95 [doi:10.1021/cb400754f]. With permission. Copyright 2014 American Chemical Society.)

The range of compounds that have a positive effect on reprogramming is impressive and essentially reflects cell processes that either promote **self-renewal** or inhibit differentiation. These compounds can act at various levels of cell signaling from cell surface receptors, through intermediate signaling components, to the transcriptional and epigenetic regulation of the genome. For example, inhibition of TGFβ family receptors enables reprogramming with Oct3/4 alone, inhibition of the Wnt or PI3 kinase signaling pathways increases the efficiency of reprogramming by the combination of Oct3/4 and Klf4, inhibition of Src kinase can replace the requirement for Sox2, and inhibition of a histone modifying arginine methyltransferase makes possible reprogramming in the presence of Oct3/4 alone (Figure 5.13). The ultimate extension of these developments was to attempt reprogramming solely with chemicals, which was achieved in the laboratory of Hongkui Deng at Peking University (Figure 5.14), the "chemical iPS cells" (CiPSC) that were generated being able to contribute to the germ line and importantly produce adults with a normal lifespan and no increased risk of developing tumors (Figure 5.15).

## The mechanism of reprogramming is complex, involving shifts in epigenetic and transcriptional control

The reprogramming process is slow, taking around 2 weeks for mouse cells and considerably longer in the case of human cells. It is also very inefficient, and the resulting iPS cells can vary widely in their developmental potential. During the decade since the first isolation of iPS cells by Yamanaka, researchers have focused intensively on the molecular mechanisms that might explain these features of reprogramming, which are quite distinct from what is seen when a nucleus is cloned following transfer into an enucleated egg. These have been explored at both population and single cell levels, using a range of approaches including transcriptome, proteome, and epigenome analyses, combined with controlled expression of reprogramming factors and selective knock down of what appeared to be key genes. Reprogramming is now generally represented as a three-phase process,

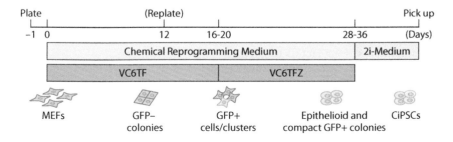

Figure 5.14 **Pluripotent reprogramming with small molecules: chemically induced pluripotent stem cells (CiPSC).** Following the identification of compounds that enhance reprogramming to generate iPS cells, investigators sought to achieve reprogramming solely through the use of small molecules. The first published strategy for chemical induction of iPS cells is illustrated. Mouse embryonic fibroblasts (MEF) were treated with chemical reprogramming medium (yellow) followed by a period in two-inhibitor (2i) medium (green). The MEF were derived from and Oct3/4-GFP mouse strain, facilitating identification of successful pluripotent reprogramming. The chemical reprogramming medium was initially composed of valproate (V), CHIR99021 (C), 616452 (6), tranylcypromine (T), and forskolin (F), and then was additionally supplemented with DZNep (Z) during the second half of the process. (From Hou P, Li Y, Zhang X et al. (2013) *Science* 341: 651–654 [doi: 10.1126/science.1239278]. Reprinted with permission from AAAS.)

involving initiation, maturation, and stabilization (Figure 5.16). These three phases include two waves of resetting of gene and protein expression, which occur during the initiation and stabilization stages. During initiation, after introduction of OSKM and over the course of the first 5 days of reprogramming, cells assume one of several states (apoptosis, senescence, transdifferentiation, reprogramming). Genes are activated that are responsible for cell proliferation, metabolism, and cytoskeletal organization, while genes associated with development and proteins involved in the electron transport system, vesicle transport, extracellular matrix, and adhesion are down regulated. These changes are related to a process known as **mesenchymal-epithelial transition** (MET). Single cell analysis reveals that although cells with ES cell-like morphology appear early, there is considerable heterogeneity of gene expression among the cells during initiation. This high degree of variation in gene expression indicates that in addition to the global alterations in the expression of genes that control proliferation, metabolism, and MET, there is a large "stochastic" element to the process at this stage.

Although it contributes the majority of the time required for reprogramming, the maturation phase is the least understood part of the process, but for sure it culminates in a transition to a more homogenous state as stable clones emerge driven by a hierarchy of core pluripotency factors. Transition from maturation to stabilization requires repression of the exogenous reprogramming factors, which is accompanied by the second wave of gene expression changes. In addition to genes encoding proteins of the core pluripotency network, the late wave of gene expression includes proteins involved in embryonic development and stem cell maintenance as well as glycolysis and the reappearance of vesicle transport, extracellular matrix, and adhesion molecules. In the transition from maturation to stabilization, genes that can promote the journey toward pluripotency are activated on a stochastic basis in a small number of cells. This promotes expression of Sox2, which defines entry into the final phase. Variation at the single cell level becomes noticeably reduced once cells start to express Nanog, and the cells enter a shorter "deterministic" phase that is governed by the hierarchical interaction between factors that stabilize the pluripotent state, including fixing the epigenome and establishing the self-maintaining core pluripotency circuitry.

As in cloning of an adult somatic nucleus, reprogramming by the Yamanaka factors must involve erasure of the somatic cell epigenetic signature and its replacement with a stem cell-like epigenome. The epigenetic changes are complex and by no means fully understood, involving both positive and negative influences on gene expression that can either enhance or hinder the reprogramming process. They include DNA methylation and X-chromosome reactivation, which occur late in the reprogramming process, and a vast array of histone post-translational modifications that start early following introduction of the four factors. For example, immediately after factor induction, *de novo* dimethylation of lysine 4 on histone H3 (H3K4me2) accumulates at the promoters of many pluripotency genes, while in parallel, the promoters of somatic genes begin to lose H3K4me2, which leads at a later stage to hypermethylation and gene silencing.

Little is known about how ectopic expression of the Yamanaka factors drives the conversion of somatic cells to the pluripotent state. The expression levels and relative stoichiometry of Oct3/4, Sox2, Klf4, and Myc does not mimic what is seen in ES cells. Immediately after factor induction, the four factors bind to genes that are actively being expressed or repressed in regions of accessible chromatin. Oct3/4, Sox2, and Klf4 also bind to distal sites throughout the genome that are not normally bound in ES cells, acting as "pioneer" factors to open up chromatin of genes that will promote reprogramming, something that can be achieved by only transient expression of the factors. Oct3/4, Sox2, Klf4, and Myc bind together at gene regions that initiate and support conversion to pluripotency, such as GLIS family zinc finger 1 (Glis1), the micro RNA-encoding mir-302-367 cluster, F box only protein 15 (Fbx15, the gene that Yamanaka originally used in his selection strategy for reprogrammed MEF), FGF4, LIN28, and factors that promote MET. Myc facilitates this process and is only essential in the first 5 days of reprogramming, acting as a general amplifier of gene expression. Although the four factors often jointly bind to their targets, subsets and different combinations of the factors frequently occupy non-overlapping genomic regions. Stoichiometry can profoundly influence reprogramming, as can a number of other factors related to mechanism, such as the presence of serum, the levels of oxygen, the availability of Vitamin C, and, as we have seen already, manipulation of various signaling pathways through the action of small molecules (Table 5.1).

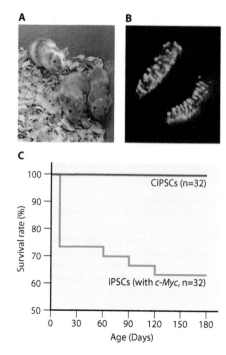

**A**   **B**

**C**

Survival rate (%) vs Age (Days)

CiPSCs (n=32)

iPSCs (with *c-Myc*, n=32)

**Figure 5.15 Mouse CiPSC can give rise to germ line cells and healthy adults.** The strongest indication that iPS cells are fully pluripotent is the demonstration that they can contribute to the germ line if introduced into a blastocyst stage embryo and that the chimeric animals have a normal lifespan without an undue occurrence of tumors. This figure shows how CiPSC derived from an Oct3/4-GFP transgenic mouse can satisfy both of these criteria. (**A**) Photograph of the chimeric mice generated using CiPSC. (**B**) Fluorescent microscopy image of testes from chimeric male mice showing CiPSC contribution to the gonads as revealed by expression of the Oct3/4-GFP reporter. (**C**) Survival curve of resulting chimeric mice produced with conventional (Myc-containing) iPS cells compared to those derived from CiPSC. (From Hou P, Li Y, Zhang X et al. (2013) *Science* 341: 651–654 [doi:10.1126/science.1239278]. Reprinted with permission from AAAS.)

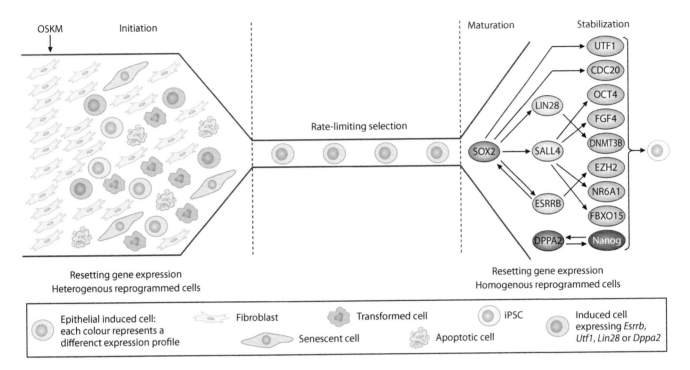

Figure 5.16 **Model for the mechanism of pluripotent reprogramming.** The length of time taken to reprogram to the pluripotent state indicates that the process is complex. Accumulating evidence suggests that there are several stages involving resetting of gene expression and selection. The diagram summarizes a current model for the phases of reprogramming, which divides the process into three phases (initiation, maturation, and stabilization). Induction of a fibroblast with Oct3/4, Sox2, Klf4, and Myc (OSKM) resets gene expression resulting in apoptosis, senescence, transformation, transdifferentiation, or reprogramming. Cells that have entered a reprogrammable state progress into a rate-limiting step delaying the conversion toward iPS cells; in this phase, pluripotency markers are activated, allowing subsequent maturation. Once Sox2 is activated, the cell enters the final phase, which entails another resetting of gene expression to establish the pluripotency gene regulatory network, resulting ultimately in the formation of the iPS cell. (From Buganim Y, Faddah D & Jaenisch R. (2013) *Nat Genet* 14: 427–439 [doi:10.1038/nrg3473]. Reprinted by permission from Springer Nature.)

## TABLE 5.1  FACTORS THAT INFLUENCE THE EFFICIENCY OF PLURIPOTENT REPROGRAMMING

| Parameter | Specific variable | Effect |
|---|---|---|
| Four factor stoichiometry | High Oct3/4 and Klf4, low Sox2 and Myc | Low reprogramming efficiency |
| | High Sox2 and Myc, low Oct3/4 and Klf4 | High reprogramming efficiency |
| Culture medium | Serum-free medium | Improved reprogramming efficiency |
| | Low oxygen | Improved reprogramming efficiency |
| Supplements | Vitamin C | Improved iPS cell quality |

*Note:* Many factors can be modulated in order to improve the efficiency of reprogramming or the quality of the resulting iPS cells. This table summarizes some of these factors and the effect that their manipulation has on the iPS cells generated.

## Molecular characterization of embryonic stem cells and induced pluripotent stem cells reveals their close similarity

Once iPS cells were being fairly regularly derived, especially from human cells, it did not take long before detailed comparisons were made with ES cell lines to determine whether factor-induced reprogramming can truly substitute for pluripotent cells derived from natural or nuclear transfer-derived embryos. The immediate conclusions, based largely on transcriptome and DNA methylation analyses, were that there are considerable differences between iPS cells and ES cells. The originating somatic cell for iPS cell generation seemed to play a particularly significant role in both gene expression differences and function, differentiation capability being somewhat skewed toward cell types related to the reprogrammed cell source. This suggests that elements of the epigenetic makeup of the starting cell are not completely removed during the reprogramming process. However, it was subsequently realized that the comparisons being made were flawed because of the very fact of the inherent genetic differences between individual humans and epigenetic variations in cells of different tissue origins. Recent studies to resolve the issue have used an elegant combination of technologies to ensure that the comparisons between iPS cells and ES cells are not influenced by these pre-existing differences. Dieter Egli's team at the New York Stem Cell Foundation Research Institute conclusively showed that iPS cells and ES cells derived from cells with equivalent genetic and epigenetic profiles are extremely similar. They produced both iPS cells and nuclear transfer ES cells from the same somatic cell culture and then compared their profiles of gene expression and DNA methylation as well as the frequencies of coding mutations and loss of imprinting. By all criteria, the iPS cells and NT-ES cell lines were extremely similar (Figure 5.17).

## Induced pluripotent stem cells have been created from several mammalian species other than mouse and human

Following the isolation of iPS cells from mouse and human cells, the same approach has been taken with a wide variety of mammalian species with a number of specific objectives in mind. There are many examples of domesticated animals for which iPS cells have been generated, usually with the intention that this will lead at some point to germ line transmission, possibly after engineering of specific desirable traits, such as disease resistance or productivity (see Chapter 19). Some initial progress toward such an ambition has been made with iPS cells derived from sheep and pigs, in particular in the latter case, for which offspring have been obtained following germ line transmission from chimeras. Several primate species have also yielded functional iPS cells that have all of the expected characteristics when assayed for differentiation potential *in vitro* or *in vivo* but, as for the human equivalents, there is

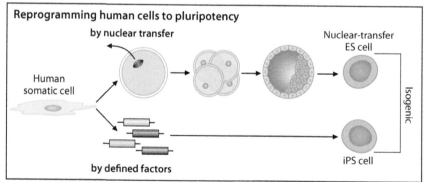

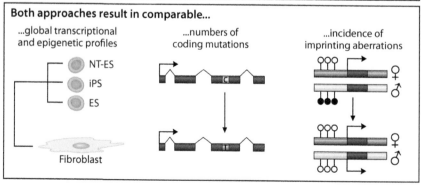

**Figure 5.17 Comparison of the molecular properties of ES cells and iPS cells.** Although the properties of ES cells and iPS cells are by definition very similar, detailed molecular analysis has been undertaken to determine if there are any underlying differences that might make iPS cells an unsuitable alternative to ES cells. In making such an assessment, it was important that the genetic background of the cells being compared be identical. The upper part of the diagram illustrates the strategy for comparison between ES cells (generated through nuclear transfer cloning) and iPS cells derived from the same human cell source to ensure that they are isogenic. The lower part of the diagram summarizes the parameters compared, which led to the conclusion that NT-ES cells and iPS cells are essentially identical when they are obtained from exactly the same genetic background. The comparisons encompassed full transcriptome and epigenome analyses, genome-wide assessment of the frequency of mutations, and determination of any aberrations in imprinting. (From Johannesson B, Sagi I, Gore A et al. (2014) *Cell Stem Cell*, 15: 634–642 [doi: 10.1016/j.stem.2014.10.002]. Copyright 2014 with permission from Elsevier.)

no intention to explore germ line transmission for reasons of ethical considerations. Nevertheless, the availability of primate iPS cells creates an excellent opportunity to investigate through comparison to identify critical mechanisms for pluripotency and reprogramming have evolved. Last, iPS cell technology is seen as a potential means to rescue endangered species from the brink of extinction (Chapter 19). To this end, iPS cells have been produced from the northern white rhinoceros, but since only two rather old individuals remain, it looks as though if these cells are to be of any use, then a strategy will be necessary involving cross-species chimeras and surrogacy involving the southern white rhinoceros.

## 5.3    STEM CELL PLASTICITY AND LINEAGE REPROGRAMMING REVISITED

In the final part of this chapter, we return to transdifferentiation, but this time in the specific context first of its relevance to **adult stem cells** and second how pluripotent reprogramming has illuminated a greater potential for reprogramming adult somatic cells from one phenotype to another.

### Adult stem cell plasticity is either a rare event or an experimental artifact

Around the turn of the millennium, there was a veritable flurry of publications and general excitement in the stem cell world around the possibility that adult stem cells with supposedly restricted potential could in fact contribute to a much wider range of tissues than expected, crossing the boundaries defined by the embryonic germ layers. Such presumed capacity, which could be considered transdifferentiation, was put under a banner of stem cell **"plasticity."** The principal observations that led to this concentration of interest came from experiments in mice in which whole or partially fractionated bone marrow was transplanted into animals that had been lethally irradiated or had been injured in specific tissues, such as the heart or skeletal muscle. Reports abounded of contributions of the donor cells, usually suspected to be **hematopoietic stem cells** (HSC), to cell lineages ranging from epithelia in skin, lung, kidney, and intestine, to liver parenchyma, skeletal muscle, cardiac muscle, and neurons in the brain. Out of these apparently amazing findings emerged a general hypothesis that differentiation of a particular adult stem cell is not hard wired but rather that the tissue microenvironment can dictate their fate. Reports were not restricted to findings of unexpected roles for hematopoietic stem cells, similar claims soon arising for the ability of **skeletal muscle stem cells** to give rise to blood cells and for **neural stem cells** being able to contribute to multiple tissues if injected into the **inner cell mass** of a blastocyst stage embryo. However, the levels of contribution of donor cells to host tissues were very low, usually on the order of 0.1%–1% or less, and the excitement was very soon dampened when reports started to come in from other research laboratories of failure to reproduce the same outcome and when more rigorous approaches were adopted, especially making use of genetic reporters that could definitively distinguish donor and host cell origin. Particularly conclusive, especially to discount a contribution of HSC to other tissue cell types, were experiments involving **parabiosis** in which two animals shared their blood circulation for up to 9 months. One of the animals of the parabiotic pair carried a transgenic fluorescent protein reporter that was expressed in all tissues, yet there was no evidence for the transfer of cells to the other partner, even when a severe injury was created that would have been expected to stimulate stem cell plasticity (Figure 5.18).

**Figure 5.18 No contribution of bone marrow cells to cardiomyocyte regeneration following injury.** Many claims have been made for the contribution of stem cells from one adult tissue in the repair and regeneration of distinct tissues, even those that differ in terms of the germ layer origin. One such set of claims were made about the ability of stem cells from the bone marrow to contribute to cardiomyocyte replacement following damage to the heart after oxygen deprivation. The definitive experiment refuting the initial claims entailed the use of parabiosis to test if marked cells in one animal could be mobilized from its bone marrow and then contribute to the repair of heart tissue in a second animal linked to its circulation. The diagram shows how two mice were joined, the cells from one animal being marked by a ubiquitously expressed GFP reporter. Injury to cardiac muscle mimicking a myocardial infarction (purple shaded zone at the apex of the heart) was induced by coronary artery ligature in the GFP-negative partner. Subsequent examination of the heart demonstrated that some GFP-positive cells entered the damaged tissue, but these did not become cardiomyocytes and instead were largely restricted to mature blood cell types. (From Chien K (2004) *Nature*, 428: 607–608 [doi: 10.1038/nature02500]. Reprinted by permission from Springer Nature.)

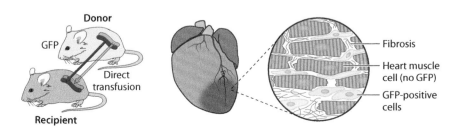

Donor

GFP

Direct transfusion

Recipient

Fibrosis

Heart muscle cell (no GFP)

GFP-positive cells

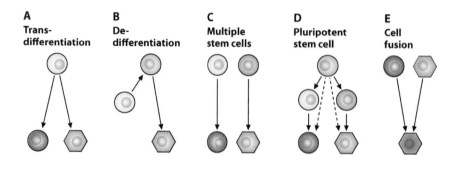

**Figure 5.19 Possible mechanisms underlying apparent stem cell plasticity.** As it became clear that adult stem cell plasticity was either a very rare event or the consequence of experimental artifact, explanations were sought for the observations that had been made. The schematic summarizes the potential mechanisms thought to account for these observations. Yellow/blue circles, tissue-specific stem cells; green circles, pluripotent stem cells; red circle/blue hexagon, differentiated cells from the yellow and blue lineages, respectively. (From Wagers A & Weissman I. (2004) *Cell* 116: 639–648 [doi: 10.1016/s0092–8674(04)00208-9]. Copyright 2004 with permission from Elsevier.)

So what was the explanation for the apparent transdifferentiation seen in some experiments, but the overall low efficiency and frequent lack of reproducibility? The discussions that followed revolved around a number of possible models (**Figure 5.19**), the current state of thinking being that one or two of these are consistent with what was reported and that there is no need to evoke a concept of transdifferentiation or plasticity. The original experiments purporting to show plasticity suffered from two main flaws; first, they often utilized highly mixed populations of cells, and second, the assays of the donor cells in the recipient tissues were either insufficiently rigorous or open to misinterpretation. Lack of precision in defining the donor cell phenotype was particularly an issue when bone marrow was the source tissue, since this contains multiple adult stem cell types in addition to HSC, including **endothelial progenitor cells** and **mesenchymal stem cells**. Likewise, some reports of muscle stem cells contributing to hematopoiesis can definitely be explained by the presence of HSC in muscle tissue, it being well known that they continuously circulate the body.

The principal explanation for apparent stem cell plasticity appears, however, to be cell fusion, which we have already described in this chapter in the context of both natural differentiation processes and experimental transdifferentiation. The contribution of bone marrow cells to liver hepatocytes, cardiac myocytes, and Purkinje neurons in the brain was eventually shown to be the consequence of cell fusion, which was best demonstrated using a Cre-activated reporter approach (Chapter 2) in which the donor and host separately contribute the Cre and the Cre-activated reporter so that only through cell fusion can the two be brought together (**Figure 5.20**). So cell fusion is a reality, but it is very rare when it does occur and can lead to what is essentially reprogramming of the host fusion partner cell nucleus. This may have some physiological relevance in the case of Purkinje neurons but is likely to have little or no therapeutic potential, which of course was the original driver for the studies that led to the misguided claims in the first place.

## Transcription factors can be used directly to reprogram adult cells from one type to another avoiding a stem cell intermediate

As we have already seen, the era of pluripotent reprogramming using combinations of transcription factors was predated by experiments that showed that single transcription factors associated with a particular differentiated lineage could result in transdifferentiation within the lineage itself or to developmentally closely related lineages. The fact that Yamanaka's four factor combination could have such a profound effect as to switch an adult fibroblast or blood cell back to a pluripotent embryonic state inspired the search for a means to convert cells such as fibroblasts directly to fully functional adult cell types. Another consideration worth mentioning is that although pluripotent stem cells can be differentiated to almost all cell types of the body, it has often proved difficult to achieve a fully adult functional profile. Differentiation can result in cells that have characteristics of fetal tissues, as is seen, for example, with hepatocytes and cardiomyocytes (Chapter 3), so that is appealing to have the prospect of reprogramming directly to an adult phenotype without the requirement for an iPS cell stage.

Informed by the extensive knowledge concerning how patterns of specific transcription factors define the development and functioning of individual cell lineages, researchers adopted a strategy akin to Yamanaka's and screened for transcription factor combinations that could yield particular adult cell types. Using this approach, mature cells with an adult phenotype have been generated for several

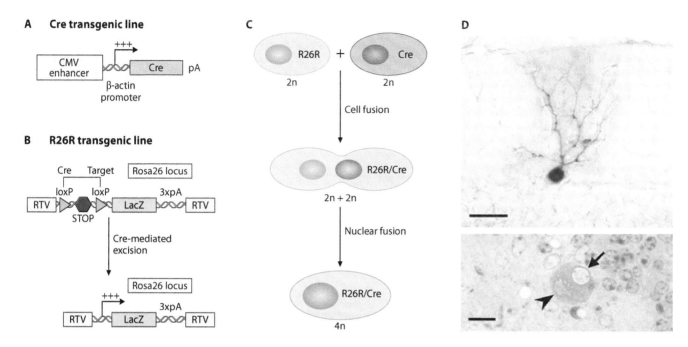

**A    Cre transgenic line**

**B    R26R transgenic line**

**Figure 5.20 Definitive proof that cell fusion underlies many reports of transdifferentiation.** One of the most likely explanations for apparent adult stem cell plasticity appears to be that cell fusion occurs between the adult stem cell and cells in the tissue under investigation. Lineage tracing using fluorescent protein expressing transgenic reporters has provided definitive proof that cell fusion does indeed occur in such circumstances. In the example illustrated, transgenic mouse strains that (**A**) constitutively express Cre recombinase and (**B**) contain a Cre-activated reporter gene (ROSA26 locus containing the LacZ gene that can only be expressed if Cre recombinase removes an upstream transcriptional stop sequence surrounded by loxP sites) were used as the source of donor cells and the host for transplantation, respectively. Cell fusion following transplantation of Cre-expressing bone marrow cells into the LacZ reporter recipient (**C**) was specifically revealed leading to the presence of β-galactosidase-positive heterokaryons that stain blue in the presence of the appropriate substrate. This is illustrated for a Purkinje neuron (**D**), the upper image showing staining in the whole neuron and the lower image showing the cell body and the presence of both donor and host nuclei. (From Alvarez-Dolado M, Pardal R, Garcia-Verdugo J et al. (2003) *Nature* 425: 968–973 [doi:10.1038/nature02069]. Reprinted by permission from Springer Nature.)

**TABLE 5.2  TRANSCRIPTION FACTOR-DEPENDENT REPROGRAMMING OF FIBROBLASTS**

| Adult cell type | Reprogramming transcription factors |
| --- | --- |
| Hepatocyte | GATA4, Hnf1a, Foxa3 |
| Cardiomyocyte | GATA4, MEF2c, Tbx5 |
| Neuron | Ascl1, Brn2, Mytl1 |
| Hematopoietic progenitors | Erg, GATA2, Lmo2, Runx1c, SCL |

*Note*: In addition to the reprogramming to pluripotent stem cells, it has become increasingly apparent that expression of lineage-associated transcription factors can be used to experimentally induce transdifferentiation of one cell type into another without the requirement to pass through a stem cell stage. The table details some transcription factor combinations that have successfully been used to convert fibroblasts into specific adult cell types·

different tissues (Table 5.2), including neural cells, cardiomyocytes, liver hepatocytes, and hematopoietic progenitor cells. Impressively, hepatocytes produced this way ("iHep") are fully able to regenerate a liver in an experimental model of liver damage (Figure 5.21). Aside from the advantage of ensuring generation of a functional adult cell, direct lineage reprogramming avoids potential problems arising from the potentially tumorigenic nature of pluripotent stem cells; however, the advantage of potentially indefinite expansion is lost. It is probably too early to know which route will prevail or whether there is a place for both, or even a combination involving partial pluripotent reprogramming, allowing expansion, followed by transcription factor-directed terminal differentiation.

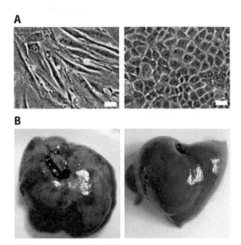

Figure 5.21 **Induced hepatocytes derived from fibroblasts are fully functional.** One adult cell type that has been produced directly from fibroblasts is the hepatocyte. This was first achieved by introducing the transcription factors Gata4, Hnf1a, and Foxa3 into fibroblasts while simultaneously inactivating the cell cycle inhibitor p19^Arf. (**A**) Phase-contrast microscopy images of mouse fibroblasts (left) and the morphological change they undergo after being transduced with the transcription factors to form induced hepatocytes (iHep) (right). (**B**) Photographs of a whole liver from a mouse with necrotic liver disease (left) compared to one from the same mouse strain that has received an injection of iHep (right). (From Huang P, He Z, Ji S et al. (2011) *Nature* 475: 386–391[doi:10.1038/nature10116]. Reprinted by permission from Springer Nature.)

## 5.4  SUMMARY

In this chapter, we have shown that the phenotype of an adult somatic cell remains flexible and in certain circumstances, either natural or experimental, can change to that characteristic of a completely different cell type. We describe some examples of circumstances in which a cell naturally converts to another lineage type, and then focus on how an adult somatic cell nucleus can be reprogrammed either by nuclear transfer (cloning) or by fusion to a distinct cell type. We highlight how enforced shifts in the function of a given cell nucleus depend upon specific transcription factors and the fact that their manipulation could in itself alter cell phenotype. In the second part of the chapter, we discuss how this latter line of thinking led to the discovery that the introduction of transcription factors associated with the pluripotent state of ES cells can be used as a way of reprogramming adult cells to stem cells with very similar properties. Finally, we return to the concept of **transdifferentiation**, first as it relates to the possibility that this might occur at the level of adult stem cells and second how it might be artificially induced by transcription factor expression directly to change one mature cell type into another functioning cell without the need to transition via a pluripotent stem cell intermediate.

## KEY POINTS

- In general, adult cells are not irreversibly fixed in a particular differentiated state, but under natural or experimental conditions can be "reprogrammed" to a different phenotype.

- Reprogramming can be relatively minor, for example, to a proliferating progenitor state (dedifferentiation) or more extensive to another cell lineage (transdifferentiation) or even back to the stage of pluripotency.

- De- and transdifferentiation occur naturally in a few specialized circumstances, usually connected with regenerative processes.

- Pluripotent reprogramming can be achieved through nuclear transfer into an egg, the underlying process required for cloning.

- Transdifferentiation, also known as lineage reprogramming, can be enforced by fusion between cells of distinct differentiated phenotypes, which in essence is akin to transferring a nucleus into a different cytoplasmic milieu.

- Reprogramming to a pluripotent state can be achieved through enforced expression of specific transcription factors associated with the maintenance of pluripotency in ES cells.

- Pluripotent reprogramming by transcription factors was first achieved using the four proteins Oct3/4, Sox2, Klf4, and Myc. These proteins are commonly referred to as the Yamanaka factors and the cells generated as induced pluripotent stem (iPS) cells.

- Factors other than Klf4 and Myc can cooperate with Oct3/4 and Sox2 in bringing about reprogramming, and one or more of the Yamanaka factors can be replaced by specific chemicals that interfere with signaling pathways and gene expression.

- iPS cells were first generated from mouse fibroblasts and were subsequently produced from human cells and from a wide range of mammalian species.

- Most adult cell types seem to be capable of being reprogrammed to iPS cells.

- The mechanism of reprogramming involves the three phases of initiation, maturation, and stabilization, which entail changes in gene expression, resetting of the epigenetic landscape, and establishment of a network of interacting pluripotency transcription factors.

- When derived from a genetically identical cell source, iPS cells are essentially indistinguishable from ES cells produced from nuclear transfer cloned embryos.

- What appeared to be transdifferentiation of adult stem cells when they were transplanted, especially in the context of particular tissue injury, turned out to be due to either extremely rare or artifactual events. The most common explanation for this so-called adult stem cell plasticity seems to be fusion with cells in the injured tissue.

Reprogramming directly from one type of mature adult cell to another can be achieved through enforced expression of specific transcription factors associated with a particular differentiated lineage.

# FURTHER READING

## Cell plasticity and transdifferentiation

Blau, H.M., Pavlath, G.K., Hardeman, E.C., Chiu, C.P., Silberstein, L., Webster, S.G., Miller, S.C. and Webster, C. (1985) Plasticity of the differentiated state. *Science*, 230: 758–766.

Davis, R.L., Weintraub, H. and Lassar, A.B. 1987 Expression of a single transfected cDNA converts fibroblasts to myoblasts. *Cell*, 51: 987–1000.

Graf, T. (2011) Historical origins of transdifferentiation and reprogramming. *Cell Stem Cell*, 9: 504–516.

Sánchez Alvarado, A. and Yamanaka, S. (2014) Rethinking differentiation: Stem cells, regeneration, and plasticity. *Cell*, 157: 110–119.

Wagers, A.J. and Weissman, I.L. (2004) Plasticity of adult stem cells. *Cell*, 116: 639–648.

## Nuclear cloning

Gurdon, J.B., Elsdale, T.R. and Fischberg, M. (1958) Sexually mature individuals of *Xenopus laevis* from the transplantation of single somatic nuclei. *Nature*, 182: 64–65.

## Induced pluripotent stem cells

Buganim, Y., Faddah, D.A. and Jaenisch, R. (2013) Mechanisms and models of somatic cell reprogramming. *Nat. Rev. Genet.*, 14: 427–439.

Hou, P., Li, Y., Zhang, X., Liu, C., Guan, J., Li, H., Zhao, T. et al. (2013) Pluripotent stem cells induced from mouse somatic cells by small-molecule compounds. *Science*, 341: 651–654.

Okita, K., Ichisaka, T. and Yamanaka, S. (2007) Generation of germline-competent induced pluripotent stem cells. *Nature*, 448: 313–317.

Takahashi, K. and Yamanaka, S. (2006) Induction of pluripotent stem cells from mouse embryonic and adult fibroblast cultures by defined factors. *Cell*, 126: 663–676.

Takahashi, K., Tanabe, K., Ohnuki, M., Narita, M., Ichisaka, T., Tomoda, K. and Yamanaka, S. (2007) Induction of pluripotent stem cells from adult human fibroblasts by defined factors. *Cell*, 131: 861–872.

Yu, J., Vodyanik, M.A., Smuga-Otto, K., Antosiewicz-Bourget, J., Frane, J.L., Tian, S., Nie, J. et al.. (2007) Induced pluripotent stem cell lines derived from human somatic cells. *Science*, 318: 1917–1920.

## Transcription factor induced conversion between adult somatic cell types

Ieda, M., Fu, J.D., Delgado-Olguin, P., Vedantham, V., Hayashi, Y., Bruneau, B.G. and Srivastava, D. (2010) Direct reprogramming of fibroblasts into functional cardiomyocytes by defined factors. *Cell*, 142: 375–386.

Vierbuchen, T., Ostermeier, A., Pang, Z.P., Kokubu, Y., Südhof, T.C. and Wernig, M. (2010) Direct conversion of fibroblasts to functional neurons by defined factors. *Nature*, 463: 1035–1041.

# The stem cell niche

**6**

We should probably not start a chapter with a question, but from what we have described up to this point, can we really say with certainty what defines a stem cell? The salient features of a stem cell are without doubt, that is, they are cells that have either a temporary role during development or a life-long role in individual tissues of an organism and have the unifying characteristics of being able to self-renew and have a potential for **differentiation** to one or more functional cell types (Chapter 1). The importance of stem cells is underlined by the fact that they have been with us since multicellular organisms evolved, originating on at least two separate occasions in the branches that gave rise ultimately to the plant and animal kingdoms (Chapter 7). Even though most stem cell biologists would agree that we have so far just skimmed the surface with regard to a complete understanding of the molecular mechanisms that control **self-renewal**, stem cell **potency**, and **commitment** to differentiation, we nevertheless have some clear insights into the **epigenetic**, transcriptional, and signaling processes that operate (Chapters 3, 4, and 5). Notwithstanding details yet to be discovered, can we be sure that these processes, whether generic or specific to individual stem cell types, will give us the ultimate answer, or is there something missing? This is a leading question, as there is one major factor that we have not yet taken account of, namely the environment, which for sure plays a permissive role but might even be deterministic and the ultimate reason a cell will behave as a stem cell.

In this chapter, we will therefore describe what is known about the environment of a stem cell and the extent to which these are similar or differ for individual stem cell types. Having looked into the details of different stem cell environments, we will then consider how these might be affected during aging and in disease, especially in relation to cancer.

## 6.1  STEM CELLS HAVE A DEFINED ENVIRONMENT

There are plenty of circumstantial indicators of the importance of a stem cell's particular environment, or **"niche"** as it is usually referred to. Most, perhaps all, stem cells are found in a specific tissue location, whether these are **pluripotent stem cells** in the developing embryo, germ cells in reproductive organs, or **adult stem cells** associated with the homeostasis of individual tissues. In some cases, the specific tissue location of stem cells is extremely precise (e.g., in fruit fly [*Drosophila*] gonads, mammalian hair follicles, or growing shoots of plants), while in others, the location seems to exhibit a degree of flexibility (e.g., **hematopoietic stem cells** [HSC] in the bone marrow). Tissue-restricted location could be a consequence of the developmental origin of a given organ and the stem cells associated with it; however, stem cell transplantation experiments, which are the "gold standard" for assessing stem cell potential (Chapter 2), reveal that there is often a requirement for homing back to the tissue of origin for a stem cell to be able to function correctly. Although many stem cell types can be isolated with a high degree of purity and subsequently maintained in culture, the retention of full stem cell potential is a problem or relies upon the cells adopting a three-dimensional (3D) level of organization, either spontaneously (e.g., intestinal stem cell **organoids**) or through provision of biocompatible matrices.

## Investigations of hematopoietic stem cells set the stage for the stem cell niche concept

As we repeat so often in this book, studies on HSC have to a large extent paved the way in stem cell science, and this is for sure also the case for our current thinking on the stem cell niche. The starting point was the realization that cultures of HSC would only retain stem cell potential in the presence of non-hematopoietic bone marrow **stromal cells**. The key experiments were performed in Manchester between 1975 and 1977 by Brian Lord and Mike Dexter, leading to the pioneering theory of the stem cell niche proposed in 1978 by their colleague Raymond Schofield (Figure 6.1). The theory that Schofield proposed viewed the niche as the ultimate driver of stemness, to the extent that it could impose the stem cell state upon a more committed cell, as exemplified by the way in which he had observed committed **progenitors** (**colony forming units in the spleen, or CFU-S**) becoming stem cells (HSC) when a vacant niche was available (Figure 6.2). Schofield championed the idea that one of the purposes of the niche was to prevent the excessive proliferation of the stem cell and thus reduce the number of genetic mutations accumulating in the mother cell. Despite the detailed description of the stem cell niche theory, Schofield had no clear evidence that this was the case, but research since then on a wide variety of stem cell scenarios in tissues from plants and animals, and ranging from the most primitive invertebrates to humans, has fully supported his vision.

Advances in the characterization of stem cell niches have been very much influenced by the accessibility of a particular tissue environment to experimental observation. For this reason, many of the most significant findings have involved small invertebrates, including the gonads of the nematode worm *Caenorhabditis elegans* and the fruit fly *Drosophila melanogaster*. Plants have also proved to be revealing subjects for the study of stem cell niches, while in animals higher up the evolutionary scale than invertebrates, some tissues that are easily observed, such as the skin, have also yielded insight into the diversity of stem cell environments and the common themes that underpin them. The model organisms used in studying stem cell niches have also generally had the added advantage of being amenable to genetic manipulation, enabling the roles of specific molecular components of the niche to be investigated.

## A stem cell niche may have multiple cell and non-cellular components

Based on Schofield's original idea and subsequent observations of diverse stem cell types, the essential components of an idealized niche have been defined (Figure 6.3). In broad terms, these components of the niche adopt a specific architecture that includes one or more cell types that are not stem cells and the **extracellular matrix** (ECM). The niche is also influenced by hormones and other signaling molecules, which derive from sources within (paracrine) or external to (humoral) it, as well as neural signals and the general metabolic status. In the rest of this chapter, we will explore each of these components through examples from across the world of multicellular organisms, highlighting both the common principles and some specific adaptations where they have been defined. We will also consider how it is that the stem cell niche can dictate

Figure 6.1 **Raymond Schofield.** (From Scadden D. (2014) *Cell* 157: 41–50 [doi: 10.1016/j.cell.2014.02.013]. Copyright 2014 with permission from Elsevier.)

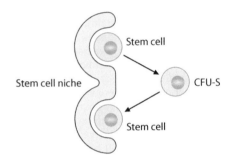

Figure 6.2 **Schofield's vision of the stem cell niche.** Simplified view of a stem cell niche as proposed by Schofield. In the context of *in vitro* cultured bone marrow, the daughter of a blood stem cell is a colony forming unit-spleen cell (CFU-S), which Schofield proposed could re-enter a stem cell state if it could find and occupy the niche. (From Schofield R. *Blood Cells* 1978;4:7–25. Copyright 1978 with permission from Elsevier.)

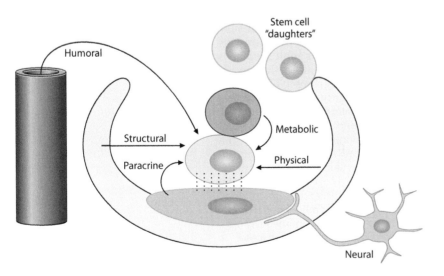

Figure 6.3 **Essential elements of a stem cell niche.** The key niche elements acting upon a stem cell within its niche include the influences of its structural architecture, physical interactions between the stem cell and its surrounding cells, signaling interactions between neighboring cells, and metabolic and neural inputs. These elements and interactions (arrows) are illustrated schematically. (From Scadden D. (2006) *Nature* 441: 1075–1079 [doi:10.1038/nature04957]. Reprinted by permission from Springer Nature.)

the fate of those cells within it, whether that is to become a stem cell, to remain as a stem cell, or to differentiate. Finally, we will consider the dynamic nature of some stem cell niches and how this can impact areas such as aging or the development of cancer and ultimately how knowledge of this might give rise to new therapeutic strategies.

## 6.2 ARCHITECTURE OF STEM CELL NICHES: SIMILARITIES ACROSS THE ANIMAL AND PLANT KINGDOMS

A number of common features are seen in many of the stem cell niches studied, including the presence of only one or a small number of actual stem cells. In addition, the stem cells are generally in direct contact with one or more cells that serve a support role, organizing their orientation, often in a limited space that likely helps to restrict stem cell number (Figure 6.3).

### Niche support cells vary between different stem cell types but often show similarities in their overall spatial arrangement

Although there are differences in the type and range of cells that support stem cells in different contexts (Table 6.1), comparison of stem cell niches from plants, invertebrate animals, and vertebrates reveals many similarities in their architecture (Figure 6.4). Orientation of the stem cell and the restricted space for expansion in proximity to the **organizer cell(s)** means that cell division often occurs such that one daughter moves in a direction away from the niche, effectively propelling it through the stage of transient amplification and eventual terminal differentiation.

Seminal to our understanding of the characteristics of stem cell niches has been the research on *Drosophila* **germ line stem cells** (GSC), especially that emanating from the laboratories of Allan Spradling at the Carnegie Institution in Baltimore and Margaret Fuller at Stanford University. Their collective findings and those of others are described in Chapter 12. Gonads of both sexes exhibit similar features, and here we will illustrate using the testis (Figure 6.4A). The male GSC sits in the testis in close contact with the so-called **hub cell**, the most important niche component. When the stem cell divides, the daughter that will maintain the stem cell pool remains firmly attached to

### TABLE 6.1 DIVERSITY OF STEM CELL NICHES IN INVERTEBRATE AND VERTEBRATE MODEL ORGANISMS

| Tissue | Stem cell | Support cells |
|---|---|---|
| Nematode (*C. elegans*) gonad | GSC | Distal tip cell |
| *Drosophila* testis | GSC | Hub cells |
| *Drosophila* ovary | GSC | Cap cells |
| Mouse bone marrow | HSC | Osteoblasts, vascular cells, megakaryocytes, MSC |
| Mouse small intestine | ISC | Paneth cells, crypt fibroblasts |
| Mouse skin | Follicular bulge stem cell | Dermal fibroblasts |
| Mouse brain (lateral ventricle) | NSC | Ependymal cells, vascular cells |
| Mouse testis | SSC | Sertoli cells, vasculature, interstitial cells |
| Mouse skeletal muscle | Satellite cell | Myofibers |

*Note:* Although different stem cell niches share a requirement for one or more support cells, the nature of these cells can vary considerably. The table lists for three commonly studied model species some of the adult stem cell types associated with particular tissues and the corresponding niche support cells identified.

*Abbreviations:* GSC, germline stem cell; HSC, hematopoietic stem cell; ISC, intestinal stem cell, NSC, neural stem cell; and SSC, spermatogonial stem cell.

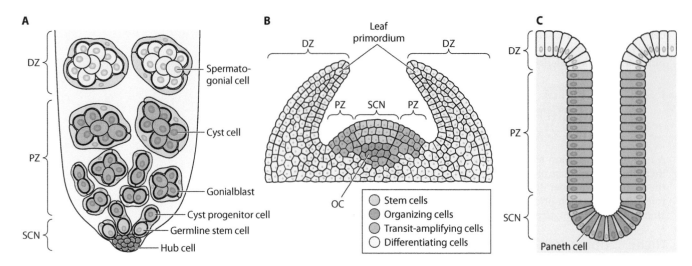

Figure 6.4 **Common organization of stem cell niches in animals and plants.** Common features of the organization of stem cell niches can be seen by comparisons between invertebrate and vertebrate animal tissues and even with plants, in spite of the fact that plant and animals appear to have evolved independently with respect to their stem cells. The three cross-sectional diagrams illustrate representative and well-studied niches: (**A**) *Drosophila* testis; (**B**) plant apical meristem; and (**C**) mouse intestinal crypt. In each case, the stem cell population (blue) resides near specialized organizing cells (purple) in the stem cell niche (SCN). Stem cells give rise to populations of transit-amplifying cells (orange) in the proliferation zone (PZ), which give rise to differentiated cells (cream) in the differentiation zone (DZ), each moving progressively away from the organizing cells of the niche. (From Heidstra R & Sabatini S. (2014) *Nat Rev Mol Cell Biol* 15: 301–312 [doi: 10.1038/nrm3790]. Adapted by permission from Springer Nature.)

the hub cell, while the other daughter cell moves away, differentiating first into **transit amplifying progenitor** cells before eventually becoming fully mature sperm.

In a highly analogous, although evolutionarily convergent, way, the stem cells in the growth zone ("**meristem,**" Chapter 8) of plant shoots lie in a small region that is immediately adjacent to cells of the organizing center (**Figure 6.4B**). Movement of the stem cells sideways away from the organizing center initiates proliferation, which continues until they reach a differentiation zone and begin to form structures such as the leaf. The shoot meristem was the first to evolve when plants became multicellular; only later did roots come on the scene, but they too have the same sort of stem cell niche, although the precise molecular mechanisms at play differ to some extent.

In vertebrates, many stem cell niches have been quite extensively examined, especially in rodents and humans, achieving a detailed picture in some cases that is comparable to that obtained from studies on *Drosophila* GSC. Epithelial tissues supported by stem cells have been particularly helpful in revealing the nature of stem cell niches, ranging from those that continuously give rise to the lining of the gastrointestinal tract (Chapter 11) to those that generate hair follicles in the skin (Chapter 15). The stem cell niche in the small intestine (**Figure 6.4C**) is another good example of the importance of a direct interaction between the stem cells and a supportive cell, in this case **Paneth cells**, and the way in which the geometry of the niche constrains **intestinal stem cell** expansion and following **asymmetric** division directs differentiation away from the niche, located in the crypts, toward the tip of the villi.

The hair follicle niche is particularly fascinating in the way that it supports more than one stem cell type in a structure about half way up the follicle known as the "**bulge**" (**Figure 6.5**). Cell contacts are important here too, but when required, stem cells migrate out of the bulge, heading downward to form a new follicle. A distinct stem cell, the melanocyte stem cell, which is the precursor of pigment-producing melanocytes, also resides in the bulge and, like the follicle stem cells, migrates downward when required as new hairs grow. The juxtaposition of different stem cells in the restricted space provided by the bulge means that there are contacts between stem cells, which appear to be important for some aspects of regulation, in particular the behavior of the melanocyte stem cells.

Another example of a highly organized stem cell niche is the one that supports adult **neural stem cells** (NSC) in the brain, as exemplified by those located in the subventricular zone (Chapter 14). Again, the stem cells are in close contact with niche cells, and the orientation and progression of cells through the structure

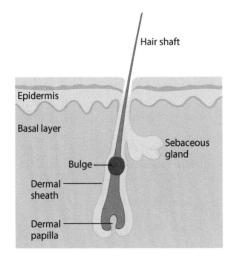

Figure 6.5 **The hair follicle stem cell niche.** The niche in the hair follicle is very well characterized and illustrates many of the features common to stem cell niches. The cross-sectional schematic of the hair follicle shows the location of the key structural elements and the position of the so-called bulge region (red). The bulge contains stem cells that migrate downward toward the base of the follicle, from where they generate a new hair. The bulge also contains melanocyte stem cells, which likewise migrate to the base of the hair.

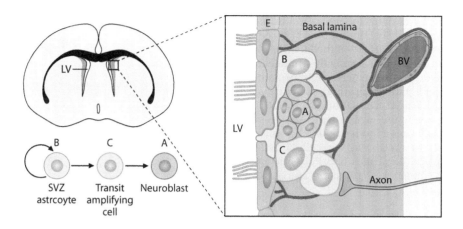

Figure 6.6 **The neural stem cell niche.** Stem cells in the brain are located adjacent to the ventricles and are organized in discrete niches that provide cell contacts, many of which are from cell types that are themselves derivatives of the stem cells. The cross-sectional diagram on the left illustrates the lateral ventricle (LV) region of the brain that contains the stem cell-generating subventricular zone (SVZ). The schematic on the right shows the cell types and anatomy of the adult SVZ niche located next to the LV. The SVZ astrocytes (B, blue) represent the stem cells, which give rise to transit-amplifying cells (C, green) that in turn generate the migrating neuroblasts (A, orange) destined to populate the olfactory bulb. The three cell types of the SVZ niche are in contact with ependymal cells (E, brown). BV: Blood vessel. (From Riquelme P, Drapeau E & Doetsch F. (2008) *Philos Trans R Soc Lond B Biol Sci* 363:123–137 [doi: 10.1098/rstb.2006.2016]. Permission conveyed through Copyright Clearance Center, Inc.)

dictates differentiation (Figure 6.6). The NSC sit in the center of the niche surrounded by multi-ciliated ependymal cells and move around the niche as they give rise to intermediate progenitor cells that divide to generate neuroblasts. The NSC remain in close contact with all of their progeny and interestingly possess a long extension that touches blood vessels.

Although the *in vitro* culture of HSC with bone marrow stromal cells gave some of the first hints of the importance of the niche and led to Schofield's theory, the real-life situation in the bone marrow is by no means fully resolved. One of the reasons for this is the anatomical location of the HSC within the bone marrow that makes *in situ* observation difficult, although modern imaging strategies are beginning to yield definitive insight. Another feature of the HSC environment is that it is constituted of multiple cell types, including osteoblasts, vascular cells, megakaryocytes, macrophages, various hematopoietic progenitors, adipocytes, **mesenchymal stem cells** (MSC), and neural cells (Figure 6.7). One accepted current view is that there are two HSC niches within the bone marrow; namely the osteoblastic niche and the vascular niche. However, this is undoubtedly oversimplistic and it is likely that there are a number of sub-niches supporting different levels of stem cell behavior and whether they are **quiescent**, expanding, or able to circulate.

## The niche probably arose early in the evolution of multicellular organisms once stem cells became essential

It is worth considering the early evolution of the stem cell niche as represented in modern day invertebrates that probably correspond quite well to the first multicellular

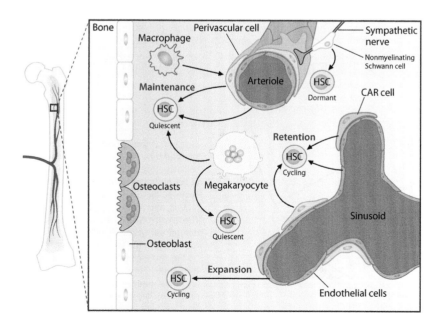

Figure 6.7 **Multiple niches for hematopoietic stem cells in the bone marrow.** Cross section through a mammalian femur showing the components of the hematopoietic stem cell (HSC) niche. The two proposed HSC niches are the osteoblastic niche and the vascular niche. An array of different cell types constitutes these niches, including osteoblasts, osteoclasts, macrophages, megakaryocytes, endothelial cells, CXCL12-abundant reticular (CAR) cells, and perivascular cells, which together interact to maintain the balance between quiescence, proliferation, and differentiation of the HSC. (From Boulais P & Frenette P. (2015) *Blood* 125: 2621-2629 [doi: 10.1182/blood-2014-09-570192]. Permission conveyed through Copyright Clearance Center, Inc.)

organisms around 500 million years ago. The best-studied organism in this respect is Hydra, which has an impressive regenerative capacity underpinned by three stem cell lineages (Chapter 7). Two of the stem cell types support epithelial layers, while the third type, known as the interstitial stem cell, resides among the epithelial cells of both layers and is **multipotent**, being able to give rise to nerve cells, nematocysts (specialized cells in the tentacles that deliver a paralyzing sting), gland cells, and gametes. The interstitial stem cells are distributed throughout the whole body column, although marking with a nanos-green fluorescent protein (GFP) reporter has shown that they are invariably located adjacent to both epithelial cells and an ECM structure known as the mesoglea, which is a translucent, non-cellular, jelly-like substance found between the two epithelial cell layers (Figure 6.8). Pairs of stem cells are always apparent, both cells being next to epithelial cells, but only one is in direct contact with the ECM.

## 6.3    HOW DOES THE NICHE CONTACT AND COMMUNICATE WITH STEM CELLS?

Irrespective of the type of niche or its complexity, it is possible to group the influences that impinge on a stem cell into four categories, namely cells, soluble growth factors, the ECM, and small molecules reflecting metabolic status. It is important not to view these categories in isolation; rather think of them as intimately interconnected. For instance, the cells constituting the niche produce a variety of growth factors and ECM molecules and can affect the metabolic balance of the environment. Likewise, the overall milieu of the niche can affect the niche cells to change their metabolism and secretion of growth factors and ECM. Ultimately, all of the factors that the stem cell is exposed to in its niche are sensed by specific cell surface or intracellular receptors (Figure 6.9), which in turn relay signals that change the behavior of the stem cell through changes in protein activity or gene expression. Although the functions of support cells and the associated ECM in different tissues are similar, they are tuned to the specific needs of the individual stem cell so that, for example, an intestinal niche Paneth cell cannot maintain a HSC, and likewise a bone marrow osteoblast would not support an intestinal stem cell.

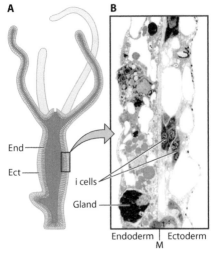

**Figure 6.8 The interstitial stem cell niche in *Hydra*.** Even simple invertebrates have evolved a specific microenvironment for stem cells. (**A**) Longitudinal cross section through *Hydra*, showing the endodermal (end) and ectodermal (ect) epithelial (epi) cell layers that constitute the body wall. (**B**) The stained section of the body wall shows how the stem, or interstitial (i), cells are positioned relative to the mesoglea (m). (From Bosch T. (2014) *Trends Immunol* 35: 495–502 [doi:10.1016/j.it.2014.07.008]. Copyright 2014 with permission from Elsevier.)

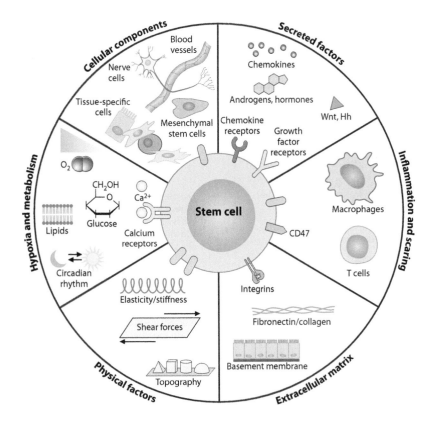

**Figure 6.9 Components and factors experienced by a stem cell in its niche.** The stem cell niche can be composed of a wide range of cell and non-cell components, including secreted factors, extracellular matrix, physical factors, and nutrients and small molecules. These components are summarized in the pie diagram, with some of the relevant receptors for receiving the various inputs illustrated on the surface of the schematic stem cell at the center. (From Lane S, Williams D & Watt F. (2014) *Nat Biotech* 32: 795–803 [doi: 10.1038/nbt.2978]. Reprinted by permission from Springer Nature.)

## Many distinct cell types can contribute to a stem cell niche

As we have seen, cells constituting a niche can come in very many different flavors, and these can have either very specific roles or more generic contributions. Some cells contact the stem cell directly, while others perform their role at a distance through release of soluble factors or ECM. Direct cell-cell interactions include the already mentioned organizer cells such as the hub cell of the *Drosophila* testis GSC niche, Paneth cells in the intestinal villus niche, or the ependymal cells of the NSC niche in the brain. A stem cell may also contact other stem cells or the committed progeny of a stem cell, including the transit amplifying population. Niche cells that directly interact with a stem cell express surface molecules that may function in a physical sense, restraining the position or orientation of the stem cell or by providing a signal that elicits a response in the stem cell such as promoting self-renewal or differentiation.

Typical cells found in the niche environment, which do not necessarily have direct contact with the stem cells, are fibroblasts, MSC and **pericytes**, endothelial cells (including those that are part of blood vessels), and nerve cells. Another important group of cells, which very much reflect changes in the health status of an organism and possibly relay a need for enhanced cell production from stem cells, are cells involved in immune and inflammatory responses, including both myeloid cells like macrophages and neutrophils and various B- and T-lymphoid cells.

## Non-cellular components of the stem cell niche include signaling proteins and extracellular matrix

One of the key components of almost all stem cell niches is the extracellular matrix, which derives from the cell components of the niche, including the stem cells. The composition of the ECM varies between niches, but the broad function remains the same (Figure 6.10). The 3D structure created by the macromolecules that constitute the ECM provides the architecture of the stem cell niche and an anchor for the cellular components. Cell-matrix anchorage also provides signaling inputs to the stem cells that influence their fate decisions, ligands contained within the ECM molecules being recognized by specific receptors on the stem cells. The ECM also largely determines the biomechanical properties of the niche that are perceived by the stem cells. Hence, in some circumstances, the elasticity of the microenvironment can dictate the degree and type of stem cell differentiation, as seen for MSC (Chapter 10, Box 10.3).

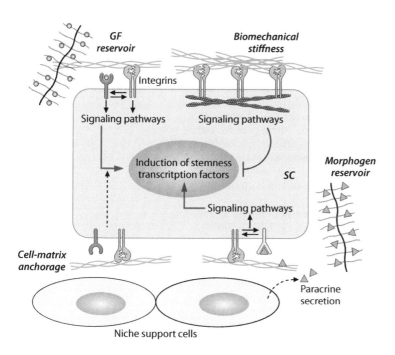

Figure 6.10 **Interactions of stem cells with extracellular matrix in the niche.** The extracellular matrix (ECM) plays a key role in providing the microenvironment for stem cells in their niche, both in structural terms and through its ability to bind to and present soluble factors produced both locally or systemically. The schematic diagram illustrates some of the interactions in the niche between the stem cell (SC) and the ECM. Stem cells express integrins that mediate adhesion of the stem cell to ECM proteins. The ECM acts as a reservoir for growth factors (GF) and morphogens, while the molecular composition of the ECM determines its stiffness and thereby the mechanical forces acting upon the stem cell. All of these interactions regulate the stem cell by means of induction of signaling pathways. (From Brizzi M, Tarone G & Defilippi P. (2012) *Curr Opin Cell Biol* 24: 645–651 [doi: 10.1016/j.ceb.2012.07.001]. Copyright 2012 with permission from Elsevier.)

Another role for the ECM in the stem cell niche is that it binds specific growth factors and **morphogens**, regulating their local availability and establishing a biochemical gradient, thereby providing an opportunity to integrate different inputs from adjacent cells and more distant sites in an organism. ECM proteins, including fibronectin, vitronectin, collagens, and proteoglycans, avidly bind many soluble growth factors such as the FGFs, HGF, and VEGFs. Similarly, collagens II and IV are capable of specifically binding BMPs and TGF-β. Some ECM proteins also present ligands that are an integral part of the macromolecule, for example, the EGF-like motifs found in laminins and several other proteins.

The wide variety of cell types associated with stem cell niches is reflected in an even wider range of secreted factors that they are capable of producing, both to act directly on the stem cells and to influence the niche. Whereas secreted ECM acts locally, growth factors can either be derived locally (within one to two cells distant) from neighboring stem cells and niche cell components or can act diffusely. Although some of these factors are unique to a particular niche, many are remarkably conserved across various tissues, although the consequence of their activation may differ between stem cell types. A classic example is the Wnt pathway, which crops up in many stem cell scenarios ranging from *Drosophila* gonads to the mammalian intestine and hematopoiesis. For example, signaling initiated by the binding of Wnt proteins to their receptors (Chapter 4) functions to maintain stem cell self-renewal in intestinal and hematopoietic tissue but directs differentiation in other stem cell types. In the GSC niche of *Drosophila* ovaries, the Wnt pathway in somatic cells helps to regulate stem cell differentiation by restricting BMP signal production to the **cap cells**. Other frequently encountered soluble factors in stem cell niches include Hedgehog, various chemokines, cytokines, peptide hormones, and the aforementioned growth factors, such as TGF-β and FGFs.

## A variety of specific molecular interactions link the stem cell to either niche cells or the extracellular matrix

There are numerous cell surface molecules that are crucial to the adherence of stem cells within their niche, and in many instances, these interactions also play a part in signaling that changes the cell's behavior (Table 6.2). The cadherins and integrins have emerged as principal links between the stem cell and its niche and serve as good examples to illustrate the various consequences of such interactions, although other classes of surface protein receptor molecules are also involved (Figure 6.11).

When expressed on two different cells, cadherins bring about adhesion through interaction between their extracellular domains (a "homophilic" interaction). The intracellular portion of the cadherin molecule interacts with cytoskeleton-associated proteins, especially the catenins, thereby providing a link to intracellular processes (Figure 6.11A). The most studied of the cadherin proteins in the context of the stem cell niche is Epithelial (E) cadherin. In the GSC niches in the gonads of *Drosophila*, E-cadherin is expressed on the niche cells (cap cells in the ovary, hub cells in the testis) and on the GSC, enabling formation of tight adherens junctions. E-cadherin is clearly important for the maintenance of the GSCs since its experimentally induced loss results in a rapid depletion of GSC from within the niche. These homophilic interactions between the niche and the GSC have been shown to be essential for the polarization of the stem cell, providing cues for the orientation of stem cell divisions and thus specifying stem cell differentiation (see Chapter 12, Figure 12.11). Cadherins are also expressed in at least some mammalian stem cell niches, where they play similar roles to that seen in the *Drosophila* gonads. For example, in the NSC niche in the brain, E-cadherin is expressed on both the NSC and ependymal cells.

The other main group of proteins involved in stem cell adhesion in the niche are the integrins, which are dimers composed of α and β subunits (a "heterodimer") that interact with proteins of the ECM, such as collagen, laminin, and fibronectin, as well as other cell surface adhesion molecules like intercellular adhesion molecule (ICAM) and vascular cell adhesion molecule (VCAM) (Figure 6.11B). Integrins are widely expressed on stem cells, and the stem cell-ECM interaction they mediate has been shown to play a part in diverse niche settings in stem cell adhesion, anchorage, and homing and in maintaining the balance of self-renewal versus differentiation. Interestingly, β1 integrin can influence the balance between symmetric and

**TABLE 6.2  SIGNALING PATHWAYS AND ADHESION MOLECULES IN STEM CELL NICHES**

| Tissue | Stem cell | Signaling pathways | Adhesion |
|---|---|---|---|
| Nematode (*C. elegans*) gonad | GSC | Notch | NI |
| *Drosophila* testis | GSC | JAK-STAT | DE-cadherin, β-catenin |
| *Drosophila* ovary | GSC | DPP-BMP | DE-cadherin, β-catenin |
| Mouse bone marrow | HSC | SCF, Wnt, Notch, ANG1... | β1 integrin |
| Mouse small intestine | ISC | Wnt, BMP | β-catenin |
| Mouse skin | Follicular bulge stem cell | Wnt, BMP | β-catenin, β1 integrin |
| Mouse brain (lateral ventricle) | NSC | Shh, BMP | N-cadherin, β-catenin |
| Mouse testis | SSC | GDNF, SCF | α6 integrin, β1 integrin |
| Mouse skeletal muscle | Satellite cell | Notch | β1 integrin |

*Note:* The stem cell niche provides a multitude of signal inputs, either from direct cell-cell contact or from soluble factors. The table lists for three commonly studied model species some of the adult stem cell types and the key signaling pathways and adhesion molecules that are involved in their interaction with the niche. The stem cell abbreviations are as indicated in Table 6.1.

*Abbreviations:* DPP, JAK, Janus kinase; STAT, signal transducer and activator of transcription; decapentaplegic; BMP, bone morphogenetic protein; SCF, stem cell factor; Wnt, wingless; ANG1, angiopoietin 1; Shh, Sonic Hedgehog; GDNF, glial cell derived neurotropic factor; NI, not investigated.

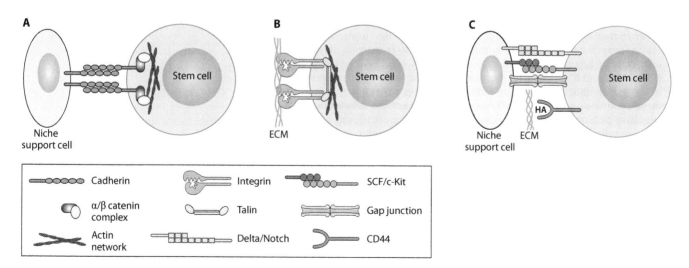

**Figure 6.11 Receptor proteins linking a stem cell to its niche.** Cell-cell or cell-extracellular matrix (ECM) interactions of a stem cell in its niche involve a number of families of transmembrane proteins expressed by the stem cell. The diagrams illustrate some of the typical molecular protein interactions that link stem cells to elements of the niche. (**A**) Classical cadherin-mediated cell-cell adhesion acting via the intracellular α/β-Catenin complex. (**B**) Integrin-mediated cell-ECM interaction. The intracellular domains of integrins interact with the actin cytoskeleton network through talin proteins. (**C**) Other molecules such as Delta/Notch, stem cell factor (SCF)/c-Kit, CD44/hyaluronic acid (HA), and gap junction components are also involved in stem cell-niche interactions.

asymmetric division of stem cells in their niche by controlling the mitotic spindle orientation, much in the same way that E-cadherin orientates GSC division in the *Drosophila* gonad. This role is revealed in the mammary epithelium if β1 integrin is conditionally ablated, whereupon the normal pattern of symmetric stem cell division is perturbed. Another example involves the NSC niche, which is likely to rely on this form of interaction. Hence, α6β1 integrin is highly expressed on NSC, helping them to adhere to the endothelial cells in the niche, an interaction that probably involves α laminin, which is abundantly expressed in the subventricular zone. Integrins can signal directly, for instance through complexes incorporating focal adhesion kinase (FAK) or integrin-linked kinase (ILK), but are also able to modulate signaling through growth factor receptors, such as those that recognize EGF, PDGF, or VEGF, which can in turn affect the activation status of the integrin heterodimer.

Some stem cells, including GSC, NSC, and HSC, appear to use both cadherin and integrin adhesion molecules to interact with their respective niches. Furthermore, many other families of adhesion molecules are utilized by stem cells for anchorage and communication with their niche environment (Figure 6.11C). One that is encountered in many different scenarios involves the "heterotypic" interaction between Notch family receptors on stem cells and their cognate ligands ("Delta-like ligands") that are expressed on niche cells. This type of interaction is important, for example, in epidermal stem cells in the skin and for HSC in the bone marrow. Another receptor-ligand interaction crucial to the maintenance of HSC is that involving the transmembrane receptor tyrosine kinase KIT and its ligand stem cell factor (SCF), which is most often active in its soluble form but can be expressed as a membrane-bound molecule that has been shown to be particularly important for stem cell maintenance. The cell adhesion molecule CD44, through its ability to bind to the glycopolymer hyaluronic acid, has the potential to influence many stem cells in their niche, again as exemplified by HSC. Last, it is worth mentioning that gap junctions, which form firm bridges between many cell types, facilitating electrical, chemical, and small molecule exchange, play some specific roles with respect to stem cells in their niche. Gap junction proteins have been found to anchor stem cells such as the HSC but also are involved in removal of reactive oxygen species from HSC, the transfer of calcium into NSC, and for the maintenance of GSC in *Drosophila* gonad niches.

## Small molecules in the niche environment link stem cells to metabolic status

The concentration of oxygen in the niche, as well as its pH, ionic strength, and the availability of nutrients such as glucose and amino acids, all play a part in the proper maintenance of stem cells within their niche. Perhaps best understood at the moment is the importance of the concentration of molecular oxygen, sensed through the transcription factor hypoxia inducible factor 1α (HIF1α). Various stem cells, including HSC, MSC, and NSC, reside in hypoxic niches, the low oxygen tensions contributing to the maintenance of an undifferentiated state. As a consequence, stem cells under these conditions must rely heavily on anaerobic glycolysis, rather than mitochondrial oxidative phosphorylation, to support ATP production. The levels of nutrients critical for stem cell metabolism, including glucose for energy production and amino acids for protein synthesis, are also sensed by stem cells, in this case through the mammalian target of rapamycin (mTOR) kinase. Finally, specific ion concentration can be detected eliciting a response by the stem cell. The best example is the effect of calcium ion concentration in the HSC niche, where it acts as a chemotactic signal to the stem cells via the calcium-sensing receptor (CaSR). A high calcium ion concentration is found in the endosteal niche due to the activity of bone-degrading macrophage-like cells known as osteoclasts.

## 6.4 CHANGES IN THE STEM CELL NICHE CAN CONTRIBUTE TO TISSUE FAILURE DURING AGING

The overall decline in tissue function as an organism ages has many underlying causes, most of which are still not fully understood. One contributory factor is a gradual loss of adult stem cell capability, resulting in loss of tissue homeostasis and organ failure. Broadly speaking, the effects of aging on stem cells are manifest in two ways, that is, either as changes in the stem cell itself, or as a consequence of the local or systemic environment (Figure 6.12 and Chapter 18). Tissue aging may be a consequence of age-related decline in: (i) the specific stem cells, (ii) the niche leading to decreased stem cell capacity, or (iii) the tissue or whole organism leading to compromise of the niche and then the stem cells. The relative contribution of these intrinsic and extrinsic changes varies depending on the stem cell system, but both can involve the interaction of the stem cell with its niche. For example, intrinsic age-related changes may affect receptors on the stem cell that interact with cells or the ECM in the niche, while alterations in extrinsic factors can include the cellular or non-cellular components of the niche structure or systemic factors delivered through the blood stream.

As *Drosophila* gonad stem cell niches have proved illuminating in terms of defining the structure and cell interactions that typify stem cell-niche interactions, so they have also highlighted the sorts of damaging changes that can accumulate with age. Most obviously, the number of cap or hub cells decline in the ovaries or testes, respectively. The association of the stem cells with the cap or hub cells is also weakened by a decrease in E-cadherin expression. In the mammalian equivalent,

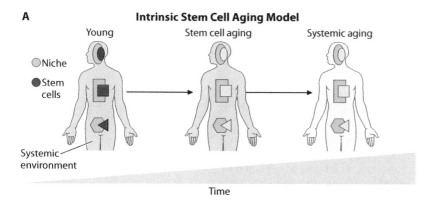

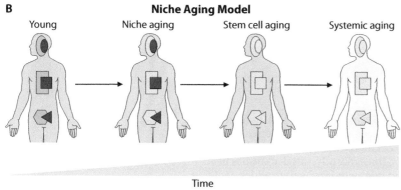

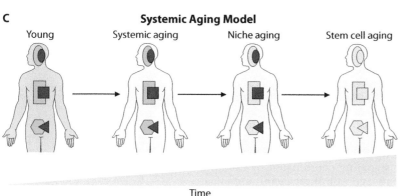

Figure 6.12 **The place of the niche in stem cell aging.** One component of the aging process in animals appears to involve specific tissue stem cells, and at least in some circumstances, changes in the niche are expected to contribute to reduced stem cell capability. Three models showing how stem cell function affects and is affected by aging; in each example, three different niche-stem cell interactions are illustrated, and aging is indicated by grey shading. (**A**) Intrinsic stem cell aging independent of the niche contributes to organism aging. (**B**) Niche aging leads to stem cell aging, which in turn leads to aging of the organism. (**C**) Systemic aging drives the aging of the niche, which in turn leads to stem cell aging. (From Drummond-Barbosa D. (2008) *Genetics* 180: 1787–1797 [doi:10.1534/genetics.108.098244]. Permission conveyed through Copyright Clearance Center, Inc.)

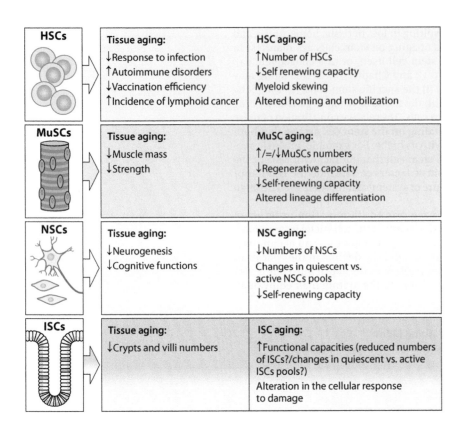

**Figure 6.13 Changes in stem cells in relation to tissue aging.** Several mammalian adult stem cells have been investigated in some detail with respect to age-associated changes in their function in relation to the overall status of the tissue they underpin. Four stem cell types are illustrated in the schematic and are shown next to the aging phenotype of their respective tissue of origin and the changes that have been defined in the stem cell. Abbreviations: HSC, hematopoietic stem cell; MuSc, skeletal muscle satellite stem cell; NSC, neural stem cell; ISC, intestinal stem cell. (From Ortells M & Keyes W. (2014) *Biochem Soc Trans* 42: 663–669 [doi:10.1042/BST20140045]. With permission from Portland Press.)

the levels of glial cell-derived neurotrophic factor in Sertoli cells, normally required for the regulation of self-renewal and differentiation of **spermatogonial stem cells**, are reduced, leading to a lessening of stem cell function, which can be regained by transplantation into a younger environment. Several mammalian adult stem cells have been investigated for the effects of aging upon them and the impact that this has on the respective tissue homeostasis, notable among which are HSC, **muscle satellite cells** (MuSC), NSC, and intestinal stem cells (**Figure 6.13**). Both stem cell intrinsic and extrinsic components have been identified to contribute to the aging processes.

## Both intrinsic and extrinsic influences contribute to the aging of muscle satellite stem cells

Perhaps the best-studied example of an age-related decline in cell-niche interactions is that of muscle satellite stem cells, which are very much affected by their environment and their ability to interact with it. Aging is associated with a progressive decline in skeletal muscle mass and regenerative capacity. Until recently, this decline has been largely attributed to extrinsic alterations in the microenvironment to which MuSC are exposed. However, recent reports show that MuSC also progressively undergo cell-intrinsic alterations that profoundly affect stem cell regenerative function with aging (**Figure 6.14**). Genetic and molecular analysis has also pinpointed decreased integrin β1 expressed on MuSC as a contributing factor in aging, leading to weaker interactions with the ECM and impaired response to environmental growth factor cues, like that from the powerful mitogen FGF2. Compounding this deficit, the ECM component fibronectin, derived from the stem cells, is not up regulated sufficiently in aged muscles, again decreasing the stem cell-ECM interaction. Regarding the niche itself, it seems that this is potentially adversely affected during aging by a number of cell-based factors, including the role played by the muscle fibers that serve as an important source of signals to regulate satellite cell function with age. In young muscles, the stem cells are transiently activated after tissue damage by inflammatory cytokines and growth factor signaling coordinated by immune cells, fibroblasts, and fibro-adipogenic progenitors (FAP). These factors stimulate regulatory pathways in the MuSC, including Delta-Notch, governing cell cycle and stem cell fate, resulting in temporary activation and self-renewal, with homeostasis

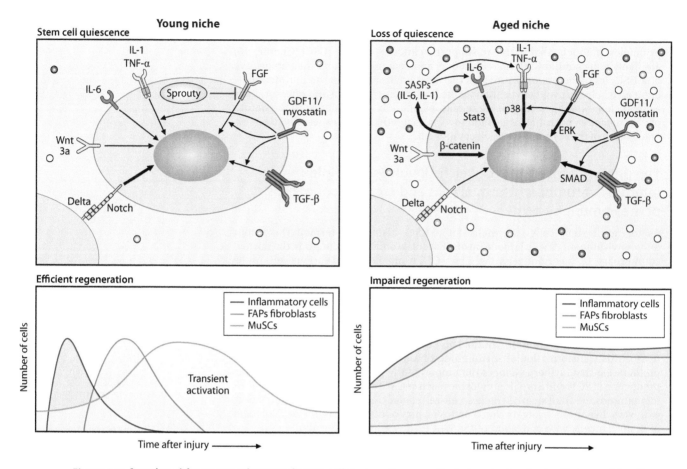

**Figure 6.14 An aging niche compromises muscle stem cell regenerative capacity.** Perhaps best understood in terms of the effects of aging on a particular stem cell and its niche are muscle satellite stem cells (MuSC). The schematic diagram shows signaling pathways involved in the maintenance of MuSC together with the response to injury and the changes in these associated with aging. The upper panels summarize signaling inputs from the niche that maintain the quiescent state. In young muscle, these are largely through Notch-Delta1 interactions. In aged muscle, the response to Notch is limited, whereas several other pathways become dominant, as indicated by the thickness of the arrows. The response to injury is represented in the lower panels. Damage to young muscle results in a transient activation of the stem cells brought about by the release of inflammatory cytokines and growth factor signaling coordinated by immune cells, fibroblasts, and fibro-adipogenic cells (FAPs). In aged muscle, the damage responses are prolonged and the activation of the stem cells is much reduced. (From Blau H, Cosgrove B & Ho A. (2015) *Nat Med* 21: 854–862 [doi: 10.1038/nm.3918]. Reprinted by permission from Springer Nature.)

reached upon completion of myofiber repair through re-induction of stem cell quiescence, in part via activation of Sprouty, an inhibitor of FGF signaling. In aged muscles, self-renewal signals such as Delta diminish and inflammatory and fibrogenic signals become stronger and longer lasting, resulting in aberrant stem cell activation, including through Wnt/β-catenin, cytokine, and growth factor signaling pathways. Last, studies involving the surgical joining of a young and old mouse (**parabiosis** – Chapter 10) provide strong evidence for a systemic, blood-borne factor that can rescue old stem cells. In such experiments, a positive effect is observed of the linked young animals on stem cells in the old animals. At least part of this rejuvenating effect appears to be due to higher levels of circulating growth differentiation factor 11 (GDF11) and oxytocin. Interestingly, GDF11 administered directly to aged mice can restore the regenerative capacity of aged muscle, and at the same time has a similar positive effect on cardiac regeneration and neural stem cell function.

## 6.5 THE ROLE OF STEM CELL NICHES IN MALIGNANCY

As much as the niche is essential for the proper maintenance of adult stem cells, there is accumulating evidence to suggest that the establishment of tumors and their propensity to metastasize depends very much on the hijacking of, or *de novo* creation of, a suitable niche environment. Although the details may differ, the fundamental principles of a niche that is able to support a **cancer stem cell** (CSC) (Chapter 18) are likely to be similar to those required by adult stem cells. From the perspective of a CSC, its niche not only provides the ideal environment for its maintenance, but it also helps to avoid eradication by the host immune system and often protects against chemotherapeutics. For these reasons, there is great interest in the nature of CSC niches as potential targets for novel targeted therapies, especially to enable eradication of the resistant stem cell component.

## The primary cancer stem cell niche involves a complex interplay of tumor and normal cells and recruitment of cells of the immune system

As for adult stem cells, the niche for a CSC, also often termed the tumor microenvironment (TME), is made up of adjacent stromal cells alongside the normal counterparts of the tumorigenic cells, non-CSC tumor cells, and various cells of the immune system (Figure 6.15). These different cell types within the TME produce an assortment of factors that not only stimulate the self-renewal of the CSC but also induce angiogenesis and recruit immune and other stromal cells to the niche to aid the progression of the tumor and eventually result in its metastasis. The CSC itself also plays a major role in the remodeling of its own niche through the production of factors that have the ability to transform normal fibroblasts into cancer-associated fibroblasts (CAF). This transformation of surrounding fibroblasts causes an increase in their proliferation that further enhances the support of the CSC because this elevates the production of ECM and a profile of cytokines such as CXCL12, VEGF, PDGF, and HGF. As for many normal adult stem cell niches, cell-cell contacts are important, and signaling pathways, like those involving Notch and Wnt, play a major part in the development of many cancers, with activation of the Wnt pathways having the ability to induce stemness toward a CSC state.

The recruitment of cells that influence the immune response to the TME is a critical factor in the ability of the CSC to evade immune surveillance. Chemokines and cytokines secreted by the tumor cells attract numerous immune cell types that contribute to the immunosuppressive environment of the CSC niche. Such cells include MSC, which have an immune suppressive effect (Chapter 16) but can also

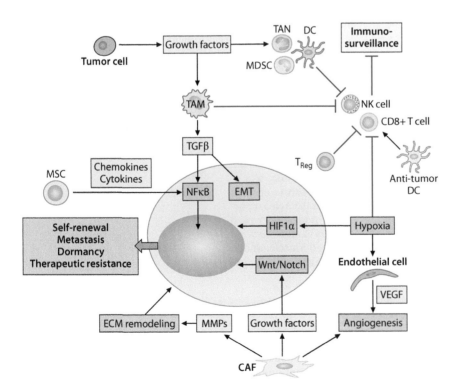

**Figure 6.15 The role of cell components of the cancer stem cell niche.** As in any adult stem cell niche, cancer stem cells (CSC) encounter and rely upon a host of interactions with normal tissue cells, other tumor cells, immune cells, and soluble factors. The diagram illustrates some representative potential interactions between CSC and cells and other components within the niche. Mesenchymal stem cells (MSC) promote cancer stem cell characteristics by up regulating nuclear factor κB (NK-κB). Surrounding non-CSC tumor cells produce growth factors that recruit and expand tumor associated macrophages (TAM), myeloid-derived suppressor cells (MDSC), tumor-associated neutrophils (TAN), and dendritic cells (DC). TAM in turn release transforming growth factor β (TGF-β), which also promotes NF-κB signaling. Accumulation of TGF-β by TAM results in the recruitment of regulatory T cells (T$_{reg}$) cells, which in addition to the hypoxic niche environment inhibit immunosurveillance by CD8+ T-cells and natural killer (NK) cells. Hypoxia promotes CSC survival by the induction of expression of the transcription factor hypoxia induced factor 1α (HIF1α). Hypoxia also induces endothelial cells to induce angiogenesis through the production of vascular endothelial growth factor (VEGF). Cancer-associated fibroblasts (CAF) also promote angiogenesis and stimulate Wingless (Wnt)/Notch signaling that supports CSC maintenance. The production of matrix metalloproteinases (MMPs) by CAF enhances extracellular matrix (ECM) remodeling and thereby the survival of CSC. (From Plaks V, Kong N & Werb Z. (2015) *Cell Stem Cell* 16: 225–238 [doi: 10.1016/j.stem.2015.02.015]. Copyright 2015 with permission from Elsevier.)

influence stem cell properties by, for instance, blocking differentiation through production of the bone morphogenetic protein (BMP) antagonist Gremlin-1. Other key players in immune modulation that are recruited to the CSC niche include tumor-associated macrophages (TAM), tumor-associated neutrophils (TAN), and populations of cells identified as myeloid-derived suppressor cells. These latter cells help in the recruitment of cells including regulatory T cells or T helper cells through their secretion of growth factors and cytokines such as TGFβ and IL6, but may also perform additional functions that, for instance, promote metastasis by eliciting ECM breakdown.

## Tumor spread and the formation of metastases can also involve the hijacking or *de novo* formation of a stem cell niche

The metastatic spread of cancer cells, and the establishment of CSC at sites distant to the primary tumor, also necessitates the creation of a suitable niche. This niche might be pre-existing, for example, around blood vessels in perivascular niches, or may be induced *de novo*. Intriguingly, it seems that some tumors, as seen experimentally for cancers of the breast, lung, and gut, stimulate a premetastatic niche at a distance through the release of factors that act systemically. These factors can, for instance, lead to the activation of fibroblasts and to the recruitment of immune system modulating cells, which together cause remodeling of the specific tissue to create a niche. Once the metastatic cells lodge in the new niche, the CSC can also continue the remodeling process.

It is probably worth reflecting at this point on the very assays that are often utilized to define CSC, namely transplantation assays (Chapters 2 and 18). Such assays, as exemplified in their use in the initial discovery of the CSC underpinning acute myeloid leukemia, require that the tumor initiating stem cells find an appropriate niche. Given what we have just said about the induction of a premetastatic niche, it is possible that the perfect location does not actually exist in the host animal receiving the test cells, and therefore the assay at best underestimates the frequency of CSC or at worst fails to demonstrate their existence. One solution that has been successfully applied to improve the assay has been to include stromal cells together with the tumor-derived cells being tested when they are injected into the experimental host.

## 6.6 TARGETING STEM CELL NICHES AS A THERAPEUTIC OPTION

Knowledge of adult stem cell niches offers a number of possibilities for intervention in a clinical context, ranging from regenerative therapies and means to combat aging processes, to the treatment of cancer (Figure 6.16).

## The hematopoietic stem cell-niche interaction serves as a paradigm for the way in which intervention might have benefits for regenerative therapies

The most likely impact on clinical practice lies in the area of transplantation, especially that involving HSC. One of the primary challenges in HSC transplantation, as well as other tissue-specific transplants, is to achieve sufficient regeneration in a short enough time to be an effective defense against disease. In order to maximize engraftment efficiency, it is important to gain sufficient donor stem cell numbers with the fitness to be able to efficiently migrate to the target tissue following transplantation and engraft and proliferate to regenerate the diseased tissue, failure to achieve any of these points increasing the risk of graft failure. The niche can be and has been an effective target to increase the number of cells available for transplantation. For example, HSC mobilization is routinely used to harvest donor cells for bone marrow transplantation. To achieve HSC mobilization, the donors are treated with G-CSF that actively remodels the bone marrow niche and causes the disruption of the normal interactions between the HSC and the niche cells and allows the migration of the HSC out of the niche and into the peripheral circulation. Alternatively, treatments can be initiated in the stem cell recipient that prime the niche environment to make it optimally receptive to the engraftment of donor stem cells and thus enhance the regenerative capacity. The efficacy of such an approach can be demonstrated in animal models, as seen, for example, when

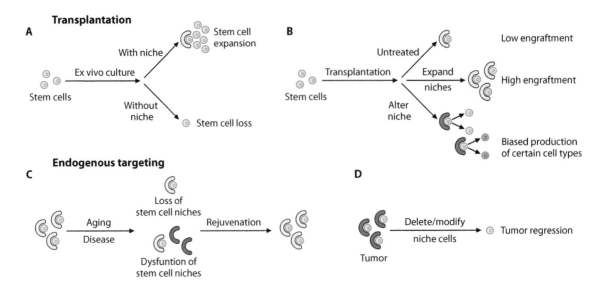

**Figure 6.16 Therapeutic manipulation of the stem cell niche.** Targeting of the stem cell niche could have therapeutic value in a number of contexts related to human health. First, for therapies involving transplantation, recapitulation of niche-derived signals could be used to expand stem cells *ex vivo* (**A**) or to expand or alter the available niches *in vivo* (**B**). Second, targeting of stem cell niches *in vivo* could be a way to rejuvenate aged or diseased tissues (**C**) or to eliminate tumors by limiting the potential for cancer stem cells to be maintained by their niche (**D**). (From Wagers A. (2012) *Cell Stem Cell* 10: 362–369 [doi:10.1016/j.stem.2012.02.018]. Copyright 2012 with permission from Elsevier.)

mice are treated with parathyroid hormone (PTH), which improves post-bone marrow transplantation recovery by promoting the proliferation of Nestin+ MSC and encouraging the differentiation of osteolineage cells in the bone marrow that ultimately support the donor HSC.

## Therapeutic manipulation of the stem cell niche may represent a way to combat some aspects of age-related tissue decline

As we discuss in Chapter 18, stem cell function declines during aging in most adult tissues, contributing to the onset and accelerated progression of age-related disease. In this chapter, we have seen that the decline in adult stem cell capability with age can be partly attributable to a deficit in the function of the relevant niche or the ability of the stem cell to interact with it. The classic experiments involving parabiotic joining of the circulatory system of aged mice with young mice have demonstrated that the age-related stem cell function decline can effectively be reversed, at least in some cases, as a result of the action of niche-derived factors originating from the younger animals acting upon the aged stem cells. Therapies are being investigated to identify a means to enhance the endogenous potential of **tissue stem cells** by manipulation of the niche housing them. For example, targeting defects in signaling pathways between the niche and the stem cell such as inactivated Notch and hyperactivated Wnt signaling.

## Targeting of stem cell niches has the potential to enhance the treatment of cancer

Finally, the fact that tumor growth and metastasis are supported by stem cells that have specialized niches raises the possibility of novel treatments for cancer. One key problem in eradicating a disease like cancer is the fact that it has a stem cell basis and that often the disease-initiating CSC is innately quiescent and therefore is resistant to therapies such as ionizing radiation and chemotherapeutics. Another problem with directly targeting the disease-initiating stem cell is that it is often difficult to distinguish from its normal stem cell counterpart. Therapies are now being employed that exploit the properties of stem cells and their requirement to reside within a specific niche microenvironment. Using the CSC as an example, the TME supports and initiates the stem cell-like program required to transform cancerous cells; therefore, targeting the TME is an indirect means to eradicate the CSC. Although this method has shown promise, such as in the treatment of colon tumors and neural gliomas, there have been some instances where the CSC have evolved mechanisms to evade the constraints of the niche and essentially become independent of niche-derived signals.

## 6.7    SUMMARY

This chapter has given an overview of the concept of the stem cell niche, and we have provided examples from diverse species and tissues for which there is quite a clear picture of the way in which a microenvironment can work in partnership with the stem cells to control their specific properties. We have also looked at the way in which the stem cell niche might contribute to disease and tissue decline during aging.

Stem cell niches share many components in their overall makeup from both cells and non-cellular factors, as well as some common architectural features. These similarities are particularly striking when comparing the stem cell niches between plants and animals, for which the evolution of their stem cells is thought to have occurred independently.

Knowledge of the stem cell niche constituents and the ways in which these can change, as in aging, or be subverted, as is seen in the development and spread of cancer, will certainly educate thinking about novel therapeutic approaches in these different clinical scenarios.

## KEY POINTS

- Stem cells in multicellular organisms throughout the plant and animal kingdoms reside in a tissue microenvironment known as the niche, which supports self-renewal and regulates commitment to differentiation.

- The stem cell niche is constituted of a variety of cells and extracellular matrix proteins and exposes the stem cell to soluble factors and molecules derived both locally and systemically.

- Specific cell contacts ensure maintenance in a self-renewing state and can dictate the orientation of cell division.

- Protein-protein interactions between surface molecules on stem cells and niche cells contribute to the positioning of cells in the niche.

- Surface proteins on stem cells also serve as receptors recognizing ligands in the form of either soluble factors or moieties in extracellular matrix proteins. The receptor-ligand interactions initiate signals that influence stem cell behavior.

- The niche also affects the stem cell through the local concentration of oxygen, ions, and nutrients.

- The function of the stem cell niche can be a contributory factor in the decline of tissues with age, either through accumulated defects in the niche or systemic environment or because of a decline in the ability of the stem cell to interact effectively with the niche.

- Cancer stem cells exist in a specialized niche, which shares many features with those occupied by normal stem cells.

- The metastatic spread of cancer cells to a distant site in the body is often preceded by the formation of a pre-metastatic stem cell niche through the action of systemic factors released by the primary tumor.

- Understanding of the nature of normal or tumor stem cell niches offers novel therapeutic opportunities for the treatment of age-related tissue decline and cancer.

## FURTHER READING

### General

Boulais, P.E. and Frenette, P.S. (2015) Making sense of hematopoietic stem cell niches. *Blood*, 25: 2621–2629.

Chen, S., Lewallen, M. and Xie, T. (2013) Adhesion in the stem cell niche: Biological roles and regulation. *Development*, 140: 255–265.

Heidstra, R. and Sabatini, S. (2014) Plant and animal stem cells: Similar yet different. *Nat. Rev. Mol. Cell Biol.*, 15: 301–312.

Janocha, D. and Lohmann, J.U. (2018) From signals to stem cells and back again. *Curr. Opin. Plant Biol.*, 45: 136–142.

Losick, V.P., Morris, L.X., Fox, D.T. and Spradling, A. (2011) *Drosophila* stem cell niches: A decade of discovery suggests a unified view of stem cell regulation. *Dev. Cell*, 21: 159–171.

Morrison, S.J. and Spradling, A.C. (2008) Stem cells and niches: Mechanisms that promote stem cell maintenance throughout life. *Cell*, 132: 598–611.

Scadden, D.T. (2006) The stem-cell niche as an entity of action. *Nature*, 441: 1075–1079.

Turksen, K. (Ed) (2015) *Tissue-Specific Stem Cell Niche.* Humana Press.

### The stem cell niche in disease

Plaks, V. et al., (2015) The cancer stem cell niche: How essential is the niche in regulating stemness of tumor cells? *Cell Stem Cell*, 16: 225–238.

Wagers, A.J. (2012) The stem cell niche in regenerative medicine. *Cell Stem Cell*, 10: 362–369.

# STEM CELLS ACROSS THE ANIMAL AND PLANT KINGDOMS

# PART THREE

# 7 Regeneration in animals: The evolution of stem cells

In the past decade or so, the word "**regeneration**" has become inextricably linked to stem cell biology, mainly in association with potential medical applications deriving from this relatively new area, but we should first clarify what is actually meant by the term regeneration in different contexts relevant to stem cell science. The goal of "**regenerative medicine,**" which is taken by many to be synonymous with "stem cell-based medicine," is to repair or replace damaged or diseased organs using cells. Essentially, there are three strategies that can be employed using cells to bring about regeneration in medicine: implantation of stem cells to build new structures, implantation of cells pre-primed to **differentiate** to a given lineage, and stimulation of endogenous stem cells to replace missing structures. Such regenerative therapeutic approaches that involve stem cells in one way or another, although currently limited in their efficacy or purely conceptual, will feature throughout subsequent chapters, in particular in Chapter 17.

In this chapter, we will look at regeneration as it occurs as a natural process across the animal kingdom, ranging from the complete rebuilding of the body of some invertebrates through to limb, heart, and skin regeneration in some vertebrates. We will pay particular attention to the extent to which stem cells play a role in the regenerative processes. The potential of regenerative medicine is usually emphasized by pointing out that certain amphibians of the order Urodela, which have much the same molecular, cellular, and structural makeup as ourselves, are able to replace a severed limb within 70 days (Figure 7.1). Although "regeneration" of human limbs following amputation presently remains in the hands of prosthetic surgeons and skilled medical engineers, it is widely anticipated that the lessons learned from understanding regeneration will eventually have an impact on regenerative medicine.

## 7.1 THE NATURE OF REGENERATION

Regeneration is widespread in animal biology, although nature has been somewhat capricious in the way that this ability has been gained or lost during evolution, even between closely related species. Generalizations are not always easy to make because of the variation that is seen in where regeneration plays a significant role, however, by and large, the capacity to regenerate is observed less in complex multicellular animals that branched off the evolutionary tree more recently (Figure 7.2). The more restricted role for regeneration in vertebrates compared to invertebrates also tends to be limited to specific body structures or organs (Figure 7.3). Invertebrates, like certain flatworms, can totally regenerate their body from a small fragment, but this only applies to some species, while other more complex invertebrates like echinoderms (such as starfish, sea urchins, and sea cucumbers) and crustaceans (such as crabs and lobsters) can replace lost appendages. Although regenerative capacity is more limited among the vertebrates, there are examples like lizards, salamanders, and ray-finned fish that can replace tails, limbs, and fins, respectively. The selective pressure that has conferred regenerative capacity in some animal species appears most often to be a way of evading predators or recovering from the consequences of encounters that were a little too close for comfort. Often, in fact, the intentional loss of a tail, limb, or

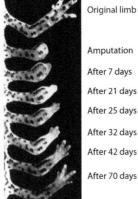

Original limb

Amputation

After 7 days

After 21 days

After 25 days

After 32 days

After 42 days

After 70 days

**Figure 7.1 Limb regeneration in salamanders.** Urodele amphibians, as exemplified by salamanders, demonstrate an amazing capability to regenerate damaged or amputated limbs in a matter of days. On the left is a photograph of an eastern red spotted newt (*Notophthalmus viridescens*) in its immature terrestrial "red eft" stage of the lifecycle. On the right is a photograph showing the stages of limb development in the newt from the original limb at the top, immediately following amputation and then the stages of blastema formation through to the newly regenerated limb by day 70 toward the bottom of the image. (From http://www.sharpschapelliving.com/tag/red-spottednewt and From Goss RJ. (1969) *Principles of Regeneration.* New York, NY: Academic Press. Copyright 1969 with permission from Elsevier.)

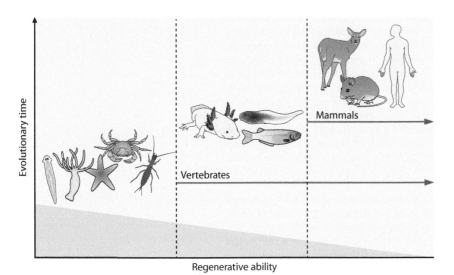

Mammals

Vertebrates

Evolutionary time

Regenerative ability

**Figure 7.2 Regenerative ability across the multi-cellular animal kingdom.** During evolution, multicellular animals diversified into species with differing complexity. There is a general inverse correlation between complexity and the ability of the organism to regenerate. The diagram depicts the correlation between regenerative capacity and the time of emergence of current extant species. Invertebrates such as planarian flatworms and *Hydra* display whole body regeneration, while starfish and arthropods, including crabs and insects, are able to regenerate appendages. Vertebrates evolved to have a much-reduced regenerative ability, although some present-day species are capable of regeneration of appendages, including amphibian tadpoles (limbs), axolotl (limbs), and zebrafish (fins). As mammals evolved, regenerative ability was almost lost, with just a few examples of limited regeneration such as the mouse and human (digit tip) and deer (antler). (From Streelman JT, editor. (2013) *Advances in Evolutionary Developmental Biology.* Hoboken, New Jersey: John Wiley & Sons, Inc. With permission from Wiley.)

| Biological level | Examples | Pre-amputation | Post-amputation | Regenerate |
|---|---|---|---|---|
| Whole body | Regeneration from a small body fragment | | | |
| Structure | Limb, fin, tail, head, tentacle, siphon, arm, stalk | | | |
| Internal organ | Heart, liver, lens | | | |
| Tissue | Epidermis, gut lining | | | |

**Figure 7.3 Differing scales of regeneration across the animal kingdom.** Regeneration occurs to varying degrees ranging from whole body regeneration to that of just one specific tissue. The figure depicts the types of regeneration (blue) that can be observed following amputation (red dashed lines) in various animals. Whole body regeneration can occur from small body fragments, as is observed in *Planaria*. Some species are capable of regenerating specific structures such as appendages following blastema formation, the example shown being limb regeneration in axolotl. Some vertebrates are able to regenerate internal organs such as the heart, as seen in the zebrafish. Replacement of tissue is a more common regenerative phenomenon, being observed for example in wound healing of the epidermis and gut lining. (From Bely A & Nyberg K. (2009) *Trends Ecol Evol* 25: 161–170 [doi: 10.1016/j.tree.2009.08.005]. Copyright 2009 with permission from Elsevier.)

Many of the animals that are able to regenerate have the need to do so due to the process of autotomy, which literally means self-amputation, usually as way to evade predation (Figure 1). Most animals that display such behavior are able fully to regenerate the missing appendage. The classic examples of autotomy are shown as loss of appendages such as limbs and tails in salamander and lizards, respectively; however, other examples exist across the animal kingdom, with at least one known species of mammal, the African spiny mouse (*Acomys*), autotomizing their skin upon predation. Autotomy also exists among invertebrates, such as crustacea and insects, and much of what is known about the regenerative abilities following autotomy has been gained from studies on such creatures.

Box 7.1 Figure 1 **Autotomy.** Survival is an innate instinct of all living creatures. A defense mechanism known as autotomy (self-amputation) has evolved across the animal kingdom in order to evade predation. The images show examples of autotomy. On the left is the sunflower starfish (*Pycnopodia helianthoides*) with a missing arm. In the center is a northern alligator lizard (*Elgaria coerulea*) with a freshly autotomized tail. On the right is an African spiny mouse (*Acomys kempi*) following autotomized loss of its dorsal skin. All of these animals are capable of regenerating the lost tissue. (Left: From SIMoN Photo Library. Photo credit: Chad King/NOAA MBNMS. Center: From http://www.californiaherps.com/lizard/pages/e.c.principis.html. Used with permission. © Gary Nafis. Right: From Seifert et al (2012) *Nature* 489: 561–565 [doi:10.1038/nature11499]. Reprinted by permission from Springer Nature.)

even the skin is facilitated as a way of escaping the clutches of a predator through a process known as **autotomy**, literally meaning self-amputation (Box 7.1).

In one sense, regeneration could be considered widespread if we consider an animal's ability to maintain organs that require regular replacement of cells, such as the skin and intestine. However, the daily turnover of cells seen in adult organs of animals relies predominantly on resident stem cell populations and will not be considered further here, as this is discussed in several chapters throughout this book.

## Regeneration can be achieved by several means, some of which involve stem cells

As we will see when we take a look at individual examples of regeneration, the cellular mechanisms involved are quite varied (Figure 7.4). There are some clear instances of the involvement of stem cells, especially in regeneration in the invertebrates. Research into the potential for the total regeneration from specialized cells seen in some of the animals arising earlier in the phylogenetic tree, which include the extant animal representation of the transition from single- to multi-cellular life, is beginning to give tantalizing clues about the evolution of the mechanisms that dictate stem cell characteristics. However, most of the regenerative scenarios in vertebrates, excluding the daily replacement of cells in adult organs, appear to involve a process that is usually

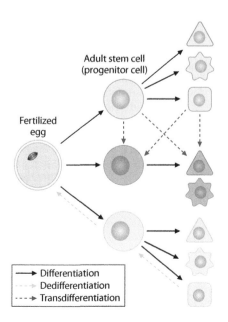

Figure 7.4 **Alternative cell fates involved in regeneration.** The cell mechanisms governing regeneration are variable depending on the species and tissue in question. This schematic depicts the routes that cells might follow during regeneration. A fertilized egg and the pluripotent stages of the embryo ultimately generate multiple cell lineages, depicted by the different colors (blue, red, and yellow), each of which gives rise to numerous cell types depicted by different shapes within each lineage. This cell fate determination occurs though several stages of differentiation (solid black arrows). Dedifferentiation (green dashed arrows) is when a cell reverts to a more immature state, whereas transdifferentiation (red dashed arrows) means that a cell from one lineage transforms directly into a cell type of a different lineage. (From Sugimoto K, Gordon S & Meyerowitz E. (2011) *Trends Cell Biol* 21:212–218 [doi: 10.1016/j.tcb.2010.12.004]. Copyright 2011 with permission from Elsevier.)

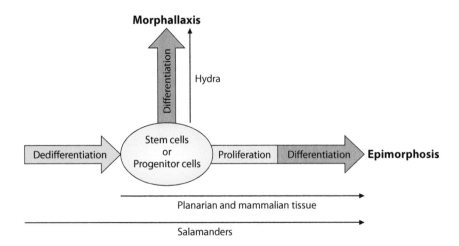

**Figure 7.5 Modes of regeneration.** Natural regeneration pathways can be grouped into two classes, as depicted in this diagram. Stem or progenitor cells (yellow) are central to nearly all types of regeneration. Cells can undergo dedifferentiation back toward the stem or progenitor cell state before undergoing direct differentiation through the process of morphallaxis, for example, as seen in regeneration in *Hydra*. Stem cells can also undergo epimorphosis in which stem or progenitor cells enter a phase of proliferation before differentiating to regenerate the tissue. Examples of epimorphosis include regeneration in planarian flatworms, salamander limbs, and mammalian tissues. (From Odelberg S (2002) *Semin Cell Dev Biol* 13: 335–343 [10.1016/s1084952102000897]. Copyright 2002 with permission from Elsevier.)

known as **dedifferentiation**. Although there has been some controversy and confusion as to exactly what the term dedifferentiation means and how it can be determined, it is probably best to stick with a quite simple definition, namely that it describes a reversal, however small, in the phenotype of a cell to an earlier state in the path that it followed to its current differentiated form. This reversal could be quite extensive, and certainly is what is seen when a somatic cell is artificially **reprogrammed** toward a **pluripotent** state (Chapter 5), but can also be subtle, involving no more than the reacquisition of the capacity to proliferate. Dedifferentiation should not be confused with **transdifferentiation**, that is, switching to a cell type that is normally on a distinct pathway of development and differentiation. Transdifferentiation may be relevant in a few restricted instances of regeneration, and is certainly a useful term when describing reprogramming from one somatic cell type to another without the necessary requirement to go via a stem cell stage.

Natural regeneration can be grouped into two classes that are distinguished by the way in which stem cells or **progenitors** are involved (Figure 7.5). Regeneration involving morphallaxis relies on the recruitment of endogenous stem cells or progenitors that undergo differentiation without the requirement for cell proliferation. In contrast, in epimorphosis, the stem cells or progenitors, which are either resident within the tissue or resulting from a process of dedifferentiation, undergo a proliferation phase before differentiating to the final cell type.

## 7.2 ANIMALS THAT REGENERATE USING STEM CELLS

Within the animal kingdom, the invertebrates sit at the apex of the hierarchy of regenerative potential. In particular, members of the phyla Porifera (sponges); Cnidaria, for example, the *Hydra* genus; some Platyhelminthes, such as members of the *Planaria* genus (flatworms); and Annelida (segmented worms) (Figure 7.6) are masters of regeneration with the ability to renew the whole animal from sections of tissue or small numbers of cells. In essence, this sort of regenerative potential makes these organisms effectively immortal and of considerable interest therefore in trying to understand aging processes and how these might be ameliorated (Chapter 18). The fact that quite closely related species can differ in their capacity for regeneration provides a golden opportunity for researchers to find the critical molecular differences that dictate whether a given species can regenerate. One example where this is apparent is among the planarian flatworms, with species such as *Schmidtea mediterranea* being able fully to regenerate, while *Dendrocoelum lacteum* have a very much reduced capacity in this respect.

### Sponges probably shed light on the evolution of the first stem cells

Sponges are phylogenetically the earliest extant multicellular animals (metazoans) and probably reflect a common ancestor that is likely to have arisen through colonial behavior of single-cell organisms. Sponges are filter feeders with an array of internal chambers linked by a canal system and are constituted from about ten

Figure 7.6 **Invertebrates capable of whole body regeneration.** Invertebrates as a group exhibit the greatest potential for regeneration, many species being able fully to regenerate their entire body from just a few cells. The photographs illustrate invertebrate species that are capable of undergoing whole body regeneration. Top left: Freshwater sponge (*Ephydatia fluviatilis*). Top right: *Hydra (Hydra viridis)*. Bottom left: Planarian flatworms (*Schmidtea mediterranea*). Bottom right: Earthworm (*Lumbricus* terrestris). (Top left: From http://www.naturephoto-cz.com/fresh-water-spongephoto-14145.html. Copyright © Jiri Bohdal. Used with permission. Top right: "Green Hydra, *Hydra viridissima*" by Frank Fox; http://www.mikro-foto.de. Licensed under Creative Commons Attribution-Share Alike 3.0 Germany via Wikimedia (https://creative-commons.org/licenses/by-sa/3.0/de/deed.en). Bottom left: From https://planosphere.stowers.org/worms. Used with permission from Stowers Group of Companies. Bottom right: "*Eisenia foetida*" by Rob Hille - Own work. Licensed under Creative Commons Attribution-Share Alike 3.0 Unported via Wikimedia. (http://creativecommons.org/licenses/bysa/3.0/deed.en).)

different cell types (Figure 7.7A). One of these cell types, known as archeocytes in the demosponges (95% of sponge species), are highly mobile and appear to function as stem cells underpinning the remarkable regenerative ability of these animals. Experimental proof for the stem cell properties of archeocytes is not dissimilar to that more commonly seen when investigating mammalian stem cells (Chapter 2). Purely on morphological grounds, archeocytes can be seen to have the potential expected of stem cells in that they are the main component of gemmules (dormant, cyst-like structures that can reconstitute the animal if most of the sponge dies) and are observed to differentiate into all other cell types, including gametes (Figure 7.7B). Further, dissociated sponge cells can be aggregated and go on to develop into a mature animal, a feature that can be prevented if archeocytes are depleted from the mix or can be perfectly recapitulated by purified archeocytes alone. A second cell type with some features that imply stem cell-like potential is the choanocyte. This cell, which has a flagellum and a collar of microvilli (reminiscent of the similarly named single-cell metazoan ancestor, the choanoflagellates), derives from the archeocyte and is the direct precursor of sperm.

Little is known so far about how these sponge stem cells are regulated, either intrinsically or through interactions akin to those seen in the **niches** required by many **adult stem cells** (Chapter 6); however, there are some tantalizing hints that factors seen in higher organisms that are involved in the **totipotency** of germ cells may have evolved early, enabling the beginnings of stem cell characteristics in the animal kingdom. For example, both archeocytes and choanocytes express a homolog

Figure 7.7 **Stem cells in sponges.** Sponges are multicellular, sessile aquatic organisms. Their body structure is adapted to maximize water flow through the cavities for the capture of nutrients. (**A**) This schematic view of a vertical section through a fresh water sponge (*E. fluviatilis*) shows the structure of internal cavities and the various cell components, including choanocytes (green) and the archeocyte stem cells (red). (**B**) A model of the stem cell system in sponges. Archeocytes are active stem cells that undergo self-renewal and are able to differentiate into various types of cells, including choanocytes. In sexual reproduction, gametes are derived from archeocytes (oocytes) or choanocytes (sperm). Choanocytes may transdifferentiate into archeocytes under specific circumstances. (From Funayama N. (2013) *Dev Genes Evol* 223:23–28 [doi: 10.1007/s00427-012-0417-5]. Reprinted by permission from Springer Nature.)

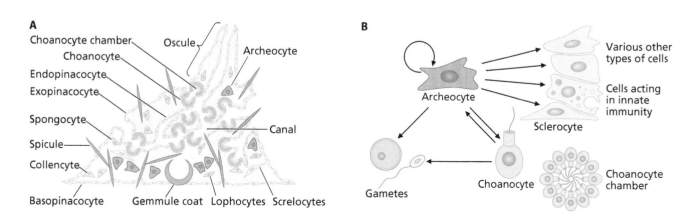

of Piwi, a protein known to function in various animals to maintain potency, mostly through binding to small RNAs (piRNAs) that serve to guide Piwi to degrade specific mRNAs. Other RNA-binding proteins associated with germ and stem cell behavior in higher animals are also expressed in both archeocytes and choanocytes (e.g., Nanos) or just in archeocytes (e.g., Vasa). Characterizing the stem cells of the earliest branching metazoans is clearly of much wider interest than just an understanding of the individual organism's processes of regeneration, in all likelihood giving important clues as to the origins of the fundamental mechanisms regulating these crucial cells.

## *Hydra*, the first experimental model of regeneration, has three distinct stem cell populations

*Hydra* is a genus of small fresh water tubular polyps belonging to the phylum Cnidaria, which also includes animals such as corals, sea anemones, and jellyfish. Although cnidarians are simple, they are more complex than the sponges, being classed as Eumetazoa because during embryogenesis they have true tissues arising from specific germ layers. Unlike the other major branch of Eumetazoa, the bilaterians, cnidarians such as *Hydra* generally have radial (rather than bilateral) symmetry and possess just two (rather than three) germ layers, that is, **endoderm** and **ectoderm** (Figure 7.8). The basal disc within the foot anchors it to surfaces, enabling the mouth opening (hypostome) at the opposite end to catch prey using the tentacles coated in specialized stinging cells. If cut in half transversely, the severed sections are able to regenerate the lost half, whether this is the head or the tail, regaining the appropriate amount of tissue in relation to the site of the cut (Figure 7.9).

*Hydra* has been a key organism in the study of regeneration for over 250 years and continues to be a very important experimental model system in trying to understand fundamental mechanisms of stem cell biology. In 1744, Abraham Trembley (Box 7.2) noted the incredible regenerative capacity of *Hydra*, leading some to suggest much later that he was among the first to study stem cells. Regeneration of *Hydra*, which makes it essentially immortal (single clones of some laboratory strains have been maintained continuously for over 50 years with no signs of aging), can be attributed to three distinct stem cell types in the central gastric region. The ectodermal and endodermal layers are generated from specialized epithelio-muscular cells that can only give rise to their respective lineage, forming unicellular sheets. The two layers are separated by an **extracellular matrix** and are primarily responsible for the shape of the polyp. Between these layers resides a **multipotent stem cell type**, termed the interstitial cell (I-cell), that can give rise to neurons, nematocytes (stinging cells), secretory cells, and gametes (Figure 7.10). Interstitial cells are characterized by their round undifferentiated morphology and large nucleus. Proliferation and differentiation of these stem cells causes a continual movement of cells through the body, resulting in excess older cells being lost from the end of the animal and enabling regeneration of either end of the animal.

As for the sponges, studying stem cells in *Hydra* provides an incredible window through which to observe how evolution established the principles of stem cells and set up fundamental regulatory mechanisms that went forward into both pluripotent embryo-derived and restricted potency adult stem cells. The cnidarians, as represented by *Hydra*, are of particular interest because they possess most of the

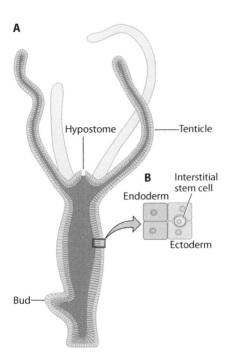

**Figure 7.8 Location of stem cells in the body wall of *Hydra*.** The freshwater tubular polyp *Hydra* is a cnidarian with remarkable regenerative capability. *Hydra* possess two germ layers, namely endoderm and ectoderm. (**A**) The cross-sectional diagram through the body of *Hydra* shows the head (hypostome) at one end and the foot at the other. The body wall contains two epithelial cell layers (**B**), endodermal (orange) and ectodermal (grey). Within the ectodermal layer exists the interstitial stem cell (blue). (From Tanaka E & Reddien P (2011) *Dev Cell* 21: 172–185 [doi: 10.1016/j.devcel.2011.06.016]. Copyright 2011 with permission from Elsevier.)

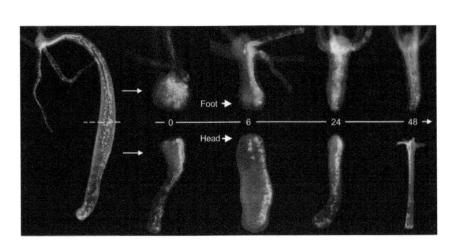

**Figure 7.9 Tracking of cells during regeneration in *Hydra*.** Transgenic *Hydra vulgaris* expressing green fluorescent protein (GFP) in its endodermal epithelial cells allows the tracking of cells during head and foot regeneration by morphallaxis following amputation (white dashed line). The numbers indicate the hours post-amputation. (From Wittlieb J, Khalturin K, Lohmann J et al. (2006) *Proc Natl Acad Sci* 103: 6208–6211 [doi:10.1073/pnas.0510163103]. With permission. Copyright 2006 National Academy of Sciences, U.S.A.)

Abraham Trembley (Figure 1) was born in Geneva in 1710 into an officer's family. An experimental zoologist, he was the first experimentalist to study *Hydra*, which he did while acting as the tutor to the 3- and 6-year-old sons of the Dutch politician Count Willem Bentinck in his summer residence near The Hague. Writing in 1740, Trembley described his first experiment on *Hydra* polyps: "The first operation I performed on the polyps was to cut them transversely.... I place it with a little water in the hollow of my left hand.... When I have

it as I want it, I delicately pass one blade of the scissors, which I hold in my right hand, under the part of the polyp's body where it is to be severed. Then I close the scissors." The appearance of the polyps and their ability to regenerate (Figure 2) reminded Trembley of the mythical Hydra, and he adopted this Greek name to refer to them. His findings on *Hydra* and its regenerative abilities were published in 1744 in the landmark book *Mémoires pour servir à l'histoire d'un genre de polypes d'eau douce*, leading to his recognition, including membership of the Royal Academy of London.

Box 7.2 Figure 1 **Abraham Trembley.** (From Wikimedia.)

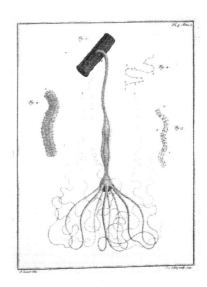

Box 7.2 Figure 2 **Abraham Trembley's drawing of *Hydra*.** (From Wikimedia.)

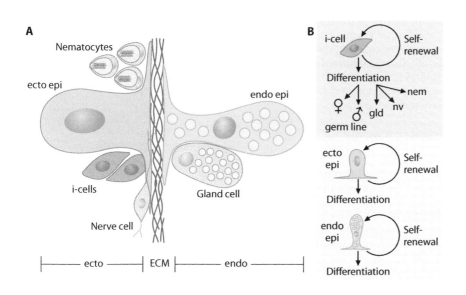

Figure 7.10 **Three distinct stem cell types in *Hydra*.** The impressive regeneration observed in *Hydra* polyps can be attributed to the presence of three distinct stem cells present in the ectodermal and endodermal cell layers. (**A**) Diagram showing the location of the three stem cell types within the body wall of the *Hydra* polyp. The ectodermal (ecto) and endodermal (endo) cell layers are separated by extracellular matrix (ECM). The stem cells are shown in green (endodermal epithelial, endo epi), blue (ectodermal epithelial, ecto epi), and pink (interstitial cells, i-cells). Differentiated derivatives of the i-cell (nematocyte, nerve cell, and gland cell) are shown in grey. (**B**) Schematic representation of the self-renewal (circular arrow) and differentiation (straight arrow) fates of the three stem cells. i-cells are multipotent and capable of giving rise to cells of the male and female germ line, gland cells (gld), nerve cells (nv), and nematocytes (nem), whereas the ecto epi and endo epi cells are unipotent stem cells capable of generating just endodermal and ectodermal layer cells, respectively. (From Hemmrich G, Khalturin K, Boehm A-M et al. (2012) *Mol Biol Evol* 29: 3267–3280 [doi:10.1093/molbev/mss134]. By permission of Oxford University Press on behalf of the Society for Molecular Biology and Evolution.)

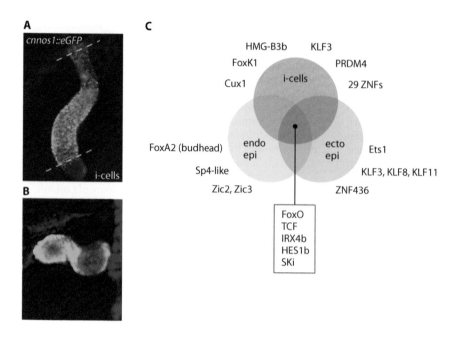

**Figure 7.11 Molecular signatures of *Hydra* stem cells.** Transgenic and lineage tracing technology has enabled the isolation of the specific stem cell types found in the *Hydra* polyp. (A) Image of a transgenic *Hydra* polyp carrying enhanced green fluorescent protein (eGFP) under the control of the interstitial cell-specific *cnnos1* promoter. (B) Confocal image of eGFP expressing interstitial cells (i-cells, green) within the *Hydra* body wall adjacent to endodermal epithelial cells and the ECM (both stained red). (C) Venn diagram illustrating the cell type-specific signature genes encoding transcription factors that are expressed in the three stem cell types. The signature genes are orthologs of transcription factors genes that encode proteins with stem cell functions in vertebrates. Genes enclosed in the box are expressed in common between all three stem cells. (A: From Boehm A-M & Bosch T. (2012) *Zoology* 115:275–282 [doi: 10.1016/j.zool.2012.03.004]. Copyright 2012 with permission from Elsevier. B and C: From Hemmrich G, Khalturin K, Boehm A-M et al. (2012) *Mol Biol Evol* 29:3267–3280 [doi:10.1093/molbev/mss134]. By permission of Oxford University Press on behalf of the Society for Molecular Biology and Evolution.)

gene families found in bilaterians. The central role of *Hydra* as a model organism has been helped by the development of the sorts of transgenic and **lineage tracing** technologies that have proven so useful in studies on phylogenetically more advanced species (Chapter 2). A good example is the work of Thomas Bosch and colleagues, who have used transgenic strains of *Hydra* in which a fluorescent protein is selectively expressed in each of the three stem cell types as a way of isolating them to obtain a molecular "signature" of the key genes that define their cell type-specific function (Figure 7.11). Each of the three stem cell types expresses genes encoding transcription factors, the orthologs of which exhibit stem cell functions in vertebrates. This points to the early evolution in metazoans of stem cell gene function controlling principal features such as **self-renewal** and differentiation. A small subset of these transcription factor genes is expressed in all three stem cell types, and most notable among these are TCF (T-cell factor) and FoxO (a member of the "Forkhead" family).

As will be seen in several chapters throughout this book, the Wnt/β-catenin signaling pathway leading to TCF activation is a central component of many regulatory mechanisms affecting vertebrate stem cells. Similarly, FoxO proteins have emerged as crucial determinants of aging (and presumably stem cell function) in species as divergent as *Drosophila* and humans (Chapter 18), FoxO3a and FoxO1 having been shown to play a role in maintenance of adult **hematopoietic stem cells** (Chapter 9) and **embryonic stem cells** (Chapter 4), respectively. The fact that such stem cell-related functions indeed evolved early in multi-cellular animals is nicely illustrated by the experiments that were performed on FoxO in *Hydra* subsequent to it being highlighted through gene expression analysis. Hence, over expression of FoxO increased interstitial stem cell and progenitor cell proliferation and activated stem cell genes in terminally differentiated somatic cells. In contrast, down regulation of the protein led to an increase in terminal differentiation and caused down regulation of stem cell-associated genes. These sorts of studies are, of course, just the beginning, and undoubtedly the future will see far more genes being revealed as being important in the overall network of regulators that control stem cell behavior.

## Animals with bilateral symmetry exhibit a range of regenerative potentials

Moving up the phylogenetic tree and onto the branch of the bilaterians (Figure 7.12), there continue to be examples of regeneration with an underpinning stem cell basis, but the situation becomes a little more complex. Not only is there an increase in the tissue complexity and functional specification of cells, there are also some big differences in the regenerative capacities of quite closely related species, a fact that is both intriguing and a golden opportunity to dissect what molecular and cellular differences might bring this about.

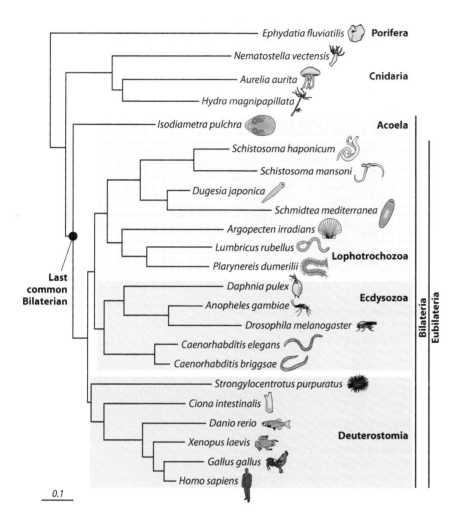

**Figure 7.12 Bilaterian evolution.** Simplified phylogenetic tree representing the evolution of the multicellular animal kingdom. At the top is the phylum Porifera (sponges), which is the link to the first multicellular animals. Of the four superphyla in the animal kingdom, the Radiata is represented by the dominant phylum, the Cnidaria. The three superphyla (Lophotrochozoa, Ecdysozoa, and Deuterostomia) constituting the Eubilaterians, which are animals characterized by a one-way intestine, are shown grouped with the Acoela phylum, representing the potential extant remnant of the basal Bilaterian. (From Egger B, Steinke D, Tarui H et al (2009) *PLOS ONE* 4: e5502 [doi: 10.1371/journal.pone.0005502]. Published under the terms of the Creative Commons Attribution License http://creativecommons.org/licenses/).

Thinking about regeneration in the early bilaterians also gives us the opportunity to introduce the concept of the **blastema**, the region of reorganization and growth that occurs at the site of amputation or wounding involving a complex tissue with multiple cell types. A blastema, whether this is associated with a severed flatworm, an amputated amphibian limb, or a damaged zebrafish heart, involves cell proliferation and differentiation of a number of functional cell types. Some blastemas, especially those in the more primitive animals, utilize stem cells, while most situations in the more complex invertebrates, such as arthropods, and the vertebrates are characterized by dedifferentiation as a prerequisite for cell expansion followed by specific differentiation.

## Flatworms set the scene for adult stem cells in higher animals

The experimental organism that has provided some of the deepest insight into regeneration and the evolution of stem cells is the pretty unassuming flatworm, which belongs to the genus *Planaria* that includes species living in freshwater, seawater, or on land. These humble animals are far more complex than those we have considered until now in this chapter, having tissues and cell types that are seen in most organisms higher up the phylogenetic tree, including central and peripheral nervous systems, eyes, an intestine, excretory organs, musculature, and an epidermis.

Similar to *Hydra*, some planarians have the ability to regenerate an entire animal from a small section of tissue containing only a few thousand cells, any region of the flatworm's body being able to achieve this apart from the tip of the head and the pharynx. Following amputation, anterior- and posterior-facing wounds form head and tail blastemas, respectively, leading to complete regeneration of the anterior-posterior pattern within a week or so (Figure 7.13). Flatworms vary, even between distinct strains of a given species, depending on whether they can totally regenerate their body, making them capable of asexual reproduction and conferring immortality, or

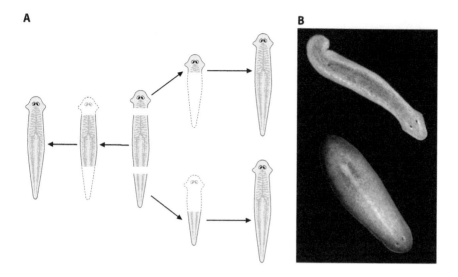

**Figure 7.13 Regeneration in planarian flatworms.** Some planarian flatworms are able completely to regenerate their entire body from a small section of tissue from any region of the body. (**A**) The diagram illustrates how complete body regeneration can occur in some species of flatworm from a fragment taken from any part of the body. (**B**) Two species of flatworm that have been used extensively in experimental studies on regeneration: *Dugesia sp.* (top) and *Schmidtea mediterranea* (bottom). (A: adapted from http://burgscience. edublogs.org/files/2010/03/planaria-regen-eration.jpg, and B: "Dugesia subtentaculata" by Eduard Sola—Own work. Licensed under Creative Commons Attribution-Share Alike 3.0 Unported via Wikimedia. (http://creativecom-mons.org/licenses/by-sa/3.0/deed.en).)

have limited regenerative potential and a strict dependence on a sexual reproductive cycle. Much of the research into the impressive regenerative capability of planarians has arisen from work on the freshwater species *Dugesia japonica* and *Schmidtea mediterranea* (Figure 7.13).

## Neoblasts are responsible for the impressive regenerative potential seen in planarians

The regenerative potential of planarians derives from a population of resident cells termed **neoblasts**, which are distributed throughout the body. Neoblasts include all dividing adult somatic cells, making up approximately 20%–30% of all cells, and are characterized by a high nuclear to cytoplasmic ratio and sensitivity to γ-irradiation. Neoblasts are distributed throughout the body with the exception that they are completely absent within the tip of the head and the pharynx, correlating with the lack of regenerative ability of these regions.

The sensitivity of neoblasts to irradiation and the ability to adjust the dose so that differing numbers of cells remained, theoretically right down to just a single cell, made it possible to show through statistical analysis that a single **clonogenic neoblast** (cNeoblast) can expand and differentiate sufficiently to regenerate the whole. Even more definitively, irradiation could be increased to eliminate all dividing cells followed by transplantation of a sorted, single cNeoblast cell (defined by their small size and compact, round morphology, and the uptake of a DNA-binding dye), again leading to complete repopulation of the host and regeneration of a new organism (Figure 7.14). cNeoblasts are therefore by definition true stem cells, although the broader population of neoblasts are likely to be partially committed. The single cell experiments argue strongly for the pluripotent nature of the cNeoblast, and two models can be envisaged for the way in which it supports regeneration. Either the cNeoblast gives rise directly to all lineages, or it yields committed progenitors that proliferate and undergo restricted lineage differentiation, which is an option more consistent with presence of a broad neoblast population.

## Regulatory mechanisms in neoblasts seem to reflect what has been defined in vertebrate adult stem cells

The ability to isolate and assay neoblasts from flatworms also opens up the possibility to work out what defines these cells in molecular terms and to ask how they compare to adult stem cells from earlier metazoans as well as higher vertebrates. Transcriptomic and proteomic screens of purified neoblasts combined with knockdown experiments of candidate regulators point to a number of features highly reminiscent of pluripotent stem cells and germ cells in other animals (Chapters 3 and 12). These features include a set of post-transcriptional regulators, especially RNA-binding proteins such as a homolog of Piwi, which also crops up in what may be earliest stem cells in evolution, represented by the modern-day sponges, through to pluripotent cells in vertebrates. Protein complexes known to control **epigenetic** modifications linked to pluripotency

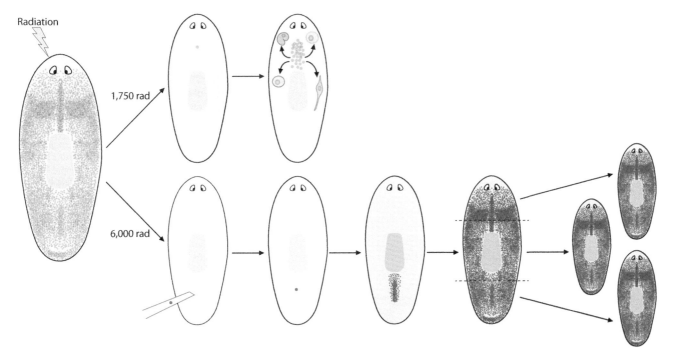

Figure 7.14 **Isolation and functional testing of neoblasts.** Irradiation of planarian (*Schmidtea mediterranea*) flatworms at a dose of 1750 rad can result in animals with just a single surviving neoblast (blue). This single cell can divide and produce a colony of neoblasts, which in turn can repopulate the differentiated cells spanning all germ layers. Irradiation with a higher dose of 6000 rad can eliminate all neoblasts. Transplantation of a single cNeoblast, isolated from a donor strain (red), results in clonogenic growth within the host and complete repopulation to give rise to an animal that can subsequently regenerate any part of its body. (From Tanaka E & Reddien P (2011) *Dev Cell* 21: 172–185 [doi: 10.1016/j.devcel.2011.06.016]. Copyright 2011 with permission from Elsevier.)

as well as direct targets and upstream regulators of the pluripotency factors Oct4, Sox2, and Nanog are also features of the neoblasts.

Another regulatory mechanism that is frequently encountered in the context of stem cells, that is, Wnt/β-catenin signaling, also appears to have been established early in multi-cellular animals, playing a key part in the realization of neoblast regenerative potential. In the planarians, this is seen as a gradient of activity from tail to head, and if the activity of β-catenin is experimentally reduced, then the tail region is seen to transform into a head-like region. The effect of β-catenin in this context is mediated by repression of extracellular-signal-regulated kinase (ERK) signaling, which itself is a feature of head formation ("anteriorization"). In the planarian *Phagocata kawakatsui*, excessive posterior β-catenin signaling interferes with the ERK activation so that the tail region shows markedly reduced head-regenerative ability, whereas tail regeneration is normal.

The ancient origins of the neoblast stem cell regulatory mechanisms seen in the planarians have been strengthened by recent findings in acoel worms (Figure 7.15), combined with a reassessment of their place in the evolutionary tree. Acoel worms were thought to be closely related to planarians, but phylogenetic analysis of gene sequences now implies that the Acoela are basal bilaterians, separated from the branch that gave rise to the planarians by some 500 million years (Figure 7.12). Like planarians, acoels are capable of whole body regeneration, implying similar underpinning stem cells/neoblasts and blastema formation. The Acoela neoblasts also express a Piwi homolog, and again Wnt/β-catenin signaling is crucial for head and tail specification.

**A**

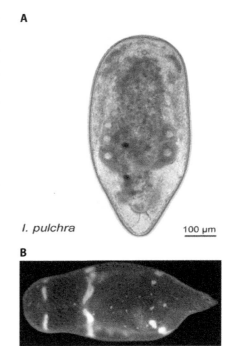

*I. pulchra*                    100 µm

**B**

Figure 7.15 **Experimental model Acoela worms.** Acoel worms perhaps most closely resemble the original form of the last common Bilaterian. Acoela are capable of whole body regeneration. (**A**) *Isodiametra pulchra*. (**B**) The panther worm, *Hofstenia miamia*. (A: From Dittman IL, Zauchner T, Nevard L et al. (2018) *J Morphol* 279:589–597 [doi: 10.1002/jmor.20794] and B: Used with permission from Matt Hooge. Copyright © Hooge 2007.)

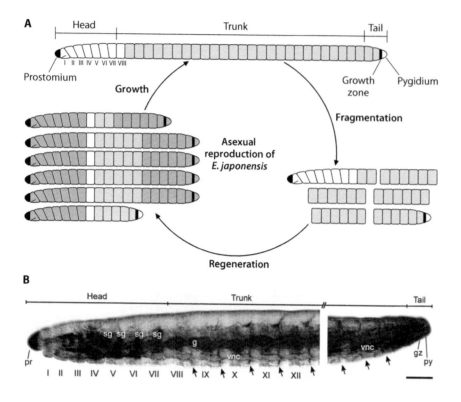

**Figure 7.16 Regeneration in segmented Annelid worms.** Annelid worms are thought to be the most recently evolved group that have retained the capability to regenerate their whole body through the presence of neoblast stem cells. (**A**) Schematic illustration of the regeneration pattern during asexual reproduction in *Enchytraeus japonensis*. The worms harbor neoblasts in all segments except for the anterior-most eight segments, that is, the seven head-specific segments (segments I–VII) and the first trunk segment (segment VIII). Following fragmentation, each fragment regenerates a complete head or tail and grows into a normal worm, irrespective of the region of the body from which the fragment was originally derived. Neoblast-bearing segments and regenerated segments are indicated in grey and orange, respectively. (**B**) Distribution of neoblasts in *E. japonensis*. Intact worms were stained with methyl green-pyronin. Neoblast pairs (arrows) are localized on the intersegmental septa along the ventral nerve cord. g, gut; gz, growth zone; pr, prostomium; py, pygidium; sg, septal (pharyngeal) gland; vnc, ventral nerve cord. Scale bar = 100 μm. (From Myohara M (2012) *PLOS ONE* 7: e37319 [doi: org/10.1371/journal.pone.0037319]. Published under the terms of the Creative Commons Attribution License http://creativecommons.org/licenses/.)

## Annelids were the first organisms in which neoblasts were identified

The most evolved animals that retain the capacity for complete regeneration of the body, brought about through neoblast stem cells, are certain species of segmented worms, or annelids. Actually, the term neoblast was first used in 1891 by Harriet Randolph to describe small, undifferentiated, embryonic-like cells involved in the regeneration of annelids of the genus *Lumbriculus*. Just as we have seen for lower invertebrates, the regenerative capacity and the contribution of neoblasts to blastemas in the annelids can vary between species. Even when whole body regenerative capacity involving neoblasts is seen, the way in which the blastema brings this about has some features of tissue remodeling that are more similar to what is seen in vertebrate limb regeneration. An example of an annelid with complete regenerative potential is the oligochaete worm, *Enchytraeus japonesis*, which is able to reproduce asexually by dividing into several fragments that can regenerate the whole animal in 5 days or so but are also able to induce sexual reproduction. In contrast to planarians, annelid neoblasts are localized at the intersegmental septa along the ventral nerve cord (**Figure 7.16**), and all segments can regenerate into a complete head or tail irrespective of which part of the body they are from. The related species *E. buchholzi*, which can reproduce sexually, lacks neoblasts but can substantially regenerate; however, complete heads are never regenerated.

## 7.3    REGENERATION OF ANIMAL STRUCTURES THAT DO NOT REQUIRE STEM CELLS

In invertebrates that branched later than the annelids in the phylogenetic tree, regeneration becomes more restricted to specific structures, such as the heart, and is especially important for the replacement of appendages of various types. The common theme, extending from what is seen at the wound site in regenerating invertebrates, is that regeneration in higher invertebrates and vertebrates is achieved through the formation of a blastema. However, unlike planarians and annelids that either fully or partially utilize neoblast stem cells in the development of the blastema, those of higher organisms generally do not require stem cell involvement but rather involve some degree of dedifferentiation and "positional memory" of those cells adjacent to the site of damage.

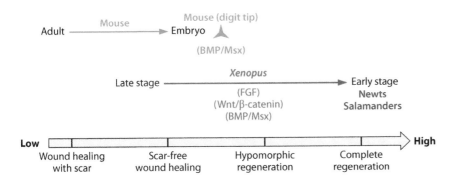

Figure 7.17 **Range of regenerative responses of vertebrates to limb amputation.** Regeneration across the animal kingdom is highly variable. This diagram illustrates the degree of regenerative responses among tetrapod vertebrates, ranging from scar-forming regeneration in adult mice through to complete limb regeneration in newts and salamanders. Between these two extremes are different degrees of regeneration in embryonic mice and *Xenopus* tadpoles. Common, evolutionarily conserved signaling pathways play an important role in the various regenerative responses: Wnt/β-catenin; BMP, bone morphogenetic protein; Msx, Msh homeobox; and FGF, fibroblast growth factor. (From Yokoyama H. (2008) *Develop Growth Differ* 50:13–22 [doi: 10.1111/j.1440-169x.2007.00973.x]. With permission from Wiley.)

Across the animal kingdom, many organisms can replace lost limbs, although the capability to do so varies with species and the stage of development (Figure 7.17). The need for new limb growth is often the consequence of intentional shedding through the process of autotomy and is most often discussed in the context of amphibian limb replacement but applies as much to arthropods like crabs and insects as it does to some vertebrates. There are also scattered examples throughout the vertebrates of structures other than limbs, such as heart and skin, which show some degree of regenerative repair and collectively give insight into some of the more exotic possibilities for regenerative medicine.

## Amphibian limb regeneration points to the possibility that this may be feasible in mammals, including humans

The study of limb regeneration has a long history, dating back to the eighteenth century polymath Lazzaro Spallanzani, who, from among many areas of biology that fascinated him, observed regeneration in a wide range of animals, including flatworms, snails, and amphibians. In 1768, he concluded that, as we have seen already, lower animals have greater regenerative power than those higher in the phylogenetic tree; that young individuals have a greater capacity for regeneration than the adults of the same species; and that with the exception of some lower animals, it is only superficial parts that can regenerate.

Rather than describe limb regeneration in an order relating to the phylogenetic positioning of different animals, we will start with those species that have provided the greatest insight into the mechanisms involved and then go on to consider how similar or otherwise limb regeneration, as well as other types of structural repair, compares in vastly different animals. Amphibians, or more precisely salamanders and newts (the Urodela), are the masters of limb regeneration and include several much-studied experimental models, such as the axolotl, or Mexican salamander (*Ambystoma mexicanum*) (Figure 7.18).

Many researchers have built on the foundations of Spallanzani, prominent among them and leading the way through studies on amphibian limb regeneration have been Jeremy Brockes and his protégé Elly Tanaka (Box 7.3). Through their work, and that of many other developmental biologists, we now have a clear picture of the process of blastema formation at the site of injury and subsequent rebuilding of the lost segment of limb that applies not just to the amphibian limb but seems to be relevant in most other related contexts.

## Amphibian limb regeneration is preceded by the formation of a blastema

The blastema (Figure 7.19) is effectively a large zone of progenitor cells that proliferate before undergoing differentiation and could be viewed as analogous to the **callus** formed in plants upon damage or pathogenic infection (Chapter 8). Immediately following limb amputation, epithelial cells migrate to form a wound epithelium, which covers the stump and thickens through distant epithelial cell proliferation and continued migration to form the apical ectodermal cap (AEC). The AEC is similar to the apical ectodermal ridge (AER) that forms during limb bud formation in embryogenesis (Box 7.4) and plays an important part by providing signals, together with those emanating from the stump, which collectively control proliferation and lineage specification. Tissue lysis and fragmentation occurs at the wound site, the induction

Figure 7.18 **Axolotl.** ("An axolotl in captivity" by th1098 – Own work. Licensed under Creative Commons Attribution-Share Alike 3.0 Unported via Wikimedia. (http://creativecommons.org/licenses/by-sa/3.0/deed.en).)

Jeremy Brockes and Elly Tanaka (Figure 1) have both contributed hugely to our current understanding of regeneration, in particular through their studies on amphibian limb regeneration. Jeremy Brockes is research professor at University College London (UCL), and for most of his career, his primary research focus has been the study of limb regeneration and pattern formation in urodeles. He began his career as a biochemist during his undergraduate studies at Cambridge University, before moving on to study for his Ph.D. in molecular biology at Edinburgh University. He then undertook post-doctoral research at both Harvard Medical School and University College London before going on to lead his own team around regenerative biology. His pioneering work led to the discovery of the nAG protein

during limb regeneration and how its interaction with Prod1 was essential for tissue patterning.

Elly Tanaka is internationally recognized for her ground-breaking successes in developing molecular genetics and imaging techniques to study limb regeneration in the salamander and is now a well-known name in the regenerative field. Having gained her first degree from Harvard University and a Ph.D. from the University of California at San Francisco, Elly started her studies on salamander limb regeneration as a post-doctoral fellow in Jeremy Brockes' laboratory in UCL. Later, as an independent investigator at the Max Planck Institute of Molecular Cell Biology and Genetics, Dresden, Germany, Elly initiated her live imaging and molecular genetics approaches to studying spinal cord and limb regeneration.

Box 7.3 Figure 1 **Jeremy Brockes and Elly Tanaka.** (Used with permission from University College London and from https://www.imp.ac.at/news/detail/article/elly-tanakaawarded-ernst-schering-prize-2017/. Copyright © IMP.)

of specific proteases being crucial for this to happen, before cells migrating from the stump accumulate to form the blastema, which then gives rise to differentiated cells that adopt the original pattern of the missing limb fragment. The blastema has the remarkable property of retaining a positional identity that depends on the point at which the limb was severed (Figure 7.20). This effective autonomy of the blastema can be seen in grafting experiments. For example, an upper arm blastema grafted onto a wrist-level stump regenerates a whole arm from the level of the wrist. Likewise, grafting of a proximal blastema of a missing leg onto another proximal blastema of another amputated leg results in the formation of two complete legs.

## The blastema is composed of distinct cells that individually remember the cell type from which they originated

The origin of the cells that constitute the blastema has only recently been definitively demonstrated, thanks largely to cell tracking experiments. The conventional view had been that somatic cells dedifferentiate into pluripotent blastema cells that are capable of generating muscle, cartilage and nerve tissues. Elly Tanaka and her colleagues used

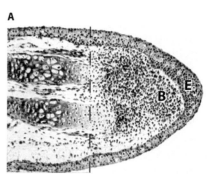

Figure 7.19 **Blastema formation.** Following amputation, the blastema quickly forms as a zone of proliferating progenitor cells that later undergo differentiation. (**A**) Cross section of a regenerating salamander limb 18 days post-amputation. The plane of the amputation is indicated by the dashed line. The blastema (B) is quite prominent, and cell proliferation is ongoing. There is close contact between the epidermis (E) and the underlying blastema. (**B**) Positional memory of the blastema. An upper arm blastema (blue) transplanted onto a lower arm stump (grey) results in duplications along the proximal-distal axis. The blastema forms the structures it would have formed *in situ*, the distal part of upper arm, the lower arm, and the hand. U, upper arm; L, lower arm; H, hand. (A: From Simon A & Tanaka E. (2013) *Wiley Interdiscip RevDev Biol* 2:291–300 [doi: 10.1002/wdev.73]. With permission from Wiley. B: From Nacu E & Tanaka E (2011) *Annu Rev Cell Dev Biol* 27:409–440 [doi:10.1146/annurev-cellbio-092910-154115]. Permission conveyed through Copyright Clearance Center, Inc.)

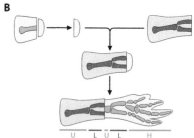

BOX 7.4 BACKGROUND: LIMB BUD DEVELOPMENT DURING EMBRYOGENESIS

The development of the vertebrate limb has been a major model for studying pattern formation in animals. The development of the mouse limb from lateral plate **mesoderm** is rapid, with expansion of the limb bud by approximately 1300 μm between embryonic days 9.5 and 12. The developing limb is patterned along three axes (Figure 1): anterior-posterior (thumb to little finger), dorsal-ventral (back of the hand to the palm), and proximal-distal (shoulder to fingertips). At the distal end of the growing limb bud is a thickened ridge called the apical ectodermal ridge (AER), a region of undifferentiated cells that have been shown to be essential for the continued outgrowth of the limb. The AER maintains outgrowth by producing signals in the form of fibroblast growth factors (FGF). Patterning across the anterior-posterior axis and the formation of the digits has been shown to involve a concentration gradient of a **morphogen** emanating from a defined zone at the posterior margin of the limb bud (Figure 2). This zone, termed the zone of polarizing activity (ZPA), was first described in the developing chick wing. The developing chick wing has three digits, but when the ZPA from an early wing bud was grafted to the opposite margin of another wing bud, the result was the formation of three additional digits that developed as a mirror image. This classic experiment demonstrated that the identity of the digit was dependent upon the distance from the ZPA. The signal deriving from the ZPA, in both the chick and mammals, is Sonic Hedgehog (Shh). Where Shh concentration is at the lowest, at the anterior margin furthest away from the ZPA, digit 1 forms, while the posterior digit 5 arises from the ZPA itself. During vertebrate limb development, proximal-to-distal segment specification is dictated by the progressively more distal expression boundaries of the highly conserved homeodomain-containing transcription factors HoxA9, HoxA11, and HoxA13. The same order of expression is seen in blastema cells and the distal tip that arises during the regeneration of the axolotl limb.

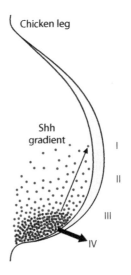

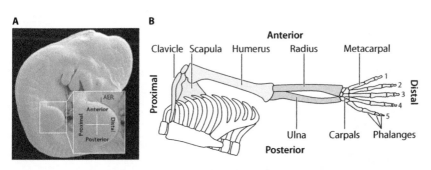

Box 7.4 Figure 1 **Limb development in the vertebrate embryo.** Analysis of vertebrate limb development has been crucial in understanding the regenerative patterns that occur following limb amputation. (**A**) Scanning electron microscopy image of a mouse embryo at gestational day 10.5. Inset: the forelimb bud forms at the level of the heart and is shown here with the two main limb bud axes indicated. The apical ectodermal ridge (AER) is highlighted in green. (**B**) Diagram to show the skeletal elements of an upper limb in the same orientation as the depicted developing embryonic limb bud in (A) (A: Used with permission from Odyssé Michos, PhD. B: From Zeller R, Lopez-Rios J & Zuniga A. (2009) *Nat Rev Genet* 10: 845–858 [doi:10.1038/nrg2681]. Reprinted by permission from Springer Nature.)

Box 7.4 Figure 2 **Limb development in the vertebrate embryo.** Anterio-posterior patterning in the developing limb bud is determined by a gradient of Sonic Hedgehog (Shh) emanating from the polarizing region (purple) at the posterior margin of the AER (beige). The numbers relate to the individual digits as shown in Figure 2 and show how these relate to the specific point in the Shh concentration gradient. (From Tickle C & Barker H (2013) *Wiley Interdiscip Rev Dev Biol* 2: 275–290 [doi:10.1002/wdev.70]. With permission from Wiley.)

transplantation of green fluorescent protein (GFP)-labeled tissues in axolotl to trace the source and fate of particular cells in the blastema and the derived regenerated tissues. They found that each tissue produces progenitor cells with restricted potential. Their conclusion was that regenerating tissues actually consist of diverse lineage-restricted progenitors and can "remember" the tissue from which they originated. For example, transplantation of a segment of skeleton, followed by amputation across the engrafted tissue, led to the appearance of GFP-positive cells in the blastema and then the regenerated limb, but only within the skeletal elements. This pattern of behavior is akin to the unipotent nature of many adult mammalian stem cells.

These transplantation experiments were also able to show that the cells have positional identity, but that this has a degree of cell type specificity (Figure 7.21). So, if the graft of skeletal tissue was derived from the proximal (upper arm) skeleton, then GFP-positive cells were ultimately seen in all parts of the regenerated skeleton. In contrast, if the graft was generated from distal (digit) cartilage, then the GFP-positive

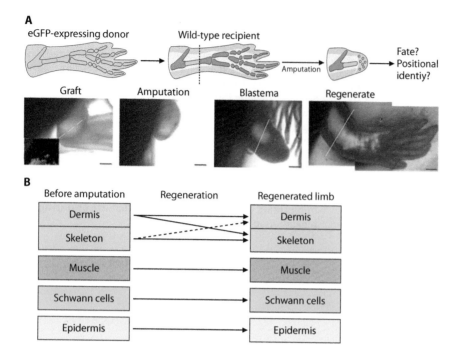

**A**

eGFP-expressing donor → Wild-type recipient → Amputation → Fate? Positional identiy?

Graft    Amputation    Blastema    Regenerate

**B**

| Before amputation | Regeneration | Regenerated limb |
|---|---|---|
| Dermis | | Dermis |
| Skeleton | | Skeleton |
| Muscle | → | Muscle |
| Schwann cells | → | Schwann cells |
| Epidermis | → | Epidermis |

**Figure 7.20 Origin of cells within the blastema.** Transgenic expression of enhanced green fluorescent protein (eGFP) in different cell types of the limb enable the tracking of cell fate. (**A**) Cartilage grafting from a transgenic eGFP-expressing (green) axolotl to a wild type recipient. After subsequent amputation, the contribution of the eGFP-positive cells can be assessed in the developing blastema and regenerated limb. Progression of eGFP-labeled donor tissue through the regeneration experiment. The amputation was made through an area containing the stable cartilage graft (dotted line), and its derivatives were then visualized by whole-mount imaging through stages of regeneration. The inset in the first image indicates that the eGFP-labeled cartilage does include skeletal muscle (red). (**B**) Summary of tissue grafting experiments using various source tissues (muscle, cartilage, Schwann cells, and epidermis). The labeled cells remain restricted to their developmental origin and compartmentalization during blastema formation and limb regeneration. There is little lineage switching, other than dermal cells contributing to the skeleton. (From Poss D. (2010) *Nat Rev Genet* 11:710–722 [doi:10.1038/nrg2879]. Reprinted by permission from Springer Nature.)

cells forming in the blastema were only able to contribute to the regenerated hand but not the upper part of the arm. This retention of positional identity by cartilage-derived blastema cells was not seen, though, for nervous tissue Schwann cells. The positional memory is thought to be the result of a gradient of expression of the Prod1 protein on the surface of the cells along the length of the limb whose effect is mediated by retinoic acid (RA). The level of amputation determines the degree of Prod1 expression in the stump cells and hence defines the regenerative memory of the blastema.

The role of Prod1 is only one example of the many signals and regulatory mechanisms that play a part in the limb regenerative process. Perhaps not surprisingly, these mechanisms are often recapitulations of what normally happens as a limb develops during embryogenesis. Other examples include various epithelial/mesenchymal interactions, such as those mediated by fibroblast growth factors. Both FGF-8 and FGF-10 are expressed within the limb blastema, and similarly to the developing limb bud AER, FGF-8 is expressed in the basal layer of the AEC. Inhibition of the evolutionarily conserved Wnt/β-catenin signaling pathway, which we have already encountered in this chapter in the context of whole body regeneration in several invertebrates, results in no regeneration, whereas activation of β-catenin in non-regenerative limbs of a frog (genus *Xenopus*) results in a regenerative response. In 2007, Jeremy Brockes described nerve dependence as one of the key aspects to regeneration, loss of a nerve input from the limb stump causing a failure in regeneration. Various neurotrophic factors (e.g., glial growth factor and FGF-2) are secreted by axons into the amputated limb, leading to mitotic activity in the blastema and up regulation of genes important for regeneration. A key factor discovered by Brockes is the secreted newt anterior gradient (nAG) protein, which peaks at around 5–7 days after amputation and appears to be crucial for the proliferation of the cells within the blastema.

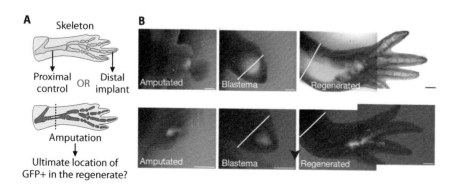

**A**

Skeleton

Proximal control OR Distal implant

Amputation

Ultimate location of GFP+ in the regenerate?

**B**

Amputated    Blastema    Regenerated

Amputated    Blastema    Regenerated

**Figure 7.21 Positional identity.** Lineage tracing experiments also showed that the cells of the blastema have positional memory. (**A**) Cartilage grafting from a transgenic enhanced GFP (eGFP) expressing axolotl to a wild type recipient as described in **Figure 7.20**. In this figure, the experiment entailed comparison of grafts of cartilage from either the proximal (upper arm) or distal (digit) skeleton. (**B**) Whole-mount images of the regenerating limbs show that proximal cartilage contributes to the development of all limb skeletal elements, whereas the distal tissue retained the memory of its site of origin and only gave rise to digits. (From Kragl M, Knapp D, Nacu E et al. (2009) *Nature* 460:60–65 [doi: 10.1038/nature08152]. Reprinted by permission from Springer Nature.)

We have already mentioned the large differences in regenerative capacity that exist between quite closely related species and the fact that this helps reveal crucial features of the involvement of stem cells in lower invertebrates such as planarians and annelids. As we have just described, the vertebrate blastema underlying limb regeneration generally seems to involve dedifferentiation with little or no transdifferentiation and certainly no requirement for resident adult stem cells. However, recent experiments, once more from the laboratory of Elly Tanaka, which compared the origin of regenerated muscle fibers in two different salamanders, a newt (*Notophtalmus viridescens*—see Figure 7.1) and the axolotl, found that muscle dedifferentiation is an integral part of limb regeneration in the newt, but not in axolotl. Very clever use was made of Cre-loxP-based genetic fate mapping (Chapter 2), labeling mature muscle cells with yellow fluorescent protein (YFP) and cells derived from Pax7-positive stem cells with GFP. They found that in the newt, muscle fiber (myofiber) fragmentation results in proliferating, Pax7-negative mononuclear cells in the blastema (YFP+/GFP-) that give rise to the skeletal muscle in the new limb. In contrast, myofibers in axolotl do not generate proliferating cells and do not contribute to newly regenerated muscle; instead, resident Pax7+ cells (muscle **satellite cells** – see Chapter 10) provide the regeneration activity (YFP-/GFP+).

## Mammals show limited regeneration of appendages

Mammals show little if no capacity for regeneration of complete appendages in the way that is seen for the amphibian limb. However, an element of regenerative repair has been documented at the tip of the digits. Following amputation of the digit tip in neonatal mice, blastema formation leading to repair takes place. However, the point at which regeneration can occur is very specific in that it is only possible if amputation occurs distal to the last finger (interphalangeal) joint. This might seem like an encouraging observation if we wish to mimic clinically what we have seen is possible for the amphibian limb, but it seems that digit regeneration in mammals is a distinct and limited process. The source of cells utilized for the regenerative process of the digit tip is actually from the remnant nail bed and requires the expression of bone morphogenetic protein 4 (BMP4) and muscle segment homeobox-1 (msx1). The Msx-1 **knock-out** mouse has a defect in regeneration of the digit tip, which can be rescued with BMP4 expression. During early limb development, BMP signaling and its target Msx-1 are involved in induction of the AER.

## Arthropod invertebrates ranging from crabs to crickets demonstrate a variety of blastemas involved in regeneration

There are many more examples in the animal kingdom of limb replacement involving the formation of a blastema. Among the invertebrates, the arthropods are particularly reliant on autotomy and regeneration, and animals as diverse as crickets and crabs have been used as experimental models to investigate how the blastema is established and generates the new limb. The details of the process may vary (the blastema in the crab is stimulated by signaling involving the ecdysteroid receptor, while that in the cricket limb requires the highly conserved JAK/STAT pathway) or are yet to be explored, and it is beyond the scope of this chapter or book to deal in any more detail with the phylogenetic comparison of regenerative blastemas. Nevertheless, it is clear that the findings in amphibians will likely define common principles that apply more broadly.

## 7.4   REGENERATION IN VERTEBRATES IS NOT RESTRICTED TO LIMBS

Given the stunning capacity of certain amphibians to regenerate near-perfect replacement limbs, it is perhaps not surprising that the vast bulk of research into natural regenerative processes in vertebrates has focused on the mechanisms underpinning this. The other key question following from this, with all its implications for potential clinical application, is why the same is not seen in other vertebrates such as mammals. However, there are a growing number of other circumstances in specific vertebrates of regenerative processes that are both fascinating from an evolutionary and mechanistic perspective and also have possible implications for ways in which biological processes might be manipulated in humans for clinical benefit. One of the most striking examples is the regeneration that can be elicited in the heart of fish following damage, with obvious implications for ischemic cardiac damage, which remains one of the largest causes of premature death in many societies. Another example is the process of skin replacement that accompanies the specialized autotomy response occurring in a few mammals, which again has tantalizing implications if this process could be harnessed for skin repair, especially following traumatic injury or burning.

## The zebrafish is a highly tractable model system for the study of regeneration in certain vertebrate tissues

As the urodele amphibians have been the models of choice in working out how limbs can regenerate, so the zebrafish (*Danio rerio*) has led the way around other structures such as fins, the heart, and the retina. The zebrafish has become a popular laboratory model largely because of the relative ease with which mutagenic and genetic studies can be performed, including the potential to generate temperature-sensitive mutants, thereby avoiding potential developmental lethality, and lineage tracing capabilities that match what is possible in the mouse.

We will take a look at how zebrafish studies have been informative in understanding of retinal regeneration when we consider the wider field of stem cells in the nervous system in Chapter 14. We will also not deal with the considerable volume of work that has been carried out on fin regeneration since for our purposes here, this does not add too much to the concepts of blastema formation and differentiation that we have already elaborated upon around limb regeneration. However, it is worth stressing that the ability to induce and then screen for mutants that affect the ability fully to regenerate fins, both in terms of form and size, has added an extra dimension to the identification of key genes that influence the process.

## Zebrafish exhibits impressive cardiac regeneration following injury to the heart

Where the zebrafish has proven to be the model of choice and has been game changing in terms of our thinking about organ regeneration is with respect to the heart. The findings around the mechanisms used in the fish to replace a lost section of the heart, especially those deriving from Cre-loxP lineage tracing, have considerable implications in the debate about the role of resident cardiac stem cells (Chapter 13) and have generated huge interest because loss of heart tissue contributes significantly to the morbidity of major diseases such as myocardial infarction, which can lead to large reductions in left ventricular muscle mass with no apparent capacity for regeneration.

Zebrafish are able to regenerate their hearts following surgical amputation of the cardiac apex (Figure 7.22), which represents about 20% of the total ventricular mass. The wound is initially sealed by a fibrin clot, which is gradually replaced by regenerating heart tissue rather than the scar tissue that is observed in mammalian cardiac injury. This regeneration relies on the proliferation of cardiomyocytes, which in the normal zebrafish heart are already proliferating at a rate of 3% of the total population, far higher than is seen in mammals, and following injury increases tenfold. The source of these regenerating cardiomyocytes was initially thought to be a resident pool of undifferentiated cardiomyogenic progenitors. However, through use of elegant Cre-loxP-based fate-mapping studies, it has become clear that they derive from pre-existing differentiated cardiomyocytes that dedifferentiate to a more immature proliferative state (expressing the transcription factor GATA4) and, once expanded, differentiate back to fully functional cardiomyocytes, which become fully incorporated into the structure of the heart, including becoming vascularized and electrically coupled.

**Figure 7.22 Heart regeneration in zebrafish.** Zebrafish are able to regenerate their hearts following surgical dissection of the cardiac apex. This diagram shows the regenerative response in the zebrafish heart following partial resection injury (red dashed line). Within a few hours post-amputation (hpa), the retinoic acid (RA) synthesizing enzyme raldh2 (light blue) is induced throughout the endocardium and later the epicardium before these responses localize to the wound by 3 days post amputation (dpa). By 7 dpa, gata4 (green) gene regulatory sequences are activated throughout the cortical muscle layer of the ventricle. Cardiomyocyte proliferation is stimulated and by 14 dpa, vascularization occurs at the wound site. By 30 dpa, a new wall of cardiac muscle has been formed and becomes electrically coupled to the existing cardiac muscle. (From Gemberling M, Bailey T, Hyde D et al. (2013) *Trends Genet* 29: 611–618 [doi: 10.1016/j.tig.2013.07.003]. Copyright 2013 with permission from Elsevier.)

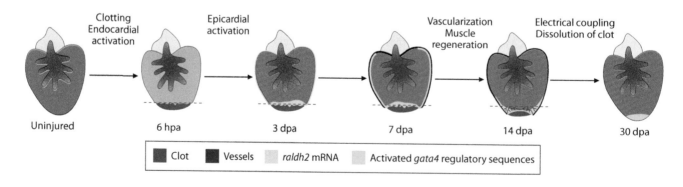

Uninjured    Clotting Endocardial activation — 6 hpa    Epicardial activation — 3 dpa    7 dpa    Vascularization Muscle regeneration — 14 dpa    Electrical coupling Dissolution of clot — 30 dpa

■ Clot    ■ Vessels    *raldh2* mRNA    Activated *gata4* regulatory sequences

## Skin regeneration in mammals: unlocking latent potential

Evidence that regenerative mechanisms have sometimes remained unlocked within mammals, not just fish, comes from observations on the African spiny mouse of the genus *Acomys*. This small mammal sheds large areas of its skin as an autotomy response to attack, which it is able to completely and rapidly regenerate in a scar-less manner. *Acomys* represents one of the only genuses of mammal that display autotomy as a means to evade predation similar to the shedding of the lizard tail, and quite incredibly, it can replace up to 60% of its skin, including new hair follicles, within 30 days (Figure 7.23).

Analysis of the skin of *Acomys* reveals that it is considerably more brittle than the skin of the house mouse (*Mus musculus*) and has a distinct underlying structure that consists mostly of the **extracellular matrix** protein collagen III rather than the aligned fibrils of collagen I that leads to the formation of scars in most mammals. The ability of *Acomys* not only rapidly to regenerate new skin but at the same time form new hair follicles led to the suggestion that the epidermal cells covering the wound tissue established interactions with underlying cells to also promote hair-follicle formation. The signals involved have been shown to be similar to those established during embryonic hair development, and the rapid epithelial crawling and hair follicle regeneration resembled the hallmark of limb regeneration in salamanders.

The mechanisms governing the regenerative ability in *Acomys* have actually been deciphered through analysis of the regenerative response observed following ear puncture. This is achieved in a manner similar to that seen in limb regeneration in the salamander involving the formation of a blastema (Figure 7.24). Following an ear hole punch, there is rapid formation of the non-proliferative epidermal layer through cells crawling (cells crawl through protrusion of the leading edge driven by actin polymerization interacting with the substrate), beneath which the mesenchymal cells proliferate forming the blastema. This mesenchymal tissue has been shown to be rich in fibronectin, tenascin, and collagen III and leads to the hole being filled in by cartilage, adipose, dermal, and epidermal tissue but not by muscle. In comparison, the non-regenerative *Mus musculus* only shows initial, but not persistent, mesenchymal cell proliferation but lacks the presence of collagen III, with cells instead laying down collagen I, leading to the formation of scar tissue. However, interestingly, a mutant strain of *Mus musculus*, known as the Murphy Roths large (MRL) strain, displays a similar capacity for wound closure without scarring. This feature is a genetically definable trait, suggesting that a single gene change might be sufficient to unlock a degree of regenerative ability.

These last two examples of regeneration in fish and mammals suggest that the differences between structures capable or incapable of complete repair are not so enormous and give some hope that it might be possible artificially to manipulate damaged human tissues to release latent mechanisms that could achieve at least a degree of self-healing.

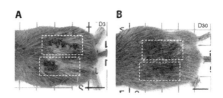

Figure 7.23 **Wound repair in the African spiny mouse.** The African spiny mouse is able to autotomize its skin following predation and then fully to regenerate scar-free skin including new hair follicles. The photographs show an African spiny mouse following autotomized loss of dorsal skin after 3 days (A), and the same region 27 days later (B), revealing that the wounds are no longer visible. Scale bar = 1 cm. (From Seifert A, Kiama S, Seifert M, et al. (2012) *Nature* 489:561–565 [doi:10.1038/nature11499]. Reprinted by permission from Springer Nature.)

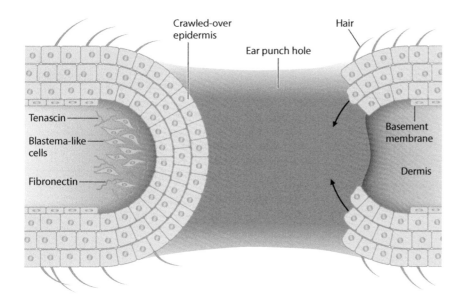

Figure 7.24 *Acomys* **mechanism of regeneration.** The mechanism of regeneration of scar-free wound repair in *Acomys* has been deciphered through analysis of responses in the ear following ear puncture wounding. The diagram illustrates the mechanism of wound repair following ear hole punching. The hole is filled in by regenerating cartilage, adipose tissue, and skin. Rapid reformation of the skin outer layer occurs by crawling epidermal cells, underneath which there is an induction of blastema-like cells together and expression of fibronectin and tenascin. (From Tanaka E (2012) *Nature* 489: 508–510 [doi:10.1038/489508a]. Reprinted by permission from Springer Nature.)

## 7.5    SUMMARY

In this chapter, we have surveyed the occurrence of natural regeneration throughout the animal kingdom. Generally, invertebrate species that arose early in the evolution of animals have the most impressive regenerative capacity, often being able to replace their whole body from just one or a few cells. In such cases, the cells capable of regeneration can be considered to have stem cell characteristics. Species arising later in evolution, leading to modern-day more complex invertebrates and vertebrates, have more limited or even non-existent capacity to regenerate beyond the normal day-to-day replacement of cells in tissues. When regeneration is seen these latter species, it is most often restricted to a specific organ, which is most notably exemplified by the amphibian limb. The cellular and molecular mechanisms of such regeneration appear not to require stem cells but rather involve lineage-specific progenitors or dedifferentiation of mature cells. Understanding the mechanisms of regeneration, especially in the specialized circumstances seen in vertebrates, and defining why this occurs in some species but not others that are closely related, has huge implications for the future development of regenerative therapies in humans.

---

### KEY POINTS

- Regeneration is a natural biological process that spans the animal kingdom, although the extent of its use is less widespread higher up the evolutionary tree.

- Some invertebrates are able to regenerate their whole body using a variety of stem cell types, making them effectively immortal.

- Many of the mechanisms that regulate the invertebrate stem cells responsible for regeneration evolved early in multi-cellular animals and persist in mammalian pluripotent stem cells.

- More complex animals can only regenerate certain structures such as limbs and fins, but rather than utilizing stem cells, regeneration requires the formation of a zone of dedifferentiation, cell proliferation, and differentiation known as a blastema.

- The blastema has been most studied in relation to limb regeneration, particularly in the urodele amphibians (salamanders and newts).

- Lineage tracing, especially using constitutive or tissue-specific activated expression of fluorescent proteins reporters, has provided definitive proof of the origin and fate of cells in the blastema.

- The blastema is formed by dedifferentiation of mature cells to a progenitor state, each cell lineage doing this independently.

- The lineage-restricted cells making up the limb blastema retain a memory of the position of the cell from which they derived.

- Some vertebrate structures other than limbs can regenerate, notable among which is the heart in zebrafish, especially since this suggests that there may be a latent potential to regenerate in animals that do not normally exhibit such a characteristic.

- Mammals show limited ability to regenerate, including the tips of digits in neonates and the skin in a few species or strains of mice.

---

## FURTHER READING

### General

Poss, K.D. (2010) Advances in understanding tissue regenerative capacity and mechanisms in animals. *Nat. Rev. Genet.*, 11: 710–722.

Reddien, P.W. (2013) Specialized progenitors and regeneration. *Development*, 140: 951–957.

Sugimoto, K., Gordon, S.P., Meyerowitz, E.M. (2011) Regeneration in plants and animals: Dedifferentiation, transdifferentiation, or just differentiation? *Trends Cell Biol.*, 21: 212–218.

### Limb regeneration

Choi, Y., Meng, F., Cox, C.S., Lally, K.P., Huard, J. and Li, Y. (2017) Regeneration and regrowth potentials of digit tips in amphibians and mammals. *Int. J. Cell. Biol.*, 2017: 5312951.

Haas, B.J. and Whited, J.L. (2017) Advances in decoding axolotl limb regeneration. *Trends Genet.*, 33: 553–565.

Kragl, M., Knapp, D., Nacu, E., Khattak, S., Maden, M., Epperlein, H.H. and Tanaka, E.M. (2009) Cells keep a memory of their tissue origin during axolotl limb regeneration. *Nature*, 460: 60–65.

Seifert, A.W. and Muneoka, K. (2018) The blastema and epimorphic regeneration in mammals. *Dev. Biol.*, 433: 190–199.

Tanaka, E.M. and Reddien, P.W. (2011) The cellular basis for animal regeneration. *Dev. Cell*, 21:172–185.

### Zebrafish regeneration

Sehring, I.M., Jahn, C. and Weidinger, G. (2016) Zebrafish fin and heart: What's special about regeneration? *Curr. Opin. Genet. Dev.*, 40: 48–56.

# Plant stem cells

# 8

It would be quite easy for a student learning about stem cells to think that everything we know has been gleaned from studies on animal models and humans, and in fact up to this point in the book, we have indeed restricted our discussion to the animal kingdom. However, it would be very remiss of us if we did not now introduce the plant kingdom because an understanding of how stem cells contribute to the life of a plant is both amazing in terms of the scale of what stem cells can achieve as well as being a highly interesting comparative exercise to help define the mechanisms that determine stem cell characteristics and how these may have evolved. Knowledge about the nature of plant stem cells and the ways in which they determine and support plant tissues will undoubtedly have ramifications for horticulture and industrial application (Chapter 19), including modification of plant features such as flower shape and color or seed yields.

We will start this chapter by looking at the need for stem cells in plants and show that the variety is much more limited compared to animals, being essentially restricted to three types, at least in the flowering plants. Then we will detail some of the molecular features of stem cell regulation in plants, which are on the face of it completely different from those seen in animal stem cells (Chapter 4), probably having arisen independently during evolution, but nevertheless revealing some fundamental principles about the mechanisms required to establish the basic characteristics of **self-renewal** and control of **commitment**. We will discuss the **plasticity** of plant cells and the extent to which they can be **reprogrammed** to a stem cell state, reminiscent of the situation we have described for **induced pluripotent stem (iPS) cells** (Chapter 5) but suggesting that there may be an inherent flexibility in plants dictated by the relatively greater influence of the **niche** compared to what is seen in animal tissues (Chapter 6). Finally, we will reflect on the comparison between plant and animal stem cells from a variety of perspectives.

## 8.1 THE NATURE OF STEM CELLS IN PLANTS

We have seen in Chapter 7 that it is likely that during evolution, the origin of multicellularity and specialization of cells with distinct properties required for the functioning of primitive tissues was accompanied soon after by the evolution of stem cells. The very fact that an analogous step of specialization accompanying a multicellular existence occurred in the evolution of plants, which had diverged from those branches that would lead to the animal kingdom way back in geological time at a stage of single-cell life forms, implies that stem cells also became necessary at that point. In this section, we will look at the way in which stem cells are involved in the life of plants, how they can be defined, and if the same rigor can be applied as is expected in the study of animal stem cells.

### Plants have a life-long requirement for stem cells

Plants are essential for all life through their ability to use the energy from sunlight to generate carbohydrates from water and carbon dioxide. The evolution of plants (Figure 8.1) has culminated in three broad categories that are characterized by apical growth, namely the simple, rootless non-vascular plants (the bryophytes, which

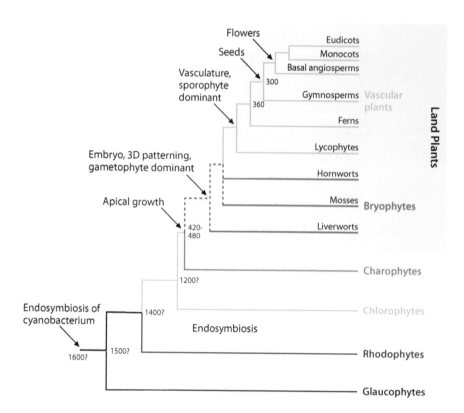

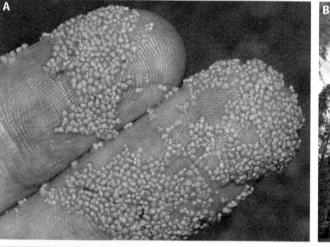

Figure 8.2 **The diversity and extremes
of the plant world.** There are claims and
counter-claims for which plants beat the
records, but irrespective of who is right, the
size and age extremes that can be achieved
are impressive. (**A**) The smallest flowering
plants, belonging to the genus *Wolffia*,
are minute rootless plants that float at the
surface of water, an average individual plant
being 0.6 mm long and 0.3 mm wide. (**B**) The
giant sequoia (*Sequoiadendron giganteum*)
grows on the western slopes of the Sierra
Nevada and can certainly claim to be the
most massive plant; the largest specimen,
nicknamed General Sherman, is 83 m tall with
a trunk 11 m in diameter and weighs nearly
1400 tons. (**C**) The bristlecone pine (*Pinus
longaeva*), living high in the mountains of
California, can live for nearly 5000 years. (**D**)
Many plants can grow as clones, and these
add a whole new dimension to the question
of longevity. One such plant that grows
as a clone is the quaking aspen (*Populus
tremuloides*) in North America, which can
spread over hundreds of hectares and, in
the case of the clone illustrated (the trees
with yellow foliage), can have been growing
for up to 80,000 years (A: "WolffiaArrhiza2"
by Christian Fischer – Own work. Licensed
under Creative Commons Attribution-
ShareAlike 3.0 Unported via Wikimedia.
(http://creativecommons.org/licenses/
by-sa/3.0/deed.en). B: "General Sherman tree"
by Kimon Berlin. Licensed under Creative
Commons Attribution-ShareAlike 2.0 Generic
via Wikimedia. (https://creativecommons.org/
licenses/bysa/2.0/deed.en). C: "Big bristlecone
pine" by Dcrjsr – Own work. Licensed under
Creative Commons Attribution-ShareAlike
3.0 Unported via Wikimedia. (http://
creativecommons.org/licenses/by-sa/3.0/
deed.en). D: Photo credit: Mark Muir, USDA.)

include liverworts, hornworts, and mosses); vascular seedless plants; and vascular seed producing plants, which are divided into gymnosperms (e.g., conifers) and angiosperms (flowering plants). Plants are amazingly diverse, with something like 300,000 distinct species, and their range of sizes equals or exceeds that seen in the animal kingdom, from the diminutive duckweed (*Wolffia arrhiza*), measuring about 600 μm, to the gargantuan giant redwood (*Sequoiadendron giganteum*), which grows up to 85 m in height and 7 m in diameter (Figure 8.2). Many plants also have incredible life spans, way in excess of any animal. The record holder is the great bristlecone pine from California and Nevada, some specimens of which are in excess of 5000 years old, although even these are veritably youthful compared to some plants that grow as clonal colonies, that is, by vegetative growth in the absence of sexual reproduction. The age of such clonal plants can be mind boggling, one claim for the record being a colony of 47,000 quaking aspens in the Fishlake National Forest in the United States, which has an estimated age of 80,000 years.

Plant cells share features common to all eukaryotes but also have characteristics related to their plant-specific functions (Box 8.1). Although in many ways different at

## BOX 8.1  BACKGROUND: PLANT CELLS

The three categories of land plants (bryophytes, seedless plants, and seeded plants) range in complexity from relatively simple mosses through to the most evolved structures that characterize the flowering plants. As in animals, cell specialization underpins the development of the various tissues of a plant, including the cells that make up the leaves and flowers, shoots, roots, and a complex network of transport vessels, constituted by the xylem and phloem. Each of these tissues has its own complex architecture. Reflecting their specialized functions, many of which are plant-specific, plant cells differ in several key aspects compared to the cells of other eukaryotic organisms (Figure 1). Generally, plant cells have a more geometric, usually rectangular shape that is supported by a cell wall constituted of polysaccharides, the most important of which is cellulose. The cell wall lies outside of the cell membrane but contains pores known as plasmadesmata

that allow cell-cell communication through shared plasma membranes and endoplasmic reticulum. The central region of a plant cell commonly contains a large vacuole that maintains cellular rigidity through the pressure of water within, as well as serving to traffic and store various molecules, although stem cells in plants differ by having small vacuoles. Like other eukaryotes, the cytoplasm of the plant cell contains mitochondria, organelles with their own genome that are presumed to be of prokaryotic origin that became incorporated through a symbiotic association. Plant cells also contain other specialized organelles (so-called "plastids") for storage and synthesis (e.g., of starch and pigments) and most importantly the chloroplast, which contains chlorophyll, a green-colored pigment that absorbs sunlight and allows the plant to make its own food through the process of photosynthesis. Plastids are also thought to have been incorporated through a symbiotic association.

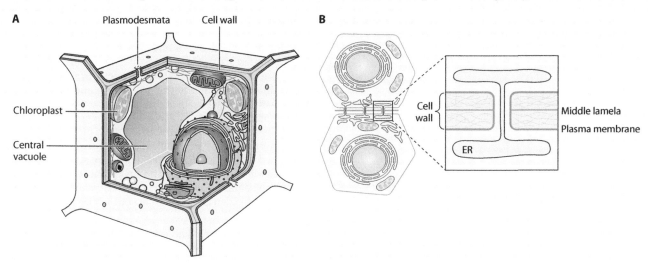

Box 8.1 Figure 1 **Plant cells.** Plant cells share all of the features with their animal counterparts that define a eukaryotic cell but have a number of specific adaptations. (**A**) Diagram of the structure of a plant cell showing features that are not seen in animal cells. These include the central vacuole, chloroplasts, cell wall, and plasmodesmata that are formed by plasma membrane connections between adjacent cells, which penetrate the cell wall. (**B**) The figure illustrates how two plant cells can be connected through their plasmodesmata. Plasmodesmata contain a shared portion of the endomembrane system (yellow), which is thought to be important for the transmission of intercellular signals. The inset shows details of the plasmodesmata connection. The cell wall is made up of polysaccharides such as cellulose (indicated by blue lines), with a central pectin-based cell plate, called a middle lamella, which connects the cell walls of adjacent cells (brown line). Molecules can pass between adjacent cells either through the joined cell walls or through the plasmodesmata link. (B: From Sparks E, Wachsman G & Benfey P. (2013) *Nat Rev Genet* 14: 631–644 [doi: 10.1038/nrg3541]. Reprinted by permission from Springer Nature.)

the cellular and organismal levels, plants and animals have the same requirements for growth, tissue maintenance, and repair and the ability to respond to environmental stresses. As in animals, these needs in plants are often underpinned by stem cells. Many of the demands placed on plant stem cells are the same in terms of ability to self-renew and **differentiate** in a controlled manner, but the scale and longevity of some species pose extra challenges.

The range of distinct differentiated cell types and their organization into functional structures varies between the three categories of plants and has a bearing on their requirements for specialized cells, analogous to stem cells in animals, which have the potential to maintain growth throughout a plant's lifetime. Depending on the complexity of the plant, ranging from the more primitive bryophytes to the most highly evolved flowering plants, it requires one or more growth and differentiation tissues, the so-called **meristems**. Cells within the meristem are able to proliferate and are relatively undifferentiated, including the most immature cells that are the ones often likened to stem cells in animals. As we will see later in this chapter, meristems present in regions of growth and differentiation share common features but also have location-specific characteristics, the most complex plants requiring meristems for the development of roots, the vasculature and stems, shoots, leaves, and flowers. Among those plants that lack any vascular system and do not produce flowers for their reproduction, there is still a reliance on the existence of stem cell populations for survival and growth, although the nature of the stem cells that exist within these is not as well characterized as for the flowering plants.

## The meristem is a structure incorporating both the stem cells and their niche

Meristems characterize the growth and development of tissues such as leaves, flowers, stems, and roots. Based predominantly on research performed on the model organism *Arabidopsis,* which is an unimpressive flowering plant that many of us regularly pull out from our gardens as a weed (Box 8.2), the nature of meristems has

---

**BOX 8.2 ENABLING TECHNOLOGY: *ARABIDOPSIS THALIANA* AS A MODEL ORGANISM**

Similar to the role that model organisms play in research on animal biology, a flowering cress related to cabbages and mustard has emerged as the organism of choice for genetic studies in plants. Thale cress, better known by its Latin name *Arabidopsis thaliana* (Figure 1), was advocated as a suitable

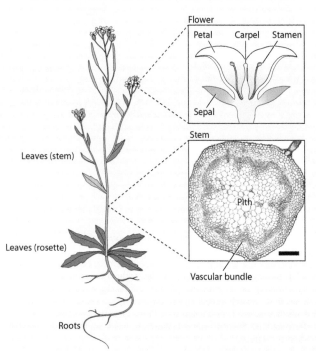

Box 8.2 Figure 1 **Arabidopsis thaliana.** An illustrative diagram of an adult *Arabidopsis thaliana* plant depicting essential organs common to most flowering land plants, including roots, stem, leaves (rosette and stem), and flowers. The top inset depicts the general structure of the flower, showing the location of the stamen, carpel, petal, and sepal. The lower inset shows a cross section through a stem to highlight the central pith and vascular bundles. (From Clark S (2001) *Nat Rev Mol Cell Biol* 2: 276–284 [doi:10.1038/35067079]. Reprinted by permission from Springer Nature and Barra-Jimenez and Ragni (2017). *Curr Opin Plant Biol* 35: 145–151. With permission from Elsevier. Photo: From Barra-Jiménez A & Ragni L. (2017) *Curr Opin Plant Biol* 35: 145–151 [doi: 10.1016/j.pbi.2016.12.092]. Copyright 2017 with permission from Elsevier.)

model for plant developmental studies by Friedrich Laibach, who in 1945 published a collection of X-ray-induced genetic mutants. Akin to the use of the fruit fly in animal research, *Arabidopsis* meets the requirements as a suitable model organism due to its small size and simple growth requirements within the laboratory. Adding to these useful characteristics, it is a self-fertilizing plant, producing thousands of seeds from just one individual. It takes just 6–8 weeks for a seed to go from germination through to the production and maturation of the plants' first seeds. In addition to these desirable features for a model organism, its role as the preferred model plant was sealed in 1986 by the publication of a method for genetic transformation, which enables easy, targeted alterations that have led to the creation of over 300,000 insertion mutants. In 2000, *Arabidopsis* became the first plant to have its entire genome sequenced (which is relatively short at about 150 megabases), identifying in excess of 25,000 genes.

Figure 8.3 **The major stem cell-containing meristems of *Arabidopsis thaliana*.** The image on the left is of an *Arabidopsis thaliana* plant 25 days after germination. The location and histology of the major meristems are shown on the right as close-up microscopical views. (From Greb T & Lohmann J. (2016) *Curr Biol* 26: 816–821 [doi: 10.1016/j.cub.2016.07.070]. Copyright 2016 with permission from Elsevier. And From Gaillochet C & Lohmann J. (2015) *Development* 142: 2237–2249 [doi: 10.1242/dev.117614]. Published under the terms of the Creative Commons Attribution License (https://creativecommons.org/licenses/by/3.0.)

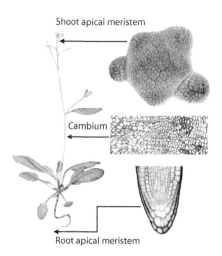

Shoot apical meristem

Cambium

Root apical meristem

been defined in considerable detail both in terms of the arrangement of the stem cells and the niche components and also, as we will see later, from the perspective of the precise molecular mechanisms at play.

Even in the most highly evolved plant species, as typified by *Arabidopsis*, stem cells appear to be restricted to three principal types of meristem: two supporting the growth at the apical tips, that is the shoots and roots, and one enabling the radial growth of stems in the tissue known as the cambium (Figure 8.3). We will summarize what is known about each of the three meristems, which demonstrate both similarities and distinct features. **Apical meristems** are the most undifferentiated structures present in the buds and growing tips of roots and can in turn give rise to other meristem types that we will look at briefly later. The basic structure of an apical meristem consists of layers and zones. The stem cells in a meristem are small and lack many of the features that characterize differentiated plant cells, including having only a very thin cell wall and extremely small vacuoles and lacking fully formed plastids in their cytoplasm. Comparative morphological studies of apical meristems of species across the plant kingdom have revealed distinct homologies that imply that the more complex apical meristems evolved from the simpler forms (Figure 8.4), although gene expression comparisons between these analogous regions suggest that there are both shared and species-specific features in these stem cell zones.

## The shoot apical meristem supports the growth and development of shoots, leaves, and flowers

In vascular plants, an apical meristem is associated with the growth of shoots (the **shoot apical meristem, SAM**). The SAM gives rise directly to shoots but can also transform to produce detached meristems that support the development of leaves and flowers. The outermost cell layers of the SAM are called the tunica, which cover the innermost layer constituting the corpus. The structure can be further divided into a central zone (CZ) that contains slowly dividing cells essential for meristem maintenance and peripheral (PZ) and rib (RZ) zones that usually have considerably greater rates of proliferation. The precise number and organization of layers and zones vary depending on the particular meristem and type of plant.

In flowering plants like *Arabidopsis*, the SAM is made up of three layers that constitute epidermis (L1), subepidermus (L2), and internal mass (L3) (Figures 8.5 and 8.6). Monocotyledons (e.g., grasses) have two layers, while gymnosperms have only a

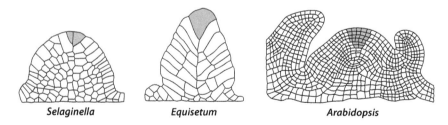

*Selaginella*      *Equisetum*      *Arabidopsis*

Figure 8.4 **Apical growth structures in land plants.** Apical meristems are present in a wide range of land plants, from simple mosses through to the most complex flowering species. Shown are diagrammatic representations of longitudinal sections through the shoot apical growth regions in three species illustrating the difference in structure of the basic apical growth structures from a seedless spikemoss (*Selaginella*) compared to the meristems in more complex land plants, including a horsetail (*Equisetum*) and a seed-forming cress (*Arabidopsis*). The stem cell-containing regions are shaded in pink. (From Banks J. (2015) *New Phytologist* 207:486–487 [doi:10.1111/nph.13525]. With permission from Wiley.)

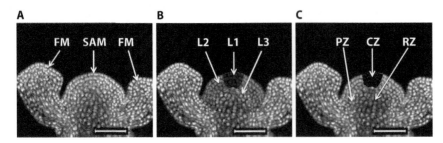

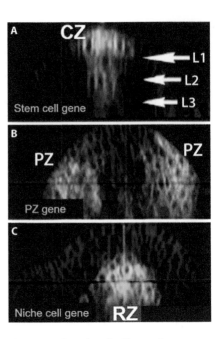

**Figure 8.5 Structure of a shoot apical meristem.** Laser scanning confocal microscope optical cross section through an apical meristem of *Arabidopsis thaliana* stained with propidium iodide to highlight the nuclei of individual cells. (**A**) The position of the shoot apical meristem (SAM) and adjacent floral meristems (FM) are highlighted. (**B**) The arrangement of the apical meristem into three layers is shown. The epidermal layer (L1, red) and subepidermal layer (L2, orange) overlay the innermost layer, also known as the corpus (L3, yellow). The L1 and L2 are together called the tunica. (**C**) The SAM can be divided into three functional zones, namely the central zone containing the stem cells (CZ, purple); the peripheral zone (PZ, green), where organ primordia are initiated; and the rib zone (RZ, blue) that gives rise to the internal part of the growing stem. (From Meyerowitz E. (1997) *Cell* 88:299–308 [doi:10.1016/s0092-8674(00)81868-1]. Copyright 1997 with permission from Elsevier.)

single layer, and in more primitive species, such as the mosses, all cells derive from a single apical cell (Figure 8.4). The individual characteristics of the layers in the SAM appears to be necessary to impose stem cell activity giving rise to the specific layers of plant tissue, namely epidermal **progenitor** cells (L1), subepidermal tissues and gametes (L2), and the majority of the ground tissue (L3). The stem cells located within each layer exhibit distinct properties with regard to their mode of division. Stem cells located within L1 and L2 generally undergo anticlinal division (division is perpendicular to the cell surface) meaning that the daughter cells remain in the same layer as the parent cell. In contrast, stem cells within L3 tend to divide in all directions but do not usually invade L2. The importance of the layered structure in the SAM has been demonstrated by layer removal followed by assessment of the effect on the stem cells within the other layers. For example, removal of L1 by laser ablation in the tomato plant SAM had profound effects on the orientation of division of the stem cells in the layers beneath, causing L2 stem cells to shift from anticlinal to periclinal (at right angles to anticlinal) division and leading to local tissue outgrowth and failure to initiate organogenesis. These observations imply that the presence of L1 provides essential signals for the maintenance of stem cell function and differentiation throughout the apical meristem.

In addition to the layered organization of the SAM, stem cell activity is also located in distinct zones of the SAM (Figure 8.7). The CZ harbors all of the stem cell activity of the SAM and consists of slowly cycling stem cells whose progeny become displaced into the neighboring PZ, where they can then differentiate in sequence to generate lateral organs. Progeny of the CZ that are displaced around the OC into the RZ differentiate and integrate as part of the growing stem. Although the RZ does not house stem cell activity, there is a subset of cells within it that make up the organizing center (OC), which is essential in the provision of cues for stem cell specification.

**Figure 8.6 Functional cell types in a meristem.** Gene expression characterizes the different cell components of a meristem. Shown are fluorescent microscopy images of cross sections through the shoot apical meristem in which specific transgenic fluorescent reporters have been used to reveal the location of cells with a particular function. Each of these reporters is controlled by the regulatory elements for a gene that is characteristic of the region-specific cells (green). The central zone (CZ) is marked by cells expressing *CLAVATA 3* (**A**), the peripheral zone (PZ) by cell expression of *UNUSUAL FLORAL ORGANS* (**B**), and the rib meristem (RM) by cells expressing *WUSCHEL* (**C**). Cell outlines are highlighted in red by staining with the lipophilic dye FM4-64, the arrows highlighting the position of the three structural layers (L1-3). (From Xie M, Tataw M & Venugopala Reddy G (2009) *Semin Cell Dev Biol* 20: 1126–1133 [doi: 10.1016/j.semcdb.2009.09.014]. Copyright 2009 with permission from Elsevier.)

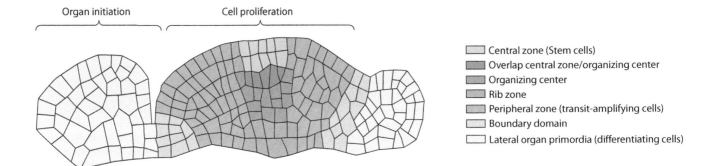

Central zone (Stem cells)
Overlap central zone/organizing center
Organizing center
Rib zone
Peripheral zone (transit-amplifying cells)
Boundary domain
Lateral organ primordia (differentiating cells)

**Figure 8.7 The shoot apical meristem.** Diagrammatic representation of a longitudinal section through the shoot apical meristem, showing the different functional domains. (From Gaillochet C & Lohmann J. (2015) *Development* 142: 2237–2249 [doi: 10.1242/dev.117614]. Published under the terms of the Creative Commons Attribution License (https://creativecommons.org/licenses/by/3.0).

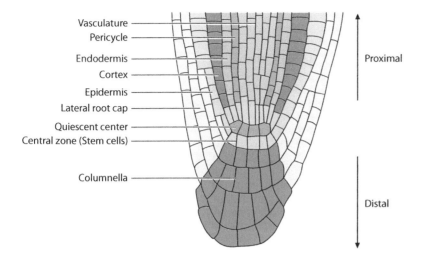

Vasculature
Pericycle
Endodermis
Cortex
Epidermis
Lateral root cap
Quiescent center
Central zone (Stem cells)
Columnella

Proximal

Distal

Figure 8.8 **The root apical meristem.**
Diagrammatic representation of a longitudinal section through the root apical meristem, showing the different functional domains. (From Efroni I, Mello A, Nawy T et al. (2016) *Cell* 165: 1721–1733 [doi:10.1016/j.cell.2016.04.046]. Copyright 2016 with permission from Elsevier.)

## The root apical meristem type supports the growth and development of roots

The root apical meristem (RAM) serves an analogous function to the SAM and has a similar organization of stem cells (Figure 8.8). In addition to producing roots, the RAM gives rise to meristems that develop the epidermis of the stem as well as the tissues of the xylem and phloem constituting the vascular transport system and elements that provide the structural support of the plant. The center of the root tip contains a small group of cells called the **quiescent** center (QC) that provides functions comparable to that of the OC in the SAM. Stem cells of the RAM (also called initials) surround the QC and give rise to differentiated cells that make up the stele (the central vascular tissue and surrounding cells), ground tissue (comprising endodermal and cortex components), epidermis, lateral root cap, and columella (the very tip of the root). Through **asymmetrical** divisions, RAM stem cells produce daughter stem cells that become progressively distant from the stem cell niche and differentiate. There are two identifiable pools of stem cells: the proximal stem cell daughters and the distal stem cell daughters. The proximal stem cells give rise to a **transit amplifying** population in which extra rounds of cell division take place, forming the meristem. These cells make up the epidermal, ground tissue, and vascular cell types, which, during differentiation, expand along the main axis carrying the stem cell niche forward with the root tip. In contrast, distal stem cells do not divide again but instead differentiate into starch-containing cap cells that eventually detach from the root tip as a new layer is formed. Comparison of stem cells between the two apical meristems shows that unlike SAM stem cells that are able to give rise to all aerial tissues, those in the RAM appear to be of limited potential, with a specific stem cell for each tissue type making up the root.

## Specialist structures including leaves and flowers have independent meristems derived from the shoot apical meristem

In addition to maintaining growth of a plant, stem cells are equally important in the establishment of peripheral organs, such as the leaves and flowers. Daughter stem cells within the SAM that are displaced into the peripheral zone undergo differentiation into leaves, side shoots, or flowers (Figure 8.9). Floral meristems contain transient stem cell populations and are therefore termed determinate meristems, as opposed to a SAM, which is described as being indeterminate. Floral meristems are maintained for just a limited period of time, as they are required only for as long as the flowers take to form and soon after are consumed in the production of the innermost floral constituents, such as the carpels. In *Arabidopsis*, stem cell activity within the floral meristem terminates after the production of 16 floral organs in four whorls. During plant growth, multiple secondary growth axes can also develop through formation of additional meristems, for example, to give lateral roots or branches, which are derived from either the root pericycle (a layer of cells encircling the vascular tissue) or the leaf axil, respectively.

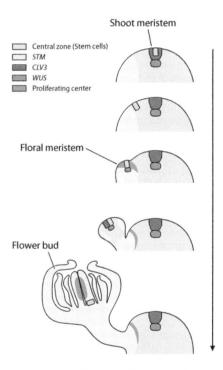

Central zone (Stem cells)
STM
CLV3
WUS
Proliferating center

Shoot meristem

Floral meristem

Flower bud

Figure 8.9 **Meristem termination and flower development.** A floral meristem develops from a cell (blue) deriving of the shoot apical meristem, which is displaced sideways to the periphery of the shoot. Here it becomes part of a newly forming floral meristem and is incorporated into a flower. Expression of the meristem marker *STM* is shown in bright green, expression of the stem-cell marker *CLV3* in red, and expression of *WUS* in purple. The stem cell derivative that goes to form the floral meristem first loses expression of *CLV3* as it leaves the shoot meristem but reacquires it in the floral meristem, where it eventually once more loses its stem cell identity as it is displaced from the proliferating center (bottom, indicated in green) of the developing flower. (From Weigel D & Jürgens G. (2002) *Nature* 415: 751–754 [doi:10.1038/415751a]. Reprinted by permission from Springer Nature.)

## Plant vasculature located in the stems is derived from stem cells in the cambium meristem

Following the discovery of stem cells within the SAM and RAM, it was proposed that there is a population of stem cells within a specified niche that are able to produce and maintain the xylem and phloem in those plants with a vasculature. Cross sections of the *Arabidopsis* stem show a distinct ring patterning of vascular tissue with phloem on the outer side of the ring and xylem on the inside. Plants that exhibit continual lateral growth form stems with a ring of cells called the cambium, which lies between the phloem and xylem (Figure 8.10). Studies on the western balsam-poplar (*Populus trichocarpa*, the first tree to have its genome sequenced) revealed that the stem cells capable of giving rise to cells of both the phloem and xylem reside within the cambium regions of the vascular niche. The location of the stem cells within the cambium between the two types of descendant tissues (Figure 8.11) is critical for the production and precisely organized structure of the phloem and xylem and is achieved by regulating the orientation of stem cell divisions and feeding of the newly derived progeny into differentiation pathways.

## Species originating at earlier points in evolution than the vascular plants might reveal how stem cells emerged in plants

Although much of the knowledge of plant stem cell character and function has come from *Arabidopsis*, studies on the stem cells from other subgroups of land plants can also give insight into the evolutionary emergence of stem cells within the plant kingdom. A basic apical growth mechanism probably evolved in the algal ancestors of land plants and was elaborated into the apical cell type meristems seen in the bryophytes (Figure 8.4). Almost all stem cell research on non-flowering land plants has been carried out using the moss *Physcomitrella patens* (Box 8.3). This moss forms a hypha-like body (protonema) and a shoot-like body (gametophore) from a protonema apical cell and a gametophore apical cell, respectively. These apical cells have stem cell characteristics. In the haploid generation, seven stem cell types have been described that produce the different tissues and organs. The first division of a spore forms the chloronema apical stem cell that contributes toward tip growth and forms the chloronema cells that constitute the filamentous body of the moss. After several days of cultivation, these chloronema apical stem cells transform into caulonema apical stem cells. Gametophore apical stem cells arise from approximately 5% of the caulonema apical stem cells to produce the cells that form the stem, ultimately producing leaf apical stem cells and rhizoid apical stem cells. At the tip of the gametophores, the reproductive organs archegonia (egg) and antheridia (sperm) are formed from a single stem cell.

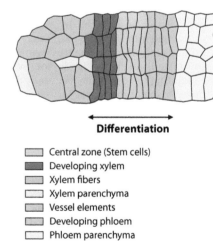

**Differentiation**

- Central zone (Stem cells)
- Developing xylem
- Xylem fibers
- Xylem parenchyma
- Vessel elements
- Developing phloem
- Phloem parenchyma

**Figure 8.11 Cambium meristem.** A schematic cross section of the cambium area in the *Arabidopsis* lower stem. The different cell types forming the functional domains found in the cambium are indicated, including the cells of the phloem toward the center of the stem and xylem on the outer circumference. (From Greb T & Lohmann J. (2016) *Curr Biol* 26: 816–821 [doi: 10.1016/j. cub.2016.07.070]. Copyright 2016 with permission from Elsevier.)

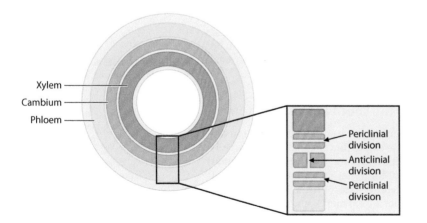

**Figure 8.10 Stem cells in the *Arabidopsis* vasculature.** Schematic illustration of a cross section through the stem of *Arabidopsis*. Following the development of the primary stem, the initially discrete vascular bundles merge during the secondary developmental phase, cells between the vascular bundles forming concentric rings of phloem (yellow) and xylem (orange) on either side of the cambium (purple). Inset: stem cells are located within the cambium and undergo both anticlinal and periclinal divisions, respectively, to self-renew or differentiate toward either the phloem or xylem. (From Miyashima S, Sebastian J, Lee J et al. (2013) *EMBO J* 32: 178–193 [doi: 10.1038/emboj.2012.301]. With permission from Wiley. © 2013 European Molecular Biology Organization.)

## BOX 8.3 BACKGROUND: THE MOSS *PSYCOMITRELLA PATENS*

The mosses have an interesting phylogenetic position halfway between algae and seed plants, originating roughly 500 million years ago, and can be regarded as a basal lineage of land plants, having diverged before the acquisition of well-developed vasculature. As an experimental model, the moss *Psycomitrella* is small and easy to cultivate and has a dominant haploid phase. Uniquely among plants, it exhibits efficient homologous recombination, enabling reverse genetics in a way that is more familiar perhaps in mouse embryonic stem cells. Genetic approaches were greatly facilitated by the determination of the genome sequence in 2006, which revealed a possible 36,000 genes within the 500 Mbp making up the 27 chromosomes of the moss. Comparison with the genome sequence of the algal ancestors of the moss showed that there were significant changes concomitant with the evolutionary movement to land,

including a general increase in gene family complexity, loss of genes associated with aquatic environments, acquisition of genes for tolerating terrestrial stresses (e.g., variation in temperature and water availability), and the development of the auxin and abscisic acid signaling pathways for coordinating multicellular growth and dehydration response. As well as the advantageous genetic approaches resulting from the dominance of the gametophyte in its life cycle (Figure 1) and the ease of genetic modification, the usefulness of *Psycomitrella* as a model organism, especially with respect to stem cell biology, is also facilitated by its relatively simple developmental pattern, its suitability for cell lineage analysis, and similar responses to plant growth factors and environmental stimuli to those observed in other land plants.

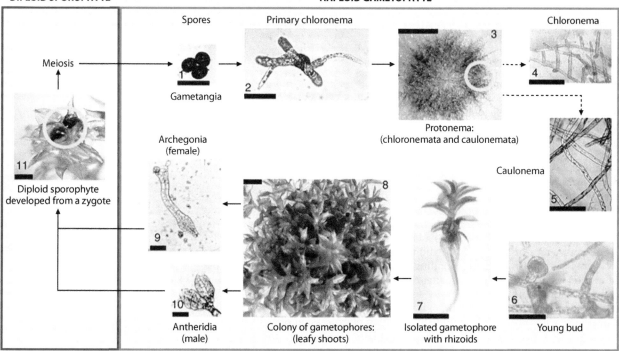

Box 8.3 Figure 1 **Life cycle of the moss *Physcomitrella patens*.** The lifecycle of *Physcomitrella* is completed under optimal conditions in less than 12 weeks. During the haploid gametophyte phase, spores (1) germinate in the presence of light into primary chloronemata (2), and development continues to the protonema phase (3), depicted here as a 15-day-old colony. Tip growth (from the region within the yellow circle) occurs from protonema yielding branching chloronema (4) and caulonema (5) cells. Young buds (6) forming from the filamentous two-dimensional structures produce a leafy shoot with developing rhizoids of the gametophores (7) and ultimately a fully developed moss (8) at 4 weeks. From the mature moss, the gametophores can give rise to female archegonium (9) and male antheridia (10), both of which are present on the same plant after exposure to reduced temperatures. After fertilization, the diploid sporophyte (11, yellow circle) develops from the zygote, undergoing meiosis within its capsule, leading to the formation of spores and the beginning of a new cycle. (From Schaefer D & Zrÿd J-P (2001) *Plant Physiol* 127: 1430–1438 [doi: 10.1104/pp.010786]. Permission conveyed through Copyright Clearance Center, Inc.)

## 8.2    THE REGULATION OF PLANT STEM CELLS

We have seen already that meristems are highly organized structures, containing a hierarchy of stem cells that essentially create their own niche through the generation of differentiated progeny. Such an organized structure of cells requires that each cell be able to assess its current position, monitor any changes, and adjust its behavior accordingly. This is no different than the many examples described throughout this book in which **adult stem cells** in animal tissues rely on their niche for communication with both their immediate and more distant environments to be able to respond as demand for cell production dictates.

Understanding the mechanisms that maintain stem cells in plants is of considerable interest in its own right but also has wider implications for understanding of stem cells in animals. Several advantages in terms of relative simplicity and highly defined location, combined with good genetics models, have meant that it has been possible over the last three decades to make great advances in the understanding of the mechanisms that control stem cell number, position, and fate in plant meristems. If anything, the definition of these mechanisms, including details of the signaling pathways and niche components, is far more advanced than for most adult stem cells in animals.

Key to the regulation of plant meristematic stem cells is the interplay between soluble signaling molecules and transcription factors, many of which are part of transcription factor families that have homologs in animals. These factors, again as is seen in animals, are often within feedback loops and part of a network of signaling molecules that is modulated by **epigenetics**, micro RNAs, and hormone action. The degree of complexity of stem cell regulation that has been worked out in plant meristems, especially in the SAM, is quite astonishing and far beyond what we can cover in this chapter. We will illustrate some key features of the types of mechanisms employed simply by focusing on two proteins, one a soluble hormone and the other a transcription factor, which interact closely to dictate stem cell behavior in the SAM.

## Genetic analysis of the shoot apical meristem in *Arabidopsis* has revealed how hormones and transcription factors produced in the niche influence stem cell activity

The best-studied plant stem cell niche from the perspective of the underpinning molecular mechanisms is that in the *Arabidopsis* SAM, for which the crucial component is a negative feedback loop involving a transcription factor expressed in the niche OC and a secreted peptide that is produced by the stem cells of the three layers in the CZ of the meristem. The discovery of this control mechanism provides a perfect illustration of the way in which an elegant genetic approach using a very tractable model organism can be used to identify the key molecular players and deduce how they interact with one another.

The main protagonists in the regulation of the SAM are *CLAVATA3* (*CLV3*), a 96 amino acid protein that is the precursor of a 12 amino acid secreted peptide hormone, and *WUSCHEL* (*WUS*), which is a homeodomain transcription factor. Mutation in *CLV3* reducing its activity leads to a larger meristem and increased flower size, while plants over expressing *CLV3* exhibit the opposite phenotype, that is, smaller meristems and flowers and correspondingly fewer stem cells (**Figure 8.12**). *WUS* loss-of-function mutants, like the consequence of *CLV3* over expression, have a reduced stem cell compartment in the SAM (**Figure 8.13**). A combination of phenotypic analysis of loss of function and gain of function mutants and over-expressing lines, respectively, with observations on the distribution of expression of *CLV3* and *WUS* has led to a clear picture of their part in the regulatory interactions that control the SAM stem cell maintenance and differentiation. *WUS* transcript expression has only been observed within the OC, although recently it has been shown to act as a mobile signal and has also been observed in the overlying stem cells. *WUS* acts in the overlying stem cells by promoting the expression of *CLV3*. The *CLV3* peptide binds to the *CLAVATA1* (*CLV1*) receptor kinase, promoting assembly of a signaling complex (including a second *CLAVATA* transmembrane protein, *CLV2*), one action of which is to negatively regulate expression of *WUS*. *CLV3* expression is restricted to the three layers of CZ stem cells, from which it is secreted, and limits its own domain of expression by preventing differentiation of PZ cells, which surround the CZ, into CZ cells and restricts overall SAM size by a separate, long-range effect on cell division (**Figure 8.14**).

Findings from studies on the SAM stem cells also illustrate perfectly how the core mechanism can be modulated through additional factors. As an example, the modulating role played by micro RNAs in stem cell maintenance can be seen in miR394, which acts in the SAM to restrict stem cell competence to the distal layer, essentially anchoring the stem cells within the tip of the apical meristem. Hence, miR394 is produced from L1 and can move into L2 and L3, where it represses its target protein *LEAF CURLING RESPONSIVENESS* (*LCR*), which normally acts to promote leaf morphogenesis through inhibition of *CLV3* activation by *WUS* (**Figure 8.15**).

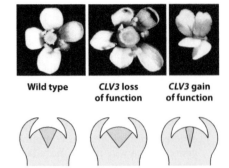

**Wild type**     **CLV3 loss of function**     **CLV3 gain of function**

**Figure 8.12 Negative feedback signaling in the SAM: *CLAVATA3* limits meristem size.** Photographs of *Arabidopsis* flowers from wild type and *CLAVATA3* (*CLV3*) mutants. On the left is the wild type flower showing normal development, in the middle a loss of function *CLV3* mutant (reduced expression) exhibiting enlarged floral features such as extra petals and stamens and an enlarged pistil. On the right is a flower from a *CLV3* transgene (over expression) showing reduced flower formation. The lower diagrams illustrate the relative sizes of the stem cell-containing central zone regions (purple) in the shoot apical meristem. (From Brand U, Fletcher J, Hobe M et al. (2000) *Science* 289:617–619 [doi: 10.1126/science.289.5479.617]. Reprinted with permission from AAAS.)

**Wild type**     **WUS loss of function**

**Figure 8.13 Negative feedback signaling in the SAM: *WUSCHEL* enhances meristem size.** Photographs of *Arabidopsis* flowers from wild type and *WUSCHEL* (*WUS*) mutants. On the left is the wild type flower showing normal development, and on the right is a flower from a loss of function *WUS* mutant (reduced expression) exhibiting reduced flower formation. The wild type flower has six stamens and a central pistil. The mutant flower has only one stamen and no central pistil. The lower diagrams illustrate the relative sizes of the stem cell-containing central zone regions (purple) in the shoot apical meristem. (From Mayer KFX, Schoof H, Haecker A et al. (1998) *Cell* 95: 805–815 [doi: 10.1016/s0092-8674(00)81703-1]. Copyright 1998 with permission from Elsevier.)

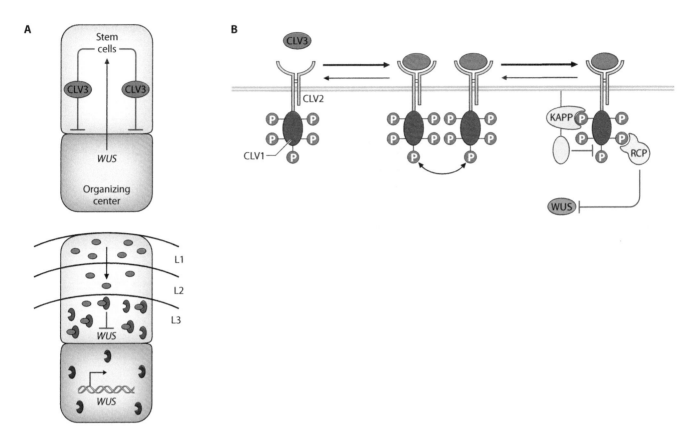

Figure 8.14 **CLV3 and *WUS* negative feedback signaling mechanism in the SAM. (A)** The upper diagram depicts some of the key signals involved in the interaction between the stem apical meristem (SAM) stem cells in the central zone (blue) and the cells of the organizing center (purple). The transcription factor *WUSCHEL* (*WUS*) is expressed by the cells of the organizing center and promotes the expression of *CLAVATA3* (*CLV3*) in the overlying stem cells that in turn represses *WUS* and regulates the size of the organizing center. The lower panel shows how *CLV3* (red circles) produced by stem cells within L1 and L2 of the SAM interacts with its receptor *CLAVATA1* (*CLV1*) (dark blue crescents) on cells of L3, resulting in a repression in *WUS* expression in these cells. Cells of the organizing center continue to express *WUS* because excess *CLV1* receptors in L3 prevent *CLV3* from entering the organizing center. **(B)** *CLV3* ligand (red oval) binds to the extracellular domain of the putative *CLV1/CLV2* heterodimeric receptor (represented in blue). Ligand binding drives *CLV1* phosphorylation, which leads to the binding of downstream effector molecules and to repression (T-bar) of the expression of the *WUS* gene. P, phosphate. (A: From Haecker A & Laux T (2001) *Curr Opin Plant Biol* 4: 441–446 [doi:10.1016/s1369-5266(00)00198-9]. Copyright 2001 with permission from Elsevier. B: From Clark S. (2001) *Nat Rev Mol Cell Biol* 2: 276–284 [doi:10.1038/35067079]. Reprinted by permission from Springer Nature.)

Figure 8.15 **Potentiation of *WUS* activity in the shoot apical meristem by miRNA.** Diagram of the shoot apical meristem (SAM) showing the positions of the stem cells in the central zone (blue), the organizing center (purple), and the peripheral zones where leaf morphogenesis occurs (orange). Activity of the transcription factor *WUSCHEL* (*WUS*), which controls the size of the SAM, is regulated by *CLAVATA3* (*CLV3*) but additionally involves modulation through the micro RNA miR394. Stem cells within Layer 1 of the SAM (L1) produce miR394, which moves into the lower layers, where it represses its target protein *LEAF CURLING RESPONSIVENESS* (*LCR*). *LCR* normally acts to promote leaf morphogenesis through the inhibition of *CLV3* activation by *WUS*. Activation and inhibition are indicated by arrows and T-bars, respectively. (From Holt A, van Haperen J, Groot E et al. (2014) *Curr Opin Plant Biol* 17:96–102 [doi:10.1016/j.pbi.2013.11.011] Copyright 2014 with permission from Elsevier.)

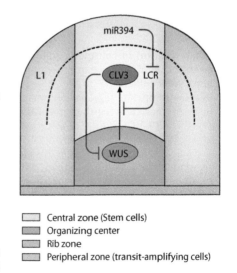

Central zone (Stem cells)
Organizing center
Rib zone
Peripheral zone (transit-amplifying cells)

## The regulation of stem cell function in the root apical meristem and the cambium meristem is analogous to that seen in the shoot apical meristem

The maintenance of RAM stem cells involves similar signaling pathways to those within the SAM (Figure 8.16). Cells in the QC of the RAM, the equivalent of the OC in the SAM, are the source of a short-range, non-cell autonomous signal that prevents neighboring stem cells from differentiating and therefore maintains stem cell characteristics. Cells of the QC express the *WUS* homolog *WUS-RELATED HOMEOBOX5* (*WOX5*), loss of function mutants of which cause premature

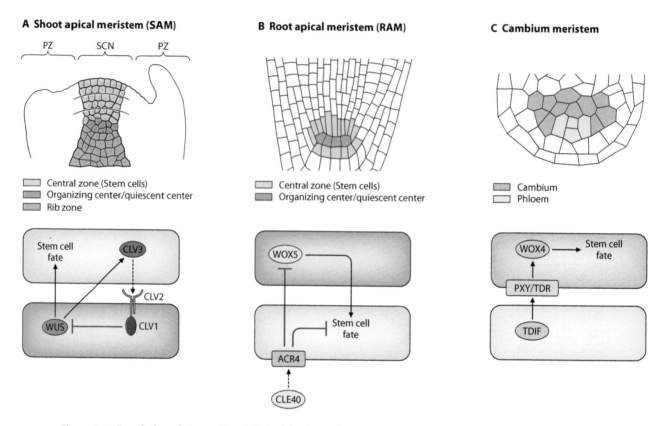

**A Shoot apical meristem (SAM)**

PZ    SCN    PZ

Central zone (Stem cells)
Organizing center/quiescent center
Rib zone

Stem cell fate

CLV3

CLV2

WUS    CLV1

**B Root apical meristem (RAM)**

Central zone (Stem cells)
Organizing center/quiescent center

WOX5

Stem cell fate

ACR4

CLE40

**C Cambium meristem**

Cambium
Phloem

WOX4    Stem cell fate

PXY/TDR

TDIF

**Figure 8.16 Regulation of stem cells and their niches in *Arabidopsis*.** Summary diagrams comparing the principal regulatory mechanisms controlling stem cell location and number in the three primary meristems of *Arabidopsis*. (**A**) The shoot apical meristem (SAM) is organized into three distinct zones; the central zone (CZ) containing the stem cells, the peripheral zone (PZ), and the rib zone (RZ). The organizing center (OC) lies in the region between the CZ and the RZ. The lower part shows the negative feedback loop between the transcription factor *WUSCHEL* (*WUS*) produced by the OC and its stimulation of *CLAVATA3* (*CLV3*) production in the stem cells, which in turn represses *WUS* expression through the action of the *CLV1/CLV2* receptor complex. (**B**) In the root apical meristem (RAM), the stem cells surround the quiescent center (QC). The lower part shows the regulatory mechanism acting on the stem cells from the QC. *WUS-RELATED HOMEOBOX5* (*WOX5*) is produced by the QC and regulates stem cell fate. *CLAVATA3/ENDOSPERM SURROUNDING REGION 40* (*CLE40*) acts via the receptor *ARABIDOPSIS CRINKLY4* (*ACR4*) to repress *WOX5* expression. (**C**) In the cambium meristem, the stem cells are located within the procambium and give rise to phloem cells (yellow). Phloem cells secrete *TRACHEARY ELEMENT DIFFERENTIATION INHIBITORY FACTOR* (*TDIF*) that binds the receptor *PXY/TDR* on the surface of the procambial stem cells, promoting the expression of the transcription factor *WOX4*, which ultimately controls stem cell fate. Stem cells are indicated in purple and the OC/QC in blue. Activation and inhibition are indicated by arrows and T-bars, respectively. (From Miyashima S, Sebastian J, Lee J et al. (2013) *EMBO J* 32: 178–193 [doi: 10.1038/emboj.2012.301]. With permission from Wiley. © 2013 European Molecular Biology Organization.)

differentiation of columella stem cells (visible by their accumulation of starch granules), while over expression effectively blocks differentiation and generates more stem cells. *WUS* and *WOX5* can functionally replace each other if expressed in the appropriate regions, demonstrating their close homology and the parallels in apical meristem regulation. Similarly, a peptide derived from the *CLV3* homolog *CLAVATA3/ENDOSPERM SURROUNDING REGION 40* (*CLE40*) acts via the receptor kinase *ARABIDOPSIS CRINKLY4* (*ACR4*) to repress *WOX5*.

The cambium adds a third example to the pattern of molecular mechanisms that regulate stem cells in meristems, which relies on specific factors but is analogous to what is seen in the apical meristems (Figure 8.16). The predominant mechanism acting upon the cambial stem cells relies on peptide-receptor signaling similar to the *CLV3-WUS* pathway in the SAM. In *Arabidopsis,* phloem cells secrete a peptide called *TRACHEARY ELEMENT DIFFERENTIATION INHIBITORY FACTOR* (*TDIF*), which interacts with the membrane kinase receptor *PHLOEM INTERCALATED WITH XYLEM/TDIF receptor* (*PXY/TDR*) expressed on the surface of cambial stem cells (Figure 8.17). This interaction results in three independent responses, namely (i) stimulation of cell division in the cambium, a cascade that requires the activity of the WUS-related transcription factor *WOX4*; (ii) prevention of stem cell differentiation toward xylem cells; and (iii) control of the orientation of stem cell division. Genetic analyses showed that *WOX4* is only required for the effect on proliferation, so that

Figure 8.17 **Control of stem cell fate in the vascular niche.** Schematic representation of the interactions taking place in the cambium meristem that control differentiation of the stem cells toward phloem or xylem. Phloem cells (yellow) secrete *TRACHEARY ELEMENT DIFFERENTIATION INHIBITORY FACTOR* (*TDIF*), which interacts with the *TDIF* receptor/*PHLOEM INTERCALATED WITH XYLEM* (*TDR/PXY*) on the surface of the cambial stem cells, promoting cell division within the stem cells themselves, differentiation toward phloem cells, and inhibition of differentiation toward xylem cells. The concentration of *TDIF* (depicted by the green triangle) decreases with distance from the phloem so that differentiation toward xylem can eventually occur once this dips below a threshold at a specific distance from the phloem. (Adapted from Hirakawa Y, Kondo Y & Fukuda H. (2010) *J Integrat Plant Biol* 52: 8–16 [doi: 10.1111/j.1744-7909.2010.00904.x].)

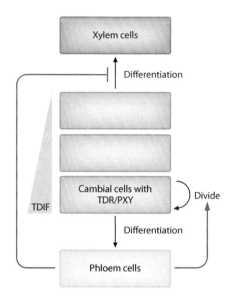

at least two intracellular signaling pathways must diverge after *TDIF* recognition by *TDR/PXY* to regulate the behavior of vascular stem cells.

## Determinate meristems are characterized by the transient nature of their stem cell component

In the determinate meristems of a plant, such as those giving rise to flowers, there is a reliance on transient stem cells whose finite existence limits flower size and form. Floral meristems are organized in a similar manner to the SAM and also depend on signaling between *CLV* and *WUS*. The transition from SAM to floral meristem requires genes that both specify the floral organs and cause the termination of the production of stem cells. The floral regulatory gene *AGAMOUS* (*AG*) is turned on in floral meristems by *WUS* in combination with the gene that actually defines floral identity, namely the *LEAFY* (*LFY*) gene. *AG* is expressed in the sexual organ primordium and remains specific to the carpel and stamens. Loss of function of *AG* leads to an absence of the sexual organs, which are replaced by whorls of sepals and petals. Once *AG* has been activated, it then negatively regulates *WUS* expression, thereby terminating the floral meristem, which it achieves indirectly by activating the transcription factor *KNUCKLES* (*KNU*) that in turn represses *WUS* expression, and directly by recruiting PcG to *WUS* and thereby bringing about its repression (Figure 8.18).

Similarly, in *Arabidopsis*, the transit amplification of meristem populations within the PZ of the SAM, during the specification of leaf outgrowths, is dependent upon the expression of another homeodomain protein, *SHOOT MERISTEMLESS* (*STM*), which is expressed throughout the meristem but is down regulated upon leaf organogenesis, an important negative effect being through the action of the Myb transcription factor *ASYMMETRIC LEAVES1* (*AS1*). The expression of *STM* throughout the PZ delays premature differentiation, but it also functions to render stem cells of the CZ responsive to the *WUS* signal from the OC (Figure 8.19).

## 8.3 REGENERATION IN PLANTS AND ITS LINK TO STEM CELLS

In Chapter 5, we have seen that most, if not all, animal cells can be reprogrammed either through nuclear transfer into an egg or through expression of transcription factors that define the **pluripotent** state, while in Chapter 7, we describe instances in which certain tissues, depending on the animal species, are able to regenerate through reprogramming-like processes. Plants generally exhibit an incredible capacity for regeneration, which in some ways might be equated with what is seen in animals but certainly provides further insight into the way in which stem cells and reprogramming are either connected or work in concert to maintain the post-embryonic organism.

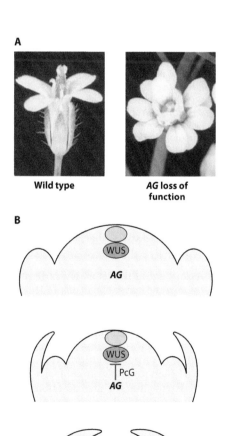

Figure 8.18 **Meristem termination and flower development.** (A) A wild type *Arabidopsis* flower (left) and *AGAMOUS* (*AG*) loss of function mutant flower (right) showing the loss of the sexual organs within the flower and their substitution by extra petals and sepals. (B) Schematic showing the regulatory interactions between *AG* and *WUSCHEL* (*WUS*). *AG* is turned on in floral meristems and directly represses the expression of *WUS* by recruitment of Polycomb Group proteins (PcG). *WUS* repression is necessary to terminate the induction of the floral meristem. (A: Used with permission from John L. Bowman, Monash University. B: From Liu X, Kim Y, Müller R et al. (2011) *Plant Cell* 23:3654–3670 [doi: 10.1105/tpc.111.091538]. Permission conveyed through Copyright Clearance Center, Inc.)

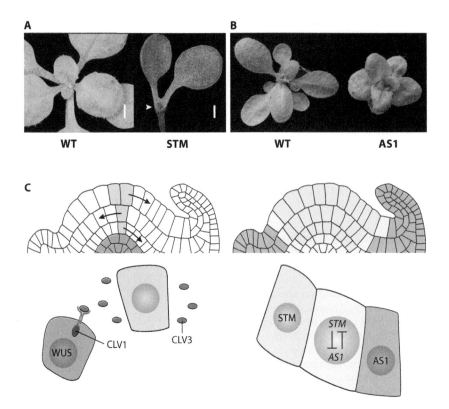

**Figure 8.19 Meristem termination and leaf development. (A)** Photographs of *Arabidopsis* seedlings from wild type and *SHOOT MERISTEMLESS* (*STM*) mutants. The mutant seedling is clearly missing the apical shoot (arrow head). (**B**) Photographs of *Arabidopsis* plants from wild type and *ASYMMETRIC LEAVES1* (*AS1*) mutants. (**C**) Transverse section through the shoot apical meristem (SAM) showing the direction of production (arrows) of daughter stem cells (blue) toward the periphery. The lower illustration depicts the *CLAVATA3* (*CLV3*, red), *CLAVATA1* (*CLV1*, dark blue), and *WUSCHEL* (*WUS*) signaling between stem cells and the organizing center (purple). The right-hand illustration shows the expression of *STM* (green) within the SAM and the inhibitory effect it has on the Myb transcription factor *AS1* (lower panel). Organogenesis coincides with expression of *AS1* (orange) in the organ primordia and loss of *STM* expression. (A: From Roth O, Alvarez J, Levy M et al. (2018) *Plant Cell* 30: 1309–1321 [doi: 10.1105/tpc.18.00222]. Permission conveyed through Copyright Clearance Center, Inc. B: From Garcia D, Collier S, Byrne M et al (2006) *Curr Biol* 16: 933–938 [doi: 10.1016/j.cub.2006.03.064]. Copyright 2006 with permission from Elsevier. C: From Scheres B. (2007) *Nat Rev Mol Cell Biol* 8: 345–354 [doi: 10.1038/nrm2164]. Reprinted by permission from Springer Nature.)

## Regeneration in plants reveals an inherent plasticity in somatic cells and the ability to generate *de novo* stem cells

As long ago as the 1890s, it was discovered that complete plants could be grown from the severed leaves of pansies and begonias, while in 1958, the group of Kathryn Mears at Cornell University showed that single cells from the vascular phloem of carrot plants could be cultured to generate a whole plant, demonstrating therefore that some adult cells at least can be regarded as being **totipotent**. Such early observations are representative of the many examples of regeneration that can be achieved in plants, either *in vivo* or in culture (Figure 8.20). Overall, plants can regenerate through two broad principles that involve somatic cells but do not require the stem cells in meristems that we have described previously. Hence, regeneration can be achieved either from cells that are relatively undifferentiated or through the reprogramming of differentiated somatic cells, although not all cells may be equal in this regard.

## Regeneration of the plant root tip can occur in the absence of stem cells and can recreate the lost root apical meristem stem cells and their niche

The regeneration of the root tip following its removal has been compared to the limb regeneration that is possible in some amphibians (Figure 8.21). Like the

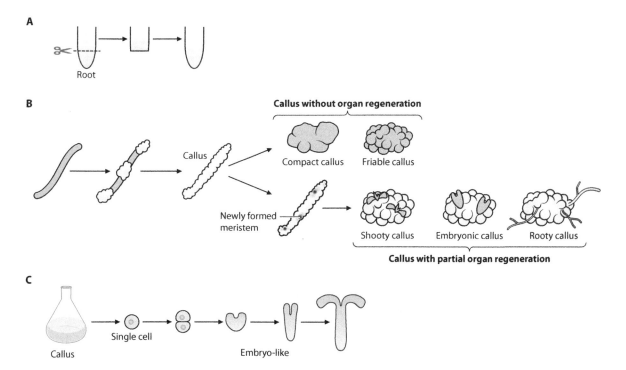

**Figure 8.20 Plant regeneration and callus formation.** Schematic diagrams illustrating possible mechanisms of regeneration in plants. (**A**) Regeneration that replaces the excised tip of a root. (**B**) Callus formation in which a piece of plant tissue (green) develops callus (cream) under the influence of plant hormones as part of the response to wounding. The callus can either be non-regenerative (compact or friable) or regenerative (shooty, embryonic, or rooty). The plant structures developing from regenerative callus arise from newly formed meristems (purple dots). (**C**) Callus growth can be induced in medium that includes plant hormones. In this case, it is possible to observe a single callus cell undergoing normal embryogenesis and development into a plant. (A and C: From Sugimoto K, Gordon S & Meyerowitz E (2011) *Trends Cell Biol* 21: 212–218 [doi: 10.1016/j.tcb.2010.12.004]. Copyright 2011 with permission from Elsevier. B: From Ikeuchi M, Sugimoto K & Iwase A (2013) *Plant Cell* 25: 3159–3173 [doi: 10.1105/tpc.113.116053]. Permission conveyed through Copyright Clearance Center, Inc and Sugimoto K, Gordon S & Meyerowitz E (2011) *Trends Cell Biol* 21: 212–218 [doi: 10.1016/j.tcb.2010.12.004]. Copyright 2011 with permission from Elsevier.)

amphibian limb, root tip regrowth does not require stem cells, but in contrast to the animal situation, this process of regeneration also restores the stem cell component and its niche in the reformed RAM. The cells responsible for root tip regeneration are contained in the central cylinder of cells constituting the stele, including vascular cells and the pericycle (see Figure 8.8). Recent studies involving single cell sequencing of *Arabidopsis* root tip cells during regeneration have given a quite clear picture of the sequence of events at the molecular level and, importantly, have revealed how the processes involved recapitulate the events that occur during the development of the plant embryo (Figure 8.22). Immediately following removal of the root tip, stele cells begin to reprogram to a state akin to the cells of the hypophysis, which is the embryonic cluster of cells that gives rise to the root. Under the influence of the asymmetrically distributed plant hormones cytokinin and auxin, the hypophysis-like cells take on the identity of quiescent center and columella, respectively. The arrangement of these two latter tissues and changing distribution of the hormones subsequently induces the generation of stem cell activity.

## Plants exhibit a considerable potential for reprogramming of somatic cells to generate meristems and even the whole organism

Regeneration from fully differentiated cells occurs through distinct mechanisms but again can lead to the formation of embryo-like structures that lead to the creation of new centers of stem cell activity. Plants are able to generate an entire plant from a piece of tissue in culture through a **callus** as an intermediate growing cell mass (Figure 8.20). Plants naturally generate callus in response to stresses such as wounding and pathogen infiltration. Many of the cells within a callus are totipotent with the ability to regenerate entire plants. Such regeneration always involves new centers of plant growth through the formation of meristems. Callus formation is not a uniform process, with a great deal of diversity in the regenerative capacity within it. Some calluses have no organ regenerative capacity at all, and these are known as friable or compact callus. Callus with some degree of regenerative potential are termed rooty, shooty, or embryonic callus depending upon the organs they can produce. Callus can also be induced in culture through the use of a carefully controlled balance of plant hormones auxins and cytokinins to form somatic embryos, leading to the formation of a whole plant being elicited by transfer to media containing no hormone or altered ratios of hormones.

The mechanisms involved in the formation of callus have been somewhat controversial, with plant biologists debating whether this involves **dedifferentiation** or activation of nascent meristems. Recent gene expression profiling suggests that callus formation reflects to a degree the process of lateral root formation, even when derived from aerial somatic tissues such as the stem. Furthermore, the study of wounding responses, which can naturally involve callus formation, has given some insight into the molecular processes involved. As the result of loss of cell-cell contacts and release of plant hormones, a wound stimulates the expression of a succession of transcription factors that collectively lead to the formation of new meristems and outgrowth of tissues. Many factors appear to be involved, which are far beyond what we will be able to cover here. As an example, wounding in *Arabidopsis* induces the aptly named transcription factor *WOUND-INDUCED DEDIFFERENTIATION1*

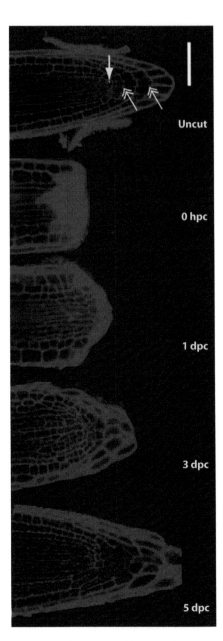

**Figure 8.21 Regeneration of a root tip.** The root tip along with its associated meristem is able completely to reform after being excised, including regeneration of the stem cells and their niche. The series of confocal microscope images show the uncut tip of an *Arabidopsis* root at the top, with the location of the quiescent center cells indicated by the arrow and the double arrow showing the position of the columella cells, and then subsequent to cutting (hpc, hours post cut; dpc, days post cut). Scale bar = 50 μm. (From Sena G & Birnbaum K. (2010) *Curr Opin Genet Dev* 20: 460–465 [doi: 10.1016/j.gde.2010.04.011] Copyright 2010 with permission from Elsevier.)

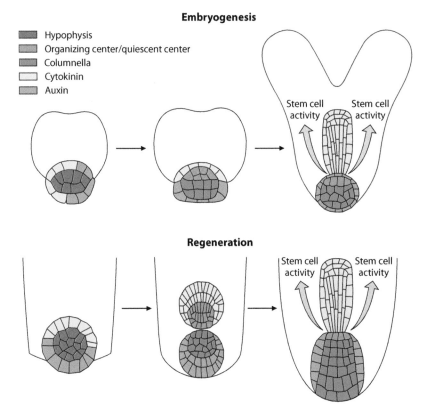

**Figure 8.22 Model for root regeneration.** Single cell RNA sequencing of cells in a regenerating *Arabidopsis* root tip has revealed gene expression changes that indicate that reprogramming creates a series of events closely mimicking the process of early embryo development in plants. The schematic diagram compares the formation of the stem cell domain in a root tip during embryogenesis (top) and regeneration following tip excision (bottom). Both processes initiate with the expression and segregation into domains of the plant hormones auxin (green) and cytokinin (yellow), which together surround the hypophysis in the embryo (purple) and the region that becomes the organizing center (pink) and columella (blue). The hormone distribution provides spatial cues for the cellular domains of the niche and root cap and initiate the formation of stem cells (arrows). (From Efroni I, Mello A, Nawy T et al (2016) *Cell* 165: 1721–1733 [doi:10.1016/j. cell.2016.04.046]. Copyright 2016 with permission from Elsevier.)

(*WIND1*). Transient over expression of *WIND1* can induce callus without wounding, which can then regenerate shoots and roots in appropriate media or, when continuously expressed, can lead to shoot formation without wounding, implying that it reprograms cells to a pluripotent state. Sequential activation of *WIND1* and an embryonic regulator, *LEAFY COTYLEDON2* (*LEC2*), induces somatic embryogenesis at both wound and non-wound sites, whereas single activation of *LEC2* permits embryogenesis only at wound sites.

Epigenetic modifications appear to play an important part in the reprogramming processes, as exemplified by *WIND1/LEC2*. An important epigenetic regulator in this context is the Polycomb Group (PcG) factor PRC2, which gives rise to the trimethylation of histone H3 on lysine residue 27 (H3K27me3). The role of PRC2 is to maintain the differentiated state, which is achieved at least in part by repression of transcription factor gene expression, including that of *WIND* and *LEC* family proteins. Mutation of PRC2 in *Arabidopsis* leads to elevated *WIND* and *LEC* expression and a consequent formation of callus and embryo-like structures.

## 8.4   SIMILARITIES AND DIFFERENCES BETWEEN PLANT AND ANIMAL STEM CELLS

Throughout this chapter, we have alluded to the possibility of similarities between plant and animal stem cells. We would argue that, strictly speaking, plant and animal stem cells both adhere to the central principles of what defines stemness, that is, the combined ability to self-renew and to commit to specific pathway(s) of differentiation. It is also clear that some of the features that characterize stem cells in the adult organism are quite similar, including the need for a structured niche and the role of downstream transit amplification of progenitors as a means to expand the population of differentiating cells. That said, it is also apparent that there are a host of differences between plants and animals in terms of the origin and role of stem cells and the mechanisms that determine and regulate them.

### The developmental origin of plant and animal stem cells differs

It is generally accepted that adult **tissue stem cells** in animals originate during embryogenesis and are destined to perpetuate throughout life, serving as a reserve to replenish cell loss and maintain homeostatic balance. Plants, on the other hand, are not so restricted in their supply of stem cells. Embryogenesis for sure generates stem cells that serve to support the various meristems, but we have seen that new stem cells can be generated *de novo* in post-embryonic plants, for instance, during regeneration of the root tip or in the formation of lateral or axial meristems. The origin of the germline also marks another distinction between stem cells in plants and animals. In plants, this arises as the result of differentiation initiated in the SAM, whereas in animals, a specific group of stem cells give rise to gametes with no replenishment during post-embryonic life. The fact that plant germ cells can be generated from somatic tissues (including via the SAM) that may have accumulated mutations might be seen as disadvantageous; however, a recent study of a long-lived plant (an oak tree) involving whole genome sequencing of tissues at the ends of temporally distant branch tips (of the order of hundreds of years since having a common cell of origin) actually revealed very little fixation of mutations.

### Regeneration involving reprogramming exhibits some similarities between plants and animals

We have described previously that differentiated cell reprogramming occurs in plants, as seen in the regrowth of the root tip or through the formation of callus tissue *in vivo* or *in vitro*. The similarities are perhaps superficial rather than mechanistic. Certainly, they differ greatly in that animals do not regenerate stem cells in the circumstances of regrowth of an amputated limb. Nevertheless, plant and animal somatic cells can both exhibit reprogramming to a pluripotent or totipotent state, as seen naturally in plants in terms of callus formation or somatic embryogenesis and artificially in animals through the generation of induced pluripotent stem cells (Chapter 5).

## The stem cell niche and regulatory networks supporting plant and animal stem cells show some common generic features

On the face of it, comparison of the available information on plant and animal stem cells suggests that there are some significant analogies in the overall architecture of stem cell niches and some clear parallels in the signaling and gene regulatory mechanisms that control stem cell maintenance and fate. A simple side-by-side look at the *Drosophila* ovary and the *Arabidopsis* RAM reveals a lot of similarities in terms of the presence and relative orientation of the organizer cells, the stem cells, and the downstream transient amplifying and more mature cells (Figure 4.20). Likewise, the networks of transcription factors and signaling inputs controlling stem cell function, spatial identity, and commitment to differentiation also look remarkably alike in plants and animals, with feedback loops and additional modulation through miRNAs. However, there is little or no homology between any of the proteins involved. In animals, the signals that derive from a stem cells' neighbor in the niche include molecules such as Notch, Wingless, and Hedgehog, ranging from invertebrates to humans, but no such proteins are encoded in the plant genome.

One regulatory protein that is shared between plants and animals and has a role in stem cell fate determination is the retinoblastoma (Rb) protein, which appears to promote exit from the stem cell state followed by differentiation. Another example is the PcG proteins, which regulate chromatin through generation of the epigenetic H3K27me3 modification. In animals, this modification is important in the maintenance of pluripotency in stem cells, whereas in plants, PcG proteins regulate the transition between the pluripotent and differentiated states, but by repressing genes in the differentiated cells that are normally expressed in the meristem, as we have described, for instance, in the case of the action of *AG* on expression of the *WUS* gene and in the control of wound response genes like *WIND1*.

## Plant and animal stem cells arose independently during evolution

Aside from the superficial, structural similarities in stem cell location and behavior, and a few examples of proteins with related stem cell regulatory functions in plants and animals, it seems almost certain that stem cells evolved independently in the two kingdoms, which is perhaps not surprising given that plants and animals evolved as multicellular organisms from single-cell ancestors that were clearly on distinct branches of the evolutionary tree. It is also feasible that within each Kingdom, stem cell characteristics evolved multiple times, although the similarities in regulatory mechanisms, whether comparing among animal or plant cells, could be compatible with common ancestors or a conserved limited set of molecular options that have been adopted multiple times. The few common mechanisms are therefore most likely to have been ones that existed already to control cell division and differentiation in unicellular organisms, and the broader similarities that are observed help us to focus on the most fundamental principles of stem cell biology.

## 8.5   SUMMARY

This is the only chapter in the book that deals with plant stem cells, but we hope that we have conveyed the great interest in and degree of understanding that there is around the role that these cells play in supporting plants. Some of the unique features of plant stem cells and their highly defined location in meristems, combined with elegant use of cell and molecular biology techniques in very amenable models for genetic manipulation and phenotypic characterization, as typified by *Arabidopsis*, has led to a very detailed understanding. Although the discovery techniques and the language used to describe plant stem cells have both taken much from the animal stem cell world, it seems from the comparisons that can be made now between plant and animal stem cells that they are analogs rather than homologs, having arisen separately during evolution. Nevertheless, there has been much gained from the shared approaches to investigating stem cells in the two kingdoms, and, most interestingly, the conclusion about parallel (convergent) evolution suggest that stemness is a fundamental principle of multicellular life.

## KEY POINTS

- Plants have an essential need for stem cells in order to maintain growth, which can be extensive and very long term.

- Plant stem cells are located in meristems, which are present at the sites of growth (shoots, roots, and in the vasculature) but are also formed during the development of organs such as flowers.

- Apical meristems constitute the growth points of shoots and roots and have an organized structure of layers and zones of cells in which the stem cells occupy a discrete region surrounded by cells forming the niche.

- Other meristems derive from the apical meristems, but, like the type that produce flowers, may only be transient.

- The majority of findings about plant stem cells have come from *Arabidopsis*, a model for flowering plants, but stem cells have also been studied in more primitive land plants like moss.

- The regulation of stem cells in the apical and vascular meristems is well defined, mainly through the study of specific *Arabidopsis* mutants.

- Stem cell maintenance and commitment are restricted to defined domains through cross-talk between the various regulatory components, intercellular communication being facilitated by transfer between individual cytoplasms through the plasmadesmata.

- The best defined regulatory mechanisms operating in a meristem are those in the *Arabidopsis* shoot apical meristem (SAM), which involve the transcription factor *WUSCHEL* that is produced from the niche and acts to stimulate expression in the stem cell zone of the receptor ligand *CLAVATA3*, which in turn negatively regulates *WUSCHEL* expression.

- Although each meristem has a distinct mechanism of regulation, these have parallels in the involvement of transcription factors and receptor-ligand interactions.

- Plants can regenerate, including the ability to produce a totipotent callus.

- Plant regeneration does not require stem cells but leads to the *de novo* formation of stem cells in the newly generated meristems.

- Plant and animal stem cells and their respective niches share many functional features but are mechanistically distinct and seem to have arisen independently during evolution.

## FURTHER READING

### General

Koornneef, M. and Meinke, D. (2010) The development of *Arabidopsis* as a model plant. *Plant J.*, 61: 909–921.

Smith, A.M., Coupland, G., Dolan, L., Harberd, N., Jones, J., Martin, C., Sablowski, R. and Amey, A. (2009) *Plant Biology*. Garland Science, New York & London.

### Plant stem cells

Efroni, I., Mello, A., Nawy, T., Ip, P.L., Rahni, R., DelRose, N., Powers, A. et al. (2016) Root regeneration triggers an embryo-like sequence guided by hormonal interactions. *Cell*, 165: 1721–1733.

Gaillochet, C., Daum, G. and Lohmann, J.U. (2015) O cell, where art thou? The mechanisms of shoot meristem patterning. *Curr. Opin. Plant. Biol.*, 23: 91–97.

Miyashima, S., Sebastian, J., Lee, J.-Y. and Helariutta, Y. (2013) Stem cell function during plant vascular development. *EMBO J.*, 32: 178–193.

Sena, G., Wang, X., Liu, H.Y., Hofhuis, H. and Birnbaum, K.D. (2009) Organ regeneration does not require a functional stem cell niche in plants. *Nature*, 457: 1150–1153.

Sugimoto, K., Gordon, S.P. and Meyerowitz, E.M. (2011) Regeneration in plants and animals: dedifferentiation, transdifferentiation, or just differentiation? *Trends Cell Biol.*, 21: 212–218.

### Plant regeneration

Ikeuchi, M., Ogawa, Y., Iwase, A. and Sugimoto, K. (2016) Plant regeneration: Cellular origins and molecular mechanisms. *Development*, 143: 1442–1451.

### Comparison between plant and animal stem cells

Birnbaum, K.D. and Sánchez Alvarado, A. (2008) Slicing across kingdoms: Regeneration in plants and animals. *Cell*, 132: 697–710.

Sugimoto, K., Gordon, S.P. and Meyerowitz, E.M. (2011) Regeneration in plants and animals: dedifferentiation, transdifferentiation, or just differentiation? *Trends Cell Biol.*, 21: 212–218.

# STEM CELLS IN SPECIALIZED ADULT ANIMAL TISSUES

# PART FOUR

# Hematopoietic stem cells

9

From the earliest stages of the evolution of multicellular animals, there has been a requirement for specialized cells that serve to defend and repair the individual. In one sense, such cells function like any other specialized cells that constitute a particular organ, yet in their own right, they do not make up a tissue. As evolution introduced progressively greater complexity, the way in which these cells reached the appropriate location went from one of essentially random migration to constraint within a system of vessels and pumps, effectively giving the reason to define blood as a tissue. From the definition of blood arises the consequent terminology that is applied widely to cells from animals with or without a defined blood system, such as the adjective "hematopoietic." Evolution also led to a diversification of hematopoietic cell function, for example, incorporating gas exchange, and ultimately leading to the full complement of cells that circulate as blood in vertebrates. The assumption is usually that the expanding range of hematopoietic cells was a process of divergence, as evidenced by the fact that all vertebrate blood lineages derive from a common stem cell; however, this need not be the case in all animal species.

Whether it is the complex blood system of vertebrates or the far more straightforward cell-based defense mechanisms in invertebrates, the life-long requirement for such functional cells necessitates a stem cell source. The understanding of these stem cells, and in fact of **adult stem cells** in general, has been dominated by studies on mammalian bone marrow blood stem cells, in particular in the laboratory mouse. Identifying, isolating, and investigating the properties of these so-called **hematopoietic stem cells** (HSC), largely driven by a therapeutic imperative, provided the impetus for technology development and a conceptual framework for stem cells that has paved the way for research on stem cells in most tissues and in many different species.

The majority of this chapter will be devoted to consideration of the mammalian HSC, not only because of the detailed insight that we have into the nature of these cells but also because the history of their investigation provides a wealth of lessons about how stem cell science has developed and the sorts of findings that we can expect to see in the future for currently less well understood adult stem cells. We will start and finish this chapter, however, with a wider view of the ways in which stem cells are known or expected to underpin evolutionarily more primitive cells that are functionally, and most likely directly, related to vertebrate blood cells. Although presently less well understood, it is quite likely that future studies on invertebrate animal models with respect to their hematopoietic cells and their associated stem cells will shed new light on the vertebrate equivalents.

## 9.1 EVOLUTIONARY DIVERSIFICATION OF BLOOD CELLS AND THEIR STEM CELLS

Our understanding of hematopoiesis and the stem cells required to support it has without doubt been built up from studies on mammals, in particular the laboratory mouse and with significant parallel effort on the human system. As we describe throughout this book, organisms that arose earlier in evolution often serve as a more tractable way to investigate complex biological systems that rely on a **differentiation** hierarchy that is maintained by stem cells, but in the case of hematopoiesis, the history of the field is somewhat reversed in this sense in that studies on invertebrates such as crustacea and flies often take a lead from knowledge obtained from vertebrates. However, as for other scenarios, work on invertebrates and the experimental advantages that this can have is likely to shed new light on fundamental mechanisms and principles.

## The vertebrate hematopoietic system is made up of several specialized cell types

Vertebrates have the largest number of blood cell types with the most diverse range of functions (Box 9.1), including red cells for the transport of oxygen and carbon dioxide, thrombocytes involved in blood clotting and vessel repair, immune cells that adapt to elicit a specific response to pathogens and foreign antigens (lymphoid cells), and the various cells of the innate immune system, including monocyte-macrophages and granulocytes that serve many functions ranging from clearance of bacteria to the generation of an inflammatory response to injury.

## Although generally simpler, the invertebrate hematopoietic system shares many features with that of vertebrates

It seems that the origins of the array of vertebrate blood cells may lie in an early adaptation in animals that led to the development of cell types whose function was to defend against invading pathogens, and freely moving cells with structural and functional properties of at least some of the vertebrate blood cells are found in most multicellular animals. The evolutionarily most primitive extant organisms that exhibit such hematopoietic cells include animals without a body cavity (coelom) such as sponges and jellyfish. In animals that have a coelom and a vascular system, whether open as in many invertebrates or closed as in some invertebrates and all vertebrates, the hematopoietic cells are usually referred to as hemocytes. The hematopoietic cells present in invertebrates can be defined structurally into a number of types, which come with very descriptive and not always consistent

---

**BOX 9.1 BACKGROUND: CELLS OF THE VERTEBRATE HEMATOPOIETIC SYSTEM**

The vertebrate hematopoietic system is made up of a variety of terminally differentiated cell types (Figure 1) that constitute either the myeloid or lymphoid compartments, which function together to exchange gases between tissues and the outside, to repair damage to the vascular system, and to bring about effective immune responses to fight off pathogenic infection or other foreign bodies. "Myeloid" literally means "associated with the bone marrow" and includes those white cells (or "leukocytes") that constitute the first line of immune defense (the "innate" response). Myeloid leukocytes include the granulocytes (neutrophils, eosinophils, basophils) and the monocytes and their more mature phagocytic form, macrophages. Cells of the myeloid compartment also include red blood cells (erythrocytes) and platelets (thrombocytes), which are respectively necessary for the oxygenation of tissues and for efficient blood clotting. In the majority of vertebrates, excluding mammals, erythrocytes and thrombocytes are nucleated

cells, whereas in mammals, the equivalent cells lack a nucleus. In the case of thrombocytes, mammals have a specialized nucleated cell called the megakaryocyte, which is the largest cell within the bone marrow and releases vast numbers of "platelets" into circulation that function to bring about blood clotting at sites of injury. Lymphoid cells ultimately originate in the bone marrow but depend heavily on the lymphoid tissues (thymus, spleen, and lymph nodes) for aspects of their maturation and function. Lymphoid cells constitute the "adaptive" arm of the immune response and are divided into B-lymphocytes and T-lymphocytes (also known as B- and T-cells), which are selected and expanded to act against individual foreign antigens, especially those present on invading pathogens. B-cells are responsible for the production of antibodies, whereas T-cells act both to destroy cells (cytotoxic or killer T-cells) and to enhance and regulate the adaptive immune response (helper and regulatory T-cells).

| Monocyte | Neutrophil | Eosinophil | Basophil | Lymphocyte | Macrophage | Erythrocyte | Platelets |

Box 9.1 Figure 1 **Mammalian blood cell types.** Illustrations of the eight principal blood cells of the mammalian hematopoietic system as seen under the microscope when a blood smear is stained with hematoxylin and eosin.

**TABLE 9.1 METAZOAN BLOOD CELL TYPES**

| | | | Vertebrate blood cell type | | | | |
|---|---|---|---|---|---|---|---|
| | | | Innate immune cells | | T- and B-cells (adaptive immune cells) | Platelets/ thrombocytes (vessel repair and clotting) | Red blood cells (gaseous exchange) |
| Invertebrate animal group | | Stem cell | Phagocytes (macrophage) | Granulocytes | | | |
| No coelom | Sponges jellyfish, hydra flatworms | Archaeocytes, interstitial cells, neoblasts | – | – | – | – | – |
| Coelom | Arthropods (insects and crustacea) | Prohemocytes | Hyaline hemocytes | Granular hemocytes, eleocytes | – | Hemocytes (aggregation and release of granules) | Acellular (hemolymph) |
| | Annelids (worms) | Prohemocytes | Hyaline hemocytes | Granular hemocytes, eleocytes | – | – | Acellular (hemolymph) or plasmatocytes |
| | Protochordates | Prohemocytes | Hyaline hemocytes | Granular hemocytes, eleocytes | – | – | – |

*Note:* The table shows the vertebrate blood cell types and the presumed correspondence of these with hematopoietic cells found in different categories of invertebrate.

names, such as prohemocytes, hyaline hemocytes, granular hemocytes, and so on. Most importantly, there is often some functional correspondence to vertebrate hematopoietic cells (Table 9.1). Notably absent in invertebrates are lymphoid cells and thrombocytes, although some functions of the latter can be seen in certain hemocytes. Cells equivalent to red cells, in the sense of being carriers of gases, first appeared in invertebrates around the time that the annelids (worms) developed the polychaete branch, more primitive animals being able to gain sufficient oxygen through passive diffusion or the presence of oxygen-carrying proteins in the fluid filling the coelom.

As for the many examples of adult tissues and their associated stem cells, the hematopoietic cells of vertebrates and invertebrates have limited lifespans and need to be replaced on a continual basis, especially when there is a high demand such as following infection or injury. Many invertebrates have lifespans that are comparable to or exceed those of vertebrates; for example, lobsters can live up to 100 years, and at least one species of mollusc, the ocean quahog clam (*Arctica islandica*), can reach the ripe old age of 500! Most invertebrates that have been investigated in any detail have revealed evidence of hematopoietic progenitor or stem cell-like cells, often those cells that are morphologically referred to as prohemocytes. Up until recently, it was believed that relatively short-lived invertebrates such as insects, best represented by the fruit fly *Drosophila melanogaster*, only require hematopoietic progenitors as part of the process of development; however, recent findings, which we will discuss further subsequently, indicate that they too have adult stem cells supporting hematopoiesis.

## 9.2 MAMMALIAN HEMATOPOIETIC STEM CELLS

Following our brief survey of the occurrence of hematopoietic cells throughout the animal kingdom and the need in probably all cases for stem cells to maintain their numbers in adult life, we will now look in some detail at vertebrate HSC, in particular those present in the bone marrow of the mouse. The study of and consequent understanding of the murine HSC can without doubt be regarded as paradigmatic, not just for mammalian HSC but also for the whole field of adult stem cell biology. Although we will start by summarizing the wealth of knowledge around the mammalian HSC, we will later go back to look at the extent of equivalence in the blood stem cells of invertebrates and how understanding both vertebrate and invertebrate cells can cross-inform experimental investigation.

## Understanding the basis of bone marrow transplantation was a strong driver toward characterization of hematopoietic stem cells

The original driver for the study of HSC was very practical, that is, the attempt to understand and improve upon bone marrow transplantation as a means to treat leukemia and other hematological disorders (Chapter 17). However, the consequence of more than half a century of research on HSC, predominantly in the mouse, has established fundamental principles of stem cell biology such as the differentiation hierarchy, stem cell **quiescence** and heterogeneity, the role of intrinsic and extrinsic factors in the regulation of stem cell renewal and differentiation (Chapters 4 and 6), and the involvement of stem cells in cancer and aging (Chapter 18). The need to find ways to identify, isolate, and characterize rare stem cells as exemplified by HSC has had a massive impact on technology innovation (Chapter 2), including flow cytometry and cell sorting, *in vivo* observation of cells, and single cell analysis of molecular properties like RNA expression.

The game changing nature of the technological, biological, and therapeutic advances resulting from the investigation of HSC not surprisingly has brought to prominence a number of individuals (Box 9.2). In 1957, E. Donnall Thomas performed the first successful HSC transplantation for the treatment of leukemia using bone

---

### BOX 9.2 PIONEERS: KEY FIGURES IN THE DISCOVERY OF HEMATOPOIETIC STEM CELLS

**E. (Edward) Donnall Thomas** (Figure 1): The HSC was discovered following pioneering bone marrow transplantations performed by E. Donnall Thomas, research for which he gained full recognition by being awarded the Nobel Prize for Physiology or Medicine in 1990. His career began in 1943 when he graduated from the University of Texas with a masters degree in chemistry and chemical engineering. Three years later, he completed his medical degree at Harvard Medical School. Following some time serving in the army, in 1955, he was appointed physician in chief at Mary Imogene Bassett Hospital in Cooperstown, NY, where he began experimenting with bone marrow transplantations in dogs. Following poor results in patients, his research into matching tissue types in dogs together with other researchers' work on humans led in 1969 to the first successful treatment of leukemia by transplantation between matched siblings. In 1974, E. Donnall Thomas was appointed director of medical oncology at the Fred Hutchinson Cancer Research Center, which went on to become the world's leading center for bone marrow transplantations.

**James Till and Ernest McCulloch** (Figure 2): The pair's pioneering work began in 1957 when, following his PhD at Yale University, James Till joined the group of Ernest McCulloch at the University of Toronto as a post-doctoral researcher. The amalgamation of the knowledge and expertise of biophysicist Till and cellular biologist McCulloch produced one of the great cornerstones of stem cell biology when they demonstrated in 1961 (published in *Nature* in 1963) that injecting mouse bone marrow cells into irradiated animals led to the formation of colonies on the recipient spleen whose number correlated proportionally with the number of cells injected. Despite never earning them a Nobel Prize, this principle paved the way for the subsequent isolation of the HSC from mouse bone marrow.

Box 9.2 Figure 1 **E. Donnall Thomas.** Pictured alongside his wife, Dotty, who was his research partner. (Used with permission from the Fred Hutchinson Cancer Research Center).

Box 9.2 Figure 2 **James Till and Ernest McCulloch.** Biophysicist James Till and hematologist Ernest McCulloch, almost 50 years after the first isolation and finding of evidence of stem cells (Used with permission from the Lasker Foundation).

marrow from one twin to treat their sibling twin, a feat for which he ultimately received a Nobel Prize. The first real documented knowledge of the existence of the HSC came from studies by James Till and Ernest McCulloch in 1961. Often credited as the fathers of stem cell science, Till and McCulloch were the first to show that colonies of myeloid cells forming on the spleens of lethally irradiated mice following bone marrow transplantation were clonal and therefore derived from one parent cell.

## The hematopoietic stem cell is at the apex of a hierarchy of progenitor expansion and terminal differentiation

The process of hematopoiesis produces approximately $10^{11}$ to $10^{12}$ new blood cells every day in a human adult. Similar to other adult **tissue stem cells**, the HSC is a rare cell type, with the frequency in mouse bone marrow being less than 0.01% of the total number of cells. Generation of such a large number of blood cells from such a small number of stem cells is made possible because of the hierarchical nature of hematopoietic differentiation involving transient but extensive amplification of progenitors on the pathway toward the fully mature functional cell types of the blood (Figure 9.1). The hierarchical organization of the hematopoietic system, with the HSC sitting at the apex, facilitates expansion while at the same time serving a protective function, that is, it minimizes the number of cell division-associated genetic mutations arising within the stem cell population that have the potential to be carried through to daughter progeny. By having a distinct but short-lived population of progenitor cells that undergo rapid expansion, any mutation occurring within this population should be rapidly eradicated, unless it confers a selective advantage such as enhanced proliferation and survival or reduced differentiation (Chapter 18).

The characterization of HSC has depended heavily on cell sorting combined with *in vivo* and *in vitro* assays of self-renewal and differentiation potential.

Definition of the stem cell and progenitor components of the hierarchy, a process that is still very much ongoing even after decades of research, relies heavily on two key technologies, namely (i) flow cytometry and **fluorescence-activated cell sorting** using antibodies to recognize combinations of cell surface antigens that characterize HSC and (ii) *in vivo* assay of **stem cell potential** by transplantation (Chapter 2). Most of what we know about mammalian HSC comes from studies on the mouse, which is at least partly due to the relative ease of performing transplantation assays with mouse cells. Although this assay is the "gold standard" for demonstrating and quantifying HSC activity, it is both time consuming and costly and so researchers have developed less stringent assays that can indicate with some degree of accuracy the stem cell function of a given HSC subtype. Most usually, the **colony forming unit (CFU)** assay in the presence of defined cocktails of cytokines is performed to test the differentiation potential of a given HSC population. CFU assays developed to assess myeloid differentiation from HSC can accurately determine the myeloid cell output by means of colony morphology (Figure 9.2). Don Metcalf at the Walter and Eliza Hall research center in Melbourne was one of the principal researchers who helped to elaborate the CFU assay, especially as a means of defining hematopoietic growth factors, which then proved so useful to

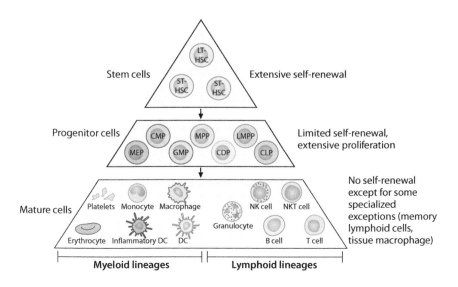

Figure 9.1 **The hematopoietic hierarchy.** Hematopoietic stem cells (HSC), like many stem cells, sit at the top of a hierarchy of cell expansion and progressive differentiation. The diagrammatic pyramid illustrates how a small number of hematopoietic stem cells, which are maintained by self-renewal, can give rise to a large number of differentiated, functional blood cell types through expansion driven by committed progenitor cells. Abbreviations: LT-HSC, long-term HSC; ST-HSC, short-term HSC; MPP, multipotent progenitor; LMPP, lymphoid-primed multipotent progenitor; CMP, common myeloid progenitor; MEP, megakaryocyte-erythroid progenitor; GMP, granulocyte-macrophage progenitor; CLP, common lymphoid progenitor; DC, dendritic cell; NK, natural killer cell. (From Manz M & Boettcher S. (2014) *Nat Rev Immunol* 14:303–314 [doi: 10.1038/nri3660]. Reprinted by permission from Springer Nature.)

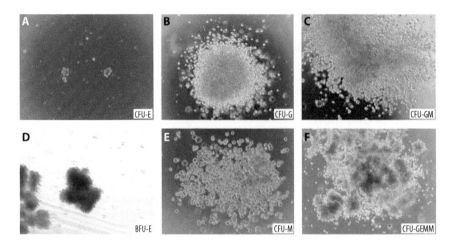

Figure 9.2 **Myeloid colony formation** *in vitro.* Hematopoietic stem cells and progenitors can be assayed for their ability to proliferate and differentiate using semi-solid media containing defined hematopoietic growth factors. Phase-contrast microscope images of hematopoietic colonies derived by the seeding of committed progenitors in semi-solid media containing a cocktail of myeloid cytokines. Abbreviations: (**A**) CFU-E, colony forming unit erythroid; (**B**) CFU-G, colony forming unit granulocyte; (**C**) CFU-GM, colony forming unit granulocyte-macrophage; (**D**) BFU-E, burst forming unit erythroid; (**E**) CFU-M, colony forming unit macrophage; and (**F**) CFU-GEMM, colony forming unit granulocyte-erythroid- megakaryocyte-macrophage. (From https://www.rndsystems. com/products/humanmethylcellulose-enriched-media_hsc005. With permission from Bio-Techne.)

the field in enabling characterization of individual cell differentiation potentials. Assessment of lymphoid potential is more difficult by CFU assay, although this can be achieved for B-cells. Demonstration of T-cell differentiation potential requires an alternative cell culture strategy using the support of a specific bone marrow **stromal cell** line (OP9) expressing the Notch ligand Delta-like 1.

## The hematopoietic stem cell population in mice can be resolved into several subtypes based on distinct patterns of surface antigen expression

It is almost three decades since one of the first descriptions of the identity of mouse HSC by Irving Weissman (Box 9.3) and colleagues. The original key strategy for HSC isolation developed by the Weissman laboratory relies on the selection of bone marrow cells for the absence of cell surface proteins that characterize the various

---

### BOX 9.3 PIONEER: IRVING WEISSMAN

Irving ("Irv") Weissman (Figure 1) has made a huge impact during his long career, both as the person who finally found a way to identify and isolate hematopoietic stem cells and as a thought leader in the field of stem cell biology, but also as an advocate and campaigner who has helped to shape public and political thinking and action. He grew up in Montana and was very early captivated by science, from the age of 16 spending a few summers in a local pathologist's laboratory helping with experiments on transplantation immunology and getting two co-authorships into the bargain. He went on to study for his BS at Montana State University in 1961, followed by his MD at Stanford University in 1965. Around this time, Weissman also showed how willing he is to challenge authority, organizing a protest against the role of doctors in the Vietnam war, as a result of which he was denied a position previously offered to him in the US National Institutes of Health. Aside from his seminal discoveries, key being the purification of HSC from mice in 1987, as professor of pathology and developmental biology and director of the Stanford Institute of Stem Cell Biology and **Regenerative Medicine**, Weissman trained many scientists who themselves became influential in hematopoietic biology, including Owen Witte, Sean Morrison, and Christa Muller-Sieburg. He has also had a huge impact on the community by championing the $3 billion California State initiative to fund stem cell research, which was very much part of a reaction to the restrictive policies on the use of

human embryos in stem cell research introduced by George W. Bush during his presidency. Passion and dedication are words that definitely describe Weissman's career and life, whether this be his scientific achievements or his devotion to fishing, the other great legacy of his upbringing in Montana.

Box 9.3 Figure 1 **Irv Weissman.** Photo credit: Norbert von der Groeben/Stanford Medicine. Used with permission.

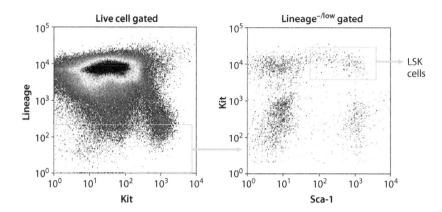

differentiated hematopoietic lineages. These cells, termed the lineage-negative (Lin⁻) population, can be subdivided on the basis of their expression of the stem cell markers Sca1 and Kit. Following the description of the Lin-Sca1+Kit+ population, or "LSK" for short (Figure 9.3), it was subsequently fractionated further using expression of **cluster of differentiation** 34 (CD34) and Flt3. Hiromitsu Nakauchi in Tokyo and Sten Erik Jacobsen at the Karolinska Institute in Stockholm showed how the surface proteins CD34 and Flt3 could be used further to restrict the definition of the murine HSC. These distinct stem cell types have differing potentials and can also be arranged hierarchically (Figure 9.4). Those LSK cells having low levels of both CD34 and Flt3 surface proteins have the greatest long-term repopulating potential, defined by sustained hematopoietic engraftment for more than 3 months in lethally irradiated recipient mice. In the simplest model of the HSC hierarchy, the stem cell residing at the apex of the HSC population has the greatest potential for expansion and differentiation. Termed the long-term HSC (LT-HSC), this CD34-negative, Flt3-negative LSK cell exists predominantly in a quiescent or "dormant" state. A gain in expression of CD34 accompanies transition to the short-term HSC (ST-HSC), which has a lesser engraftment potential. ST-HSC in turn give rise to multipotent **progenitors** (MPP), which express both CD34 and Flt3 and have essentially lost stem cell potential.

The more surface markers that became available, and the more of these that could be investigated simultaneously by flow cytometry and cell sorting, the more refined and sophisticated the scheme of the HSC population has become. For example, use of the surface marker VCAM-1 enabled discrimination of a multilineage progenitor downstream of the MPP that has distinct **commitment** bias toward lymphoid cells. This lymphoid primed multipotential progenitor (LMPP) retains full lymphoid potential and an element of myeloid potential with the exception of megakaryocyte-erythroid differentiation (Figure 9.5).

The most highly resolving strategy to date utilizing surface markers for the characterization and isolation of HSC came about through the identification using RNA microarray screening of a family of proteins that are expressed in specific combinations on different HSC subpopulations. Members of the family known as signaling lymphocyte activation molecules (SLAM) were first described over a decade ago in the laboratory of Sean Morrison and include the proteins CD48, CD150, CD229, and CD244. Using these markers, it has been possible to subdivide the LSK population into at least seven fractions (Figure 9.6).

**Figure 9.3 Murine hematopoietic stem cell isolation.** Hematopoietic stem cells are generally purified by combining immunofluorescent staining of surface antigens with passage through a fluorescence activated cell sorter. The figure is a representation of two-dimensional plots of mouse bone marrow cells that have passed through a cell sorter and been detected on the basis of fluorescent antibodies against various surface antigens. The panel on the left shows the total population of cells discriminated on the basis of their staining with a cocktail of antibodies targeted against lineage-specific antigens ("Lineage" on the Y-axis) and one against the Kit receptor ("Kit" on the X-axis). The panel on the right is an analysis of the cells defined within the yellow rectangle in the left-hand panel ("Lineage⁻/low") and represents the staining for Kit (Y-axis) and the stem cell antigen Sca-1 (X-axis). The region in the yellow rectangle includes HSC and multipotent progenitors and is usually referred to as Lineage⁻Kit⁺Sca-1⁺ ("LSK").

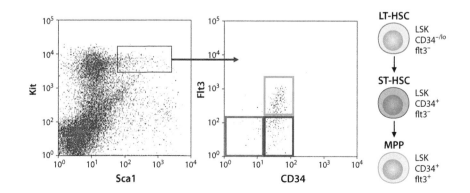

**Figure 9.4 Discrimination of hematopoietic stem cell subpopulations using CD34 and Flt3.** The heterogeneous hematopoietic stem cell (HSC) population can be subdivided using a number of additional surface markers. The panels represent cell-sorted mouse bone marrow cells as described in Figure 9.3. The LSK cells defined in the panel on the left are further resolved in the panel on the right based on their expression of Flt3 (Y-axis) and CD34 (X-axis). The colored rectangles define the surface antigen characteristics of the HSC subtypes that are illustrated in the hierarchical lineage shown on the right, which shows long-term (LT) HSC (blue), short-term (ST) HSC (red), and multipotent progenitors (MPP) (green).

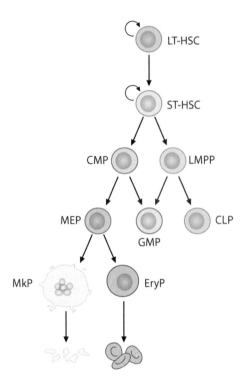

**Figure 9.5 A modified hematopoietic hierarchy.** The diagram shows the relationship between the self-renewing (circular arrows) hematopoietic stem cell (HSC) and the committed progenitors, with the inclusion of two branches of multipotent progenitors that give rise either solely to myeloid cells or predominantly to lymphoid cells. Abbreviations: LT-HSC, long-term HSC; ST-HSC, short-term HSC; LMPP, lymphoid-primed multipotent progenitor; CMP, common myeloid progenitor; MEP, megakaryocyte-erythroid progenitor; GMP, granulocyte-macrophage progenitor; CLP, common lymphoid progenitor; MkP, megakaryocyte progenitor; EryP, erythroid progenitor.

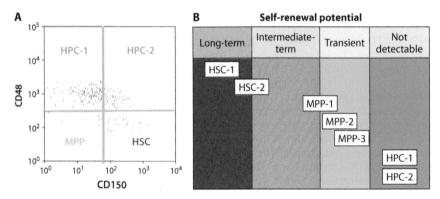

**C**

| Fraction | Surface marker expression | % of WBM |
|---|---|---|
| **HSC-1** | CD150$^+$CD48$^{-/low}$CD229$^{-/low}$CD244$^-$Lineage$^{-/low}$Sca-1$^+$c-Kit$^+$ | 0.0020±0.0005 |
| **HSC-2** | CD150$^+$CD48$^{-/low}$CD229$^+$CD244$^-$Lineage$^{-/low}$Sca-1$^+$c-Kit$^+$ | 0.0065±0.0021 |
| **MPP-1** | CD150$^-$CD48$^{-/low}$CD229$^{-/low}$CD244$^-$Lineage$^{-/low}$Sca-1$^+$c-Kit$^+$ | 0.0006±0.0001 |
| **MPP-2** | CD150$^-$CD48$^{-/low}$CD229$^+$CD244$^-$Lineage$^{-/low}$Sca-1$^+$c-Kit$^+$ | 0.0159±0.0038 |
| **MPP-3** | CD150$^-$CD48$^{-/low}$CD229$^+$CD244$^+$Lineage$^{-/low}$Sca-1$^+$c-Kit$^+$ | 0.0237±0.0056 |
| **HPC-1** | CD150$^-$CD48$^+$Lineage$^{-/low}$Sca-1$^+$c-Kit$^+$ | 0.1132±0.0271 |
| **HPC-2** | CD150$^+$CD48$^+$Lineage$^{-/low}$Sca-1$^+$c-Kit$^+$ | 0.0032±0.0014 |

**Figure 9.6 SLAM markers distinguish distinct hematopoietic stem cell populations.** Ever greater refinement of hematopoietic stem cell (HSC) purification became possible with the use of antibodies against the "signaling lymphocyte activation molecule" (SLAM) family of cell surface receptors. (**A**) The panels represent cell-sorted mouse bone marrow cells as described in **Figure 3.4**, the LSK cells being further resolved based on their expression of the SLAM markers CD48 (Y-axis) and CD150 (X-axis). The right-hand panel illustrates the surface antigen characteristics of the defined populations of stem cells and progenitors: HSC, hematopoietic stem cells; MPP, multipotent progenitor cells; HPC, hematopoietic progenitor cells. (**B**) Summary of the self-renewal properties of the defined HSC, MPP, and HPC populations. (**C**) Corresponding surface marker characteristics of each of the defined stem cell and progenitor populations and an indication of their frequency in whole bone marrow (WBM). (B and C: From Oguro H, Ding L & Morrison S. (2013) *Cell Stem Cell* 13:102–116 [doi: 10.1016/j.stem.2013.005.014]. Copyright 2013 with permission from Elsevier.)

## Properties of hematopoietic stem cells other than surface antigen expression can be utilized in their identification and purification

There are some means to identify and isolate HSC other than through the use of antibodies against cell surface proteins. Most notably, it is possible to exploit the ability of quiescent stem cells to selectively expel small molecules from their cytoplasm, potentially protecting them from cytotoxic insult. There is a family of transmembrane proteins known as ABC transporters that are expressed by stem cells and whose function it is to actively pump out cytotoxic chemicals. The

**TABLE 9.2  COMPARISON OF SURFACE ANTIGENS USED TO CHARACTERIZE AND PURIFY MOUSE AND HUMAN HSC**

| Surface antigen | Mouse | | | Human | |
|---|---|---|---|---|---|
| | LT-HSC | ST-HSC | MPP | HSC | MPP |
| Lineage | – | – | – | – | – |
| Sca-1 | + | + | + | NR | NR |
| CD117 (kit) | + | + | + | + | + |
| CD90 (Thy1) | + (low) | + (high) | – | + | – |
| Flt3 | – | – | + | + | – |
| CD34 | – | – | + | + | + |
| CD38 | + | + | + | – | – |
| CD150 | + | + | + | – | – |

*Note:* The table lists a number of key cell surface antigens that are used to identify either mouse or human HSC.
*Abbreviations:* LT-HSC, long-term HSC; ST-HSC, short-term HSC; MPP, multipotent progenitor; NR, not relevant.

family member ABC-G2 is highly expressed in HSC and has the ability to expel certain chemical dyes from the cytoplasm by passive diffusion. One such dye is the DNA binding compound Hoechst 33342, which, following excitation by a UV laser, can be visualized on a flow cytometer by measuring red and blue fluorescent light emissions. Selective exclusion of the dye from stem cells can be seen in the resulting characteristic profile in which the stem cells are found in the so-called **side population** (Chapter 2).

On top of the increasingly fine resolution of HSC subpopulations using ever more surface marker combinations, functional experiments have revealed other layers of complexity in terms of differences in cell cycle behavior or differentiation bias of the cells. Bromodeoxyuridine (BrdU) label retention (Chapter 2) reveals that there is heterogeneity in CD34-CD150+CD48- LSK HSC in terms of cell cycling, which has been equated to an active population responsible for homeostasis that cycles once every 30 days or so and dormant HSC that divide once every 150–200 days and serve an emergency function following injury.

## The characterization of human hematopoietic stem cells lags significantly behind that of mouse hematopoietic stem cells

Although we have gained significant knowledge about the nature of the various HSC subpopulations within mouse bone marrow, an equivalent level of understanding is severely lacking for human HSC. This is largely a result of a lack of analogous surface markers in humans; for example, the SLAM protein CD150 used in the identification of mouse LT-HSC is not expressed on human HSC. Also, many crucial HSC markers on murine cells, such as CD34 and Flt3, exhibit a different pattern of expression on human HSC, the majority of human HSC being CD34⁺, whereas the mouse equivalent cells are CD34⁻ (Table 9.2).

CD34 was the first marker of human HSC identified in the early 1990s but, as we have pointed out, its expression is counter to what is seen on mouse HSC, which only become CD34-positive when they have committed to the ST-HSC stage. The majority of human CD34⁺ bone marrow cells are actually lineage restricted progenitors and do not meet the precise definition of a stem cell. Incorporating antibodies against the surface markers CD38, CD45RA, and Thy1 enabled a more precise definition of HSC that are clearly distinct from MPP. In 2011, John Dick and colleagues identified the expression of CD49f (integrin α6) as another feature of human HSC and showed that transplantation of a single CD34⁺CD38⁻CD45RA⁻Thy1⁺CD49f⁺ bone marrow cell was sufficient to establish hematopoietic reconstitution in immune compromised mice (Figure 9.7).

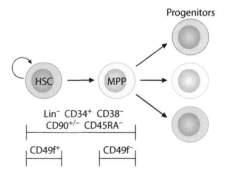

**Figure 9.7 The human hematopoietic stem cell hierarchy.** Human hematopoietic stem cells (HSC) are less well characterized and resolved compared to their mouse equivalents. At present, only two stages have been clearly defined based on their surface antigen characteristics. The schematic illustrates the surface marker differences that distinguish the most immature human HSC, which has the ability to repopulate all blood cell types, from the multipotent progenitor (MPP).

Another factor that has impeded the characterization of human HSC is the difficulty of performing the gold standard transplantation assay. A variant technique is used, involving transplantation of human cells into immune compromised mouse strains ("**xenotransplantation**," Chapter 2), thereby avoiding their rejection, but the environment for engraftment is by no means ideal, mainly because the potential **niches** for HSC are not able to provide all of the correct support and signals for effective **self-renewal** (Figure 9.8). This niche deficiency can be overcome to some extent, for example, by co-transplantation of stromal cell components or by direct injection of the test cells into the bone marrow, but the efficiency is often low, and there can be considerable variability.

Similar to studies in mice, alternative strategies for the isolation of human HSC have been explored. For example, human HSC express high levels of cytosolic aldehyde dehydrogenase (ALDH) to protect them from oxidative insult. Using a substrate for ALDH that yields a fluorescent product, it is possible to isolate ALDH$^{high}$ HSC. These cells are also enriched for the expression of CD133, a surface marker that has been shown to be expressed on several adult stem cells and could serve as a useful alternative to CD34 selection.

## 9.3    CELL HETEROGENEITY AND THE NICHE HAVE EFFECTS ON CLINICALLY RELEVANT ASPECTS OF HEMATOPOIETIC STEM CELL BEHAVIOR

Beyond reliable methods for the identification and isolation of HSC from the mouse and humans, the study of the stem cell basis of the hematopoietic hierarchy has enabled advances in our understanding of the intrinsic dynamics of adult stem cell populations as well as the requirements of the stem cell's environment, or niche. The niche plays a huge role in determining the maintenance and differentiation of HSC but also regulates stem cell mobilization from the bone marrow into the body as a whole. Both HSC heterogeneity, and how this can change with age, and the control elicited by the niche are of considerable importance in a clinical context.

### Hematopoietic stem cells are heterogeneous in terms of their likelihood of giving rise to myeloid or lymphoid lineage restricted progenitors

Using single cell transplantation experiments, Christa Muller-Sieburg, who in 1986 was the first in the world to prospectively purify HSC when working with Irving Weissman, provided findings in 2002 from her own laboratory in La Jolla showing that programmed behavior leads to biased differentiation of individual stem cells. She used limiting dilution analysis of bone marrow cultures to classify HSC into three types: myeloid biased, lymphoid biased, and balanced HSC (Figure 9.9). These lineage-biased phenotypes appear to be hard-wired in that secondary transplantation of HSC derived from primary transplants of myeloid, lymphoid, or balanced HSC demonstrates a similar lineage output among the engrafted cells. The relative frequency of the lineage-biased HSC changes with age, favoring myeloid-biased cells over lymphoid-biased or balanced HSC, perhaps explaining the shift away from lymphoid differentiation in older mice and humans (see also Chapter 18).

Other examples of functional bias among HSC have subsequently been identified, most notably through association with expression of the platelet-associated glycoprotein von Willebrand factor (vWF). Using a fluorescent protein reporter driven by the *Vwf* gene, Sten Erik Jacobsen and colleagues in Oxford identified a subpopulation of LT-HSC that are either platelet or platelet myeloid biased. Yet another functionally defined cell also serves to maintain the supply of platelets, but specifically under conditions of stress when they can become severely depleted, for example, as a result of infection and inflammation. Hence, stem-like megakaryocyte-committed progenitors (SL-MkP) can be identified in the HSC population. Under normal circumstances, these cells are quiescent, although they already contain platelet-specific RNAs, priming them for megakaryocyte differentiation. In response to acute inflammation, SL-MkP become activated, resulting in megakaryocyte protein production and efficient replenishment of platelets.

Many questions remain about heterogeneity among the HSC population, and it is probably fair to say that there is some confusion about the correspondence or otherwise of cells defined in individual experiments using different criteria. For instance, it is not totally clear what the relationship is between platelet-biased HSC

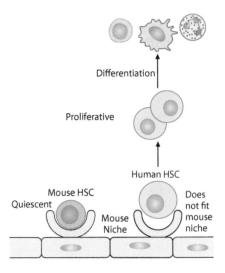

**Figure 9.8 Niche compatibility differences when engrafting mouse or human hematopoietic stem cells.** Grafting of human hematopoietic stem cells (HSC) into immune compromised mice to generate a so-called xenograft model provides the best means to assess human stem cell numbers and function. The schematic illustrates how mouse HSC engage with elements of the mouse bone marrow niche and can thereby persist in a quiescent state following transplantation. In contrast, transplanted human HSC in a xenograft model engage less well and as a consequence tend toward proliferation and differentiation. (Adapted from Goyama S, Wunderlich M & Mulloy J. (2015) *Blood* 125:2630–2640 [doi: 10.1182/blood-2014-11-570218].)

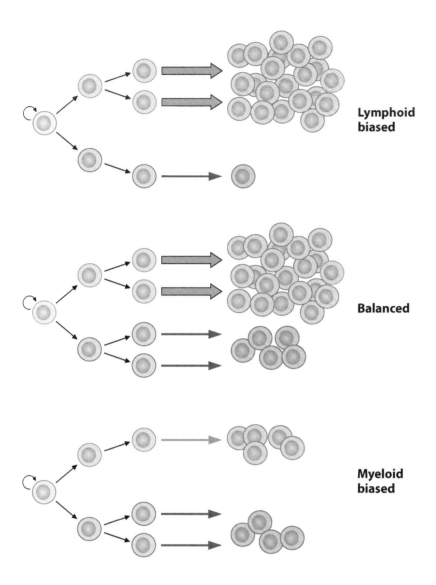

**Lymphoid biased**

**Balanced**

**Myeloid biased**

Figure 9.9 **Heterogeneity in the hematopoietic stem cell population.** The hematopoietic stem cell (HSC) population is heterogeneous in a number of ways, one of which is their propensity for commitment toward myeloid or lymphoid lineages. The diagram illustrates the types of stem cells that can be defined based on this commitment preference. HSC (blue) each exhibit distinct, stable differentiation programs. The thickness of the arrows corresponds to the ability of the HSC to generate progeny of the indicated type (blue arrows for lymphoid and red arrows for myeloid). Lymphoid-biased HSC generate few myeloid cells (orange) but standard levels of lymphocytes (purple). Myeloid-biased HCS produce few lymphocytes but standard levels of myeloid cells. Balanced HSC generate more lymphocytes than myeloid cells (at least in mouse) and are named because their output resembles the average output of mature cells from all HSC subsets together. Each HSC subtype has self-renewal capacity (circular arrow) and gives rise to intermediate progenitors. (Adapted from Muller-Sieburg C, Sieburg H, Bernitz J et al. (2012) *Blood* 119:3900–3907 [doi: 10.1182/blood-2011-12-376749].)

and SL-MkP. Also, how do the HSC subtypes defined by cell cycle behavior or self-renewal activity overlap with those based on lineage potential? Lastly, what are the mechanisms that determine the various aspects of the observed heterogeneity? The serial transplantation studies already mentioned point to a stable cell autonomous explanation, such as epigenetic modification, which might reflect a developmental origin, but there could also be a contribution from the stem cell environment, especially in the context of changes related to aging.

## Hematopoietic stem cell heterogeneity can become restricted during aging

Irrespective of the functional bias of individual HSC, it is also now accepted that aging introduces a change in the profile of the overall population of stem cells through a progressive reduction in its diversity. Hence, so-called "age-related clonal hematopoiesis" (ARCH) results from the effects of selective advantage or disadvantage in the self-renewal of individual HSC, which ultimately leads to a predominance of a more restricted number of stem cell clones. The health impact of ARCH is thought to be a higher predisposition to disease involving hematopoietic insufficiency or malignancy (Chapter 18). The underlying mechanisms leading to ARCH are likely to involve epigenetic modification or random somatic mutation, the latter actually being the basis of several genome-wide studies that have mapped the occurrence of gene copy number variations in people of different ages. Two such studies, both conducted in Harvard Medical School, but independently in the laboratories of Benjamin Ebert and Steven McCarroll, illustrate the extent and the potential effects of ARCH. These studies found that somatic mutations were rare in blood cells from people younger than 40–50 years of age but rose appreciably in frequency with age, the Ebert study concluding that individuals in their 70s, 80s, and 90s carry a clonal mutation burden of 10%, 12%, and 18%, respectively, with a corresponding increased risk of hematological malignancy, which is probably at least partially linked to the fact that a large number of the mutations occurred in three leukemia-associated genes.

## Hematopoietic stem cells reside in the bone marrow in distinct niches that regulate their self-renewal and differentiation

All stem cells are dependent upon their environment, the various components of which, whether these are cells, **extracellular matrix**, or soluble factors generated locally or at a distance in the organism, constitute the niche (Chapter 6). In the case of HSC, the niche is not only a crucial feature of their existence and ability to function correctly, but it is also the determinant of the ability of the *in vivo* transplantation assays to yield a meaningful result. Hence, as we have mentioned, the engraftment of human HSC in immune compromised mice is less than ideal because of mouse-human differences, many of which relate to the composition of the stem cell niche.

Ironically perhaps, although the HSC is by far the most studied adult stem cell, there is still considerable uncertainty about the precise nature of its niche, or indeed if there are multiple niches suited to cells with distinct potentials. Part of the reason for the relative lag in understanding HSC niches is the fact that they are within the bone marrow, which poses challenges for the visualization of cells either in sections or in the *in vivo* context. The other main problem around direct observation is the fact that, for all that we have some extremely refined definitions of the surface antigen phenotype of HSC, the uniqueness of this phenotype relies upon a combination of markers, rather than one specific HSC-restricted molecule, and each of these defining markers is expressed on one or more of the non-stem cell types.

## The bone marrow niche is far more complex than was first suspected

HSC have long been thought to reside at the interface of the bone and bone marrow (the "endosteum") and that this represents the stem cell niche (Figure 9.10). HSC certainly reside in close proximity to osteoblasts (Figure 9.11), the cells that produce new bone at the interface with the bone marrow, and experimental manipulation of osteoblasts, by genetic or pharmacological means, influences HSC number and function. The interaction between HSC and osteoblasts has been proposed to involve the adhesion molecule N-cadherin; however, studies involving the **knockout** of its gene in either cell type failed to reveal any effect, casting doubt on this particular conclusion. The current view is rather that osteoblasts support HSC by a paracrine effect, producing growth factors including Angiopoietin-1 and Thrombopoietin (TPO), which respectively bind to the receptors Tie-2 and MPL present on HSC, thereby helping to maintain their normal quiescent state.

Direct observation by confocal microscopy of LT-HSC, highlighted by fluorescent antibody staining for the SLAM markers, revealed their location to be in close proximity to sinusoidal blood vessels (Figure 9.10), and it is now believed that this identifies a "perivascular niche" as a crucial microenvironment responsible for the maintenance of HSC (Figure 9.11). The chemokine CXCL12, also known as SDF-1 (Stromal derived factor-1), is the ligand for CXCR4, is crucial for HSC maintenance in the bone marrow, and is produced in large amounts by a specific cell type in the perivascular niche. These cells, termed CXCL12 abundant reticular (CAR) cells, are in close contact with HSC near blood vessels and are uniformly scattered throughout the bone marrow. Similar cells expressing CXCL12 can be seen in human bone marrow, and their genetic ablation in the mouse results in a loss of HSC. Changes in the interaction between CXCR4 and CXCL12 result in altered dynamics between HSC and the bone marrow niche.

Adding to the complexity of cell types involved in the HSC bone marrow niche, a **mesenchymal stem cell (MSC)** related cell, which specifically expresses the protein Nestin, appears to play an important role. These Nestin-positive MSC are also distributed in the perivascular domain, express high levels of CXCL12 and stem cell factor (SCF) and have been seen associating with CD150$^+$CD48$^-$ HSC. Depletion of the Nestin-positive MSC correlates with a reduction in HSC number in the bone marrow.

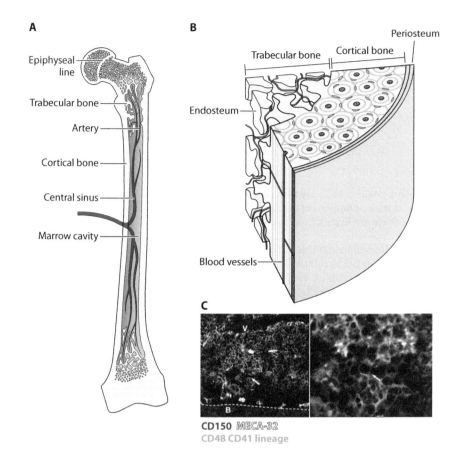

**A**

Epiphyseal line

Trabecular bone

Artery

Cortical bone

Central sinus

Marrow cavity

**B**

Periosteum

Trabecular bone

Cortical bone

Endosteum

Blood vessels

**C**

**CD150** MECA-32
CD48 CD41 lineage

**Figure 9.10 The bone marrow niche.** Hematopoietic stem cells (HSC) reside mainly in the bone marrow. (**A**) Diagram of a femur illustrating the bone structure, especially the minute projections of bone known as trabeculae, which provide a large surface for contact with cells. (**B**) Segment of bone from the trabecular region showing the endosteum, which acts as the interface between bone and the marrow. The endosteal region is supplied with a rich blood supply through a network of arteries and arterioles (red). (**C**) Section through the endosteal region showing the interface of bone and marrow. Cells have been stained with specific antibodies linked to fluorescent labels that are either green (CD48, CD41, and Lineage antigens) or red (CD150). Low (left) and high (right) magnification views show an HSC (arrow) with the phenotype CD150$^+$CD48$^-$CD41$^-$Lin$^-$. This HSC is close to the endosteum (dotted line) but not detectably in contact with cells lining bone (B, bone; V, blood vessel). (A and B: From Morrison S & Scadden D. (2014) *Nature* 505: 327–334 [doi: 10.1038/nature12984]. Reprinted by permission from Springer Nature. C: From Morrison S & Spradling A. (2008) *Cell* 132:598–611 [doi: 10.1016/j.cell.2008.01.038]. Copyright 2008 with permission from Elsevier.)

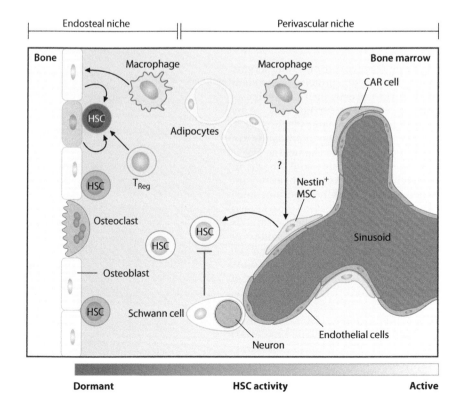

Endosteal niche | Perivascular niche

Bone

Macrophage

Macrophage

Bone marrow

CAR cell

HSC

Adipocytes

T$_{Reg}$

HSC

?

Nestin$^+$ MSC

Osteoclast

HSC

HSC

Sinusoid

Osteoblast

HSC

Schwann cell

Endothelial cells

Neuron

**Dormant** — **HSC activity** — **Active**

**Figure 9.11 Bone marrow hematopoietic stem cell niches.** Most hematopoietic stem cells (HSC) reside in one of two locations in the bone marrow, that is, the endosteal and perivascular niches. HSC located in the endosteum tend to be more quiescent, whereas HSC located in the perivascular region are more active. In the endosteal niche, HSC are located near osteoblast progenitors (preosteoblasts). All of the indicated cell types present in the bone marrow (with the exception of osteoclasts, which remodel bone) are known to modulate HSC behavior (pointed and blunt arrows indicating positive and negative influences, respectively). Two of the downstream HSC progeny cells, regulatory T cells (T$_{Reg}$) and macrophages, participate in HSC regulation. Macrophages seem to maintain HSC in their niche through an indirect mechanism mediated by Nestin+ mesenchymal stem cells (MSC) and possibly by osteoblasts. (From Hsu Y-C & Fuchs E. (2012) *Nat Rev Mol Cell Biol* 13:103–114 [doi: 10.1038/nrm3272]. Reprinted by permission from Springer Nature.)

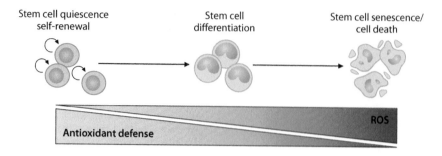

**Figure 9.12 The influence of ROS levels on hematopoietic stem cells.** Quiescent or self-renewing hematopoietic stem cells (HSC) have low reactive oxygen species (ROS) levels due to their strong antioxidant machinery, maintained by proteins such as FOXO3. Intermediate ROS levels prime HSC for differentiation, while high ROS levels cause stem cell senescence and death. (Adapted from Bigarella C, Liang R & Ghaffari S. (2014) *Development* 141: 4206–4218 [doi: 10.1242/dev.107086].)

Cell types such as osteoblasts, CAR cells, and Nestin-positive MSC are undoubtedly just the tip of the iceberg in terms of the variety of niche components that impact HSC. For example, it has recently been shown that there are inputs from nerves that influence HSC.

Importantly, the bone marrow environment exhibits considerable variation in the availability of oxygen, regions of low vascularization being relatively hypoxic. Interestingly, HSC residing in hypoxic conditions have a higher reconstitution potential than those that do not. *In vitro* cultures of HSC under hypoxic rather than normal oxygen conditions contain fewer cells due to exit into a cell cycle G0 state, but these have a higher engraftment capability. The levels of reactive oxygen species (ROS) play a part in the maintenance of HSC. ROS are produced by the partial reduction of oxygen and include superoxide ($O_2^-$), hydrogen peroxide ($H_2O_2$), and the hydroxyl radical ($OH^-$). Genetic reduction of proteins responsible for monitoring and regulating cellular ROS levels, such as the FOXO transcription factor, leads to a loss of HSC quiescence and self-renewal capacity, an observation that correlates with the enhanced engraftment of HSC grown under hypoxic conditions (**Figure 9.12**).

## Hematopoietic stem cell mobilization from the niche and homing back to it are key both for its normal function and in a therapeutic context

The interaction of HSC with the niche is by no means static, and under normal conditions there is a continual, if somewhat limited, outward flow of stem cells from the bone marrow. These "mobilized" HSC circulate around the body but do not return to the bone marrow. The return of HSC to the bone marrow niche, which is seen during stem cell transplantation, is referred to as "homing," a process that is crucial to the success of therapeutic bone marrow donation.

The mechanisms of mobilization and homing involve various interactions between HSC and the niche, including receptors and adhesion proteins on the stem cell and corresponding ligands expressed on the surface of cells such as osteoblasts or associated with the extracellular matrix, which can be further affected by locally or systemically released growth factors, lipids, or proteases (**Figure 9.13**). Perhaps the best understood factors that influence mobilization and homing are components of the SDF-1-CXCR4 chemokine-receptor axis. The chemokine SDF-1 is produced by bone marrow stromal cells, such as osteoblasts and CAR cells in the endosteal and perivascular niches, respectively, and acts as a potent attractant for HSC through binding to its receptor CXCR4. This is clinically important since disruption of the binding of SDF-1 to CXCR4 using the drug AMD3100 (also known as Plerixafor) is an effective means to artificially mobilize HSC for collection of cells to use in transplantation. SDF-1 and CXCR4 are also important in the context of the mode of action of granulocyte colony stimulating factor (G-CSF), which is the most frequently used clinical HSC mobilizing agent. G-CSF acts at multiple points (**Figure 9.13**), including down regulation of SDF-1 expression and inactivation of both SDF-1 and CXCR4 through the action of proteases that are released from bone marrow stromal cells.

Aside from SDF-1-CXCR4, several other interactions between HSC and the niche are important in mobilization and homing. Notable examples include Kit (CD117) on HSC binding to its ligand SCF and the integrin α4β1 (VLA-4) on HSC that binds to the adhesion molecule VCAM-1. The importance of the latter dimeric pair is apparent when either molecule is inhibited using specific antibodies or genetically

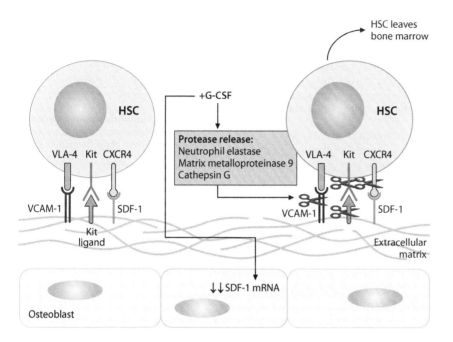

**Figure 9.13 Interactions with niche components determine the mobilization of hematopoietic stem cells.** A variety of receptors expressed by hematopoietic stem cells (HSC) bind to ligands in the extracellular matrix and are susceptible to pharmacological treatments that lead to mobilization of the stem cells. Addition of G-CSF induces protease release that cleaves VCAM-1 and SDF-1 in the niche, as well as CXCR4 and VLA-4 on the HSC, thereby allowing the HSC to exit the bone marrow. In addition, G-CSF causes a reduction in SDF-1 mRNA expression in SDF-1 osteoblasts. (Adapted from https://www.pparesearch.com/whitepaper/stem-cell-mobilization-protocols-filgrastim-vsmozobil/.)

deleted, which leads to HSC mobilization, and, interestingly, VCAM-1 is also a target of the proteolytic effects of G-CSF. Last, an important additional factor that synergizes with the molecular interactions just described involves the existence of gradients of soluble lipids, most notably of sphingosine-1-phosphate, which act as chemoattractants for HSC when the lipid binds to its cognate receptor on the stem cell.

## Until recently, the requirement for the niche has limited the potential for expansion of hematopoietic stem cells *ex vivo*

Although HSC have been under intense investigation for well over half a century and as a consequence can be identified and isolated with considerable precision, it has proven very difficult to maintain them once removed from their natural environment. There would be obvious advantages if HSC could be cultured *in vitro*, both from an experimental perspective and importantly so that more cells could be available for therapeutic application. However, it has proved to be very difficult to maintain and expand fully functional HSC, loss of self-renewal and engraftment potential being an inevitable consequence within a short space of time. Such difficulties attest to the importance of the HSC niche, which many have tried to replicate. Recently, Satoshi Yamazaki and colleagues in the University of Tokyo appear to have cracked the problem and have shown that transplantable mouse HSC can be expanded *ex vivo*. Their solution has been to utilize high levels of the growth factor TPO to synergize with low levels of SCF and fibronectin in the presence of polyvinyl alcohol. The degree of expansion over a month is impressive, up to 900 cell doublings being achieved, and it will be important now to see how this methodology translates into the human context.

## 9.4 STEM CELLS SUPPORTING BLOOD CELLS IN INVERTEBRATES

As we have already discussed, there are clear descriptions of hematopoietic cells in most multicellular invertebrate animals, certainly back to the evolutionary stage of the sponges, although the precise relationship between these invertebrate cells and those that are so well characterized in vertebrates is not completely clear. There are instances of functional homology, but it is not necessarily the case that this corresponds to an evolutionary relationship; that is, convergent evolution may have played as much a part as divergence. Nevertheless, a requirement for a stem cell source for these cells can be confidently assumed in many cases because of the longevity of the organism. Divergent evolution is of course a prerequisite if lessons learned from lower-order animals can be applied to the understanding of the mechanisms operating in vertebrate scenarios. There are many examples, some of which we discuss in this book, where the study of invertebrate stem cells and their niche can directly inform the study of the vertebrate equivalent. The classic example is the fruit fly *Drosophila* testis, in which the cells and signals involved in the maintenance and differentiation of the **spermatogonial stem cells** are quite closely mirrored in the mammalian situation (Chapter 12).

From annelids to arthropods, including crustaceans and insects, there are examples of invertebrate animals in which hematopoietic cells are continuously produced from progenitor clusters. However, in only two organisms have these centers of hematopoiesis been studied in detail in a way that allows comparison to what we know about hematopoiesis and HSC in vertebrates. One is the crayfish, and the other, perhaps unsurprisingly since it is the invertebrate model organism par excellence, is the fruit fly. Of the two, the understanding of hematopoiesis and the stem cells supporting it is significantly more advanced for *Drosophila*, so much so that you will rightly leave this chapter with the view that HSC in the animal kingdom are presently predominantly described by a comparison of the fruit fly and the mouse! What this means, of course, is that the groundwork is there to allow informed and more efficient investigation of other animals, whether that be human compared to the mouse, or a crustacean or worm compared to the fruit fly.

**Figure 9.14 Crayfish hematopoietic tissue.** The freshwater signal crayfish, showing the location (red oval) of the hematopoietic tissue (HPT).

## A hierarchy of hematopoiesis with a hematopoietic stem cell-like cell at the apex has been defined in crayfish

The crustacean model of choice to investigate hematopoiesis has been the signal crayfish (Figure 9.14), *Pacifastacus leniusculus*, which is a native of North America but is a pest in Europe, having escaped from collectors into the river systems with disastrous consequences for the local ecology. Studies in the laboratory of Irene Söderhäll in Uppsala have defined the hematopoietic tissue (HPT) of the crayfish as a distinct organ. The HPT contains five distinct types of hemocytes that develop along two lineages, namely semigranular cells (SGC) and granular cells (GC), which fit into a differentiation hierarchy starting from a presumed stem cell, the Type 1 cell (Figure 9.15). The most significant advance based on the crayfish model has been the identification of two novel cytokines known as astakines, which are related to vertebrate prokineticins, a group of small, secreted proteins that are thought to be involved in several important physiological processes like neurogenesis, tissue development, and angiogenesis. Astakine 1 stimulates proliferation of HPT cells, the presumed stem cells, but also induces their differentiation along the SGC lineage, whereas Astakine 2 plays a role in GC differentiation. If parallels indeed exist between invertebrates and vertebrates with respect to hematopoietic hierarchy regulation, then it might be expected that astakine homologues in vertebrates, possibly the prokineticins, might be discovered.

## Although evolutionarily distant from vertebrates, hematopoiesis and its stem cell components in *Drosophila* exhibit some shared features

The extent of parallels between invertebrate and vertebrate hematopoiesis are somewhat more obvious in the fruit fly. For a long time, it was thought that adult hemocytes are only generated during the larval stages of *Drosophila* development (Figure 9.16), but very recently this has been found to be a simplistic view and that actually, as in vertebrates, blood cells are continuously produced from stem cells throughout life. Without doubt, though, the larval source of adult hemocytes, the lymph gland (LG), has been most thoroughly studied. The LG yields adult hemocytes from a pool of **multipotent** precursors upon metamorphosis or in response to immune challenge. The largest lobe of the LG is subdivided into the posterior signaling center

Figure 9.15 **The hematopoietic hierarchy in crayfish.** Within the crayfish hematopoietic tissue (HPT), Type 1 pluripotent stem cells, or prohemocytes, can give rise to Type 2 cells, which are the precursors of both semigranular cells (SGC) and granular cells (GC). Type 3 and Type 4 cells represent GC precursors, and Type 5 cells are the precursors of SGC. Astakine 1 (Ast1) acts on HPT cells, promoting their proliferation (yellow arrow), blocking apoptosis, pushing their differentiation toward Type 2 cells, and then sending cells along the SGC lineage. Ast2 plays an important role in the hematuration of the GC lineage. (Adapted from Lin X & Söderhäll I. (2011) *Blood* 117: 6417–6424 [doi: 10.1182/blood-2010-11-320614].)

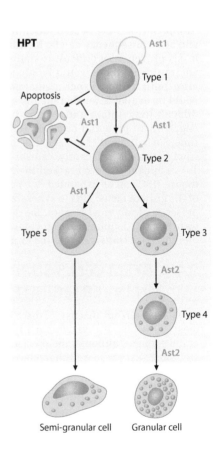

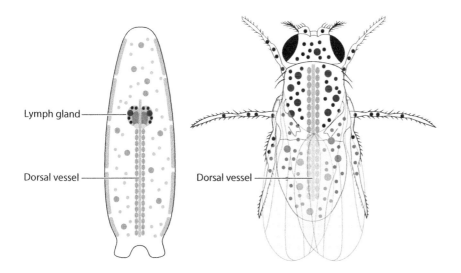

Figure 9.16 *Drosophila* **hematopoietic cells of larval origin.** The schematic diagram illustrates the location of the lymph gland (purple) surrounding the anterior end of the dorsal vessel in the *Drosophila* larval stage (left). Hematopoietic cells in the larva emanating from the lymph gland are indicated by small circles (plasmatocytes) or large circles (crystal cells), and these are colored to depict purely larval cells (pink) or cells that will populate the adult (red and purple). The representation of the adult fly similarly shows the presence of the adult type plasmatocytes and crystal cells distributed throughout the body. (From Evans C, Sinenko S, Mandal L et al. (2007) *Adv Dev Biol* 18:259–299 [doi: 10.1016/S1574-3349(07)18011-X]. Copyright 2007 with permission from Elsevier.)

(PSC), the medullary zone (MZ), and the cortical zone (CZ) (Figure 9.17). In much the same way as **lineage tracing** has been utilized in mouse models to confirm stem cell self-renewal and differentiation, clonal marking has shown that cells in the MZ are multipotent stem cell-like cells known as prohemocytes, which migrate into the CZ as they differentiate into the three types of functional hemocytes (plasmatocytes, crystal cells, and lamellocytes). The PSC controls this commitment process and can be regarded as the homolog of the vertebrate HSC niche. Furthermore, many transcriptional regulators and signaling pathways involved in hematopoiesis in both the niche and HSC are conserved between mammals and *Drosophila* (Table 9.3). Regulators such as Wingless (Wg), Hedgehog (Hh), and Notch emanate from or are influenced by ligands produced by the PSC and lead, for instance, to effects on prohemocyte signaling involving JAK-STAT (Figure 9.18). All of these molecules have mammalian homologs so that the relative ease of genetic studies in *Drosophila* provides a good opportunity to gain insight into signal integration and crosstalk in HSC and their environment. Another example of evolutionary conservation of regulatory mechanisms between prohemocytes and HSC concerns the effects of molecular oxygen in the stem cell niche. As we have seen already, variation in oxygen levels throughout the bone marrow environment plays an important part in the regulation of vertebrate bone marrow HSC. Low oxygen levels characterize the osteoblastic niche and are probably crucial for the maintenance of LT-HSC, while higher oxygen concentrations in the vascular niche favor proliferation of more committed cells. Low oxygen levels limit accumulation of ROS, thereby limiting damage and helping to maintain quiescence and enhance survival. In *Drosophila*, ROS levels in prohemocytes sensitize them to differentiation, and just as the Forkhead transcription factor FOXO plays a role in regulating vertebrate HSC responses to oxidative stress, so too is the homologous regulator dFOXO in the fruit fly involved in lamellocyte differentiation.

Finally, as we already alluded to, true adult stem cells have recently been discovered in *Drosophila* with the finding of active blood cell formation in the dorsal part of the abdomen (Figure 9.19). These active hematopoietic hubs are capable of hemocyte specification and can respond to bacterial challenge to produce increased numbers of functional cells to eradicate the infection. The adult fruit fly hemocyte progenitors and differentiated hemocytes appear to be embedded in a functional network of extracellular matrix proteins, so it is quite possible that, just as the stem cell niche of the *Drosophila* testis has given clues about the nature and function of the mammalian male **germ stem cell** niche (Chapter 12), there may be a good chance that an understanding of this simple invertebrate HSC niche might provide fresh insight into the mammalian bone marrow niche. Interestingly, the precursor cells within the hub are not homogeneous and seemingly mirror some of the lineage-bias heterogeneity of HSC present in the vertebrate bone marrow. While some prohemocytes are capable of differentiating solely into plasmatocytes, there exist a few bipotent precursors. Notch appears to play a key role in determining the fate of these bipotent HSC in that cells that maintain signaling adopt a crystal cell fate, while those in which Notch signaling declines are triggered to become plasmatocytes.

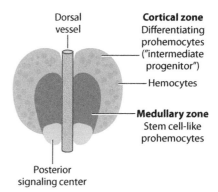

Figure 9.17 *Drosophila* **lymph gland hematopoietic niche.** Schematic drawing of the *Drosophila* larval lymph gland (primary lobe), showing the three major zones, which are symmetrically positioned on either side of the dorsal vessel, and the presence of hematopoietic cells. The posterior signaling center effectively serves as the niche. The medullary zone contains the stem cell-like prohemocytes, while the outer cortical zone contains intermediate progenitors and fully differentiated hemocytes. (From Grigorian M & Hartenstein, V. (2013) *Dev Genes Evol* 223:103–115 [doi: 10.1007/s00427-012-0428-2]. Reprinted by permission from Springer Nature.)

## TABLE 9.3  CONSERVED PROTEINS WITH IMPORTANT ROLES IN HEMATOPOIETIC STEM CELL REGULATION

|  | *Drosophila* function | Mammalian function |
|---|---|---|
| **Transcription factor** | | |
| Serpent (GATA) | Specification of hemocytes; progenitor maintenance | Maintenance of HSC; lineage commitment; differentiation |
| Lozenge (RUNX) | Specification of hemocytes | Maintenance of HSC; lineage commitment; differentiation |
| U-shaped (FOG) | Specification of hemocytes; differentiation of lamellocytes | Lineage commitment; regulation of differentiation |
| Collier (EBF) | Maintenance of PSC cells | B-cell differentiation; maintenance of the niche |
| **Signaling pathway** | | |
| Notch | Differentiation of crystal cells | Maintenance of HSC; hematopoietic regeneration; regulation of differentiation |
| Hedgehog | Maintenance of prohemocytes | Lineage commitment; regulation of differentiation |
| Wingless (WNT) | Proliferation of PSC cells; maintenance of prohemocytes; differentiation of crystal cells | Maintenance and proliferation of HSC |
| JAK-STAT | Maintenance of prohemocytes; differentiation of lamellocytes | HSC renewal during regeneration; lineage commitment |
| ROS/FoxO | Maintenance of prohemocytes | Maintenance of HSC; differentiation of erythroid cells |
| Pvf (PDGF/VEGF) | Differentiation of plasmatocytes | Maintenance of HSC |

*Source:*  From Crozatier M & Vincent A. (2011) *Dis Mod Mech* 4: 439–445 [doi: 10.1242/dmm.007351]. Published under the Creative Commons Attribution (CC BY) license (https://creativecommons.org/licenses/by/4.0/).

*Note:*  The table summarizes the involvement in hematopoietic cell specification and differentiation of transcription factors and signaling components that are conserved between *Drosophila* and mammals. When the gene name differs between *Drosophila* and mammalian homologs, the mammalian name is shown in parentheses.

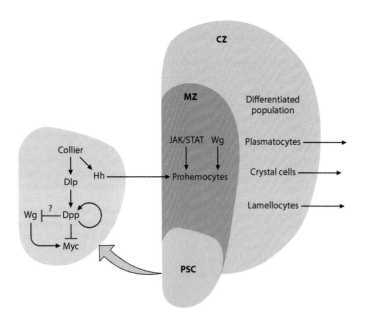

**Figure 9.18 Signaling pathways controlling hematopoietic progenitors in the lymph gland.** Signaling within and between the three zones in the lymph gland combine to maintain the stem cell population (prohemocytes) and regulate differentiation to the three mature cell types (plasmatocytes, crystal cells, and lamellocytes). The posterior signaling center (PSC) produces signaling molecules such as Hedgehog (Hh) and Wingless (Wg), ultimately under the control of the transcription factor Collier (homolog of the mammalian protein EBF). Signaling from the PSC impacts prohemocytes in the medullary zone (MZ), which are also acted on by further signals from Wg and the JAK/STAT pathway before differentiated cells enter the cortical zone (CZ). (From Lee J, Cranna N, Chahal A et al. (2012) *Cells* 1:1182–1196 [doi: 10.3390/cells1041182]. This article is distributed under the terms and conditions of the Creative Commons Attribution license (http://creativecommons.org/licenses/by/3.0/).)

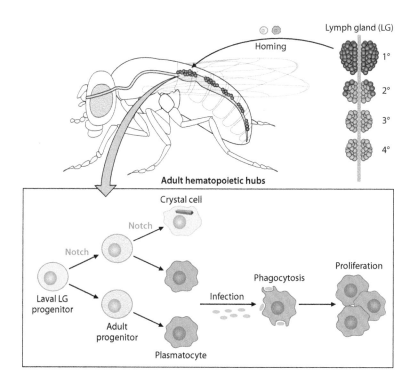

Figure 9.19 **Active hematopoiesis in adult *Drosophila*.** Hemocytes from the larval lymph gland (LG) populate the adult hematopoietic hubs, which lie along the gut. Within the adult hubs, the larval LG prohemocytes can adopt two progenitor states, one that is monopotent and gives rise to plasmatocytes and one that is bipotent and able to give rise to both plasmatocytes and crystal cells. Further expansion of plasmatocytes, the homolog of the mammalian macrophage, is stimulated following bacterial infection. Notch signaling is important in controlling the fate of prohemocytes and the bipotent progenitor, favoring the pathway leading to crystal cells. (From Ghosh S, Singh A, Mandal S et al. (2016) *Dev Cell* 33:478–488 [doi: 10.1016/j.devcel.2015.03.014]. Copyright 2016 with permission from Elsevier.)

## 9.5 SUMMARY

In this chapter, we have aimed not only to describe HSC as another example of an adult stem cell type but have used the story of its discovery and characterization to show how this has been a long journey involving a host of laboratories around the world. The history of the progressive increase in our understanding of HSC illustrates many of the principles and technologies that underpin thinking about and research into stem cell biology. Although the mouse has been the primary experimental model for investigation of HSC, and for sure the degree of detailed knowledge is nothing short of impressive, the imperative of therapeutic potential is ever more rapidly pushing forward research on human HSC. We also take the opportunity to step back in evolutionary terms to invertebrates because, with the stage set by studies on mammals, these provide further potential as model systems in which to gain insight into how the HSC has evolved and establish the underlying molecular and cellular mechanisms controlling its properties.

## KEY POINTS

- The vertebrate blood system consists of specialized cells that perform diverse functions, including protection against various foreign bodies and infections, repair of injuries, and transport of gases to and from tissues. The blood system is maintained in the adult by a hierarchy of cells at the top of which sit stem cells, which reside in the bone marrow of mammals or related tissues in other vertebrates.

- Cells related to some of the vertebrate blood cell types, usually termed hemocytes, can be found in many invertebrate species back in evolutionary terms as far as the sponges. In most species with hemocytes, a self-renewing cell has been identified that is believed to maintain their production throughout life.

- Driven by the desire to perform and to understand the cellular basis of therapeutic transplantation of hematopoietic (blood) stem cells (HSC), a vast research effort over more than 50 years has given us a detailed picture of their phenotype, location, and functional properties.

- Studies on HSC have been most extensive in the mouse, and our understanding of human HSC is far less complete.

- Mouse HSC, and presumably those in humans, are a complex mix of distinct cell types with different functional properties, including their ability to engraft, their role in homeostasis versus emergency regeneration, and their potential to give rise to the full range of hematopoietic cells or only a restricted subset of lineages.

- Mammalian HSC reside in discrete niches in the bone marrow, at least two being recognized, that is, the endosteal niche next to the bone that maintains quiescent cells in a low oxygen environment and a perivascular niche adjacent to blood vessels, which is exposed to higher levels of oxygen and supports more proliferative HSC.

- The presence of HSC in the niche is not static in that they are able to both leave and return, either naturally or as part of a clinical intervention, through processes that involve modulation of one or more of the molecular interactions linking the stem cell to the bone marrow environment.

- Invertebrate blood cell systems are simpler than their vertebrate counterparts, but studies on model organisms are revealing some interesting parallels with vertebrates in the way that invertebrate stem cells, or prohemocytes, are maintained and regulated.

- The two best studied invertebrate models are both arthropods, namely a crustacean (the crayfish) and an insect (the fruit fly *Drosophila*). *Drosophila* hematopoiesis is particularly well studied, especially that involving stem cells in the larval lymph gland, although an adult site of hematopoiesis has also now been identified.

- Key regulators of *Drosophila* hematopoiesis, including transcription factors and signal transduction molecules, often have close homologs in mammals that perform similar roles in either the maintenance of the stem cell component or in determining differentiation fate.

- Lessons learned from the study of adult stem cells giving rise to blood cells, especially in the mouse, have provided a technological and intellectual impetus that has facilitated studies aimed at understanding many other tissue stem cells.

## FURTHER READING

### General

Eaves, C.J. (2015) Hematopoietic stem cells: Concepts, definitions, and the new reality. *Blood*, 125: 2605–2613.

Wilkinson, A.C., Ishida, R., Kikuchi, M., Sudo, K., Morita, M., Crisostomo, R.V., Yamamoto, R. et al. (2019) Long-term *ex vivo* haematopoietic-stem-cell expansion allows nonconditioned transplantation. *Nature*, doi: 10.1038/s41586-019-1244-x.

### Hematopoietic stem cell heterogeneity

Crisan, M. and Dzierzak, E. (2016) The many faces of hematopoietic stem cell heterogeneity. *Development*, 143: 4571–4580.

Jaiswal, S., Fontanillas, P., Flannick, J., Manning, A., Grauman, P.V., Mar, B.G., Lindsley, R.C. et al. (2014) Age-related clonal hematopoiesis associated with adverse outcomes. *N Engl J Med*, 371: 2488–2498.

Muller-Sieburg, C.E., Sieburg, H.B., Bernitz, J.M. and Cattarossi, G. (2012) Stem cell heterogeneity: Implications for aging and regenerative medicine. *Blood*, 119: 3900–3907.

### Invertebrate hematopoietic stem cells

Dey, N.S., Ramesh, P., Chugh, M., Mandal, S. and Mandal, L. (2016) Dpp dependent hematopoietic stem cells give rise to Hh dependent blood progenitors in larval lymph gland of *Drosophila*. *Elife*, 5. pii: e18295.

Ghosh, S., Singh, A., Mandal, S. and Mandal, L. (2015) Active hematopoietic hubs in *Drosophila* adults generate hemocytes and contribute to immune response. *Dev Cell*, 33: 478–488.

Grigorian, M. and Hartenstein, V. (2013) Hematopoiesis and hematopoietic organs in arthropods. *Dev Genes Evol*, 223: doi:10.1007/s00427-012-0428-2.

Minakhina, S. and Steward, R. (2010) Hematopoietic stem cells in *Drosophila*. *Development*, 137: 27–31.

# Musculo-skeletal and dental stem cells in mammals

10

Many multicellular organisms rely on a solid support structure that serves multiple functions, ranging from protection of soft tissues through to an anchor point for muscles. Such structures are broadly of two types, that is, exoskeletons or endoskeletons. Exoskeletons are external and are found in many invertebrates. Endoskeletons, in contrast, are internal and vary in complexity from simple siliceous spicules in sponges to the true endoskeletons composed of mineralized tissue derived from the **mesoderm** that are present in echinoderms and chordates.

In this chapter, we are going to focus on the stem cell basis of the mammalian musculo-skeleton, considering both the skeletal components, including bones, cartilage, and teeth, as well as the skeletal muscle, the function of which is intimately linked with the structural elements of the skeleton. We will cover the biology of individual musculo-skeletal stem cells in relation to the tissue of which they are a part, including their diversity, function, and what is known about regulatory mechanisms involved in their maintenance and progression toward **differentiated** cells. Where relevant, some examples will be given of any current or prospective clinical applications of these stem cells.

Several properties and functions of **stem cells of the musculo-skeleton** have been described in addition to being the precursors for differentiated cell types, and in particular their ability to modulate the behavior of immune cells in various normal and disease contexts is attracting a huge amount of research interest and stimulating many avenues of potential clinical application. We will discuss the involvement of MSC in the immune system in Chapter 16.

## 10.1 THE FUNDAMENTAL COMPONENTS OF THE MAMMALIAN SKELETON, DENTITION, AND MUSCULATURE

Considering the musculo-skeleton as being made up essentially of two components, that is, structural or mechanical elements and force-generating muscles, nicely parallels the underpinning stem cell basis. Hence, the mechano-structural components are all derived from a group of stem cells known collectively as mesenchymal stem cells, while skeletal muscle cells are generated from a specific monopotent stem cell, generally known as the **satellite cell**.

### The cells that give rise to mineralized structures or to tissue components with high tensile strength are derived from a related group of mesenchymal stem cells

The structures of individual bones have evolved to provide strength but also to support specific function. Bones are overlaid by or connected at points with cartilage and act as the framework and anchor point for tendons and muscles (Figure 10.1), enabling complex movement but also providing a protective structure for organs such as the brain and a contribution to the functioning of tissues such as the lungs and the blood cell system. Although bones contain calcium, phosphorus, sodium, and other minerals,

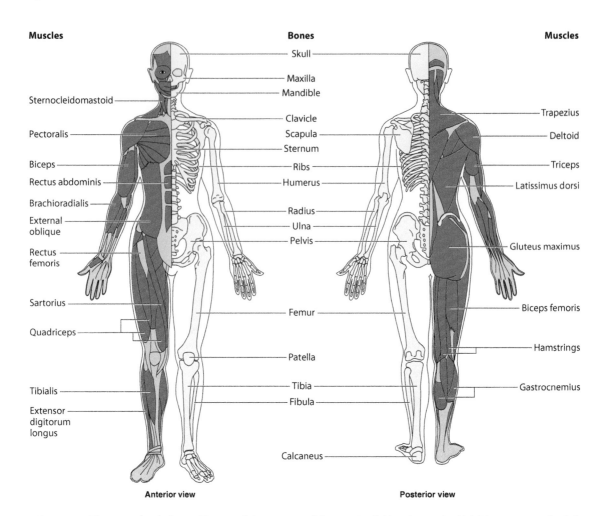

**Figure 10.1 The musculo-skeleton.** Diagram of the anatomy of the anterior (left) and posterior (right) human musculo-skeleton, the muscle components being highlighted in pink. (From the Merck Manual Consumer Version (Known as the Merck Manual in the US and Canada and the MSD Manual in the rest of the world), edited by Robert Porter. Copyright (2019) by Merck Sharp & Dohme Corp., a subsidiary of Merck & Co., Inc., Kenilworth, NJ. Available at http://www/merckmanuals.com/consumer. Accessed 11 December 2019.)

they are live structures with a blood supply and innervation and mostly consist of the protein collagen, which gives them their flexible framework, allowing them to withstand pressure, while the mineralization around the framework gives strength (Figure 10.2).

Contained within or built around the rigid framework of the skeleton, the precise orientation and correct functioning of most organs also relies upon the "mesenchyme," often referred to as "connective tissue." Mesenchyme, a generic term for masses of cells that fill the spaces, act as anchors, protect against damage, and probably create specific environments for a host of biological processes occurring outside of defined organs.

Although not strictly part of the skeleton, teeth have clear parallels with bone in terms of their structure and stem cell origins, and for these reasons, they will also be discussed in this chapter. Like bone, teeth are composed of calcium, phosphorus, and other minerals, but as the hardest part of the body not requiring the flexibility of bones, they mostly consist of a calcified tissue called dentin (Figure 10.3).

All of these cell types that define the structure of bones, cartilage, tendons, connective tissue, and teeth are derived from MSC, although, as we will describe in the next sections, these stem cells are probably members of a family of related cell types that each have distinct features relevant to the tissue to which they contribute.

## The muscles associated with the skeleton are maintained by a specific, monopotent stem cell

Skeletal muscle is the most abundant muscle type in the human body, constituting almost half of body mass, being responsible for movement, maintenance of posture, body shape, and stabilization of bones and joints, and is a source of heat as a by-product of its activity. Skeletal muscle is made up of bundles of cells called myofibers, which are held together by connective tissue (Figure 10.4). Each fiber is composed of a

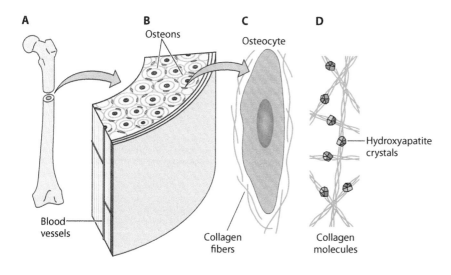

Figure 10.2 **The structure of bone.** Bone is a highly structured tissue, built for both strength and flexibility based largely on a combination of collagen fibers and a matrix of calcium minerals. The schematic diagram illustrates the different levels of organization of the structure. (**A**) Long bone. (**B**) The internal structure of a bone consists of cylindrical units running lengthwise. These "osteons," as they are known, are built from osteocytes and have a centrally located blood supply. (**C**) The microstructure of the osteon is seen with the constituents of bone extracellular matrix (ECM) surrounding the osteocytes. (**D**). The molecular structure of the ECM consists of parallel collagen molecules built around crystals of the calcium phosphate mineral hydroxyapatite. (Adapted from Stevens M, George J. *Science* (2005) 310:1135–1138 [doi: 10.1126/science.1106587].)

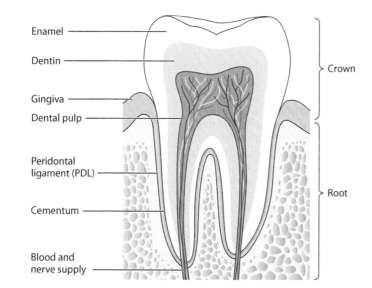

Figure 10.3 **Structure of a tooth.** The tooth is complex combination of mineralized matrices and soft tissues. Three distinct mineralized tissues (enamel, dentin, and cementum) and soft tissues, including the dental pulp and the periodontal ligament. The pulp is a vital tissue with its own blood and nerve supply. The diagram illustrates the major components of the tooth. Enamel and dentin encase the dental pulp. Cementum and the periodontal ligament help secure the tooth roots in the jaw.

membrane, many scattered nuclei, and thousands of inner strands termed myofibrils. The myofibrils are the same length as the myofiber they constitute, and they make up the contractile unit of skeletal muscle. Myofibrils accomplish contraction upon nerve stimulation through the action of smaller components called sarcomeres that attach end-to-end throughout the myofibril. Sarcomeres are made up of two contractile

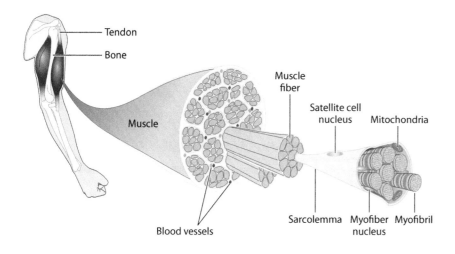

Figure 10.4 **Structural organization of muscles.** Like bone, muscle is a highly organized structure built up from smaller units. The schematic diagram shows how bundles of muscle fibrils (myofibrils) are constructed into a complete muscle, which links to the skeletal bones through a tendon connection. Myocyte nuclei are located peripherally, and the satellite stem cells lie in close proximity to the myofibers, being sandwiched between them and the enclosing sarcolemma.

proteins, namely myosin and actin, and it is the interaction between these proteins, specifically the sliding of the actin filaments at each end of the central myosin toward the center of the myosin, that ultimately results in shortening of the myofiber and force generation.

The stem cell that maintains skeletal muscle during adult life, the so-called satellite cell, is monopotent, and we will describe how it has been extensively characterized in terms of its molecular control and the way in which it interacts with its **niche**. Furthermore, skeletal muscle is a good example of a tissue prone to the effects of aging, and our understanding of the muscle satellite cell has given considerable insight into the way in which stem cells and their environment can influence the changing ability of a tissue to maintain homeostasis throughout life.

## 10.2 MESENCHYMAL STEM CELLS

With the exception of the **hematopoietic stem cell** (HSC), MSC are probably the most intensively studied **adult stem cell** from the perspective of potential clinical application, although they remain surprisingly little understood in terms of their origins and biology. As of 2019, over 1000 clinical trials have been registered that involve the use of MSC, although a majority of these are not related to their ability to generate cells involved in the structural tissues and relate instead to their effects on the immune system (Chapter 16).

### The discovery and characterization of mesenchymal stem cells dates back over 50 years

Although not named as such until the 1990s, the discovery of the MSC is usually credited to the Russian biologist Alexander Friedenstein (Box 10.1). Following from earlier observations that whole pieces of bone marrow transplanted to different anatomical sites could give rise to ectopic bone, Friedenstein showed that a small population of rapidly adhering cells from the bone marrow could achieve the same (Figure 10.5). When seeded sufficiently sparsely on plastic culture dishes, these cells formed discrete colonies (**"colony-forming unit (CFU) fibroblast"** or CFU-F), which when transplanted could give rise not only to bone but also other elements of skeletal tissues, including fat, cartilage, and fibrous tissue.

Since Friedenstein's seminal findings, MSC have been the subject of considerable research effort, but quite strangely, they remain relatively poorly defined and understood, especially if we compare our state of knowledge of them with adult stem cells such as the HSC. Fundamental questions remain about the diversity of MSC within the body and their heterogeneity of function; in fact, doubt has been expressed that they are indeed true stem cells, so much so that some people prefer to use the MSC acronym to mean "mesenchymal **stromal cells**." The definition of MSC as true

---

**BOX 10.1 PIONEER: ALEXANDER FRIEDENSTEIN (1924–1998)**

Alexander Friedenstein (Figure 1) was a true visionary, who made predictions on the clinical application of mesenchymal stem cells several decades in advance of the present-day avalanche of activity seeking to demonstrate their therapeutic efficacy. He started his medical career in the Military Medical Academy in St. Petersburg, and early in his research, he was interested in the regeneration of skin and bones. From 1950 onward in the Gamaleya Institute of Epidemiology and Microbiology in Moscow, Friedenstein began to focus on interactions between bone and hematopoietic tissues, and as a result of this work is generally credited with the discovery of the mesenchymal stem cell. His critical observation in the 1960s was that bone marrow contains stem cells for skeletal tissues that he was able to isolate as adherent, fibroblast-like, colony forming cells that he could differentiate into adipocytes, chondrocytes, and osteoblasts *in vivo*.

Box 10.1 Figure 1 **Alexander Friedenstein.** (From Bianco P. (2015) *Bone* 70:2–9 [doi: 10.1016/j.bone.2014.08.011]. Copyright 2015 with permission from Elsevier.)

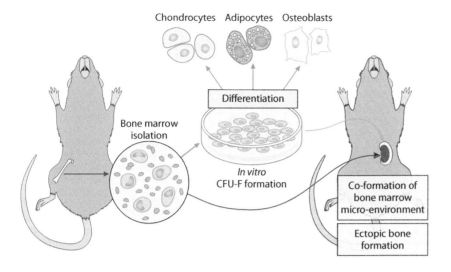

Chondrocytes    Adipocytes    Osteoblasts

Differentiation

Bone marrow
isolation

*In vitro*
CFU-F formation

Co-formation of
bone marrow
micro-environment

Ectopic bone
formation

**Figure 10.5 Experiments by Friedenstein that defined the mesenchymal stem cells.** Friedenstein conducted three key sets of experiments that enabled him to define mesenchymal stem cells and point toward their stem cell characteristics. In 1968 (red arrows), he discovered that rodent bone marrow could be transplanted under the kidney capsule, leading to the formation of ectopic bones. Two years later (blue arrows), he showed that non-hematopoietic cells within the bone marrow can form fibroblast-like colonies (CFU-F) when cultured *in vitro* and that these could be differentiated into cartilage-, fat-, and bone-forming cells. Last, in 1974 (green arrow), Friedenstein proved that the CFU-F cells are responsible for the formation of the ectopic bone seen when bone marrow is transplanted under the kidney capsule. (From Andrezejewska A, Lukomska B & Janowski M. (2019) *Stem Cells* 37:1–10 [doi:10.1002/stem.3016]. With permission from Wiley.)

stem cells has long been presumed and only recently have *in vivo* experiments been performed that demonstrate that at least some cells in a cultured population are capable of **self-renewal** while retaining their **multipotentiality**. The history leading to the definitive statement that MSC are indeed stem cells progressed from the initial observations of Friedenstein and others that whole mouse bone marrow could be transplanted and form ectopic bone structures. Similar experiments performed first with cultured non-hematopoietic stromal cells, isolated CFU-F cells, and ultimately single isolated perivascular cells from the bone marrow showed the formation of miniature bone structures or "ossicles," establishing that stem cell activity does exist within the MSC population (Figure 10.6).

## Simple procedures and a small set of cell surface markers are sufficient for the isolation of mesenchymal stem cells

Given the background to MSC biology and the general advances in stem cell biology over the last two decades, it is perhaps quite surprising that the usual method of obtaining MSC has not changed significantly since Friedenstein's original discoveries. Bone marrow, usually obtained by aspiration from the iliac crest of the

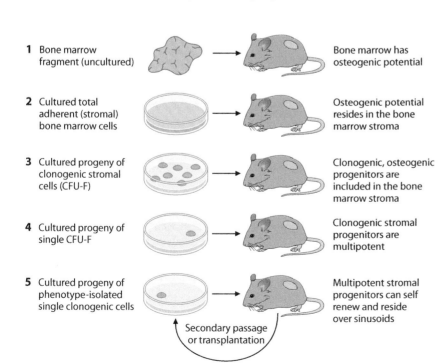

**1** Bone marrow fragment (uncultured)

Bone marrow has osteogenic potential

**2** Cultured total adherent (stromal) bone marrow cells

Osteogenic potential resides in the bone marrow stroma

**3** Cultured progeny of clonogenic stromal cells (CFU-F)

Clonogenic, osteogenic progenitors are included in the bone marrow stroma

**4** Cultured progeny of single CFU-F

Clonogenic stromal progenitors are multipotent

**5** Cultured progeny of phenotype-isolated single clonogenic cells

Multipotent stromal progenitors can self renew and reside over sinusoids

Secondary passage or transplantation

**Figure 10.6 Identification of mesenchymal stem cell activity.** Cartoon summarizing the methods that have been used to determine mesenchymal stem cell (MSC) activity in various populations of cells derived from bone marrow by demonstration of the formation of heterotopic ossicles (miniature bone organs that include the hematopoietic microenvironment). (1) Transplantation of bone marrow fragments lacking any bone structures, which was the seminal experiment in the discovery of MSC. (2) Transplantation of adherent, cultured bone marrow stromal cells revealed the role of the non-hematopoietic population. (3) Transplantation of pooled cells expanded from fibroblastic colony-forming units (CFU-F) showed that there are discrete stromal progenitors in the bone marrow that can be cloned *in vitro*. (4) Transplantation of cells from a CFU-F progenitor clone proved that a single multipotent progenitor is able to give rise to all skeletal tissues represented in the ossicle, that is, bone, adipocytes, and fibroblasts. (5) Serial transplantation of cells derived from an ossicle formed following primary transplantation of clonally expanded bone marrow CFU-F progenitors established that the skeletal multipotent progenitors are able to self-renew. (From Bianco P, Cao X, Frenette P et al. (2013) *Nat Med* 19:35–42 [doi: 10.1038/nm.3028]. Reprinted by permission from Springer Nature.)

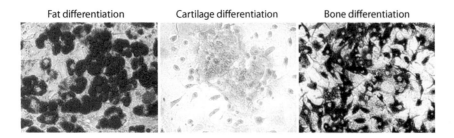

Fat differentiation    Cartilage differentiation    Bone differentiation

Figure 10.7 **Trilineage differentiation of mesenchymal stem cells.** Mesenchymal stem cells grown as a monolayer on tissue culture plastic can be differentiated toward the three mesenchymal lineages. On the left is a culture differentiated toward fat (adipocytes) verified by staining with Oil Red-O, which stains lipid droplets red. The middle image shows cartilage (chondrocyte) differentiation revealed by staining with Toluidine blue. The image on the right shows bone (osteogenic) differentiation as defined by a positive staining for alkaline phosphatase, the osteogenic cells staining a deep purple color. (From Placencio V, Li X, Sherrill T et al. (2010) *PLOS ONE* 5: e12920 [doi: 10.1371/journal.pone.0012920]. Published under the terms of the Creative Commons Attribution License (http://creativecommons.org/licenses/).)

pelvis or the large bones of the leg, remains the most utilized source of MSC. These non-hematopoietic stem cells are present in bone marrow at a frequency of about 0.01%–0.0001% of nucleated cells, roughly similar to the abundance of HSC.

Simple adherence to plastic is the standard approach to MSC isolation and expansion, although combinations of positive and negative selection based on the presence or absence of specific surface antigens in combination with **fluorescence-activated cell sorting** (Chapter 2) has helped to refine the process. The International Society for Cellular Therapy (ISCT) has proposed a set of criteria that need to be satisfied in order to define a population of human cells as MSC, namely adherence to tissue culture plastic; expression of the surface antigens **cluster of differentiation** 73 (CD73), CD90, and CD105; absence of the markers CD14/CD11b, CD34, CD45, CD79a/CD19, and HLA-DR; and the ability to differentiate into fat (adipocytes), cartilage (chondrocytes), and bone (osteoblasts/osteocytes) (Figure 10.7). Once established in culture, MSC grow as a monolayer and adopt a fibroblast-like morphology. The tri-lineage differentiation potential of MSC can be quite simply tested by seeding the cells in defined chemical and growth factor conditions that selectively favor differentiation into fat, cartilage, and bone (Box 10.2).

The culture conditions for MSC can be modified to enhance growth or the specific differentiation potential of the cells by the selective addition of growth factors such as platelet-derived growth factor (PDGF), fibroblast growth factor (FGF-2), or transforming growth factor β (TGF-β). MSC are prone to phenotypic change in culture, as exemplified by their surface antigen profile, and there has been concern that prolonged growth might lead to chromosomal mutations and possible selection for cells with the potential to grow as a tumor, although failure to demonstrate **teratoma** formation upon transplantation of human MSC into immunocompromised mice (Chapter 2) argues against this concern.

Using very similar protocols, MSC have been successfully derived from mouse, rat, dog, baboon, pig, sheep, goat, and rabbit, among other species, although there are some species-specific differences, notably as observed in the most popular experimental model for studies on MSC, that is, the mouse. For reasons that are not clear, obtaining MSC from mouse bone marrow has proven to be more challenging, although recent use of antibodies against platelet-derived growth factor receptor α (PDGFRα) and stem cell antigen-1 (Sca-1) as positive selection of cells in the non-hematopoietic fraction of ground-up bones, rather than the marrow, has resulted in a dramatic improvement in the efficiency of their isolation. Likewise, a number of more recently developed cell sorting strategies involving additional surface antigens such

---

**BOX 10.2 BACKGROUND: CONDITIONS FOR THE TRILINEAGE DIFFERENTIATION OF MESENCHYMAL STEM CELLS**

MSC growing as monolayers on tissue culture plastic can be readily differentiated into fat (adipocytes), cartilage (chondrocytes), and bone (osteoblasts/osteocytes) by simple changes to the growth conditions:

• **Fat**—Addition of dexamethasone, methyl isobutylxanthine, insulin, and indomethacin results in differentiation that is recognized by the appearance of lipid rich vacuoles (stained with Oil Red O) and expression of proteins such as peroxisome proliferation-activated receptor γ2, lipoprotein lipase, and fatty acid-binding protein aβ2.

• **Cartilage**—When cells are pelleted and cultured in the presence of TGF-β, this leads to the development of a multi-layered, matrix-rich mass staining positive for Toluidine Blue due to the abundance of glycosaminoglycans within the **extracellular matrix** (ECM).

• **Bone**—Addition of ascorbic acid, β-glycerophosphate, and dexamethasone results in differentiation that is recognized by the presence of alkaline phosphatase and the accumulation of calcium, which can be seen by staining with Alizarin Red or the Von Kossa method.

as CD271 and CD56 have led not only to improvements in the isolation from various tissues of human MSC with CFU-F capability but have also helped to define the extent of heterogeneity of these cells.

## Mesenchymal stem cells can be derived from multiple tissue locations, although it remains unclear if this is reflected in a significant degree of heterogeneity in these cells

MSC grown according to the ISCT guidelines represent a heterogeneous population of cells, the majority of which do not have stem cell properties; rather they are a mixture of stem cells, **progenitors**, and stromal cell components. Only when the cells are plated at a density that allows for clonal expansion can progenitors and stem cells be recognized through the formation of CFU-F, which can be further discriminated following re-plating and definition of their differentiation potential. As we have already mentioned, only CFU-F that can be re-plated to generate clones from which single cells can be transplanted to produce miniature bone-like structures known as ossicles can be regarded as true stem cells.

Based on the criteria defined by the ISCT, it has proven possible to grow MSC from most tissues, with the indication that each derived MSC population shares a common origin, albeit exhibiting differences in differentiation potential. Apart from bone marrow, commonly studied sources of MSC include adipose tissue, joint tissue (synovium), lung, teeth, the placenta, and Wharton's jelly from umbilical cord (Figure 10.8). MSC from birth-associated tissues, such as placenta and umbilical cord, offer certain advantages, including ease of availability and the fact that they possess greater proliferative capacity, especially under hypoxic conditions, in comparison to MSC populations obtained from adult tissues.

## A highly defined population of mesenchymal stem cell-like cells isolated from bones have been specifically termed skeletal stem cells

The care that needs to be taken over using the term mesenchymal stem cells in a way that fails to discriminate the diversity of this related group of stem cells is nicely illustrated by recent detailed investigation of those cells in bones that give rise to the non-hematopoietic components, that is, bone, cartilage, and the stromal matrix within which hematopoiesis occurs. Using the "rainbow" reporter mouse (Chapter 2), the laboratories of Michael Lineaker and Irving Weissman at Stanford University showed that clones could be labeled showing that a stem cell present at the ends

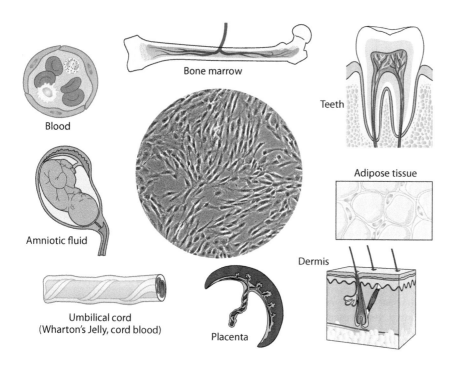

**Figure 10.8 Mesenchymal stem cells can be derived from a wide range of tissues.** Mesenchymal stem cells (MSC) are widely distributed in the body, although their origins and precise function in each location are not entirely clear. The central image shows a phase contrast image of rat MSC growing as a monolayer on a plastic surface; this is surrounded by representations of some of the tissues that have been used to derive these cells. A number of tissues associated with a developing fetus are rich sources of MSC; they can be identified in the amniotic fluid, the Wharton's jelly from the umbilical cord, the umbilical cord blood, and the placenta. In adult tissues, MSC are present and readily isolated from, for example, bone marrow, adipose tissue, the skin, and teeth. (Adapted from Merino-González C, Zuñiga F, Escudero C et al. (2016) *Front Physiol* 7:24 [doi:10.3389/fphys.2016.00024]. This is an open-access article distributed under the terms of the Creative Commons Attribution License (CC BY). Photo: From https://www.sciencellonline.com/human-bone-marrowderived-mesenchymal-stem-cells.html. With permission from ScienCell Research Laboratories, Carlsbad, CA, USA.)

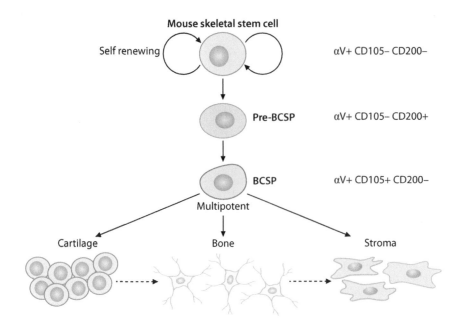

Mouse skeletal stem cell

Self renewing    αV+ CD105– CD200–

Pre-BCSP    αV+ CD105– CD200+

BCSP    αV+ CD105+ CD200–

Multipotent

Cartilage    Bone    Stroma

Figure 10.9 **Skeletal stem cells.** One subtype of mesenchymal stem cells (MSC) present in the bone marrow appears to be largely responsible for the bone structure and the stroma of the marrow cavity. These MSC, which have been termed skeletal stem cells, are positive for the expression of αV integrin and are at the top of a hierarchy of cells that can be discriminated on the basis of their expression of the surface markers CD105 and CD200. As illustrated in the diagram, the skeletal stem cell is capable of self-renewal and differentiation into multilineage progenitor cells that give rise to bone, cartilage, and hematopoietic-supportive stroma (BCSP). (From Chan C, Seo E, Chen J et al. (2015) *Cell* 160:285–298 [doi: 10.1016/j.cell.2014.12.002]. Copyright 2015 with permission from Elsevier.)

of long bones could give rise to bone, cartilage, and the bone stroma but did not produce adipocytes or muscle. Following from this observation, they used surface markers and both *in vitro* and *in vivo* assays of self-renewal and differentiation to define a hierarchy of differentiation from what they termed the **skeletal stem cell (SSC)**, which is characterized by the expression of alpha V integrin and CD200 (Figure 10.9). When transplanted to ectopic sites, SSC could give rise to ossicles that become populated with hematopoietic cells, mimicking normal bone marrow. Gene expression analysis of SSC and their downstream progenitors suggested the importance of bone morphogenetic protein (BMP) and Wnt signaling, probably acting in either an autocrine or paracrine fashion. At about the same time, the laboratory of Timothy Wang at Columbia University in New York also identified the SSC, but in their case, they used expression of the Gremlin 1 protein to selectively identify and isolate the cells.

The conclusion that can be reached from these sorts of studies is that, although sharing characteristics, "mesenchymal stem cells" from different sources have specific differentiation capacities, such that those from bone marrow are preferentially skeletal progenitors, and likewise those from adipose tissue are geared toward adipogenesis. An interesting extension to the experiments of Lineaker and Weissman showed that there is likely flexibility in the fate of some tissue MSC such that they can be forced to behave like SSC. When they placed a sponge containing BMP2, which is one of the key factors they identified as being important for SSC **commitment** to bone differentiation, subcutaneously in the mouse, then ossicles developed locally, presumably from MSC, and these too became populated with hematopoietic stem cells and progenitors from distant sites.

## There is some controversy about the tissue location of mesenchymal stem cells

The location of MSC in different tissues is in most instances not entirely obvious, or actually unknown, although it has been suggested that MSC are closely associated with blood vessels. A cell type that is defined by its location and could be a good candidate for the MSC in multiple tissues is the so-called "**pericyte**," a perivascular cell found at the surface of endothelial cells on the outer face of vessels in the microvasculature of every vascularized connective tissue. It is thought that pericytes are recruited from the surrounding tissue during organ development. It has been shown that isolated bone marrow perivascular cells behave like MSC, growing proficiently in culture and exhibiting the morphology and surface antigens (CD44, CD73, CD90, CD105) of conventionally isolated MSC. However, recent **lineage tracing** studies using lineage tracing driven by a gene selectively expressed in pericytes in multiple adult organs has pretty much dispelled the idea that these cells are **tissue stem cells** involved in repair and regeneration. Hence, CreERT2 recombinase driven by the *tbx18* gene allows highly efficient labeling of pericytes *in vivo*. Cell surface antigen profiling of Tbx18-expressing cells of multiple tissues showed them consistently to express CD29, CD140b, CD146, and the proteoglycan NG2. However, aging of the labeled mice or induction of injury failed to reveal any sign that the pericytes contribute to homeostatic or post traumatic repair processes, although the same labeled cells isolated and cultured *in vitro* were able to give rise to osteocytes, chondrocytes, and adipocytes.

## Some mesenchymal stem cells may have the capacity to differentiate into lineages other than fat, cartilage, and bone

In addition to providing a source of musculo-skeletal cells, MSC have been proposed to have other possible fates and may also perform a wide variety of functions in their uncommitted state. Many studies have claimed that cells as diverse as neurons, hepatocytes (liver cells), and cardiomyocytes can be derived from MSC, but this remains a controversial issue, the evidence often being weak. In 2001, the group of Catherine Verfaillie reported the discovery of **multipotent adult progenitor cells (MAPC),** which were co-purified along with the MSC fraction from bone marrow. These MAPC had a greater expansion capability than MSC and additionally appeared to differentiate *in vitro* toward cells belonging to all three germ layers (mesoderm, **ectoderm**, and **endoderm**) and were able to form chimeric mice following injection into the **blastocyst**-staged embryo. However, the concept of the MAPC remains controversial due to difficulties of reproducibility outside of the original group. Reports of neural differentiation of MSC (and similar properties attributed to stem cells in teeth) may relate to a likely dual origin for these cells during development. Hence, the majority of tissue MSC probably arise from the mesoderm; however, it is known that a minor proportion in the bone marrow are of neural crest origin, reflected in their expression of the neural-associated *Nestin* gene. These Nestin-positive cells could well be the precursors of neural differentiation, especially if they are a component of a mixed population derived solely on the grounds of the ISCT criteria.

## Specific signaling pathways and transcription regulators control the self-renewal and differentiation of mesenchymal stem cells

As we see in many circumstances with adult stem cells, the regulatory mechanisms that determine the balance between stem cell maintenance and differentiation have some shared principles and components as well as specific aspects (Chapter 4). Several signaling pathways and transcription factors have been investigated and shown to play critical roles in MSC differentiation, including those involving Wnt, Hedgehog, and TGF-β/BMP, and in respect to the three principal lineages contributing to the musculo-skeleton, factors promoting one particular lineage often repress the alternative pathways of differentiation. For example, the transcription factor Runx2, which is regulated by signaling pathways such as the Wnt, BMP, and Notch, is key for osteoblast differentiation but inhibits adipogenesis and chondrogenesis. Chondrogenic differentiation, on the other hand, is driven by the transcription factor Sox9, which in turn suppresses osteogenic differentiation, partly at least through its physical interaction with Runx2. Adipogenesis is predominantly under the control of the transcription factor PPARγ, and likewise commitment to this lineage blocks the differentiation of MSC along other pathways.

## Tissue elasticity directs mesenchymal stem cell lineage differentiation

Reflecting the *in vivo* microenvironments in which MSC differentiate into elements of the musculo-skeleton and how these are composed of radically different matrices, ranging from the hardness of bone to the relative softness of fat, it has been shown that the stiffness of the substrate on which cells are induced to differentiate *in vitro* can have a defining effect on the outcome. An *in vitro* gel system can be utilized to control stiffness through cross-linking and the thickness of the gel and by control of cell adhesion through covalent attachment of Collagen-I, thereby mimicking specific tissue types (Figure 10.10). When MSC are grown on matrices with an elastic modulus (a measure of stiffness) in the range typical of the brain (0.1–1 kPa), muscle

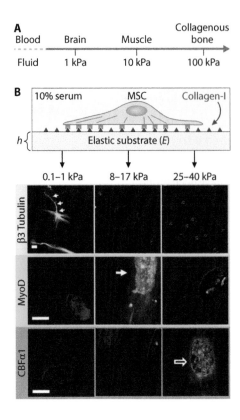

**A**

| Blood | Brain | Muscle | Collagenous bone |
|-------|-------|--------|------------------|
| Fluid | 1 kPa | 10 kPa | 100 kPa |

**B**

10% serum    MSC    Collagen-I

$h${    Elastic substrate (*E*)

0.1–1 kPa    8–17 kPa    25–40 kPa

β3 Tubulin

MyoD

CBFα1

**Figure 10.10 Matrix elasticity directs mesenchymal stem cell lineage differentiation.** Many cells are able to sense and respond to the mechanical properties of their environment. The differentiation of mesenchymal stem cells (MSC) is an example of such sensing in that their fate can be determined by the stiffness of the tissue of which they are a part. The diagram illustrates one set of experiments using bone marrow MSC growing on artificial substrates that are adjusted to mimic different degrees of stiffness. (**A**) Solid tissues stiffness can be quantified by the elastic modulus (E) measured in Pascals (Pa). (**B**) The effect of elasticity on MSC differentiation can be modeled *in vitro* using a gel system that allows control of E through cross-linking, manipulation of cell adhesion by covalent attachment of Collagen-I, and variation of gel thickness (h). When plated on the gel, the MSC are initially small and round but develop increasingly branched, spindle, or polygonal shapes when grown on matrices, respectively, in the range typical of brain (0.1–1 kPa), muscle (8–17 kPa), or stiff cross-linked collagen matrices (25–40 kPa). These changes in morphology are paralleled by the expression of lineage-specific markers. The neuronal cytoskeletal marker β3 tubulin is expressed in branches (arrows) of MSC grown on the soft, neurogenic matrices. The muscle transcription factor MyoD1 is up regulated and nuclear localized (arrow) only in MSC on myogenic matrices. The osteoblast transcription factor CBFα1 (arrow) is likewise expressed only on stiff, osteogenic gels. Scale bar is 5 μm. (From Engler A, Sen S, Sweeney H et al. (2006) *Cell* 126:677–689 [doi: 10.1016/j.cell.2006.06.044]. Copyright 2006 with permission from Elsevier.)

(8–17 kPa), or stiff cross linked-collagen matrices typical of bone (25–40 kPa), the cells are initially small and round but develop increasingly branched, spindle, or polygonal shapes over the course of a few hours to days and begin to express molecular markers characteristic of neural cells, muscle cells, or collagenous bone.

## Knowledge of the mesenchymal stem cell and its contribution to cells composing the musculo-skeleton might be translated for therapeutic benefit

The capacity of MSC to differentiate into fat, cartilage, and bone is attracting massive interest with the prospect of them becoming a source of these tissues for regenerative therapies, including replacement of damaged cartilage and bone to soft tissue reconstruction. Apart from traumatic damage or tissue loss through disease or associated clinical procedures, there is a rapidly growing demand on the horizon given the increasing proportion of older people in the population. For example, aging is associated with problems such as reduced fracture repair, reduced skeletal bone mass, and conditions such as osteoporosis, which leads to increased bone fragility. Aging is also associated with cartilage damage, for example, in the joints or intravertebral discs. MSC have already been used to generate cartilage for use in clinical trials of novel treatments for a range of conditions, including joint repair and replacement of damaged bronchi; we will come back to this when we focus on **regenerative medicine** (Chapter 17).

In addition to direct application of MSC and their downstream differentiated progeny for the purpose of regeneration, some interesting insight into how their function in tissues can be compromised in a disease state and how this might be circumvented has come from recent studies on the activity of SSC in diabetes. Bone fracture repair is known to be compromised in diabetes, a fact that encouraged Lineaker and colleagues to look into the way in which SSC, which as we described previously are the source of the cells of the bone structure, might be affected. Their evidence gained from a mouse model of diabetes revealed that decreased expression of one of the Hedgehog signaling proteins caused by the inflammatory state in diabetes negatively impacts SSC function. Delivering the deficient protein (Indian Hedgehog, Ihh) to a fracture site using a slow-release hydrogel restored fracture repair by enhancing SSC expansion, survival, and osteogenic potential.

## 10.3 DENTAL STEM CELLS

The tooth is a durable hard structure, but it is subject to everyday wear and tear and is highly vulnerable to damage as a result of mechanical trauma, chemical decay, and bacterial infections. Traditional treatment for such damage revolves around the use of synthetic implants and structural substitutions composed of inert compounds. MSC-related cells resident in or around teeth are crucial to the maintenance of this highly specialized tissue and could serve as the basis for novel approaches to dental repair and replacement.

### Several mesenchymal stem cell-related stem cells are associated with teeth

A number of distinct MSC-like cells have been identified in association with teeth and are of considerable interest due to their normal role in maintaining tooth integrity as well as their clinical potential. These cells are also intriguing because of their similarities and differences compared to other MSC, especially those present in the bone marrow, shedding light on their potential origins and context dependent mechanisms regulating fate determination and differentiation.

MSC in the tooth were first described as dental pulp stem cells (DPSC) in 2000 by Stan Gronthos and colleagues in post-natal human pulp tissue and were shown, upon transplantation into immunocompromised mice, to be able to regenerate dentin pulp-like complexes composed of a mineralized matrix with tubules lined with **odontoblasts** (analogous to osteoblasts) and fibrous tissue containing blood vessels similar to the arrangement seen in adult human teeth (Figure 10.3). Odontoblasts are long-living post-mitotic cells that align along the dentin-pulp interface, where they maintain pre-dentin and dentin apposition throughout the whole life of a tooth. Dentin produced by odontoblasts acts as a physiological barrier that protects the dental pulp following damage such as caries lesions and exposure as a result of traumatic damage. Minor repair can be achieved by pre-existing odontoblasts, but if these are destroyed, then DPSC or other MSC-like cells can likely differentiate into replacement odontoblasts.

DPSC are in many ways analogous to bone marrow-derived MSC. Both cell types are responsible for the generation of mineralizing cells, that is, odontoblasts in teeth and osteoblasts in bone. Unlike bone marrow MSC, dental associated stem cells are derived from the neural crest (NC), although a small subpopulation of Nestin-positive MSC in the bone marrow are of a similar developmental origin. Isolated DPSC and bone marrow MSC have a similar morphology and share expression of a number of surface proteins such as CD73, CD90, and CD105. However, DPSC are distinct from bone marrow MSC in that their greatest propensity is to give rise to odontoblasts. Mechanisms controlling odontoblast differentiation remain unclear, although again there are parallels with osteoblast differentiation from MSC in the involvement of BMP, Wnt, and Notch signaling and the transcription factor Runx2. Interestingly, they appear to exhibit a tendency for neurogenic differentiation, perhaps reflecting their neural crest origins, which is illustrated, for instance, by their expression of neurotrophins.

Following the identification of DPSC, five other MSC-like populations were obtained from various dental tissue sources (Figure 10.11), namely stem cells from human exfoliated deciduous teeth (SHED), periodontal ligament stem cells (PDLSC), stem cells from apical papilla (SCAP), the dental follicle precursor cell (DFPC), and gingival fibroblastic stem cells (GFSC). The precise relationship between these stem cells remains unclear, as they exhibit differences in their gene expression, proliferation, and differentiation properties. For example, SHED proliferate much faster than DPSC. Whether because of differences in their individual environmental exposure or to an inherent difference in terms of their developmental origins or position in a stem cell hierarchy, the different dental MSC have clearly distinct properties, which is best illustrated by a comparison that was made in terms of their natural deposition of minerals. Using Raman spectroscopy, the mineral composition of odontoblasts differentiated from the various dental MSC was compared to that of the "pure" dental minerals (dentin, enamel, and cementum), revealing a considerable variation, which presumably relates in some way to their specific role in the tooth (Figure 10.12).

Localization of DPSC and MSC in the dental pulp tissues is not clear, largely because of the lack of discriminating surface markers, although lineage tracing

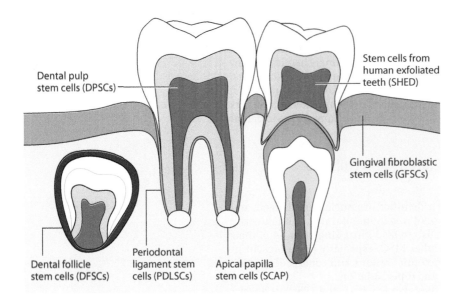

**Figure 10.11 Dental stem cells.** A number of distinct mesenchymal stem cell (MSC)-like cells have been isolated from emergent teeth, decidual first teeth, or adult teeth and their surrounding gingiva or periodontal ligaments. Pulp stem cells can be derived from adult teeth (dental pulp stem cells, DPSC—red) and exfoliated teeth (SHED cells—dark pink), while similar MSC-like cells can be found in the apical papilla (yellow). From the supporting tissues, dental MSC can be derived from the dental follicle (brown), the periodontal ligament (blue), and the gingiva (pale pink). (Adapted from Raza SS, Wagner A, Hussain Y et al. (2018) *Stem Cell Res Ther* 9:245 [doi:10.1186/s13287–018-1005-z]. This article is distributed under the terms of the Creative Commons Attribution 4.0 International License (http://creativecommons.org/licenses/by4.0).)

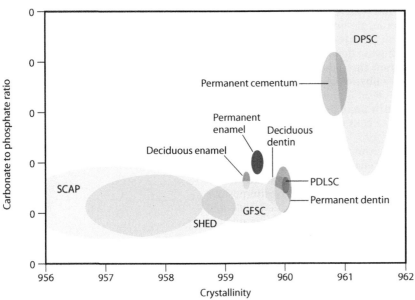

**Figure 10.12 Dental stem cell heterogeneity.** Mesenchymal stem cells (MSC) from dental or other tissue sources often appear very similar once cultured *in vitro*. The different dental MSC subtypes can be distinguished based on the type of mineral that derived odontoblasts produce *in vitro*. The diagram is a representation of a Raman microspectroscopic analysis of the mineral content in different differentiated dental MSC populations compared to the pure forms of the dental minerals. The measurements define the ratio of carbonate to phosphate (vertical axis) and the degree of crystallinity (horizontal axis). Permanent dentin—dark gray; deciduous dentine—pale gray; permanent enamel—red; deciduous enamel—orange; permanent cementum—purple. Abbreviations: DPSC, dental pulp stem cells; GFSC—gingival fibroblastic stem cells; PDLSC—periodontal ligament stem cells; SCAP—stem cells from the apical papilla; and SHED—stem cells from human deciduous teeth. (From Volponi A, Gentleman E, Fatscher R et al. (2015) *J Dental Res* 94:1568–1574 [doi:10.1177/0022034515599765]. With permission from Sage Publications.)

experiments have given some useful insight. We have seen earlier in this chapter that pericytes have been effectively excluded as tissue MSC, whereas the use of Ng2 as a characteristic marker of NC-derived pericytes to drive Cre recombinase expression revealed that some odontoblasts can be generated from pulp pericytes following damage. However, this strategy also showed that a proportion of odontoblasts must derive from an alternative MSC source.

## Incisors in rodents continue to grow in adults

Although adult teeth in mammals have a contingent of resident stem cells that can bring about a degree of repair when necessary, they generally do not continue growing throughout life. An exception that has been characterized are rodent incisors, which have a specialized stem cell-based mechanism to enable life-long growth in the face of the continual process of abrasion between the upper and lower teeth during eating. The stem cells responsible for the maintenance of incisor growth are located at the proximal end of the tooth, in the region between the lingual and labial parts of the epithelial cervical loop (Figure 10.13). Pulse-chase labeling experiments using radio-labeled DNA nucleosides showed there to be slow- and rapid-cycling mesenchymal cells that are now known to correspond to stem cells and **transit amplifying** progenitors involved in incisor growth. Lineage tracing using a number of different Cre recombinase lines identified both CD90+ MSC and glial cells as the stem cell precursors of about 50% of the incisor odontoblasts, and it is assumed that the remainder derive from another population of MSC-like cells, possibly pericytes (thereby distinguishing dental MSC from other tissue MSC of non-NC origin). Reminiscent of hair follicle **bulge stem cells** that can give rise to the epithelium of the skin (Chapter 15), lineage tracing using the transcription factor gene *Gli1* shows that all odontoblast and pulp cells in the incisor arise from Gli1+ cells, which are located in neural-vascular bundles and suggests a responsiveness to nerve-derived Sonic Hedgehog (Shh).

## Dental stem cells have considerable potential for the repair and restoration of damaged teeth

The role of odontoblasts in repair and regeneration of dentin, in the latter case requiring new cell generation from resident stem cells, opens up some avenues for therapeutic intervention. Although presently poorly understood, eventual manipulation of signals that mobilize dental MSC to differentiate could provide a useful intervention. The growth factor TGF-β, which is known to be involved in MSC differentiation, is sequestered in dentin and interestingly can be released by exposure to low energy laser light, providing an example of how dentin-producing odontoblasts might be increased.

Pulp regeneration would be a highly valuable addition to dental treatment options, to improve upon the current approaches to root canal therapy, and it is for sure going to be the case that there will eventually be application of DPSC in this context. A bigger challenge is the generation of a whole tooth. This has been achieved from DPSC in animal models *in vivo* but remains difficult in the clinical setting due to the lack of control over tooth shape, size, and availability of dental epithelium. SHED cells can be seeded onto a synthetic biodegradable scaffold, whereupon they differentiate into odontoblast-like cells. Another method being explored is the use of SCAP cells seeded onto a root-shaped block, which is coated with a gel foam containing PDLSC and transplanted into the tooth socket of a recipient, leading, after 3 months, to mineralized root-like growth within the tooth socket.

Knowledge of tooth development, in particular the crucial role played by the epithelium (Box 10.3), is helping researchers define improved strategies for the production of stem cell-derived dental structures and serves as a good example of how whole organs might be engineered in the future (see Chapter 17). For regenerative purposes, the identification of the first differentiation signals and the epithelial cell population responsible for sending these signals will facilitate the design of strategies for the engineered stimulation of tooth development (Figure 10.14). Whole tooth regeneration will require the formation of the solid structure of the crown from dental stem cells but also the roots, periodontal ligament, and nerve and blood supplies from other progenitors, possibly mimicking the normal developmental process.

## 10.4 SKELETAL MUSCLE STEM CELLS

Skeletal muscle is subject to constant injury through weight-bearing exercise and trauma and can increase its mass as the result of training. Repair of such damage or training-related increases in muscle mass requires a renewable source of replacement cells. Since the nuclei of the myofibers cannot divide, increased muscle mass and muscle regeneration has to be achieved through either an increase in myofibril mass or the production of cells from a specialized skeletal muscle stem cell.

Although it was documented during the nineteenth century that skeletal muscle had the ability to regenerate following injury, the underlying mechanism was not elucidated until the 1960s. The skeletal muscle stem cell was first discovered in 1961 by Alexander Mauro (Box 10.4) based on its anatomical location between the basal lamina and the sarcolemma as visualized by electron microscopy. At that time, the cell's function as a stem cell was not apparent, only later was it demonstrated that these cells, which became known as "**satellite cells**" because of their location, could give rise to muscle cells.

Satellite cells, or more recently sometimes referred to as "**muscle stem cells**" (MuSC), are now considered the major source of regeneration of skeletal muscle

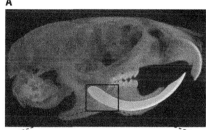

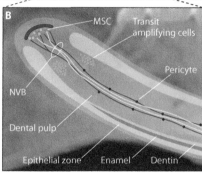

Figure 10.13 **Mouse incisor stem cells.** Unlike most adult teeth that do not grow throughout life, rodent incisors are continuously extending to overcome the effects of wear. (**A**) Micro CT scan of a mouse incisor showing the extent of the incisor (white solid line). The box represents the region schematized in (**B**). (**B**) Diagrammatic representation of the stem cell and transit amplifying areas responsible for maintaining incisor growth. The dental pulp (gray), dentin (pink), enamel (purple), and epithelial zone (blue) are shown. Within the pulp on the left can be seen slow-cycling mesenchymal stem cells (magenta and green dots, signifying Gli1+ and Glia1+ cells) associated with the neurovascular bundle (NVB) (nerves—yellow lines; blood vessels—red lines). Fast-cycling transit amplifying cells are represented as clusters (orange dots in clusters) located between the lingual cervical loop and the labial cervical loop at the proximal end of the incisor. Pericytes (dark blue dots) are shown along the blood vessels. (A: From Sarper S, Inubushi T, Kurosaka H et al. (2018) *Sci Rep* 8:10906 [doi: 10.1038/s41598-08-29317-6]. Licensed under a Creative Commons Attribution 4.0 International License (http://creativecommons.org/licenses/by4.0).). B: Adapted from Cui D, Li H, Wan M et al. (2018) 13:39–45 [doi:10.2174/157488X12666170913150403].

BOX 10.3 BACKGROUND: TOOTH DEVELOPMENT

Tooth development (Figure 1) starts from initial inductive signals between the epithelium and the mesencyhme and proceeds by proliferation of epithelial cells that form a bud around which mesenchymal cells condense. The signals exchanged between epithelial and mesenchymal cells are key at the induction and bud formation stages; mesenchymal cells first receive signals and then send reciprocal signals back to the epithelial cells. The cells begin to differentiate and the proliferating epithelial cells in the bud lead to the cap stage of development. Crown morphogenesis is initiated by epithelial signaling, and at the bell stage, precursors of specialized tooth cells (ameloblasts) coordinate enamel deposition and odontoblast formation.

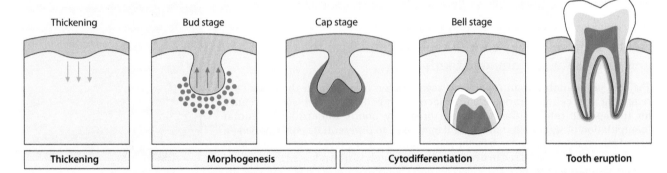

Box 10.3 Figure 1 **Tooth development.** Schematic to show the stages of tooth development initiating first with thickening of the oral epithelium (pink). The epithelium signals to the neural crest-derived mesenchyme (pink arrows). During morphogenesis, the bud stage develops with condensing mesenchyme (dark blue) forming around and signaling (blue arrows) to the bud. Development progresses through the cap stage and bell stage in the process of cytodifferentiation, finally resulting in eruption of the developed tooth through the oral epithelium. (From Volponi A, Pang Y & Sharpe P. (2010) *Trends Cell Biol* 20:715–722 [doi:10.1016/j.tcb.2010.09.012]. Licensed under a Creative Commons Attribution 3.0. Unported license (https://creativecommons.org/licenses/by/3.0/).)

Figure 10.14 **Tooth engineering from stem cells.** Understanding the processes that lead from epithelial and mesenchymal stem cells to the development of a tooth could help to define a strategy for the generation of replacement teeth from stem cells. The diagram illustrates a possible means of generating replacement teeth through creation of a tooth primordium *in vitro* flowed by transplantation into the desired location. Sources of epithelial and mesenchymal cells are expanded in culture before being combined to initiate direct contact, mimicking the *in vivo* situation. A tooth primordium, equivalent to a tooth bud or cap, begins to form following the interaction between epithelial (pink) and mesenchyme (blue) regions. Mesenchymal cells (dark blue) begin to condense around the tooth primordium. Following formation, the tooth primordium can be surgically transplanted into the mouth and left to develop into a bioengineered tooth. (From Volponi A, Pang Y & Sharpe P. (2010) *Trends Cell Biol* 20:715–722 [doi: 10.1016/j.tcb.2010.09.012]. Licensed under a Creative Commons Attribution 3.0 Unported license (https://creativecommons.org/licenses/by/3.0/).)

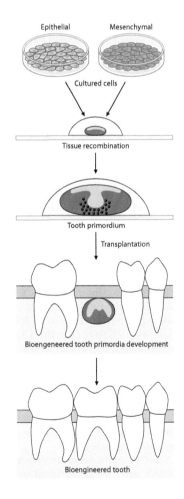

and fulfill the criteria for a stem cell. Their highly defined location and an extensive experimental effort, especially using mouse models and lineage tracing, have made the satellite among the best understood stem cells, in particular from the perspective of its niche interactions and the way in which aging impacts its functional capacity.

## Aside from their location, satellite cells can be identified through their expression of specific transcription factors and a combination of cell surface proteins

Satellite cells have a wedge-shaped appearance, with a large nuclear-to-cytoplasm ratio and few organelles (Figure 10.15), making up 2%–7% of all nuclei in the myofiber in healthy adults. Satellite cell frequency varies in different muscles as a result of variation in the composition of muscle fiber type (slow oxidative, fast oxidative, or fast glycolytic). The distribution of satellite cells along the myofibers at defined locations enhances longitudinal growth and regeneration of the muscle, their density being greatest toward the end of the myofibers, usually in close proximity to capillaries.

Alexander Mauro (Figure 1) was a biophysicist studying the physiology of muscle cells at the Rockefeller Institute and in 1961 discovered the existence of the skeletal muscle stem cell, which researchers now know to be the stem cell responsible for muscle regeneration. Mauro identified the presence of a quiescent cell lying on the surface of the myofiber, beneath the basement membrane, and coined the name "satellite cell," the name by which they are still frequently known. Mauro's original observation was made during an electron microscopic study of the skeletal muscle fiber of the frog, but it was later shown that the location of the satellite cell is conserved among all vertebrates. Although Mauro was unable to demonstrate the stem cell function of the satellite cell, his discovery evolved in the same year that Till and McCulloch discovered the HSC (Chapter 9), and he has been quoted as saying that his electron microscopy images "might be of interest to students of muscle histology and furthermore, as we shall suggest, might be pertinent to the vexing problem of skeletal muscle regeneration," paving the way for the research that followed over the next half century.

Box 10.4 Figure 1 **Alex Mauro.** Used with permission from The Rockefeller University.

The ultimate experimental proof that satellite cells are indeed myogenic precursors was obtained by showing that they are able to self-renew, proliferate, and differentiate into myogenic progeny *in vitro* and *in vivo*. Satellite cell characterization *in situ* in myofibers and isolation for *ex vivo* study relies upon a fairly limited set of markers. As for other stem cell types, satellite cells can be defined to some extent by the proteins present on their surface. Cells can be identified by immunofluorescence *in situ* but also following enzymatic disruption of myofibers, for example, using collagenase, they can be analyzed by flow cytometry or purified by cell sorting. As for MSC, discussed earlier in this chapter, the absence of expression of a specific marker can be as useful as the presence of a particular antigen. Markers not expressed on mouse satellite

**A**

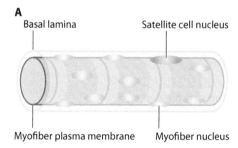

Basal lamina    Satellite cell nucleus

Myofiber plasma membrane    Myofiber nucleus

**B**

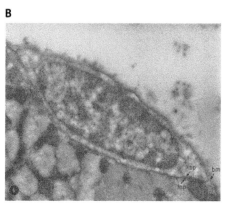

Figure 10.15 **Muscle satellite cells.** Satellite cells, the stem cells of skeletal muscle, have a characteristic appearance and a highly defined location in close contact with the myofibers. (**A**) Schematic diagram of the location of the muscle satellite cell. A satellite cell (purple) resides on the plasma membrane of the myofiber beneath the basal lamina (red). It is a rare cell along the myofiber length, with the majority of nuclei belonging to other myofiber cells (blue). (**B**) The image is an electron micrograph of a satellite cell lying just below the basal lamina (bm) and sitting on the plasma membrane of the myofiber. (A: From Yablonka-Reuveni Z & Day K. (2011) Skeletal muscle stem cells in the spotlight: the satellite cell. In: Cohen I, Gaudette G (eds) *Regenerating the Heart: Stem Cells and the Cardiovascular System*. New York: Springer, Humana Press. Reprinted with permission from Springer Nature. B: Reproduced from Scharner J, Zammit P. (2011) *Skeletal Muscle* 1:28 [doi:10.1186/2044-5040-1-28]. Originally published in Mauro R. (1961) *J Biophys Biochem Cytol* 9:493–495. Photo was furnished courtesy of Dr. G. Palade.)

cells include CD31, CD45, and Sca1, while positive selection can be achieved using antibodies against molecules such as CD34, M-Cadherin (M-Cad), and α7 integrin and by studying the expression of muscle-defining transcription factors such as Pax7. Much of the research into satellite cell characterization has been performed in amphibians and rodents, but many of the markers of satellite cells are not expressed in the human counterparts. However, human satellite cells can be reliably identified by their expression of Pax7 and M-Cad, akin to mouse satellite cells, and as well through expression of β1 integrin (CD29) and neural cell adhesion molecule (NCAM/CD56), which are not present on the mouse cells.

Once isolated, the myogenic capability of the satellite cells can be assayed by their ability to form colonies *in vitro*. These are myogenic colonies of more committed cells with a proportion retaining the ability to differentiate. More compelling definitive evidence for the stem cell nature of satellite cells comes from the transplantation of single isolated myofibers (containing approximately seven satellite cells) from normal mice into immunocompromised muscular dystrophic animals (*mdx*—Box 10.5), which demonstrates not only the production of dystrophin-positive myofibers but also the generation of more donor-derived satellite cells (Figure 10.16).

Most notably, considerable use has been made of the fact that satellite cells and their committed progeny express specific transcription factors, in particular a member of the Pax Family called Pax 7. These transcription factors can be detected within cells using specific antibodies, but more importantly, the genes encoding them have been utilized for the purpose of genetic marking (Chapter 2), usually through insertion of coding sequences downstream of a muscle-specific promoter for a fluorescent protein or an enzyme that can be used to label cells (e.g., β-galactosidase, or in more complex strategies the Cre recombinase enzyme, which then activates a silent transgenic reporter gene—Figure 10.17).

As we introduced in Chapter 2, a common phenotypic property of many adult stem cells is their occurrence within the so-called **side population** (SP) when cells stained with the DNA-binding dye Hoechst are illuminated with a UV laser in a flow cytometer. Best exemplified in the characterization of HSC as first reported by Margaret Goodell, skeletal muscle has also been shown to contain SP cells. The muscle SP is heterogeneous, containing an abundant sub-population (~80%) composed of CD31+ endothelial cells associated with blood vessels. At least two minor sub-populations can be defined, one being CD45+ hematopoietic cells (2%-10%) and the other, which accounts for the myogenic activity of the SP is CD31− CD45− and makes up about 5% of the total. The myogenic SP cells (Lin− SP) are roughly 0.1% of muscle mononuclear cells; express Pax7, Sca1, and Syndecan-4, which is a transmembrane proteoglycan that acts as a receptor for several niche components; and can generate both satellite cells and their myogenic progeny in the *mdx* transplantation assay (Figure 10.18). The relationship between the Lin− SP cells and satellite cells therefore appears to involve a progenitor/progeny component, but it also seems that both can directly generate myogenic cells.

## Muscle satellite cells are a heterogeneous population

Heterogeneity within the population of satellite cells, which seems to reflect their relative position in a hierarchy from **quiescent** stem cells to committed progenitors, has been deduced using fluorescent reporters or Cre recombinase linked to myogenic regulatory transcription factor genes. Hence, high versus low expression of Pax7

---

**BOX 10.5  ENABLING TECHNOLOGY: X-LINKED MUSCULAR DYSTROPHY (*MDX*) MOUSE MODEL**

Animal models have been crucial in developing an understanding of the mechanisms of regeneration of damaged muscle fibers. One mouse model in particular has been important in contributing knowledge of stem cell function in muscle regeneration and in addition has served as a potent model for Duchenne muscular dystrophy. The mouse model X-linked muscular dystrophy is a result of a single base substitution within an exon of the gene encoding dystrophin (termed the *mdx* gene).

The *mdx* mutant mice exhibit a less severe phenotype than that seen in human muscular dystrophy, although they display signs of muscular degeneration from three weeks of age onward in the form of muscle necrosis and muscle weakness. The muscles of *mdx* mice have reduced elasticity and are more susceptible to injury, but nevertheless, the *mdx* mutant mice have a normal lifespan, making them highly useful models for the study of muscle satellite cell function in the regeneration of muscle tissue. Importantly, the *mdx* mouse can be used as the host to test the stem cell properties of satellite cells in transplanted myofibers.

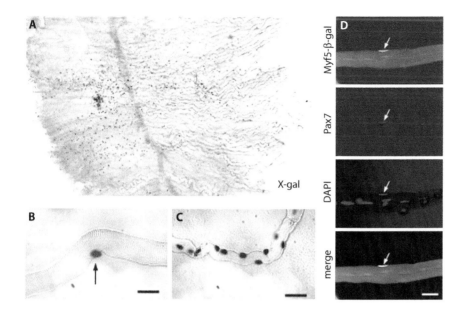

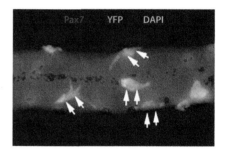

Figure 10.16 **Myofiber transplantation assay of satellite stem cells.** The *mdx* mutant mouse provides a useful model for transplantation studies of satellite cell function. In the experiment illustrated, *mdx* mice were irradiated and grafted with a single *Myf5*^LacZ+ soleus myofiber containing approximately seven satellite cells into their tibialis anterior muscle. After three weeks, the recipient mice were analyzed for the presence of grafted cells. (**A**) Staining of a segment of muscle with X-gal (substrate for β-galactosidase that is the product of *LacZ*) reveals numerous positive cells (dark blue) that have populated the muscle. (**B**) Recipient *mdx* mutant myofiber bearing a single Myf5-β-gal+ satellite cell (arrow). (**C**) A chain of newly formed donor β-gal+ myonuclei in a newly formed donor myofiber. (**D**) Fluorescent images of single myofibers from recipient mice showing co-expression of Myf5 and Pax7 in a graft-derived satellite cell (arrows). Cell nuclei are stained with DAPI. Scale bar = 25 μm. (From Collins C, Olsen I, Zammit P et al. (2005) *Cell* 122:289–301 [doi:10.1016/j.cell.2005.05.010]. Copyright 2005with permission from Elsevier.)

correlated with slower entry into cell division following injury, while tracing the history of the expression of Myf5, another transcription factor that regulates myogenesis downstream of Pax7, showed that some Pax7 expressing cells have never expressed Myf5 and are more able to engraft into the satellite cell niche, whereas those expressing Pax7 that have expressed Myf5 give rise to new myofibers but do not home to the niche.

## The control of satellite cell self-renewal and differentiation is dependent on a number of intrinsic and extrinsic factors

The investigation of processes that regulate satellite cells is complicated by the fact that the cell's highly specialized niche location makes it difficult to recreate the correct conditions for growth and differentiation once they are isolated, but some success has been achieved through experiments using intact single myofibers. When placed in culture, the satellite cell within a myofiber remains in a quiescent state until an extract from crushed muscle is added, where the satellite cells begin to divide but stop upon withdrawal of the extract, indicating a rapid activation response to signals emanating from within the muscle tissue itself.

Although satellite cells are largely quiescent, they become activated upon injury and exercise. This activation is a complex process initiated by inflammatory responses and the release of critical growth factors from the microvasculature and the muscle niche. Such factors include hepatocyte growth factor (HGF), fibroblast growth factor (FGF), insulin-like growth factor (IGF), and nitric oxide. HGF is a particularly strong candidate for the activation of satellite cells since, when in the quiescent state, they express the HGF receptor c-Met, and the expression of HGF in regenerating muscle is high. HGF is also a potent mitogen and chemotactic agent, promoting the migration of activated satellite cells to sites of injury facilitating efficient regeneration.

As we have already described, quiescent satellite cells express Pax7 and have an active Myf5 locus (although there is no evidence of Myf5 protein), but do not express MyoD, one of the earliest markers of myogenic commitment. Pax7 is absolutely essential for the specification and survival of murine satellite cells, made evident by the complete absence of satellite cells in *Pax7* **knockout** mice. The progression from the quiescent stem cell state toward self-renewal and differentiation is characterized by sequential increases and decreases in the level of expression of myogenic transcription factors (Figure 10.19). Upon activation, satellite cells start to express the primary transcriptional regulators Myf5 and MyoD, and continue to express Pax7. They then start to divide and progress through the myogenic program, including expression of the transcription factors Myogenin and Mrf4, which regulate terminal differentiation.

Reflecting the more immature stem cell phenotype of the cells expressing high Pax7, they asymmetrically segregate their DNA so that the daughter cells receiving the template DNA strand are the ones that maintain expression of stem cell markers (see Chapter 1). The asymmetry of DNA strand segregation is paralleled by components of

Figure 10.17 **Genetic marking of satellite cells.** Lineage tracing using fluorescent reporters activated by Cre recombinase driven from satellite cell-specific gene elements have proved invaluable in understanding the behavior of these stem cells in muscle. The fluorescent microscope images show an isolated myofiber from the tibialis anterior muscle of a mouse carrying the Pax7CreERT2 allele and the yellow fluorescent protein (YFP) reporter driven from the Rosa 26 (R26R) locus. The mouse had been treated each of the preceding 5 days by injection of tamoxifen to activate Cre recombinase. Satellite cells can be seen highlighted by immunostaining for Pax7 (pink) and through the activated YFP expression. Myofiber cell nuclei are stained blue with DAPI. (From Kuang S, Kuroda K, Le Grand F et al. (2007) Cell 129:999–1010 [doi: 10.1016/j.cell.2007.03.044]. Copyright 2007 with permission from Elsevier.)

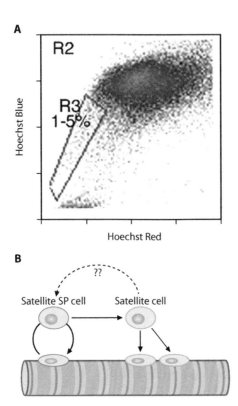

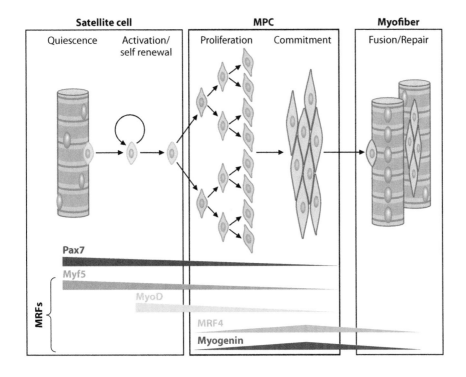

Figure 10.18 **Side population cells in skeletal muscle.** Side population (SP) characteristics following staining of a cell population with a DNA binding dye is a way to enrich for stem cells from many tissues. A proportion of muscle satellite cells can be isolated on the basis of their SP characteristics. (**A**) Flow cytometry profile of skeletal muscle-derived cells stained with Hoechst 33342, revealing the SP representing 1%–5% of whole muscle cells. (**B**) The myogenic activity of the muscle SP is in a CD31-CD45- population that is roughly 0.1% of muscle mononuclear cells and can generate both satellite cells and their myogenic progeny in the *mdx* transplantation assay. The satellite SP cells (blue) may divide asymmetrically to produce a satellite SP cell and a satellite cell. Satellite cells are capable of symmetric division to produce satellite cell progeny. It is not known if satellite SP cells constitute a lineage or arise from the interaction of a satellite cell with a specialized niche. (From Tanaka K, Hall J, Troy A et al. (2009) *Cell Stem Cell* 4:217–225 [doi: 10.1016/j. stem.2009.01.016]. Copyright 2009 with permission from Elsevier.)

Figure 10.19 **Regulation of satellite cell self-renewal and differentiation.** As for the regulation of many stem cells, the role of lineage-specific transcription factors in satellite cells is crucial. The diagram illustrates how changes in the expression of key transcription factors (myogenic regulatory factors—MRF) correspond to the transition from the quiescent stem cell state through to myocyte differentiation. Quiescent satellite cells (green) express high levels of Pax7 and Myf5. Upon activation to proliferate, MyoD is up regulated as they differentiate toward muscle precursor cells (MPC). Satellite cell progeny then follow one of two decisions: they either down regulate MyoD and undergo self-renewal to produce Pax7+ satellite cells or further differentiate by up regulation of MRF4 and Myogenin to become committed to undergo fusion to form new myofibers. (From Boldrin L, Muntoni F & Morgan J. (2010) *J Histochem Cytochem* 58:941–955 [doi:10.1369/ jhc.2010.956201]. With permission from Sage Publications.)

Notch signaling, which is known to be a major determinant of satellite cell fate. The high Pax7 cells express the Notch ligand Delta 1 and the Notch signaling inhibitor Numb, while the more committed cells that have expressed Myf5 express the Notch receptor Notch 3.

As in many scenarios of cell differentiation from stem cells, micro RNAs (miRNAs) appear to play an important modulating role in gene expression. Over 350 miRNAs are differentially expressed when comparing quiescent and activated satellite cells. A good example of the role that miRNAs can play concerns miR-31, which is highly expressed in quiescent cells and rapidly down regulated following muscle injury.

miR-31 targets *myf5* RNA, preventing its translation so that, once down regulated, there is a rapid increase in translation of the protein.

## The satellite cell niche is crucial for normal stem cell function

The concept of the stem cell niche (Chapter 6) was first proposed by Ray Schofield in 1978 in the context of the HSC, and, as for the majority of stem cell niches, the myofiber niche maintains satellite cells progressing from quiescence to an activated cycling state and back again. This dynamic interchange between quiescence and activity is in response to a diverse range of external stimuli triggered by exercise, injury, and disease. Removal of satellite cells from their niche within the myofiber and culture *in vitro* results in a reduction in self-renewal and myogenic potential. Similar to other stem cell types described throughout this book, some degree of maintenance of stem cell properties can be achieved by culturing on cell-specific substrates, and in the case of the satellite cells, Helen Blau and colleagues showed that this can be achieved by growing them on a more elastic cell substrate.

The satellite cell in its niche is in intimate contact with both the myofibers and the basal lamina, with closely associated blood vessels (Figure 10.20). Direct physical links involve surface receptors such as integrins, M-cadherin, and Notch, and various growth factors such as FGF, among several others, have been identified as being important in satellite cell maintenance within the niche. The interplay between Wnt and Notch signaling, with the receptors being expressed on the satellite cell and the ligands being presented by the niche, is perhaps the most important determinant of muscle stem cell fate, as it is in several adult stem cell scenarios. Canonical Wnt signaling (Chapter 4), mainly through the ligand Wnt3a, drives differentiation of satellite cells, while non-canonical Wnt7a-dependent signaling promotes symmetric satellite cell divisions, migration of satellite cells, and the growth of myofibers. A switch from Notch to canonical Wnt signaling is necessary for satellite cell differentiation. Satellite cells express high levels of Notch to retain them in a quiescent state, the Notch ligands being sequestered to the surface of the myofibers. Upon activation, canonical Wnt signaling increases, antagonizing the effects of Notch, but once the satellite cells return to quiescence, they switch back to Notch signaling.

## Changes in the interaction between the satellite cell and its niche are a major factor in the decline of stem cell capacity during aging

During aging, there is a progressive loss of skeletal muscle mass, strength, and endurance, typically termed sarcopenia. The regenerative capacity of aged muscle is reduced compared to that seen in younger adults, with rapid myofiber formation

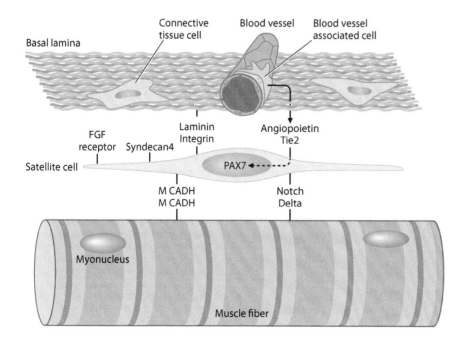

Figure 10.20 **The satellite cell niche.** The satellite cell interacts with its environment through a very defined niche in which it is in direct contact with the muscle fiber and the adjacent basal laminal layer. The diagram shows a Pax7+ satellite cell contacting the muscle fiber and the basal lamina through specific receptor-ligand interactions. Abbreviations: MCADH—M Cadherin; FGF—fibroblast growth factor. (From Lander A, Kimble J, Clevers H et al. (2012) *BMC Biology* 10:19 [doi: 10.1186/1741-7007-10-19]. Licensed under the terms of the Creative Commons Attribution License (http://creativecommons.org/licenses/by/2.0).)

being replaced by a delayed and partial myogenic response, often accompanied by deposition of fibrous scar tissue where normal muscle should have formed. At birth, satellite cells make up approximately 30% of the mouse myofiber nuclei, declining to 4% in adulthood and in aged individuals representing only 2%.

The decline in muscle function with age was originally thought to be the consequence of loss of the **potency** of satellite cells after it was observed that when derived from older tissue, they exhibit a delayed response to activating stimuli and show reduced proliferative potential. However, this theory was challenged by the discovery that the decline in function of the satellite cell is more likely a result of a defective niche environment. Experiments performed in the 1980s showed that when muscular tissue from either young or aged donors was transplanted into young or aged recipients, it was obvious that the host environment was the determining factor in the regenerative capacity of the satellite cells from the donor tissue. Further support for the importance of the loss of niche efficacy with age came from classic experiments making use of parabiotically paired mice (Figure 10.21). In these key experiments, two mice would be surgically connected through flaps of skin, a process termed **parabiosis**, so that they developed a shared vascular system, enabling the transfer of circulating factors from one mouse to the other. When an old mouse was paired with a young mouse (**heterochronic** pairing) but not with another old mouse (**isochronic** pairing), the regenerative capacity of the aged damaged muscle was enhanced as a result of increased activation of the resident aged satellite cell by factors released from the young mouse.

## Changes in a number of receptor ligands and growth factors contribute significantly to the decline in the satellite cell niche

As for many adult stem cells and their interaction with their niche, the satellite cell's dependence on active Notch signaling is believed to be one of several features of the

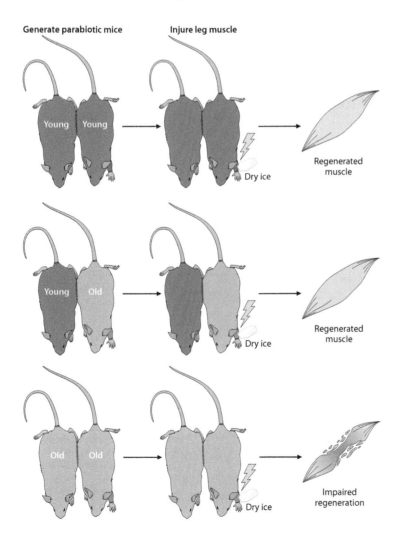

Figure 10.21 **Proof that the niche rather than the satellite cell declines in function with age.** Parabiotic mice can be used to assess the influence of the environment of aged animals on muscle regeneration. At the top of the figure, two young mice are surgically joined. Injury to one mouse limb is achieved by subjecting to dry ice treatment, but this will subsequently regenerate. In the middle, a young mouse (dark) is joined to an aged mouse (light), which is consequently seen to be able to regenerate its injured muscle. At the bottom, two aged mice are joined, but in this case, the injured muscle does not regenerate. (From Snoeck H. (2005) *Nature Biotechnol* 23:434–435 [doi: 10.1038/nbt0405-434]. Reprinted by permission from Springer Nature.)

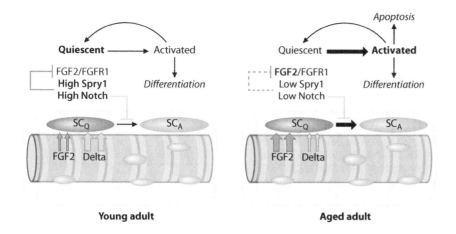

Figure 10.22 **Age-related changes in the satellite cell niche.** Schematic illustration of the satellite cell niche in young adult and aged skeletal muscle. One myofiber (MF) is represented with multiple myonuclei (blue). The satellite cell is shown progressing from a quiescent state (purple, $SC_Q$) to the activated state (pink, $SC_A$). In the young adult, the niche produces FGF2 (orange arrows), although its action on proliferation is limited by the inhibitory action of Sprouty 1 (Spry1) on the FGF receptor 1 (FGFR1) pathway leading to a quiescent state. The myofiber is normally able to respond to requirements for cell replacement through production of the Notch ligand Delta (green arrows). In aged muscle, FGF2 is produced at higher levels, leading to a greater proliferative stimulus through FGFR1 (and inhibition of Spry1), and at the same time, the ability to express Delta is restricted. Cumulatively, these altered responses lead to overproliferation, apoptosis, and a decline in satellite cell numbers. (From Tajbakhsh S. (2013) *Cell Res* 23:455–457 [doi:10.1038/cr.2013.3]. Reprinted by permission from Springer Nature.)

myofiber niche that is impaired during aging (Figure 10.22). With increasing age, the induction of Delta within the niche is defective, resulting in attenuated Notch signaling and a consequent reduction in satellite cell proliferation and impaired regenerative capacity. Given the reciprocal relationship between Notch and Wnt signaling in satellite cells, the up regulation of Wnt3a signaling in aged skeletal muscle further impairs satellite cell function. Wnt3a can be counteracted by the soluble protein Klotho, which can presumably at least partially explain the results of the parabiosis experiments, but this too shows a tendency of age-related decline. In addition, there are reports of age-associated changes in the level of signaling through the FGF receptor on the satellite cell. FGF growth factors are potent stimulators of mitogenic activity, and it has been noted that there is an accumulation of FGF-2 around the satellite cells during aging, although the precise cause for the increase in FGF-2 is not known. This age-related increase in FGF-2 has been linked to a loss of satellite cell quiescence through an increased inhibition of the signal transduction modulator Sprouty1.

There are likely to be many other factors at play, including some that are secondary to such deficits but serve to accentuate the decline in muscle regenerative capacity; reduced regeneration capacity within the aged muscle results in the deposition of fibrous tissue and adipocytes, which in turn alters the secretion of growth factors within the microenvironment and ultimately hinders the migratory behavior of infiltrating macrophages during regeneration.

## Understanding of the satellite cell and its niche offers new avenues for novel therapeutic strategies to tackle disease and the effects of aging

A major purpose of expanding knowledge on the muscle satellite cell population is ultimately to gain an insight into ways in which these cells might be exploited for therapeutic purposes. As we have noted, there are several ways in which muscle function can be lost, either as the result of acquired or inherited defects. Atrophic muscle losses include traumatic damage, sudden loss of body fat and muscle as a consequence of diseases such as cancer and AIDS (a condition known as cachexia), and age-associated sarcopenia, as we have just described. In contrast, dystrophic muscle loss is the result of a genetic defect.

The experimental testing of satellite cells in models of muscle loss and damage, such as transplantation into the *mdx* mouse, clearly demonstrate that it is feasible to enhance the normal process of regeneration or the restoration of full function in individuals with muscle dystrophies, which is a group of diseases characterized by a progressive weakening of the skeletal muscle, resulting in limitations to mobility, heart and lung failure, and eventually death. The most common and severe form, Duchenne muscular dystrophy, is an X-linked recessive disease caused by mutations in the Dystrophin protein, which is found in the membrane of muscle fibers and links the intracellular cytoskeleton with the extracellular membrane, providing mechanical stabilization. Potential therapy can be envisaged around the use of satellite cells, either from the affected individual and used as a target for genetic correction or through transplantation from a matched donor with a normal Dystrophin gene, but there are many problems that would have to be resolved in order to make this a realistic prospect. As we will discuss in Chapter 17, there are many hurdles that need to be crossed to go from proof-of-principle experiments to actual application of stem cells, including satellite cells, in patients.

Importantly, and quite likely to be of much wider benefit in the treatment of atrophic muscle loss caused by disease and aging, increasing our understanding of the intrinsic and niche-related regulatory mechanisms and how these might be manipulated, perhaps through pharmacological means, are realistically tractable goals for the foreseeable future. Indeed, we have described two examples of proteins that might be used to combat muscle atrophy in that both Wnt7a and Klotho, either produced in the niche or systemically available, can influence the Wnt-Notch signaling balance in a beneficial manner. An intrinsic age-related change in satellite cells limiting their effectiveness has been identified as elevated activity of the mitogen-activated kinase (MAPK) signaling pathway, which cannot be alleviated by transplantation into the microenvironment of young recipient muscles. However, treatment of the aged satellite cell population with MAPK inhibitors can expand residual functional cells, illustrating that it might be feasible to tackle muscle wasting in the elderly from a number of directions.

## 10.5   SUMMARY

In this chapter, we have described two stem cell types that are required for the generation and maintenance of the major components of the musculo-skeleton. One of these, the mesenchymal stem cell, is often quite loosely defined and exhibits considerable heterogeneity at a number of levels. Heterogeneity in *in vitro* culture has to some extent masked the stem cell nature of MSC, leading some to believe that MSC are not stem cells. Heterogeneity of MSC is also seen across different tissues, particularly with respect to the relative propensity for differentiation into fat, cartilage, or bone. Some of the heterogeneity observed in MSC is connected to their developmental origins; although most adult tissue MSC are of mesodermal origin, others are derived from the embryonic neural crest. A particular specialization of MSC-like cells is seen in teeth, in which at least six subtypes of neural crest-derived stem cells have been defined that each predominantly give rise to dentin-producing odontoblasts. In contrast to the nature of MSC, we have described how a monopotent stem cell known as the satellite cell is responsible for the production of muscle cells. Helped by the ease of observing satellite cells *in situ*, their niche has been very well characterized, and a picture has emerged of the mechanisms that control self-renewal and differentiation as an interplay between stem cell intrinsic factors and the extrinsic contribution of the niche. We have described how loss of muscle function in older age can be at least partially attributed to changes in both the stem cell and its niche. Throughout the chapter, we have highlighted how the understanding of the stem cells of the musculo-skeleton is opening up a number of clinical opportunities for tissue regeneration and pharmacological interventions to alleviate the decline in stem cell function during aging.

## KEY POINTS

- The musculo-skeleton is maintained largely through two adult stem cell types, that is, a family of related mesenchymal stem cells (MSC) and skeletal muscle stem (satellite) cells.

- MSC are generally quite loosely defined based on adherence to plastic culture dishes; some shared cell surface antigens; and *in vitro* capacity to differentiate into fat, cartilage, and bone.

- MSC were first identified from the bone marrow but are also present in many other tissues.

- True stem cell behavior of MSC can be demonstrated by transplantation and the formation of bone-like structures termed ossicles.

- *In vivo* differentiation of MSC is influenced by the tissue type in which they reside, and are particularly affected by the stiffness of the tissue.

- The assumption that vessel-associated pericytes are MSC appears not to be correct, although they do share some characteristics and *in vitro* assume a similar phenotype.

- MSC-like stem cells play an important part in the maintenance and repair of teeth.

- Dental MSC are particularly geared to generating dentin-producing odontoblasts and the dental pulp.

- The stem cells that maintain adult skeletal muscle, usually known as satellite cells, are closely associated with muscle fibers and are only able to differentiate into muscle cells.

- The transcription factor Pax7 defines the stem cell state of the satellite cell.

- The satellite cell niche is well defined and known to communicate with the satellite cell through several key ligand-receptor interactions, notably Delta-Notch and FGF2-FGFR.

- During aging, the decline in skeletal muscle regenerative capacity seems largely to be caused by changes in the niche, including decreased Delta signaling and increased FGF2 production, although stem cell intrinsic effects and systemic mediators also play a part.

# FURTHER READING

## Mesenchymal stem cells

Bianco, P., Cao, X., Frenette, P.S., Mao, J.J., Robey, P.G., Simmons, P.J. and Wang, C.Y. (2013) The meaning, the sense and the significance: Translating the science of mesenchymal stem cells into medicine. *Nat. Med.*, 19: 35–42.

Engler, A.J., Sen, S., Sweeney, H.L. and Discher, D.E. (2006) Matrix elasticity directs stem cell lineage specification. *Cell*, 126: 677–689.

Guimarães-Camboa, N., Cattaneo, P., Sun, Y., Moore-Morris, T., Gu, Y., Dalton, N.D., Rockenstein, E. et al. (2017) Pericytes of multiple organs do not behave as mesenchymal stem cells *in vivo*. *Cell Stem Cell*, 20: 345–359.

Sacchetti, B., Funari, A., Michienzi, S., Di Cesare, S., Piersanti, S., Saggio, I., Tagliafico, E. et al. (2007) Self-renewing osteoprogenitors in bone marrow sinusoids can organize a hematopoietic microenvironment. *Cell*, 131: 324–336.

## Skeletal stem cells

Bianco, P. and Robey, P.G. (2015) Skeletal stem cells. *Development*, 142: 1023–1027.

Chan, C.K.F., Seo, E.Y., Chen, J.Y., Lo, D., McArdle, A., Sinha, R., Tevlin, R. et al. (2015) Identification and specification of the mouse skeletal stem cell. *Cell*, 160: 285–298.

Tevlin, R., Seo, E.Y., Marecic, O., McArdle, A., Tong, X., Zimdahl, B., Malkovskiy, A. et al. (2017) Pharmacological rescue of diabetic skeletal stem cell niches. *Sci. Transl. Med.*, 9. doi: 10.1126/scitranslmed.aag2809.

Worthley, D.L., Churchill, M., Compton, J.T., Tailor, Y., Rao, M., Si, Y., Daniel Levin, D. et al. (2015) Gremlin 1 identifies a skeletal stem cell with bone, cartilage, and reticular stromal potential. *Cell*, 160: 269–284.

## Dental stem cells

An, Z., Sabalic, M., Bloomquist, R.F., Fowler, T.E., Streelman, T. and Sharpe, P.T. (2018) A quiescent cell population replenishes mesenchymal stem cells to drive accelerated growth in mouse incisors. *Nat. Commun.*, 9: 378.

Feng, J., Mantesso, A., De Bari, C., Nishiyama, A. and Sharpe, P.T. (2011) Dual origin of mesenchymal stem cells contributing to organ growth and repair. *Proc. Natl. Acad. Sci. USA*, 108: 6503–6508.

Sharpe, P.T. (2016) Dental mesenchymal stem cells. *Development*, 143: 2273–2280.

## Muscle satellite stem cells

Collins, C.A., Olsen, I., Zammit, P.S., Heslop, L., Petrie, A., Partridge, T.A. and Morgan, J.E. (2005) Stem cell function, self-renewal, and behavioral heterogeneity of cells from the adult muscle satellite cell niche. *Cell*, 122: 289–301.

Conboy, I.M., Conboy, M.J., Wagers, A.J., Girma, E.R., Weissman, I.L. and Rando, T.A. (2005) Rejuvenation of aged progenitor cells by exposure to a young systemic environment. *Nature*, 433: 760–764.

Cosgrove, B.D., Gilbert, P.M., Porpiglia, E., Mourkioti, F., Lee, S.P., Corbel, S.Y., Llewellyn, M.E., Delp, S.L. and Blau, H.M. (2014) Rejuvenation of the muscle stem cell population restores strength to injured aged muscles. *Nat. Med.*, 20: 255–264.

Schmidt, M., Schüler, S.C., Hüttner, S.S., von Eyss, B. and von Maltzahn, J. (2019) Adult stem cells at work: Regenerating skeletal muscle. *Cell. Mol. Life Sci.*, https://doi.org/10.1007/s00018-019-03093-6.

Yin, H., Price, F. and Rudnicki, M.A. (2013) Satellite cells and the muscle stem cell niche. *Physiol. Rev.*, 93: 23–67.

# Stem cells in endoderm-derived organs

11

Consuming food and water and breathing oxygen are the critical inputs that keep our bodies functioning and necessarily require an ability to dispose of the waste products generated. Two body systems, the gastrointestinal (GI) and respiratory tracts (Figure 11.1), can be thought of as elaborate tubes that deal with these inputs and, with the exception of the role played by the kidneys, also handle most of the waste issues. The linings of both the gastrointestinal and respiratory systems are largely products of the embryonic **endoderm**, that is, the innermost layer formed following gastrulation in the developing embryo (Box 11.1). Beginning at the mouth and ending at the anus, the human GI tract is approximately nine meters long. The upper reaches of the GI tract consist of the esophagus and stomach, while the lower part contains the small intestine (duodenum, jejunum, and ileum) and the large intestine (colon and rectum). Likewise, the respiratory tract includes an upper airway, a lower airway, and the lungs. The upper airway, which is not endoderm derived, ranges from the nose to the pharynx. The lower airway includes the larynx, trachea, bronchi, and bronchioles. The lungs include the respiratory bronchioles, alveolar ducts, alveolar sacs, and alveoli. From the bronchi, the dividing tubes become progressively smaller, with about 20 subdivisions in humans, before ending at an alveolus. The endoderm is not only the source of most of the tube linings of the GI and respiratory tracts but also gives rise to associated organs, including the liver, gall bladder, pancreas, and thyroid, which facilitate the processing of nutrients, the handling of metabolic waste products, and the rate of metabolism.

As we will see in this chapter, the stem cells associated with the GI and respiratory tracts serve well to illustrate the demands that need to be satisfied by tissue-restricted stem cells in organs whose cell turnover can range from low, as for the respiratory system, to very high in the GI tract. Complete renewal of the intestinal villus epithelium occurs every 2–6 days in most adult mammals, meaning that enterocytes have the highest turnover rate of any fixed-cell population in the body. Mathematical modeling suggests that $10^{11}$ cells are shed from the human small intestine per day. The study of the GI tract epithelial stem cells, which underpin such enormous replacement, has also given insight into several of the fundamental principles of stem cells. In particular, studies of GI tract stem cells have shed light on the nature of the **niche** and importantly have revealed how a **tissue stem cell** can have distinct forms that act as either the source of everyday replacement or perform a reserve role following damage. Although to date less intensively studied, and therefore less well defined, stem cells with a relatively low rate of cell turnover in the GI tract-associated organs and respiratory tract also appear to perform a reserve role. At least in some of these tissues, the role of the **adult stem cells** also has to be put in context with the ability of the differentiated cells derived from them to proliferate and bring about extensive regeneration (see Chapter 7).

## 11.1  GASTROINTESTINAL TRACT LINING STEM CELLS

The epithelial lining of the tube constituting the GI tract has high demands placed on it both in terms of the need for continuous replacement and the fact that the

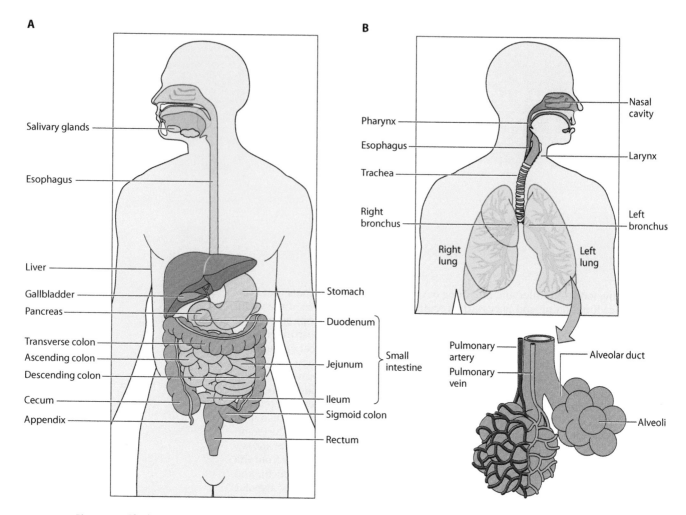

**Figure 11.1 The human gastrointestinal and respiratory tracts.** Diagram showing the structural organization of (**A**) the gastrointestinal tract and (**B**) the respiratory system. (B: from http://training.seer.cancer.gov/anatomy/respiratory/passages/bronchi.html).

## BOX 11.1  BACKGROUND: DEVELOPMENT OF THE EMBRYONIC ENDODERM INTO THE GASTROINTESTINAL AND RESPIRATORY TRACTS

The definitive endoderm is one of the three embryonic germ layers, and, among other organs, gives rise to many of the component tissues of the gastrointestinal (GI) and respiratory tracts (**Figure 1**). The endoderm forms the innermost layer of the embryo and as gastrulation proceeds, morphogenetic movements (beginning at E7.5 in mouse) transform it into the primitive gut tube (**Figure 2**), which is surrounded by the mesodermal layer. The endodermal cells generate only the lining of the GI tract and its glands, the mesodermal mesenchyme cells surrounding this tube providing the muscles and supporting tissue. As the endodermal tubes form, the epithelium responds to the different mesenchymal environments, enabling the digestive and respiratory tubes to develop specific structures. The primitive gut becomes divided into the foregut, midgut, and hindgut. The foregut is eventually specified to give rise to the esophagus, trachea, stomach, lungs, thyroid, liver, gall bladder, and pancreas, whereas the midgut gives rise to the small intestine and the hindgut to the large intestine. The liver and pancreas develop

by budding of the foregut just below the level of the stomach. An outgrowth of endoderm extends from the foregut into the surrounding mesenchyme, where induction of proliferation and branching leads to the formation of the glandular epithelium of the liver. The section of the tube forming the liver that is closest to the digestive tube continues to function as the drainage duct of the liver, and a branch from this creates the gall bladder. The lungs are also a derivative of the foregut, even though they serve no role in digestion. As in the digestive tube, the regional mesenchyme dictates the differentiation of the developing respiratory tube as it extends from the floor of the pharyngeal area. In the region of the neck, the respiratory tube grows straight, forming the trachea, and after entering the thorax, it branches, forming the two bronchi and then the lungs. The esophagus, stomach, and duodenum are specified from the remaining foregut, while the jejunum, ileum, and small intestine are induced from the midgut, and the large intestine and portions of the urogenital system arise from the hindgut.

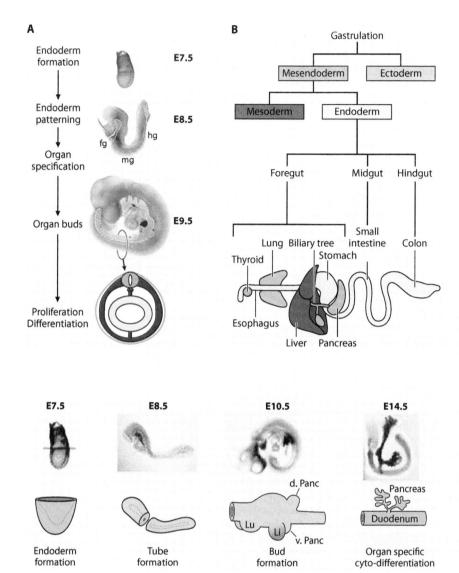

**A**

Endoderm
formation

↓

Endoderm
patterning

↓

Organ
specification

↓

Organ buds

↓

Proliferation
Differentiation

E7.5

E8.5

fg   hg

mg

E9.5

**B**

Gastrulation

Mesendoderm        Ectoderm

Mesoderm        Endoderm

Foregut        Midgut        Hindgut

Lung  Biliary tree    Small
intestine    Colon

Thyroid        Stomach

Esophagus

Liver   Pancreas

**E7.5**

Endoderm
formation

**E8.5**

Tube
formation

**E10.5**

d. Panc

Lu   Li

v. Panc

Bud
formation

**E14.5**

Pancreas

Duodenum

Organ specific
cyto-differentiation

Box 11.1 Figure 1 **Formation of the gastrointestinal and respiratory tracts from the endoderm.** (**A**) Depiction of endoderm organ specification during development of the mammalian embryo. The images show mouse embryos at embryonic (E) day 7.5 (top), 8.5 (middle), and 9.5 (bottom). Endodermal regions are shaded in yellow, with the regions of the foregut (fg), mid gut (mg), and hindgut (hg) labeled. The cross section at the bottom is through an E9.5 embryo showing the arrangement of the three germ layers, with the endoderm lining the gut tube (yellow), surrounded by mesoderm (red) and ectoderm (blue). (**B**) Distribution of endoderm-derived tissues along the gastrointestinal tract. (Adapted from Zorn A & Wells J (2009) *Annu Rev Cell Dev Biol* 25: 221–251 [doi: 10.1146/annurev.cellbio.042308.113344]. Images: From Zorn AM. Liver development. (2008) In: *StemBook* [Internet]. Cambridge (MA): Harvard Stem Cell Institute. [doi:10.3824/stembrook.1.25.1]. This is an open-access article distributed under the terms of the Creative Commons Attribution License.)

Box 11.1 Figure 2 **Stages of endoderm development.** The GI tract and associated organs develop from the endoderm, which in the mouse commences around E7.5, organs being specified by E14.5. The upper images are of embryos, and the schematic diagrams below show how the gut tube forms and organs arise by budding around E10.5. Abbreviations: Lu, lung; Li, liver; St, stomach; d. Panc., dorsal pancreatic bud; and v. Panc., ventral pancreatic bud. (From Wells J, Melton D. (1999) *Annu Rev Cell Dev Biol* 15:393–410 [doi:10.1146/annurev.cellbio.15.1.393]. Permission conveyed through Copyright Clearance Center, Inc.)

environment faced by the cells can be very harsh, for example, the acidic contents of the stomach. Within the human small and large intestine, up to $10^{11}$ epithelial cells are lost every day, the small intestinal epithelium having the highest **self-renewal** rate of any tissue, with an approximate turnover time of around 3–5 days. The most studied stem cells in the GI tract are those that underpin the maintenance of the small intestine; in fact, it is probably fair to say that our understanding of these cells and the environment with which they interact is probably greater even than that of the **hematopoietic stem cell** (Chapter 9). Although there are many similarities between small **intestinal stem cells (ISC)** and those of other sections of the GI tract epithelium, we will also see that there are some subtle differences in the equivalent cells in the stomach and colon, probably reflecting the respective environmental demands.

## Stem cells reside in the crypts of the small intestinal epithelium

The distinctive epithelial lining of the small intestine is adapted to maximize surface area for efficient absorption through the presence of finger-like protrusions called villi, which are about 0.5–1.6 mm in length (Figure 11.2). Each villus is encircled by at least six crypts of Lieberkühn, which are invaginations of the epithelial surface and the continual source of replacement cells that move in a stream from their

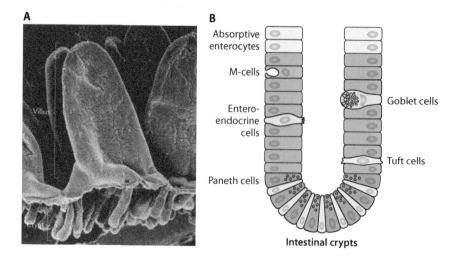

**A**

**B**

Absorptive enterocytes

M-cells

Entero-endocrine cells

Goblet cells

Tuft cells

Paneth cells

Villus

**Intestinal crypts**

Figure 11.2 **Architecture of the epithelial surface in the small intestine.** (**A**) The small intestinal epithelium is constituted of villi that project into the lumen. The scanning electron micrograph shows such villi and the multiple crypts that lie at their base. (**B**) Several terminally differentiated cells that arise from the crypts characterize the epithelium of the intestinal villi. Absorptive cells have a brush border (a dense array of microvilli) on their apical border. Goblet cells and enteroendocrine and Paneth cells are all secretory cells. Goblet cells secrete mucus, and their apical cytoplasm is distended with mucus-filled secretory granules. Enteroendocrine cells secrete various gut hormones. Paneth cells secrete antibacterial proteins. Tuft cells are chemosensory cells, while M ("membranous") cells function as an interface between the contents of the gut lumen and the underlying immune cells. (A: From Magney J, Erlandsen SL, Bjerknes M et al. (1986) Amer J Anat 177:43–53 [doi: 10.1002/ aja.1001770106]. With permission from Wiley. B: From Gerbe F, Legraverend C & Jay P. (2012) Cell Mol Life Sci 69: 2907–2917 [doi: 10.1007/ s00018-012-0984-7]. Published under the terms of the Creative Commons Attribution License http://creativecommons.org/licenses/.)

base through to the tip of the villus. The intestinal epithelium is made up of several terminally differentiated cell types, namely absorptive enterocytes, mucus secreting goblet cells, hormone secreting enteroendocrine cells, tuft cells that are thought to sense the contents of the gut, **Paneth cells** that secrete bactericidal products such as lysozyme and defensins, and microfold (M) cells that overlay the Peyer's patches and play a key role in mucosal immunity.

Given the harsh luminal environment at the villus surface, it was obvious to begin the search for the intestinal stem cell within the crypts, where cells are protected from the lumen, and the results of early experiments verified that they indeed house a self-renewing cell population. These experiments included the mapping of heritable genetic markers, which had been introduced at a low frequency, and their appearance in clones containing all of the epithelial lineages of the intestine. During the 1970s, experiments were performed using radioisotope and bromodeoxyuridine (BrdU) to label DNA in dividing intestinal cells. The epithelial cell migration pattern indicated that the cells had a common origin close to the crypt base. Further ablation of the crypt base itself by targeted irradiation completely abolished any epithelial cell renewal, presumably due to the loss of the resident stem cell population.

We are going to devote a large fraction of this chapter to consideration of the stem cells in the small intestine because the understanding that has been gained through their study is fundamental to the field of adult stem cell biology. Compared to other adult stem cells, in the case of the small intestine, we know more about the precise nature of the stem cells and their location, how their population dynamics are regulated, and how they interact with specific components of their niche. Their study also represents a paradigm in the sense that **lineage tracing** (Chapter 2) has been crucially important and allowed definitive conclusions to be made about self-renewal and **commitment** as well as the nature of distinct subtypes of stem cells.

## Although the intestinal stem cell is well characterized, there remains some doubt about the precise details of the hierarchy

Although the location of the intestinal stem cell to the crypts was revealed by the early experiments, its identity and characteristics remained unknown. Over the last decade or so, advances in stem cell labeling and **lineage tracing** methods, combined with specific culture assays for intestinal epithelial cells, allowed very precise definition of the position and properties of what turned out to be more than one stem cell type. In spite of such detailed characterization, there remains some debate as to what exact cell type is responsible for the regenerative capacity of the intestinal epithelium. Two models have been proposed (Figure 11.3), and although neither has since been accepted as the definitive model, it is likely that aspects of both are relevant and we are close to understanding how stem cells maintain the intestinal epithelium.

The first model was proposed by Chris Potten (Box 11.2) and was termed the "+4 position model" since the intestinal stem cell population was proposed to reside at position 4 from the base of the crypt as a ring of approximately 16 cells lying just above the Paneth cells. These +4 cells were first identified as DNA-**label retaining cells** (LRC), although the observation that these cells readily incorporated BrdU, and were therefore considered to be actively cycling, apparently conflicts with the view that

**A** "+4 position" model

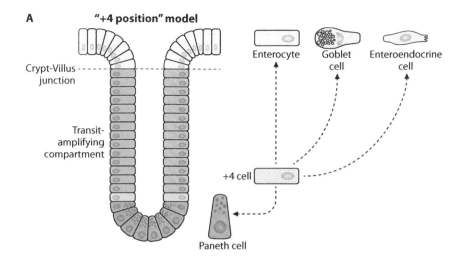

Crypt-Villus junction

Transit-amplifying compartment

+4 cell

Paneth cell

Enterocyte

Goblet cell

Enteroendocrine cell

**B** "Stem cell zone" model

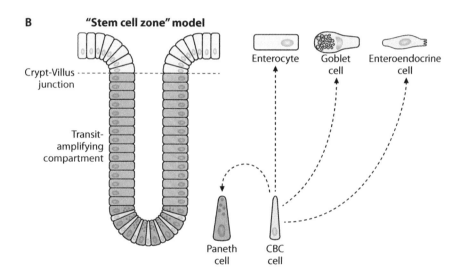

Crypt-Villus junction

Transit-amplifying compartment

Paneth cell

CBC cell

Enterocyte

Goblet cell

Enteroendocrine cell

Figure 11.3 **Models of small intestine stem cell identity.** Two models have been proposed for the nature of the small intestinal stem cells that lie in the base of the villi crypts. (**A**). The "+4" model: the exact position of the "+4" stem cell (blue) can be seen directly above the Paneth cells (purple). (**B**) The "stem cell zone" model: the crypt base columnar (CBC) cells (beige) are located at the base of the crypt interspersed between the Paneth cells. (From Barker N, van de Wetering M & Clevers H. (2008) *Genes* Dev 22: 1856–1864 [doi: 10.1101/gad.1674008]. Used with permission from Cold Spring Harbor Laboratory Press.)

---

**BOX 11.2    PIONEERS: CHRIS POTTEN AND HANS CLEVERS**

Chris Potten and Hans Clevers (Figure 1) have been central figures in the thinking and investigation of the nature of intestinal epithelial stem cells spanning over half a century, and through the insight they have achieved in this system have also contributed hugely to our overall understanding of adult stem cells. Chris Potten died in 2012, but Hans Clevers continues to push the boundaries, using ever more elegant genetic marking strategies to define the precise location and behavior of the stem cells of the GI tract.

Chris Potten was always a pioneer within the field of epithelial and stem cell biology. His work focused mainly around the identification and characterization of stem cells

Box 11.2 Figure 1 **Christopher Potten (left) and Hans Clevers (right).** (From http://static.guim. co.uk/sysimages/Guardian/Pix/ pictures/2012/9/5/1346845484687/ Professor-Christopher-Pot-008.jpg, and Henk Thomas/KNAW (https://www. flickr.com/photos/79173061@N08/ albums/72157631675823928).)

within the GI tract, and he was the first to demonstrate that the few stem cells were located at cell position 4 from the base in the small intestinal crypts and at the base of the large intestine. His career began in Manchester, England, with his PhD studies on mouse epidermal cells, using radiation as a tool to unravel the mechanisms of tissue renewal and response to injury. This work during the 1960s led to a change in the view that all cycling cells in epithelia were regenerative but rather that there is a defined hierarchy of cell lineages. Following his switch to the study of stem cells within the GI tract, carried out in the Paterson Institute for Cancer Research in Manchester, where he was to lead an epithelial cell biology group for 29 years, he was the first to investigate and provide substantial evidence to support John Cairns's immortal DNA strand hypothesis. Upon his retirement in 2000, he co-founded Epistem Ltd, a highly successful company based at the University of Manchester. In recognition of his contribution to the field, he was awarded a life fellowship from Cancer Research UK and also received International Marie Curie and Weiss medals for services to radiation research.

Hans Clevers obtained his MD degree in 1984 and his PhD degree in 1985 from the University of Utrecht. After doing postdoctoral work in the Dana-Farber Cancer Institute of the Harvard University, Boston, USA, Hans was first professor of immunology and then, since 2002,

professor in molecular genetics and director of the Hubrecht Institute at Utrecht University. Speaking of his career path, Hans said: "There was a bit of an awkward route. I actually studied biology first, and then took up medical school at about the same time ... Did two separate studies, graduated from both, was going to be a pediatrician, then decided to spend a year in science, liked it so much more that I realized I didn't—I shouldn't become a real doctor. I was not good with—I liked patients, but I was a little bit impatient with them. I then decided to go for a post-doc at Boston to Dana Farber, where I really learned the trade." Well he certainly learned his trade well—his work on intestinal epithelium and intestinal cancer have been nothing short of definitive in terms of the specific understanding of the particular tissues and fundamental with regard to so many of the principles that underpin adult stem cell biology. His findings include characterization of the transcription factor TCF-1, defining details of Wnt signaling in normal and cancerous cells, and the discovery of the Wnt target gene Lgr5, peaking in his detailed studies on stem cells in the small intestine and subsequently in other epithelia, including the stomach. Reflecting his achievements, Hans has been the President of the Royal Netherlands Academy of Arts and Sciences since 2012 and has been the recipient of many prizes and awards from around the world, culminating perhaps in becoming a Knight in the Order of the Netherlands Lion in 2012.

label retention is a mark of **quiescence**. The discrepancy can be accounted for by the concept of the "immortal strand" hypothesis (Chapter 1), which results by **asymmetric** segregation of the template DNA strand during mitosis, a feature that is thought to serve to limit the accumulation of DNA damage during stem cell replication.

The second model was proposed by Hazel Cheng and Charles Phillipe Leblond (Chapter 12) in 1974 and was called the "stem cell zone model." This model proposes that adult intestinal stem cells reside in a specialized niche termed the "stem cell permissive zone" at the base of the crypt. Cells proliferate to generate daughter cells that exit the niche and commit to **differentiation** at the "common origin of differentiation" located around position +5 from the crypt base. Paneth cells mature as they migrate back down toward the crypt base, and non-Paneth cells mature as they migrate toward the villus tip. Their model arose from electron microscopy observations of the crypt from which they noted that the crypt base is not exclusively populated by Paneth cells but rather has another cell type with a distinct morphology intercalated among the Paneth cells. They termed these crypt base columnar (CBC) cells and showed that they are able to give rise to all four major epithelial cell lineages of the intestine. Following incorporation of tritiated thymidine into mouse crypts, Cheng and Leblond saw that CBC cells that survived the irradiation contained radiolabeled phagosomes after engulfing neighboring dead CBC cells. Over time, all cell lineages of the crypt became radiolabeled, highlighting the role of the CBC cell as the common ancestor stem cell. Additional evidence for the CBC stem cell came two decades later when Matthew Bjerknes and Hazel Cheng used heritable somatic mutations within the crypt to demonstrate that a small proportion of these resulted in epithelial clones containing all major lineages as well as CBC cells.

## Lgr5 is a defining stem cell marker in the small intestine

The stem cell zone model failed to gain acceptance until the discovery in 2007 of intestinal stem cell-specific markers that enabled researchers to perform lineage-tracing studies on individual cell types within the crypt. The single most important marker of intestinal stem cells that allowed a step change in understanding of the role of each cell in the crypt is the protein encoded by the Wnt target gene Lgr5. Many of the most significant of these advances, including the discovery of Lgr5, were the result of work in the laboratory of Hans Clevers (Box 11.2) at Utrecht University. Using reporters (e.g., lacZ or eGFP—see Chapter 2), either driven directly by Lgr5 gene regulatory elements or irreversibly activated by Lgr5-driven **Cre recombinase**, it was possible to

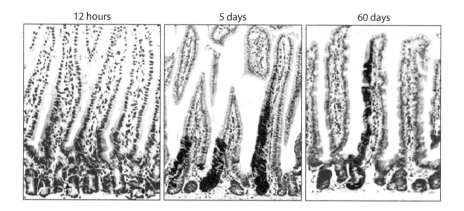

12 hours　　　　　5 days　　　　　60 days

Figure 11.4 **Lineage tracing studies as evidence of CBC stem cell function.** Lineage tracing has been very effective in highlighting the location of stem cells in the intestinal crypt. The histological sections are through the small intestine of a mouse derived from a cross between the Lgr5-EGFPIRES-creERT2 knock-in strain and the Rosa26-lacZ reporter mouse. The panels represent various time points after tamoxifen injection to induce the activity of Cre recombinase and consequent expression of lacZ. Left panel: 12 hours after induction. The blue stain shows those cells that are positive for β-galactosidase activity. The middle and right panels represent 5 and 60 days after induction, showing that over time, ribbons of lacZ expression develop along the length of the villus deriving from the initial Lgr5+ stem cell. (From Barker N, van Es J, Kuipers J et al. (2007) Nature 449:1003–1007 [doi: 10.1038/nature06196] Reprinted by permission from Springer Nature.)

label CBC cells and highlight the presence of an average of 14 CBC cells at positions 1–4 in the base of the crypt lodged between the Paneth cells. Lineage tracing following irreversible reporter activation revealed clones expressing the reporter expanding to form the epithelium spanning from the crypt through to the villus tip and representing all epithelial cell lineages of the intestine (Figure 11.4), thus providing definitive evidence that the Lgr5+ CBC cells are self-renewing, **multipotent stem cells**.

## Stem cells isolated from the intestinal epithelium can be grown in culture to produce organoids resembling the small intestine lining

In addition to genetic lineage tracing studies, further evidence of intestinal stem cell activity has been highlighted by the development of an elegant *ex vivo* technique termed **organoid** culture (Chapter 2). In such cultures, isolated intestinal stem cells can be cultured in a 3D matrix to form "mini-guts" that can be used to study the biology of the stem cells and their direct progeny. Culturing of Lgr5+ cells with the Wnt agonist R-spondin, EGF, Notch ligand, and the BMP inhibitor Noggin leads to the formation of the mini-gut, which fully recapitulates the central features of the normal gut epithelium (Figure 11.5). The mini-gut is composed of crypts housing resident Lgr5+ CBC cells, Paneth cells, and **transit amplifying cells** that feed into the central lumen. Studies of the Lgr5+ stem cells within these mini-gut cultures reveal that

**A**

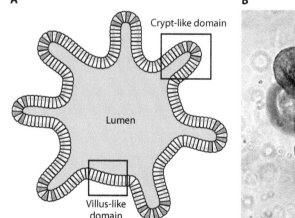

Crypt-like domain

Lumen

Villus-like domain

**B**

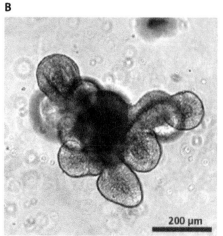

200 μm

**C**

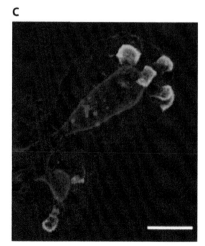

Figure 11.5 **Organoid culture of intestinal stem cells.** Stem cells from many adult tissues can be cultured *in vitro* under conditions that generate 3D structures that mimic the organization of the tissue of origin, containing all differentiated cell types and a self-renewing component of stem cells. This has been most extensively studied using intestinal epithelium stem cells. (**A**) Schematic of a mature intestinal organoid. The central lumen is surrounded by an epithelial monolayer with budding crypt-like domains that harbor stem cells (purple) at their base. (**B**) Brightfield image of a mature (day 5) mouse intestinal organoid. (**C**) Immunofluorescent staining of a mouse intestinal organoid. Paneth cells located at crypt bases are visualized by staining for lysozyme (green). F-Actin (red) reveals the crypt structure at the apical surface of the epithelium, and DAPI (blue) reveals cell nuclei. Scale bar is 25 μm. (B: From *Intestinal Epithelial* Organoid Culture with IntestiCult Organoid Growth Medium (Mouse). Version 2.0.1. Technical Bulletin. STEMCELL Technologies Inc. Cat No. 28223. Copyright © 2016 STEMCELL Technologies Inc. C: From O'Rourke K, Dow L & Lowe S. (2016) *Bio-protocol* 6: e1732 [doi:10.21769/BioProtoc.1732]. With permission from Bio-Protocol.)

their self-renewal resembles *in vivo* observations and shows that stem cell activity is completely restricted to the Lgr5$^+$ fraction since Lgr5$^-$ crypt cells failed to generate organoids. Mini-gut organoids can be cultured for well over a year without any change in either phenotype or karyotype. The capacity of fully formed organoids to bring about epithelial regeneration is most dramatically seen following transplantation into recipient mice that have undergone epithelial damage following chemical treatment. Labeling of the mini-gut organoids (e.g., with red fluorescent protein) allows the grafts to be tracked and reveals that they persist and remain healthy for up to 6 months.

## The expression signature of intestinal stem cells identifies Wnt as a critical regulator

Following the identification of markers that can be used to enrich intestinal stem cells, transcriptomic and proteomic techniques (Chapter 2) were applied to identify key genes important in the function of intestinal stem cells and their direct progeny. Microarray analysis of Lgr5$^+$ CBC cells identified approximately 500 genes whose expression appeared to be associated with these cells, which, following gene ontology categorization, were predominantly linked to the Wnt signaling pathway. These Wnt target genes include Sox9, Ascl2, Ephb2, Troy, and Axin2, the importance of which can be demonstrated by ablating their expression. For instance, loss of the transcription factor Ascl2 results in a rapid and selective loss of Lgr5$^+$ stem cells from the crypt base. Loss of Lgr5 expression itself results in a suppression of Wnt signaling and ultimately in the death of the intestinal stem cell population. Analysis of Lgr5$^+$ stem cells highlighted the crucial requirement for Wnt levels to be maintained at an optimal level for proper intestinal stem cell function. Interestingly, many of the genes that have been identified in the +4 stem cells, such as Bmi1, Lng1, Tert, and Hopx, are also expressed in the Lgr5$^+$ CBC cells. This observation led to some doubt being cast on whether the +4 and the Lgr5$^+$ CBC cells are truly independent populations or whether they are in fact representatives of a broader intestinal stem cell population. Like Lgr5$^+$ CBC cells, the +4 cells are able to form mini-gut organoids.

## Human crypts contain a stem cell population analogous to mouse Lgr5$^+$ stem cells

Definition of the equivalent human Lgr5$^+$ CBC stem cells in the intestinal crypt has been more challenging due to the lack of antibodies against Lgr5 that can be utilized to identify and isolate the cells. However, human intestinal stem cell research has not been completely grounded by this deficit. Since Wnt signaling in mouse intestinal stem cells has proven so important in the regulation of their function, the rational approach was to try to identify alternative Wnt target genes that could be utilized to isolate human intestinal stem cells. Ephrin Type B Receptor 2 (EPHB2) proved to be the solution, antibodies against this protein being effective in the isolation stem cells from the human crypt base. Cells isolated this way express high levels of the Lgr5 RNA and have organoid forming ability analogous to the Lgr5$^+$ CBC cells isolated from mouse crypts.

## The current model of the intestinal stem cell involves a dynamic balance between crypt base columnar cells and the +4 cells

A unifying model for the identity of the intestinal stem cell arose following the detailed characterization of both Lgr5$^+$ CBC cells and the +4 cells, including their overlap in gene expression profiles, taken together with the realization that the intestine can survive the targeted ablation of Lgr5$^+$ CBC cells because they are rapidly replaced (Figure 11.6). The currently favored model for the intestinal stem cell is that the Lgr5$^+$ CBC cells are the engines of the everyday regenerative ability in the intestinal epithelium, which generate transit amplifying cells that go on to differentiate as they proceed toward the villus. In addition, the CBC cells give rise to long-lived Paneth cells and to the LRC cells that co-express Lgr5, thought to represent the +4 stem cell that acts as the reserve intestinal stem cell.

A very elegant variation of the Lgr5-driven lineage tracing has added yet another dimension to the contemporary model, which again might serve as a paradigm for the way in which adult stem cells in some circumstances behave as a population rather than individuals. The concept of **"neutral drift"** in the intestinal stem cell population has derived from a couple of experimental approaches but most vividly using activation of multicolor lineage tracing (see Chapter 2, Figure 2.7). Using mice carrying four

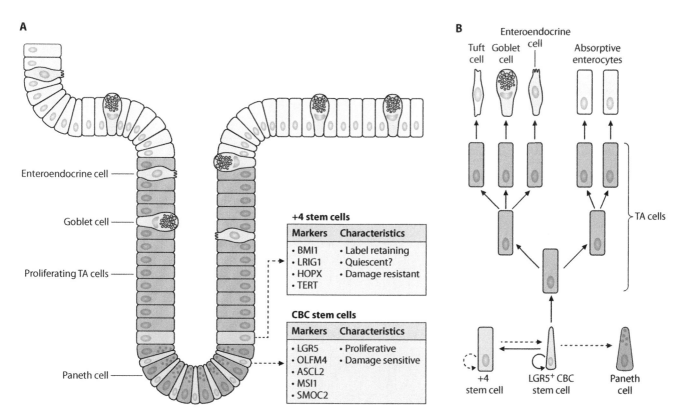

**Figure 11.6 Intestinal stem cell markers and features.** The intestinal crypt is very well defined in terms of the locations of stem cells and their differentiated progeny, which are each characterized by specific patterns of gene expression. (**A**) Diagram of the intestinal crypt showing the position of the crypt base columnar (CBC) cells (gray), "+4" stem cells (blue), Paneth cells (purple), transit amplifying (TA) cells (orange), enteroendocrine cells (green), and goblet cells (yellow). Inset: lists of the known markers and characteristics of "+4" and CBC stem cells. (**B**) Tree diagram showing the differentiation hierarchy originating with the Lgr5+ CBC stem cell that gives rise to Paneth cells and +4 stem cells, which act as reserve stem cells following epithelial injury. (From Barker N. (2014) *Nat Rev Mol Cell Biol* 15:19–33 [doi: 10.1038/nrm3721]. Reprinted by permission from Springer Nature.)

distinct fluorescent protein reporters, any one of which could be activated in a given CBC stem cell, crypts can be produced containing a population of distinctly colored cells. Each cell generates differentiated progeny or stem cells, resulting in "rainbow" crypts over the course of two weeks. However, over the subsequent months, some crypts gradually convert to a single color as a result of neutral competition between symmetrically dividing cells (Figure 11.7).

## The intestinal stem cell niche is defined by the juxtaposition of the stem cells to Paneth cells

As in other adult tissues, the precise regulation of the number and function of intestinal stem cells is determined by the cells with which they are in direct contact. The close association of Lgr5+ CBC stem cells with Paneth cells highlights the likely role that Paneth cells play in the formation of the intestinal stem cell niche (Figure 11.8). Experimental proof of the importance of Paneth cells in the support of the stem cells comes from the observation that their genetic ablation in mice results in the loss of stem cells from the crypt. The organoid culture of stem cells also points in this direction in that single Lgr5+ CBC cells rarely survive *in vitro*, whereas sorted doublets consisting of one Lgr5+ CBC stem cell and one Paneth cell resulted in the formation of a "mini-gut." As for many of the scenarios of regulation of stem cells that we describe throughout this book, the intestinal stem cells encounter signals within their niche that maintain stemness, which are counterbalanced by signals inhibiting this or promoting differentiation. Not surprisingly, key signaling components commonly involved in stem cell maintenance and differentiation (Chapter 4) serve as critical players, with Wnt, EGF, and Notch signals emanating from the Paneth cells maintaining the stem cell characteristics of the CBC, while an opposing gradient of BMP that increases away from the crypt base helps to support differentiation.

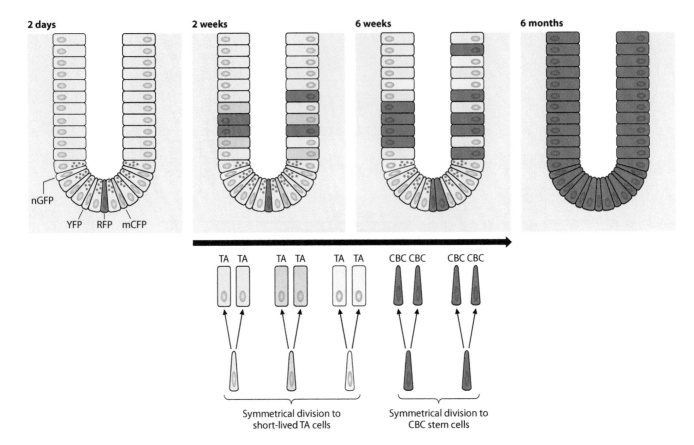

**Figure 11.7 Neutral drift within the stem cell pool.** The dynamics of stem cell division can lead to the evolution of a clone if a stem cell undergoes symmetrical self-renewal as opposed to immediate production of transit amplifying cells, which eventually are lost from the crypt through differentiation. The upper diagrams illustrate cross sections though a crypt showing a time course of the progression of stem cells labeled by activation of fluorescent reporters in crypt base columnar (CBC) stem cells. The mouse strain used for this experiment contains multiple fluorescent reporters (nuclear green fluorescent protein [nGFP], yellow fluorescent protein [YFP], red fluorescent protein [RFP], and membrane-bound cyan fluorescent protein [mCFP]), but only one is activated per cell. Each labeled stem cell generates progeny of the corresponding fluorescence, resulting in the appearance of "rainbow" crypts by 2 weeks. As the months progress, a combination of unbiased symmetrical division leading to either self-renewal or commitment to differentiation leads to the gradual dominance of the descendants of one of the originally labeled CBC cells and the cells that derive from it, in the case illustrated this being the cell that had the RFP reporter activated. The lower diagram shows how the symmetrical division toward either transit amplifying (TA) cells or CBC cells causes the neutral drift within the stem cell pool. (From Barker N. (2014) *Nat Rev Mol Cell Biol* 15:19–33 [doi: 10.1038/nrm3721]. Reprinted by permission from Springer Nature.)

## Stem cells in the colon are also Lgr5⁺ but are of one type only and do not associate with Paneth cells in their niche

Perhaps not unexpectedly, the epithelial lining of the colon has many features in common with the small intestine and correspondingly contains similar stem cells; however, there are some differences. Unlike the small intestine, the colon has a smooth luminal face without villi, reflecting its specialized role in the compaction of feces, and contains crypts that have a higher frequency of goblet cells (Figure 11.9). The base of the crypts is also distinct, containing Lgr5⁺ cells but no Paneth or +4 cells. Presumably this means that there are differences in the way that the stem cells support replacement of the epithelium, which, although a little slower than in the small intestine, is still required every 5–7 days. The absence of +4 cells and Paneth cells implies that both the dynamics of stem cell use and the niche interactions supporting them are different, even if only subtly. Indeed, one study of the colonic epithelium has identified goblet cells positive for expression of the Kit, receptor for stem cell factor (SCF), which may perform a similar function to Paneth cells in the small intestine. These cells express factors implicated in stem cell maintenance (such as Dll1, Dll4, and EGF) and *in vitro* can promote organoid formation by Lgr5⁺ colon cells.

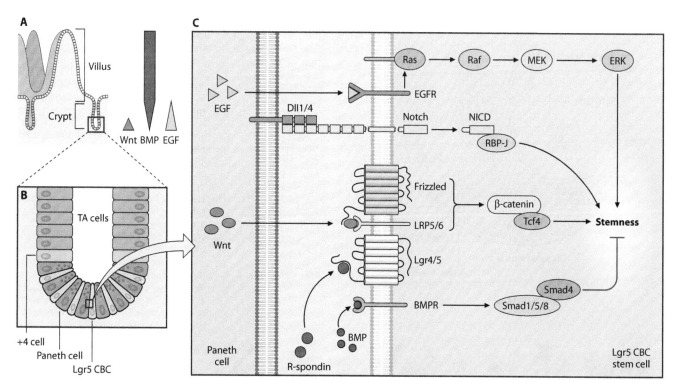

**Figure 11.8 Molecular interactions of intestinal stem cells with their niche.** As for any stem cell in its niche, intestinal stem cells communicate with adjacent cells through a number of receptor-ligand interactions. (**A**) Cross section of an intestinal villus with a magnification (**B**) of the crypt base showing the location of the +4 cell, Lgr5+ crypt base columnar (CBC) cell, and Paneth cells. (**C**) Diagram showing the interactions that take place between the Lgr5+ CBC cell and the Paneth cells. Three signals (EGF, Notch, and Wnt) are essential for intestinal epithelial stemness, while bone morphogenic protein (BMP) negatively regulates stemness and promotes differentiation. Full Wnt activation requires the presence of R-spondin, the ligand for LGR5. (From Sato T & Clevers H. (2013) *Science*, 340:1190–1194 [doi:10.1126/science.1234852]. Reprinted with permission from AAAS.)

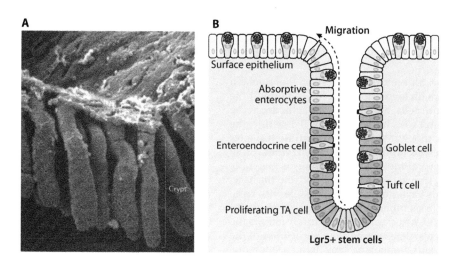

**Figure 11.9 Stem cells in the colon.** The structure of the epithelium in the colon is very similar to that of the small intestine. Reflecting this similarity, stem cells and their downstream progeny are located in analogous positions, although there are some subtle differences. (**A**) Scanning electron micrograph of the colon, showing the colonic crypt with flat luminal surface lacking the villi that are characteristic of the small intestine. (**B**) Cross-sectional diagram though a colonic crypt showing the positions of the Lgr5+ stem cells (beige) in relation to the differentiating lineages of cells. (A: From Magney J, Erlandsen SL, Bjerknes M et al. (1986) *Amer J Anat* 177:43–53 [doi: 10.1002/aja.1001770106]. With permission from Wiley. B: From Barker N. (2014) *Nat Rev Mol Cell Biol* 15:19–33 [doi: 10.1038/nrm3721]. Reprinted by permission from Springer Nature.)

## Stem cells in the stomach exhibit distinct features in different regions of the organ

Upstream of the small intestine in the GI tract is another organ with a rapidly self-renewing epithelium, namely the stomach. Like the other epithelia we have already looked at, that of the stomach has crypt-like structures, in this case referred to as gastric units or glands. From top to bottom, the stomach is divided into three regions: the fore stomach or cardia region, the main body or corpus, and the pylorus (Figure 11.10). Like the small intestine and colon, Lgr5 expression has been found to mark adult stem cells in the pyloric region, which are located at the bottom of the glands and capable of long-term renewal of the epithelium.

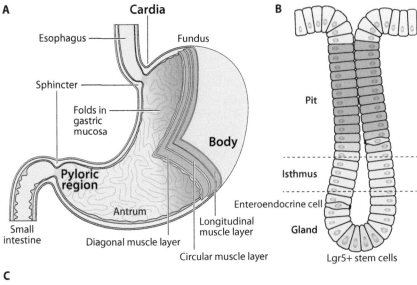

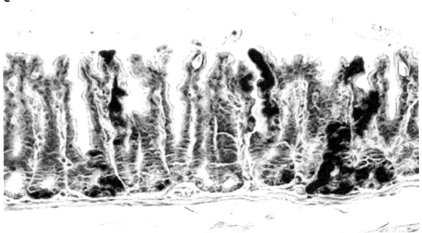

**Figure 11.10 Lgr5 expression marks stem cells in the pyloric region epithelium of the stomach.** The stomach epithelium shares many features with the equivalent structures in the intestine, both in terms of organization and the nature of the supporting stem cells. **(A)** Schematic diagram of the structure of the stomach illustrating the major regions (cardia, body or corpus, and pylorus), muscular and epithelial components, and the linkage to the rest of the GI tract at the esophagus and small intestine. **(B)** Cross section through a gastric pit or gland from the pyloric region showing the principal zones and the location of the Lgr5+ cells at the base. **(C)** Lineage tracing of Lgr5+ cells in the base of the pyloric glands using a mouse derived from a cross between the Lgr5-EGFPIRES-creERT2 knock-in strain and the Rosa26-lacZ reporter. The section, which is stained for expression of β-galactosidase (blue), is from the pyloric region of the stomach 10 days after tamoxifen injection to induce the activity of Cre recombinase and consequent expression of lacZ. (B and C: From Barker N, Huch M, Kujala P et al. (2010) *Cell Stem Cell*, 6:25–36 [doi: 10.1016/j.stem.2009.11.013]. Copyright 2010 with permission from Elsevier.)

Intriguingly, the situation seems to be different in the rest of the stomach (Figure 11.11). In the corpus, the glands are divided into (from top to bottom) the pit, isthmus, neck, and base. The most mature cells, mainly mucous cells, are at the top, with rapidly dividing (transit amplifying) cells in the isthmus. Below this, the neck region contains mucous cells that are thought to **transdifferentiate** (Chapter 5) into chief cells, which populate the base and produce digestive enzymes. Chief cells turn over slowly over the course of several months. The isthmus is generally viewed as the stem cell zone, cells migrating from it toward both the pit and the base, but neither it nor any other part of the glands have stem cells identifiable with Lgr5. However, using the stem cell marker Troy (a member of the tumor necrosis factor receptor superfamily), which is also a target of Wnt signaling and closely follows the expression pattern of *Lgr5* in intestinal crypts, cells can be seen at the base of the gland that are the fully differentiated chief cells. Employing Troy in lineage tracing, some single marked chief cells can be seen to label complete gastric units over several months, demonstrating that they are in fact acting as the stem cells (Figure 11.11). Furthermore, as for Lgr5+ cells in the intestine, these Troy+ cells are capable of organoid formation. Troy therefore seems to mark a specific set of chief cells that are akin to the quiescent +4 reserve stem cells in the small intestine.

## 11.2 GASTROINTESTINAL TRACT ASSOCIATED ORGAN STEM CELLS

Apart from the cells that constitute the alimentary canal from the pharynx to the colon, many major organs are closely associated with the GI tract by virtue of their common developmental origin from the endoderm. These organs are diverse in their

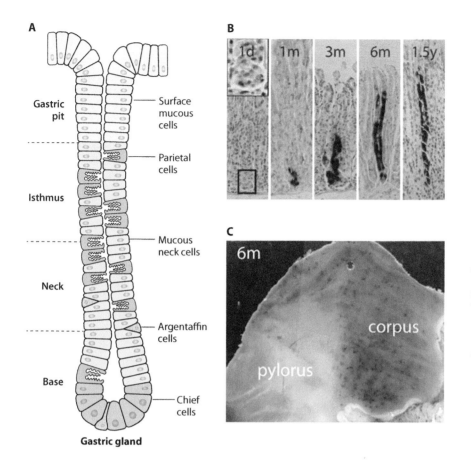

**Figure 11.11 Troy⁺ stem cells in the stomach epithelium.** Like the lower pyloric region, the upper part of the stomach (corpus) contains glands that contain stem cells at their base. These stem cells are distinct from those in the pyloric glands, as exemplified by the expression of distinct stem cell-specific genes. (**A**) Cross section through a gland in the corpus region of the stomach illustrating the principal cell types and the location of the stem cells ("chief cells"). (**B**) Expression of a Rosa26-lacZ reporter gene activated following tamoxifen induction of Cre recombinase expressed from a Troy-eGFP-IRES-CreERT2 knock in (Troy-ki) mouse in Troy⁺ cells at the base of the glands in the corpus. Sections stained for β-galactosidase expression (blue) are shown for 1 day (Inset at top: expanded image of the gland base); 1, 3, and 6 months; and 1.5 years post induction. (**C**) Whole-mount β-galactosidase staining of the Troy-ki stomach 6 months post induction. Lineage tracing is evident in the corpus, whereas the pylorus is negative. (B and C: From Stange D, Koo B-K, Huch M et al. (2013) *Cell* 155: 357–368 [doi:10.1016/j.cell.2013.09.008]. Copyright 2013 with permission from Elsevier.)

organization, function, and the nature of their associated stem cell components. The precise character of the stem cells, or whether they even exist, in the liver, pancreas, and thyroid gland remains fully to be resolved.

## The adult liver has huge regenerative capability, but this probably has little or no reliance on stem cells

The liver is the largest organ in the body and performs many functions critical to the absorption and processing of food taken up through the GI tract. These functions include detoxification, glycogen storage, the secretion of bile, and the synthesis of various serum proteins. The highly organized structure of the liver facilitates efficient delivery and screening of substances absorbed from the gut and excretion of waste and bile salts back into the gut through the biliary system. It is constructed around the building block of the lobule and its functional subunits that consist of a parallel arrangement of a sinusoid and a bile canaliculus on either side of a row of hepatocytes (Box 11.3).

A defining feature of the liver is that it is capable of huge feats of regeneration, being able fully to recover after the surgical removal of large portions. Mature hepatocytes turn over slowly in a period of several months, and maintenance of their numbers has been thought to be largely due to their proliferation within the liver parenchyma. Much or all of the regenerative capacity of the liver seems to be the result of the stem cell-like self-renewal capacity of the hepatocytes, which is an extension of the normal homeostatic mechanisms, and there has been some controversy about any possible involvement of resident stem cells.

Some experiments in rats implied a maintenance mechanism in the liver involving stem cells since radiolabeled nucleotide incorporation showed new hepatocytes appearing in the periportal (bile duct) area, which then moved toward the pericentral region. Lineage tracing also revealed how cholangiocytes might contribute to hepatocyte replacement. Hence, using Sox9 to drive expression of Cre recombinase to in turn activate a reporter, labeled cholangiocytes were seen gradually to spread to hepatocytes incorporating the whole parenchyma after 1 year, but at the same time

BOX 11.3 BACKGROUND: STRUCTURAL ORGANIZATION OF THE LIVER

The adult liver (**Figure 1**) consists of two major lobes, which are divided into lobules arranged around triads of vessels that both deliver and remove blood and collect bile. In the functional unit of the lobule (**Figure 2**), blood flows from the portal vein through the sinusoidal capillaries toward the central vein. Polarized hepatocytes are arranged between the sinusoid and a parallel bile canaliculus.

Juxtaposition of one face with sinusoidal endothelial cells aids in the transfer of materials between hepatocyte and blood stream. The other face of the hepatocyte excretes bile salts into the bile canaliculus, which is connected to the bile duct by the canals of Hering. A second epithelial cell type within the liver, namely the cholangiocyte, lines the bile duct.

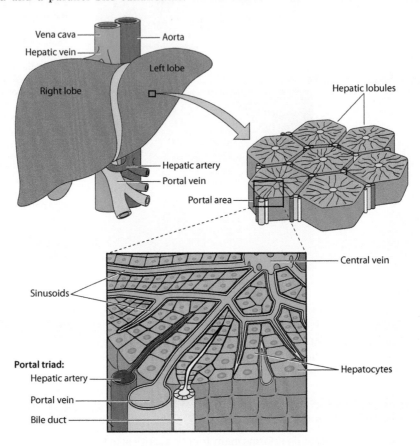

Box 11.3 Figure 1 **Architecture of the liver.** Diagram of the structure of the liver with magnification to show the individual hepatic lobules and their detailed architecture.

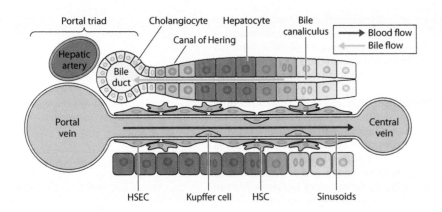

Box 11.3 Figure 2 **Structure of the fundamental unit that constitutes a liver lobule.** The building blocks of liver lobules are structured around three vessels, usually referred to as the portal triad, consisting of the portal vein, the hepatic artery, and a bile duct. Blood flows from the portal vein (red arrow) toward the central vein through the sinusoids, which are flanked by hepatocytes on either side. The sinusoids are lined by fenestrated hepatic sinusoidal endothelial cells (HSEC), hepatic stellate cells (HSC), and Kupffer cells. Bile produced from the hepatocytes flows toward the bile duct along the bile canaliculus (green arrow). Cholangiocytes line the bile ducts, and here they meet the hepatocytes forming a region called the canal of Hering. (From Miyajima A, Tanaka M & Itoh T. (2014) *Cell Stem Cell* 14: 561–574 [doi: 10.1016/j.stem.2014.04.010]. Copyright 2014 with permission from Elsevier.)

remaining in the bile duct, implying both commitment to differentiation and self-renewal. As we will see shortly, there is good evidence for a liver stem cell, although there is some doubt over the validity of these particular lineage tracing experiments, which have not been repeated using related, but distinct strategies.

## Cells with stem cell properties can be isolated from the liver, but the *in vivo* relevance remains controversial

Injury to the liver that inhibits the proliferative capacity of the hepatocytes, for instance, in chronic liver diseases such as viral hepatitis and alcoholic liver disease, is accompanied by considerable repair. In the case of the rat, at least, this involves the induction and emergence of a unique epithelial cell population. This activated epithelial cell type, which emerges from the periportal area in what is known as the "ductular reaction," was first described in 1956 by Emmanuel Farber at the University of Toronto when he injured rat livers by partial hepatectomy with prior treatment with a compound (2-acetylaminofluorene) to block hepatocyte proliferation. Due to the shape of the nucleus of the activated epithelial cells, Farber coined the term **oval cell**. Such experiments enabled researchers extensively to characterize the stem cell features of these activated oval cells, including their source.

Although it became clear from different injury models in rodents that the activated population of oval cells was different depending on the nature of the liver insult, these populations are all encompassed by the term liver **progenitor** cells (LPC). An analogous cell type has also been identified in human chronic liver diseases, where they are known as hepatic progenitor cells or intermediate hepatobiliary cells. LPC have therefore been proposed to be facultative stem cells that become activated under conditions of injury to the liver. Following liver insult, the LPC almost always emerge from the periportal area termed the canals of Hering, which are situated between the bile canaliculus and the bile duct (Box 11.3 and Figure 11.12). Utilizing the expression of specific markers, such as proteins associated with cholangiocytes (e.g., cytokeratin 19) or hepatocytes (e.g., albumin), and also the epithelial cell adhesion molecule (EpCAM) and **cluster of differentiation** 133 (CD133) (also known as Prominin1), enabled the sorting of cells from the liver that could be shown to be clonogenic and to behave as bipotent progenitors, differentiating to both cholangiocytes and hepatocytes *in vitro*.

More recent lineage tracing experiments in the mouse have, however, found no evidence of the existence of stem cells in the adult liver. Using various injury models

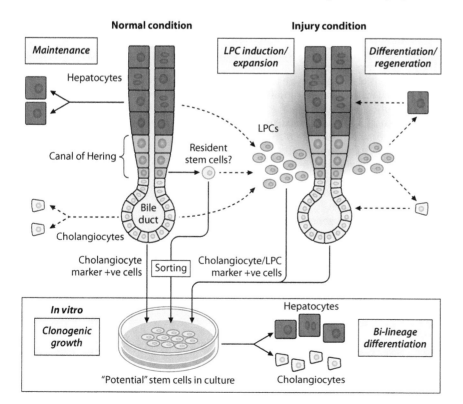

Figure 11.12 **Speculative model of liver progenitor cell activation.** Representations of postulated liver stem cell activation under normal (left) and injury (right) conditions and how the isolated cells can be cultured *in vitro* to give rise to hepatocytes (red) and cholangiocytes (green). Under normal homeostatic conditions, both hepatocytes and cholangiocytes are maintained by their own proliferative ability. Certain types of injury inhibit this proliferation, resulting in the expansion of a population of liver progenitor cells (LPC, purple), which emerge from the canal of Hering and have the ability to differentiate toward both hepatocytes and cholangiocytes. (From Miyajima A, Tanaka M & Itoh T. (2014) *Cell Stem Cell* 14: 561–574 [doi: 10.1016/j.stem.2014.04.010]. Copyright 2014 with permission from Elsevier.)

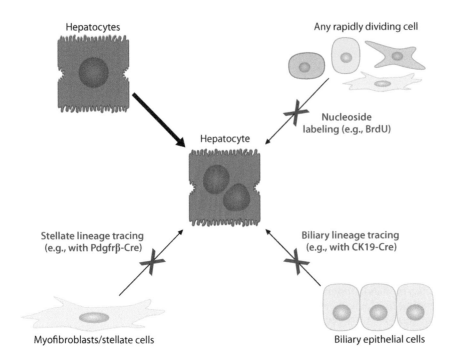

**Figure 11.13 Evidence against hepatocyte replacement from stem cells.** There have been conflicting views about the role of stem cells in the replacement of hepatocytes in the liver. Comparison of several lineage tracing strategies has now led to the conclusion that the only significant origin for replacement hepatocytes is through proliferation of the hepatocytes themselves. The diagram summarizes the various approaches, red crosses and the adjacent annotation indicating those tracing methods that showed that non-hepatocyte sources do not generate hepatocytes under circumstances of oval cell injury. The methods included: nucleoside (e.g., BrdU) labeling of rapidly dividing cells; lineage tracing of biliary cells using Cre recombinase driven from CK19, Sox9, and Hnf1b gene elements; and lineage tracing of stellate/fibroblastic cells with Pdgfrβ-driven CreERT. Proof of the contribution of hepatocytes to their own regeneration (bold black arrow) came, for example, from cell marking using virally expressed hepatocyte-specific Cre recombinase. (From Grompe M. (2014) *Cell Stem Cell* 15:257–258 [doi:10.1016/j.stem.2014.08.004]. Copyright 2014 with permission from Elsevier.)

combined with lineage tracing, it has now proven impossible to demonstrate labeling of new hepatocytes originating from biliary epithelial cells, hepatic stellate cells, or any other rapidly dividing cells in the liver, whereas marking cells with hepatocyte-specific viruses (adeno-associated viruses, or AAVs) did show that hepatocytes themselves can regenerate new parenchymal cells (**Figure 11.13**).

## Like the liver, the existence of stem cells in the pancreas has been an issue of debate

Originally thought to be a non-dividing organ, it has now been shown that both the endocrine and exocrine tissues (**Box 11.4**) of the pancreas can regenerate. As for hepatocytes in the liver, regeneration of β-cells within islets has been described to be as a result of the endogenous ability of the cells to replicate and not as a result of a resident stem cell population. This has become a contentious issue, though, following the discovery of a progenitor-like cell residing in the ductal regions similar to the oval cells described within the liver. Although able to show an element of regeneration, this cell is insufficient to recover the lost β-cell mass that is observed in type I diabetes mellitus. Nevertheless, interest in the possibility of a β-cell stem cell remains high because of its huge potential for regenerative therapeutic application.

The understanding that the β-cell may have capacity for self-renewal arose from lineage tracing experiments depending on expression from the insulin gene to achieve heritable labeling with a tamoxifen-inducible Cre-Lox system. Any β-cell generated after induction with tamoxifen would be labeled only if they were the direct progeny of pre-existing β-cells, while any new β-cell from a source such as a stem or progenitor cell population would not be labeled. The results from this experiment implied that during normal growth, all β-cells arise from pre-existing β-cells. However, this conclusion was later challenged in that the tracing studies using the insulin gene may have been based on an incorrect assumption. This assumption was that all insulin+ cells within the adult pancreas are fully mature β-cells, neglecting the possibility that a putative pancreatic stem cell might in fact itself express insulin. Indeed, it proved possible to identify a pancreatic multipotential progenitor (PMP) within the adult pancreas that was positive for the expression of insulin, although the stringency of the promoter used has been brought into question, making it impossible to draw firm conclusions.

## The thyroid has to date been insufficiently investigated to reach a definitive conclusion about the presence of stem cells

The thyroid is another organ that only requires a slow rate of turnover (about five times in the life of a human), yet at the same time retains an ability for large-scale regeneration. It is not unreasonable to expect that similar means are employed in the thyroid to those we have seen in the liver in order to deal with these requirements, although the experimental evidence is too limited at present to enable any definite conclusions. Modest damage to the thyroid, involving removal of a small portion of the tissue, may involve immature, quiescent cells that reside in the "solid cell nest" (SCN), a structure believed to be a developmental remnant. In contrast, more massive damage, such as removal of a whole lobe of the organ, could involve resetting of mature cells (e.g., follicular cells or calcitonin-producing cells) to an immature proliferative state followed by subsequent maturation. Some experimental evidence does point to the presence of adult stem cells, but the link to the SCN is not clear. For example, cells can be isolated from human goiter that expresses endoderm markers such as GATA4 and can be maintained in culture as spheres, which under appropriate conditions can be differentiated toward thyrocyte-like cells expressing markers like thyroid stimulating hormone (TSH) receptor and thyroglobulin and ultimately to cells that exhibit TSH-dependent uptake of $^{125}$iodine.

## 11.3  STEM CELLS IN THE RESPIRATORY TRACT

The mammalian lung is a complex organ that exhibits low cell turnover. It is composed of multiple cell types, including cells of **mesodermal** and **ectodermal** origin, but it is the endoderm-derived epithelia that form the branching structure

---

### BOX 11.4  BACKGROUND: DEVELOPMENT AND STRUCTURAL ORGANIZATION OF THE PANCREAS

The pancreas consists of both exocrine and endocrine tissue secreting digestive enzymes and hormones, respectively (Figure 1). The exocrine functions are carried out by the acinar cells, which secrete at least 15 different digestive enzymes, and the ductal cells that produce bicarbonates and mucins, both of which are secreted via the pancreatic ductal system to the duodenum for aiding digestion. The endocrine function of the adult pancreas originates from the islets of Langerhans, which are distributed throughout the pancreas. The islets are made up of five cell types, each with specific endocrine properties: alpha ($\alpha$) cells that secrete glucagon; beta ($\beta$) cells that secrete insulin, C-peptide, and amylin that make up 70%–90% of the population of islet cells; delta ($\delta$) cells that secrete somatostatin; epsilon ($\varepsilon$) cells that secrete ghrelin; and PP-cells that secrete pancreatic polypeptide. The pancreas

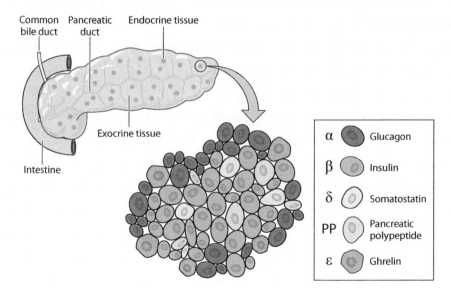

Box 11.4 Figure 1 **Architecture of the pancreas and the islets of Langerhans.** The pancreas, illustrated schematically on the left, is composed of both endocrine (brown dots) and exocrine (yellow) tissues that respectively secrete enzymes directly into the intestine, to which the pancreas is directly connected, and release various hormones into the bloodstream. The exocrine tissue is contained in the islets of Langerhans (center diagram), which are constituted of five cell types that are identified in the expanded view on the right. Each islet is made up of a central mass of insulin producing β-cells (green) surrounded by glucagon secreting α-cells (red). Somatostatin secreting δ-cells (blue) and pancreatic polypeptide (PP) producing cells (yellow) are scattered throughout the islet, while the most recently discovered ghrelin-producing ε-cells (purple) are located on the periphery. (Left: From MacDonald P & Rorsman P (2006) PLoS Biol 4:e49 [doi: doi: 10.1371/journal. pbio.0040049]. Published under the terms of the Creative Commons Attribution License http://creativecommons.org/licenses/.)

(like other organs such as the lung and mammary gland) is a branching organ. During pancreatic development, the pancreatic epithelium contains multipotent progenitors that are able to give rise to all cell types of the pancreas (Figure 2). Initially, these progenitors are located at the tips of the branches. Eventually, cells at the tips of the branches will develop into acinar cells, while endocrine progenitors form from the trunk of the branches.

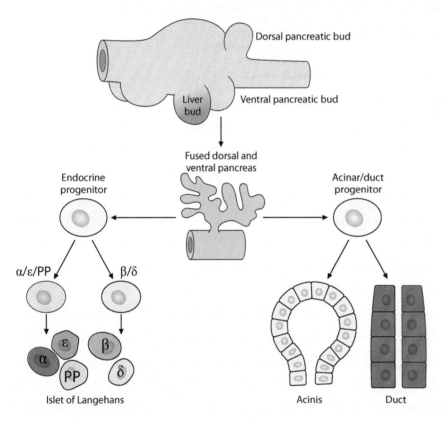

**Box 11.4 Figure 2 Developmental origins of the pancreas.** Following the formation of the primordial buds out of the gut tube, the dorsal and ventral components fuse, initiating the differentiation of progenitors that go on to give rise to either the exocrine components of the pancreas or the endocrine cell types present in the islets. As in any differentiation process, the expression of different combinations of transcription factors determines the sequential lineage segregation. The endocrine progenitor first commits to differentiation along two pathways, one of which ultimately produces the $\alpha$, $\varepsilon$, and PP cells, and the other the $\beta$ and $\delta$ cells.

and are the critical determinant of gas exchange with the blood. Various endoderm-derived epithelia characterize the different levels of the airway from trachea to alveolus (Figure 11.14).

The trachea and proximal airways are lined by a pseudostratified epithelium consisting of basal cells, secretory cells (goblet and club cells), ciliated cells, and neuroendocrine cells. The distal conducting airways are lined by simple columnar epithelium consisting of secretory cells, ciliated cells, and intermittent clusters of pulmonary neuroendocrine cells. The alveolar gaseous exchange surfaces are lined by cuboidal surfactant secreting alveolar epithelial type II (ATII) pneumocytes and squamous gas-exchanging alveolar epithelial type I (ATI) pneumocytes. The mouse and human respiratory systems differ in several respects, most notably in that in human lungs, cartilage rings are located in the extrapulmonary airways and extend for several bronchial extensions. When comparing experimental studies, generally the mouse trachea and mainstem bronchi are a better reflection of the human lung cellular structure. A combination of principally lineage tracing studies and single cell isolation and analysis has enabled discrimination of the relationship between stem cells, progenitors, and their differentiated progeny in the two broad categories of lung epithelium.

## Normal homeostasis requires minimal cell replacement achieved through self-renewal of basal stem cells and some of their differentiated progeny

During normal homeostasis of the pseudostratified epithelium (Figure 11.15A), the basal cells can act as progenitor cells, becoming luminal precursor cells (Cytokeratin 5 positive) before further differentiation into secretory cells and neuroendocrine

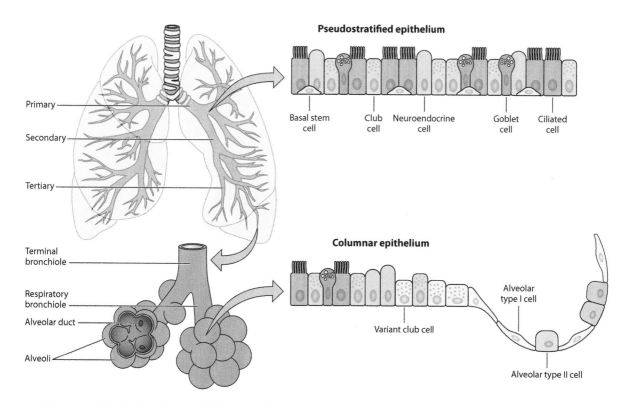

**Figure 11.14 Distribution of lung epithelial cells.** The structure of the epithelium in the airways, from the trachea through to the branching bronchioles terminating in the alveoli, involves many of the same cell types throughout, although the precise stem cell contribution varies. The branching structure of the airways in the lung is represented on the left, and the two principal epithelial types, namely pseudostratified epithelium in the branching airways and columnar epithelium in the small airways, are depicted to the right, showing the various constituent cell types (secretory goblet and club cells, ciliated cells, neuroendocrine cells, and variant club cells). The basal cells are progenitor cells of the pseudostratified epithelium (yellow). The alveolar epithelium consists of alveolar type I (gray, flat shape) and type II (beige) cells and alveolar progenitors (orange and gray cuboidal). (Right: From Schilders K, Eenjes E, van Riet S et al. (2016) *Resp Res* 17:44 [doi: 10.1186/s12931-016-0358-z. Published under the terms of the Creative Commons Attribution License (http://creativecommons.org/licenses/by/4.0/).)

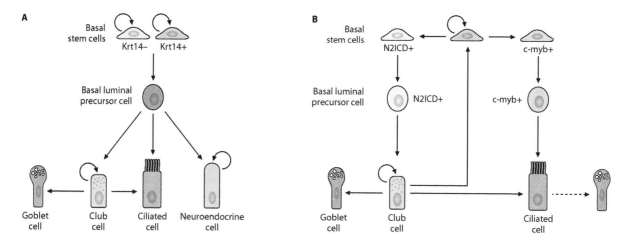

**Figure 11.15 Bronchiolar stem cells.** The replacement of airway epithelial cells from basal stem cells appears to involve related but distinct pathways under normal homeostatic conditions compared to circumstances of injury. (**A**) During normal homeostasis, the self-renewing (circular arrows) basal cells are progenitor cells of the pseudostratified epithelium, either subtype (distinguished by the expression of Cytokeratin 14, Krt14) giving rise to a luminal precursor cell before further differentiation into ciliated cells, neuroendocrine cells, and club cells. The latter two differentiated cells are themselves capable to self-renewal, the club cells also being able then to produce goblet cells and ciliated cells, a process that under the low replacement demands of homeostasis is likely to be sufficient to regenerate ciliated cells and goblet cells. (**B**) Under conditions of epithelial loss caused by injury, basal stem cells proliferate and differentiate along two parallel paths defined by the expression of the Notch2 intercellular domain (N2ICD) or the transcription factor c-myb that respectively give rise to secretory cells and ciliated cells. Loss of the basal stem cells can be compensated by the dedifferentiation of secretory cells. Lineage tracing studies have shown that depending on the injury conditions, goblet cells can arise from the secretory cells or ciliated cells. (From Schilders K, Eenjes E, van Riet S et al. (2016) *Resp Res* 17:44 [doi: 10.1186/s12931-016-0358-z. Published under the terms of the Creative Commons Attribution License (http://creativecommons.org/licenses/by/4.0/).)

cells. Lineage tracing of these precursors using Krt5-Cre showed that they can rise to labeled ciliated cells and club cells that are maintained for at least 14 weeks without being replaced by unlabeled progeny. Both neuroendocrine cells and club cells are also capable of self-renewal, the latter being able to give rise to goblet cells and ciliated cells. Since there is a very low turnover of cells in the homeostatic epithelium, it is likely that the dividing secretory cell population is sufficient to regenerate ciliated cells, although their generation from basal cells is not excluded. Basal cells make up approximately 30% of the total epithelial cell population of the adult mouse lung, and it is expected that there is a degree of heterogeneity within the population, but there are insufficient markers available at the moment to resolve these, although they do at least fall into populations that are positive and negative for the expression of Cytokeratin 14 (Krt14).

As for the gut, further evidence in support of airway basal cells containing a stem cell population comes from *in vitro* culture. Sorted cells can form colonies that are able to regenerate differentiated secretory and ciliated epithelial cells when seeded into tracheas stripped of epithelial cells and transplanted subcutaneously into immunocompromised rats. Similar experiments utilizing mouse Krt5+ basal cells, but this time in a 3D *in vitro* culture system, enabled the formation of clonal "tracheospheres" containing both ciliated and secretory cells, which could be subcultured for multiple passages, implying an ability to self-renew. Cells with similar properties have also been isolated from human lungs, in which they represent approximately 0.01% of total airway epithelial cells.

Following studies of bromodeoxyuridine (BrdU) incorporation, one area of the mouse airway proposed to be enriched in quiescent basal cells is that of the ducts of the submucosal glands and the intercartilage regions. The cells in these locations retaining BrdU for long periods following pulse-chase labeling, indicating that they cycle and divide at a much lower rate than other basal cell populations.

## The pattern of basal stem cell activity changes following injury in the lung

Upon depletion of luminal cells (Figure 11.15B), for example, by exposure to sulfur dioxide, basal cells proliferate and subdivide into two populations, positive for either the Notch2 intracellular domain (N2ICD) or the c-Myb transcription factor, which differentiate into secretory or ciliated cells, respectively. Also, after loss of basal cells following injury, club cells can dedifferentiate into basal stem cells, while recent lineage tracing studies have shown that goblet cells can derive from both club cells and ciliated cells.

## The basal stem cell niche is an undefined structure constituted of the surrounding epithelial neighbors

Many of the stem cell systems described throughout this book rely on specific niche components in order for the stem cells to function. Basal cells do not have a defined anatomical niche; rather they are distributed throughout the airways, forming a continuous monolayer in the upper levels and becoming sparser as clusters or individual cells in the distal sections. The basal cell niche is likely composed of surrounding epithelial neighbors (other basal cells, ciliated cells, goblet cells, and club cells) and interactions with the subjacent smooth muscle, dendritic cells, and neuroendocrine cells.

## The alveolus appears to be a relatively simple structure but exhibits a complex dependence on interconnected stem cells and progenitors

The relative simplicity of the structure of the alveolus should lead to greater clarity in our understanding of the involvement of stem cells in the maintenance of this terminal component of the airways, but there are nonetheless several distinct interconnecting **stem cell potentials** that are called upon under various circumstances (Figure 11.16). As described previously, alveoli are predominantly lined by two cell types: ATI cells and cuboidal ATII cells. Stem cells can be isolated from the junction between the bronchioles and the alveolus, and these will differentiate into bronchiolar (club cell) and alveolar (ATI and ATII) lineages in culture. Self-renewing club cells have the

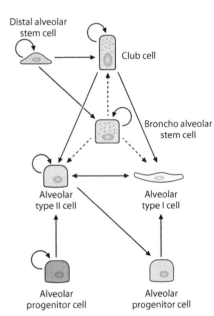

**Figure 11.16 Putative lung epithelial stem and progenitor cells in the alveolar region.** In spite of the simpler nature of the alveolar epithelium, the situation regarding stem cells at the junction between the terminal bronchioles and the alveolus is more complicated than is seen in the airway epithelium. Stem cells isolated from the broncho-alveolar duct junction (blue/beige) can differentiate into bronchiolar and alveolar lineages in culture (dashed lines). The bronchiolar club cells (blue) have the potential to form alveolar type I (ATI) and type II (ATII) cells. ATII cells (beige) can self-renew and differentiate to ATI cells (flat, gray). After removal of a portion of lung, ATI cells can also regenerate ATII cells. An alveolar progenitor cell (orange) expressing α6-β4 integrins can also regenerate ATII cells after injury. Yet another cell type (gray, columnar) expressing Sca1+ can arise from ATII cells and regenerate ATI cells. Distal alveolar stem cells (purple) appear after severe injury and give rise to secretory club cells and alveolar cells. (From Schilders K, Eenjes E, van Riet S et al. (2016) Resp Res 17:44 [doi: 10.1186/s12931-016-0358-z. Published under the terms of the Creative Commons Attribution License (http://creativecommons.org/licenses/by/4.0/).)

potential to form ATI and ATII cells after some types of injury. ATII cells themselves can self-renew and differentiate into ATI cells, but this can occur too in the reverse direction following removal of part of a lung. Other progenitor cells can play a part following injury; for example, an alveolar progenitor cell with self-renewal potential can regenerate ATII cells, while another progenitor that is itself generated from ATII cells can regenerate ATI cells. Yet more complexity is seen after severe injury, when a distal alveolar stem cell appears that can give rise to club cells and broncho-alveolar stem cells.

## 11.4  SUMMARY

In this chapter, we have seen how quite diverse organs that develop from a common origin, namely endoderm, giving rise to both the GI tract and its associated organs as well as the respiratory system, exhibit quite distinct requirements and characteristics of their resident stem cells. The GI tract itself has very clearly defined stem cells that are quite well understood in terms of their niche location and relationship to progenitors and differentiated cell types. With some subtle differences, the stem cells that populate the GI tract from the stomach to the colon are relatively similar. In contrast, the GI tract associated organs are less well understood and do not conform to a simple model of stem cell-based homeostatic or regenerative cell replacement. In fact, at least in the case of the liver, a role for adult stem cells is much in doubt, and the same might apply in other organs such as the pancreas and thyroid. Unlike the GI tract, the respiratory system has a low requirement for cell replacement under normal conditions but requires additional capability in order to deal with injury. In both circumstances, stem cells play a role, although this varies between the airways and the terminal alveoli, involving defined stem cells but also an ability of some differentiated cell types in the epithelium to self-renew and on occasion dedifferentiate.

## KEY POINTS

- The gastrointestinal (GI) and respiratory tracts are built around endoderm-derived tubes of epithelial cells.

- The GI and respiratory tract epithelia have quite different rates of turnover, ranging from very frequent cell replacement in the gut to relatively little homeostatic demand for new cells in the respiratory tract.

- The GI tract is associated with several organs that co-develop from the endoderm, including the liver, gall bladder, pancreas, and thyroid gland.

- The GI tract-associated organs exhibit a low rate of cell turnover but can often regenerate if extensively damaged. Homeostatic replacement and regeneration appear to be achieved differently, but any substantial role for adult stem cells remains controversial.

- The study of stem cells in the small intestine, which reside in crypts surrounding the base of the villi, serves as a paradigm for the use of lineage tracing and *in vitro* and *in vivo* functional assays. The findings from these studies have established many of the basic principles of adult stem cell biology.

- The current model for stem cells in the small intestine involves two cell types: the Lgr5+ crypt base cell, which is highly active, and a relatively quiescent cell derived from

it located four cells above the base. This +4 cell acts as a reserve in case of injury.

- Stem cells in other parts of the GI tract, including the stomach and colon, show features in common with those in the small intestine but also have distinct characteristics.

- The small intestine stem cell niche is very precisely defined based on the location of the two stem cell types and also includes the adjacent Paneth cells.

- The liver is maintained by a slow turnover of cells, largely involving the proliferation of mature hepatocytes.

- A defining feature of the liver, which can also be seen in other GI tract associated organs, is its ability to regenerate after damage. This can occur through hepatocyte proliferation, but in circumstances when this is compromised, as in liver disease, an adult stem cell has been proposed to come into play, although recent evidence argues against this.

- Repair and regeneration of the pancreas and thyroid gland seem to show similarities to the liver.

- The epithelium of the respiratory tract differs significantly between the brochiolar airways and the alveoli. This difference is reflected in distinct, but nevertheless overlapping, utilization of stem cells and progenitors.

## FURTHER READING

Barker, N. (2014) Adult intestinal stem cells: Critical drivers of epithelial homeostasis and regeneration. *Nat Rev Mol Cell Biol.*, 15: 19–33.

Grompe, M. (2014) Liver stem cells, where art thou? *Cell Stem Cell*, 15: 257–258.

Miyajima A., Tanaka, M. and Itoh, T. (2014) Stem/progenitor cells in liver development, homeostasis, regeneration, and reprogramming. *Cell Stem Cell*, 14: 561–574.

Schilders, K.A., Eenjes, E., van Riet, S., Poot, A.A., Stamatialis, D., Truckenmüller, R., Hiemstra, P.S. et al. (2016) Regeneration of the lung: Lung stem cells and the development of lung-mimicking devices. *Resp Res.*, 17: 44.

Smukler, S.R., Arntfield, M.E., Razavi, R., Bikopoulos, G., Karpowicz, P., Seaberg, R., Dai, F. et al. (2011) The adult mouse and human pancreas contain rare multipotent stem cells that express insulin. *Cell Stem Cell*, 8: 281–293.

Stange, D.E., Koo, B.K., Huch, M., Sibbel, G., Basak, O., Lyubimova, A., Kujala, P. et al. (2013) Differentiated Troy[+] chief cells act as reserve stem cells to generate all lineages of the stomach epithelium. *Cell*, 155: 357–368.

# Stem cells underpinning the generation of germ cells

# 12

The life of a multicellular organism is finite, but the aim of life is to continue forever, persisting in or extending into any available environment that will support its existence. For this reason, all life forms have a means of reproducing themselves and continuing their genetic characteristics. Broadly speaking, the transfer of an organism's genetic makeup into the next generation can involve exact copying (asexual reproduction) or, as in the majority of multicellular plants and animals, the combination of genetic material derived from two individuals (sexual reproduction). Asexual reproduction has the benefit that it can enable rapid expansion of the numbers of an organism but lacks the enormous potential provided in sexual reproduction through the bringing together of genetic material from two individuals, which serves to protect against the dangers of mutational damage and at the same time to introduce variation that might confer selective advantage, enabling adaption to the environment.

Stem cells play a role in both asexual and sexual reproduction, although in very different ways. We will not discuss asexually reproducing organisms here, but examples will be found in other chapters in which, for example, specialized stem cells or the acquisition of stem cell properties enable a complete plant or animal to be produced from a small fragment of tissue (e.g., vegetative growth in plants [Chapter 8] or reproduction in some species of flatworms [Chapter 7]). Here, we will focus on the involvement of stem cells in sexual reproduction in animals and will limit our discussion to those stem cells that are directly part of the formation and early development of a new organism. We could also talk about those stem cells that support the reproductive process through their role in the creation of the organs required for reproduction, but we will leave any reference to these to other relevant chapters (e.g., **mesenchymal stem cells** [MSC], which are associated with the cyclical development of the endometrium, are discussed in detail in Chapter 10).

Our starting point in this chapter will be a brief overview of early development to highlight how stem cells, in an essentially transient manner, underpin the way in which the embryo is established (this is also covered in Chapter 3 in relation to the generation of **pluripotent stem cell** lines). An important transient population of cells in all animals that have defined organs producing male and female gametes are the **primordial germ cells (PGC)**. These latter cells can be viewed as stem cells because of their pluripotent characteristics, and we will show how they give rise to the **germ stem cells** within the gonads. Although focusing on vertebrates, including the continuing debate as to whether mammals retain female germ stem cells into adulthood, we also pay considerable attention to the germ stem cells of invertebrates, especially in *Drosophila*. Not only have we learned a huge amount about the role of stem cells in the gonads through such work (also considered in Chapter 7), the findings have provided paradigms for the **niche** structures and regulatory mechanisms that operate more widely in the control of **adult stem cells** and will be referred to throughout this book.

## 12.1 STEM CELLS THROUGHOUT THE REPRODUCTIVE CYCLE

The ultimate goal of an organism is to perpetuate its genes into the next generations. In vertebrates, this is achieved through the union between a spermatozoa and oocyte through the process of fertilization, resulting in the formation and development of the fetus. The reproductive capacity of vertebrates is vast, and hence there is a requirement for the production of large quantities of sperm and eggs throughout the reproductive lifetime of an individual. In this section, we will introduce **spermatogonial stem cells (SSC),** which lie at the heart of male reproductive capacity, and the controversy surrounding the existence of an equivalent stem cell in the female gonads. First, though, we will introduce the embryonic ancestor to the adult germ line stem cells, the primordial germ cell.

## Embryonic development highlights a weakness in the definition of what constitutes a stem cell

If we stick with the simple definition that a stem cell is able to **self-renew** and has the potential to **differentiate** into one or more cell types with specific functions (Chapter 1), then is there an argument for certain cells in the developing embryo being classified as stem cells. For sure, there are cells from several stages or tissues at specific stages that can be used to generate pluripotent stem cell lines (Chapter 3), but does their transient nature preclude them from the stem cell definition? We are probably in the realms of semantics, and there is almost certainly no right answer! So, what do we mean? The fertilized egg, **or zygote,** is one cell, which of course has the potential to give rise to a whole organism, and in so doing undergoes repeated cell division and progressive fate determination. The first divisions give rise to cells, all of which have the full potential (they are **totipotent**) to yield a complete embryo, up to the morula stage, in fact (Figure 12.1). Does this make these cells stem cells? Further divisions beyond the morula stage lead to loss of totipotency, but some cells remain pluripotent, including the cells of the **inner cell mass** of the **blastocyst** (the source of **embryonic stem [ES] cells** in mice), right up to the **epiblast** stage (the source of similar stem cells from human embryos). The argument around **potency** and ability to give rise to stem cell lines in culture could also be extended to the primordial germ cells, which will be discussed in the next section, since these are both ultimately the source of germ stem cells and gametes as well as being capable of culture as **embryonal germ (EG) cell** lines (Chapter 3). Of course, the (absurd?) progression of this line of argument, given that the definition of stem cell in this context is reliant on intervention in the sense of isolation from the organism and cell culture in a laboratory, is that, as we have seen in Chapters 3, 4, and 5, all cells have stem cell potential since they can be **reprogrammed** to a pluripotent state!

## Primordial germ cells establish during embryogenesis and go on to produce spermatozoa and oocytes in the adult

The developmental origins of germ stem cells have been characterized in both invertebrates and vertebrates, and there are a remarkable number of similarities, as there are also in the structures and mechanisms that maintain them in the adult.

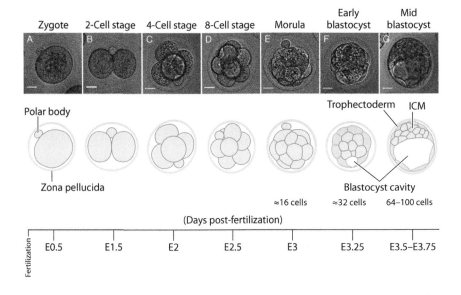

| Zygote | 2-Cell stage | 4-Cell stage | 8-Cell stage | Morula | Early blastocyst | Mid blastocyst |

Polar body

Zona pellucida

Trophectoderm    ICM

Blastocyst cavity

≈16 cells    ≈32 cells    64–100 cells

(Days post-fertilization)

Fertilization    E0.5    E1.5    E2    E2.5    E3    E3.25    E3.5–E3.75

**Figure 12.1 Early development of the mouse embryo.** Fertilization of an oocyte by a sperm to form the zygote is followed by a series of cell divisions. Depicted here is a schematic representation of the stages of early embryonic development in the mouse embryo together with phase-contrast photos from each stage (scale bars = 20 μm) and the corresponding time line showing days post-fertilization (**E**). (**A**) zygote (fertilized egg) highlighting the polar body and zona pellucida, (**B**) 2-cell stage, (**C**) 4-cell stage, (**D**) 8-cell stage, (**E**) morula, (**F**) early blastocyst, and (**G**) mid blastocyst showing the blastocyst cavity, trophectoderm, and inner cell mass (ICM). (From Saiz N & Plusa B. (2013) *Reproduction* 145:R65-R80 [doi: 10.1530/REP-12-0381]. © Society for Reproduction and Fertility. Permission conveyed through Copyright Clearance Center, Inc.)

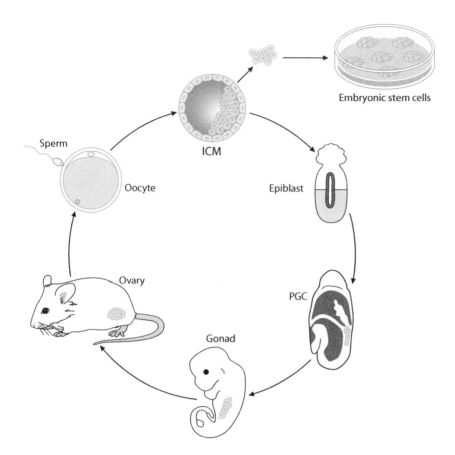

Figure 12.2 **Identification of primordial germ cells through Oct4-GFP expression.** The expression of Oct4 can be used as an indicator of potential pluripotent cells in the embryo. The use of an Oct4-GFP expressing transgenic mouse line enables the identification and tracking of Oct4 positive cells (green) and reveals that expression is highest in the oocyte and becomes gradually restricted to the primordial germ cells as embryonic development proceeds, these cells eventually migrating to the gonads. (From Wu G & Schöler H. (2014) *Cell Regen* 3:7 [doi: 10.1186/2045-9769-3-7]. Published under the terms of the Creative Commons Attribution License (http://creativecommons.org/licenses/by/2.0).)

Early in development, at the time of gastrulation, a small group of cells arise that will go on to form the stem cells that produce spermatozoa and oocytes in the adult. These cells, termed the primordial germ cells, at first migrate into the **endodermal** tissues that will form the gut and from there follow paths toward the bilaterally symmetrical genital ridges, the somatic precursors of what will eventually become the gonads.

Studies of primordial germ cell origins and development have been facilitated by the production of a **transgenic** mouse in which reporter green fluorescent protein (GFP) expression is driven from the promoter of the pluripotency transcription factor Oct4 (Chapter 4). During embryonic development, the pattern of Oct4 expression becomes progressively restricted to the PGC (Figure 12.2) and can therefore be used to image these cells during development.

Two basic mechanisms are used in animals to induce the formation of PGC. In most organisms, including invertebrates and non-mammalian vertebrates, this is achieved through exposure to germ plasma containing inductive cytoplasmic RNAs, RNA-binding proteins, and various organelles that have been accumulated in the mature oocyte. During the first divisions of the embryo, these inductive agents are segregated to the cells fated to become PGC. In contrast, PGC arise in mammals through a process of inductive signaling from extra-embryonic tissues. In spite of the evolutionary distance, the origin and migration of the PGC in invertebrates such as *Drosophila* is quite similar to what is seen in mammals (Figure 12.3).

In the mouse embryo at day 6.25 of gestation, six PGC precursors are set aside in the proximal epiblast near where the primitive streak will eventually form (Figures 12.3 **and** 12.4), their determination being the result of signals in the form of bone morphogenetic proteins (BMP) secreted by the extraembryonic **ectoderm** and visceral endoderm. Induction of PGC fate is characteristically marked by the expression of transcriptional repressor proteins of the PRDM family, namely PR domain zinc finger protein 1(Prdm1), also known as B lymphocyte-induced maturation protein 1 (Blimp-1), and Prdm14. The Prdm1 positive PGC precursor cells move out of the embryo into the extraembryonic **mesoderm** at the base of the allantois, probably to avoid other signals that would induce somatic cell lineage differentiation. During this period, PGC are specified, and expression of the protein Stella marks the event (around day 7.25 after fertilization), along with the expression of other PGC markers such as Kit. PGC initially exist as a cluster of about 40 cells but then re-enter the embryo proper during

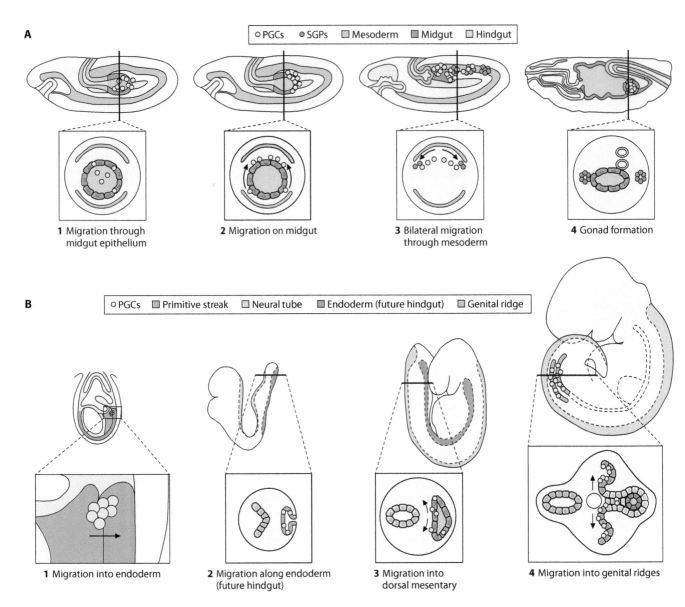

**Figure 12.3 Evolutionary conservation of primordial germ cell development and migration.** Studies in both *Drosophila* and mouse have revealed an evolutionary conserved pattern of primordial germ cell (PGC, yellow) development and migration in the embryo. Lateral views (top panels) and transverse sections (insets) showing PGC migration in (**A**) *Drosophila* and (**B**) mouse. (**A**) PGC in *Drosophila* migrate through the midgut epithelium following specification (1), where they reorient themselves on the midgut (2). PGC migrate bilaterally toward the somatic gonadal precursors (SGP) (3), where they become associated to form the embryonic gonad (4). (**B**) In the mouse embryo, PGC migrate from the primitive streak to the endoderm on embryonic day 7.5 (E7.5) following their specification (1). By E8, PGC begin their migration along the endoderm (2). At E9.5, PGC migrate bilaterally into the dorsal mesentery (3), and by E10.5, they reach the genital ridges, where they are specified to become either gonocytes (male) or oocytes (female). (From Richardson B & Lehmann R. (2010) *Nat Rev Mol Cell Biol* 11:37–49 [doi:10.1038/nrm2815]. Reprinted by permission from Springer Nature.)

early gastrulation and multiply to approximately 20,000 cells while they migrate toward the genital ridges. During their journey, the PGC undergo extensive nuclear reprogramming (including activation of pluripotency genes and changes in the **epigenetic** modifications to the genome). In humans, the PGC migrate from the yolk sac epithelium to the hindgut by approximately 4 weeks post-fertilization and move through the dorsal mesentery, finally reaching the genital ridge around 6 weeks of gestation. Once at the genital ridges, PGC adopt different fates depending on their sex. In males, the PGC enter the seminiferous tubules and become gonocytes, where they cease dividing but retain stem cell potential by differentiating into spermatogonial stem cells. In females, PGC undergo meiosis to become oocytes, seemingly ending their stem cell potential, although this is now matter of debate that we will touch upon in more detail later.

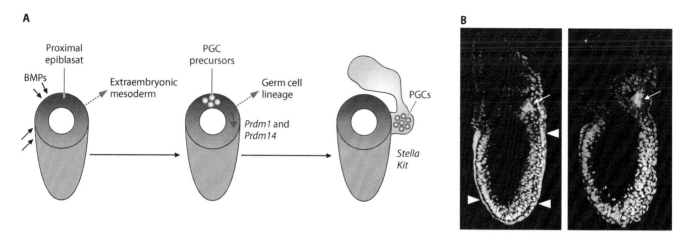

**Figure 12.4 The earliest committed primordial germ cell precursors in the mouse.** At embryonic day 6.25 during mouse development, primordial germ cell (PGC) precursors are set aside in the proximal epiblast. (**A**) The signals responsible for the retention of PGC at the proximal epiblast are bone morphogenic proteins (BMP) secreted by the extraembryonic ectoderm and visceral endoderm. The PGC precursors begin to express PR domain zinc finger protein 1 (*Prdm1*) and *Prdm14* and begin to migrate into the extraembryonic mesoderm, where PGC are specified and begin to express the PGC markers *Stella* and *c-Kit*. The schematic depicts these crucial stages of PGC specification in the mouse embryo. The embryo is shown as an orange cone, with the epiblast as a layer at the top. (**B**) Mice transgenic for green fluorescent protein in which expression is driven by either *Prdm1* (left) or *Prdm14* (right) expression show that *Prdm1* is expressed in the nascent PGC precursors emerging from the most proximal part of the posterior epiblast (arrow) as well as in the visceral endoderm (arrow heads). *Prdm14* is exclusively expressed in the germ cell lineage (arrow). (A: From De Felici M. (2009) *Int J Dev Biol* 53: 891–894 [doi:10.1387/ijdb.082815mf]. Copyright © 2009 UBC Press. Used with permission. B: From Saitou M & Yamaji M. (2010) *Reproduction* 139: 931–942 [doi: 10.1530/REP-10-0043]. © 2010 Society for Reproduction and Fertility. Permission conveyed through Copyright Clearance Center, Inc.)

## Distinct signaling cues guide primordial germ cells to migrate to the gonads

The molecular basis of how PGC give rise to gametes is understood well in flies (*Drosophila melanogaster*) and nematode worms (*Caenorhabditis elegans*), and a number of the genes responsible for the process have been identified. Investigation of the human equivalents has been technically and ethically more challenging, and therefore the mouse model has served as an alternative to study mammalian germ cell origins. PGC require an intrinsic motility program and external guidance cues to survive and successfully migrate. Proper guidance involves both attractive and repulsive cues and is mediated by protein and lipid signaling. PGC obtain directional signals from cells along their migration path. One such signal, the chemokine stem cell derived factor 1 (SDF-1), is critical for proper PGC migration in zebrafish and in mice. Similarly, in *Drosophila*, a molecule that is structurally related to chemokine receptors is important for PGC migration. Programmed cell death plays an important part in PGC migration by eliminating those cells that are in danger of colonizing locations other than the gonads, leading potentially to the later occurrence of germ cell tumors (Chapter 18). Signaling involving stem cell factor (SCF, or Steel Factor), which is the ligand for the receptor tyrosine kinase c-Kit, is often seen in the context of stem cells (e.g., in **hematopoietic stem cells** [HSC], Chapter 9) and is important in PGC, being required for their proliferation and migration but also needing to be reduced to ensure that misplaced PGC are eliminated.

## Spermatogonial stem cells maintain spermatogenesis throughout the life of an adult male

Mature spermatozoa (sperm cells) are produced through the process known as spermatogenesis (Box 12.1). In man, spermatogenesis initiates during the years of puberty around the age of 10–15 years old. Spermatogenesis is a complex process involving several stages of proliferation, differentiation, and maturation, all of which are coordinated in synchronous cycles, each taking around 70 days to complete, yielding something like 5.5 million sperm per gram of testis tissue per day. In mammals, the adult SSC that support spermatogenesis are rare cells ($2–3 \times 10^4$ per adult mouse testis) that reside upon the basement membrane of the seminiferous tubules in the testis.

The identification and characterization of SSC in adult mammalian testis has been challenging. Over 50 years ago, work by Charles Phillipe Leblond and Yves Clermont (Box 12.2) on human testis identified two Type A spermatogonial subtypes based on their morphological characteristics, which they referred to as Dark Type A ($A_{dark}$) and Pale Type A ($A_{pale}$) spermatogonia (Figure 12.5). According to the model proposed by Leblond and Clermont, both $A_{dark}$ and $A_{pale}$ spermatogonia (a distinction that is restricted to primates) can act as stem cells, with the $A_{dark}$ spermatogonia acting as a reserve stem cell pool, dividing infrequently but being able to be triggered to self-renew in case of injury or disease. In contrast, the $A_{pale}$ spermatogonia represent the active SSC population that can give rise to Type B spermatogonia and ultimately spermatocytes, spermatids, and spermatozoa.

Following the initial findings of Leblond and Clermont, numerous research groups continued to try to deduce the precise scheme of spermatogenesis and define the true SSC. Several schemes of spermatogonial multiplication and self-renewal have been proposed during the last 50 years. In 1971, Huckins and Oakberg proposed one scheme termed the $A_s$ ($A_{single}$) model, which has been largely accepted as dogma for the last 40 years. However, during the last decade, two alternative models have been proposed (Figure 12.6): the fragmentation model and more recently the hierarchical $A_s$ model, which determined that the $A_s$ SSC at the apex of the hierarchy are in fact heterogeneous in their stem cell capability, as dictated by the level of expression of the protein Inhibitor of Differentiation 4 (ID4).

## Spermatogonial stem cells can be isolated based on antigen expression on their surface and successfully cultured *in vitro*

Although the first observations by Leblond and Clermont identified the presence of the Type A spermatogonial cell as stem cells, they were not able to prove their function. The evolution of the differentiation schemes described previously came about through a rapid acceleration in the techniques available to test their ability as stem cells. As described in Chapter 2, a process, often iterative, of prospective isolation using surface markers followed by *in vitro* and *in vivo* bioassays is required if true stem cell identity is to be confirmed. Most of the information surrounding the characteristics of SSC has been obtained from studies in mouse and to a lesser extent rats and pigs, with primate and human SSC research lagging behind due to difficulties in obtaining testicular

### BOX 12.1 BACKGROUND: SPERMATOGENESIS

Spermatogenesis takes place in the seminiferous tubules in the adult testis, producing mature spermatozoa that are stored in the epididymis. The structural organization of the testis is largely conserved between rodents and man, a fact that has enabled studies of spermatogenesis relevant to humans using animal models. A seminiferous tubule is composed of both somatic cells and spermatogenic cells, derived from the primordial germ cells. The main constituents of the spermatogonial stem cell (SSC) niche are Sertoli cells, which reside within the seminiferous epithelium and provide essential cues for the maintenance and coordination of spermatogenesis. In addition to Sertoli cells, the space between seminiferous tubules contains Leydig cells, which are responsible for androgen production, and myoid cells, which have a smooth muscle-like appearance and are thought to be responsible for the rhythmic contractile motion of the tubules. The spermatogenic cells constitute a hierarchy, with SSC at the top and a series of generations of dividing and progressively more differentiated cells below, including the spermatogonia, spermatocytes, and spermatids. This hierarchy of expansion and differentiation is reflected in the layered organization of the cells within the seminiferous tubules, with spermatogonia residing at the

basement membrane, spermatocytes toward the middle of the seminiferous epithelium, and spermatids within the adluminal region (Figure 1).

As in other stem cell hierarchies, the SSC, as the most immature cell, is not the only cell that self-renews. The SSC is only one of several defined cell types among a population of self-renewing spermatogonia (Figure 2). Spermatogonia are grouped into three classes; Type A, Intermediate, and Type B. Type A spermatogonia are the most undifferentiated and are divided into $A_{single}$ ($A_s$), $A_{paired}$ ($A_{pr}$), and $A_{aligned}$ ($A_{al}$), with SSC activity residing within the $A_s$ subpopulation (0.02%–0.03% of all germ cells). Type A undifferentiated spermatogonia differentiate into $A_{1-4}$ spermatogonia, the $A_4$ spermatogonia then further differentiating into intermediate (In) and Type B spermatogonia. Each of these stages of spermatogonial differentiation involves mitotic division, maintaining the diploid DNA content. In order to become a functional germ cell, the DNA content of the terminally differentiated spermatozoa must become haploid. Meiotic divisions begin to take place when Type B spermatogonia differentiate into primary and secondary spermatocytes (meiosis I/II). Secondary spermatocytes finally give rise to spermatids during meiosis II that in turn undergo spermiogenesis to fully differentiated spermatozoa.

**A**

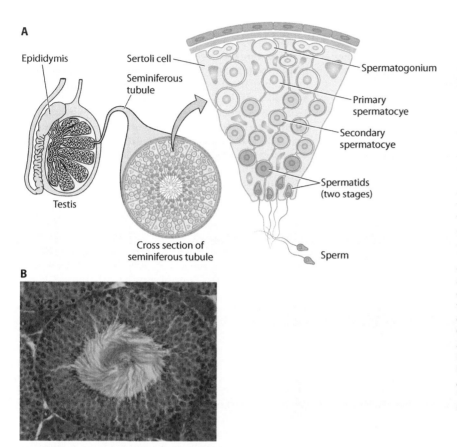

Epididymis

Sertoli cell

Seminiferous tubule

Spermatogonium

Primary spermatocye

Secondary spermatocye

Spermatids (two stages)

Sperm

Testis

Cross section of seminiferous tubule

**B**

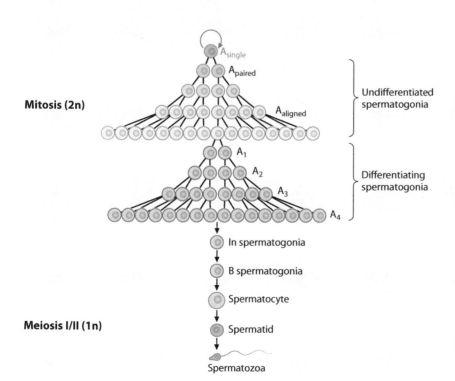

**Mitosis (2n)**

A_single

A_paired

A_aligned

Undifferentiated spermatogonia

A_1
A_2
A_3
A_4

Differentiating spermatogonia

In spermatogonia

B spermatogonia

Spermatocyte

**Meiosis I/II (1n)**

Spermatid

Spermatozoa

Box 12.1 Figure 1 **Spermatogenesis.** Spermatogenesis is the process that produces millions of mature spermatozoa every day beginning at the time of puberty in males. (**A**) Spermatogenesis takes place in the testes in specialized seminiferous tubules, which are a network of tubules with spermatogenesis taking place along its entire length. A cross section though the seminiferous tubule reveals a distinct hierarchy of differentiation with the immature spermatogonium residing at the basement membrane in contact with Sertoli cells. As cells progress toward the seminiferous tubule lumen, the spermatocytes become more mature, eventually forming mature sperm cells that are released into the lumen for storage in the epididymis. This process can be distinguished in a histological staining of a section through the seminiferous tubule of an adult mouse (**B**), where the tails of the spermatids and mature spermatozoa can be clearly defined. (A: Adapted from Campbell N, Reece J, Mitchell L et al. (2003) *Biology: Concepts & Connections*, 4th ed. San Francisco: Pearson Benjamin Cummings. B: From Xie B, Li J & Zhu W-J. (2014) *Exp Ther Med* 7:654–656 [10.3892/etm/2014.1481]. Used with permission from Spandidos Publications.)

Box 12.1 Figure 2 **Transient amplification and differentiation in spermatogenesis.** The process of spermatogenesis originates with the spermatogonial stem cell that resides at the apex of a stem cell hierarchy. The diagram depicts this hierarchy in a mouse testis, which initiates at the apex with $A_{single}$ spermatogonia (dark blue) and has self-renewal activity (indicated by the circular arrow). $A_{single}$ spermatogonia give rise to more undifferentiated spermatogonia, $A_{paired}$ and $A_{aligned}$, which function as transit amplifying cells. The $A_{aligned}$ spermatogonia give rise to differentiating spermatogonia A1-A4, which serve further to expand cell numbers. The A4 spermatogonia yield intermediate (In) spermatogonia, which in turn produce B spermatogonia. These stages each involve mitotic division, maintaining the genome in a 2n state. Production of spermatocytes from B spermatogonia and progressive differentiation toward spermatids and spermatozoa require meiotic divisions, resulting in haploid (1n) cells ready for fertilization of an oocyte. (From Kanatsu-Shinohara M & Shinohara T. (2013) *Annu Rev Cell Dev Biol* 29:163–187 [doi:10.1146/annurevcellbio-101512–122353]. Permission conveyed through Copyright Clearance Center, Inc.)

## BOX 12.2 PIONEERS: CHARLES PHILLIPE LEBLOND AND YVES CLERMONT

It is widely acknowledged that the foundations for the principles of stem cell behavior were laid down during the 1960s by Till and McCulloch following their pioneering bone marrow transplantation experiments (Chapter 9). However, earlier work by the researchers Charles Phillipe Leblond and Yves Clermont (**Figure 1**) had already led to the proposal in 1953 that adult tissues can be maintained from a resident stem cell population.

Leblond, who was born in Lille in 1910 and educated in Paris, qualified as a doctor in 1934 and specialized in anatomy. Leblond moved to McGill University in Montreal in 1941, where he eventually became the chair of the Department of Anatomy, which he held until 1974. He is

**Box 12.2 Figure 1 Charles Phillipe Leblond and Yves Clermont.** Charles Phillipe Leblond and Yves Clermont are shown on the left and right, respectively (Left: Scanned and released by the family of CP Leblond. Published under Creative Commons Attribution 2.0 Generic Licensed (https://creativecommons.org/licenses/by/2.0/) via Wikipedia. Right: Used with permission from McGill University Archives, PR050559.)

widely considered to have been a pioneer in cell biology, being among the first to use synthetic radioactive isotopes in the localization of labeled molecules within tissues, developing the now universally used technique known as autoradiography. Leblond used autoradiography to track radioactive precursors of DNA and thereby examine the renewal and fate of cells in many different tissues. Through the understanding gained about the turnover and mitotic rates of cells, Leblond and his colleagues paved the way for our current understanding of the cell cycle, asymmetric cell divisions, and stem cells.

In 1950, the Canadian Yves Clermont joined Leblond's research team while he undertook his PhD studies on the cell associations found in the seminiferous epithelium of three different species. Clermont's thesis findings became the building blocks for the modern understanding of spermatogenesis, including the spatial and temporal organization of the seminiferous epithelium. Clermont identified the progressive stages of differentiation of germ cells from spermatogonia through to spermatozoa and described the kinetics of spermatogenesis in many species, including humans. In their seminal publication in 1953, Clermont and Leblond are the first to designate groups of dividing cells in an adult organ as stem cells and proposed their "Stem Cell Renewal Theory." Leblond and his colleagues went on to define stem cells in other tissues, including **satellite cells** in skeletal muscle (Chapter 10).

Leblond did not stop at stem cells. He continued research well into his 90s, and through his focus on the application of radioactive tracing of biomolecules made major discoveries in areas as diverse as protein glycosylation, bone growth, and DNA replication.

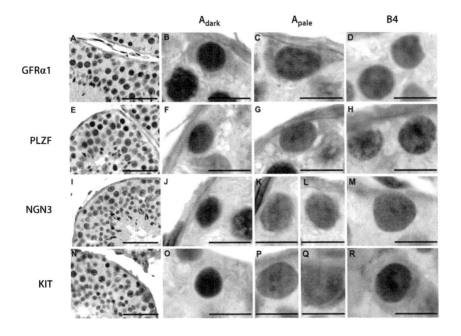

**Figure 12.5 Heterogeneity in the spermatogonial stem cell pool.** Pioneering work by Leblond and Clermont on mammalian testis identified two Type A spermatogonial subtypes based on their morphology, $A_{dark}$ and $A_{pale}$. The images are of sections of the adult rhesus monkey testis stained by immunohistochemistry for the characteristic markers of spermatogenesis, GFRα1, PLZF, NGN3, and KIT (brown staining), and then counterstained (blue) to reveal nuclear morphology. In this way, it is possible to identify $A_{dark}$ and $A_{pale}$ spermatogonia, as well as differentiating B spermatogonia. The first image in each row shows part of one seminiferous tubule cross section (scale bar: 50 μm). Enlargements are also shown of representative $A_{dark}$, $A_{pale}$, and B4 spermatogonia (scale bar: 10 μm). (From Herman B, Sukhwani M, Simorangkir D et al. (2009) *Hum Reprod* 24:1704–1716 [doi:10.1093/humrep/dep073]. By permission of Oxford University Press.)

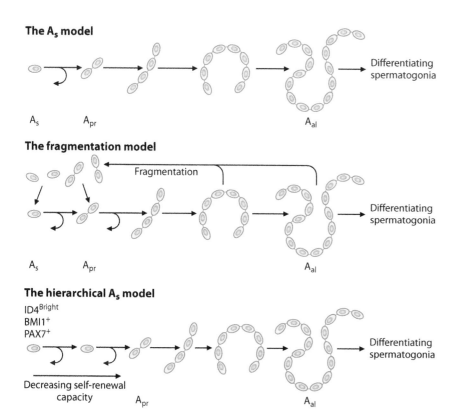

**The A$_s$ model**

A$_s$     A$_{pr}$     A$_{al}$     Differentiating spermatogonia

**The fragmentation model**

Fragmentation

A$_s$     A$_{pr}$     A$_{al}$     Differentiating spermatogonia

**The hierarchical A$_s$ model**

ID4$^{Bright}$
BMI1$^+$
PAX7$^+$

Decreasing self-renewal capacity     A$_{pr}$     A$_{al}$     Differentiating spermatogonia

**Figure 12.6 Alternative models of spermatogenesis.** Following Leblond and Clermont's seminal findings, several models of spermatogonial self-renewal have been proposed. First, the A$_s$ model, which has been largely accepted for the last 40 years, depicts a simple hierarchy where A$_s$ spermatognia give rise to A$_{pr}$, which in turn form chains of A$_{al}$ spermatogonia. The later proposed fragmentation model proposes that A$_s$ spermatogonia give rise to A$_{pr}$ and A$_{al}$ chains, which fragment to give rise to A$_s$ and A$_{pr}$ again. More recently suggested, the hierarchical A$_s$ model envisages the A$_s$ spermatogonial stem cell pool as being heterogeneous, giving rise to further A$_s$ cells that have decreased self-renewal capacity and in turn produce A$_{pr}$ and A$_{al}$ as originally proposed. The most immature A$_s$ spermatogonium is thought to express ID4, BMI, and PAX7. (Adapted from de Rooij D. (2017) *Development* 144: 3022–3030 [doi: 10.1242/dev.146571].)

tissue for research, although within the last decade, it has been shown that human SSC share many, if not all, of the phenotypic markers defined in mice.

Mouse spermatogonia and their path through differentiation can be defined through a combination of surface markers. An important marker is the stem cell factor receptor Kit, which is only weakly present on immature spermatogonia, but as SSC transit toward differentiating spermatogonia (A$_{1-4}$), they begin to express high levels on their surface (Figure 12.5). Other defining markers include α6-integrin, GFRα1, Plzf, and Vasa. The identification of such markers has facilitated the isolation of SSC from testicular tissue samples by **fluorescence activated cell sorting** (FACS) for both research and clinical applications. As for several stem cell types, mouse SSC have been shown to exclude the DNA binding dye Hoechst, enabling their isolation by FACS with the **side population** sorting technique (Chapter 2). Until recently, however, there was no unique surface marker that could be used to isolate SSC, most of the surface markers described labeling other male germ cells or somatic cells from testicular tissue. The identification of the expression of the protein ID4 in A$_s$ spermatogonia has enabled the isolation of various subtypes of A$_s$ cells and the identification of the most immature SSC with maximum self-renewal capacity. In 2017, through the generation of an ID4-eGFP transgenic mouse, the group of Oatley identified A$_s$ spermatogonia cells with distinct levels of ID4 (20% bright, 40% dim, and 40% intermediate levels of GFP expression). Upon transplantation into recipient mice testes that lacked endogenous spermatogenesis, the ID4$^{Bright}$ A$_s$ cells transplanted better than the ID4$^{Dim}$ cells (Figure 12.7). In addition to the identification of ID4, two other proteins have been shown to mark the most immature SSC, namely Paired Box 7 (Pax7) and Bmi1. A$_s$ cells expressing either Pax7 or Bmi1 also co-express ID4 and in the case of Pax7$^+$ID4$^+$ cells make up the rarest of A$_s$ spermatogonia subtypes (0.25% of total testicular cells) and are therefore thought to represent the ultimate SSC. Interestingly, despite these proteins seemingly marking the most immature SSC, they do not appear to play an essential role in the function of SSC given the fact that their genetic ablation in mice does not result in any significant decline in spermatogenesis.

As for other stem cell types discussed throughout this book, maintenance of stem cell characteristics such as self-renewal during and after culturing the cells is crucial if SSC are to be used within the clinic and for further research

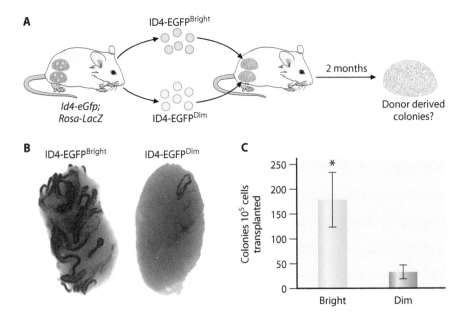

**Figure 12.7 Identification of the most immature spermatogonial stem cell through its expression of ID4.** Identification of a unique spermatogonial stem cell marker called Inhibitor of Differentiation 4 (ID4) has enabled the isolation of the most immature SSC. Through the generation of a transgenic mouse expressing GFP driven by *Id4* gene elements that control its expression, together with the LacZ reporter, it was possible to use FACS to isolate either ID4-EGFP^Bright or ID4-EGFP^Dim SSC (**A**) that were transplanted into recipient mice testis and analyzed after 2 months. Following activation of the LacZ reporter, it was evident that maximum engraftment was achieved from the ID4-EGFP^Bright SSC compared to ID4-EGFP^Dim SSC (**B** and **C**). (From Helsel A, Yang Q, Oatley J et al. (2017) *Development* 144(4): 624–634 [doi: 10.1242/dev.146928]. Reproduced with permission from The Company of Biologists Ltd.)

studies. SSC are tightly controlled through their niche, and hence removal from this environment can have detrimental consequences on the overall function of the SSC, resulting in apoptosis. Despite the complexity of the spermatogenesis process, progress has been made in the ability to culture SSC *in vitro*. Mouse SSC have been isolated from pre-pubertal mouse testes and have been propagated in long-term culture with confirmation of the presence of SSC within the cultures by transplantation into recipient mouse testes lacking endogenous spermatogenesis. Successful culture of both mouse and human SSC relies on three factors: a basic conditioning medium; feeder cells; and growth factors such as glial cell-derived neurotrophic factor (GDNF), a name based on its original discovery, epidermal growth factor (EGF), fibroblast growth factor 2 (FGF2), and leukemia inhibitory factor (LIF). With the recent identification of the "ultimate" ID4+Pax7+ SSC, culture conditions have been increasingly better defined, and FGF2 is the one crucial factor required to induce self-renewal. In spite of the progress made in culture conditions for SSC, scientists have yet to achieve full transition through spermatogenesis to mature spermatozoa from SSC *in vitro*, a feat that has only been achieved with organ explant culture.

## Transplantation of spermatogonial stem cells enables determination of their stem cell potential *in vivo*

The most definitive test of a stem cell is to demonstrate its ability to self-renew and yield daughter cells with the capability of differentiating into mature cell lineages. Transplantation assays have been developed that enable a true test of stem cell capacity for several stem cell types (Chapter 2), including those from testicular tissue. The testis cell transplantation assay was pioneered by Ralph Brinster (Box 12.3) and has revolutionized the field of spermatogonial stem cell research. In addition to facilitating SSC research, the prospect of utilizing testis cell transplantation as a therapeutic intervention for the treatment of male factor infertility is a realistic prospect, in much the same way that transplantation of bone marrow HSC is widely used in the treatment of many different hematological disorders.

The simplicity of the SSC transplant procedure does not at all reflect the complexity of the demands on the transplanted cells. Testis cells are harvested from fertile male donor mice and microinjected as a cell suspension into the seminiferous tubules of infertile recipient mice. Following microinjection, the SSC must migrate through layers of differentiating germ cells and Sertoli cell tight junctions to find and occupy their niche on the basement membrane of the seminiferous tubule. By labeling the donor cells with a reporter transgene, donor-derived spermatogenic colonies within the recipient testes can easily be identified 3 to 5 months post-transplantation (Figure 12.8). The recipient male mice are then mated to females, and the production of viable offspring carrying the reporter transgene serves

BOX 12.3 PIONEER: RALPH BRINSTER

Raised on a small farm in New Jersey, Ralph Brinster (Figure 1) developed a keen interest in fertility, reproduction, and germ cells that sparked half a century of ground-breaking research in embryonic cell differentiation and stem cell physiology, with him being one of the principal founders of the field of mammalian transgenesis. Brinster's career began in 1953 when he completed his undergraduate studies in animal science at the Cook School of Agriculture at Rutgers University. Following a 3-year tour as an officer in the United States Air Force, where he served in the Korean War, he studied for an MD and then in 1964 completed a PhD in physiology at the University of Pennsylvania. Brinster's work initially focused on the development of culture and manipulation strategies for mouse eggs. These methods were adopted by the scientific community and remain in force to the present day with little modification. He also demonstrated that **teratocarcinoma** stem cells would colonize a blastocyst, an observation that led to the development of embryonic stem cells and their use in genetic engineering. Brinster's research also led to successful modifications to the germ line and the generation of transgenic animals. Importantly, and most applicable to the field of stem cell biology, Brinster pioneered techniques to assay spermatogonial stem cell (SSC) function by testis cell transplantation, first performing such an assay in 1994, and was able fully to demonstrate the stem cell potential at a single cell level. This pioneering work on SSC enabled scientists to gain a greater understanding of the process

of spermatogenesis and provide a potential technique to preserve the fertility of men following chemotherapeutic treatment for cancer.

Box 12.3 Figure 1 **Ralph Brinster.** (Photo credit: Scott H. Spitzer. Used with permission from University of Pennsylvania School of Veterinary Medicine.)

as proof that fertility has been restored by the SSC transplant. A colony arising within the testis of a recipient mouse following SSC transplantation represents the functionality of a single SSC. Additionally, the transplantation assay can be used in conjunction with experimental manipulation of gene expression in SSC, or somatic cells (Sertoli, Leydig, or myoid) potentially constituting elements of the niche, to determine the importance of key molecules in the maintenance and function of the stem cells (Figure 12.9).

## Xenogeneic spermatogonial stem cell transplantation indicates conservation of the niche throughout evolution

Transplantation of SSC provides researchers the perfect opportunity to study the role that the stem cell niche plays in the regulation and maintenance of the stem cells. By performing **xenogeneic** SSC transplants, such that rat spermatogenesis can take place within the seminiferous tubules of a mouse, it has been possible to show that timing of the cell divisions from SSC through to spermatozoa is intrinsic to the stem cell and is not controlled by the niche cells. This was evident from the production of mature rat spermatozoa in the lumen of the mouse seminiferous tubules after 52 days, rather than the 35 days that is normally observed in mice. Xenogeneic SSC transplantations have also highlighted the evolutionary conservation or otherwise of the signals that regulate spermatogenesis. The fact that mouse Sertoli cells are capable of supporting rat spermatogenesis is indicative of conserved functioning of surface recognition molecules and growth factors. Likewise, the ability of transplanted rat SSC to move from the seminiferous tubule lumen through to the basement membrane, where they lodge and give rise to differentiating spermatogonia through and mature spermatozoa, shows that the signals used in these two species have been conserved since their evolutionary divergence 10 million years ago. Species that diverged away from rodents between 50 and 100 million years ago, such as dog, cow, and baboon, are also able to colonize the basement membrane of the seminiferous tubule in mice but are unable to differentiate toward spermatozoa. This highlights that the signals maintaining the

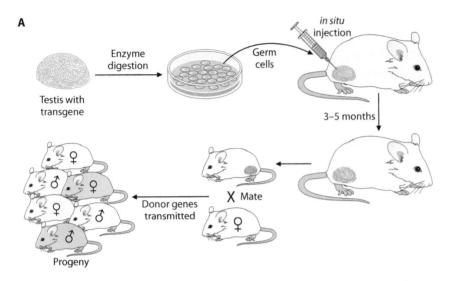

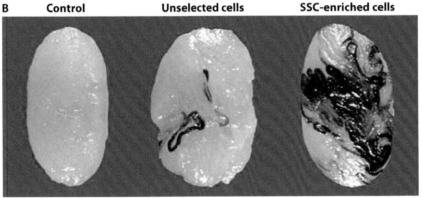

Figure 12.8 **Spermatogonial stem cell transplantation assay.** Ultimately, stem cells are defined based on function, and one test of this is their ability to transplant into recipients, engraft, and produce terminally differentiated cells. Spermatogonial stem cell (SSC) transplantation enables the true determination of prospective SSC populations following their isolation. (**A**) This cartoon illustrates the testis transplantation method. A single cell suspension is produced from a fertile donor testis by enzyme digestion. These cells can then either be used immediately or cultured. Cells are injected into the lumen of seminiferous tubules of an infertile male mouse, so that only cells with SSC potential are able to engraft and form a clone of spermatogenesis in the recipient testis. When the donor cells incorporate a transgenic reporter, it is possible to highlight spermatogonial cells arising from them. Using a β-galactosidase reporter, cells can easily be seen as blue stretches of tubule. Mating of an engrafted recipient male to a wild type female produces progeny, some males of which carry the transgene. (**B**) Images of testes engrafted with SSC expressing a ubiquitously expressed β-galactosidase reporter and stained blue to reveal expression of the enzyme. On the left is a control testis showing no SSC engraftment. In the middle is a testis transplanted with unselected testis-derived cells showing minimal engraftment. On the right is a testis in which SSC-enriched cells were transplanted, giving rise to extensive engraftment. (A: From Brinster R. (2002) *Science* 296:2174–2176 [doi:10.1126/science.1071607]. Reprinted with permission from AAAS. B: From Oatley J & Brinster R. (2008) *Annu Rev Cell Dev Biol* 24:263–286 [doi: 10.1146/annurev. cellbio.24.110707.175355]. Permission conveyed through Copyright Clearance Center, Inc.)

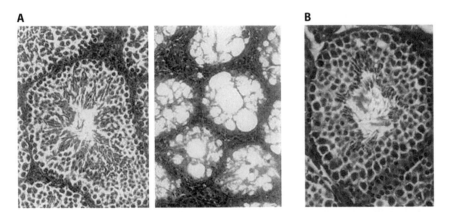

Figure 12.9 **Determining crucial spermatogonial stem cell niche components through transplantation.** Spermatogonial stem cell (SSC) transplantation can be used to determine which of the components of the SSC niche are the most crucial to their function. Transplantation of wild type SSC into a defective environment can restore spermatogenic function to the testes. (**A**) This figure shows hematoxylin and eosin stained sections of 12 week-old mouse testis from a wild type animal (left) and a mutant (right) that is genetically deficient in Kit (W/W$^v$), the receptor for stem cell factor (SCF) that plays a critical role in spermatogenesis. Note the absence of spermatogenesis in the mutant testis. (**B**) This image is from the testis of a W/W$^v$ host that has been engrafted with SSC from a wild type mouse. It can be seen that the normal architecture of the seminiferous tubule is restored, including the presence of cells at various stages of spermatogenesis. (From Brinster R & Zimmerman J (1994) *Proc Nat Acad Sci* 91:11298–11302 [doi: 10.1073/pnas.91.24.11298]. Copyright (1994) National Academy of Sciences, U.S.A.)

stem cells in the short term are conserved, but the ability for SSC to respond to differentiation factors has changed during evolution.

## The occurrence of female germ cells in vertebrates is a matter of some debate

As we will describe later, in at least some invertebrates exemplified by *Drosophila*, both males and females have adult stem cells within their gonads that maintain the production of gametes. In contrast, the central dogma of human reproductive biology has been that women are born with all the oocytes they will ever possess and become reproductively senescent (menopause) later in life. This foundation was laid down by Sir Soloman Zuckerman in 1951 and of course implies that there is no requirement for an adult stem cell in human females and presumably at least other mammals among the vertebrates. The fact that a woman is born with about a million follicles but will only ovulate around 400 times in her lifetime suggests that replenishment from stem cells is anyway unnecessary.

In recent years, the dogma has been challenged following a publication in 2004 from Jonathan Tilly (Figure 12.10) and colleagues that reported the existence of oocyte stem cells (OSC) in the female mouse ovary. These researchers were investigating follicular atresia, that is, the process responsible for the loss of the majority of follicles throughout life. They reported that follicles were dying at a rate such that the ovary would be devoid of oocytes far earlier than is actually seen and therefore concluded that an OSC must be present to maintain the follicle number. The report sparked a huge debate over the existence of OSC that has divided opinions, many being in support, while others believed Tilly's data had been misinterpreted.

Tilly's conclusion stemmed from analysis of the mouse ovary, where he discovered the presence of oval-shaped cells that stained positive for the germ cell-specific marker DEAD box polypeptide 4 (Ddx4, also called Vasa homolog) that is expressed in the germ cells of all animals. They also observed incorporation of bromodeoxyuridine (BrdU), a DNA precursor analogue that can be detected once incorporated and is used as a marker of cell proliferation, and positive staining for the meiosis-specific protein, Synaptonemal Complex Protein 3 (SCP3). The apparently definitive proof of their belief that they had detected OSC was transplantation of ovarian tissue from normal mice into the ovaries of mice that ubiquitously express GFP. They observed oocytes expressing GFP in the recipient ovary surrounded by recipient (not green) granulosa cells, indicating that, by their reasoning, OSC from the donor mouse had initiated folliculogenesis. In a similar study 5 years later, another research group described the existence of female germ stem cells (FGSC) that could be isolated and cultured from neonatal and adult mouse ovaries. These FGSC were isolated from transgenic mice expressing GFP by FACS following antibody staining against Ddx4. Following isolation, the FGSC were transplanted into the ovaries of infertile female mice that later were able to produce viable GFP-positive offspring. These latter experiments are considered not to be definitive because of the possibility that reprogramming to a pluripotent state occurred during the culture process. A more recent publication from Tilly and his colleagues has addressed many of the criticisms, including the markers and sorting strategy employed, and using genetically marked cells, they have been able to engraft both mouse and human OSC into recipient mouse ovaries, giving rise to follicle-like structures (Figure 12.11).

Despite these studies, stem cell and reproductive biologists remain skeptical about the existence of mammalian FGCS. The controversy continues because it seems strange that the existence of OSC should have remained elusive for so long, and only a few research

**Figure 12.10 Jonathan Tilly.** (From Gura T. (2012) *Nature* 491: 318–320 [doi:10.1038/491318a]. Reprinted by permission from Springer Nature.)

**Figure 12.11 Mouse oocyte stem cells generate functional eggs after intraovarian transplantation.** The existence of mammalian female oocyte stem cells (OSC) remains controversial, with most believing that females lack any OSC activity. This dogma has been challenged recently by evidence suggesting that OSC can be transplanted into the ovaries of infertile mice and produce viable offspring. This figure shows examples of growing follicles, ovulated eggs, and the resultant embryos deriving from the ovaries of wild type mice that had been injected with GFP-expressing presumed OSC 5–6 months before analysis. The top and bottom rows of images are respectively derived from the host (GFP-negative) or the donor (GFP-positive). The images on the left of each row are immunohistochemically stained for GFP (brown; with a blue hematoxylin counterstain), showing oocytes *in situ* in the ovary. The rest of the images are fluorescent microscopy views of eggs, in either cumulus-oocyte complexes (left) or ovulated (right), and embryos generated by IVF after induced ovulation of the recipient female mice. Embryos are shown at the 2-cell, 4-cell, 8-cell, compacted morula (CM), and blastocyst (B) stages. Scale bars, 30 μm. (From White Y, Woods D, Takai Y et al. (2012) *Nat Med* 18:413–422 [doi:10.1038/nm.2669]. Reprinted by permission from Springer Nature.)

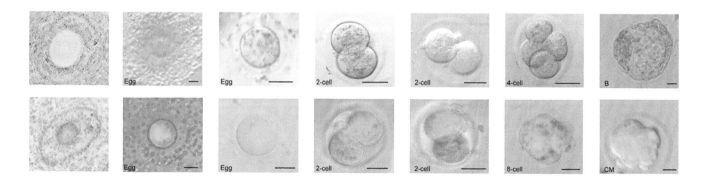

groups claim to have successfully isolated OSC from ovarian tissue. Most investigators in the field question both the experimental technique and interpretation of the results obtained. For example, some are critical of the use of Ddx4 as a cell surface marker for sorting OSC since it is described as being localized in the cytoplasm of oocytes. Then again, questions both for and against the OSC have to be posed. For instance, why would there not be a female germ line stem cell equivalent to the spermatogonial stem cell? Conversely, why do females enter menopause if the OSC does exist? Ultimately, proof that FGSC persist in the adult requires the same sorts of evidence that are sought for any other tissue-associated stem cell; that is, putative GSC must be traceable *in vivo* and shown to be able both to give rise to fertile oocytes and to self-renew.

## Some non-mammalian vertebrates definitely have female germ stem cells, but do all invertebrates?

It is generally accepted that the presence of female GSC is the norm in invertebrates and "lower" vertebrates. However, in spite of the findings about female GSC in model organisms like *Caenorhabditis elegans* and *Drosophila*, as well as what we have seen in Chapter 7 about flatworms and their ability to produce germ cells from pluripotent adult **neoblasts**, it appears that the majority of invertebrate species restrict mitotic divisions of germ cells, and hence the requirement for stem cells, to preadult stages.

Contrary to the still prevalent view that female mammals lack OSC, there are some vertebrate species that have the ability to generate oocytes from resident GSC. These vertebrates tend to have high egg-producing capacity, such as is seen in fish, and it could be this crucial difference in reproductive capacity that has accounted for the presumed loss of female GSC activity in mammals. Female GSC in fish have been best characterized in ray-finned fish such as medaka (*Oryzias latipes*) and zebrafish (*Danio rerio*). Medaka ovaries contain two compartments: the dorsal cavity and ventral stromal compartment. Between these compartments exists the germinal epithelium that constitutes the ovarian GSC niche. The germinal epithelium contains cords of **stromal cells** expressing Sox9b that associate closely with GSC expressing stem cell markers such as the homologue of Ddx4 (Vasa) (**Figure 12.12**). These cords very closely resemble the germarium described in *Drosophila* ovaries, the importance of the association between niche cells and GSC made evident by differentiation of those GSC furthest away from the niche signals.

The fact that the basic elements of germ cell development have been conserved throughout evolution and that male GSC are spread widely across the animal kingdom put together with the observation that some species utilize female GSC into adulthood begs the question why such a strategy is not more widely adopted in females. One possible answer is down to essential differences between male and female gametes. Sperm are simple, throw-away devices that if damaged can easily be eliminated and not impact fertilization, whereas oocytes are generally produced in small numbers and contain a significant reserve of cell components that contribute to the first stages of development of the next generation. Evolution may have decided that it is easier to generate the complex cytoplasm of an oocyte by establishing a reserve of primordial follicles that can then be progressively supported to undergo this final maturation rather than repeatedly go through a process of germ cell **commitment** and differentiation.

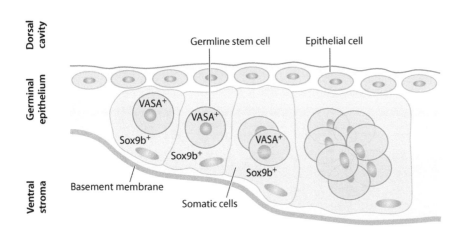

**Figure 12.12 The female germline stem cell niche in medaka.** Despite the dogma that female mammals lack germline stem cell (GSC) activity, this is not necessarily the case for all vertebrates, fish being an example. This schematic representation of the GSC niche in the medaka fish shows the division of the ovary into two compartments, the dorsal cavity and the ventral stroma. GSC reside in the germinal epithelium, a region that lies in between the basement membrane and the epithelial cells. Cords of Sox9b-expressing stromal cells lie in close association with the VASA-expressing GSC. (From Hanna C & Hennebold J. (2014) *Fertil Steril* 101: 20–30 [doi: 10.1016/j.fertnstert.2013.11.009]. Copyright 2014 with permission from American Society for Reproductive Medicine. Published by Elsevier.)

## 12.2  MECHANISMS THAT REGULATE GERM STEM CELLS

Arguments about the degree to which adult females rely upon a pool of adult GSC aside, there is no doubt that males and females in some species have stem cells dedicated to the germ line. As is the case for other tissues, the study of the regulatory mechanisms governing the development of GSC in humans is difficult due to access to tissue and ethical constraints. Studies on model organisms, chosen not because they are representative of animals in general or have particular relevance to reproduction, but rather because these are the ones adopted generally in biological research, suggest that there are some common themes around the developmental origins of GSC and the way in which these are maintained and gamete production from them is regulated. These organisms include the *Caenorhabditis elegans* gonad, the *Drosophila* ovary, and testes from *Drosophila* and mice. We have already given some thought to common aspects to the origin of GSC from primordial germ cells. In this section, we are going to take a closer look at what studies on model organisms, in particular *Drosophila*, have told us about the GSC niche and the molecular and cellular mechanisms, both cell intrinsic and between cells, which control stem cell self-renewal and differentiation.

### *Drosophila* is an essential model for the study of germ stem cells

On the face of it, the male and female gonads in *Drosophila* are very similar structures, and although there are distinguishing features, there are many parallels in the way in which the respective GSC are arranged and controlled. A common feature in the testis and ovary, which is seen in the gonads of other species, too, is that the mitotic divisions downstream of the GSC are incomplete, leading to permanent cytoplasmic bridges between mitotic sisters and to the formation of a cyst of germ cells that develop in synchrony. As seen for other adult stem cells, GSC produce progeny that proliferate through several rounds of divisions before terminally differentiating, in this case marked by entry into meiosis. This pattern conforms to the general strategy of transient amplification (Chapter 1) in the case of male GSC, but in females, there is no such expansion of the number of oocytes, pointing to a specific modification in the female allowing most of the mitotic sisters to contribute to the creation of the specialized oocyte. Studies on GSC in both sexes are facilitated by the morphology and easy access to the gonads combined with genetic approaches that readily enable selective manipulation of gene expression in the stem cells or the niche constituents. Such studies have highlighted the importance of the niche structure and its cell contacts and signaling input as well as the role played by changes in the GSC at the level of the regulation of transcription, RNA processing, and translation. Much of what has been learned from studies of GSC regulation within the niche in *Drosophila* gonads has informed our understanding of stem cell-niche interactions in other tissues

### There is a defined germ line stem cell niche in female *Drosophila*

The *Drosophila* ovary is organized in a similar way to the male testis (Figure 12.13). The niche is composed of terminal filament cells, **cap cells** that directly contact the GSC, and escort cells, which wrap around the GSC, preventing direct contact between the stem cells. **Asymmetric division** is less orchestrated in the female compared to the male, available space in the ovary being the primary determinant of which of the daughters resulting from GSC division is able to remain in contact with the cap cells. Nonetheless, adhesion between a GSC and a cap cell involving E-cadherin is important in ensuring the correct degree of stem cell retention. Premature differentiation of the GSC is to a large extent controlled by repressors of translation, including the proteins Nanos and Pumilio, and the stimulus for germ cell development involves modulation of the expression of a differentiation inducer that is rather prosaically known as Bag-of-Marbles (Bam) – a whole host of these somewhat colorful *Drosophila* gene names will follow! A growth factor called Unpaired (Upd), which emanates from the terminal filament cells, engages with its cognate receptor called Domeless (Dom) on the cap and proximal escort cells, eliciting a JAK-STAT signal that results in the production of the BMP ligands Decapentaplegic (Dpp) and Glass bottom boat (Gbb), which then act on the adjacent GSC to maintain repression of Bam. Only when the GSC move away from the niche following their asymmetric division do they escape the influence of BMP signaling and consequently up regulate Bam, which in turn

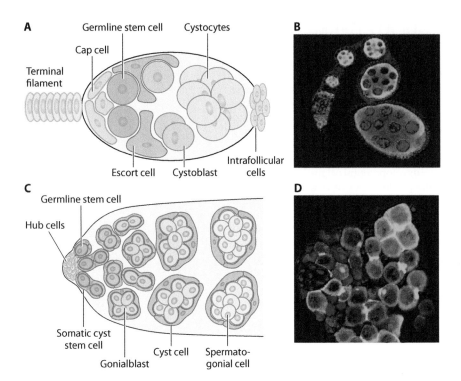

**Figure 12.13 Comparison of the germline stem cell niches in *Drosophila* ovary and testis.** The *Drosophila* ovary and testis are both organized in a similar manner, comprising supporting niche cells and germline stem cells (GSC) that progressively differentiate as they migrate away from the niche cells. (**A**) Schematic diagram of a single germarium in the *Drosophila* ovary, illustrating the location of GSC, their maturing derivatives, and supporting niche cells. The terminal filament, cap cells, and escort cells that make up the niche lie against a basement membrane. The cystoblast and developing cysts move away from the anterior tip while in contact with escort cells. The escort cells receive signals from the terminal filament and cap cells that control stem cell number and proliferation rate. (**B**) Fluorescent confocal microscope image of an ovary germarium, showing stem cells at the tip leading to the assembly of egg chambers connected to one another like a string of different sized beads. Germline cells, including GSC and differentiated germ cells, are labeled green for VASA, while germline specific structures and surrounding follicle cells are labeled (red) for a specific protein involved in regulating cytoskeletal structure and function during oogenesis (the Adducin-like protein Hu-li tai shao). Nuclei are shown in blue. (**C**) Schematic diagram of a *Drosophila* testis, illustrating the location of GSC and somatic cyst stem cells and their respective maturing derivatives, the gonialblasts/spermatogonial cells and their encasing cyst cells. The niche hub cells lie at the apical tip. Daughter cells of the GSC and somatic cyst stem cells eventually differentiate and leave the niche. (**D**) Fluorescent confocal microscope image of a whole *Drosophila* testis. A nuclear marker stains somatic cyst stem cells bright red, whereas it diminishes in their daughters. GSC and their progeny are marked green with cortically localized GFP. DNA is blue. (A: From Hanna C & Hennebold J. (2014) *Fertil Steril* 101: 20–30 [doi:10.1016/j.fertnstert.2013.11.009]. Copyright 2014 with permission from American Society for Reproductive Medicine. Published by Elsevier. B: From http://www.stowers.org/faculty/xie-lab. Used with permission from Stowers Group of Companies and Dr. Ting Xie. D: From https://www.cell.com/pictureshow/reproduction. Picture by Erika Matunis and Becca Sheng, Johns Hopkins University. Copyright 2019, with permission from Elsevier.)

promotes differentiation (**Figures 12.14 and 12.15**). Other refinements, beyond the scope of this description, help to focus the extent of the BMP signal and ensure that Bam is activated immediately outside of the niche. Through its interaction with translation factors such as eIF4A, Bam promotes the translation of differentiation-inducing transcripts and reduces expression of E-cadherin, thereby relaxing the association of GSC with cap cells and preventing cytoblasts from entering the niche.

## The germ line stem cell niche in male *Drosophila* is also well characterized

An adult male *Drosophila* has a pair of testes, each being a blind-ended tube coiled around a seminal vesicle (**Figure 12.13**). The male GSC niche is located at the apical tip of the testis, consisting of a small closely packed dome-shaped arrangement of stromal cells (approximately 10–15 cells in total) termed the **hub** that protrudes into the testis. Surrounding the hub cells, which are equivalent to the cap cells in the ovary, are 6–12 GSC, each surrounded by somatic cyst stem cells (CySC) that give rise to the somatic cyst cells that nurture the developing germ cells called gonialblasts. Gonialblasts are produced by asymmetric division of GSC and are surrounded by two somatic cyst cells. The gonialblasts undergo four rounds of divisions to produce a cluster of around 16 spermatogonial cells, each remaining connected to one through the intercellular bridges. These spermatogonial cells later progress through the

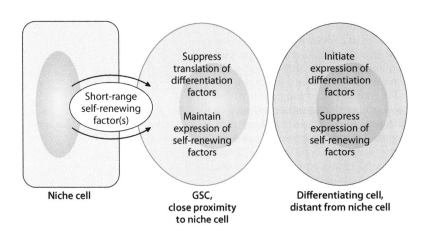

**Figure 12.14 General features of stem cell regulation in *Drosophila* gonads.** Maintenance of the GSC within *Drosophila* gonads is tightly controlled by means of short-range self-renewing factors emanating from the niche cell. The further away the GSC is from these signals, the more likely they are to enter a program of differentiation, as illustrated here in this simplified schematic.

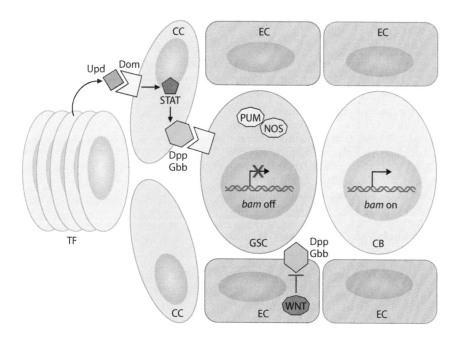

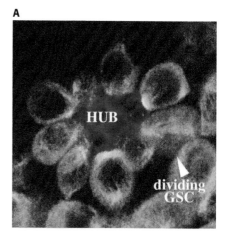

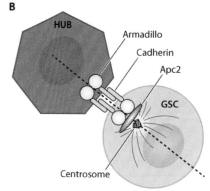

Figure 12.15 **Key signaling pathways regulating germline stem cells in the *Drosophila* ovary.** The regulatory mechanisms governing the maintenance of self-renewing germline stem cells (GSC) in the *Drosophila* ovary have been largely identified, providing an insight into the interactions between stem cells and their niche. This schematic diagram illustrates the regulatory mechanisms that influence GSC in the *Drosophila* ovary. Terminal filament cells (TF) produce the growth factor Unpaired (Upd) that interacts with its receptor Domeless (Dom) on the cap cells (CC). This interaction initiates JAK/STAT signaling, resulting in the production of the bone morphogenetic proteins Decapentaplegic (Dpp) and Glass bottom boat (Gbb), which then act on the adjacent GSC to maintain the repression of Bag of marbles (Bam) by Nanos (Nos) and Pumilio (Pum). When GSC move away from the cap cells during differentiation into cystoblasts (CB), they move out of the range of the Dpp and Gbb signals, releasing the repression of Bam and promoting their differentiation. The focus of the BMP signaling and the repression of Bam is further restricted in EC cells through Wnt inhibition.

premeiotic S phase and end spermatogenesis by their development into spermatids, which lose the intercellular bridges and are released following maturity into the seminal vesicle, where they are stored.

Although there are both structural parallels between the female and male GSC niches in *Drosophila*, the molecular mechanisms supporting the way they work have both similarities and differences. For example, the inhibition of translation of existing differentiation-associated RNA transcripts is not required in the same way, and although the factor Bam is likewise involved in controlling differentiation, it does not come into play until much later in the process, around the time when mitosis must stop and meiosis is initiated.

As seen in the ovary, the regulation of asymmetric division of the male GSC is critically dependent on the physical orientation of the stem cell in relation to the niche cell it contacts. Understanding of the details of this particular mechanism undoubtedly has implications in many other stem cell settings given the importance of maintaining tight control of asymmetric division. The mitotic spindle of the *Drosophila* male GSC is set up perpendicular to its interface with the hub cell. As in females, this interaction involves adherens junctions, interactions between E-cadherin at the hub and GSC membranes linking the cells and providing the orientation to the cell division machinery through binding to Armadillo (the *Drosophila* name for β-catenin) and other components such as Apc2 linking to the centrosome (Figure 12.16). When the duplicated centrosomes separate during cell division, the one anchored through Armadillo and E-cadherin stays next to the hub, while the other migrates to the other side of the cell, thereby dictating the orientation of the plane of daughter cell separation.

Figure 12.16 **Asymmetric division of the male germline stem cell is determined by oriented adherens junctions between the hub and the stem cell.** Similar to the *Drosophila* ovary, regulation of asymmetric division of the male germline stem cell (GSC) is dependent upon interactions with surrounding niche cells that regulate the orientation of the mitotic spindle of the GSC. (**A**) This fluorescent confocal microscope image shows the association of GSC (green, transgenically labeled with GFP-tagged α-tubulin) around the hub cells (red, stained with a specific antibody against E-cadherin). The mitotic spindle visible in one GSC (arrow) is oriented away from the center of the hub. (**B**) This schematic diagram illustrates how the asymmetric division of the male GSC is determined by its orientation in relation to the hub cell. The mitotic spindle is set up perpendicular to the GSC's interface with the hub cell. Cadherins bind to Armadillo on both the hub cell and the GSC and are anchored to the centrosome in the GSC through Apc2. When the cell divides, the duplicated centrosome anchored to Armadillo and Cadherins remains within the cell associated with the niche hub cell. (A: From Yamashita M, Fuller M & Jones D. (2005) *J Cell Sci* 118: 665–672 [doi: 10.1242/jcs.01680]. Reproduced with permission from The Company of Biologists Ltd.)

Another feature of male compared to female GSC in their niche is that they are surrounded by the somatic cyst stem cells. The CySC cell bodies are displaced from the hub but maintain contacts with it through cytoplasmic processes that extend between the GSC. CySC, like GSC, orient their mitotic spindles and divide so that one daughter is attached to the hub and maintains stem cell identity and the other daughter is displaced and differentiates as a cyst cell. Two cyst cells interact with the gonialblast daughter derivative of a GSC to form the functional unit of differentiation, the germ line cyst. The cyst cells do not divide again but enclose the developing germ cells as they undergo transient amplification and differentiation throughout the rest of spermatogenesis. Interestingly, the **transit amplifying cells** are not irreversibly committed to differentiation and can break away and reoccupy the GSC niche and revert to being stem cells. Such dedifferentiation of transit amplifying cells may be a mechanism to replace lost stem cells and can occur in other scenarios, including not just the testes of other species such as mammals but also in other stem cell niches like the gut (Chapter 11).

JAK-STAT signaling plays an equally important key role in the male GSC and the supportive CySC to the one it performs in the female GSC niche (**Figure 12.17**). As in the ovary, the Unpaired ligand is produced by a niche hub cell and interacts with the receptor Domeless on the GSC, leading to activation of the *Drosophila* JAK, known as Hopscotch, which in turn phosphorylates the transcription factor STAT. In the case of the GSC, the activation of STAT is not required to mediate their self-renewal but rather is required to maintain their attachment to the hub. In contrast, genetic manipulation of STAT signaling in the CySC has shown that its activation determines the stem cell state of these somatic support cell precursors. So, the combined effect of JAK-STAT signaling in the male is to maintain the specialized niche through the CySC and at the same time ensure that the GSCs remain within that environment. The balance of self-renewal versus differentiation of the GSC itself is via signals received from the CySC. In much the same way as female GSC are influenced by BMP ligands from the cap and escort cells, themselves elicited through JAK-STAT signaling, so the molecule Gbb produced from CySC as the result of JAK-STAT activation acts on the male GSC, although it is not clear if this is an instructive signal for self-renewal or is permissive for the stem cell state, perhaps by repressing the expression of Bam.

So, a very detailed picture has been created from studies on the *Drosophila* gonads of the ways in which a few molecular pathways can be utilized in the GSC and their adjacent niche components, sometimes in subtly different stem cell-specific combinations, to bring about stem cell maintenance and differentiation. As we will see in a moment, this picture can be of relevance for mammalian male GSC, but more importantly provides probably the most complete view so far of the fundamental components of stem cell regulation that can be a good starting point when thinking about any of the other adult stem cells described throughout this book.

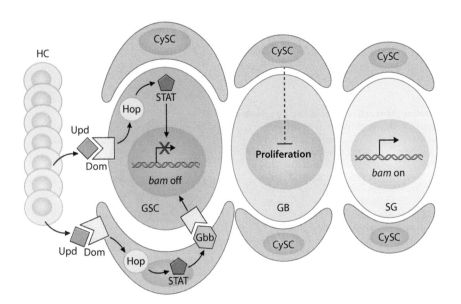

Figure 12.17 **Key signaling pathways regulating germline stem cells in the *Drosophila* testis.** Schematic diagram illustrating the regulatory mechanisms that influence germline stem cells (GSC) in the *Drosophila* testis. Hub cells (HC) produce the growth factor Unpaired (Upd) that interacts with its receptor Domeless (Dom) on the GSC and the somatic cyst stem cells (CySC). This interaction initiates JAK/STAT signaling, which in the GSC leads to repression of Bag of marbles (Bam), whereas in the CySC, it leads to the production of the bone morphogenetic protein Glass bottom boat (Gbb), which then acts on the adjacent GSC to maintain the repression of Bag of marbles (Bam). When the GSC and CySC move away from the apical niche cells, they differentiate into gonialblasts (GB), which eventually activate Bam expression when the spermatogonial cells (SG) are no longer under the influence of the Upd signals.

## The mammalian male germ line stem cell niche shows some similarities to that supporting the equivalent cells in *Drosophila*

Compared to the situation in *Drosophila*, we know relatively little about the male GSC and its niche interactions in mammals. The hierarchy of differentiation underlying spermatogenesis has obvious parallels comparing *Drosophila* and mammals, including the formation of intercellular cytoplasmic bridges. However, no equivalent discrete niche has been defined, although the stem cells appear to have a tendency to reside in certain regions of the seminiferous epithelium, near areas of the so-called interstitium that contain branching blood vessels, and they are influenced by factors secreted by adjacent somatic cells, including Sertoli, myoid, and Leydig cells (Figure 12.18). Since the discovery of the ID4$^+$ SSC in the mouse, which is thought to represent the ultimate SSC population, the precise definition of the male mammalian germ stem cell niche has changed. The ID4$^+$ SSC localize outside of the areas originally thought to be the SSC niche, raising the possibility that there are in fact two distinct stem cell niches in the testes. The thinking behind this is that the ID4$^+$ SSC sits in its own niche and produces transitory SSC that then migrate to another niche where A$_s$ SSC reside, but the precise reason for a second distinct niche is as yet unclear.

Genetic studies suggest that both transcriptional and translational regulation is important for GSC self-renewal in mice, and although a few candidate regulators operating in the stem cell have been identified (e.g., the Plzf transcription factor and Nanos-2 translational regulator), little is known about how these work together or the nature of the downstream targets. Although the GSC niche is not so easily defined, some detail is emerging about the way in which it supports the stem cells. For example, the growth factor GDNF is produced from Sertoli cells and in binding to its receptor and co-receptors GFRα1 and c-Ret is able to control GSC self-renewal (Figure 12.19). GDNF is also produced by the peritubular myoid cells together with colony-stimulating factor 1 (CSF1), which has also been shown to enhance SSC self-renewal.

Unlike their insect equivalents, mouse male GSC are highly mobile, live imaging of GFP-labeled stem cells showing that they move extensively across the basal lamina. This may explain the absence of an obvious niche and may be an adaptation to allow stem cells to respond to changes in the density of production of germ cell production throughout the testis. Once in the more mature state of interconnected spermatogonia, mouse and *Drosophila* seem to be similar in that cells in the transient amplifying stage can revert to stem cells, a phenomenon in mouse that possibly reflects the observed expression of self-renewal related regulators such as Plzf and Nanos-2 across the broad range of spermatogonial cells. Putting together the mobility

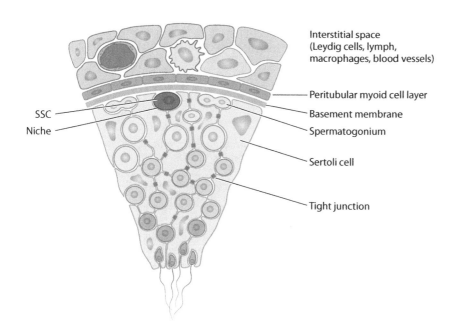

Interstitial space
(Leydig cells, lymph, macrophages, blood vessels)

Peritubular myoid cell layer
Basement membrane
Spermatogonium

Sertoli cell

Tight junction

SSC
Niche

**Figure 12.18 Mouse male germline stem cell niche.** The hierarchical organization of spermatogenesis within the mouse testes is illustrated in the schematic diagram. The precise mechanisms governing the mammalian male germline stem cell (GSC) are not as well understood as those operating in *Drosophila*. No discrete niche has been formerly identified in mammals, but GSC, also known as A$_{single}$ (A$_s$) spermatogonia or spermatogonial stem cells (SSC, red), reside in close association with the basement membrane of the seminiferous tubule. Other developing spermatogonial cells (blue) exist as cysts connected by cytoplasmic bridges and are enveloped by somatic Sertoli cells (beige), which are linked through tight junctions. As spermatogonia develop, they move toward the lumen of the seminiferous tubule. Although not precisely defined, the mammalian male GSC niche also has possible contributions from myoid cells (brown) and cells of the interstitial space, including Leydig cells (pink) and blood vessels.

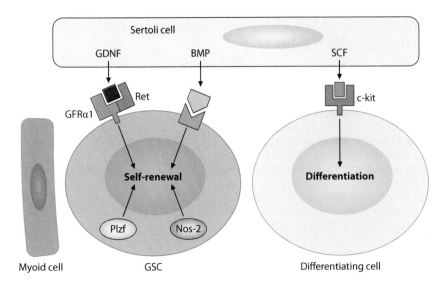

Figure 12.19 **Key signaling pathways regulating germline stem cells in the mammalian testis.** Despite the lack of a precisely defined niche, a few candidate regulators of male germline stem cells (GSC) in mice have been proposed. GSC within the adult male testes reside in close proximity to both myoid cells and Sertoli cells. Self-renewal of the GSC is thought to involve regulation by Plzf and Nanos2 (Nos-2). Sertoli cells surrounding the GSC produce glial cell-derived neurotrophic factor (GDNF), which binds to its receptor GFRα1 and co-receptor Ret on the surface of the GSC and is known to regulate self-renewal. The role of bone morphogenetic protein (BMP) produced by Sertoli cells is not yet known. As GSC differentiate, they up regulate the expression of c-Kit on their surface, enabling them to respond to the stem cell factor (SCF) produced by the Sertoli cells as they move toward the lumen of the seminiferous tubule.

of mouse GSC and this capacity to revert from the transient amplifying stage might mean that the elegant mechanisms controlling asymmetric division, which are certainly likely to apply in some higher organism adult stem cell niches, are not actually necessary in this circumstance.

## 12.3  SUMMARY

In some ways, this chapter represents a cornerstone of the book as a whole in that the study of germ line stem cells has provided many of the paradigms of stem cell biology. We could be so bold as to say that the founding work of Charles Phillipe Leblond on sperm was instrumental in establishing the basic principles upon which we still rely. Certainly, his approach is a shining example of hypothesis-driven science. The more recent controversy associated with the continuing debate about female germ line stem cells, and the atmosphere of claim and counter-claim that this creates, is also a wonderful reflection of the process of scientific discovery and the competitive nature of those involved, who are often dedicating their reputations and lives to what they believe in so strongly. Perhaps better than any other **tissue stem cell**, with the possible exception of hematopoietic stem cells (Chapter 9), we understand the precise developmental origins of germ line stem cells, have a picture from comparisons of diverse model systems how evolution has shaped their form and function, and have some very detailed definitions of the cell-cell interactions and molecular mechanisms that dictate the basic features of self-renewal and commitment that define a stem cell. Last, but not least, our growing understanding of mammalian male germline stem cells (and perhaps the female equivalent if their proponents are right) will be invaluable in future regenerative therapies for various genetic and acquired infertility problems (Chapter 17).

## KEY POINTS

- Specialized stem cells enable the production of germ cells (gametes) in adults.

- Males of most, if not all, animal species that reproduce sexually possess germ stem cells (GSC) in their gonads.

- Female GSC may be a specialized adaptation in a limited number of species, most animals relying instead on predetermined precursors of oocytes being laid down during embryogenesis.

- Male and female germ cells originate from primordial germ cells (PGC), which arise early in development by reprogramming cells to a pluripotent state, followed by their migration from their extra-embryonic location toward the developing gonads.

- Embryogenesis is characterized by a number of transient cell types that have stem cell-like properties, especially if placed in cell culture.

- Male GSC, also known as spermatogonial stem cells, give rise to a hierarchy of differentiation toward mature sperm, which includes a phase of transient amplification of interconnected cells in the form of a cyst. Individual cells within the cyst can revert to the stem cell state if required.

- Female GSC have been defined in a few species, including *Drosophila*, *C. elegans*, and certain fish, but their existence in other species, including mammals, is controversial.

- Studies on the *Drosophila* gonads and their respective GSC have described in great detail the molecular mechanisms that control stem cell self-renewal and differentiation.

- Basic principles for GSC regulation deduced from studies on *Drosophila* include the importance of spatial restriction and direct attachment to the niche cells, oriented division determining asymmetric division and daughter cell fate, and exposure to and progression through gradients of signaling growth factors produced by niche cells that shift the balance of regulation from promotion of self-renewal and inhibition of differentiation toward enhancement of differentiation.

- The male GSC niche in mammals appears to be less defined because of the different requirements for maintenance of gamete production within the context of the testis.

## FURTHER READING

De Felici, M. (2009) Primordial germ cell biology at the beginning of the XXI century. *Int. J. Dev. Biol.*, 53: 891–894.

Clermont, Y. and Leblond, C.P. (1953) Renewal of spermatogonia in the rat testis. *Am. J. Anat.*, 93: 475–502.

Evron, A., Blumenfeld, Z. (2013) Ovarian stem cells—The pros and cons. *Clin Med Insights Reprod Health*, 7: 43–47.

Spradling, A., Fuller, M.T., Braun, R.E. and Yoshida, S. (2011) Germline stem cells. *Cold Spring Harb Perspect Biol.*, 3: a002642.

White, Y.A., Woods, D.C., Takai, Y., Ishihara, O., Seki, H. and Tilly, J.L. (2012) Oocyte formation by mitotically active germ cells purified from ovaries of reproductive-age women. *Nat. Med.*, 8: 413–421.

# Stem cells in the cardiovascular system

13

In order for multicellular organisms to function, their cells need water, oxygen, and nutrients and require the removal of waste products of metabolism, including carbon dioxide. Throughout the evolutionary tree, it is evident that an increasing complexity in body structure and function demands an increasingly complex mechanism to sustain the transport of these essentials of life. In the simplest of multicellular animals, such as the sponges, this cycle of uptake and expulsion is adequately achieved through diffusion. However, as complexity and size increased during evolution, it soon became necessary for animals to develop a means to enhance passive diffusion. The solution, which is seen in complex invertebrates such as squid and is a hallmark characteristic of vertebrates, is a circulatory system that can move nutrients and waste from sites where they are encountered to where they need to be.

## 13.1  CIRCULATORY SYSTEMS IN COMPLEX ANIMALS

Just like an air conditioning system, the circulatory systems of animals rely on pipes and a pump. Analogous to a radiator or heat exchanger, the interface where nutrient and waste exchange occurs must be sufficiently permeable and large enough to maximize the rate of exchange. In animals, these uptake and release interfaces are the gills or lungs for gases and the gastrointestinal tract and urinary system for nutrients and their ultimate waste products. The precise arrangement and complexity of the circulatory plumbing systems, especially the pump, or heart, varies from relatively simple in squid (which actually has three hearts!) and fish, to the four-chambered structure seen in advanced reptiles, birds, and mammals (Box 13.1).

The magnitude of what the circulatory ("cardiovascular") system must perform throughout life is astonishing. A human heart has to beat consistently and effectively on the order of three billion times throughout our lifetime, shifting about 30 million gallons of blood, and if the flow is blocked for more than a few minutes, we die. What plumbing system in our homes could work that well? This vulnerability to failure is reflected in many life-threatening diseases of the heart and vasculature, especially if the function of the heart is partially or completely compromised when its own blood supply is shut off due to a blockage in the cardiac arteries. Such "myocardial infarction," or MI, is one of the leading causes of mortality in the western world, either as a result of its immediate consequences or the longer-term failing of the heart due to the loss of cardiomyocytes and their replacement by fibroblastic scar tissue, leading to so-called congestive heart failure. World Health Organization (WHO) figures reveal that globally, over 17 million people die annually from cardiovascular diseases.

The fact that the cardiovascular system does function so well most of the time and throughout life for the majority of people (or animals) suggests that there must be a degree of continual repair, presumably involving stem cells, although the potential for regeneration following damage, as is seen in tissues with a rapid cell turnover, is less clear. Unlike tissues such as the blood, gut, and skin (Chapters 9, 11, and 15), the rate of turnover of cells in the cardiovascular system is very low, so much so that evidence is only now emerging that homeostatic replacement of cells actually

Complex solutions became necessary once organisms evolved to the point where simple diffusion was not sufficient for the cycling of gases, nutrients, and waste through the body. Vertebrates developed closed circulatory systems that share common essential features, including (i) a set of vessels to distribute the fluid transport medium (i.e., blood), (ii) a pump, (iii) organs in which the blood exchanges what it is carrying (i.e., gills/lungs and intestine to pick up oxygen and nutrients and gills/lungs and kidneys to release gases and waste), and (iv) a network of fine capillaries at the exchange interfaces. In vertebrates such as fish, these features are incorporated into a single loop (Figure 1), but this is limited because the capillary

beds present considerable resistance, meaning that the force imparted by the heart is dissipated after blood passes through the gills, limiting the pressure to push oxygenated blood around the tissues and hence reducing the animal's metabolic capacity. The solution in higher vertebrates, such as reptiles, birds, and mammals, has been to evolve two circuits: one through the lungs returning to the heart and the other serving the rest of the body. The way in which the heart achieves this separation became progressively more sophisticated during evolution, culminating in the four-chambered structure of mammals, in which the two circuits are completely separated (Figure 2).

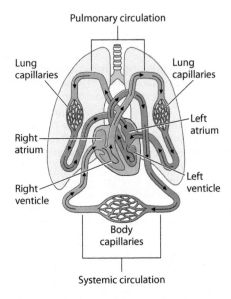

Box 13.1 Figure 1 **Circulatory systems in fish.** The circulatory system of fish is a simple single loop consisting of a two-chambered heart. Blood passes through the heart once on each circuit, meaning the fish heart must generate enough driving pressure for both passage through the gill capillaries and the body capillary beds. The direction of blood flow is indicated by the arrows; deoxygenated blood (blue) passes through the heart toward the gills, where it becomes oxygenated (red) and then passes to the systemic circulation, where gaseous exchange occurs in the tissue capillary networks.

Box 13.1 Figure 2 **Circulatory systems in mammals.** Mammalian circulatory systems are more complicated than fish, as they have a double-looped system. Mammals have a four-chambered heart, ensuring segregation of blood into the pulmonary and systemic circulatory systems. Deoxygenated blood (blue) flows from the heart to the lungs, where it becomes oxygenated (red), and is then pumped back to the heart to enter into the systemic circulation, where gaseous exchange occurs at the capillary interface before returning back to the heart as deoxygenated blood ready to re-enter the pulmonary circulation.

occurs. Perhaps reflecting the low rate of cell turnover, the identification of the stem cells that support cell replacement has been challenging and controversial to say the least, and there still remains doubt about their identity in both the vasculature and the heart.

In this chapter, we will focus separately on the stem cells that constitute the vasculature and the heart, although there is some overlap in that the cardiac endothelium is contiguous with that in vessels and both tissues require replenishment of endothelial cells and smooth muscle. There appear to be significant differences in the regenerative capacity of vertebrate hearts, those species that arose earlier in evolution, such as fish, being able to replace large portions of damaged heart (Chapter 7). This capability is not seen in adult mammals, including humans, but it suggests that it may be possible to achieve regeneration in humans if only the "missing" components could be initiated in a therapeutic context.

## 13.2   STEM CELLS IN THE VASCULATURE

The circulatory system in vertebrates is composed of a linked network of vessels that vary in terms of their dimension and structure of the vessel wall (Box 13.2). In essence, vessels diminish in size the further they are from the heart relative to the exchange interface in the capillary beds (e.g., in the lungs and gut), and the thickness of the wall reflects the pressure experienced and the need to make exchange between the blood and the tissues. Arteries, receiving the outflow from the heart, have the thickest walls, while capillaries within tissues have a thin, often single-cell layered wall. A basic artery or vein is made up of layers, which from the lumen outward include the endothelium, a layer of smooth muscle cells, and an outer, elastic layer.

### Vessels of the circulatory system require continual cell replenishment

Throughout life, blood vessels are subjected to damaging stresses that can compromise their structure and functional integrity, and therefore a capacity to repair is essential. Although the key component cells, such as endothelial cells (EC) and smooth muscle

### BOX 13.2 BACKGROUND: VESSELS THAT CONSTITUTE THE VASCULAR SYSTEM

The vascular system is composed of a complex network of blood vessels that function in the transport of blood around the body in order to exchange water, nutrients, and gases to and from the major organs. The three types of blood vessel, namely arteries, capillaries, and veins, are involved in blood transport away from the heart, exchange between blood and tissue, and transport back toward the heart, respectively (Figure 1). Arteries and veins are made up of three layers of tissue. The outer layer (tunica adventitia) is the thickest layer in veins and is

made up of connective tissue and contains nerves. The middle layer (tunica media) is the thickest in arteries and is rich in vascular smooth muscle cells and surrounded by elastic fibers arranged radially. The inner layer (tunica intima) is the thinnest layer and consists of a single layer of squamous endothelial cells that makes the interface between the vessel and the blood itself. Capillaries principally consist of just a layer of endothelial cells, allowing efficient gaseous and nutrient exchange with the surrounding tissue.

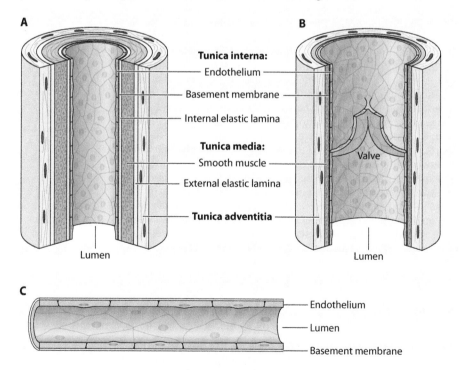

Box 13.2 Figure 1 **Structure of vertebrate blood vessels.** Cross section through an artery (**A**), a vein (**B**), and a capillary (**C**). The principal layers of the tunica intima, the tunica media, and the tunica adventitia are highlighted.

cells (SMC), could conceivably achieve this by re-entering the cell cycle, the picture that has emerged over recent years is one that involves the persistent presence of multiple **progenitors** distributed throughout the vessel walls. These progenitors include those restricted to the generation of EC or SMC, as well as cells with multipotentiality with respect to vessel components. The growth of new vessels out of the existing vasculature through the process known as "angiogenesis" continues throughout life and adds to the need for progenitor cells. Angiogenesis is driven by the demand for an increased supply of oxygen, for example, in response to exercise or in diseases such as cancer for which a rich supply of blood vessels is required to fuel their malignant nature. Many tumors promote the process themselves to sustain their high metabolic demands.

Although the basic scenario for progenitors in vessels is becoming clearer, there remains considerable uncertainty about the precise nature of these progenitors, the degree of overlap between cells defined by different researchers utilizing distinct methods of isolation and assay, and the specific locations in which individual progenitors reside. For the purposes of this chapter, we will divide our discussion of vessel wall progenitors into three categories, namely **endothelial progenitor cells** (EPC), smooth muscle progenitor cells (SPC), and **multipotent** vascular stem cells (VSC) that encompass so-called **pericytes** and **mesenchymal stem cells** (MSC) (Chapter 10).

## The identity and location of endothelial progenitor cells is contentious

The most active investigation of vessel wall progenitors has without a doubt been around the nature and origin of EPC and, as so often is the case in the pursuit of **adult stem cells,** has been quite controversial. The most contentious aspect regarding EPC relates to their origin; essentially, the debate has been whether they are resident in the vessel wall or derived from a distant site, of which the bone marrow has been a much talked about potential source.

The history of investigation into the progenitor cells that support vessel endothelia spans two decades, and yet even now, there remain many questions about their true identity and location. The field has been dominated to a large extent by the pursuit of EPC in the human circulation, starting with ideas developed from seminal research conducted in 1997 by Takayuki Asahara, who at the time was working in the Tufts University School of Medicine. However, even older experiments involving the radiolabeling of replicating cells had pointed to the possibility of resident progenitor cells in vessel walls.

Studies aimed at the identification of circulating EPC have been hampered by the relative dearth of surface markers that can be used to definitively identify and purify the cells, a fact that has been compounded by the overlap with hematopoietic cells in expression of many of the most useful markers, including **cluster of differentiation 31** (CD31) and CD34 (Table 13.1). The consequences of shared expression of several markers between EPC and hematopoietic cells has been further complicated because some of the assays used to test for endothelial cell activity have been found to be influenced by paracrine effects elicited by hematopoietic cells. Isolation procedures for putative EPC contained in the circulation have relied in part on the selection for adherent cells. One particular problem has been that this method can also select for monocytes that in culture conditions suitable for EPC, including vascular endothelial growth factor (VEGF), express surface proteins such as VEGF receptors CD31 and von Willebrand Factor (vWF), which are also characteristic of endothelial cells. What on the face of it might seem like a fairly solid assay of EPC, namely the *in vivo* repair of vessel endothelium injury (often brought about by physical abrasion or ischemic damage in rodent models), can give a positive result when falsely assigned hematopoietic cells are injected because of the latter's ability to secrete factors that assist vessel regrowth or through the process of angiogenesis, possibly by aiding recruitment of circulating or resident EPC. For this reason, such cells are often termed pro-angiogenic cells (PAC).

## Assays of proliferative potential may be a better method to isolate progenitors

What seems to be a more reliable way to isolate EPC from blood relies upon their proliferative potential, which would be expected to be considerable for any genuine stem cell. The method involves growing cells on collagen-coated culture dishes (Figure 13.1). After 2–3 weeks, colonies of cells appear, leading to them being termed "late outgrowth cells," which can be serially plated to form more colonies and will form capillary-like structures *in vitro* and can promote vascular recovery in

## TABLE 13.1 CHARACTERISTICS OF ENDOTHELIAL PROGENITOR CELLS

| Phenotypic or functional characteristic | Proangiogenic hematopoietic cells | Endothelial progenitors |
|---|:---:|:---:|
| CD31 | + | + |
| CD34 | +/− | +/− |
| CD45 | +/− | − |
| CD115 | + | − |
| CD133 | + | − |
| CD146 | +/− | + |
| VEGFR2 | + | + |
| ALDH^bright | + | + |
| acLDL uptake | + | + |
| Clonal proliferation and replating ability | − | + |
| *In vivo* vessel formation | − | + |
| Homing to ischemic sites *in vivo* | + | + |
| Paracrine effect on angiogenesis | + | + |

*Source:* Adapted from Basile & Yoder (2014) Circulating and tissue resident endothelial progenitor cells. *J Cell Physiol.,* 229: 10–16. Surface marker (gray shading) and functional characteristics (blue shading) of proangiogenic hematopoietic cells compared to endothelial progenitors.

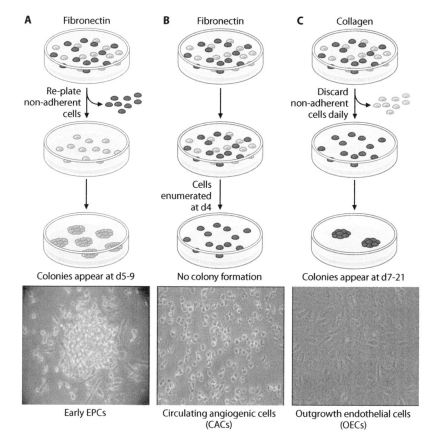

**A** Fibronectin — Re-plate non-adherent cells — Colonies appear at d5-9 — Early EPCs

**B** Fibronectin — Cells enumerated at d4 — No colony formation — Circulating angiogenic cells (CACs)

**C** Collagen — Discard non-adherent cells daily — Colonies appear at d7-21 — Outgrowth endothelial cells (OECs)

Figure 13.1 **Endothelial progenitor cell candidates isolated from blood.** The isolation and culture of endothelial progenitor cells (EPC) is highly dependent on the conditions employed. The diagram illustrates three methods that have been used to isolate EPC from the blood. (**A**) Non-adherent mononuclear cells are placed on fibronectin-coated plates and form colonies of early EPC after 5–9 day. (**B**) Mononuclear cells are plated on fibronectin-coated plates for 4 days and then non-adherent cells are removed, leaving behind the adherent circulating angiogenic cells (CAC) that lack colony forming ability. (**C**) Mononuclear cells are plated on collagen-coated plates, non-adherent cells are removed daily, and the adherent cells form colonies after 7–21 days with a cobblestone morphology and are referred to as outgrowth endothelial cells (OEC). (Line art: From Prater D, Case J, Ingram D et al. (2007) *Leukemia*, 21: 1141–1149 [doi: 10.1038/sj/leu.2404676]. Adapted by permission from Springer Nature. Part A photo: Used with permission from Dr. Sahena Haque. Part B photo: Used with permission from Dr. John Reynolds. Part C photo: From Williamson K, Stringer S & Alexander M. (2012) *Front Physiol* 3:1–7 [doi:10.3389/fphys.2012.00030]. Used with permission.)

### TABLE 13.2  VASCULAR WALL RESIDENT ENDOTHELIAL PROGENITORS

| Vessel source | Vessel wall location | EPC phenotype |
|---|---|---|
| Human thoracic artery, isolated by arterial ring assay | Between media and adventitia | CD34+VEGFR2+Tie2+, differentiate to EC |
| Human coronary arteries, isolated by immune-selection | Intima, media and adventitia | VEGFR2+Kit+ with self-renewal capacity, differentiate to EC and SMC |
| Mouse lung vasculature, isolated by immune-selection | – | CD31+CD105+Sca-1+Kit+, self-renewing and able to form vessels *in vivo* |
| Mouse limb vasculature, isolated by immune-selection | Intima | CD31+Sca-1+CD133+, clonogenic and proliferative, able to form vessels *in vivo* |

*Source:* Adapted from Psaltis P & Simari R. (2015) *Circ Res* 116:1392–1412 [doi: 10.1161/CIRCRESAHA.116.305368].

*Note:* The table summarizes the location and surface antigen phenotype of EPC isolated from a variety of vessels.

immunodeficient mouse models of ischemic vessel damage in the heart, limbs, and eyes. Although some of the same similarities to hematopoietic cells are seen in these apparently genuine EPC, they are quite distinct in other respects from hematopoietic PAC. The origin of the late outgrowth cells remains obscure, and although rare cells, they can be readily isolated from the CD34+CD133−CD146+ fraction of human umbilical cord blood and peripheral blood mononuclear cells. They are indistinguishable from mature human umbilical vein endothelial cells (HUVEC) by morphology, surface antigen expression, immunohistochemistry, proliferation, and functional assessments. Recent studies comparing the late outgrowth EPC with cells isolated directly from bone marrow using the same methodology suggests that they do not arise from this source and challenge the concept of a bone marrow-derived circulating precursor for endothelial cells.

As already mentioned, early evidence suggested that cell division can occur in vessels. Hence, in the 1970s, experiments involving the labeling of rat aortic walls with tritiated thymidine revealed discrete foci of proliferating cells in the endothelial intima, consistent with the presence of EPC. Various investigations have shown that the turnover of cells in normal circumstances is low, but increased in response to injury. A wide range of approaches have been used to try to isolate and characterize vascular wall EPC, and although the results are varied, there are some trends to the location from which the cells derive and their surface marker phenotype (Table 13.2). One might expect that tissues capable of undergoing physiological angiogenesis might be rich in EPC and exhibit clear signs of cell division. One such tissue is the lung, and investigation of the pulmonary microvasculature in the mouse lung revealed the presence of a CD31+CD105+Sca1+c-Kit+ cell able to give rise to vessels *in vivo* and with many features expected of an adult stem cell. These cells were identified as a small subpopulation within EC capable of undergoing clonal expansion *in vitro*, while other EC populations demonstrated a very limited proliferative capacity. When transplanted into another mouse, a single cell with this phenotype was shown to be able to generate functional blood vessels that connected to host circulation (Figure 13.2).

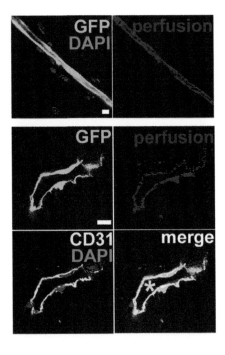

Figure 13.2  **c-Kit+ vascular endothelial progenitor cell.** Confocal microscopy image showing the formation of GFP+ blood vessels (green) 14 days after transplantation of a single c-Kit+ colony-forming endothelial progenitor cell derived from mice constitutively expressing GFP from a β-actin-driven transgene. To show that this vessel was integrated into the host vasculature, mice were perfused with fluorescent microspheres (red) that label the endothelial cells in functional blood vessels. The endothelial cells were also stained for expression of the characteristic marker CD31 (yellow) and DAPI (blue) for nuclei staining. Scale bars = 100 μm. * highlights the blood vessel lumen. (From Fang S, Wei J, Pentinmikko N et al. (2012) *PLOS Biol* 10:10e1001407 [doi: 10.1371/journal.pbio.1001407]. Published under the terms of the Creative Commons Attribution License http://creativecommons.org/licenses/.)

## Multipotent vascular stem cells are thought to be the source of smooth muscle cells

SMC provide vessels with mechanical stability and regulate vascular tone and blood flow through their ability to contract. It has been dogma for half a century that, in response to vascular injury, mature contractile SMC in the tunica media adjacent to the vascular lesion dedifferentiate into proliferative or "synthetic" SMC with both proliferative and migratory properties (Figure 13.3). Such an interchange of SMC phenotypes has been proposed to underlie the pathological process of restenosis, that is, recurrence of abnormal narrowing of an artery or valve after corrective surgery. More recently, however, an alternative source of synthetic SMC has entered the picture, namely resident multipotent vascular stem cells, either restricted to SMC **differentiation** (smooth muscle progenitor cells) or multipotent and able to produce other cell components of the vessel wall (Figure 13.4). However, the relative contributions of mature SMC dedifferentiation and resident VSC differentiation to the generation of synthetic SMC in the vasculature remain unknown. In spite of this uncertainty, similar approaches to those used to elucidate the origin of EPC in vessels, including cell grafting and **lineage tracing**, have reached very much the same sorts of conclusions regarding the existence of SPC. As for EPC, there is not yet complete clarity about the range and locations of cell types with SPC characteristics, although expression of Sca-1$^+$ appears to be a common feature in mice. Progenitor cells with differentiation potential restricted to smooth muscle cells, so true SPC, have been identified in both the medial and adventitial regions of vessel walls, but in addition, some multipotential cells, including VSC and pericytes or MSC, can certainly also serve as sources for SMC.

## Pericytes surround vessels and are most likely specialized mesenchymal stem cells

Although first described in the late 1800s and named as such in 1920, pericytes remain somewhat enigmatic, especially in terms of their developmental origins and the way that they relate to MSC. Pericytes surround vascular EC and, being contractile, share some of the roles of SMC. Location and morphology are good indicators of pericyte identity (Chapter 10), but there is a significant overlap with MSC in their expression of surface markers (Table 13.3). Cells that from the outset have been defined as MSC have also been found associated with vessels, but unlike pericytes, their location is within the wall, especially in the adventitial region. When pericytes are dissociated from their vascular location, they adopt more MSC characteristics such as the capacity to differentiate into adipocytes, chondrocytes, and osteocytes (Chapter 10) and the ability to influence

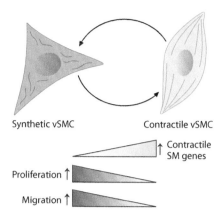

**Figure 13.3 Smooth muscle cells.**
Contractile vascular smooth muscle cells (vSMC) can switch to a synthetic vSMC phenotype in response to various stimuli. This switch results in the reduced expression of contractile smooth muscle (SM) genes with a corresponding increase in the degree of proliferation and migration in the synthetic vSMC. (From Davis-Dusenbery B, Wu C & Hata A. (2011) *ATVB*, 31: 2370–2377 [doi: 10.1161/ATVBAHA.111.226670]. Used with permission from Wolters Kluwer Health and American Heart Association.)

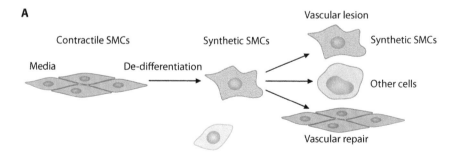

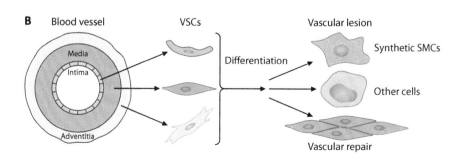

**Figure 13.4 Potential sources of smooth muscle cells in vessels.** Diagrams depicting the potential sources of smooth muscle cells (SMC) within blood vessels. (**A**) Dedifferentiation: SMC within the tunica media dedifferentiate in response to injury into synthetic SMC and then further differentiate toward contractile SMC, among other cell types. (**B**) Vascular stem cells (VSC) are located within the tunica intima, tunica media, and tunica adventitia. Following vascular injury, the SMC within the media die and the VSC differentiate into SMC or other cell types to initiate vascular repair. (From Leach D, Mathis B, Qu C et al. (2014) *Austin J Anat* 1:1007. Published under the terms of the Creative Commons Attribution License http://creativecommons.org/licenses/)

**TABLE 13.3 COMPARISON OF PERICYTE AND MSC FEATURES**

| Phenotypic or functional characteristic | Pericytes | MSC |
|---|---|---|
| Location | Perivascular | Various, including perivascular |
| CD44 | + | + |
| CD73 | + | + |
| CD90 | + | + |
| CD105 | + | + |
| CD146 | + | +/− |
| NG2 | + | − |
| αSMA | + | − |
| Multilineage differentiation | Demonstrated *in vitro* | + |
| Immune regulation | Demonstrated *in vitro* | + |

*Note:* Summary of surface antigen phenotype and functional characteristics of pericytes compared to mesenchymal stem cells (MSC).

immune system processes (Chapter 16). These behaviors beg the question whether this is an artifact of experimental manipulation or a reflection of a hierarchical relationship.

## The vessel wall acts as a stem cell niche

In spite of the continuing doubts about the precise definition of individual vascular stem/progenitor cell phenotypes and the degree of overlap, the current picture of the vessel wall is not dissimilar to other scenarios involving adult stem cells (Chapter 6). As in other tissues, **quiescent** as yet uncommitted cells reside in optimal **niches** awaiting the signal to mobilize to come to the aid of compromised areas either nearby or perhaps at some distance. The most important area of the vessel wall serving as a niche seems to be the inner adventitia, adjacent to the external elastic lamina, where progenitor cells are maintained by a diverse range of cells including fibroblasts, adipocytes, inflammatory cells, and neurons, many of which may act as injury sensors (Figure 13.5).

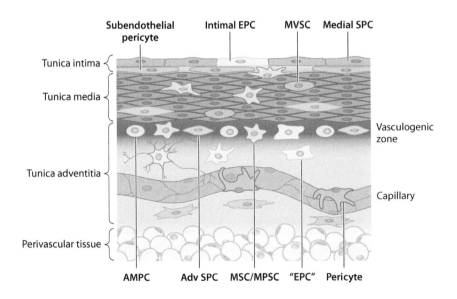

Figure 13.5 **Vessel wall stem cells.** A blood vessel wall is a complex structure that is supported by a number of different types of resident stem cells. The adventitial region is enriched in distinct yet overlapping populations, including adventitial smooth muscle progenitors (Adv SPC), adventitial macrophage progenitor cells (AMPC) in mice, and mesenchymal stromal cells (MSC) and multipotent stem cells (MPSC) in humans. Other cell types present in the vessel wall are indicated: pericytes, subendothelial pericytes, intimal endothelial progenitor cells (EPC), multipotent vascular stem cell (MVSC), and side population cells (SP). (From Psaltis P & Simari R. (2015) *Circ Res* 116:1392–1412 [doi: 10.1161/CIRCRESAHA.116.305368]. Used with permission from Wolters Kluwer Health and American Heart Association.)

## 13.3   STEM CELLS IN THE HEART

The long-held dogma surrounding the mammalian heart (Box 13.3) was that it is a terminally differentiated organ with little or no capacity for regeneration and an inability to compensate for loss of cardiomyocytes, resulting in reduced heart function with age and disease. However, two key findings strongly suggested that there is at least a limited capacity for regeneration, although the origins of replacement cells and the general applicability in mammals remain uncertain. These two findings, which may not be linked, were: (i) that there is a continual, albeit low-level, turnover of cardiomyocytes in adults and (ii) that some vertebrates, notably zebrafish and the neonatal mouse, can regenerate cardiac tissue following surgical amputation of the cardiac apex (Chapter 7). Identification of genuine cardiac stem cells has proved to be technically challenging, and the current body of evidence points to only a minor role, if any, in the maintenance of uninjured cardiac muscle and favors instead that cardiomyocyte proliferation contributes the majority of replacement cells throughout life.

### There is a great variation in the observed rates of cell cycle in the cardiomyocyte population

That there is continuous cell cycle activity within the cardiomyocyte population was first made evident by the use of carbon dating, exploiting the spike of atmospheric

---

**BOX 13.3  BACKGROUND: THE MAMMALIAN HEART**

The mammalian heart wall is separated into three layers, namely the innermost endocardium that forms a continuum with the endothelial lining of the blood vessels, the middle myocardium, and the outermost epicardium (Figure 1). The myocardium is the thickest layer of the heart wall and accounts for the contractibility of the heart muscle due to the presence of cardiac muscle cells, also known as cardiomyocytes. Cardiomyocytes make up 99% of the cells within the myocardium, with the remaining cells being modified cardiomyocytes called pacemaker cells. These pacemaker cells have a limited ability to contract but are essential for the production of action potentials that regulate the timing of contractions of the cardiomyocytes. Cardiac muscle is an involuntary striated muscle that exhibits auto-rhythmicity regulated by both the endocrine and nervous systems. In contrast to skeletal muscle (Chapter 10), cardiomyocytes form branch-like structures rather than a linear arrangement.

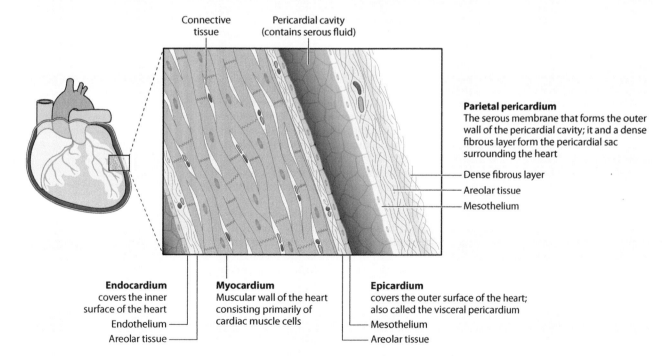

Box 13.3 Figure 1 **The mammalian heart.** Cut-away schematic of the mammalian heart depicting the major veins (blue) and arteries (red) that carry deoxygenated and oxygenated blood, respectively. Inset: cross section through the wall of the heart showing the outer pericardium, the epicardium, the myocardium (muscle layer), and the endocardium (inner lining).

**TABLE 13.4  ESTIMATED RATES OF CARDIOMYOCYTE RENEWAL**

| Deduced annual rate of cardiomyocyte renewal | Species | Method |
|---|---|---|
| 0.5%–1.9% | Human | $^{14}C$, accelerator mass spectrometry |
| 10%–40% | Human | Ki67, phospho-H3, Aurora B, IdU |
| 7%–23% | Human | $^{14}C$, accelerator mass spectrometry |
| 0.04%–4.5% | Human | Phospho-H3 |
| 1.3%–4% | Mouse | BrdU |
| 0.74% | Mouse | $^{15}N$, imaging mass spectrometry |
| 1.09% | Mouse | [$^3$H]thymidine |

*Source:* From Garbern J & Lee R. (2013) *Cell Stem Cell* 12:689–698 [doi: 10.1016/j.stem.2013.05.008]. Copyright 2013 with permission from Elsevier. Comparison of rates of annual cardiomyocyte replacement estimated using a variety of cell labeling strategies.

carbon-14 ($^{14}C$) levels during the 1950s to 1960s as a result of nuclear bomb testing. By measuring the content of nuclear $^{14}C$ in cardiomyocytes and comparing it to the atmospheric $^{14}C$ level, it could be shown that new cardiomyocytes are formed in young adult humans at a rate of about 1.5% per year, declining to just 0.45% at 75 years of age. These rates of division equate to a 45% replacement of cardiomyocytes throughout our lifetime. Although the essence of these findings is not in dispute, there remains some disagreement around the interpretation of the modeling employed and the precise rate of replacement (Table 13.4). Whichever value is correct, these sorts of studies give no insight as to the mechanism of cell replacement, and the question remains: What are the cells that can produce new cardiomyocytes? The possible answers are much the same as for the situation in the vessels of the cardiovascular system; that is, they could arise through the division of existing differentiated myocytes or be derived from progenitors resident in the heart or recruited from elsewhere in the body.

## The hunt for the elusive endogenous cardiac stem cell has been highly contentious

Although dedifferentiation of existing cardiomyocytes is feasible in vertebrates, as observed in the zebrafish heart, and for some time there was a belief that adult stem cells from other tissues, notably the hematopoietic system, could **transdifferentiate** (Chapter 5), the consensus view has become that there exist one or more progenitor populations in the adult mammalian heart. But the word "consensus" hides the fact that the identity of these progenitors is fiercely debated, to the point that some findings have been contested in the courts! What is known for definite is that there is a population of cardiac progenitors in the embryonic mammalian heart that cease their activity once the heart is formed. If these progenitor cells exist in the adult heart but in a dormant state, the thought is that they could be reactivated in order to replenish the damaged heart tissue. Even if endogenous progenitors in the heart only support replacement of 1 in 200 cells per year and are therefore inherently insufficient to mediate clinically significant repair, many believe that by understanding their properties, perhaps cardiac researchers can find a way to boost that number with a drug to a level that could have a meaningful impact for patients. The requirement of putative progenitors in the heart, just as in vessels, is not restricted to the provision of one lineage of cells. It is generally accepted that progenitors are required for at least three lineages, that is, cardiomyocytes, smooth muscle cells, and endothelial cells, and that this may be represented in one discrete cell type.

A number of **cardiac progenitor cell** (CPC) types have been described within the adult mammalian heart, but there is not general agreement on their molecular identity or specific markers with which they can be isolated or indeed whether there are one or more progenitor cell types. A common problem influencing the interpretation of studies on putative stem cell populations is that limitations on available distinguishing markers mean that most work has involved mixed populations of cells, only a small

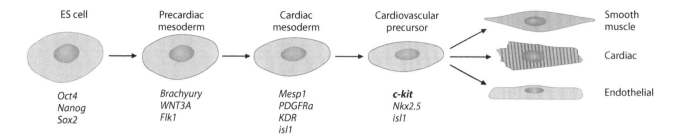

**Figure 13.6 Cardiac progenitor cells derived from ES cells.** Embryonic stem (ES) cells can be differentiated in culture through a number of stages that reflect the normal process leading toward cardiovascular precursor cells. The progression of differentiation is illustrated, highlighting key genes expressed at each stage equivalent to key steps in embryonic development. (From Hesse M, Fleischmann B & Kotlikoff M. (2014) *Stem Cells* 32:1701–1712 [doi: 10.1002/stem.1696]. With permission from Wiley.)

proportion of which actually have the actual properties that are expected of stem cells. The principal markers and phenotypic characteristics that have enabled work on CPC include c-Kit (receptor for the stem cell growth factor SCF, also designated CD117), the stem cell antigen Sca-1, **side population** (SP) behavior, and the ability to form colonies *in vitro* that are capable of differentiating into cardiac lineage cells.

## c-Kit⁺ cells have been proposed to represent the cardiac progenitor cells

Resident heart cells expressing the c-Kit antigen are considered by some to be CPC that are sufficient and necessary to repair damaged myocardium. However, genetic fate mapping experiments are not in agreement with this view, showing minimal contribution of the c-Kit⁺ cells to cardiac myogenesis, and instead suggest a rather more specific role. A clue about what form a CPC might take has come from investigations into the stages of differentiation from an **embryonic stem (ES) cell** (Chapter 3) to the three lineages that constitute the heart. Although this can only be a guide, since cardiac differentiation from mouse ES cells reflects more the embryonic development of the heart, at the stage after **commitment** to the **mesoderm** and before terminal differentiation to cardiomyocytes, smooth muscle cells, and endothelial cells, cardiovascular precursors are represented by a c-Kit⁺ progenitor, which also expresses the cardiogenic transcription factors Nkx2.5 and Isl-1 (Figure 13.6). One of the first descriptions of a putative adult mammalian CPC came in 2003 with the isolation of c-Kit⁺ cells from rodent hearts by the group of Piero Anversa working at Harvard University and the Brigham and Women's Hospital. A series of prominent papers were published by Anversa's group, paving the way even for clinical trials (e.g., the SCIPIO trial—see Chapter 17) utilizing c-Kit⁺ CPC in the treatment of patients following myocardial infarction. These c-Kit⁺ cells were seen to reside among the cardiomyocytes in the myocardium (Figure 13.7), to express characteristic markers of the cardiac lineages, and to have renewal and differentiation potential in culture once isolated. Crucially, it was claimed that these cells were able to regenerate the myocardium and blood vessels following injection into the heart of recipient animals after an experimental myocardial infarction. However, this work has elicited a great deal of skepticism because many other groups have been unable fully to reproduce the findings.

Soon after the first studies were published from Anversa's group, some parallel experiments involving the culture of cells derived from the heart implied the presence of progenitor cells that were capable of expansion, at least under laboratory conditions. The original procedure involved the culture of cells obtained from cardiac biopsies following mincing and enzymatic digestion. Eventually, these cultures develop a monolayer of cells that spontaneously migrate away from the explanted tissue (Figure 13.8), which can then be collected and placed in suspension culture, where they begin to form spherical three-dimensional aggregates. These aggregates are termed **cardiospheres** and contain a heterogeneous population of cells. Cells within cardiospheres include c-Kit⁺ cells and ones with SP behavior, cells expressing the endothelial markers CD31 and CD34, and cells expressing the mesenchymal stem cell markers CD90 and CD105. As cardiospheres develop, they become enriched in c-Kit+ cells at their center, while the outer cells become more differentiated and express sarcomeric proteins. The c-Kit⁺ cardiosphere-derived cells were shown to have cardiogenic potential when injected into injured adult pig hearts, but whether this ability was a direct effect of differentiation toward cardiomyocytes or a result of paracrine activity has not yet been determined.

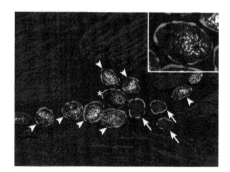

**Figure 13.7 Putative c-Kit+ cardiac progenitor cluster.** Confocal microscopy image from a murine heart section showing a cluster of c-Kit+ cells (green rim), some of which co-express Nkx2.5 (white dots, arrow heads) in their nuclei (blue, propidium iodide stained). One cell (asterisk and inset) also stains positive for α-sarcomeric actin (red) in the cytoplasm. (From Beltrami A, Barlucchi L, Torella D et al. (2003) *Cell* 114: 763–776 [doi: 10.1016/s0092-8674(03)00687-1]. Copyright 2003 with permission from Elsevier.)

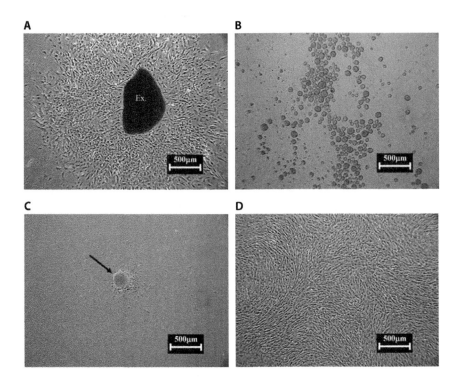

Figure 13.8 **Cardiospheres.** The presence of cardiac progenitor cells in the adult heart can be presumed from the fact that a tissue explant placed in culture results in the outgrowth of cardiospheres. The series of microscope images show an example of the stages in the formation of cardiospheres in culture. (**A**) A cardiac biopsy explant (Ex) cultured on a fibronectin-coated plate forms a monolayer cell outgrowth. (**B**) Monolayer cells are harvested and further plated onto a poly-D-lysine coated plate on which they form floating cardiospheres. (**C**) A cardiosphere (arrow) is isolated and plated on a fibronectin-coated plate to which it adheres and spreads as a monolayer of cardiosphere-derived cells that becomes confluent (**D**) and can be harvested. (From White A, Smith R, Matsushita S et al. (2013) *Eur Heart J* 34: 68–75 [doi: 10.1093/eurheartj/ehr172]. By permission of Oxford University Press on behalf of the European Society of Cardiology.)

As with most work on the potential and fate of cells suspected of being stem cells, the best that can be done experimentally is usually to perform fate-mapping studies. Using this more definitive type of approach, specifically involving **transgenic** mice expressing green fluorescent protein (GFP) under the control of gene regulatory sequences that normally control the expression of the c-Kit gene, it has been found that c-Kit is only transcribed to any significant extent in the adult heart following injury. This tracking approach has also revealed important differences between the resident cardiac c-Kit⁺ cells in the neonatal compared to adult heart. First, the number of c-Kit⁺ cells seen in a mouse heart a few days after birth is massively higher than in the adult (Figure 13.9), and when these cells were isolated on the basis of their GFP expression and placed into culture, only those from the neonatal animals could differentiate toward cardiomyocytes (Figure 13.10). Most strikingly, if the heart was experimentally injured by localized rapid freezing, mimicking a myocardial infarction, then a massive activation and localization of c-Kit⁺ progenitor cells was seen at the site of injury in the neonate, followed soon after by the appearance of GFP⁺ cardiomyocytes (Figure 13.11). Little or no such reaction was seen in the adult heart after a similar injury, and what c-Kit-GFP⁺ cells were apparent followed a path of differentiation toward restricted vascular fates, that is, endothelial and smooth muscle cells (Figure 13.12).

## Sca-1 could be a better marker of cardiac progenitors in mice

Perhaps stimulated by the controversies associated with the putative c-Kit⁺ CPC, many researchers have been looking elsewhere for alternative progenitor cells in the heart. Much of the research progressed using rodent models and predominantly focused

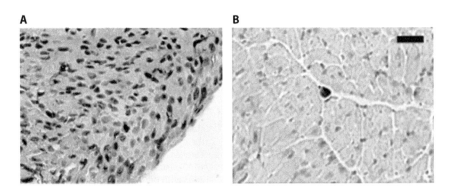

Figure 13.9 **c-Kit+ progenitor cells in neonatal and adult hearts.** Putative cardiac progenitor cells can be identified *in situ* in the mouse heart by virtue of their expression of cell surface c-Kit receptor. Immunohistochemical analysis of GFP protein expression reveals the presence of c-Kit+ cells (brown) in sections of heart from neonatal (**A**) and adult (**B**) c-Kit-GFP mice. Nuclei are stained blue/grey. (From Tallini Y, Greene K, Craven M et al. (2009) *Proc Natl Acad Sci* 106:1808–1813 [doi:10.1073/pnas.0808920106]. Used with permission.)

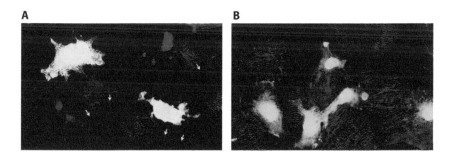

**Figure 13.10 Differentiation of c-Kit+ progenitor cells from neonatal and adult hearts.** Confocal microscopy images of cultures of fetal cardiomyocytes seeded for 7 days together with either c-Kit⁺GFP⁺ cardiac cells from neonate (**A**) or adult (**B**) hearts. The cultures were analyzed for GFP expression (green) and α-actinin (red). GFP+ cardiomyocytes differentiated from the c-Kit⁺ progenitors express mature sarcomeric structure and appear yellow due to the overlapping green and red fluorescence, whereas the adult cells do not give rise to mature cardiomyocytes and appear green. (From Zaruba M, Soonpaa M, Reuter S at al. (2010) *Circulation* 121:1992–2000 [doi:10.116 1CIRCULATIONAHA.109.909093]. Used with permission from Wolters Kluwer Health and American Heart Association.)

on another frequent marker of stem cells, stem cell antigen 1 (Sca-1), which is most commonly used in the identification of **hematopoietic stem cells** in mice (Chapter 9). A small population of Sca-1⁺ cardiac cells were identified in rodent hearts that were negative for the expression of c-Kit and that differentiated toward spontaneously beating cardiomyocytes *in vitro*. These Sca-1⁺ cells were shown to home to injured myocardium following ischemia-reperfusion injury, therefore representing a prime CPC candidate. Unfortunately, Sca-1 is not expressed on human cells and therefore does not reflect a representative human CPC population. However, continuing to work with Sca-1⁺ cells in mice enabled researchers to tease out other features of those cells that exhibit the most promising features consistent with a true role as cardiac progenitors.

As we have seen in the resolution of the relative importance of c-Kit⁺ cells in the neonate compared to the adult heart, fate mapping has given a clearer picture of the ability of Sca-1⁺ cells to contribute to cardiogenesis. In this case, lineage-tracing using **Cre recombinase** driven by Sca-1 gene sequences revealed that Sca-1⁺ cells are able to contribute to cardiac muscle in adults as they age and following induced cardiac injury. Definitive proof of the true CPC nature of Sca-1⁺ cells, or at least a fraction of the population that express this surface marker, came about by combining the presence of Sca-1 with other potential markers of progenitor cells. The side population method (Chapter 2), first used in the identification of hematopoietic stem cells, highlights a subpopulation of Sca-1⁺ cells that represent approximately 1% of the total cardiomyocyte number. These Sca-1⁺ SP cardiac cells are clonogenic; that is, they can be isolated as single cells and then expanded, enabling detailed examination of their properties. In this way, Michael Schneider and his co-workers, working in London's Imperial College, discovered that Sca-1⁺ SP cells obtained from the adult mouse heart are truly tri-potent when grafted into an injured heart, giving rise to cardiomyocytes, smooth muscle cells, and endothelial cells (Figure 13.13). Further characterization of these cells showed that the greatest potential correlated with expression of the platelet-derived growth factor receptor α (PDGFRα) on the surface, while elegant lineage-tracing experiments employing transcription factor gene-driven Cre recombinases together with a fluorescent reporter that is turned on by Cre recombinase were used to map the origin of these Sca-1⁺ CPC, showing a divergence away from cells fated to become purely endothelial progenitors (CD31⁺) (Figure 13.14). Based on single-cell gene expression analysis, the Sca-1⁺PDGFRα⁺ cells appear to be reminiscent of MSC and express transcription factors expected for an early CPC (e.g., Tcf21, Tbx20, GATA4, and Hand2). In contrast, Sca-1⁺ cells that instead express the marker CD31, express a distinct profile of genes, and have EC differentiation potential, even though they have the same ultimate development origin as the Sca-1⁺PDGFRα⁺ cells.

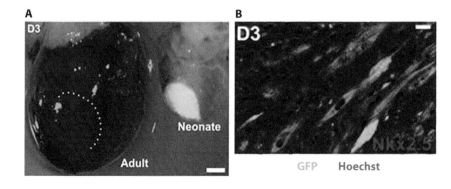

**Figure 13.11 Occurrence of c-Kit+ cells during post injury myogenesis in neonatal and adult hearts.** (**A**) Merged fluorescence and bright field microscopy images 3 days after localized cryoinfarction injury (dotted region) in adult (left) and neonatal (right) hearts. c-Kit⁺GFP⁺ cells are induced to localize to the area of damage in the neonatal heart. (**B**) Confocal image of the cryoinfarcted region of the neonatal heart showing striated c-Kit⁺GFP⁺ myocytes, some of which are Nkx2.5⁺ in their nuclei (red). (From Jesty S, Steffey M, Lee F et al. (2012) *Proc Natl Acad Sci* 109:13380–13385 [doi:10.1073/pnas.1208114109]. Used with permission.)

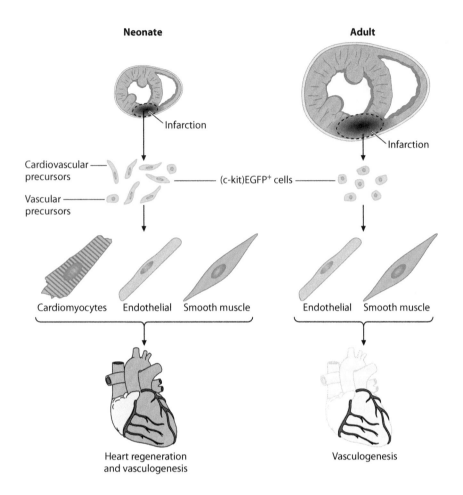

Figure 13.12 **The role of c-Kit+ progenitors during regeneration after cardiac injury in neonatal and adult hearts.** Diagram depicting the role that c-Kit+ progenitors seem to play in the hearts of neonatal and adult mice following injury (area surrounded by dotted circle). Neonatal hearts respond to injury by inducing the localization of c-Kit+ cardiovascular progenitor cells that are capable of differentiating towards all three cardiac tissue types, resulting in heart regeneration. Adult hearts activate a c-Kit+ population of vascular precursors that are only capable of differentiating toward endothelial and smooth muscle cells but not to cardiomyocytes, resulting solely in a vascular repair response (Adapted from Hesse M, Fleischmann B & Kotlikoff M. (2014) *Stem Cells* 32:1701–1712 [doi:10.1002/stem.1696]. Images modified from Patrick J. Lynch, medical illustrator; C. Carl Jaffe, MD, cardiologist; http://patricklynch.net Creative Commons Attribution 2.5 License 2006.)

## The current consensus points to cardiomyocytes rather than stem cells as the homeostatic source of replacement cells

A recent assessment of the results of studies from multiple laboratories around the world concluded that the majority of homeostatic replacement of cardiomyocytes is the result of pre-existing cardiomyocyte mitosis, which can occur more robustly in vertebrates such as fish and amphibians compared to mammals. This replacement through cardiomyocyte division may increase following cardiac injury, although definitive proof is not yet available. In contrast, although resident progenitor cells contribute to multiple cell types within the ventricle, including cardiomyocytes, their contribution to homeostasis is minimal, being perhaps <0.01% per year.

## The epicardium may be a source of cardiac progenitor cells

The prime focus in the search for the CPC has been within the myocardium, but recent evidence suggests that the lining surrounding the heart, the epicardium (**Figure 13.15**), may be able to contribute to the regeneration of both cardiomyocytes

Figure 13.13 **Clonogenic cardiac Sca-1⁺side population cells show tri-lineage potential after cardiac grafting.** Confocal microscopy images following the engraftment of clonogenic cardiac side population (SP) cells into infarct injured mouse heart. Clonogenic cardiac Sca-1⁺ SP cells were transduced prior to engraftment with an mOrange fluorescent protein expressing vector to facilitate cell tracking. (**A**) Cardiomyocyte differentiation is revealed by expression of cardiac troponin I (cTn1, green) and sarcomeric α-actin (sarc actin, purple). (**B**) Smooth muscle cell differentiation is evidenced by staining for smooth muscle myosin heavy chain (SM-MyHC, green). (**C**) Endothelial cell differentiation is highlighted by staining for von Willebrand factor (vWF, green). (From Noseda M, Harada M, McSweeney S et al. (2015) *Nat Comm* 6:6930 [doi:10.1038/ncomms7930]. Published under the terms of the Creative Commons Attribution License (http://creativecommons.org/licenses/by/4.0/).)

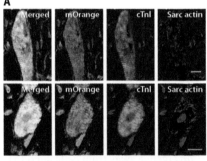

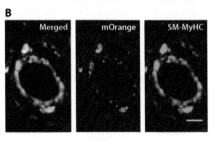

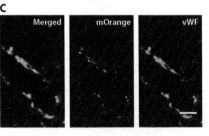

**A**

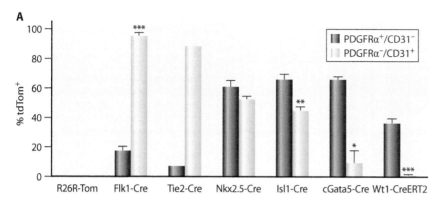

**B**

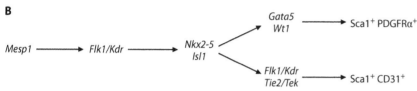

**Figure 13.14 Transcription factors determining the development of Sca-1⁺PDGFRα⁺ progenitors. (A)** Mice carrying a Cre recombinase-activated Tomato fluorescent protein (Tom) reporter were crossed to strains in which Cre was driven from gene regulatory regions of transcription factors involved at different stages of cardiac lineage development. The proportions of the fractionated Sca-1⁺ cardiac progenitors isolated from these strains positive for Tom expression are shown in the histograms distinguishing cells that are PDGFRα⁺CD31⁻ (cardiac lineage progenitors, dark blue) and PDGFRα⁻CD31⁺ (endothelial lineage progenitors, pale blue). Positivity for Tom indicates that the cells had passed through a stage of development during which the particular transcription factor was active. **(B)** The lineage diagram summarizes the conclusions reached in terms of the paths along which the precursors of each of the two progenitor cell types had progressed. (From Noseda M, Harada M, McSweeney S et al. (2015) *Nat Comm* 6:6930 [doi:10.1038/ncomms7930]. Published under the terms of the Creative Commons Attribution License (http://creativecommons.org/licenses/by/4.0/).)

and the vasculature in the form of endothelial cells. During embryogenesis, cells migrate to the developing heart to form the epicardial envelope, which then begins to contribute to the vasculature and myocardium. Once the heart is fully developed, these epicardial cells become dormant but remain within the epicardium. In 2007, a team led by Paul Riley set out to determine if these dormant progenitors, known as epicardium-derived cells (EPDC), could be reactivated. EPDC remain capable of giving rise to both heart tissue and the blood vessels that work in synergy to enable the adult heart to function properly. The starting point for Riley's laboratory was an early discovery they had made concerning the critical function of a small actin monomer binding protein called thymosin β4 (Tβ4). They had found that when Tβ4 is genetically deleted in the mouse, this causes a failure in the formation of the heart during embryogenesis as a result of a lack of coronary blood vessel formation. It had also been shown that when Tβ4 was applied to explant cultures similar to those described previously for cardiosphere propagation, then epicardial cells grew out that subsequently differentiated into fibroblasts, smooth muscle cells, and endothelial cells, essentially the components required for vessels. When Riley's team experimentally introduced Tβ4 into the adult epicardium, they found that EPDC were mobilized and then proliferated before undergoing **epithelial-to-mesenchymal transformation** (EMT—see Chapter 5) and differentiating into endothelial and smooth muscle cells constituting the coronary vasculature, interstitial fibroblasts, and Purkinje nerve fiber network.

The embryonic epicardium is characterized by expression of several genes, notable among which is Wt1, but these are down regulated postnatally as the epicardium becomes just a single-cell layer. Using a Wt1-driven Cre recombinase inducible by administration of tamoxifen in combination with an activateable fluorescent protein reporter that switches from red to green fluorescent protein when recombined (Chapter 2), it was found that only a fraction of epicardium expresses Wt1, and in the uninjured heart, there is no migration of Wt1-expressing cells into the myocardium. Applying the same strategy during injury mimicking myocardial infarction, increases in the number of WT1-positive cells have been observed in the epicardium, and some, but not all, laboratories have also seen migration of cells into the cardiac tissue. Fibroblasts seem to represent the default program in the activated epicardial layer, although there is some evidence for the generation of new endothelial and smooth muscle cells and cardiomyocytes. In addition, the epicardial cells appear able to make a paracrine contribution to repair of the heart (Figure 13.16).

Clearly, the reactivation of EPDC is still to be fully characterized, and indeed the diversity of EPDC phenotypes remains uncertain. The fact that it may well be possible to improve upon the epicardial contribution to cardiac repair through the use of growth factors and perhaps other molecules will for sure maintain interest in the epicardium for some time to come.

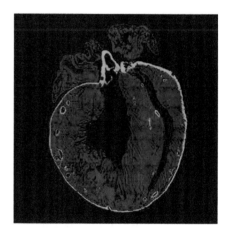

**Figure 13.15 The epicardium.** Confocal image of a cross section through an adult mouse heart showing cardiac muscle (red) and epicardium/vessels (green). (From httsp://www.dpag.ox.ac.uk/research/riley-group. Used with permission.)

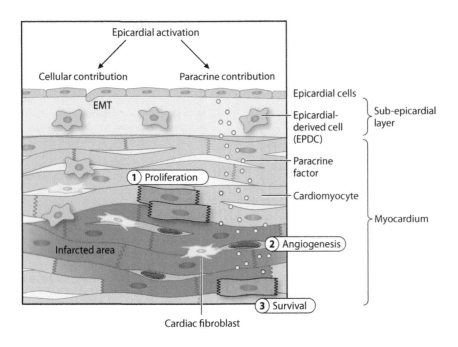

**Figure 13.16 Potential mechanisms of epicardial-derived contribution to cardiac repair.** Activation of the epicardium following injury can have both direct and indirect effects that facilitate repair of damaged cardiac tissue. A direct contribution to cells in the myocardium can occur when epicardial cells undergo EMT, migrate into the myocardium, and differentiate toward cardiac cells such as cardiac fibroblasts and smooth muscle cells. A second indirect influence involves paracrine signaling, which can induce (1) proliferation of cardiomyocytes, (2) angiogenesis, and (3) cardiomyocyte survival in the injured area. (From Smits A, Dronkers E & Goumans M-J. (2018) *Pharmacol Res* 127:129–140 [doi: 10.1016/j.phrs.2017.07.020]. Published under the terms of the Creative Commons Attribution License (http://creativecommons.org/licenses/by/4.0/).)

## 13.4  SUMMARY

In this chapter, we have looked at the potential role of stem cells in the maintenance of the two principal components of the cardiovascular system, that is, vessels and the heart. In both cases, there is controversy regarding the identity of endogenous stem cells, and for the heart whether there is actually any significant involvement of stem cells in homeostasis or repair following injury. Vessels utilize a variety of stem cells in the generation of new smooth muscle cells and endothelial cells, the former from either specialized smooth muscle progenitors or MSC-like cells, while endothelial cells arise from a dedicated progenitor, which is now thought to correspond to so-called late outgrowth cells. The involvement of stem cells in the heart seems to be minimal or nonexistent, certainly with respect to cardiomyocytes under normal conditions. Homeostatic cardiomyocyte replacement is most likely largely through their own ability to divide. Putative stem cells expressing the surface markers c-Kit of Sca-1 can be isolated and demonstrate properties of proliferation and differentiation, but their capacity to contribute to cardiomyocytes is at best limited, even following injury. A population of cells in the epicardium surrounding the heart may, however, be able to contribute to regeneration in a more significant way, giving rise to both cardiomyocytes and endothelial cells following injury and also possibly through a paracrine mechanism that leads to beneficial effects on the damaged myocardium.

---

### KEY POINTS

- The cardiovascular system, consisting of a continuous loop of vessels carrying blood propelled by the heart, is a low cell turnover tissue but nevertheless requires continual repair and regenerative capacity.

- Both the vessels and the heart undergo cell replacement to maintain homeostasis and in response to injury. Vessels additionally require new cells in order to grow during the process of angiogenesis.

- Depending on their size and structure, the blood vessels have resident monopotent and multipotent stem/progenitor cells that can give rise to endothelial cells and smooth muscle cells.

- Endothelial cells may derive from both *in situ* resident and circulating endothelial progenitor cells (EPC), although the nature and source of the cells found in the blood is controversial.

- Genuine EPC in the blood do not originate from the bone marrow, have features very similar to differentiated endothelial cells, and are known as late outgrowth cells, reflecting their distinctive behavior in culture.

- New vascular smooth muscle cells are traditionally viewed to be derived from existing cells that undergo dedifferentiation. However, both monopotent smooth muscle progenitors and multilineage pericytes/mesenchymal stem cells are now believed to play a part.

- Identification of stem cells in the heart has been particularly difficult and controversial; however, a picture is emerging that includes a variety of progenitors that become active in response to cardiac injury, although their contribution to regeneration seems to be limited.

- Cells expressing the c-C-Kit growth factor receptor have been much vaunted as tri-lineage (cardiomyocyte, smooth muscle cell, and endothelial cell) cardiac progenitors, but the consensus now is that these cells are only relevant in the neonate, whereas in adults, C-Kit⁺ cells are restricted progenitors for cells of the cardiac vasculature.

- Sca-1⁺PDGFRα⁺ cells have true stem cell properties and the potential for expansion and tri-lineage differentiation.

The layer surrounding the heart, the epicardium, contains one or dormant progenitors that can be reactivated in response to cardiac damage and may contribute cells to the repair process.

# FURTHER READING

### Vascular stem cells

Basile, D.P. and Yoder, M.C. (2014) Circulating and tissue resident endothelial progenitor cells. *J. Cell. Physiol.*, 229: 10–16.

Psaltis, P.J. and Simari, R.D. (2015) Vascular wall progenitor cells in health and disease. *Circ. Res.*, 116: 1392–1412.

### Cardiac stem cells

Beltrami, A.P., Barlucchi, L., Torella, D., Baker, M., Limana, F., Chimenti, S., Kasahara, H. et al. (2003) Adult cardiac stem cells are multipotent and support myocardial regeneration. *Cell*, 114: 763–776.

van Berlo, J.H., Kanisicak, O., Maillet, M., Vagnozzi, R.J., Karch, J., Lin, S.C., Middleton, R.C. et al. (2014) c-c-Kit+ cells minimally contribute cardiomyocytes to the heart. *Nature*, 509: 337–341.

Eschenhagen, T., Bolli, R., Braun, T., Field, L.J., Fleischmann, B.K., Frisén, J., Giacca, M. et al. (2017) Cardiomyocyte regeneration: A consensus statement. *Circulation*, 136: 680–686.

Noseda, M., Harada, M., McSweeney, S., Leja, T., Belian, E., Stuckey, D.J., Abreu Paiva, M.S. et al. (2015) PDGFRα demarcates the cardiogenic clonogenic Sca1+ stem/progenitor cell in adult murine myocardium. *Nat. Commun.*, 6: 6930.

Smits, A.M., Dronkers, E. and Goumans, M.J. (2018) The epicardium as a source of multipotent adult cardiac progenitor cells: Their origin, role and fate. *Pharmacol. Res.*, 27: 129–140.

# Stem cells in the nervous and sensory systems

# 14

As soon as multicellular organisms began to acquire complex tissues, it became essential that evolution find a way to coordinate and control both the internal bodily processes and the animal's interactions with its environment, including identification of nutrients and potentially dangerous conditions as well as recognition of friends and foes. The solution, which applies to all multicellular animals with the exception of a few simple sponges and other organisms that lack germ layers and defined tissues, is a nervous system.

## 14.1 NERVOUS SYSTEMS IN ANIMALS

The extent of an animal's nervous system can be little more than a simple network of cells, as seen, for example, in jellyfish, although most have complex structures including a brain, a central neural cord, and nerves radiating from them (Figure 14.1). In vertebrates, these structures are divided into two parts termed the central nervous

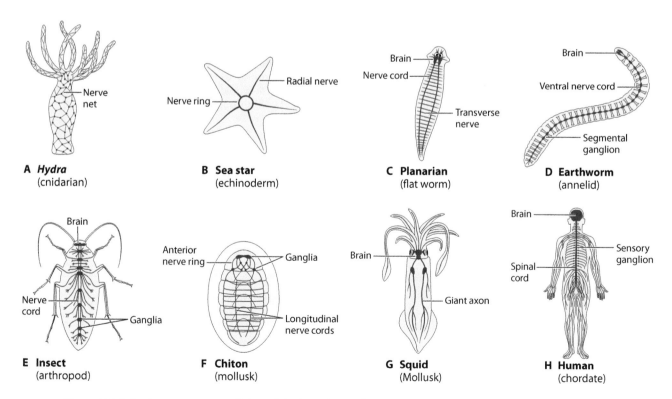

**A** *Hydra*
(cnidarian)

**B** **Sea star**
(echinoderm)

**C** **Planarian**
(flat worm)

**D** **Earthworm**
(annelid)

**E** **Insect**
(arthropod)

**F** **Chiton**
(mollusk)

**G** **Squid**
(Mollusk)

**H** **Human**
(chordate)

Figure 14.1 **Animal nervous systems.** Quite early in the evolution of multicellular animals, nerves developed as a means of allowing communication between different tissues. Examples of nervous systems within the animal kingdom range from nerve nets and nerve rings in cnidarians and echinoderms to the formation of nerve cords in flatworms and development of the brain in more complex invertebrates and the vertebrates. (Adapted from Campbell N, Reece J, Urry L et al. (2008) *Biology*, 8th ed. San Francisco: Pearson Benjamin Cummings.)

system (CNS) and the peripheral nervous system (PNS). The CNS contains the brain and spinal cord, whereas the PNS is made up of nerves that link the CNS to every other part of the body. CNS tissues and peripheral nerves are constituted by a variety of specialized cells (Box 14.1), which together serve to detect, relay, and integrate information.

As we have seen in the case of the cardiovascular system (Chapter 13), some organs in the body have a very low cell turnover and a limited ability to repair following traumatic injury or disease. This is also the case for the nervous system, and in the past, the techniques and results available suggested that the production of neurons is limited to embryonic development. However, as for the heart, logic suggests that there must be a fulfilled life-long requirement for cell replacement in the nervous

## BOX 14.1 BACKGROUND: CELL COMPONENTS OF THE NERVOUS SYSTEM

The first recognized form of nervous tissue arose 550–600 million years ago in worm-like organisms, and since that time the complexity increased massively to the point that in humans the nervous system consists of on the order of 100 billion cells, involving several key cell types, each of which exhibits a wide range of specialized morphologies and functions. The essential cell of the nervous system is the neuron, the defining feature of which is that it can rapidly send signals in the form of electrochemical waves traveling along thin fibers called axons, causing chemicals (neurotransmitters) to be released at junctions (synapses) with other neurons, thereby exciting or inhibiting the receiving cell. The nerves connecting parts of the body into the central nervous system are made up of enclosed bundles of axons, each of which is covered by an insulating sheath of myelin. In addition to a single axon, a neuron has other cytoplasmic extensions from its cell body in the form of one or more dendrites, which serve to extend the cell's surface for the receipt of signals from other neurons.

Nervous tissues contain both neural and non-neural cell types (Figure 1). There are three main functional types of neurons: (1) sensory neurons, which act as receptors of stimuli themselves or are activated by receptors to which they are connected; (2) motor neurons, which send information to the body's effectors, causing them to respond by contracting (muscle) or eliciting secretion (glands); and (3) interneurons, which connect neurons to one another.

Non-neuronal, or glial (literally "glue"), cells facilitate the functioning of neurons throughout the nervous system, providing support and protection. The myelin sheath surrounding many axons is produced by Schwann cells in the PNS or oligodendrocytes within the CNS and helps to insulate and increase the efficiency of signal transmission. The second major glial cell type is the astrocyte, literally meaning "star-like cell." Astrocytes are the most numerous and diverse glial cells in the CNS, outnumbering neurons by a factor of five. Actually, not all astrocytes are star-like cells, and they display remarkable heterogeneity in their morphology and function. For example, astrocytes contribute to the brain environment, regulate energy substrates, facilitate the formation of synapses, and provide for brain defense. Microglia, perhaps not surprising given their name, are smaller than other glial cells and have their embryonic origin in the hematopoietic system, being the brain and spinal cord equivalent of macrophages and likewise serving to remove diseased or damaged cells or pathogens that cause disease. Ependymal cells are a glial cell type, reminiscent of endothelial cells, which line the cavities (ventricles) of the brain and spinal cord, and through their contact with blood vessels absorb components of the blood to produce cerebrospinal fluid. Ependymal cells have cilia on their apical surface that help to move the cerebral spinal fluid through the ventricular space.

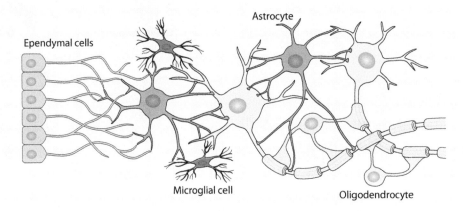

Box 14.1 Figure 1 **Neural and non-neuronal cells in the CNS.** In addition to neurons (yellow), the central nervous system is composed of several other cell types that attach to or work with the neural cells. Most important are: oligodendrocytes (pale blue), which provide the insulating myelin sheath around the axons; astrocytes (purple), the star-shaped cells that make contacts with the neurons and help maintain the neural environment; ependymal cells (brown), which line the neural cavities and link to the astrocytes; and microglial cells (res), which are hematopoietic, macrophage-like cells.

BOX 14.2 PIONEER: JOSEPH ALTMAN

Born in 1925, Altman (Figure 1) eventually became a postdoctoral researcher in 1961 in a psychology department at the Massachusetts Institute of Technology (MIT). Altman taught himself neurobiology and starting in the early 1960s, he challenged the idea of "no new neurons in the adult brain," publishing a series of papers reporting autoradiographic evidence for the continuous birth of new neurons. Unfortunately for Altman, this dogma was universally held and vigorously defended by the most powerful and leading primate developmental anatomist of his time. Although published in prestigious journals, Altman's findings were totally ignored or dismissed as unimportant for over two decades. Altman was not granted tenure at MIT and moved to Purdue University in 1968, where he conducted research for over 50 years, turning to more conventional developmental questions, perhaps because of the lack of recognition of his work on adult neurogenesis. Unable to get grants, he supported his work by producing brain atlases. Of course, history eventually proved that the essence of Altman's findings was correct.

Box 14.2 Figure 1 **Joseph Altman** (From brainmindevolution.org. With permission from Dr. Shirley Bryer-Altman).

system. Realization that the adult mammalian brain does indeed have the capacity for continuing production of neurons, and a requirement for specific stem cells, actually came about before their discovery in many other tissues. In the early 1960s, Joseph Altman (Box 14.2) and his colleagues adapted Charles Philippe Leblond's technique (Chapter 12) for the use of tritium-labeled thymidine to mark dividing **progenitors** in rodent brains. Further, Altman was able to track these progenitors in several regions of the brain as they went on to **differentiate** into neurons. These findings were highly controversial at the time, being essentially ignored by the scientific community, the widespread belief remaining that no new neurons can be created in the adult brain. In the early 1970s, Fernando Nottebohm at Rockefeller University in New York challenged this dogma by demonstrating adult neurogenesis in birds. Using similar labeling techniques to those employed by Altman, Nottebohm showed that the vocal center in the brain of adult songbirds undergoes neural expansion and death in parallel with the seasons. Although initially heavily criticized, the findings in birds were eventually accepted, but it remained much harder for people to believe that adult neurogenesis occurs in mammals. An influential publication in the mid 1980s effectively convinced many researchers that adult neurogenesis is restricted to evolutionarily lower-order animals and that it is irrelevant in humans. The field did not recover from this until almost two decades later, when the identity of the labeled progenitor cells became clear and the principle of continuing neurogenesis was fully accepted, even in higher mammals including humans. Despite this stuttering early progress, as it turned out the study of what eventually became known as **neural stem cells** (NSC) actually led the way along many avenues, providing paradigms in our understanding of **adult stem cell** biology.

The neurons of the CNS are only a part of the overall structure of a nervous system. A nervous system needs inputs and to be able to send outputs, requirements that are the essential role of the peripheral nervous system. The data the CNS receives are collected through the sensory components of the PNS, which can be broken down according to the classical view of the five senses, that is, sight, hearing, smell, taste, and touch. The nerve-linked detectors associated with each of these senses have distinct structures and specialized cells, and often, but not always, have their own dedicated stem cell component.

In this chapter, we will focus initially on the neural stem cells that contribute to continual neurogenesis in the CNS. As we will see, much of what we now understand has come from studies on rodents and humans, but studies on fish and birds have also contributed very important insights. We will then turn our attention to the specialized sensory organs in the PNS and the nature of the stem cells that maintain their integrity and which may eventually prove to be the basis for innovative therapies for the treatment of disorders such as blindness and deafness.

## 14.2  STEM CELLS IN THE CENTRAL NERVOUS SYSTEM

### Two regions in the adult brain are sites of active stem cell self-renewal and commitment

Joseph Altman's original findings, in which he had observed what he thought was neurogenesis in the dentate gyrus (DG, a simple cortical region of the hippocampus located on the floor of the lateral ventricles of the brain, functioning as the center of emotion, memory, and the autonomic nervous system), were supported by strong evidence for neurogenesis in the adult brain when he and his colleague Gopal Das subsequently identified the birth of new cells in the ventricular-subventricular zone (V-SVZ) of rodents. They were able to follow the differentiation of these progenitors into neurons in the olfactory bulb (OB), which the cells reached by migration along the rostral migratory stream (RMS) (Figure 14.2). Although also seen in primates, the presence of an analogous RMS in humans is doubtful, possibly because the OB is significantly less developed.

During the mid to late 1990s, new methods for immunofluorescence analysis, confocal microscopy, and bromodeoxyuridine (BrdU) labeling made possible *in situ* observation simultaneously of NSC division and differentiation and confirmed the importance of neurogenesis in the DG and SVZ. More recent support for the notion of adult neurogenesis came from a rather unexpected direction that made use of the fact that atmospheric testing of nuclear bombs in the 1950s and 1960s enabled carbon-14 dating of neurons from individuals born during this period, very much in the same way that cardiomyogenesis was proven to be ongoing in adults (Chapter 13).

Several laboratories led the way with *in vitro* cell culture and *in vivo* transplantation assays that enabled characterization of NSC from the adult brain. In 1992, Reynolds and Weiss were the first to isolate NSC from adult mouse brain tissue. Conditions were developed, in particular requiring the growth factors epidermal growth factor (EGF) and fibroblast growth factor-2 (FGF-2), that allow isolated adult brain cells to be grown as **multipotent** proliferative clones that form as balls of cells or "**neurospheres**" (Figure 14.3). The cells enabling the growth of neurospheres are considered NSC, although formal proof required *in vivo* transplantation (Chapter 2). These clusters are capable of giving rise to neurons, oligodendrocytes, and astrocytes. As for other cell differentiation cascades fed from a multipotent stem cell, the neural cell hierarchy initiates once the NSC undergoes **asymmetric division**, giving rise to a **transit amplifying** component, generally termed the neural progenitor cells (NPC). These NPC can be restricted in their fate potential, being either limited to neuron differentiation or serving as the precursor of both oligodendrocytes and astrocytes (Figure 14.4). A wide range of markers can be used to define the NSC, transit amplifying cells, and lineage committed progenitors and the differentiated cells deriving from them (Table 14.1).

Combining these various approaches, work over the last decade and a half has provided a picture of stem cells and their **niches** in the adult mammalian brain in which NSC are predominantly located within two distinct regions: the

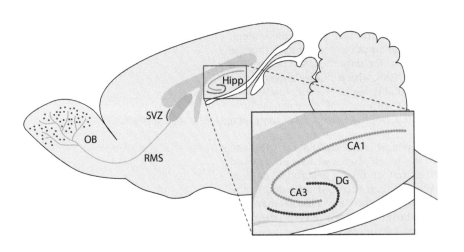

**Figure 14.2 Neurogenesis in the adult mammalian brain.** Although the turnover of cells within the brain is overall quite slow compared to many tissues, there is a continual requirement for production of specific neurons in some regions. The simplified cross section through the mammalian brain highlights the key regions of neurogenesis in the subventricular zone (SVZ) and the hippocampus (HIPP) within the subgranular zone. Inset: magnified representation of the hippocampus identifying the precise location of neural stem cells in the dentate gyrus (DG). Neuroblasts (red) derived within the SVZ migrate along the rostral migratory stream (RMS) to the olfactory bulb (OB). (Adapted from http://www.scholarpedia.org/article/File:Adult_neurog.gif.)

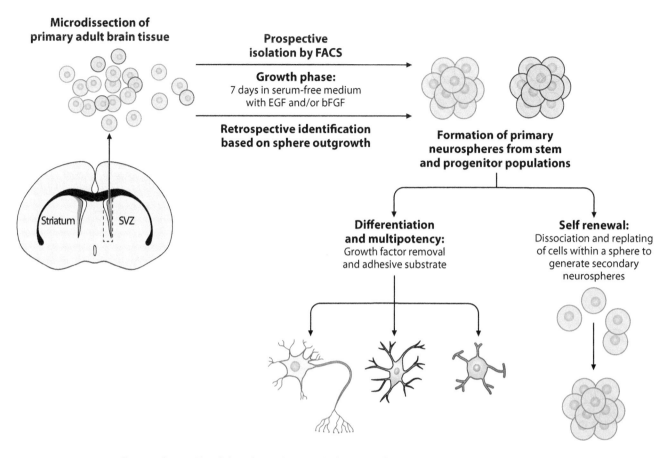

**Figure 14.3 Neurospheres.** The ability of neural stem cells from specific regions of the brain to self-renew and produce differentiated neural and glial cells can be captured in culture through formation of neurospheres. Schematic diagram showing the derivation of neurospheres from a biopsy of the subventricular zone (SVZ) of an adult mammalian brain. Following isolation by fluorescent activate cell sorting (FACS), neural stem cells are grown for 7 days in serum-free medium supplemented with growth factors epidermal growth factor (EGF) and fibroblast growth factor (FGF) before they begin to form primary neurospheres. Stem cell characteristics are assayed by secondary re-plating to test self-renewal properties and by culturing in appropriate medium to induce multi-lineage differentiation to neurons, oligodendrocytes, and astrocytes. (From Pastrana E, Silva-Vargas V & Doetsch F. (2011) *Cell Stem Cell* 8: 487–498 [doi: 10.1016/j.stem.2011.04.007]. Copyright 2011 with permission from Elsevier.)

already mentioned V-SVZ of the lateral ventricles and the subgranular zone (SGZ) of the hippocampal dentate gyrus (Figure 14.2). The V-SVZ and SGZ represent neurogenic stem cell niches, the cellular constituents of which include astrocytes, ependymal (endothelial) cells, macrophage-like microglia, and the blood vascular system. NSC from these two niches differentiate into neurons under normal physiological conditions through a process that can take several months, and when transplanted to different sites in the brain, they tend to become oligodendrocytes and astrocytes.

Although adult stem cells are generally thought to perform a role in tissue homeostasis and repair, the generation of new neurons from NSC is perhaps more geared toward expanding the **plasticity** of the brain. Through creation of new synaptic contacts with the existing circuitry, the neurons produced from the V-SVZ and the SGZ are likely to be involved in learning, the former for olfactory memory and the latter in creating spatial memory and pattern separation.

## Neural stem cells and downstream progenitors in the ventricular-subventricular zone are arranged in a hierarchical manner in a highly organized niche

The V-SVZ is the largest center of stem cell activity in the adult brain. Three types of neural progenitor cell have been identified within the V-SVZ (Figure 14.5), designated A, B, and C. The **self-renewing** NSC residing within the V-SVZ are represented by B

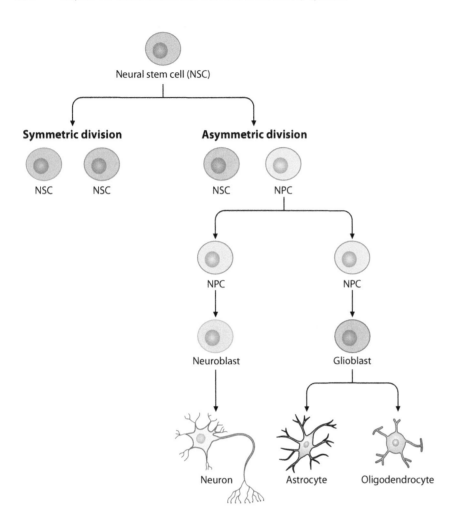

Figure 14.4 **The differentiation hierarchy from neural stem cells.** As for all adult stem cells, neural stem cells (NSC) sit at the top of a hierarchy of progenitors and differentiated cells. The diagram illustrates how NSC can undergo either symmetric division, yielding two NSC, or asymmetric division, resulting in one NSC and one committed neural progenitor cell (NPC), which can then give rise to a neuroblast that differentiates to a neuron or a glioblast that can differentiate toward an astrocyte or an oligodendrocyte. (From Vishwakarma S, Bardia A, Tiwari S et al. (2014) *J Advanced Res* 5: 277–294 [doi: 10.1016/j. jave.2013.04.005]. Copyright 2014 with permission from Elsevier.)

**TABLE 14.1  MARKERS OF NSC AND THEIR DIFFERENTIATED PROGENY**

| Type of cell | Markers |
| --- | --- |
| Neural stem cell | Prominin (CD133), NCAM (CD56), Nestin, Sox-2, Notch-2, ABCG2 |
| Neural progenitor | PSNCAM, Neurotrophin Receptor |
| Astrocyte progenitor | CD44, A2B5 |
| Oligodendrocyte progenitor | NG2, PDGFRa, Olig-2 |
| Neuron | MAP-2, Doublecortin, b-tubulin III, Neuro D, Neu N |
| Astrocyte | GFAP |
| Oligodendrocyte | Olig-1, Olig-4, Galactocerebroside |

cells, the ultimate progeny of which are able to migrate the distance along the RMS, connecting the lateral ventricles to the olfactory ventricle. When the cells reach the OB, they terminally differentiate into interneurons that integrate into the existing neural circuitry. Each B cell is able to yield as many as 10,000 differentiated neurons in young adults, the amplification in numbers being achieved at the intermediate progenitor stages. Hence, B cells give rise to C cells, which represent the transit amplifying cells in the hierarchy, dividing three to four times and differentiating into A cells, or neuroblasts. The A cells in turn divide further before migrating along the RMS toward the OB, where they then detach from the chain, migrate radially into the granule and glomerular cell layers, and undergo the final stages of differentiation.

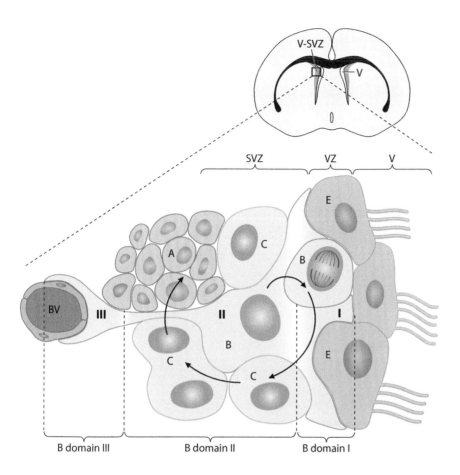

Figure 14.5 **The V-SVZ NSC niche.** Neural stem cells (NSC) occupy a very well-defined niche in the ventricular-subventricular zone (V-SVZ) of the adult mouse brain. The cross section through the mouse V-SVZ and the expanded view illustrates the location of the NSC and their differentiation pathway. The V-SVZ niche is composed of neural stem cells (B, blue) surrounded by multi-ciliated ependymal cells (E, brown). The B cells contact the ventricle (V) by a short apical process, and give rise to intermediate progenitor cells (C, green) that divide to generate neuroblasts (A, orange). The B cell remains in close contact with all of its progeny and possesses a basal process that contacts blood vessels (BV). Reflecting these multiple cell-cell contacts, the B cell structure is defined by three domains (I, II, III). (From Fuentealba L, Obernier K & Alvarez-Buylla B. (2012) *Cell Stem Cell* 10: 698–708 [doi: 10.1016/j.stem.2012.05.012]. Copyright 2012 with permission from Elsevier.)

B cells are a heterogeneous population of stem cells, which is most apparent from the fact that where they are located in the V-SVZ determines which of about ten different types of interneuron they give rise to.

Type B cells retain the apical-basal polarity that is observed within their embryonic predecessors, radial glia, which constitute the NSC of the embryo and early post-natal brain. This specialized polarity enables the B cell to contact the ventricle through small apical processes and also blood vessels via long basal processes. Strictly speaking, B cells exist in two variant forms, that is, B1 and B2, which differ in their contact with the ventricle. Unlike B1 cells, the minority B2 cells do no protrude into the ventricle. The B1 NSC have been divided into three domains (Figure 14.5). In domain I, the B1 cells contact the ventricle through a single primary cilium that is surrounded by a rosette of ependymal (E) cells, which are themselves ciliated on the surface facing the ventricle. The function of the E cells is to maintain the molecular composition of the apical compartment, their cilia propelling cerebral spinal fluid containing soluble factors crucial to the maintenance and proliferation of the NSC, including Insulin-like growth factor-2 (IGF-2), Bone morphogenetic protein (BMP), Wnt, Sonic Hedgehog (Shh), and Retinoic acid (RA). Domain II is where both B1 and B2 cells make direct contact with the surrounding transit amplifying C cells and the A cell neuroblasts, providing direct feedback from the NSC progeny. Domain III contains the basal processes of B1 and B2 cells, which make direct contact with the adjacent blood vessel endothelial cells. The B cell therefore has a well-defined niche (see Chapter 6), receiving inputs both from cells with which it is in contact and also from the blood and the cerebrospinal fluid in the ventricle.

## Stem cells and progenitors in the subgranular zone niche show both similarities and differences compared to those in the ventricular-subventricular zone

The SGZ is the other region of the brain known to have neurogenic activity and is located in the hippocampus at the interface of the hilus and the dentate gyrus. In young adults, roughly 9000 new neural cells are generated per day within the SGZ,

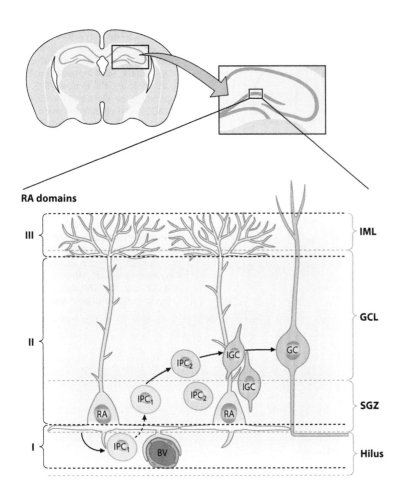

Figure 14.6 **The SGZ NSC niche.** As for the NSC niche in the V-SVZ, stem cells in the subgranular zone (SGZ) occupy a specific niche. The frontal cross section through the adult mouse brain (top left) shows the region containing the hippocampus. The dentate gyrus is shown magnified (top right) and the region depicted by the square is shown in greater detail (bottom). Within the SGZ resides the cell body of the radial astrocyte (RA, blue) that gives rise to intermediate progenitors (IPC1 and IPC2, green), which in turn differentiate into immature granule cells (IGC, orange) that remain in contact with the RA. Mature granule cells (GC, brown) send an axon parallel to the SGZ into the hilus, projecting dendrites that branch through the granule cell layer (GCL) into the inner molecular layer (IML). The RA can be divided into three domains (I, II, III). (From Fuentealba L, Obernier K & Alvarez-Buylla B. (2012) *Cell Stem Cell* 10: 698–708 [doi: 10.1016/j.stem.2012.05.012]. Copyright 2012 with permission from Elsevier.)

although only half of these go on to differentiate fully and integrate into the existing neural circuitry. Similar to the V-SVZ, three neural progenitors have been identified within the SGZ (Figure 14.6). The NSC within the SGZ has been defined as the radial glia-like cell (RA), which generates proliferative non-radial, or intermediate, precursors (IPC). The IPC in turn gives rise to neuroblasts, also known as immature granule cells (IGC) that differentiate toward immature neurons. The RA cells possess radial processes that span the entire granule cell layer into the inner molecular layer, while their cell bodies remain within the SGZ. Similar to B cells in the V-SVZ, the RA stem cells can be partitioned into three domains. The proximal domain I is where the RA cell contacts blood vessels via a single cilium. The cell body of the RA NSC lies within the intermediate domain II and forms a main shaft throughout the granule cell layer, its thin appendages being interspersed among mature granule neurons. In the distal domain III, the RA NSC is highly branched and contacts with neuronal processes, synapses, and other glial cells within the inner molecular layer.

### The spinal cord exhibits limited stem cell activity under normal conditions but can enlist quiescent stem cells for repair following an injury

Unlike the V-SVZ and SGZ of the brain, the spinal cord is generally regarded to be non-neurogenic; only myelinating oligodendrocytes (OL) are produced from OL progenitors scattered throughout the tissue. Nevertheless, experiments in mice and humans have shown that cells lining the central canal of the spinal cord, forming the ependymal layer, have stem cell properties *in vitro* and can generate multipotent neurospheres. Furthermore, studies that have combined *in vivo* labeling with **lineage tracing** have provided evidence that there is also a capacity for replacement of other cell types, which is especially prominent following injury. Using Bromodeoxyuridine (BrdU) to label the spinal cord in the mouse, it was observed that cells actively proliferating are scattered within all areas of the spinal cord. Each of these lineages

**Transgenic lines**         **Cre-dependent reporter lines**

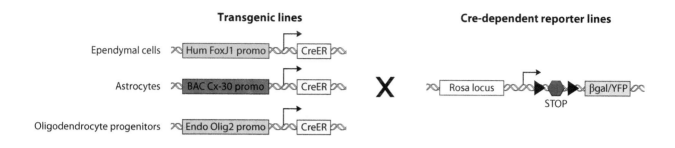

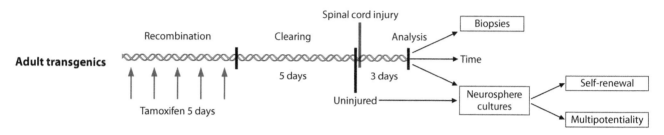

Figure 14.7 **Tracing stem cell potential in the rodent spinal cord.** Lineage tracing has proved to be effective in defining stem cell activity in the spinal cord in normal or injured animals. The upper panel illustrates three transgenic lines that express tamoxifen-inducible Cre recombinase (CreER) from gene promoters specifically expressed in ependymal cells (FoxJ1, green), astrocytes (Cx30, red), and oligodendrocyte progenitors (Olig2, blue). Each CreER transgenic line is crossed with a Cre-activated reporter (β-gal/YFP), driven from the ubiquitously expressed Rosa26 promoter that has been engineered with a loxP site (black arrow head) flanked transcriptional stop sequence. In the lower panel is shown how recombination is induced by five daily injections of tamoxifen, followed by a 5-day "clearing" period. In order to determine the contribution of the distinct cell populations after injury, a lesion is performed after the clearing period. To assess the stem cell potential of the distinct cell populations in normal animals or following spinal cord injury, observations can either be performed on cells *in situ* in brain samples or by *in vitro* culture of neurospheres, which can be serially passaged or plated under differentiating conditions to assess self-renewal and multipotency. (From Barnabé-Heider F, Göritz C, Sabelström H et al. (2010) *Cell Stem Cell* 7: 470–482 [doi: 10.1016/j.stem.2010.07.014]. Copyright 2010 with permission from Elsevier.)

were fate mapped using **transgenic** mouse lines combining a Cre-activateable reporter gene and tamoxifen-inducible **Cre recombinase** (CreER) under the control of lineage specific promoters, which after tamoxifen administration led to permanent and heritable expression of the reporter (Chapter 2). The use of promoters from the *FoxJ1*, *Connexin 30*, and *Olig2* genes enabled identification of ependymal cells, astrocytes, or oligodendrocyte lineage cells, respectively (Figure 14.7). Ependymal cells, astrocytes, and oligodendrocyte lineage cells together accounted for the vast majority of those cells labeled, with the latter being by far the most numerous, representing more than 80% of the total (Figure 14.8, **middle image**). Astrocytes and ependymal cells, which are restricted to limited self-duplication in the intact spinal cord, generate the largest number of cells after spinal cord injury. Only ependymal cells, which line the central canal of the spinal cord in the same way that they line the ventricles in the brain, generate both astrocytes and oligodendrocytes *in vivo* (Figure 14.8, **right panel**). A latent potential of the spinal cord ependymal cells as NSC is revealed *in vitro* since only cells labeled through FoxJ1-CreER can give rise to self-renewing neurospheres that yield neurons in addition to astrocytes and oligodendrocytes. This capacity is seen for cells derived both from uninjured animals and after spinal cord injury, giving some hope that it may be possible to utilize this property in a therapeutic context (Chapter 17).

The opportunities for clinical application of NSC are considerable, both due to the range of relevant neurological diseases that might be treated and also because of the potential for very significant impacts on quality of life and longevity. Consequently, a myriad of pre-clinical experiments on animal models have been carried out or are underway, and some initial early phase trials are even being undertaken for some conditions, such as spinal cord injury, stroke, and traumatic brain injury. NSC from rodents, primates, and humans are being tested, using brain- or spinal cord-derived cells as a direct source, or cells obtained by **reprogramming**, either involving an **induced pluripotent stem (iPS) cell** intermediate or direct conversion of somatic cells to NSC (Chapter 5). Partial alleviation of symptoms has been achieved in models of diseases

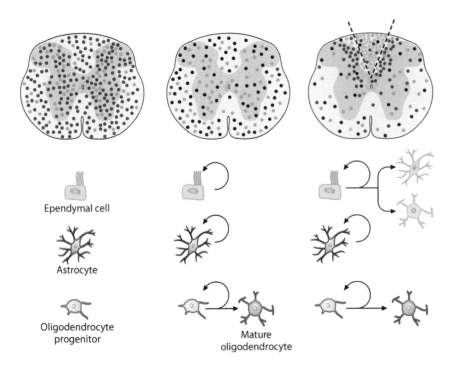

Figure 14.8 **Uni- and multi-potential stem cells in the spinal cord.** Lineage tracing has helped to define how stem cells become activated and differentiate in the spinal cord in normal conditions and following injury. The diagrammatic cross sections through the spinal cord of an adult mouse illustrate the occurrence of ependymal cells (green), astrocytes (red), and oligodendrocyte progenitors (pale blue) under normal conditions (left) and the extent to which each cell type is actively generated in normal circumstances (middle) or following injury (right), the extent of the injury being indicated by the dotted lines. Under normal conditions, astrocytes predominate but proliferate infrequently, ependymal cells line the spinal canal and also proliferate little, whereas oligodendrocyte progenitors divide extensively and give rise to mature oligodendrocytes (dark blue). Following injury, the astrocytes and ependymal cells proliferate more actively, especially around the site of the lesion, with the ependymal cells giving rise to both astrocytes and oligodendrocytes. (From Barnabé-Heider F, Göritz C, Sabelström H et al. (2010) *Cell Stem Cell* 7: 470–482 [doi: 10.1016/j. stem.2010.07.014]. Copyright 2010 with permission from Elsevier.)

Ependymal cell

Astrocyte

Oligodendrocyte progenitor

Mature oligodendrocyte

ranging from amyotrophic lateral sclerosis (ALS) and Parkinson's disease to spinal cord injury and stroke, while some efficacy has been determined in some early human trials.

## 14.3   STEM CELLS SUPPORTING THE SENSORY ORGANS IN THE PERIPHERAL NERVOUS SYSTEM

The peripheral nervous system serves to relay neural signals to and from the brain and spinal cord in the central nervous system. The input of information to the CNS comes from a variety of neurons linked to sensory receptors throughout the body. These receptors sense a wide range of stimuli originating from outside and sometimes inside the body encompassing the five accepted senses, namely vision, sound, taste, smell, and touch, the latter including the properties of surface texture, vibration, and the sensations of pain, heat, and cold. The sensory components are themselves neural, and with the exception of the nerve sensors for touch and sensation, are known to have at least some capacity for continual replacement after birth, implying the existence of specific adult stem cells.

In order to work out the identity and properties of the sensory organ stem cells, very many of the same experimental approaches have been employed as we have already seen for the CNS. Hence, *in situ* labeling of cell division combined with subsequent tracking of fate, incorporating the use of specific surface and gene markers, has commonly been used to demonstrate perpetuation of peripheral sensory neurogenesis in the adult. The ability to culture these stem cells is also important; they, like those from the CNS, grow as spheres containing both stem cells and differentiated progeny.

The continual flow of new neurons into the olfactory organ to maintain and adapt the sense of smell has already been discussed in the context of stem cells in the forebrain. We will now take a close look at what is known presently about the adult stem cells in vertebrates that can replace neurons in the eye, the inner ear, and related mechano-sensory organs and in the taste buds of the tongue. As we will see, the sensory component of each of these PNS senses incorporates a stem cell element, although the extent to which this can support regeneration is often less in mammals than it is in lower vertebrates.

### Resident stem cells serve to repair both epithelial and neural aspects of the structure of the eye

Stem cell biology research on the eye has been very active partly because it represents a highly accessible organ for experimentation but also for the obvious reason that many diseases and injuries that can lead to partial or complete loss of vision in a significant proportion of individuals may have a chance of being treated if the appropriate cells could be replaced. The relevant stem cells and the components of the eye that they could help to repair are quite diverse. Hence, the cornea at the front of the eye is an epithelial tissue that is frequently rendered opaque by disease or damage and, as we describe in Chapter 15, this is generated from specialized **limbic stem cells** that have good therapeutic potential. Likewise, the supportive layer at the back of the retina, the retinal pigmented epithelium (RPE), is frequently lost in diseases collectively known as macular degeneration, again leading to loss of vision. In this case, the best prospect for a stem cell-based therapy is not from tissue resident cells but rather utilizing pluripotent stem cells (Chapter 3) as a source of replacement RPE (Chapter 17). Here, we will describe the stem cells that support maintenance of the neural component of the vertebrate eye, which are concentrated in a layer in the retina.

The vertebrate retina (Box 14.3) is a complex structure hosting millions of photoreceptors (rods and cones) that function to convert light focused through the lens of the eye into signals that are transmitted along the optic nerve to the visual center of the brain, where an image of the outside world is interpreted. Diseases that affect retinal neural function result in irreversible damage through the specific loss of a cell population from the retina, for example, in glaucoma, a degenerative disease causing the loss of retinal ganglion cells (RGC) that make up the optic nerve, or retinitis pigmentosa, in which the photoreceptors are lost.

We will consider those cells capable of a degree of regeneration of the neural retina under two broad categories, that is, the **retinal stem cell** (RSC) and its immediate

## BOX 14.3 BACKGROUND: STRUCTURE OF THE VERTEBRATE EYE AND RETINA

The eye is a transducer that has been highly conserved throughout vertebrate evolution, whose function is to convert the energy of light into nervous impulses. The eye is built like a camera (Figure 1) to control the input of light via the iris and then focus this light as an image on the retina at the back of the eye. The retina converts the light into nervous impulses, but unlike a camera, the retina does far more in terms of signal processing and interpretation. The retina is a complex structure composed of ten layers, the innermost nine of which contain the light-sensitive elements (Figure 2). The outer, retinal pigmented epithelium (RPE) is a layer of cuboidal cells impregnated with melanin. It has the vital functions of increasing acuity of vision and ensuring that the light-sensitive neural cells are maintained in working order. The outermost layer of the neural part of the retina contains the light-sensitive rods and cones. The outer limiting membrane (OLM) is actually not a membrane but is the site of numerous junctions between the plasma membranes of the rod segments and a glial element, the Müller glial cell. The outer nuclear layer (ONL) is the location for the nuclei and cell bodies of the rod and cone cells. The outer plexiform layer (OPL) is the site of numerous synapses between the rod and cone cells and various integrator neurons, including the horizontal, bipolar, and amacrine cells and the Müller glial cells, which collectively allow processing of the initial neural signal. The inner nuclear layer (INL) contains the cell bodies and nuclei of the integrator neurons. Next inward is the inner plexiform layer (IPL), which, like the outer one, is a region of synapses. Here the bipolar cells synapse with the dendritic processes of ganglion cells, the final neuronal element of the eye itself. The ganglion cell layer (GCL) is where the ganglion cell bodies are located and has far fewer nuclei than the INL or ONL. In the nerve fiber layer (NFL), the axonal fibers from the ganglion cells are bundled together and run radially around the inner surface of the retina,

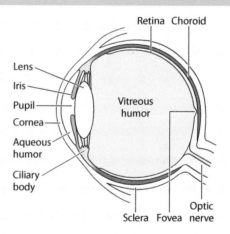

Box 14.3 Figure 1 **The vertebrate eye.** Diagram of a cross section through the vertebrate eye, showing the main structures, including the neural components of the retina and the optic nerve.

converging at the optic nerve. The final component of the retina, which is in direct contact with the vitreous humor, is the inner limiting membrane (ILM), which again is not really a membrane, but is instead a place where the foot processes of the Müller glial cells come together. Müller glial cells actually span most of the neural layers of the retina, serving as barriers and conduits for the transfer of a wide variety of molecules between different retinal cells. They also support neurons by releasing trophic factors, recycling neurotransmitters, and controlling ionic balance in the extracellular space. In addition, Müller glial cells phagocytize cone outer segments, contribute to outer segment assembly, and help to recycle the retinal chromophore for photodetection. Last, Müller glial cells directly contribute to vision by acting as optical fibers that guide light to photoreceptors.

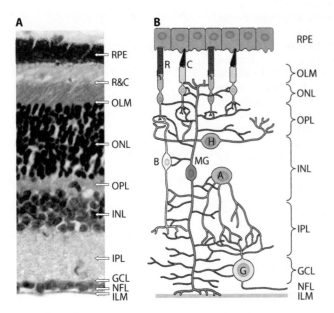

Box 14.3 Figure 2 **The cell layers of the vertebrate retina.** Stained cross section (**A**) and schematic representation (**B**) of the neural cell layers in the vertebrate eye. The retina is divided into three layers in which the neural cell bodies are concentrated: the outer nuclear layer (ONL), the inner nuclear layer (INL), and the ganglion cell layer (GCL). The cell types constituting the layers are: the rod (R, pink) and cone (C, blue) photoreceptors in the ONL; bipolar (B, yellow), horizontal (H, purple), and amacrine (A, brown) interneurons, and Müller glial cells (MG, red), in the INL; and ganglion cells (G, green) in the GCL. Ganglion cell axons run just beneath the GCL, forming the nerve fiber layer (NFL). Synapses between photoreceptors and interneurons take place in the outer plexiform layer (OPL), and synapses between interneurons and ganglion cells take place in the inner plexiform layer (IPL). Müller glial cell processes span all retinal layers and contribute to the formation of the inner limiting membrane (ILM) and outer limiting membrane (OLM). The retinal pigment epithelium (RPE) consists of pigmented cells that absorb light and contacts the photoreceptors. (A: From Willermain F, Libert S, Motulsky E et al. (2014) *Front Physiol* 30:199 [doi: 10.3389/fphys.2014.00199]. Published under the terms of the Creative Commons Attribution License http://creativecommons.org/licenses/ and B: From Goldman D. (2014) *Nat Rev Neurosci* 15: 431–442 [doi: 10.1038/nrn3723]. Adapted by permission from Springer Nature.)

progeny the **retinal progenitor cell** (RPC), and Müller glial cells. The precise location of the RSC and RPC in vertebrates varies between species, and their potential is often only revealed under circumstances of injury (Table 14.2).

## Retinal stem cells have a defined niche in the eye, the precise location of which varies between different vertebrate groups

In mammals, retinal stem cells derive from the ciliary epithelium, the inner, non-pigmented layer of which is continuous with the neural retina, while the outer, pigmented

**TABLE 14.2 SOURCES OF ADULT NEUROGENESIS IN VERTEBRATE RETINAS**

| Location in eye | Species | Proliferative capacity | Self-renewal ability | Multipotentiality |
|---|---|---|---|---|
| Ciliary Marginal Zone (CMZ) | Fish | + | + | + |
| | Birds | + | + | + |
| Inner Nucleated Layer (INL) | Fish | + | + | + (following injury) |
| Müller Cells (following injury) | Fish | + | ? | + |
| | Birds | + | ? | + |

layer that is continuous with the RPE is actually where the RSC reside. Normally, the RSC in the ciliary pigmented epithelium (CPE) are in a **quiescent** state, but when put into culture are able to proliferate, generating spheres akin to neurospheres derived from brain NSC and likewise are more effectively expanded in the presence of the growth factor FGF-2. The stem cells are rare, with only about 1 in 5000 derived from the CPE being clonal, and when transplanted into the central, vitreous cavity of the eye of a day-old mouse engraft into the distal outer layer of the retina and RPE. In fish and amphibians, the RSC are located differently, being in the ciliary marginal zone (CMZ). From this site, their immediate retinal progenitor cell derivatives move inward toward the neural retina, where they contribute new neurons and Müller glial cells. The CPE expresses proteins known or expected to be involved in the specification or maintenance of stem cells in the eye. For example, the transcription factor Pax-6 is well known for its highly conserved role at the earliest stages in the development of the eye, and this is reflected by its expression in the CPE and by the fact that conditional deletion of the *Pax6* gene in isolated cells prevents their ability to form neurospheres. In fish, retinal progenitor cells with a restricted potential for rod neuroreceptors also exist in the inner nuclear layer (INL), and only under injury conditions is a broader contribution to retinal neurons seen.

As we have seen so often for tissue-associated stem cells, fate tracing experiments utilizing fluorescent proteins driven from regulatory elements of lineage-restricted genes is a powerful way to locate stem cells and to follow their differentiation into mature cells. This is no less the case for RSC and is beautifully illustrated by studies on the CMZ stem cells of the fish retina. Pulse labeling of adult fish with BrdU, in much the same way that Joseph Altman had done over half a century ago using tritiated thymidine, reveals active proliferation in the CMZ, with the marked cells progressively moving into differentiated layers and constituting more central positions in the retina (Figure 14.9). The same fish model is also well suited to trace the precise fate of the CMZ stem cells by direct transplantation of isolated cells into the stem/progenitor cell region. By deriving the donor cells from a fish line that can express GFP in any cell type of the adult, it is possible to visualize the progeny that individual transplanted cells give rise to (Figure 14.10). The fact that a single transplanted RSC produces cells in each of the three cell-containing retinal layers (INL, GCL, and ONL) and that all differentiated cell types can be identified, including photoreceptors, amacrine cells, horizontal cells, and Müller glial cells, demonstrates the mutlipotential nature of the injected stem cells.

## Müller glial cells perform a reserve stem cell function in the injured eye

Müller glial cells are a very intriguing component of the neural retina, being well positioned to monitor retinal homeostasis and contribute to retinal structure and function since they span all of the neural layers of the retina and make synaptic contacts with each of the other cell components. Most interestingly, Müller glial cells can act as additional stem cells when the eye has suffered an injury. Retinal progenitor-like characteristics have been noted in Müller glial cells, including gene expression, proliferative responses, and the ability to generate neurons under special conditions, at least in some species. In order for Müller glial cells to participate in mammalian retinal repair, three important steps can be envisaged (Figure 14.11): (1) cell reprogramming to adopt stem cell characteristics, (2) generation of a proliferating population of multipotent progenitors, and (3) progenitor cell cycle exit and neuronal differentiation.

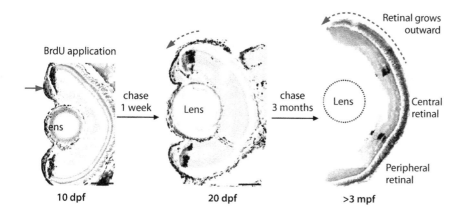

**BrdU application**

chase 1 week

chase 3 months

Retinal grows outward

Lens

Lens

Central retinal

Peripheral retinal

10 dpf          20 dpf          >3 mpf

**Figure 14.9 *In vivo* labeling of retinal stem cells in the fish eye.** Stem cells at the margins of the retina of vertebrates contribute to the development of the retina and in some species can continue to replace cells in adult life. Proliferating stem and progenitor cells within the ciliary marginal zone can be pulse labeled with bromodeoxyuridine (BrdU) at 10 days post fertilization (dpf). Once the BrdU is removed, the labeled cells can be tracked as they differentiate and become localized toward the central part of the retina. (From Centanin L, Hoeckendorf B & Wittbrodt J. (2011) *Cell Stem Cell* 9: 553–562 [doi: 10.1016/j.stem.2011.11.004]. Copyright 2011 with permission from Elsevier.)

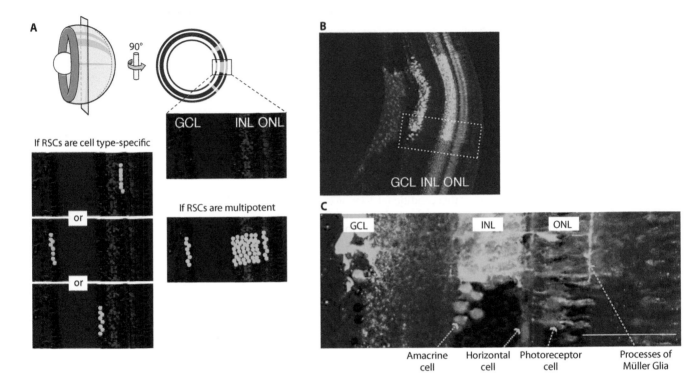

**Figure 14.10 Retinal stem cells in the fish are multipotent.** Single retinal stem cells isolated from the ciliary marginal zone (CMZ) of a transgenic medaka fish line that expresses GFP in all tissues of the adult can be injected into the CMZ of another fish. These cells incorporate into the growing retina, and their fate can be traced as they form along a strip from what becomes the back of the retina to the CMZ as it continues to move forward and the injected stem cell continues to self-renew. (**A**) Two scenarios are possible depending on the potency of the retinal stem cell injected: either it is monopotent, in which case only one cell type located in a specific retinal layer will result, or it is multipotent and can contribute to neural cell types in all or several of the layers. (**B**) The actual findings of such experiments conclusively demonstrate that a fish CMZ stem cell is multipotent. (**C**) A higher magnification view of the region marked in (B). CMZ, ciliary marginal zone; GCL, ganglion cell layer; INL, inner nuclear layer; ONL, outer nuclear layer, RSC, retinal stem cell. (From Centanin L, Hoeckendorf B & Wittbrodt J. (2011) *Cell Stem Cell 9*: 553–562 [doi: 10.1016/j.stem.2011.11.004]. Copyright 2011 with permission from Elsevier.)

Müller glial cells from fish, birds, and mammals share many aspects of their structure and function. Unlike birds and mammals, Müller glial cells from fish respond to retinal injury by undergoing a reprogramming event that allows them to acquire properties of a RSC. A variety of secreted growth factors from injured cells and Müller glial cells themselves appear to drive this reprogramming by activating a variety of signaling cascades, transcription factor expression, and changes in **epigenetic** modifications. Such stimulation of Müller glial cells is also seen in damaged retinas of birds and mice, but the proliferating cells exhibit only a limited ability to regenerate new neurons and generally do not survive.

Unraveling the mechanisms underlying Müller glial cell reprogramming and retina regeneration in fish along with studies on these cells in other species like birds and mammals may reveal novel strategies for stimulating retina regeneration in humans. For example, forced expression of Ascl1, a transcription factor involved in the specification of various neural lineages, can stimulate Müller glial cells in postnatal mouse retinal explants to reprogram and generate bipolar neurons. Human Müller glial cells share numerous characteristics with other neural stem cells and have been shown to express markers of retinal progenitors, such as *Sox2, Pax6, Chx10*, and *Notch1*, and when placed in culture are also capable of forming neurosphere-like cell aggregates. Additionally, when isolated Müller glial cells are tested *in vivo* by transplantation into the subretinal space of a rat model of retinal degeneration, the cells migrate and integrate into the photoreceptor layer and the INL and GCL.

## Specific stem cells support the maintenance of auditory, vestibular, and related sensory organs

Vibration of the media within which animals live is used to communicate and sense changes in the environment. The most obvious form that this takes is the vibration of air, which across a specific range of frequencies is what we and many other animals

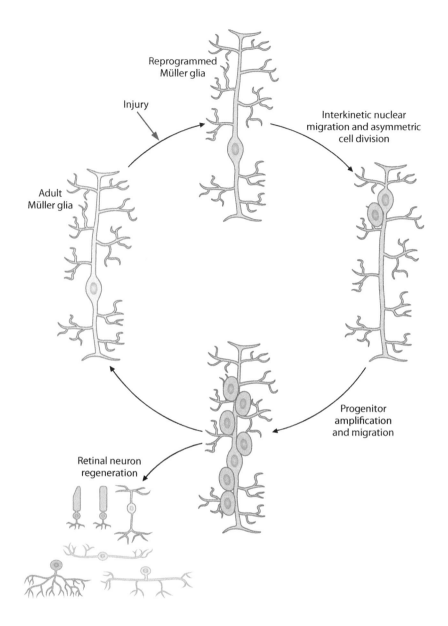

Figure 14.11 **Müller glial cell reprogramming following injury.** Studies involving injury to the retina in zebrafish revealed that Müller glial cells (yellow) are able to reprogram their genome (peach) to acquire stem cell properties. This results in nuclear migration and asymmetric division generating multipotent progenitors (pink) that transiently proliferate to restore the original Müller glial cells. The progenitors are also able to migrate to other retinal cell layers and regenerate all retinal neural cell types. (From Goldman D (2014) *Nat Rev Neurosci* 15: 431–442 [doi: 10.1038/nrn3723]. Adapted by permission from Springer Nature.)

perceive as sound through the mechano-sensory organs within the ear. Fish, and other aquatic animals, of course live in a quite different medium, but this is equally able to transmit vibrations, and sensory cells that are quite homologous to those in land animals have evolved to be able to detect this information. Although the sensory organs of the ear respond to the vibrations in the air, they actually detect movement in an enclosed, liquid-filled environment, probably reflecting the evolutionary origins of this sense in aquatic vertebrates.

As for the neural retina, the capacity for mechano-sensory organs to regenerate varies greatly between vertebrate species. The hair cell is the mechano-sensory receptor found in all vertebrates that functions to detect movement and vibrations, being found in many different mechano-sensitive organs ranging from the lateral line of fish to the inner ear of mammals. In fish, amphibians, and birds, regeneration of sensory hair cells through asymmetric cell divisions of supporting cells can contribute to recovery of hearing and balance after hair cell loss caused by trauma or toxicity. In mammals, on the other hand, the hair cells of the auditory and vestibular systems in the inner ear (Box 14.4) show at best a minimal ability to renew, explaining the permanent loss of hearing consequent to damage and associated with aging.

In fish, mechano-sensitive hair cells are found in the structure known as the lateral line, which, due to its location at the surface, has made experimental observation relatively easy. Clusters of hair cells called neuromasts have been shown to regenerate

BOX 14.4 BACKGROUND: THE MAMMALIAN INNER EAR

Ears are involved in the reception and transmission of sound to the brain but also function to control our balance. The inner ear encompasses the structures from the point at which the small bone called the stirrup transmits sound pressure onto a thin membrane, the oval window, to the exit of the nerves that take the impulses to the brain (Figure 1). The inner ear can be thought of as two organs: the vestibular apparatus, which serves as the body's balance organ, and the cochlea, which is essentially the body's microphone, converting sound pressure waves from the outer ear into nerve impulses.

The vestibular apparatus contains fluid filled semi-circular canals arranged in planes orthogonal to one another, which are joined at their ends (ampullae) to two linked chambers, the utricle and saccule. Two types of sensory epithelium are contained within these vestibular structures, both of which contain mosaics of sensory hair cells and surrounding support cells. In the ampullae, hair cells that detect rotation project their cilia into a floppy protuberance called the cupula that sways as the canal fluid shifts. The utricle and saccule each house an otolithic organ, or macula, which senses gravity and acceleration. The hair cells in the maculae are coupled to a layer of tiny floating particles (otoconia, effectively small stones), which help to transmit movement to them (Figure 2). The hair cells in each of these hair cell structures are connected to the vestibular nerve.

The cochlea is a snail-shell like structure linked to the vestibular apparatus, which is divided into three fluid-filled parts. Two are canals for the transmission of pressure and in the third is the sensitive organ of Corti, which consists

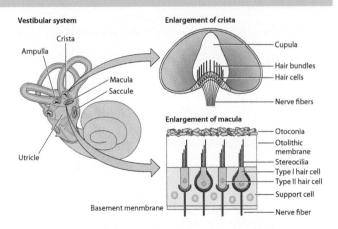

Box 14.4 Figure 2 **Sensory cells in the vestibular system.** Diagram depicting the structure of the vestibular system and its sensory organs. Sensory hair cells are found as bundles in the crista organs present in each ampulla at the end of the semi-circular canals. The cilia of these hair cells protrude into the cupula, which is a flexible structure that sways in response to movements in the fluid in the ampulla. The other sensory organ, the macula, is present in the chambers known as the utricle and saccula. Hair cells in the macula similarly protrude via their cilia (kinocilia and stereocilia) into a semi-solid substance, the movement of which in response to acceleration is enhanced by a layer of proteinized calcium carbonate granules (otoconia).

of just two specialized sensory cells, known as the outer and the inner hair cells, that function to amplify low-level sound and transmit the sound signal to the brain along the auditory nerve (Figure 3).

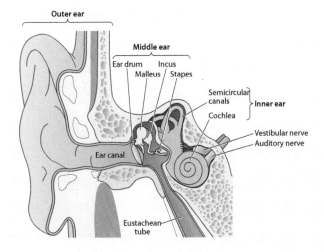

Box 14.4 Figure 1 **The structure of the human ear.** Cut-away section through the human ear illustrating the location of the inner ear (mauve) composed of the cochlea and semi-circular canals, which respectively constitute the auditory and vestibular functions of the ear. The auditory and vestibular nerves emanating from these structures are highlighted in yellow.

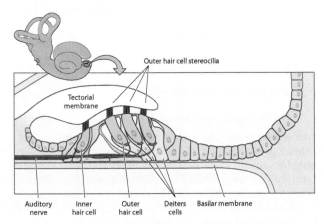

Box 14.4 Figure 3 **Sensory cells in the cochlea.** A diagrammatic cross section of the sensory structure of the mammalian cochlea showing detail of the organ of Corti. The hair cells (green), discriminated as inner and outer, project toward the tectorial membrane, which transduces fluid vibrations into shear force that it imparts to the stereocilia of each sensory cell. The hair cells are surrounded by support cells (yellow and pink), which rest on the basilar membrane.

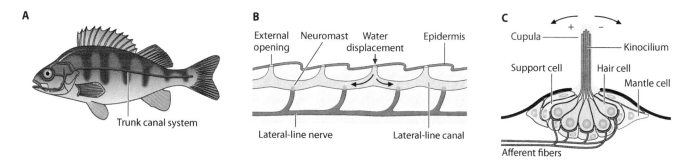

**A** Trunk canal system

**B** External opening · Neuromast · Water displacement · Epidermis · Lateral-line nerve · Lateral-line canal

**C** Cupula · Kinocilium · Support cell · Hair cell · Mantle cell · Afferent fibers

Figure 14.12 **Mechano-sensory hair cells of the fish lateral line.** Sensory neural cells in the lateral line of fish are maintained by stem cells that reside in a niche associated with each neuromast. (**A**) Location of the lateral line along the body of a fish (indicated by the red line). (**B**) Schematic diagram of a longitudinal section through the lateral line showing how hair cells (blue) in the neuromast protrude into the lateral line canal to detect any fluid movement. (**C**) Cross-sectional diagram of a neuromast illustrating the relative arrangement of the sensory hair cells and the surrounding support and mantle cells. (Adapted from http://www.britannica.com/science/lateralline-system.)

following damage. Detailed imaging of the neuromast has identified three distinct cell types: the hair cells, the support cells in direct contact with the hair cells, and mantle cells at the periphery (Figure 14.12). Studies using time-lapse videography have shown that a lost hair cell is replaced by both **transdifferentiation** of the contacting support cell and by the proliferation and differentiation of the mantle cells, indicating that the support cells and mantle cells can effectively act as stem cells of the lateral line.

In birds, both the hair cells of the utricle and of the cochlea spontaneously regenerate after damage and therefore restore lost balance and hearing. Although the mechanism of hair cell regeneration is almost identical to the situation in the fish, there is one key difference in that a distinct stem cell population is not so evident in the avian cochlea. In birds, the response to damage is for the supporting cells to divide and the daughter cells become new hair cells. In some cases, the transdifferentiation from the supporting cell to the hair cell phenotype appears to take place without mitosis. Once a support cell has entered cell cycle and replenished a lost hair cell, time-lapse imaging suggests that this same support cell rarely repeats this act. This is not to say that the support cells do not retain an element of stem cell-like activity since they are clearly capable of reactivating developmental genes such as *Pax, Notch, Atoh1,* and *Cdkn1b.*

## Vestibular and auditory ear hair cells in mammals differ in their capacity for replacement from stem cells

Hair cell regeneration in the mammalian inner ear is much more complicated. Evidence suggests that hair cells in the vestibular system have a limited capacity to regenerate, this ability is well below the level required for functional recovery following damage, and regenerative capacity in the cochlea is lost very soon after birth. Similar to other neural stem cell populations, the sphere-forming assay can be used as a means to detect stem cell activity. When treated with EGF or FGF-2, the number of spheres, in this case called **otospheres**, increases. Cells explanted from the vestibular sensory epithelia and the organ of Corti of neonatal mice give rise to spheres that express genes that are indicative of inner ear progenitor cells, including *Otx2, BMP4, BMP7,* and *Islet1.* During further differentiation, some cells grow hair bundle-like structures with hair cell markers myosin VIIA and espin. The ability for sphere formation in the cochlea decreases about 100-fold within 3 weeks of birth, which is substantially faster than the reduction of stem cells in the vestibular organs (Figure 14.13). Intriguingly, expression of the Wnt target gene *Lgr5,* which, as we discuss in Chapter 11, was initially identified as a marker of stem cells in the small intestine and other epithelial structures in the gastrointestinal tract, has been linked to hair cell regeneration in the mammalian inner ear. Using lineage tracing, it was shown that Lgr5-expressing inner pillar and third Deiters cells, which act as supporting cells in the cochlear epithelium, can differentiate into new hair cells in the neonatal mouse following damage (Figure 14.14). Recently, it has been shown that hair cell damage in the neonatal mouse utricle activates Lgr5 in supporting cells, and using lineage tracing and time-lapse microscopy, it could be seen that these cells then transdifferentiate into hair-like cells *in vitro* and *in vivo* and regenerate hair cells through mitotic and transdifferentiation pathways (Figure 14.15).

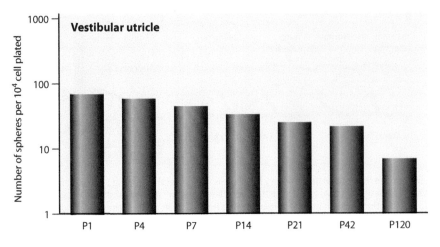

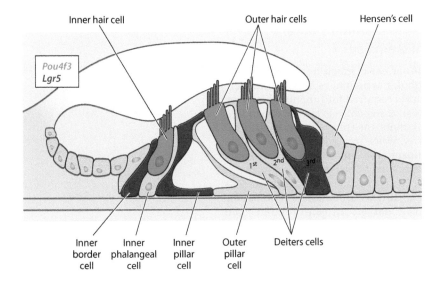

Figure 14.13 **Otosphere forming potential of cells from the vestibular and auditory epithelia of the mouse.** The formation of clusters (otospheres) of cells in culture from the inner ear epithelium can be an indication of the presence of cells with stem/progenitor characteristics. The graphs show quantification of otosphere-forming cells in populations of cells isolated from the sensory epithelia of mouse vestibular utricle and cochlear organ of Corti at different times (postnatal days, P). (Adapted from Oshima K, Grimm C, Corrales C et al. (2007) *JARO* 8:18–31 [doi: 10.1007/s10162–006–0058–3].)

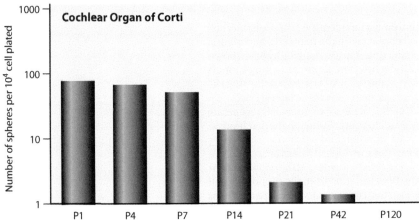

Figure 14.14 **Lgr5 expression in the stem cells of the cochlea.** The organ of Corti in the inner ear senses sound through the stimulation of hair cells. The diagram shows the structure of the organ of Corti and the location of hair cells (blue, expressing the transcription factor Pou4f3) and Lgr5-positive supporting cells (red). (Adapted from Bramhall NF (2014) *Stem Cell Reports*, 2: 311–322.)

The reasons for the difference between vestibular and auditory epithelia in mammals in terms of their stem cell properties is not well understood. It has been suggested that evolution of the auditory organ, which is more recent than vestibular peripheral sensory organs, was accompanied by a loss of stem cell capability in the cochlear supporting cells in exchange for an increased complexity allowing a greater range of hearing at both frequency and amplitude. However, the presence in the inner ear, at least in the mammalian neonate, of cells that can generate hair cells, whether these are true stem cells or work through cell division and transdifferentiation, suggests

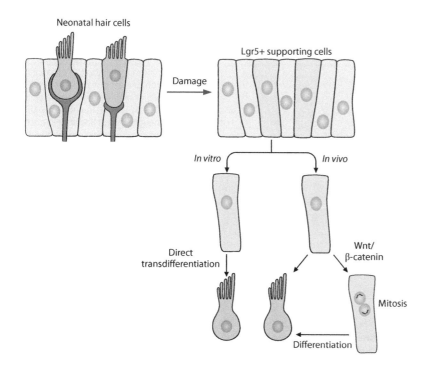

Figure 14.15 **Lgr5+ cells regenerate hair cells in the vestibular utricle by transdifferentiation.** Lineage tracing experiments have been used to show how transdifferentiation of Lgr5+ support cells in the inner ear epithelium of neonates can lead to replacement of damaged hair cells. The presence of Lgr5 in the neonatal mouse vestibular utricle sensory epithelium is revealed by expression of a Lgr5-GFP transgene. Induced toxic damage to the macula hair cells results in activation of Lgr5 expression (blue) in some of the surrounding support cells. When isolated by cell sorting, these Lgr5+ cells can be transdifferentiated *in vitro* into functional hair cells. Likewise, observations *in vivo* show a similar generation of new hair cells, but this process can also be accompanied by an expansion in cell number through mitotic division of the Lgr5+ cell. (From Wang T, Chai R, Kim G et al. (2014) *Nature Comm*, 6: 6613 [doi:10.1038/ncomms7613]. Reprinted by permission from Springer Nature.)

that it may eventually be possible to repair damaged cochlear hair cells, perhaps harnessing developmental genes, such as *Atoh1*, to enhance transdifferentiation in the mature sensory epithelium (Chapter 17).

## Stem cells for taste receptors are active throughout life

In contrast to the sensory organs in the eye and the inner ear, the sensory tissue that underpins the sense of taste is able to regenerate continuously throughout life in mammals. Based on morphological and functional characteristics, within a taste bud (Figure 14.16), there are at least three different types of mature taste bud cells (Figure 14.17), each of which is post-mitotic and short lived: Type I glial-like supporting cells; Type II receptor cells, including those responsible for sensing sweet,

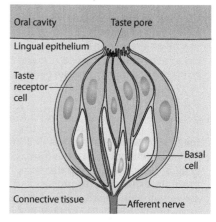

Figure 14.16 **Taste bud sensory organ.** Taste is elicited through the interaction of "tastant" molecules with specific receptors on cells in discrete organs (taste buds) in the lingual epithelium. The cross section of a taste bud in a mammalian tongue shows the arrangement of taste cells with support cells terminating in microvilli at the taste pore. (Adapted from "Schematic drawing of a taste bud" by NEUROtiker – Own work. Licensed under Creative Commons Attribution-Share Alike 3.0 Unported via Wikimedia. (http://creativecommons.org/licenses/bysa/3.0/deed.en.)

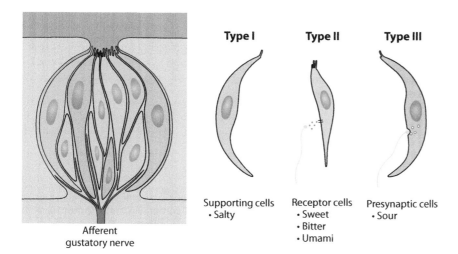

Figure 14.17 **Taste bud sensory cells.** The cells of the taste bud belong to one of three types depending upon the range of taste molecules that they detect. The diagram on the left is a cross section of a taste bud, illustrating the arrangement and diversity of taste cells and support cells and their synaptic connections to the gustatory nerve. The three types of sensory taste cells are depicted on the right, together with the ranges of tastants to which they can respond (Adapted from http://knowingneurons.com/2013/04/01/newion-channel-identified-for-the-neurotransmission-of-sweetbitter-and-umami-tastes/.)

**A**

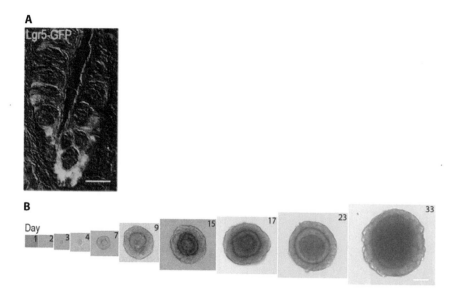

**B**

Figure 14.18 **Taste bud sensory neuron stem cells form organoids.** Lineage tracing of Lgr5 expression shows the presence of potential stem cells in taste buds and enables their isolation and culture as organoids. (**A**) Section of a tongue papilla containing taste buds, showing the presence of Lgr5 as revealed by expression of a Lgr5-GFP transgene (green). (**B**) Bright field microscopy images of cultured organoids derived from single sorted Lgr5-GFP+ taste cells at various days of the culture period. (From Ren W, Lewandowski B, Watson J et al. (2014) *Proc Natl Acad Sci* 111:16401–16406 [doi:10.1073/pnas.1409064111]. With permission.)

bitter, and umami stimuli; and Type III presynaptic cells, including sour sensors. The continual turnover of taste bud cells certainly necessitates the involvement of specific stem cells, but their identity and the nature of the regenerative process are not well characterized. Yet again, though, the *Lgr5* gene (as well as its closely related homolog *Lgr6*) is emerging as a good candidate to mark adult taste stem or progenitor cells in the tongue. Single isolated Lgr5+ cells from taste tissue can generate continuously expanding **organoids** (Figure 14.18), analogous to neurospheres, from which cells can be differentiated in culture that are functional in response to tastants in a dose-dependent manner. Genetic lineage tracing *in vivo* also shows that Lgr6+ cells gave rise to taste bud cells in taste papillae in both the anterior and posterior tongue.

## 14.4 SUMMARY

In this chapter, we have seen that the central and peripheral nervous systems, including the sensory organs that input information from the environment, have a varied dependence on resident stem cells. The role of these stem cells ranges from the generation of new neurons that serve to increase the learning capacity of the brain, for example, expanding the profile of olfactory neurons in rodents; to repair of injury, as seen in the spinal cord; and life-long replacement of sensory organ components such as vestibular hair cells and taste bud nerve cells. The understanding of key features of adult stem cells is nicely illustrated through studies on the different examples seen in the nervous system, including the importance of heterogeneity and the way in which the structure of a given niche affects the maintenance and function of a stem cell. Without doubt, the growth in knowledge of nervous system stem cells, especially the NSC, will have clinical implications in the not too distant future.

---

## KEY POINTS

- Both the central and peripheral components of the vertebrate nervous system exhibit continual cell replacement, although this varies in its extent depending on the specific neural cell type and between different species.

- Two regions of the vertebrate brain contain regions of active stem cells, namely the ventricular-subventricular zone (V-SVZ) and the subgranular zone (SGZ) in the dentate gyrus.

- The stem cells in the V-SVZ and SGZ are known as neural stem cells (NSC), and these have precisely defined niches and hierarchical progression toward their differentiation

progeny. The role of the V-SVZ is at most minimal in humans that, unlike rodents and some primates, do not have the same demand for new neurons in the olfactory bulb.

- NSC can be grown in culture as neurospheres and are able to differentiate into both neurons and non-neuronal astrocytes and oligodendrocytes.

- True NSC are not found in the spinal cord; however, ependymal cells lining the spinal canal, as well as astrocytes and oligodendrocyte progenitors, are able to proliferate under normal conditions. When the spinal cord is injured, the ependymal cells expand dramatically and can give rise to both astrocytes and oligodendrocytes.

- Specialized neural cells act as the sensory component of vertebrate organs for vision, hearing, smell, taste, and touch/sensation. All but the latter can be replaced from stem cells or by transdifferentiation, although the potential for this is very much species dependent.

- The neural cells of the eye, contained in the retina, can be replaced from retinal stem cells. In addition, Müller glial cells have some capacity under injury conditions to reprogram and give rise to retinal neural cells.

- Sound and motion are detected by specialized neurons called hair cells. These cells are located in a variety of structures in the inner ear of higher vertebrates and can be regenerated from stem cells in most species, although much of this capability has been lost in mammals.

- The vestibular apparatus in the inner ear detects motion and contains stem cells in two hair cell-containing organs. All higher vertebrates, including mammals, exhibit a degree of regenerative capacity after loss of these cells through injury.

- Hair cells in the vertebrate inner ear auditory organ of Corti, located in the cochlea, can be regenerated from surrounding support cells, although this capability is very limited in mammals after birth.

- The equivalent sensory organ to the ear in fish is the lateral line, in which a structure known as the neuromast utilizes analogous hair cells to detect vibrations in water. These hair cells can likewise be replaced from surrounding support cells.

- Three types of neural sensory cells are present in the taste buds of the human tongue, each capable of detecting different classes of taste. These sensory cells are replaced from adjacent Lgr5+ support cells.

## FURTHER READING

### Neural stem cells

Barnabé-Heider, F., Göritz, C., Sabelström, H., Takebayashi, H., Pfrieger, F.W., Meletis, K. and Frisén, J. (2010) Origin of new glial cells in intact and injured adult spinal cord. *Cell Stem Cell.*, 7: 470–482.

Bond, A.M., Ming, G.L. and Song, H. (2015) Adult mammalian neural stem cells and neurogenesis: Five decades later. *Cell Stem Cell.*, 17: 385–395.

Fuentealba, L.C., Obernier, K. and Alvarez-Buylla, A. (2012) Adult neural stem cells bridge their niche. *Cell Stem Cell.*, 10: 698–708.

Gage, F.H. and Temple, S. (2013) Neural stem cells: generating and regenerating the brain. *Neuron*, 80: 588–601.

Obernier, K. and Alvarez-Buylla, A. 2019 Neural stem cells: origin, heterogeneity and regulation in the adult mammalian brain. *Development*, 146.

Tang, Y., Yu. P. and Cheng, L. (2017) Current progress in the derivation and therapeutic application of neural stem cells. *Cell Death Dis.*, 8: e3108.

### Retinal stem cells

Centanin, L., Hoeckendorf, B. and Wittbrodt, J. (2011) Fate restriction and multipotency in retinal stem cells. *Cell Stem Cell.*, 9: 553–562.

### Sensory organ stem cells

Bramhall, N.F., Shi, F., Arnold, K., Hochedlinger, K. and Edge, A.S. (2014) Lgr5-positive supporting cells generate new hair cells in the postnatal cochlea. *Stem Cell Rep.*, 2: 311–322.

Franco, B. and Malgrange, B. (2017) Concise review: Regeneration in mammalian cochlea hair cells: Help from supporting cells transdifferentiation. *Stem Cells.*, 35: 551–556.

Ren, W., Lewandowski, B.C., Watson, J., Aihara, E., Iwatsuki, K., Bachmanov, A.A., Margolskee, R.F. and Jiang, P. (2014) Single Lgr5- or Lgr6-expressing taste stem/progenitor cells generate taste bud cells *ex vivo. Proc Natl Acad Sci U S A.*, 111: 16401–16406.

Yee, K.K., Li, Y., Redding, K.M., Iwatsuki, K., Margolskee, R.F. and Jiang, P. (2013) Lgr5-EGFP marks taste bud stem/progenitor cells in posterior tongue. *Stem Cells.*, 31: 992–1000.

# Stem cells in the skin, its appendages, and other epithelial tissues

# 15

Although it may not seem like it, the skin is our largest organ and one that has impressive demands on stem cells for its upkeep. The dust you wipe from the surfaces in your home contains skin debris, which is shed continually and therefore must be being replaced at the same rate. In fact, we effectively replace our skin about once a month. The shedding of skin reflects its principal function, that is, as a barrier fending off assaults from the environment in the shape of infectious agents, allergens, chemicals, heat and the risk of dehydration, radiation, and anything else it cares to throw at us. Skin is much more though than simply a barrier protecting our bodies, playing important roles in features as diverse as temperature regulation, camouflage, and sexual attractiveness, and as such incorporates hair, sebaceous glands, sweat glands, and pigmentation. Considering the vertebrate epidermis more widely, it has an amazing ability to produce elaborate appendages such as the scales on a snake, hooves of many animals, and the feathers on birds, which, despite varying greatly in morphology and function, all begin their development in a similar manner.

In this chapter, we will take a close look at the stem cells that underpin the skin and related epithelial surfaces, including the outer surface of the eye (the cornea) and the oral epithelium. The specialized epithelia of the lungs and the gastrointestinal (GI) tract below the oral cavity are considered in detail in Chapter 11, although where appropriate we will highlight any parallels between the stem cells in these tissues and those found in the skin and eyes. We will also discuss epithelia-associated organs (or "appendages"), including hair (or feathers in the case of birds), sebaceous glands, sweat glands in the skin, and salivary glands in the oral cavity. Last, we will focus our attention on the mammary gland, which can be regarded as an appendage of the skin, just like hair follicles and sweat glands, in that it has a similar developmental origin and relies on a secretory epithelium that has many features in common with the other epithelial tissues, in particular the sweat glands.

## 15.1 EPITHELIA: THE FRONTIER TISSUES OF THE BODY

The majority and essential component of skin, the epidermis, is constituted of epithelial cells. The epidermis is only one of several different surface epithelia throughout the body, each of which has epithelial cells serving in some way as an interface between the body and the outside world. Such interfaces include the whole length of the GI tract, from the mouth to the anus, epithelia in the genitourinary tracts, the airways and lungs, and the outer surface of the eye. This diversity of tissue epithelia is mirrored by distinctive specializations of the component epithelial cells (Box 15.1).

### Epithelial tissues and their associated appendages have a common developmental origin in the ectoderm

Of the three germ layers of the developing embryo, the **ectoderm** represents the external layer that yields the epidermis and other stratified epithelia, including that in the mouth. These tissues give rise to a range of specialized appendages such as hair

## BOX 15.1 BACKGROUND: EPITHELIAL CELLS

Epithelial cells form the layers that define surfaces that are exposed to the external environment, including the skin, the surface of the eye (cornea), the lining of the gastrointestinal and genitourinary tracts, and the surface of the airways and lungs. Epithelial cells are bound together more tightly than other cells, utilizing structures including tight junctions, desmosomes and gap junctions, but can slough off and glide in order to replace dead cells, enabling maintenance of a closed barrier. Epithelial cells are functionally polarized (Figure 1) so that the external (apical) face is in contact with either the outside world or the luminal space, while the opposite (basal) surface is supported by a basement membrane called the basal lamina, below which lies a capillary bed that provides nutrients and disposes of waste products. The cell nucleus tends to be closer to the basal surface. Epithelia can be divided into six types that relate to their distinct functions of barrier protection, absorption and excretion, lubrication, and movement of substances in a luminal space: (1) simple squamous, (2) simple cuboidal, (3) simple columnar, (4) ciliated columnar, (5) glandular, and (6) stratified (Table 1).

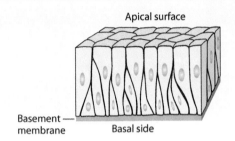

**Box 15.1 Figure 1 Polarized epithelial cell structure.** The epithelium is composed of layers of tightly bound epithelial cells that act as a protective barrier to the external environment. The schematic diagram represents a segment of epithelial membrane showing the apical-basal polarization of the epithelial cells. Epithelial cells have specialized apical membranes enhancing their absorptive and secretive abilities. Neighboring cells adhere to one another (via desmosomes, tight junctions, and adherens junctions) and to the basement membrane and extracellular matrix (ECM) through the presence of various receptors, including integrins.

### BOX 15.1 TABLE 1  DIVERSITY OF EPITHELIA TISSUES

| Epithelial type | | Example tissue |
|---|---|---|
| Simple squamous | | Lung alveoli, kidney glomerulus |
| Simple cuboidal | | Kidney tubules, respiratory bronchioles |
| Simple columnar | | Gastrointestinal tract, nasal passage |
| Ciliated columnar | | Trachea and bronchi, Fallopian tubes |
| Glandular | | Sebaceous glands, gastrointestinal linings, thyroid follicles |
| Stratified (squamous, cuboidal or columnar) | | Skin, lining of the mouth |

*Note:*  There are many types of epithelia in the body, which can be divided into at least six types that relate to their particular function as either barrier protection, absorption and excretion, lubrication, and movement of substances in the lumen. The table lists the types of epithelial cells found throughout the body, schematically represents their structure, and gives representative examples of the tissues in which they are found.

follicles, sweat glands, teeth, salivary glands, and mammary glands (Figure 15.1). A putative common stem cell population in the developing embryo's surface ectoderm is thought to give rise to stem cells in the skin and oral epithelium that further specialize into the appendages. In a similar way, and as we describe in Chapter 11, the **endoderm** is the ultimate source of those epithelial structures that are found in the GI tract from the pharynx to the anus.

In their exposed locations, the epithelial surfaces and associated structures are all at great risk of being damaged by physical, chemical, and biological stresses and as such rely heavily on the ability of **adult stem cell** populations to repair and regenerate. As we will see, a lot of progress has been made in determining the nature of these stem cells in several structures, although the picture is also far from clear in some cases.

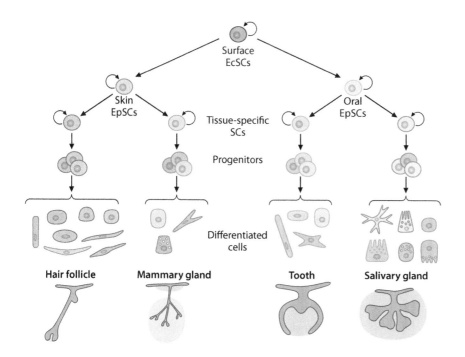

Figure 15.1 **Developmental origins of epithelial tissues and their associated appendages.** During embryonic development, the ectoderm germ layer yields the skin epidermis and oral epithelia. A common stem cell population at the apex of the developmental hierarchy gives rise to the stem cells of the skin and oral epithelia that in turn can develop into appendages such as the hair follicle, mammary gland, tooth, and salivary gland. The putative common stem cell, the surface ectodermal stem cell (EcSC), gives rise to both skin and oral epithelial stem cells (EpSC) that in turn yield tissue-specific stem cell types that ultimately produce the mature differentiated cells of the specific organ. (From Jiménez-Rojo L, Granchi Z, Graf D et al. (2012) *Front Physio* 3:107 [doi: 103389/fphys.2012.00107]. Copyright © 2012 Jimenez-Rojo, Granchi, Graf and Mitsiadis. Used with permission.)

## 15.2 STEM CELLS IN THE SKIN

The skin of vertebrates is a complex organ (Box 15.2) made up of layers of specialized epithelial cells capable of maturing into keratinocytes as they progress toward the surface, where they eventually die and are shed. These epithelial cells constitute the epidermis, which itself is underlain by dermal tissues containing blood vessels, nerves, muscles, connective tissue, and a layer of fat.

### BOX 15.2 BACKGROUND: THE SKIN AND ITS ASSOCIATED ORGANS

The skin is made up of multiple cell types arranged in a layered manner (Figure 1). The outer layer is the stratum corneum, which is constituted by fully differentiated post-mitotic keratinocytes. Below lies the epidermis, which functions primarily as a protective barrier against the environment. The skin is peppered with a number of different embedded structures, or "appendages," that are crucial for its overall tissue function. These include a multitude of hair follicles, which not only produce hairs but are also the source of protective oily secretions produced

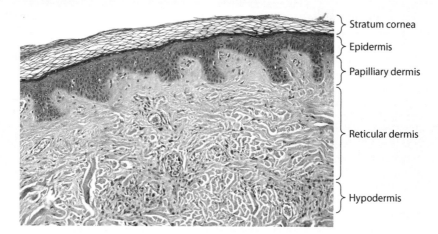

Box 15.2 Figure 1 **Structure of the skin.** The skin of vertebrates is highly complex and is made up of a range of cell types arranged in layers. Histological analysis of a cross section of mammalian skin stained with hematoxylin-eosin clearly depicts the defined layers from the stratum cornea at the surface through to the underlying hypodermis ("Normal epidermis and dermis with intradermal nevus 10xcropped" by Kilbad – Own work. Licensed under Public Domain via Wikimedia.)

by the sebaceous gland, and the sweat glands that, like hairs, contribute to temperature regulation (Figure 2). The number of these appendages varies considerably in different parts of the body as well as in a species-specific manner. The area between the hair follicles is termed the interfollicular epidermis, the thickness of which varies between different

parts of the body. Beneath the epidermis lies the dermis, which is enriched in fibroblasts that produce collagens and elastic fibers that make up the **extracellular matrix** (ECM) and provide the skin with its elasticity. The dermis is further subdivided into three layers: the papillary, which contains the highest density of dermal fibroblasts; the reticular, which maintains an abundance of collagens; and the hypodermis, which is constituted by a thick layer of adipocytes. Within the dermis, there are many other cell types, including cells of the peripheral nervous system, blood vessels, and cells of the innate and adaptive immune systems. There are two mesenchymal structures also present within the dermis, namely the dermal papilla that is thought to control hair follicle cycles and the arrector pili muscle that plays a critical role in thermoregulation by causing hairs to stand erect (causing "goose bumps"). Below the dermis is the hypodermis that has a layer of fat that provides protective padding and insulation and acts as an energy reservoir.

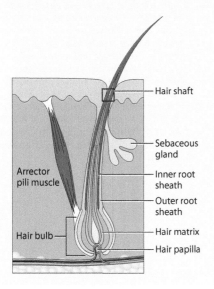

Arrector pili muscle

Hair bulb

Hair shaft

Sebaceous gland

Inner root sheath

Outer root sheath

Hair matrix

Hair papilla

**Box 15.2 Figure 2 The hair follicle.** Throughout the epithelium of the skin, there are numerous appendages such as the hair follicle. The schematic cross-sectional representation of a hair follicle embedded in the skin shows the main components from the growth point of the papilla with its associated blood supply through to the hair shaft. Additional appendages linked to the hair follicle include the sebaceous gland; nerves (not shown); and a muscle fiber, the arrector pili muscle, which can make the hair stand upright.

Definition of the nature and location of stem cells in the skin, although playing catch up to some extent compared to **hematopoietic stem cells** (HSC) (Chapter 9), has arguably become a paradigm for the application of methodology for the identification, isolation, and characterization of stem cells (Chapter 2). In fact, in many ways, our current understanding of some aspects of skin stem cell biology, in particular the details of their **niches** (Chapter 6), is unrivalled, with the possible exception of the stem cells giving rise to the specialized epithelial cells of the GI tract (Chapter 11).

### There are two models for the way that stem cells support the continual replacement of the skin

Over 40 years ago, Chris Potten, working in the Paterson Laboratory in Manchester, England, proposed that what he termed epidermal proliferative units maintain the homeostatic balance of the interfollicular epithelium (IFE). The presence of stem cells within the IFE was suggested by label retention experiments (Chapter 2) showing that 1%–10% of cells are labeled in the basal layer (Figure 15.2). Potten's model hypothesized the presence of small groups of cells in the basal layer with a central, slowly dividing stem cell, which produce columns of cells consisting of rapidly dividing **transit amplifying cells** that migrate out of the epidermal basal layer niche and **differentiate** toward keratinocytes as they move toward the surface (Figure 15.3).

The presence of stem cells and transit amplifying cells in the skin epidermis was also deduced around the same time from cell culture experiments. In the mid 1980s, Yann Barrandon and others discovered that cells in the epidermis that are able to form colonies *in vitro* could be distinguished by the morphology of the colony. They described three different colony morphologies: holoclones, which are small, round and infrequently differentiate; meroclones, which are smaller colonies with irregular edges; and paraclones, which contain cells with a limited division potential. It is now accepted that holoclones consist primarily of stem cells; meroclones contain slightly more differentiated yet highly proliferative transit amplifying cells; and paraclones are composed of **committed**, terminally differentiating cells.

Potten's model assumes that the stem cells within the basal layer behave in a hierarchal manner (Chapter 1), that is, they divide **asymmetrically,** giving rise to one stem cell and one transit amplifying cell, the latter achieving the majority of the cell expansion within the epidermis and ultimately leading to the non-dividing differentiated cells. This model is now known as the invariant asymmetry model. More

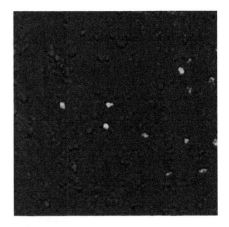

**Figure 15.2 Label retention identifying potential interfollicular stem cells.** Stem cells within the interfollicular epidermis (IFE) were first identified by label retention experiments. This fluorescent microscopy image of murine IFE illustrates label retention by putative stem cells. The IFE has been labeled by treating the animal with bromodeoxyuridine (BrdU), followed by a label-free chase period. Retention of the BrdU label (green) clearly distinguishes the stem cells from transit amplifying cells revealed by immunofluorescent staining for the proliferation marker Ki67 (red). (From Kaur P. (2006) *J Invest Dermatol* 126:1450–1458 [doi: 10.1038/sj.jid.5700184]. Copyright 2006 with permission from Elsevier. Photo credit: Dr. Kristin Braun & Dr. Fiona Watt. Used with permission.)

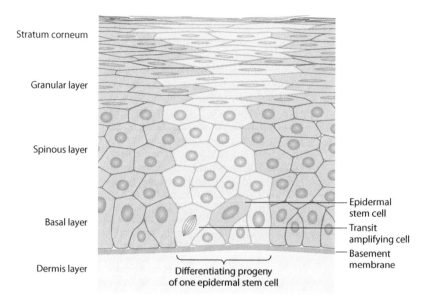

Stratum corneum

Granular layer

Spinous layer

Basal layer — Epidermal stem cell
— Transit amplifying cell
— Basement membrane

Dermis layer

Differentiating progeny of one epidermal stem cell

**Figure 15.3 The invariant asymmetry model of stem cells in the IFE.** Chris Potten first proposed the invariant asymmetry model for the behavior of stem cells in the skin epidermis over 40 years ago. His invariant asymmetry model, as it is known, assumes that within discrete cellular territories in the IFE, there are epidermal proliferation unit (dark green) stem cells that behave in a hierarchical manner, dividing to give a stem cell and a transit amplifying (TA) cell (pale green) in the plane of the epidermis. The TA cells then proliferate and differentiate in a direction perpendicular to the basement membrane, moving progressively to the skin surface as they become mature keratinocytes.

recent studies on the epidermis, and also the intestinal and esophageal epithelia, suggest that **epidermal stem cells** actually have three rather than two possible fates. Using a **lineage tracing** model based on hormone-activated **Cre recombinase** that irreversibly activates fluorescent protein expression from a **transgene**, it was shown that the distribution of labeled basal cells in distinct **clones** in these epithelial tissues exhibits dynamics that are best explained by a population asymmetry model. This model, also known as the **neutral drift model**, proposes that all basal layer cells in the epidermis are equipotent stem cells that can divide randomly in one of three ways: into two stem cells, into a stem cell and a cell marked for differentiation, or into two cells marked for differentiation (Figure 15.4). Thus, at the individual cell level, stem cells can divide symmetrically, but at the population level, these divisions are asymmetric in that they are balanced to maintain tissue homeostasis.

## The hair follicle contains several distinct stem cell populations that support skin epithelial appendages

Hairs are produced by individual hair follicles, each of which is a complex structure that not only serves the function of producing hairs and secretions but importantly also represents a major location for stem cells within the skin. These stem cells in the hair follicle act as the source of differentiated cells that make up the hair itself, the coloration of hair and skin, and the structure of the sebaceous gland. They can also contribute, under some circumstances, to the repair of the IFE. Hair development, and the activity of the stem cell populations within the hair follicle, follows a cycle of growth, regression, and **quiescence**, which is repeated with a periodicity dependent on the location of the hair follicle (Box 15.3).

During the 1990s, pulse chase experiments, using labeling with bromodeoxyuridine or tritiated thymidine, identified a region within the hair follicle that after a 4-week-long chase period contained **label-retaining cells** (LRC). This region is the hair follicle **bulge**, which is present throughout the entire hair follicle cycle (Figure 15.5). After the discovery of the putative bulge stem cells, markers were identified that could be used directly to track or isolate cells or the promoters of their encoding genes could be utilized for lineage tracing. Since then, tracing studies have led to the identification of several distinct stem cell populations that contribute to the hair follicle and the IFE. The markers of these different populations, which have distinct and sometimes overlapping patterns of expression throughout the hair follicle (Figure 15.6), include keratins (5, 14, 15 and 19), **cluster of differentiation** 34 (CD34), the transcription factor Blimp1, the leucine-rich repeat-containing G-protein coupled receptors 5 and 6 (Lgr5 and Lgr6), the leucine-rich repeats and immunoglobulin-like domains protein 1 (Lrig1), and the surface protein MTS24. Interestingly, Lgr5 was initially identified as a marker of stem cells present in the crypts of the small intestine (Chapter 11), perhaps reflecting some common principles underlying epithelial stem cells, even when they have arisen from different embryological routes.

**A**

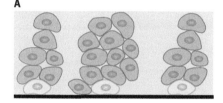

**B**

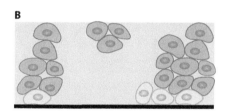

**Figure 15.4 The neutral drift model of stem cells in the IFE.** Lineage tracing experiments utilizing hormone-dependent Cre recombinase to irreversibly activate a fluorescent protein reporter was used to probe the dynamics of stem cells underpinning the IFE. Such studies revealed that the distribution of basal cells of the epidermis in distinct clones can be explained by a population asymmetry, or neutral drift, model. (**A**). The diagram illustrates Potten's original invariant asymmetry model, each stem cell (green) producing clones of more differentiated cells (red) that vary little in their size distribution. (**B**). The neutral drift model proposes that the fate of an individual stem cell can vary from asymmetric division (left, as in Potten's model), to division that can give rise to either two differentiated cells (middle) or two stem cells (right). This means that clone sizes will vary and that the overall number will decrease over time as some clones are lost. (Adapted from Klein A & Simons B. (2011) *Development* 138:3103–3111 [doi: doi:10.1242/dev.060103].)

**BOX 15.3 BACKGROUND: THE HAIR CYCLE**

Hair grows in cycles, consisting of defined phases. These are: anagen, the growth phase; catagen, the involuting or regressing phase; and telogen, the resting or quiescent phase (**Figure 1**). Normally, up to 90% of the hair follicles are in anagen, 1%–2% in catagen, and about 10% in telogen. The cycle's length varies at different locations on the body. For eyebrows, the cycle is completed in around 4 months, while it takes the scalp follicle up to 8 years to finish; this is the reason eyebrow hair has a much shorter length limit compared to hair on the head. During the active growth phase of anagen, the cells at the root of the hair divide rapidly, adding to the hair shaft so that the hair grows about 1 cm every 28 days. At the end of anagen, an unknown signal causes the follicle to go into the catagen phase, which is a short transition of about 2–3 weeks while the hair converts to a club hair as the part of the hair follicle in contact with the lower portion of the hair becomes attached to the hair shaft. This process, which takes about 2 weeks, cuts the hair off from its blood supply and from the cells that produce new hair. The hair follicle then enters the telogen, or resting phase. The club hair is the final product of a hair follicle in the telogen stage, and is a dead, fully keratinized hair.

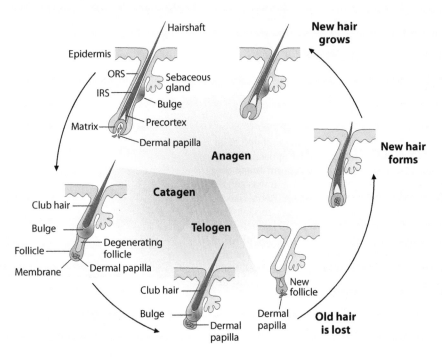

Box 15.3 Figure 1 **The hair follicle cycle.** Hair is being continually shed from the skin epidermis. The hair follicle goes through cycles of growth in order to maintain the production of hair. This diagram illustrates the hair follicle cycle, the major component parts of the follicle being labeled. At the top of the cycle is the growth phase (anagen), during which a hair is actively produced from the follicle. As the hair follicle cycle progresses toward catagen, the matrix cells become exhausted and the follicle regresses. In the rest phase (telogen), the bulge stem cells proliferate, ready to regenerate the hair follicle during the next anagen phase. ORS: outer root sheath; IRS: inner root sheath.

Labeling of bulge stem cells by lineage tracing techniques revealed that just prior to proliferation they migrate toward the hair germ at the base of the follicle and the vacated spaces are filled by the **self-renewal** of neighboring bulge stem cells (**Figure 15.7A**). Such experiments also demonstrated that stem cells in the bulge region contribute to the homeostatic maintenance of all hair follicle cells below the sebaceous gland but do not give rise to cells constituting the gland itself or the IFE, which has its own population of keratin 5/14+ stem cells (**Figure 15.7B**). However, following trauma to the IFE, bulge-derived stem cells can be seen to incorporate into the epidermis.

## The sebaceous gland has its own stem cell component as well as one that it shares from the bulge

The failure of lineage tracing studies to show anything more than occasional labeling of sebaceous glands from the bulge stem cell population pointed to the requirement for additional stem cell types in the hair follicle. Again, use of marker proteins and

Figure 15.5 **The hair follicle bulge.** Mammalian hair continuously cycles between a phase of active growth (anagen) and one of quiescence (telogen). The fluorescent image is of a murine hair follicle during anagen. During anagen growth phase, the bulge stem cells (red) are rapidly dividing. Epithelial cells express histone H2B fused to GFP, labeling them green. Actively proliferating cells are labeled red with an antibody against the proliferation protein, Ki67; DAPI stains nuclei blue. (From http://www.cell.com/pictureshow/stem-cells. Picture by Elizabeth Deschene, Valentina Greco Laboratory, Yale School of Medicine. Copyright 2019, with permission from Elsevier.)

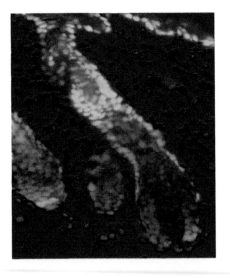

Figure 15.6 **Characteristic gene expression throughout the hair follicle.** Following the discovery of the bulge stem cell population, putative markers for stem cells within the bulge and other regions of the hair follicle were identified. A cross-sectional diagram through a representative hair follicle (HF), sebaceous gland (SG), and the surrounding interfollicular epidermis (IFE), highlighting the bulge (Bu), the isthmus (IS), and the infundibulum and junctional zone (JZ). Markers characteristic of each stem cell population are indicated: keratins (K) 5, 14, 15, and 19; CD34; the transcription factor Blimp1; the leucine-rich repeat-containing G-protein coupled receptors 5 and 6 (Lgr5 and Lgr6); the leucine-rich repeats and immunoglobulin-like domains protein 1 (Lrig1); the surface protein MTS24; and label retaining cells (LRC). (From Beck B & Blanpain C. (2012) *EMBO J* 31:2067–2075 [doi:10.1038/emboj.2012.67]. With permission from Wiley. © 2012 European Molecular Biology Organization.)

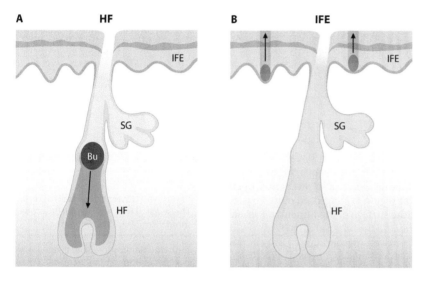

Figure 15.7 **The interfollicular epithelium and hair follicle are supported by separate stem cell populations.** Diagrams showing how defined stem cell populations support maintenance of (**A**) the hair follicle (HF) and (**B**) the interfollicular epidermis (IFE). These stem cells do not contribute to the sebaceous gland (SG). The arrows indicate the direction followed by cells differentiating from the different stem cell populations. (From Beck B & Blanpain C. (2012) *EMBO J* 31:2067–2075 [doi: 10.1038/emboj.2012.67]. With permission from Wiley. © 2012 European Molecular Biology Organization.)

genes with expression restricted to the upper regions of the hair follicle helped to resolve the situation. Immediately above the bulge, at the base of the region known as the isthmus, lies a population of Lgr6+ cells that through lineage tracing experiments have been shown to give rise to IFE and the sebaceous gland (Figure 15.8). Within the isthmus, there is also a population of cells marked by expression of MTS24, while above in the upper isthmus and junctional zone, Lrig1+ cells predominate and have been shown to reconstitute all epidermal lineages of the skin. At the mouth of the sebaceous gland is a population of unipotent **progenitor** cells expressing high levels of Blimp1 that are able to give rise to the cells of the sebaceous gland. Although multiple stem cell populations have now been identified, it is as yet unclear whether this represents a stem cell hierarchy similar to that observed within the HSC population (Chapter 9) or whether these are in fact independent stem cell populations. This will ultimately be resolved by yet more sophisticated lineage tracing experiments.

## Sweat glands are autonomous structures in the skin with their own stem cells

Sweat glands are vital for the regulation of body temperature and in some locations, such as the palms, respond to emotional stimuli. In addition to this latter role, sweat glands are essential to many animals in the attraction of mates by the release of pheromones. At least in the human body (mice, for instance, only have equivalent glands on their paws), sweat glands are the most abundant glandular structures, being present at a density of up to 700 per square centimeter. The predominant ("eccrine") sweat glands are simple coiled tubular structures, whereas apocrine glands empty directly into the hair follicle (Figure 15.9). The single coiled tubular portion of the gland, responsible

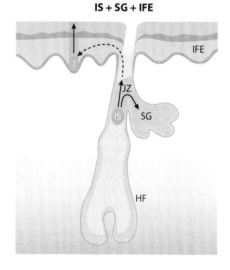

Figure 15.8 **The sebaceous gland has its own stem cell population.** Diagram showing the defined stem cell populations supporting maintenance of the interfollicular epidermis (IFE) and the sebaceous gland (SG) from cells residing in the isthmus (IS) and junctional zone (JZ). The arrows indicate the direction followed by cells differentiating from the stem cell populations. (From Beck B & Blanpain C. (2012) *EMBO J* 31:2067–2075 [doi: 10.1038/emboj.2012.67]. With permission from Wiley. © 2012 European Molecular Biology Organization.)

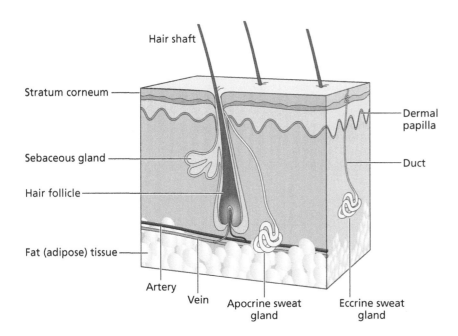

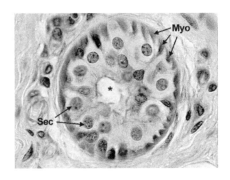

Figure 15.9 **Sweat glands.** Sweat glands are small tubular structures located in the lower dermal layers of the skin. There are two types of sweat gland, namely eccrine and apocrine. The predominant eccrine sweat glands open directly onto the surface of the skin through pores and serve as the primary source of cooling in humans. Apocrine sweat glands open into the hair follicle and are important in the secretion of pheromones and respond to both stress and sexual cues.

Figure 15.10 **Sweat glands structure.** The sweat glands are essentially a coiled tubular structure. The figure shows histological staining of a section across the glandular portion of an eccrine sweat gland, highlighting the secretory (Sec) and myoepithelial (Myo) cells surrounding the lumen (asterisk). (From Chapter 13: Skin. In Gallik S. (2013) *Histology OLM 4.0: The On-Line Lab Manual for Mammalian Histology.* Used with permission from Stephen Gallik, PhD.)

for production of the sweat secretion, is located deep in the dermis. From the coiled base, a duct continues as a relatively straight tubule upward in a spiral fashion, opening directly at the surface of the skin. The glandular region is composed of two basic cell types, namely secretory luminal cells forming the surface of the lumen and so-called myoepithelial cells, which wind around and underlie the luminal cells (Figure 15.10).

As we will see later, the repeated hormonally driven regeneration of the mammary gland in relation to pregnancy and lactation prompted the search for stem cells in that tissue, while the much less dynamic nature of sweat glands suggested to some that stem cells might not play a part in their homeostasis, although wound repair might necessitate some sort of regenerative capacity. Indeed, over half a century ago, some early injury experiments indicated that cell proliferation could be elicited in parts of the sweat gland, but it was not until 2012 that the laboratory of Elaine Fuchs (Box 15.4) in New York

Elaine Fuchs (Figure 1) grew up outside Chicago in a family of scientists, including her sister, father, and aunt, so it is hardly surprising that she followed the educational and research path that led her to become one of the most prominent figures in the field of epidermal stem cell research. Elaine earned her PhD from Princeton University in 1977 for studies on bacterial cell walls, and it was during her postdoctoral work at the Massachusetts Institute of Technology in the laboratory of Howard Green that she became interested in the mechanisms controlling epidermal stem cells and differentiation. Elaine's first academic position in 1980 was at the University of Chicago, where she progressed to be the professor of molecular genetics and cell biology, and eventually left in 2002 to become a Howard Hughes Investigator at Rockefeller University in New York.

Elaine Fuchs is a world leader in skin biology, her research focusing on the mechanisms that control the differentiation of the epidermis and its appendages from stem cells. Her approach to this research has been notable for its use of genetic approaches in defining how stem cell fate is determined and relating the function of individual genes to their role in human diseases when they are defective. Elaine's research has without doubt moved the discipline of dermatology into the realm of molecular and cell biology, and she is very active in applying her research in

the clinical context. Her achievements have been reflected in many awards and honors, including being a member of the National Academy of Sciences and a past president of the American Society of Cell Biology.

Box 15.4 Figure 1 **Elaine Fuchs.** (Used with permission from The Rockefeller University.)

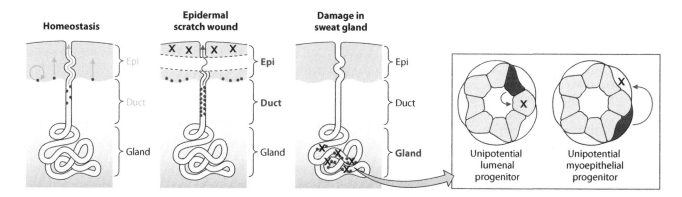

**Homeostasis**  **Epidermal scratch wound**  **Damage in sweat gland**

Epi  Duct  Gland

Epi  Duct  Gland

Epi  Duct  **Gland**

Unipotential lumenal progenitor

Unipotential myoepithelial progenitor

tackled the issue in the mouse paw, using the sorts of label-retention, lineage tracing, and cell isolation strategies that have been used successfully in many other tissues, including the skin itself. After the skin had been wounded, cells of the sweat gland duct were seen to replace lost ductal cells and to contribute to the surrounding epithelium (Figure 15.11). Using a very clever genetic approach entailing application of diptheria toxin to selectively kill cells, Fuchs's group destroyed either luminal or myoepithelial cells in the glandular coil. Similar to the results seen for the duct cells, this revealed that each cell type in the gland could only regenerate their lost counterparts, not the other cell type, implying strict unipotentiality. However, when they isolated and sorted cells from the sweat gland and assayed their ability to regenerate when transplanted, a more complicated picture emerged. When transplanted into mouse shoulder fat pads, luminal cells were not able to regenerate the gland tissue, whereas the purified myoepithelial cells consistently formed sweat glands. Intriguingly, in addition to this **multipotential** behavior of the myoepithelial cells, they show a degree of **plasticity** in terms of their possible fates. Hence, when they were instead transplanted into skin on the back of the animal or into mammary tissue of a lactating mouse, differentiation was seen to either stratified epithelium or milk-producing branching glandular structures, respectively.

Figure 15.11 **Stem cells associated with sweat glands.** Stem cells have been identified in the eccrine sweat gland using lineage tracing techniques. These stem cells were seen to contribute to both ductal cells and the epithelium following wounding. The schematic diagram illustrates the location of monopotent stem cells (red dots) in a homeostatic eccrine sweat gland (left) and their response following an epidermal scratch wound (middle) or damage to the gland itself (right). Depending on the type of injury, the monopotent stem cells respond to regenerate either luminal cells or myoepithelial cells, but not both (inset). (From Lu C, Polak L, Rocha A et al. (2012) *Cell* 150:136–150 [doi: 10.1016/j. cell.2012.04.045]. Copyright 2012 with permission from Elsevier.)

## The hair follicle bulge also contains stem cells for melanocytes

Melanocytes, the cells that give pigmentation to skin and color to hair correspondingly, reside in both locations, although they have a common origin in the melanocyte stem cells (MCSC) that occupy the hair follicle together with the other stem cells we have already described. As the hair follicle cycles, melanocyte numbers fluctuate in parallel, expanding in number and differentiating during anagen (Figure 15.12). The group of Shin-Ichi Nishikawa in Kyoto used lineage tracing in mice, employing transgenic

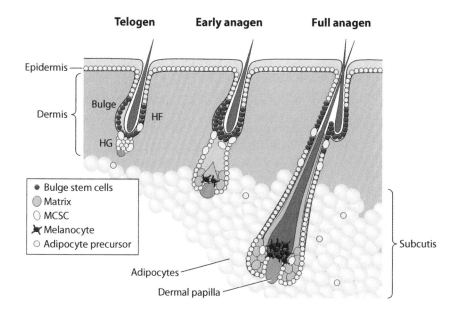

**Telogen**  **Early anagen**  **Full anagen**

Epidermis

Dermis

Bulge

HF

HG

- ● Bulge stem cells
- ● Matrix
- ○ MCSC
- ✶ Melanocyte
- ○ Adipocyte precursor

Adipocytes

Dermal papilla

Subcutis

Figure 15.12 **Melanocyte stem cells fluctuate with the hair cycle.** Melanocytes are responsible for the pigmentation of both hair and skin and are found in both the epidermis and the hair follicle (HF). Melanocytes (dark blue) are produced from melanocyte stem cells (MCSC, pale blue) that reside within the hair follicle. During the hair follicle cycle, melanocyte numbers fluctuate. This diagram shows a hair follicle during the three phases of the cycle. During the rest phase (telogen), MCSC reside within the hair germ (HG). As the cycle progresses toward anagen, the location of the stem cells shifts away from the dermal papilla and proliferation and differentiation of melanocytes occurs. (Zhang B & Hsu Y-C. (2017) *Wiley Interdiscip Rev Dev Biol* 6:e282 [doi: 10.1002/wdev.282]. With permission from Wiley.)

reporters driven by promoters of genes expressed early in melanocyte differentiation, to map the stem cell niche to the lower permanent portion of the follicle, including the bulge. Marking of cells using similar reporters facilitated sorting of a population of cells (also characterized by low expression of c-Kit, the receptor for stem cell growth factor, which is present at high levels on maturing melanocytes) that can be grown *in vitro* in the presence of keratinocytes and growth factors. Importantly, in a hair regeneration assay, purified MCSC are able to populate new hair follicles and generate pigment cells.

## Feather shape and their amazing patterns are determined by the architecture of the stem cell niche

Feathers are fundamental to a bird's existence, serving roles in aerodynamics, insulation, communication, and camouflage. Although they evolved separately, both feathers and hair are epidermal appendages that have remarkable regenerative capabilities that rely on the presence of stem cells. Feathers are more complex structures than hair (Box 15.5), much of this complexity depending on the precise location of the stem cells that maintain them. Like hair, feathers emerge from a follicle,

### BOX 15.5 BACKGROUND: FEATHER STRUCTURE AND PIGMENTATION

The primary function of feathers is to provide protection, insulation, and the capacity for flight. Feather color is important for communication and camouflage. Feathers (Figure 1) have a central shaft, the rachus, with broad vanes on either side. The calamus at the base of the feather anchors it into the follicle. Lateral branches off the brachis are called barbs, which themselves have many branches or barbules, which in turn have projections called hooklets. Feathers can be radially or bilaterally symmetrical, as for downy

and flight feathers, respectively. Feathers vary enormously in their color, with patterns formed within an individual feather and along the body axis of the bird, resulting in spectacular multicolored plumage, which can change with the seasons and from juvenile to adult (Figure 2). Feather color is three dimensional, with patterns such as stripes or spots appearing within the barbs, and their appearance can alter depending on the direction from which the feather is viewed.

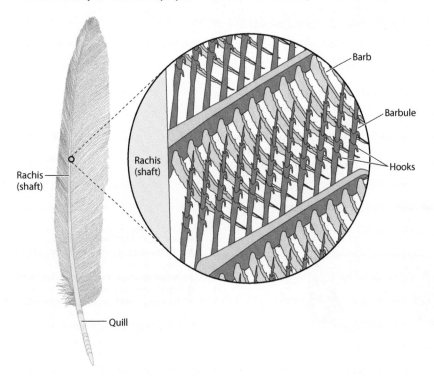

Box 15.5 Figure 1 **Feather structure.** The feather has numerous functions in birds, providing aerodynamic properties, insulation, communication, and camouflage. The diagram of a typical feather here shows the complexity of its structure. The major features of a feather are the quill (or calamus) connecting the feather to the follicle, which further up the feather becomes the rachis (shaft) to which vanes are attached on either side. The vanes consist of parallel branches called barbs (inset) of which each is made up of a series of short barbules containing tiny hooklets that tie the barbules and barbs together, creating a robust but light feather.

Box 15.5 Figure 2 **Feather shape and coloration can be exotic.** Birds are the most successful class of tetrapods, with approximately 10,000 living species. They exhibit a diverse range of feather patterning and structures that have helped in their success. The peacock is just one example of how elaborate both feather shape and color can be, all as the result of the location and activity of the stem cells within the feather follicle.

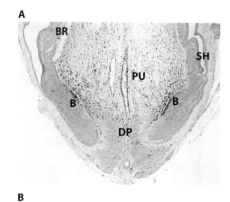

Figure 15.13 **Label retaining cells in the feather follicle bulge.** Using label retention techniques, it has been possible to identify feather follicle stem cells. The top image shows a section through a feather follicle from a chicken highlighting the regions of the follicle: dermal papilla (DP), bulge (B), pulp (PU), barb ridges (BR), and sheath (SH). The brown staining highlights cells that have incorporated and retained bromodeoxyuridine after a period without label present, therefore representing label retaining stem cells. On the bottom is a schematic representation of the same region of the feather follicle in which the bulge is highlighted in green and from which emanates transit amplifying (TA) cells (yellow) and then the barb ridges (BR, red). Also shown are the epidermis bordering the follicle (E), the follicle (F), and the papillary collar (PC). (From Alibardi L, Wu P & Chuong C. (2014) *J Morphol* 275:768–774 [doi: 10.1002/jmor.20257]. With permission from Wiley.)

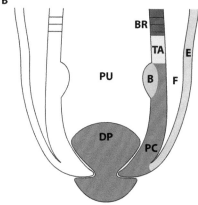

which is not only a niche for the epidermal stem cells that give rise to the feather itself but also other stem cells, including those that differentiate into melanocytes.

Those working on stem cell involvement in feather formation have so far not had the advantages of the molecular and genetic tools available for studies on mammals but nevertheless have been able to utilize some basic methods similar to those used to identify the hair follicle stem cell in mammals. Identification of feather follicle stem cells in a growing feather was initially made through observation of LRC in the region known as the collar bulge (Figure 15.13), combined with transplantation studies that demonstrated multipotency of the LRC. Definitive proof of the location of stem cells in this region came from experiments involving intricate grafting of adult quail follicles onto day 5 chick embryo limb buds, engrafted quail stem cells revealing themselves as keratinocytes within the chick follicle. As the feather follicle progresses through its cycle from initiation of the growing phase to resting through to the molting phase, the size of the collar bulge changes. During the resting phase, the collar bulge size decreases and the LRC migrate toward the lower collar, where they contact with the dermal papilla. After the old feather is molted, the LRC are able to migrate to the newly formed collar bulge region in order to initiate new feather formation.

The development of symmetry and shape of feathers is dictated by the regulation and orientation of the stem cells within the follicle. If the stem cells within the collar bulge reside on a horizontal plane, they give rise to radially symmetrical downy feathers, whereas if they are tilted by 45°, then the feathers that emerge are bilaterally symmetrical, such as seen in the flight feathers (Figure 15.14). Similarly, the arrangement of stem cells also impacts feather patterning dictated by a defined pigment stem cell population, which likewise resides in the feather follicle. Cheng-Ming Chuong, working at the University of Southern California, was instrumental in many of the studies on the location and activity of feather follicle stem cells and most notably worked out how complex patterns of coloration can be generated through a combination of the distribution of the bird equivalent of melanocyte stem cells and an extrinsic modulating activity. This modulating activity is in the form of a soluble factor provided by surrounding cells in the environment of the follicle and affects the expression of melanin in the melanocytes. Chuong determined that the pattern of feather coloration is dictated by the presence, arrangement, and differentiation of MCSC in the ring of collar bulge cells (Figure 15.15), which work together to change the pattern of melanocytes from the front to the back of the feather follicle, from the center to outside, and over time as the feather grows. In addition, the peripheral pulp layer can provide a modulating influence on color patterning by producing a peptide called agouti, which inhibits melanocytes to produce unpigmented or white regions of feather.

## A few specific signaling pathways are crucial in the regulation of interfollicular epithelium and hair follicle bulge stem cells

A number of signaling pathways, which are commonly encountered in adult stem cell scenarios (Chapters 4 and 6), have been found to be involved in both the development and maintenance of the epithelial tissues and their appendages. These include the Wnt/β-catenin, bone morphogenetic protein (BMP), and Sonic Hedgehog (Shh) pathways, among others. As the hair follicle undergoes an active cycle of anagen, catagen, and telogen, there is a requirement for very precise regulation of stem cell activation. An important component of the hair follicle environment is the dermal papilla, which forms prior to hair follicle development and remains in close contact with the base during all phases of the hair follicle cycle. The dermal papilla is a prime source of multiple signaling molecules, including Wnt and BMP.

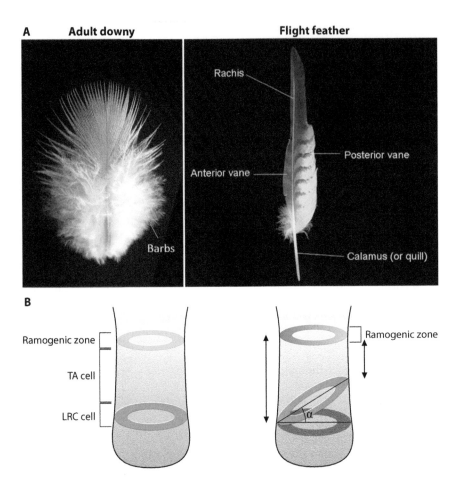

**A**    **Adult downy**

**Flight feather**

**B**

**Figure 15.14 The orientation of feather follicle stem cells affects the structure of the feather.** The complex range of feather structures on a bird's body are all produced from feather follicle stem cells. The nature of the feather shape and barb type is determined by the plane in which the stem cells reside. (**A**) Two types of feather are illustrated: adult downy (left) and a flight feather (right). (**B**) Schematic cross sections through the feather follicle at the level of the ramogenic zone (where the differentiation to barb ridges occurs). The label retaining cells (LRC, orange) lie in a ring from which emanates the transit amplifying (TA) cells that differentiate as they move toward the ramogenic zone (blue). In a downy feather follicle, the plane of the stem cell zone is horizontal, whereas in a flight feather follicle, it is tilted at an angle ($\alpha$) leading to asymmetry that breaks the radial symmetry at the ramogenic zone leading to bilateral symmetry forming there. (A: Left photo credit: Buntysmum via Pixabay. (https://pixabay.com/photos/feather-downy-bird-nature-3258267/. Right photo credit: US Fish and Wildlife Service. The Feather Atlas of North American Birds. (https://www.fws.gov/lab/featheratlas/glossary/php). B: From Yue Z, Jiang T-X, Widelitz R et al. (2005) *Nature* 438:1026–1029 [doi: 10.1038/nature04222]. Reprinted by permission from Springer Nature.)

The importance of the various signaling pathways involved in the development and cyclical regulation of the hair follicles and the IFE has taken the efforts of many laboratories, perhaps most notably those led by Elaine Fuchs (Box 15.4) and Fiona Watt (Box 15.6). Signaling through the Wnt/β-catenin pathway is important in many adult stem cells and certainly plays a central role in several different epithelial stem cell types, including those in the intestinal crypts to maintain the **intestinal stem cell** population (Chapter 11). The work of the likes of Fuchs and Watt has shown the central importance of Wnt/β-catenin signaling in the skin. Hence, the experimental loss of signaling through the Wnt/β-catenin pathway results in the absence of hair follicle formation, whereas in the right circumstances, artificially enforced expression can induce the formation of new hair follicles. BMPs are also crucial in the maintenance of hair follicle bulge stem cells through repression of their proliferation, and loss of signaling through this pathway results in stem cell activation. Another example of

**Figure 15.15 Feather patterning is affected by the orientation of the zone containing melanocyte stem cells.** Melanocytes provide the pigment to color feathers as they develop in the follicle. The diversity of colors and patterns seen is a result of the intricately positioned melanocyte stem cells and their interaction with the environment that determines their differentiation characteristics. (**A**). A feather from a barred Plymouth Rock chicken showing the alternating black and white pattern of stripes. (**B**). Schematic drawing showing a feather follicle and the location of melanocyte lineage cells within. (**C**). A three-dimensional reconstruction of melanocyte lineage cells (purple) dictated by the expression of the melanocyte lineage marker microphthalmia-associated transcription factor (Mitf) shows the presence (yellow arrows) or absence (red arrows) of progenitors in the lower bulb (LB). (Photo credit: Ting Xin Jiang, MD & Randall B. Widelitz, PhD. Used with permission. B and C: From Lin S, Foley J, Jiang T et al. (2013) *Science* 340: 1442–1444 [doi: 10.1126/science.1230374]. Reprinted with permission from AAAS.)

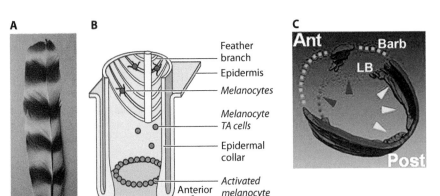

BOX 15.6 PIONEER: FIONA WATT

Fiona Watt (Figure 1) had no doubts even as a young girl that she wanted to be a scientist and believes that "being a scientist is in a sense hardwired." Perhaps because of this conviction, her early career in research was by any standards quite meteoric and rapidly demonstrated independence. Her PhD at the University of Oxford in 1979 on the role of microtubules in cancer cells was based on collecting data over just a few weeks. After a 2-year postdoctoral fellowship at the Massachusetts Institute of Technology, which like Elaine Fuchs was under the supervision of Howard Green, Fiona became head of the Molecular Cell Biology Laboratory at the Kennedy Institute of Rheumatology in London. A few years later in 1987, Fiona moved to the Cancer Research UK

**Box 15.6 Figure 1 Fiona Watt.** (From eLife Sciences Publications, Ltd. Licensed under the Creative Commons Attribution 3.0 Unported license via Wikimedia. (http://creativecommons.org/licenses/by/3.0/deed.en).)

London Research Institute, where she was the head of the Keratinocyte Laboratory for almost 20 years. Reflecting on this rapid progress, Fiona admitted that "my time from being a PhD student to running a lab was so short that I didn't have as much chance to be at the bench myself as I would have liked." Following a stint in Cambridge as the deputy director of the Cancer Research UK Cambridge Research Institute, she returned to London and is presently the director of the Centre for Stem Cell and Regenerative Medicine Research in King's College.

Throughout her research career, Fiona has focused on epidermal stem cells and their progeny and how the normal processes of growth and differentiation in these cells are controlled and can go awry in cancer. The world-leading part she has played in defining the basic principles in this important field is reflected in the many honors that Fiona has received over the years. Interestingly, though, she says that when she was first interested in stem cells "they weren't very respectable," the dominance of developmental biology at the time imposing a focus on the embryo such that "developmental biologists just couldn't understand why you would be interested in … adult tissues." Fiona has without doubt become a shining example of how to be successful in science and how to overcome the added difficulties that are often faced by women trying to juggle the work/life balance, to the extent that she has been well known for being accompanied by her eldest son on the stem cell conference circuit! Perhaps this attitude to work and life is reflected in her thoughts on how to organize a research laboratory and mentor the new generation of scientists. As she said, "Looking after a lab is sort of like tending a garden. If you forget to water it, you just can't go back in and have it look as good as it would have been if you hadn't neglected it."

a signaling molecule involved in the regulation of hair follicle stem cells is Sonic Hedgehog. Sensory neurons surrounding the hair follicle bulge (Figure 15.16) are sources of Shh, which signals to a population of cells in the upper bulge marked by the Hedgehog response gene Gli1. These Gli1-expressing bulge cells function as multipotent stem cells in their native environment and repeatedly regenerate the follicle. In response to the Shh signal, these cells can incorporate into healing skin wounds where they can change their lineage into epidermal (IFE) stem cells. Denervation of the hair follicle results in loss of the Shh-dependent regulator Gli1;

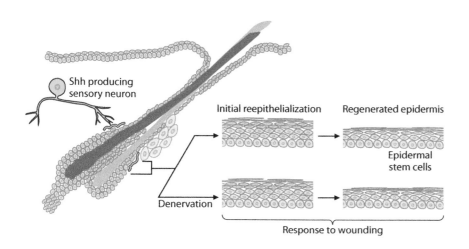

**Figure 15.16 Sensory neurons act on the hair follicle bulge.** Sensory neurons play an important role in regulating the signaling of many stem cell types, including regulating the hair follicle stem cells in the bulge region. This schematic cross section of a hair follicle shows the interaction of a sensory neuron with the bulge region. The sensory neurons produce Sonic Hedgehog (Shh) that signals to a region (green) of cells positive for expression of the Hedgehog response gene Gli1 (blue) that lie within the bulge during telogen. Experimental denervation reveals that the Shh responsive Gli1+ cells are necessary for maintaining the capability of bulge stem cells to become epidermis, for example, in response to wounding, and to provide new stem cells in the epidermis. (Adapted from Brownell I, Guevara E, Bai C et al. (2011) *Cell Stem Cell* 8: 552–565 [doi: 10.1016/j.stem.2011.02.021.)

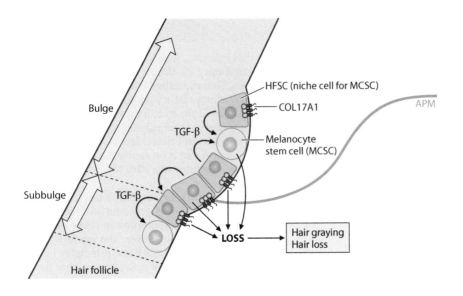

**Figure 15.17 Regulatory interactions between different hair follicle stem cells.** Our understanding of the regulation of hair follicle stem cells is progressively revealing the considerable complexity of intrinsic and extrinsic contributions including interactions between distinct stem cell types. The multiple populations of stem cells identified in the hair follicle are not only important for the production of their intended lineage but are seemingly also important for the regulation of other stem cell types present within the follicle. This diagram shows the regulatory interactions between different hair follicle stem cell populations. Collagen 17A1 (COL17A1) is highly expressed in hair follicle stem cells (HFSC) and is required for both their maintenance and that of the melanocyte stem cell (MCSC) through transforming growth factor beta (TGF-β) signaling. Loss of either stem cell type or COL17A1 leads to hair loss and hair graying. APM: Arrector pili muscle. (From Tanimura S, Tadokoro Y, Inomata K et al. (2011) *Cell Stem Cell* 8:177–187 [doi: 10.1016/j. stem.2010.11.029. Copyright 2011 with permission from Elsevier.)

however, this loss does not affect normal hair follicle regeneration but instead impacts the response to wounding and homeostasis of the sebaceous gland, preventing the bulge cells from becoming epidermal stem cells

What we have learned about the nature and role of stem cells in the skin is clearly considerable and provides clues for the understanding of other epithelial stem cells and adult stem cells more generally. Nevertheless, in a sense, the knowledge we have is just the tip of the iceberg, and the scene is set to tease apart the finer details. An example of the type of complexity that is yet to be revealed comes from studies on the interactions between some of the distinct stem cell types in the hair follicle. Hence, not only do MCSC occupy the hair follicle with the stem cells that contribute to hair, sebaceous gland, and epidermis during wounding, it appears that they are also in some sense interdependent. Genetic knock out experiments performed in the laboratory of Emi Nishimura in Tokyo showed that loss of collagen 17, which is expressed in hair follicle stem cells and fixes them in the bulge niche, leads to the expected hair loss, but also causes loss of MCSC (which do not express collagen 17) and premature hair graying. The reason seems to be that the hair follicle stem cells normally produce transforming growth factor β, which is crucial for the maintenance of the MCSC – when they are lost, so are the MCSC (Figure 15.17).

## 15.3 THE CORNEAL EPITHELIUM

The cornea is the front surface of the eye that protects it from injury and infection (Box 15.7). It is a remarkable tissue, which is uniquely transparent and essential for vision. The outer layer of the cornea is an epithelial structure that shows complete renewal in humans in 9–12 months and like other epithelia relies on stem cells to achieve homeostatic maintenance and regeneration after injury. The current status of our understanding of the specific stem cells underpinning the corneal epithelium is somewhat lagging that of the skin and intestine, but we have quite a good picture of the nature of the specialized stem cell niche, and the imperative of being able to use stem cell technology to replace damaged corneas, which is a large unmet medical need (Chapter 18), is driving progress.

### The stem cells that maintain the cornea are located in the region known as the limbus

As in the skin epithelium, the squamous epithelial layer of the cornea is subject to constant shedding, requiring regeneration from a resident stem cell population. Pioneering label retention experiments during the late 1980s led to the identification of putative corneal stem cells within the limbus, a transitional zone separating the cornea from the surrounding conjunctiva (Figure 15.18). These cells were not surprisingly called **limbal epithelial stem cells** (LSC). LRC were shown to be able to proliferate in response to wounding. Additional evidence that the limbal region houses the corneal stem cells came from experimental removal of the limbal zone,

## BOX 15.7 BACKGROUND: THE CORNEA

The cornea (Figure 1) protects the eye against the environment, forming the first defensive barrier against infection, dust, chemicals, and UV light. It also plays a very large part in the refraction of light, effectively acting as an additional lens focusing light onto the retina. The cornea is made up of multiple layers that vary in number between species. In humans, the corneal surface is composed of the epithelium, a highly regenerative tissue that itself is arranged into three layers: basal, wing, and squamous (Figure 2). The columnar basal layer secretes matrix molecules required for the basement membrane and stroma and is likely to be important in mediating cell migration in response to injury. The central "suprabasal"

wing layer is largely responsible for wound healing, and the surface squamous layer is made up of epithelial cells that form tight junctions, providing a resistant barrier to the external environment. These squamous cells have surface microvilli that serve to increase the surface area to facilitate close association with the tear film and are continuously shed from the ocular surface. Beneath the epithelium is the corneal stromal layer composed of regularly arranged collagen fibers and keratocytes required for repair and maintenance. The corneal stroma is surrounded by a membrane called Bowman's layer on the anterior side and by Descemet's membrane on the posterior side, below which is a squamous monolayer of corneal endothelium.

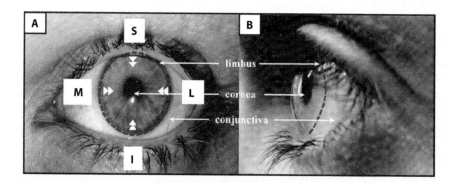

Box 15.7 Figure 1 **The cornea.** The cornea is the protective surface of the eye, preventing both injury and infection of the eye. Photographs of face (**A**) and lateral (**B**) views of the human eye showing the cornea, conjunctiva, and limbus (red dashed line). The superior (S), inferior (I), medial (M), and lateral (L) aspects of the eye are indicated. The double arrow heads depict the direction of cell migration within the cornea (centripetal). (From Ordonez P & Di Girolamo N. (2012) Stem Cells 30:100–107 [doi: 10.1002/stem.794]. With permission from Wiley.)

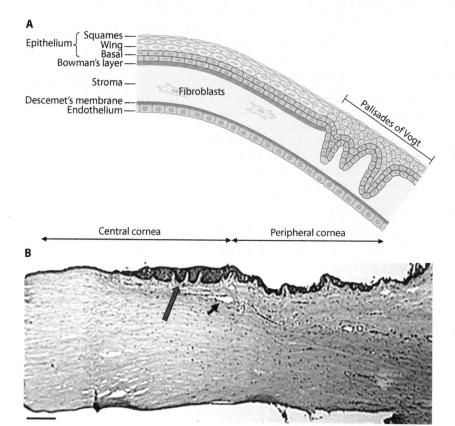

Box 15.7 Figure 2 **The corneal epithelium.** (**A**) A schematic cross section of the mammalian cornea illustrating the layered structure of the epithelium and the underlying tissue. Below the epithelium, which is made of three layers of cells termed squames, wing, and basal, is a layer of stroma that is populated with fibroblasts and collagen fibrils. The stroma is separated from the epithelium by Bowman's layer. Below the stroma lies Descemet's membrane, which itself lies on top of a single sheet of endothelial cells. In the limbus, the epithelial layers are thicker and reach deeply into the stroma in a region known as the Palisades of Vogt. (**B**) Histological section of the cornea stained with hematoxylin, which shows the undulating structures in the epithelium that constitute the Palisades of Vogt in the limbic region (red arrow). The underlying limbal stroma has high cellularity and vascularity with a blood vessel highlighted (black arrow). The scale bar represents 200 μm. (A: From Ordonez P & Di Girolamo N. (2012) Stem Cells 30:100–107 [doi: 10.1002/stem.794]. With permission from Wiley. B: From Li W, Hoyushida Y, Chen Y et al. (2007) *Cell Res* 17: 26–36 [doi:10.1038/sj.cr.7310137] Reprinted by permission from Springer Nature.)

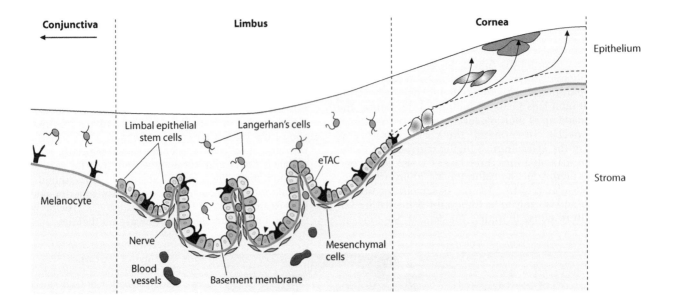

**Figure 15.18 Stem cells in the limbus.** Using label retention experiments, epithelial stem cells were identified in the cornea in a region called the limbus at the interface between the cornea and the conjunctiva. Limbal epithelial stem cells are thought to reside within the basal layer together with their progeny early transit amplifying cells (eTAC), melanocytes, and Langerhans's cells, which are a type of tissue-resident macrophage. In this hypothetical scheme, it is thought that these cell types could act as niche cells for the limbal epithelial stem cells together with an influence from blood vessels, nerves, and mesenchymal cells lying below the basement membrane. (From Li W, Hoyushida Y, Chen Y et al. (2007) *Cell Res* 17: 26–36 [doi:10.1038/sj.cr.7310137] Reprinted by permission from Springer Nature.)

which resulted in hindered epithelial cell proliferation and healing following corneal insult. LSC were subsequently shown to reside within the basal layer of the limbal epithelium.

Utilizing detailed non-invasive confocal imaging techniques, it was determined that LSC typically reside in the Palisades of Vogt (PV), which are at the periphery of the cornea at the limbal borders. These PV are fibrovascular ridges that are oriented in a radial manner and are thought to protect the LSC from physical stresses in a similar way to the crypts of Lieberkühn of the small intestine (Chapter 11). Additionally, their placement within the PV also allows a greater surface area to accommodate a greater number of stem cells in a confined region. Transmission electron microscopy has shown clusters of small primitive cells residing in direct contact with the basement membrane with heterochromatin-rich nuclei and cytoplasm sparsely populated with mitochondria and ribosomes. Other cells within these clusters are larger and more complex and are likely to represent a population of transit amplifying cells derived from the LSC.

Definitive markers of LSC have until recently been elusive, although a number of surface proteins have been utilized in order to isolate these cells, including keratins (14, 15 and 19), integrins ($\alpha$6, $\alpha$V, and $\beta$5), and the adhesion molecule N-cadherin. Putative markers useful in the identification can also be negative, that is, selectively absent from the stem cells. For example, corneal epithelium markers such as cytokeratins 3 and 12 are noticeably absent from the limbal basal cells. The transcription factor p63 is important in epithelial cell development and has been used to distinguish stem cells from transit amplifying cells. Within the limbus, one particular isoform of p63 ($\Delta$Np63$\alpha$) is preferentially expressed during the resting state, and its expression is not detected in the central cornea. *In vitro* clonal expansion of epithelia cells from the cornea, performed in much the same way as originally used by Barrandon for the skin, revealed that p63 is expressed in holoclones (i.e., those clones originating from stem cells) with little or no expression in meroclones or paraclones.

As we have seen in Chapter 2, expression of the membrane transporter protein ABCG2 gives rise to a flow cytometric profile termed the **side population** that is characteristically associated with many stem cell populations. In the same way, ABCG2+ limbal cells are enriched in epithelial stem cell activity. Recently, the related transporter protein ABCB5 has been found to mark LSC and to be required for stem cell maintenance and corneal development and repair. Abcb5 gene loss of function in **knockout** mice caused depletion of LSC due to enhanced proliferation and apoptosis and resulted in defective corneal differentiation and wound healing. Most interestingly, isolated human or murine ABCB5+ LSC could be shown to restore the cornea upon grafting to LSC-deficient mice in transplantation models.

There has been some controversy around the location of the LSC, some research groups claiming to have identified alternative crypts adjacent to the PV that are surrounded by highly cellular and vascular stroma. However, regardless of the definition of the limbal crypt, it is now well accepted that LSC are housed within multiple locations of the limbal epithelium. The sole location for epithelial stem cells in the limbus has, however, been challenged recently following the demonstration in the mouse that central corneal epithelium can be serially transplanted.

A group working in Sun Yat-sen University in the Chinese city of Guangzhou has recently made the exciting discovery that it may be possible to interconvert epithelial stem cells. Using a 3D *in vitro* culture system for LSC, these investigators firstly showed that p63 cooperates with another transcription factor PAX6 to specify LSC and that the signaling protein WNT7A controls corneal epithelium differentiation through PAX6. Loss of WNT7A or PAX6 induced the LSC to differentiate into skin-like epithelium, whereas the counter-experiment involving expression of PAX6 in skin epithelial stem cells converted them to LSC-like cells, which were effective in repairing the corneum in a rabbit corneal injury model.

## 15.4 STEM CELLS IN THE EPITHELIA OF THE GASTROINTESTINAL TRACT

The epithelia and their associated structures in the GI tract are actually derived from two embryonic germ cell layers, the oral cavity as far as the pharynx being of ectodermal origin like the skin, whereas the vast majority, from the pharynx to the anus, derives from endoderm. Given the endodermal origins of the majority of the GI tract, we cover the epithelial structures and their stem cell components of tissues, including the intestine, stomach, and lung, in Chapter 11, which leaves only the oral cavity to consider here.

### The oral epithelium and its appendages all have their own associated stem cell components

Like the skin, the surfaces of the oral cavity, including gums, palate, and tongue, are constituted of a stratified epithelium and similarly are likely to be continually renewed from adult stem cells similar to those that maintain the IFE. Although definitive experiments are as yet lacking, it should be feasible to identify stem cells in the oral epithelium through the application of lineage tracing experiments similar to those that have been used so effectively to study the skin. It is unclear whether oral epithelium stem cells (OESC) behave similarly to the IFE, given that both of these tissues are ectodermally derived, or if instead OESC behave more like the endodermally derived esophageal epithelium.

Also paralleling the skin, the oral epithelium has its own set of appendages that originate from the epithelium during embryogenesis, including the highly specialized dental and salivary epithelia. In Chapter 10, we describe how the development of teeth requires the interplay between cells of the oral epithelium and cells that form the structure of the tooth, which requires specific dental stem cells related to **mesenchymal stem cells**. The latter stem cells persist and are capable of some repair throughout life to combat damage such as dental caries. Although humans are not able to grow new teeth, some animals can, suggesting perhaps that not only can the mesenchymal stem cell components continue into adulthood but also that the specialized epithelium and its associated stem cells might persist. An example is the mouse, which has incisors that grow continually as they are worn down. Stem cells associated with the continual growth of mouse incisors have been located in a specific epithelial niche on the tongue side of the tooth. No such epithelium persists in the molar teeth of the mouse or for any teeth in humans, but there is some hope that a quiescent remnant attached to tooth ligament, represented by the so-called epithelial cell rests of Mallassez, might be capable of being reactivated to act in the same way as those in the murine incisor epithelium.

The salivary gland is analogous to the sweat gland in the skin in that it is composed of luminal secretory cells, myoepithelial cells, and cells making up the duct to the surface. Evidence for a stem cell component comes from experimental damage of the salivary glands below the tongue (in rats) brought about by tying off the duct. Release of the block triggers regeneration, implying the presence of stem cells. These types of experiments, combined with the now familiar toolkit of radioactive labeling, surface marker analysis, cell sorting, and transplantation, has pinpointed cells expressing the surface antigen EpCAM as the stem cell component in this tissue. Sorted EpCAM[high] cells can be maintained *in vitro* and generate salivary gland **organoids** in the presence of Wnt proteins, the ultimate proof of their stem cell capability being that they can regenerate salivary glands following transplantation into damaged

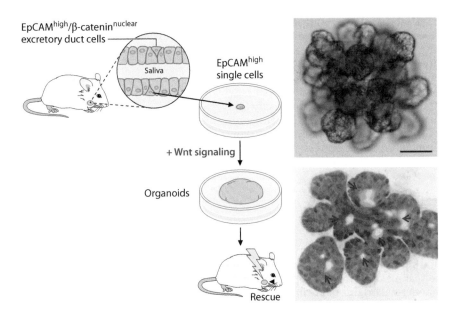

EpCAM$^{high}$/β-catenin$^{nuclear}$ excretory duct cells

Saliva

EpCAM$^{high}$ single cells

+ Wnt signaling

Organoids

Rescue

**Figure 15.19 Salivary gland stem cells can be cultured *in vitro* and used to regenerate damaged tissue.** The ductal compartment in the salivary gland (SG) has been identified as the location harboring stem cells. Rare SG ductal EpCAM$^{high}$ cells expressing nuclear β-catenin can be isolated as single cells and expanded *in vitro* in the presence of Wnt signals. Organoids develop in culture that contain all differentiated SG cell types. Cells derived from such organoids can be transplanted into submandibular glands of irradiated mice to robustly restore a function tissue and saliva secretion. The insets show a representative example of a minigland originating from a single EpCAM$^{high}$ in a 9-day-old culture. The upper differential interference contrast image shows the overall appearance of a growing minigland, while the lower section, stained with Toluidine blue, shows uniform lumen formation throughout minigland (arrows). Scale bar represents 10 μm. (From Maimets M, Rocchi C, Bron R et al. (2016) *Stem Cell Rep* 6: 150–162 [doi:10.1016/j.stemcr.2015.11.009]. Copyright 2016 The authors. With permission from Elsevier.)

tissue (Figure 15.19). It remains to be seen to what extent the stem cells are similar or different to the equivalent cells in the other secretory appendages and whether there is any flexibility in their potential fate, for example, to give rise to sweat gland or mammary gland cells.

## 15.5    STEM CELLS IN MAMMARY GLANDS

A defining feature of mammals is the ability of the females to produce milk to feed their young from the moment of birth. The tissue that makes this possible, the mammary gland (Box 15.8), is effectively an appendage of the skin having formed as an invagination of the epidermis during development. The mammary gland has some similarities to the sweat gland, although it has a much more elaborate branching structure and is able to undergo repeated cycles of extensive growth and regression ("involution") in synchrony with the need to feed successive offspring. The construction of the large mass of secretory tissue requires an increase in the number of mammary epithelial cells, a feat that relies on stem cell activity.

### Stem cells in the mammary gland reside in the basal epithelium

The ability of the mammary gland to grow cyclically in response to the need for lactation relies upon the presence of the **mammary stem cell** (MaSC). These stem cells are solely responsible for the complete development of mammary gland tissues and for the required expansion and remodeling of the mammary gland throughout pregnancy, lactation and involution and additionally act as a reserve population for repair should tissue damage occur.

In 1959, perhaps not realizing that he was a discoverer of the existence of an adult **tissue stem cell**, Kenneth B DeOme (Figure 15.20) developed a transplantation assay for the growth potential of mammary tissues. DeOme operated on the mouse mammary gland to produce a "cleared fat pad," which in essence involves the removal of the underdeveloped epithelial tree so as to create an environment that is receptive to growth of new mammary tissue. In this way, DeOme was able to show that the entire ductal epithelial tree could be regenerated following transplantation of a segment of normal mammary gland tissue. Successful engraftment of any segment of the mammary gland epithelial tree indicated that repopulating capacity is widely distributed throughout the tissue. Since the days of those early experiments, the ability of a single stem cell to transplant serially and generate complete mammary outgrowths in cleared fat pads of mice has become the gold standard assay to determine MaSC activity, in much the same way that hematopoietic stem cells have been so extensively characterized using the bone marrow transplantation assay (Chapters 2 and 9). Additionally, similar to cardiosphere and neurosphere formation (Chapters 13 and 14), single cell suspensions from mammary tissue can be incubated

BOX 15.8  BACKGROUND: THE STRUCTURE AND FUNCTION OF THE MAMMARY GLAND

Mammary gland tissue demonstrates a remarkable capacity to regenerate during repeated cycles of pregnancy, lactation, and involution. The mammary gland appears as a tree-like structure composed of hollow branches that until puberty remain immature, with the ducts branching only one or several times. At this time during mammary gland development, two distinct epithelial cell lineages are identifiable: luminal epithelial cells (surrounding the central lumen) and elongated myoepithelial cells (in a basal position adjacent to the basement membrane) – this is very much the same as seen in the sweat gland. Following the onset of puberty, the ductal system expands and differentiates rapidly at the highly proliferative terminal end buds (alveoli), structures that form at the tips of growing mammary ducts and are lined with cap cells (Figure 1). During pregnancy and lactation, secretory alveolar epithelial cells differentiate at the terminal end buds, producing milk proteins that are secreted into the alveolar lumen and further collected in the ducts ready for expulsion through the nipple during suckling. Following lactation, most of the secretory alveolar cells undergo programmed cell death, resulting in the gland involuting, ready for the development cycle to begin again in successive pregnancy.

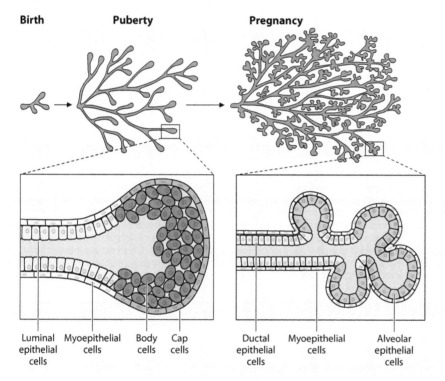

**Birth**    **Puberty**    **Pregnancy**

Luminal epithelial cells | Myoepithelial cells | Body cells | Cap cells

Ductal epithelial cells | Myoepithelial cells | Alveolar epithelial cells

Box 15.8 Figure 1 **Mammary gland development and architecture.** At birth, the mammary gland consists solely of a rudimentary bud that remains morphogenetically quiescent until puberty. During puberty, hormonal cues trigger the formation of the terminal end buds (inset, left), and then during pregnancy extensive elongation, bifurcation, and lateral branching lead to the formation of the full epithelial tree. An expanded view of a terminal end bud is shown in the right-hand inset. As the mammary gland tree matures during pregnancy, these terminal buds further subdivide, ultimately forming the alveoli, which consist of the luminal secretory cells (pink) enclosed by an epithelial layer. The lumen of the alveoli drains into the ducts, which are themselves lined by luminal epithelial cells (pale blue) that are in turn surrounded by myoepithelial cells. (Top: From Gjorevski N & Nelson C. (2011) *Nature Rev Mol Cell Biol* 12: 581–593 [doi: 10.1038/nrm3168]. Reprinted by permission from Springer Nature.)

in a non-adhesive medium, and any stem cells present give rise to cell aggregates called **mammospheres**, from which cells can be serially isolated to form more of the same (Figure 15.21). The cell content of mammospheres is heterogeneous, the dispersed cell population being able to differentiate along one or both of the epithelial or myoepithelial lineages. Three-dimensional culture of mammosphere-derived cells in the presence of ECM components yields organoids demonstrating features of mammary tissue. Of note, the fact that epithelium formed through transplantation can only be serially transplanted for about seven cycles suggests that MaSC have a limited self-renewal potential.

Although the presence of MaSC has been clearly demonstrated through assays of stem cell activity, their precise location is not yet determined. As so often happens in the adult stem cell field, isolation of MaSC has proven difficult due to the lack of any uniquely defining stem cell markers. The essential cell components of the mammary gland are the ductal and glandular luminal epithelial cells, which are underlain on the basal side by myoepithelial cells. There are numerous known surface markers to distinguish between the myoepithelial and luminal cells. For example, myoepithelial cells express keratins 5 and 14, the epithelium-associated transcription factor p63, and smooth muscle-specific contractile proteins, whereas luminal epithelial cells are distinguished by their expression of keratins 8 and 18. A subset of luminal cells (30%–50%) is positive for hormone (estrogen/progesterone) receptors and most likely represents the mature ductal cells. All mammary epithelial cells express CD24 and

Figure 15.20 **Kenneth B DeOme.** (From Cardiff R, Bern H, Faulkin L et al. (2002) *Comp Med* 52:12–31. With permission from American Association for Laboratory Animal Science.)

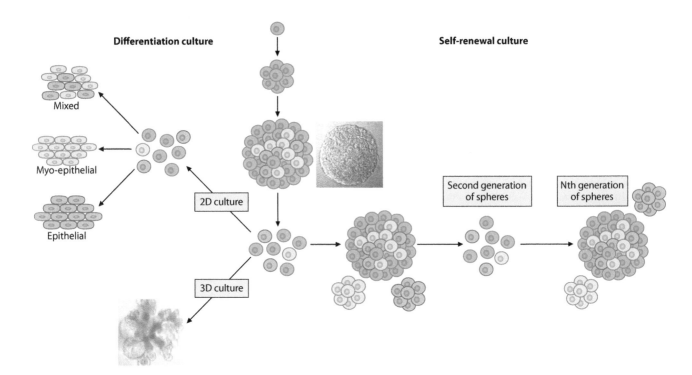

**Figure 15.21 Growth of MaSC as mammospheres.** Similar to cardiosphere and neurosphere formation, mammary stem cells (MaSC) present in single cell suspensions are capable of forming mammospheres in *in vitro* culture. The diagrammatic scheme illustrates how MaSC (blue) isolated from a mouse mammary gland can be cultured in serum-free medium containing growth factors epidermal growth factor (EGF) and fibroblast growth factor (FGF) to form clusters of cells that are termed mammospheres. These clusters contain a mixture of stem cells and cells committed to the lineages of the mammary gland, that is, epithelial (red) and myoepithelial (pale blue) cells. Serial self-renewal capacity of stem cells within the mammosphere can be demonstrated by dissociation and re-plating of cells. Alternatively, growth in media that support differentiation (e.g., containing serum and collagen) allows the committed cells fully to mature into epithelial and myoepithelial cells or the stem cells to differentiate into both cell types. Three-dimensional culture is also possible in the presence of an ECM-like matrix, leading to the formation of organoids that display the morphological features of luminal and alveolar cells. (Photos: From Dontu G & Wicha M. (2005) *J Mammary Gland Biol Neoplasia* 10:75–86 [doi: 10.1007/s10911–005-2542-5]. Reprinted by permission from Springer Nature. Line art: Adapted from Dontu G, Abdalla W, Foley J et al. (2003) Genes & Dev 17:1253–1270 [doi: 10.1101/gad.1061803].)

EpCAM; however, basal cells have a lower level of expression of CD24 and higher levels of CD29 and CD49f compared to luminal cells (**Figure 15.22**). Mammary cells sorted by FACS using combinations of these latter markers and transplanted into the cleared fat pads of mice revealed that only basal epithelial cells contained the ability completely to regenerate the mammary epithelium, indicating that fully competent stem cells resided only in this compartment of the mammary gland tissue.

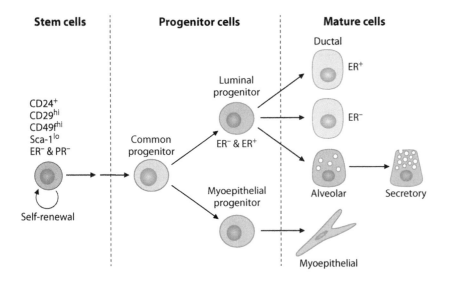

**Figure 15.22 MaSC markers and hierarchy.** The schematic illustrates the hierarchical relationship between stem cells, progenitors, and differentiated cells in the formation of the mammary gland and highlights some of the surface markers that have been used (in the mouse) to isolate and characterize them. In addition to the expression of the indicated surface markers, the presence or absence of estrogen receptor (ER) also discriminates between subpopulations of the luminal progenitors.

Several recent reports have also linked the expression of specific markers to a state of quiescence, reminiscent of what has been described so well for HSC in the bone marrow. For example, high levels of Bcl11b expression in 5% of the cells in a $CD49f^{high}CD24^{med}Lin^-$ basal cell population mark a quiescent stem cell that has potent regenerative activity and possibly lies upstream of more committed progenitors. Using a reporter mouse, it was shown that the $Bcl11b^{high}$ cells localize to the interface of basal and luminal cells throughout the adult mammary gland.

## 15.6  SUMMARY

In this chapter, we have covered what are seemingly a highly related group of tissues, linked through their function as interfaces with the environment and their fundamental building block, that is, the epithelial cell. Indeed, these common connections appear to be reflected to some extent in the underpinning stem cells. However, each tissue is also revealed as a complex of component parts requiring several different cell types, each of which is supported by its own distinct stem cell. The skin illustrates these complexities very well, being not only an interface between the body and the environment but incorporating hair, sweat glands, and a means to achieve coloration. The hair follicle and its interplay with the skin epithelium also beautifully illustrates the diverse contributions of different stem cells, their flexibility depending on circumstances, and potential for communication between different stem cell types. Hence, the hair follicle itself can be regarded as being capable of generating at least three tissues, that is, the hair, the sebaceous gland, and in some circumstances a contribution to the epithelium, each structure originating from particular stem cells, although it remains to be determined if there is a hierarchical relationship between any of these. The hair follicle also provides a perfect example of the stem cell niche, not only for its component stem cells and how they are maintained but also for the stem cells that give rise to melanocytes, which intriguingly exhibit a cross-talk that influences the potential of each stem cell. Comparison between the hair and feather follicles shows not only how evolution has crafted apparently quite different tissues using similar principles but also that the niche has evolved in some circumstances, as seen for feathers, to play yet more roles by orienting stem cells to affect shape and patterning. Some of the appendages discussed also show how stem cells can be involved in a repetitive cycling of tissue structure, as we have seen for both the hair follicle and the mammary gland, and that there may often be flexibility for stem cells to give rise to different cell types and structures depending on their environment. Without doubt, the study of skin, related tissues, and their appendages has provided and will continue to provide many of the guiding principles to understanding stem cells and form the basis for many novel therapeutic opportunities.

## KEY POINTS

- Epithelial cell layers form a barrier between the external environment and the rest of the body. Specialized structures called appendages are associated with a given epithelium to support its function.

- Epithelial cells are lost on a continual basis as part of their function and need to be replaced from stem cells that reside in specific niches particular to each epithelium.

- The skin is constituted of a stratified epithelium in which are embedded several appendages, including the hair follicle and sweat glands. Under normal conditions, the epithelium is replaced from stem cells that sit in the basal cell layer.

- The hair follicle is a complex structure containing a number of distinct stem cell types that support the repeated cycling of hair growth, the sebaceous gland, and after wounding of the skin can also produce new epithelium.

- Sweat glands have their own contingent of stem cells, some of which can give rise to all cell types of the structure. These multipotential myoepithelial stem cells have a degree of plasticity in that they can give rise to stratified epithelium or milk-producing cells when transplanted into the appropriate tissue context.

- The hair follicle also contains the niche that supports melanocyte stem cells.

- Feather follicles exhibit many similarities to hair follicles, including the presence of melanocyte stem cells, but evolved independently. The feather follicle has incorporated the additional feature of variable spatial arrangement of the stem cells in their niche, which allows for the production of different feather shapes from the feather follicle stem cells as well as a variety of coloration patters from the melanocyte stem cells.

- The front surface of the eye, the cornea, is an epithelial tissue that is supported by stem cells residing in the limbic regions on either side of the iris.

- The GI tract contains various epithelial cell-lined surfaces, most of which are derived from the endoderm, as opposed to the ectodermal origins of the skin, its appendages, and the cornea. The exception is the oral cavity, in which the epithelia arise from the ectoderm in the same way as the skin.

- The oral epithelium also has appendages, including salivary glands and in some animals a specialized epithelium that supports the continual growth of teeth through an inductive interaction with mesenchymal cells.

- The mammary gland is an epithelial appendage of the skin that has some passing similarities to sweat glands. Stem cells have been characterized and play an essential part in the cycle of growth, lactation, and eventual regression that is associated with pregnancy.

## FURTHER READING

### Skin and its appendages

Beck, B. and Blanpain, C. (2012) Mechanisms regulating epidermal stem cells. *EMBO J.*, 31: 2067–2075.

Hsu, Y.C., Li, L. and Fuchs, E. (2014) Emerging interactions between skin stem cells and their niches. *Nat. Med.*, 20: 847–856.

Tanimura, S., Tadokoro, Y., Inomata, K., Binh, N.T., Nishie, W., Yamazaki, S., Nakauchi, H. et al. (2011) Hair follicle stem cells provide a functional niche for melanocyte stem cells. *Cell Stem Cell*, 8:177–187.

Watt, F.M. (2014) Mammalian skin cell biology: At the interface between laboratory and clinic. *Science*, 346: 937–940.

### Salivary gland

Maimets, M., Rocchi, C., Bron, R., Pringle, S., Kuipers, J., Giepmans, B.N.G., Vries, R.G.J. et al. (2016) Long-term *in vitro* expansion of salivary gland stem cells driven by Wnt signals. *Stem Cell Rep.*, 6: 150–162.

### Cornea

Nowell, C.S. and Radtke, F. (2017) Corneal epithelial stem cells and their niche at a glance. *J. Cell Sci.*, 130: 1021–1025.

Secker, G.A. and Daniels, J.T. *Limbal Epithelial Stem Cells of the Cornea*. StemBook. http://www.stembook.org/node/588

### Mammary gland

Lloyd-Lewis, B., Harris, O.B., Watson, C.J., Davis, F.M. (2017) Mammary stem cells: Premise, properties, and perspectives. *Trends Cell Biol.*, 27: 556–567.

Visvader J.E. and Stingl, J. (2014) Mammary stem cells and the differentiation hierarchy: Current status and perspectives. *Genes Dev.*, 28: 1143–1158.

# 16

# Atypical stem cells that enable and modulate immune system function

At the very beginning of this book, we have given the simple, yet definitive, description of what features constitute a stem cell (Chapter 1). The two stem cell characteristics of **self-renewal** and capacity to **differentiate** along specific cell lineages are amply illustrated in many of the subsequent chapters, especially those dealing with **tissue stem cells**. Where strict compliance with the stem cell criteria has not been demonstrated, this is more often than not because the necessary "gold standard" assays have not been available to researchers. However, as with many strict definitions, there are inevitably some gray areas. The immune system in vertebrates, which is constituted of a variety of differentiated cell types that help defend against a broad range of "invaders," including microorganisms and "foreign" tissues, relies on some interesting examples of cells that have features of stem cells yet in one way or another do not fit so well with the "standard" view of what defines a stem cell.

In Chapter 9, we have already described the ultimate stem cell origin of the hematopoietic hierarchy in the form of **hematopoietic stem cells** (HSC), which give rise to the cells that make up blood including the functional effector cells. In this chapter, we will focus on four very different aspects of the immune system that are either totally independent of HSC or involve stem cell properties at the bottom rather than the top of the hematopoietic hierarchy. First, we will look at effector cells involved in aspects of the immune response that appear to arise early in development independently of HSC but have a life-long ability to self-renew. Second, we will discuss various cells of the immune response that have the capacity of memory, which at least at first sight seems to recall elements of stem cell behavior. Then we will consider what are perhaps more conventional tissue-specific stem cells that support the so-called secondary lymphoid organs in which many aspects of immune responses occur. Last, we will describe the other face of **mesenchymal stem cells** (MSC) (Chapter 10), that is, how they are able to moderate the effector cells of the immune system, which is increasingly being realized to be critical for the fine balance between an appropriate immune response and one that is overly vigorous and potentially harmful.

## 16.1   EFFECTOR CELLS OF THE IMMUNE SYSTEM

All organisms are exposed to pathogens in their environment, including parasitic, bacterial, or viral infection. To combat such infections, living organisms have evolved mechanisms of immune defense that come into play if purely physical barriers are breached (Figure 16.1). The first line of protection, which occurs within hours and is seen in both invertebrates and vertebrates, involves hematopoietic cells and is referred to as innate immunity. A slower response, provided by the adaptive immune system, evolved in vertebrates less than 500 million years ago. The purpose of the adaptive immune system is the production of a population of cells that attack the pathogen and then develop an element of memory so that the next time the particular invader is encountered, the response is much faster.

Both innate and adaptive immunity rely on hematopoietic cells whose stem cell origin is not in question in that they derive from HSC resident in the bone marrow (Chapter 9). The innate immune system is largely composed of HSC-derived bone marrow ("myeloid") cells, including granulocytes and macrophage, although one

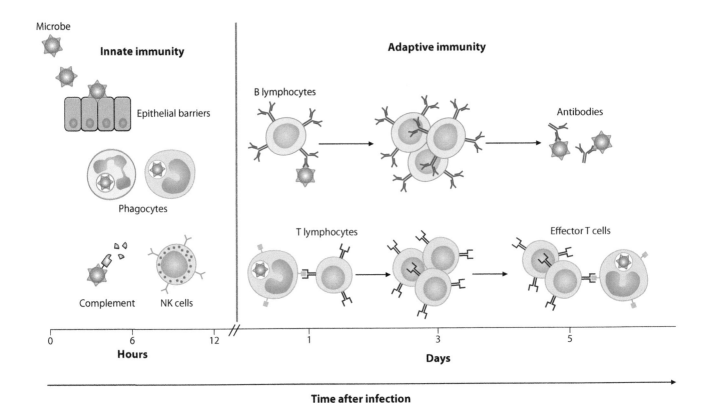

Figure 16.1 **The innate and adaptive immune responses.** The innate and adaptive immune responses provide a defense against microbes and foreign antigens. Innate immunity is ready to respond immediately, whereas the adaptive response involves the activation and expansion of specific lymphocytes that have the correct receptor for a given foreign antigen. The diagram summarizes the essential non-cellular and cell components and gives an indication of the relative time scales for induction of innate immunity and adaptive immunity. (From Abbas A & Lichtman A. (2006) *Basic Immunology: Functions and Disorders of the Immune System*, 2nd ed. Philadelphia: Saunders. Copyright 2006 with permission from Elsevier.)

cell of lymphoid origin, the natural killer cell, also plays a role. The granulocytes consist of the highly abundant neutrophil and the less abundant basophil and eosinophil. Neutrophils function as first responders, in particular engulfing pathogens by a process called phagocytosis. Basophils and eosinophils are important in inflammatory reactions, in particular during allergic reactions. Macrophage, like neutrophils, can engulf and digest microbes by phagocytosis and also clean up cell debris. In contrast, the adaptive immune response is composed of lymphoid cells, namely B- and T-cells, which respectively attack pathogens through the production of antibodies or by direct cell-mediated killing.

There is a degree of overlap between the two immune systems with some cell types such as natural killer cells, which respond rapidly upon viral infection, and γδ T-cells acting in both systems, and some macrophage and the myeloid dendritic cells serving to process and present antigens to the lymphoid cells of the adaptive response (Figure 16.2).

## 16.2 HEMATOPOIETIC STEM CELL-INDEPENDENT MYELOID CELLS IN THE INNATE IMMUNE RESPONSE

In vertebrate animals, the origins and complement of myeloid and lymphoid hematopoietic cells vary during the course of development from early stages of embryogenesis through to the adult. The source of hematopoietic cells ranges from the yolk sac and then the major blood vessel wall in early embryos, to the fetal liver around mid-gestation, and then one or more adult tissues. In birds and mammals, the adult tissue source of hematopoietic cells is the bone marrow, while in fish, it is the kidney. The profile of hematopoietic cells changes during development in a way that mirrors the different sources of cells. Early embryonic hematopoiesis tends to be

**Innate immunity**
**(rapid response)**

**Adaptive immunity**
**(slow response)**

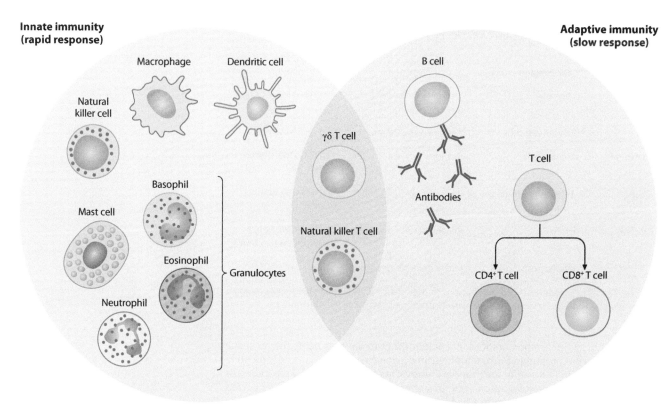

Figure 16.2 **Cells of the immune system.** Representation of the cell types that constitute the innate and adaptive immune systems and the degree of overlap between the two systems. (From Dranoff G. (2004) *Nat Rev Cancer* 4:11–22 [doi:10.1038/nrc1252]. Reprinted by permission from Springer Nature.)

restricted to the generation of myeloid cells, and only later do HSC become established, for instance, in the liver and then the bone marrow, which is able to produce the full gamut of hematopoietic cell types seen in adults. An important hematopoietic cell type during early development is the macrophage, which is initially produced from the yolk sac before HSC in the bone marrow become their predominantly life-long originating stem cell. Interestingly, it has recently been discovered that the early macrophage can persist throughout life, a capacity that seems to have a stem cell basis.

## Adults have multiple distinct populations of macrophage that have different developmental origins

Although circulating macrophage involved in the innate immune response are derived from HSC in the bone marrow, similar cells are resident in nearly every tissue of the body (Table 16.1). For example, the brain contains cells known as microglia, the skin is home to Langerhans cells, and Küpffer cells in the liver represent this organ's resident macrophage population. Despite performing broadly similar roles within each tissue, there is a great deal of heterogeneity among the different macrophage types, as seen in their patterns of surface antigen expression, implying that there are actually significant functional differences.

The role of HSC in the generation of bone marrow and circulating macrophages is well understood, but the origins of tissue-resident macrophages are somewhat more controversial. This controversy is, however, perhaps more a matter of detail rather than a profound disagreement amongst researchers. What is agreed is that a proportion, sometimes the majority, of specific tissue-resident macrophage originate not from monocytes derived from bone marrow HSC, but instead arise at an early stage of development. The issue of detail is exactly from which point of development tissue-resident, as opposed to infiltrating bone marrow-derived, macrophage arise. As often the case around scientific contentions, everybody is right, at least to a degree! Deducing the source of tissue-resident macrophage has relied principally on **lineage tracing** experiments involving activation of a fluorescent protein reporter by **Cre recombinase** (Chapter 2) expressed from regulatory elements of genes encoding proteins expressed

**TABLE 16.1 TISSUE RESIDENT MACROPHAGE TYPES**

| Adult tissue | Macrophage cell type | Function | Characteristic surface markers |
|---|---|---|---|
| Brain | Microglia | Clearing cellular debris and dead neurons | CD11b, CX3CR1, CD115 |
| Skin | Langerhans cells | Antigen presentation | CD11b, F4/80, MHC II |
| Liver | Küpffer cells | Phagocytosis of microbes | F4/80, CD169, CD64, CD68 |
| Lung | Alveolar macrophages | Phagocytosis of microbes | CD11c, F4/80, CD169, CD64 |
| Colon | F4/80$^{hi}$ | Phagocytosis of microbes | CD11c, F4/80, MHC II, CX3CR1, CD115, CD64 |
| Blood | Monocytes | Innate immune response | CD11b, Ly6C, CD115 |

*Note:* The table lists some of the principal types of macrophage found in adult tissues, describing their function and characteristic surface markers that have been defined in the mouse.

on macrophage lineage precursors at particular times in development. The presence of the fluorescent protein in the macrophage in a given tissue is taken to be indicative that the cell originated within the time period covering the expression of the indicator protein. The differences of opinion seem to stem from the selection of the gene used to drive expression of Cre recombinase and the assumptions about where this gene is first expressed in embryo development. Based on several such studies in the mouse, the current view is that brain microglia arise solely in the early embryo from the yolk sac site of hematopoiesis. Langerhans cells in the skin partly derive from the yolk sac and partly from fetal liver cells, all other tissue resident macrophage being of fetal liver origin (**Figure 16.3**). There remains some disagreement about the precise nature of the "second wave" of cells coming from the fetal liver, although for the purposes of our discussion, this is a detail that does not impact the important point that not all tissue resident macrophage can be continuously replaced from HSC-derived monocytes.

## Non-hematopoietic stem cell-derived tissue-resident macrophage appear to have a continuing requirement for stem cell activity

Why does the origin of tissue-resident macrophage matter from the perspective of stem cell biology? The answer is simply that the persistence of populations of macrophage

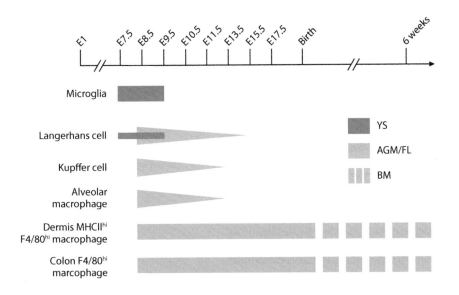

Figure 16.3 **Distinct developmental origins of adult macrophage.** Tissue macrophage originate from embryonic, fetal, and adult sources of hematopoietic cells. The schematic time line summarizes the developmental origins of some of the most important adult macrophage types. The days of embryonic development (E) are indicated up to birth and early perinatal life. Abbreviations: YS – yolk sac; AGM – aorta-gonad-mesonephros; FL – fetal liver; BM – bone marrow; Mφ – macrophage. (From Sheng J, Ruedl C & Karjalainen K. (2015) *Immunity* 43: 382–393 [doi:10.1016/j.immuni.2015.07.016]. Copyright 2015 with permission from Elsevier.)

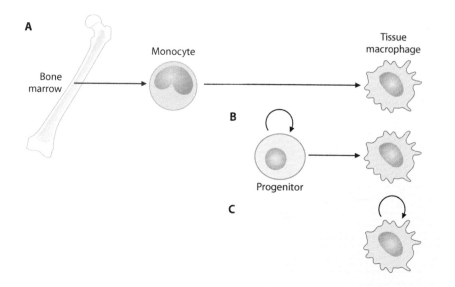

Figure 16.4 **Models of tissue-resident macrophage maintenance.** Adult tissue macrophages arise either from the bone marrow or from embryonic sites of hematopoiesis. Macrophage that persist from early development require a mechanism to allow their replacement. (**A**) Bone marrow-derived monocytes are able to differentiate into macrophage that infiltrate tissues throughout adult life. (B and C) Two models can be proposed for the way in which macrophage that are deposited in a tissue during embryonic or fetal life are able to be maintained in adult life; (**B**) a self-renewing monopotent stem cell provides differentiated macrophage as required, or (**C**) the tissue-resident mature macrophage retain an ability to replicate.

deriving from the embryo or fetus immediately implies that these cells can be maintained through proliferation in a way akin to regular tissue stem cells. Two possible scenarios can be envisaged to explain the apparent stem cell behavior of tissue macrophage (Figure 16.4). One idea is that there are **monopotent progenitors**/stem cells for tissue-resident macrophage that can expand by **asymmetric** division and then differentiate toward mature functional cells as required, in much the same way as expected of tissue stem cells. The alternative view is that tissues are seeded during development with essentially mature macrophage, but these retain an ability to re-enter a replicative cell cycle to make exact copies of themselves, again reflecting an essential feature of stem cells.

Certainly, the tissue-resident macrophage can respond to situations requiring expansion of numbers, for example, during inflammation, in much the same way as bone marrow-derived cells that infiltrate from the blood stream (Figure 16.5). Both recruited and tissue-resident macrophage respond to macrophage colony stimulate factor (M-CSF) and lymphoid cell-derived interleukin (IL) 4 by proliferating, although the expansion of the resident cells can be massive. Tissue-resident macrophage demonstrate some site-dependent differences in their cytokine responsiveness, so that those in the brain and skin are stimulated by M-CSF and IL-34, while those in the lung respond to granulocyte-macrophage (GM) CSF.

## Details are beginning to emerge about the mechanisms that underlie tissue-resident macrophage self-renewal

Studies from the laboratory of Michael Sieweke at the University of Marseille have given an important insight into the way that apparently differentiated tissue-resident macrophage can re-enter a proliferative state and thereby self-renew. By making genome-wide analyses of transcription factor binding and histone modifications associated with gene activation status and comparing **quiescent** and self-renewing macrophage, they were able to show that enhancers bound by the lineage-specifying transcription factor PU.1 are more activated in the proliferating cells. The activated enhancers are particularly associated with a network of genes encoding regulatory proteins, centered on Myc and Klf2, which are important for self-renewal. Interestingly, the same set of genes are linked to **embryonic stem (ES) cell**

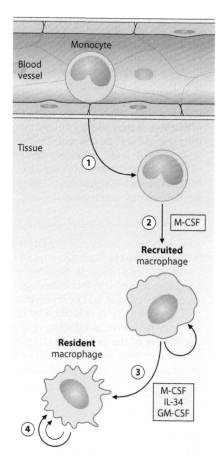

Figure 16.5 **Tissue-resident and recruited macrophage respond to inflammatory signals.** Both tissue-resident macrophage and those derived from the bone marrow contribute to the response to inflammation. HSC-derived monocytes infiltrate the tissue from the blood stream (1). Infiltrating cells differentiate (2) *in situ* in response to macrophage colony stimulating factor (M-CSF), which can also induce limited proliferation (circular arrow) of recruited macrophages, which can sometimes integrate into the resident macrophage pool (3). Self-renewal of resident macrophages (4) during homeostasis (small circular arrow) or acute inflammatory conditions (large circular arrow) can be driven by M-CSF or interleukin-34 (IL-34) in the brain and epidermis and by granulocyte-macrophage colony stimulating factor (GM-CSF) in the lung. (Adapted from Sieweke M, & Allen J. (2013) *Science* 342: 946 [doi: 10.1126/science.1242974].)

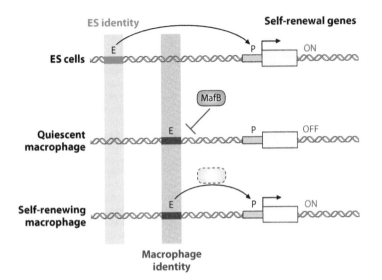

Figure 16.6 **Tissue-resident macrophage self-renewal genes are inhibited by MafB.** Genome-wide analysis of gene expression and epigenetic modification have shown that tissue-resident macrophage (MΦ) are capable of expressing a set of self-renewal genes that are also important in embryonic stem (ES) cells but through the action of distinct enhancer (E) elements. In macrophage, the transcription factors that act on the relevant enhancer are inhibited by the factor MafB when in the quiescent state, its down regulation enabling activation of the self-renewal gene promoters (P). (From Soucie E, Weng Z, Geirsdóttir L et al. (2016) *Science* 351:680 [doi: 10.1126/science.aad5510]. Reprinted with permission from AAAS.)

self-renewal genes, although in these cells, distinct enhancers are involved in their regulation (Figure 16.6). Furthermore, the macrophage lineage transcription factor MafB was found to be bound to the self-renewal genes in the quiescent state only, and *in vivo* single cell analyses on resident macrophage revealed that they exhibit transient down regulation of MafB when they become proliferative.

## 16.3    MEMORY CELLS OF THE ADAPTIVE IMMUNE RESPONSE HAVE STEM CELL PROPERTIES

The adaptive immune response represents an interesting challenge to the definition of what constitutes a stem cell and perhaps requires a view that goes beyond the apparently straightforward dogma (Chapter 1) and the features of molecular mechanisms underpinning **adult stem cells** (Chapter 4). This scenario relates to the feature of immunological memory, which is inherent in specific lymphoid cells and can persist throughout a large part of life.

An acquired response by lymphocytes to a given foreign antigen, leading either to the generation of specific antibodies by B-cells or functional T-cells, involves unique rearrangements of immunoglobulin or T-cell receptor genes, respectively. An essential feature of these responses, and the reason for the effectiveness of vaccination against a host of diseases, is the ability of the immune system to "remember" the first encounter and to then respond very rapidly and effectively should the same organism be encountered later in life. Both B- and T-cells have the ability to generate a memory of their first experience of pathogen-associated antigens. This cell-based memory is reminiscent of what stem cells have to achieve in terms of self-renewal and differentiation, but are there really any similarities in the molecular and cellular features of these two very different scenarios?

### The cellular basis for B-cell memory is little understood

Although the memory component of acquired immunity is without doubt, the precise definition of the cells involved is not completely resolved. In the case of B-cells, there are at least two routes to memory, one requiring the so-called germinal centers in lymph nodes and involving the stages during which the antibody response "matures" through expansion and selection of cells that produce antibodies with higher affinity, and a second that is independent of the germinal centers and represents more the primary response, producing IgM class antibodies. Beyond this understanding, the nature of the cells conferring B-cell memory is somewhat sketchy, and there is no conclusive insight into the cellular mechanisms that specifically support the maintenance and subsequent differentiation of these cells.

### T-cell memory can be ascribed to cells with a specific stem cell phenotype

The situation for the T-cell component of the immune system is more clearly defined than it is for B-cells and provides a better basis to test whether stem cell principles apply to memory cells in the adaptive immune response. There are a number of subtypes

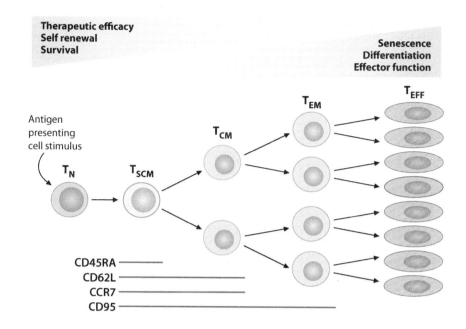

Figure 16.7 **Memory T-cell hierarchy.** The development and maintenance of memory T-cells can be shown schematically as a hierarchy with the surface markers that define each stage depicted as an expression bar below. T-cell differentiation initiates at the apex with the naïve T-cell ($T_N$), which upon activation by engagement with a stimulus from an antigen presenting cell enters into a program of proliferation and differentiation through the T memory stem cell ($T_{SCM}$), the central memory T-cell ($T_{CM}$), the effector memory T-cell ($T_{EM}$), and culminating in the generation of the short-lived effector T-cell ($T_{EFF}$). (Adapted from Gattinoni L & Restifo NP (2013) *Blood*, 121: 567–568)

of T-cells, each with a distinct function during the immune response. T-helper ($T_H$) cells, which express the **cluster of differentiation** 4 (CD4) surface marker, assist other white blood cells in mounting an immune response, such as macrophage by causing their activation or promoting the maturation of B-cells. CD8+ cytotoxic T-cells ($T_C$) act to destroy virus-infected cells. Regulatory T-cells ($T_{reg}$) are also CD4+ but are additionally characterized by their expression of CD25. The role of $T_{reg}$ is to suppress activation, proliferation, and cytokine production of $T_H$ and $T_C$, and they are also thought to suppress B-cells and dendritic cells.

The current view of the development and maintenance of memory T-cells, gleaned from experiments in mice, can be represented as a hierarchy (Figure 16.7), which starts with naïve T-cells ($T_N$) and their encounter with foreign antigen presented on specialized macrophage-like cells (antigen presenting cells; APC). Once a $T_N$ is activated, it proliferates and differentiates, ultimately becoming a short-lived effector T-cell ($T_{EFF}$), including both $T_H$ and $T_C$. Under the influence of particular cytokines, this process of differentiation can be effectively diverted to generate T memory stem cells ($T_{SCM}$). Some more **committed** intermediaries have also been defined between the $T_{SCM}$ and $T_{EFF}$ and can be distinguished by certain surface antigens, but $T_{SCM}$ have the greatest capacity for self-renewal and are able to reconstitute the full diversity of memory and effector cells, as revealed by serial transplantation assays (Chapter 2). This latter feature is reminiscent of the self-renewal and **multipotency** properties of adult stem cells, and certainly the most recent demonstration that single CD8+ $T_{SCM}$ can serially transfer a specific $T_C$ immune response argues very much in favor of a true stem cell identity. Some other characteristics of $T_{SCM}$ and their immediate central memory ($T_{CM}$) derivatives contrast with the adult stem cell paradigm. Hence, adult stem cells are laid down in individual tissues as they develop in the fetus, whereas $T_{SCM}$ progressively accumulate as an animal encounters pathogens during early life. Also, unlike the rarity of adult stem cells in relation to the overall population of hematopoietic cells in the bone marrow, memory T-cells represent the predominant subset of T-cells in many tissues, including in the bone marrow and spleen, but also locations such as mucosal surfaces and the skin.

## T-cell memory stem cells actively proliferate

$T_{SCM}$ cells have been described in both mice and humans, in the latter being defined by the surface antigen characteristics CD8+CD27[bright]CD45RO–CCR7+CD95+. These cells exhibit a life-long presence at stable frequencies, which suggests two possibilities for their behavior; that is, either $T_{SCM}$ cells remain in a relatively dormant state with prolonged survival, or the pool of these memory cells may be sustained by ongoing proliferation and cell turnover. Evidence supporting the second option was recently provided using a novel approach of *in vivo* labeling of proliferating human cells using deuterium (as $^2H_2O$) followed by a chase period and a detailed flow cytometric quantification of the T-cell subpopulations. The conclusion reached was that $T_{SCM}$

cells are maintained in a state of dynamic flux. The degree of cell turnover fits with the high levels of the proliferation associated protein Ki67 that were detected in both the CD4+ and CD8+ $T_{SCM}$ cells. In contrast, Ki67+ cells were rare in the corresponding $T_N$ populations. This study also looked at individual **telomere** lengths, which were found to be long. Moreover, $T_{SCM}$ cells displayed higher levels of telomerase activity than either $T_N$ or other memory T-cells, again attesting to their stem cell nature.

The apparent similarity between memory lymphocytes and adult stem cells with respect to stem cell characteristics has led researchers to look for cellular and molecular parallels. It would be wrong, of course, to assume that an apparent relatedness based on functional criteria between adult stem cells, such as the HSC, and memory lymphocytes like the $T_{SCM}$ necessarily means that the molecular mechanisms have to be in any way the same. Some interesting observations have been made in CD4+ $T_N$ cells in which the protein mouse target of rapamycin (mTOR) has been inhibited by rapamycin. The consequence of such treatment is the induction of $T_{SCM}$ cells, which are functionally more effective than *in vivo* derived CD4+ memory T-cells when tested by transplantation. This conversion involves a shift to fatty acid metabolism favored by the memory T-cells. Transcriptome comparisons of CD4+ $T_N$ or $T_{CM}$ cells with $T_{SCM}$ cells (natural or pharmacologically induced) revealed some genes with distinct expression levels in the stem cell subtype, but there is little yet to indicate which of these differences are functionally significant.

## Epigenetic and transcriptional regulation is important in the specification of $T_{SCM}$

Genetic evidence from mice has shown there to be some specific roles for histone modification in the establishment or maintenance of $T_{SCM}$. For example, comparison of CD8+ $T_{SCM}$ and $T_{EFF}$ cells from wild type animals and mice deficient in the enzyme Suv39h1, which methylates histone 3 on lysine reside 9 (H3K9me3) and thereby silences gene expression, revealed a set of affected stem cell-related memory genes. Suv39h1-defective CD8+ $T_{SCM}$ cells show sustained survival and increased long-term memory capacity, suggesting that Suv39h1 silences memory stem cell genes during CD8+ $T_{EFF}$ terminal differentiation (Figure 16.8).

At the level of transcription factor-dependent regulation, one protein, Myb, has recently emerged as a pivotal regulator of CD8+ $T_{SCM}$. Interestingly, Myb is known to play important roles in the control of self-renewal and differentiation in a number of adult stem cells, most notably HSC. CD8+ T-cells lacking Myb undergo terminal

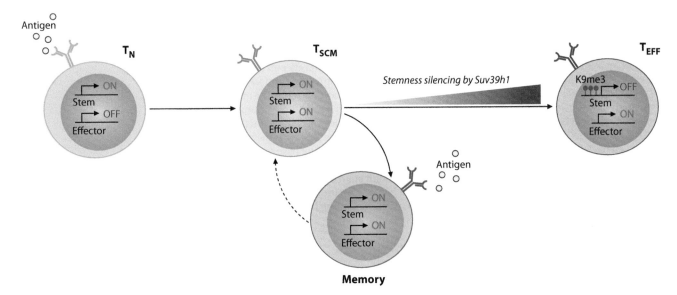

**Figure 16.8 CD8+ T memory stem cell commitment to effector cells is regulated by specific histone methylation.** Genes that control self-renewal and effector function in naïve T-cell ($T_N$), T memory stem cells ($T_{SCM}$), and T effector cells ($T_{EFF}$) appear to be subject to epigenetic regulation that modifies the control elicited by differentiation-associated transcription factors. The working model illustrated for CD8+ T-cells shows how differentiation affects the activation status ("ON" versus "OFF") of both stem cell and effector function-associated genes and further how the Suv39h1 protein can inhibit stem cell genes through methylation of histone H3 lysine 9 (K9me3), allowing the activated effector genes to dominate and promote terminal differentiation. (From Pace L, Goudot C, Zueva E et al. (2018) *Science* 359:177–186 [doi: 10.1126/science.aah6499]. Reprinted with permission from AAAS.)

differentiation and generate fewer stem cell-like memory cells than Myb-sufficient T-cells. Myb acts both as a transcriptional activator of the gene encoding the transcription factor Tcf1 to enhance memory development and as a repressor of the *Zeb2* transcription factor gene to hinder effector differentiation.

## 16.4 STEM CELLS IN THE PRIMARY AND SECONDARY LYMPHOID ORGANS

Although the effectors of the immune system are cells of hematopoietic origin, the way in which these cells are deployed, respond to foreign invaders, and then deliver their function, whether this is through antibody production or cell-mediated actions, is very much dependent on structures constituted by non-hematopoietic cells. Since these structures persist throughout life, it follows that they are likely to be maintained by cells that have a stem cell origin, either locally within the structure or deriving from an exterior source.

### Mesenchymal cells constitute the structural elements of lymphoid tissues

Tissues where lymphoid cells mature and function are divided into primary and secondary lymphoid organs (Figure 16.9). The primary lymphoid organs provide an environment for HSC and committed progenitors to divide and mature into B- and T-cells (Figure 16.10). The bone marrow serves as the ultimate source of all lymphoid cells in mammals because it harbors the HSC. On the other hand, precursors of T-cells migrate from the bone marrow to the thymus (Box 16.1) to undergo their process

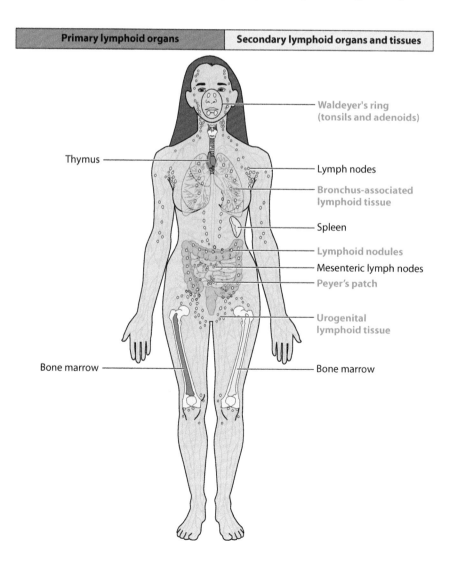

**Figure 16.9 Lymphoid organs.** Schematic cross section through the human body showing the location of primary (orange) and secondary (yellow) lymphoid organs and tissues.

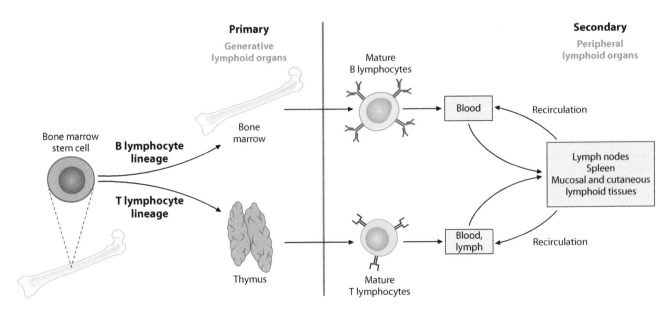

**Figure 16.10 The role of primary and secondary lymphoid organs in immune responses.** Ultimately deriving from the bone marrow hematopoietic stem cell, B- and T-cells differentiate and mature within the primary lymphoid organs (bone marrow and thymus). Upon migration to secondary lymphoid organs (lymph nodes, spleen, mucosal and cutaneous tissues), the lymphoid cells encounter foreign antigens and initiate an adaptive immune response.

## BOX 16.1  BACKGROUND: THE THYMUS AND T-CELL DEVELOPMENT

Although T-cell precursors are produced within the bone marrow, they do not become functionally mature until they have been primed by the thymus, a crucial component of the adaptive immune system. In humans, the thymus is located in the front of the chest cavity near the trachea and is composed of two identical lobes each consisting of multiple lobules (Figure 1). The thymic lobes are segregated into a central medulla, a peripheral cortex, and an outer capsule. Thymic activity, and therefore growth, is at its greatest during the neonatal and pre-adolescent years. The cells

within the thymus that perform the crucial support for T-cell differentiation are the thymic epithelial cells (TEC, Figure 2), which are specific to either the medulla (mTEC) or the cortex (cTEC). Lymphoid progenitors migrate from the bone marrow to the thymus, where they mature through progressive phases of proliferation, differentiation, and selection. The role of the cTEC is to aid in the production of a diverse repertoire of T-cells, generated at random, which then undergo selection for those that are able to recognize antigens in the presence of the MHC molecules of the individual ("self MHC"). The selected and expanded T-cell clones then progress through the medulla, where they are negatively selected to eliminate cells that react with self-antigens, thereby avoiding the possibility of autoimmunity, that is, generating inherent "tolerance" in the immune system.

Box 16.1 Figure 1 **The thymus.** Histological section (H&E stained) of a human infant thymus showing the defined cortex (dark staining) and medulla (light staining) regions and the separation into distinct lobules. (From Kerr J. (1999) *Atlas of Functional Histology*. St. Louis: Mosby. Copyright 1999 with permission from Elsevier.)

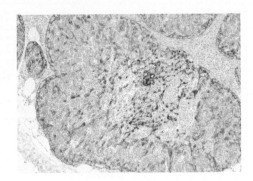

Box 16.1 Figure 2 **Thymic epithelia.** Section through a thymic lobule immunostained using an antibody against cytokeratin (brown staining), which is expressed on epithelial cells that are distributed throughout the structure. (From http://slideplayer.com/slide/8073055/.)

of differentiation. Secondary, or peripheral, lymphoid organs, which include lymph nodes and the spleen, are the sites in which lymphocytes encounter foreign antigens and adaptive immune responses initiate, cells expand, and the initial response matures and develops a memory component. The way in which the secondary lymphoid organs are connected to the blood and lymphatic systems means that they can rapidly "survey" incoming pathogens and at the same time serve as a focus in order to bring together different cell components that need to interact in order to elicit an immune response.

The structural elements of the primary and secondary lymphoid tissues are constituted largely from cells of mesenchymal and epithelial origin. The specific epithelia that are critical to the function of the thymus undergo replacement in adult life, and it is therefore no surprise that these are likewise supported by a stem cell component. As we have seen in Chapter 10, the presence of MSC in the hematopoietic environment of the bone marrow provides a source of bone and fat cells that play a big part in the generation of the HSC **niche** (Chapter 6). Related, but specialized, mesenchymal cells known as fibroblastic reticular cells (FRC) are found in the lymph nodes, spleen, thymus, and other lymphoid tissues. Both MSC and FRC express platelet-derived growth factor receptor-α, but FRC can be distinguished by their expression of podoplanin.

## A bipotent stem cell maintains the thymic epithelium in adults

The thymic epithelium is the site at which T-cells are selected, either to be **clonally** expanded or to be eliminated because they recognize self-antigens. Thymic epithelial cells (TEC) exhibit continuous turnover, those in the medulla having a half-life of just two weeks, implying the need for a source for their continual replenishment. Evidence for a bipotent TEC progenitor first emerged from studies on the embryonic thymus, and until recently, the presence of a similar cell in the postnatal thymus has remained rather elusive. Rare **stromal cells** have now been identified in the murine adult thymus, which, under special culture conditions, including the presence of epidermal growth factor and fibroblast growth factor 2, can form spheres termed "thymospheres" (Figure 16.11), very much in the same way that stem cells from tissues as varied as heart and mammary gland expand and differentiate in culture (see Chapters 13 and 15). These thymosphere-forming cells (TSFC), which have the surface marker profile CD45- EpCAM- Sca1+ CD24- and express cytokeratins, display features of stem cells in that they are slow cycling, self-renewing, and bipotent, being able to commit to both the medullary and cortical TEC lineages.

## Lymph node fibroblastic reticular cells are replaced during adulthood from two precursor cell sources

The structure of lymph nodes (Box 16.2) is laid down during embryogenesis, but there is a continuing requirement for cell replacement in adulthood, which is most apparent under conditions of inflammation. The majority component cells of lymph nodes are FRC, which create a matrix toward which lymphocytes and antigen presenting cells

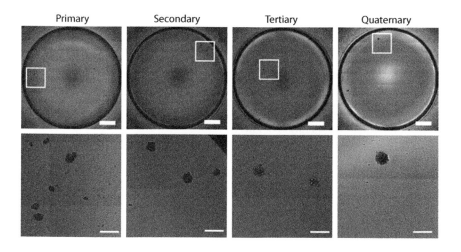

Primary | Secondary | Tertiary | Quaternary

**Figure 16.11 Thymosphere forming cells.** Adult thymic epithelium contains stem cells that are able to form thymosphere colonies *in vitro*. The figure shows representative images of thymospheres formed after sequential plating in optimal assay conditions. Colonies visualized in the low magnification images (white squares) on the upper row are shown at higher magnification at the bottom. Scale bars represent 1 mm and 200 μm in upper and lower rows, respectively. (From Ucar A, Ucar O, Klug P et al. (2014) *Immunity* 41: 257–269 [doi:10.1016/j.immuni.2014.07.005]. Copyright 2014 with permission from Elsevier.)

BOX 16.2 BACKGROUND: LYMPH NODES

The lymph nodes are abundant in number; in humans, there are between 500 and 700 distributed throughout the body (**Figure 1**). Lymph nodes make up part of the lymphatic system that is responsible for the transport of lymphocytes around the body and plays a role in the removal of excess fluid from tissues. The functional components of the lymph node consist of the outer cortex and the inner medulla (**Figure 2**). Lymph nodes are immunological meeting places, in which B-cells, T-cells, dendritic cells, plasma cells, and macrophages congregate inside an encapsulated mesenchymal sponge that is created by a network of fibroblastic reticular cells (FRC) surrounding thin strands of **extracellular matrix** and infiltrating lymphatic vessels. The structure of the lymph node is crucial to its function, as it funnels antigens and antigen-presenting cells toward rare antigen-specific lymphocytes to maximize their chance of finding each other and of initiating an adaptive immune response.

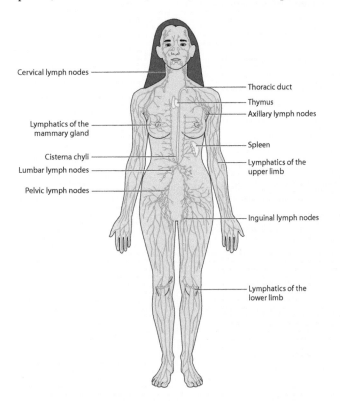

Box 16.2 Figure 1 **The lymphatic system.** Schematic cross section through the human body illustrating the location of the lymphatic network and lymph nodes.

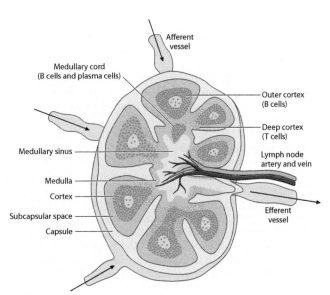

Box 16.2 Figure 2 **Lymph node.** Schematic cross section through a typical lymph node showing the key structural components and the location of the predominant lymphoid cell types.

migrate. FRC appear to have two origins that continue in adult life, one of these being located within the lymph node and the other being external (**Figure 16.12**). Within the lymph node, FRC most likely develop from specialized stromal progenitor cells, known as lymphoid tissue organizer (LTo) cells, which are themselves the descendants of MSC-like cells. More recent evidence reveals a second stream of mesenchymal precursor cells that migrate into the lymph nodes from adjacent adipose tissue. Pre-adipocytes can be recruited to lymph nodes from adipose tissues, or "fat pads," by a process that depends on lymphotoxin-β receptor (LTβR) and to a lesser extent the chemokine receptor CXCR4. Upon arrival in the lymph node, the pre-adipocytes differentiate into LTo cells and can generate up to 60% of FRC in adults. The new generation of FRC is particularly relevant in activated lymph nodes as part of an inflammatory response.

## 16.5 THE IMMUNE MODULATORY FUNCTION OF MESENCHYMAL STEM CELLS

MSC, as we have already discussed in Chapter 10, come with more than their share of controversy related to what they actually are in terms of origin, heterogeneity,

**Figure 16.12 Dual sources for lymph node reticular cell replacement.** Fibroblastic reticular cells (FRC) can be derived from two distinct cell sources, both of which lead initially to the generation of the lymphoid tissue organizer (LTo) cell that ultimately gives rise to the FRC. Within the lymph node, a mesenchymal precursor cell acts as one source of FRC. In addition, mesenchymal cells external to the lymph node, which are committed to the adipocyte lineage, can contribute to FRC production. The adipocyte-derived cells are especially important under conditions of inflammation. (From Fletcher A, Acton S & Knoblich K. (2015) *Nat Rev Immunol* 15: 350–361 [doi: 10.1038/nri3846]. Reprinted by permission from Springer Nature.)

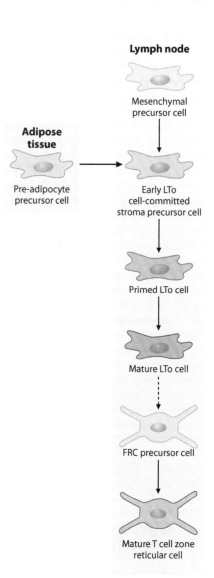

and adherence to the fundamental principles that we use to define a stem cell. These sorts of questions about MSC revolve around the apparently straightforward issue of their ability to differentiate toward osteoblasts, adipocytes, and chondrocytes, also adding to the controversy whether they can give rise to other cell types such as neural cells or hepatocytes. Here we will consider a totally different side to the character of MSC, that is, their capacity to affect the function of cells of the immune system. The way in which MSC can influence various aspects of the body's immune responses, especially when these are too vigorous or dysregulated, has raised considerable interest in possible clinical applications that, although exciting, are not without controversy.

## Mesenchymal stem cells can augment and moderate immune cell function

A number of observations have pointed toward an immunological relevant activity of MSC, including the fact that they appear to be "immune privileged," seem to induce a degree of tolerance in mismatched bone marrow transplantations, and both present at their cell surface and secrete many molecules with the potential for an effect on immune cells. In 2002, it was shown directly that human MSC can inhibit the proliferation of stimulated T-cells *in vitro* and can prolong skin engraftment when infused into non-human primates. Since then, the complexity of the interaction between MSC and the immune system has been fully realized, if perhaps not yet fully understood.

Although MSC interact with the majority of immune cells of both the innate and adaptive immune systems, they themselves are not considered an immune cell; rather, they can be thought of as safety valves or governors of an immune response, which if unchecked could get out of hand and cause serious damage to single or multiple tissues. MSC produce a wealth of growth factors, cytokines, chemokines, proteases, and adhesion molecules that have the potential to act upon immune cells in inflamed and damaged tissues, thereby modulating the inflammatory and defense reactions. *In vitro*, MSC have been shown to have effects on T-cells, B-cells, NK cells, dendritic cells, neutrophils, and macrophage. The immunosuppressive behavior of MSC appears to be absent, or at least not well developed, in MSC that have not received an inflammatory stimulus. Three factors seem to be particularly relevant for the activation of the immunosuppressive functions of MSC, namely interferon $\gamma$ (IFN-$\gamma$), tumor necrosis factor $\alpha$ (TNF-$\alpha$), and interleukin 1$\beta$ (IL-1$\beta$). For example, exposure of MSC to IFN-$\gamma$ increases their immunosuppressive function by stimulating the production of inflammatory inhibitors such as indoleamine-pyrrole 2,3-dioxygenase (IDO), prostaglandin E2 (PGE2), transforming growth factor $\beta$ (TGF$\beta$), and hepatocyte growth factor (HGF). The source of the inflammatory stimuli is at least in part those cells that the MSC, once activated, will suppress, and only following this cell-to-cell cross-talk does the release of soluble factors have an inhibitory effect. Although the overall effect of MSC is one of immune suppression, this can be elicited by both positive and negative effects on the target immune cells (Figure 16.13).

The precise mechanisms by which MSC interact with these multiple cell types have not been fully elucidated, but it is clear that they are numerous. The major ways in which MSC influence immune cells include (i) secretion of anti-inflammatory factors, (ii) expression at their cell surface of immunosuppressive molecules, (iii) secretion of immune cell attractants (chemokines), and (iv) control through expression of enzymes that metabolize compounds essential to lymphocyte proliferation (Table 16.2). MSC can also contribute to immune responses, particularly under inflammatory conditions, when they start to express MHC Class II and can present exogenous antigens to elicit activation of CD4$^+$ T$_H$ cells. Innate immune cells recognize antigens on pathogens through so-called pattern recognition receptors, which MSC can also express, especially in response to an inflammatory environment. When a microbe

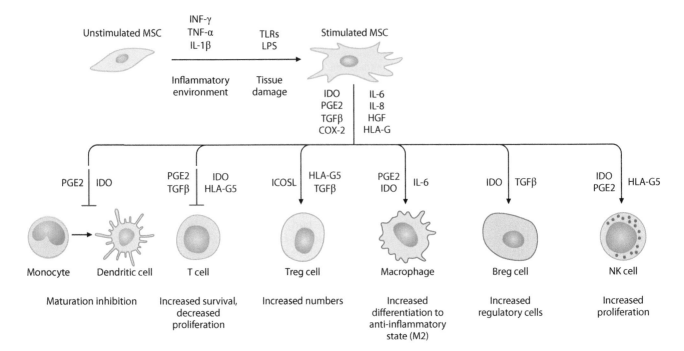

**Figure 16.13 Overview of the interactions between mesenchymal stem cells and immune cells. Mesenchymal stem cells** (MSC) respond to tissue inflammation or damage by becoming activated to produce a range of cytokines, growth factors, and other molecules that have activities against immune cells such as prostaglandins and proliferation inhibitory surface proteins and enzymes. The diagram summarizes some of the principal effects of these stimulated factors, which can be either positive (indicated by a pointed arrow) or inhibitory (indicated by a bar-ended arrow) on the cells that constitute the innate and adaptive immune systems. The cellular consequence of the released factors on each cell type is indicated below the cell. Abbreviations: INF-γ – interferon gamma; TNF-α – tumor necrosis factor α; IL-1β – interleukin-1 β; PGE-2 – prostaglandin E2; IDO – indoleamine 2,3-dioxygenase; TGF-β – transforming growth factor-β; HLA-G5 – human leukocyte antigen-G5; Treg – regulatory T-cell; Breg – regulatory B-cell. (From Lee D & Song S. (2018) *Cellular Immunol* 326: 68–76 [doi:10.1016/j.cellimm.2017.08.009]. Copyright 2018 with permission from Elsevier.)

## TABLE 16.2 IMMUNE MODULATORY FUNCTIONS OF MESENCHYMAL STEM CELLS

| Function | Examples |
| --- | --- |
| Secretion of anti-inflammatory factors | TGF-β, HGF, PGE-2 |
| Expression of surface immune suppressive molecules | PD-L1, FasL |
| Secretion of immune cell attractants | IL-8, CCL2 |
| Expression of metabolic enzymes | IDO, CD73 |

*Note:* The table summarizes some of the principal immunomodulatory functions of mesenchymal stem cells, with examples of key molecules involved in each mechanism.
*Abbreviations:* TGF-β – transforming growth factor β; HGF – hepatocyte growth factor; PGE-2 – prostaglandin E2; PD-L1 – programmed death ligand 1; FasL – Fas ligand; IL-8 – interleukin-8; CCL2 – chemokine CCL2; IDO: indoleamine 2,3 dioxygenase; CD73 – ecto-5′-nucleotidase.

is engaged by these receptors, the MSC increases expression of immunomodulatory genes, perhaps thereby acting as a negative feedback loop protecting against damaging immune reactions to the invader.

The ability of MSC to influence immune responses is multifaceted, but it is unclear how the diversity of modulatory functions relates to single cells or to the differentiation potential of those cells toward the three cell types associated with tissue structure. Another area of uncertainty concerning the effects of MSC on the immune system is the relevance this has *in vivo* given that almost all studies that have defined actions of these cells on innate or acquired immunity have been performed *in vitro* and may not reflect the actions in their normal tissue location.

## The release of exosomes from mesenchymal stem cells is an alternative mode of delivery of immune modulatory molecules

Aside from secretion of soluble factors that have an effect on immune cells or direct cell-cell interaction, MSC are also able to deliver modulatory molecules through the release of double membrane particles ranging from 40–100 nm in diameter known as exosomes. As vehicles able to transport bioactive molecules, some of which are not expected to be secreted, the possibility of exosome production from MSC is thought to be an explanation for some observations of multiple biological effects throughout the body in experimental models.

Exosomes originate from early endosomes, which as they mature undergo inward budding, leading to the accumulation of vesicles within the lumen of what is known as a multivesicular body (MVB). The MVB releases its cargo of exosomes when it fuses with the cell surface membrane (Figure 16.14). Generally, exosomes are harvested by ultracentrifugation of MSC culture supernatants and have been proposed to mimic the range of MSC immunosuppressive effects based on testing in a variety of *in vivo* models of inflammatory disorders, although the evidence for equivalent efficacy is as yet still inconclusive. The function of exosomes derived from MSC is largely the result of delivery of their content of mRNA, miRNA, lipids, and proteins to target cells, which they achieve in several ways, including receptor interaction, endocytosis, or release in the cell's vicinity. The source of a particular MSC population used to isolate exosomes, whether this be bone marrow, adipose tissue, or umbilical cord, among others, influences the profile of molecules that they contain.

## Immune response modulation by mesenchymal stem cells is of considerable interest as a novel therapy for many diseases

The relative ease of producing and expanding MSC and their immune privileged characteristics that make **allogeneic** use more possible has led to the initiation of a large number of clinical trials of efficacy against inflammatory and immune system disorders. The immunosuppressive actions relevant to particular diseases varies so that, for example, MSC are thought to be effective in treating graft-versus-host disease (GvHD), which is a condition that can occur after an allogeneic bone marrow transplant involving donated lymphoid cells seeing the recipient's body as foreign

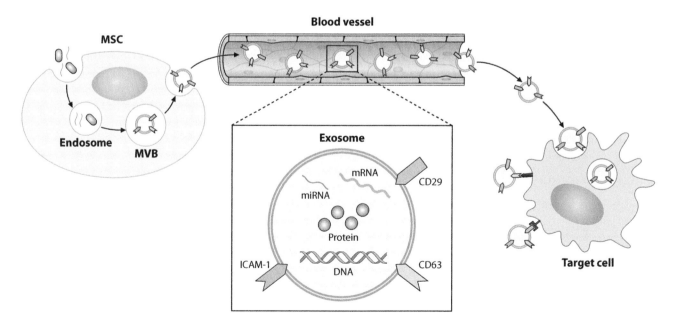

**Figure 16.14 Mesenchymal stem cell-derived exosomes.** One route for the release of immune system modulating molecules from mesenchymal stem cells (MSC), which may be the major mechanism, involves the formation of membrane-bound exosomes. These are formed within multivesicular bodies (MVB) through fusion of endosomes. The exosomes are shed from the MSC when the MVB fuses with the cell surface membrane, and they then travel through the blood stream to target cells where they can interact through receptor interactions, releasing their cargo at the surface or intracellularly following endocytosis. The expanded image of an exosome summarizes the types of molecules that they can contain and illustrates three of the commonly encountered surface proteins, that is, intercellular adhesion molecule 1 (ICAM-1), integrin β1 (CD29), and tetraspanin 30 (CD63). (From Zhou J, Tan X, Tan Y et al. (2018) *J Cancer* 9: 3129–3137 [doi: 10.7150/jca.25376]. With permission from Ivyspring International Publisher.)

and causing damaging cytotoxic reactions. The effect of MSC against GvHD involves induction of $T_{reg}$ and inhibition of the proliferation of CD4+ and CD8+ effector T-cells, while in the case of systemic lupus erythematosus, which is an autoimmune disease that results in damage to many tissues, MSC appear to act by suppressing the activation of B-cells and augmenting the action of $T_{reg}$ and $T_H$ cells.

By the end of 2017, over 700 MSC-based clinical trials had been registered worldwide, a significant proportion of which relate to modulation of the immune response (Figure 16.15). The variety of immune disorder diseases entering clinical trials of MSC is diverse, including multiple sclerosis, Type I diabetes, Crohn's disease, systemic lupus erythematosus, and GvHD (Table 16.3). Although allogeneic MSC appear to be tolerated, and therefore advantageous in acute condition requiring immediate treatment, the possibility would exist to use a patient's own MSC if the condition is subacute, for example, as in autoimmune diseases.

## 16.6 SUMMARY

We have taken the opportunity in this chapter to take a second look at a system and a stem cell that are partly explored elsewhere in this book, with the aim to understand how adult stem cell characteristics can be applicable in a context other than at the top of a differentiation hierarchy. Our focus has been on the cells that provide an immune response in mammals, which, although supported through a stem cell-based hierarchy originating with the HSC, provide some intriguing examples of the way in which stem cells can impinge on the functioning of a biological system. We have described how some tissue macrophage that contribute to innate immunity have a non-HSC origin early in development but are maintained throughout life by stem cell-like self-renewal. We also describe how differentiated hematopoietic cells in the form of B- and T-cells of the adaptive immune response can adopt stem cell characteristics as part of the establishment of immunological memory. Both of these hematopoietic scenarios are beginning to be understood in mechanistic terms, revealing that there are very different ways to achieve a stem cell state. We have also looked at the role of various non-hematopoietic cells in the functioning of the adult immune system, including the maintenance of the structures of the primary and secondary lymphoid organs through stem cells supporting mesenchymal and epithelial tissues. Last, we took a look at the intriguing ability of non-hematopoietic MSC not only to contribute to structural and other tissue cell types but also to act as immune system modulators, which is achieved through their ability to produce a whole host of protein, lipid, and nucleic acid molecules that are active against specific cellular components of the immune system. The understanding of the atypical stem cell behaviors that we have covered in this chapter has huge implications from a clinical perspective, and many outcomes of fundamental research already being applied in clinical trials for immune system disorders or to enhance cell-based immune therapies against cancer.

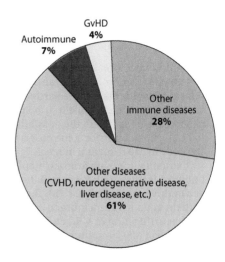

**Figure 16.15 Mesenchymal stem cell clinical trials.** The ability of mesenchymal stem cells (MSC) to moderate the immune response, especially when its excessive or inappropriate action can have damaging effects on tissues, has created huge interest in their therapeutic potential. The pie chart summarizes how the 500 or so clinical trials involving MSC that were registered at the end of 2017 were split between different clinical indications, showing that almost 40% of these related to immune system conditions. (From Rawat S, Gupta S & Mohanty S. (2018) Mesenchymal stem cells modulate the immune system in developing therapeutic interventions. In: Tyagi R & Bisen P (eds) *Immune Response Activation and Immunomodulation*. IntechOpen. [doi:10.5772/intechopen.80772]. Published under the terms of the Creative Commons Attribution License (http://creativecommons.org/licenses/by/3.0).)

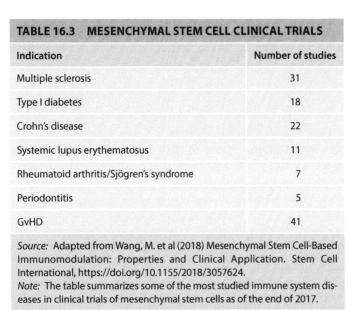

| TABLE 16.3 MESENCHYMAL STEM CELL CLINICAL TRIALS | |
|---|---|
| **Indication** | **Number of studies** |
| Multiple sclerosis | 31 |
| Type I diabetes | 18 |
| Crohn's disease | 22 |
| Systemic lupus erythematosus | 11 |
| Rheumatoid arthritis/Sjögren's syndrome | 7 |
| Periodontitis | 5 |
| GvHD | 41 |

*Source:* Adapted from Wang, M. et al (2018) Mesenchymal Stem Cell-Based Immunomodulation: Properties and Clinical Application. Stem Cell International, https://doi.org/10.1155/2018/3057624.
*Note:* The table summarizes some of the most studied immune system diseases in clinical trials of mesenchymal stem cells as of the end of 2017.

## KEY POINTS

- The adult immune system presents several examples of cell types that might be regarded as stem cells, yet these do not fit into the standard view of tissue specific adult stem cells.

- Aside from HSC, which supply the pipeline of hematopoietic cells involved in both innate and adaptive immune responses, there are examples of both hematopoietic and non-hematopoietic cells with stem cell characteristics that are involved in the immune system.

- Macrophage resident in a variety of tissues are not derived from HSC but rather have their origin in the embryo or fetus. Tissue-resident macrophage persist throughout life and can be considered to have stem cell characteristics, the mechanistic basis for which is beginning to be elucidated.

- B- and T-cells of the adaptive immune response develop a memory component, enabling more rapid and effective responses when a pathogen is encountered a second time. These memory cells have stem cell features, which have been most clearly defined for the T-cell lineages.

- Both primary and secondary lymphoid tissues require a source of replacement cells throughout life, implying the need for one or more stem cell components in the tissue or coming from an external site.

- Epithelia within the thymus are crucial for the development of the T-cell repertoire. The two component epithelial cell types in the medulla and cortex of the thymus are replaced from a bipotent progenitor/stem cell.

- MSC-related cells contribute to the structural elements of both primary (e.g., the thymus) and secondary (e.g., the lymph node) lymphoid organs.

- Lymph nodes also receive new cell input from adipose tissue MSC-related adipocyte progenitor cells, especially during inflammatory conditions.

- MSC have a profound immune regulatory function, which in essence prevents uncontrolled activity of an immune response.

- The immune regulatory function of MSC consists of multiple components ranging from secretion of anti-inflammatory molecules to the expression of enzymes that metabolize compounds that are essential to immune cell proliferation.

- The release of exosomes containing proteins, lipids, and nucleic acids may be responsible for a proportion of the immune function of MSC.

- MSC are being extensively tested in clinical trials aimed at alleviating a variety of immune disorders.

## FURTHER READING

### General background

Janeway's Immunobiology. *Kenneth Murphy, Casey Weaver,* 9th ed. Garland Science, New York.

### Adult myeloid stem cells

Gentek, R., Molawi, K. and Sieweke, M.H. (2014) Tissue macrophage identity and self-renewal. *Immunol. Rev.,* 262: 56–73.

Sieweke, M.H. and Allen, J.E. (2013) Beyond stem cells: Self-renewal of differentiated macrophages. *Science,* 342: 946.

Soucie, E.L., Weng, Z., Geirsdóttir, L., Molawi, K., Maurizio, J., Fenouil, R., Mossadegh-Keller, N. et al. (2016) Lineage-specific enhancers activate self-renewal genes in macrophages and embryonic stem cells. *Science,* 351: 680.

### Lymphoid memory stem cells

Ahmed, R., Roger, L., Costa Del Amo, P., Miners, K.L., Jones, R.E., Boelen, L., Fali, T. et al. (2016) Human stem cell-like memory T cells are maintained in a state of dynamic flux. *Cell Rep.,* 17: 2811–2818.

Gattinoni, L., Speiser, D.E., Lichterfeld, M. and Bonini, C. (2017) T memory stem cells in health and disease. *Nat. Med.,* 23: 18–27.

Graef, P., Buchholz, V.R., Stemberger, C., Flossdorf, M., Henkel, L., Schiemann, M., Drexler, I. et al. (2014) Serial transfer of single-cell-derived immunocompetence reveals stemness of CD8(+) central memory T cells. *Immunity,* 41: 116–126.

Pace, L., Goudot, C., Zueva, E., Gueguen, P., Burgdorf, N., Waterfall, J.J., Quivy, J.P. et al. (2018) The epigenetic control of stemness in CD8+ T cell fate commitment. *Science,* 359:177–186.

### Secondary lymphoid organ stem cells

Ucar, A., Ucar, O., Klug, P., Matt, S., Brunk, F., Hofmann, T.G. and Kyewski, B. (2014) Adult thymus contains FoxN1(-) epithelial stem cells that are bipotent for medullary and cortical thymic epithelial lineages. *Immunity,* 41: 257–269.

### Immunomodulation by mesenchymal stem cells

Lee, D.K. and Song, S.U. (2018) Immunomodulatory mechanisms of mesenchymal stem cells and their therapeutic applications. *Cellular Immunol.,* 326: 68–76.

Phinney, D.G. and Pittenger, M.F. (2017) Concise review: MSC-derived exosomes for cell-free therapy. *Stem Cells,* 35: 851–858.

# BIOMEDICAL AND ECONOMIC APPLICATIONS OF STEM CELLS

# PART FIVE

# Stem cells in biomedicine: Disease modeling and tissue repair

# 17

If one thing has driven the science of stem cell biology over the last two decades, then it is the prospect of a medical revolution for the treatment of a seemingly vast array of diseases and conditions. The expectation is that stem cells can become an integral part of greater understanding of diseases, the discovery of new drugs, or themselves contribute directly to the repair of replacement of organs, a concept generally referred to as "**regenerative medicine**". This simple vision of a panacea, which is more often than not over hyped by the media and a variety of vested interests, needs to be deconstructed in order to understand the challenges and true potential, both of which can be enormous.

In this chapter, we draw upon the content of many of the other chapters in this book, whether those on **pluripotent stem cells** or the descriptions of **adult stem cells** in different organs. We will start by setting the scene and explore what unmet medical need is likely to benefit from the use of stem cells. We will then look at how investigations utilizing stem cells can shed fresh light on the mechanisms underpinning specific diseases or aid in the process of drug discovery. The majority of the chapter will be used to consider the prospects for stem cell application in regenerative medicine, either through the manipulation of resident stem cells in the patient or by provision of cells or tissues that are developed *ex vivo* and then introduced into the patient. We will reflect on the challenges that have to be faced in order to turn possible strategies into a viable prospect, including the need for the science of "**tissue engineering**", and will conclude on a positive note using a few recent examples of successes to date to show that the barriers to implementation of regenerative therapies involving stem cells are by no means insurmountable and that the results can be nothing short of life changing for those affected.

## 17.1 THE UNMET NEED IN MEDICINE THAT CAN BENEFIT FROM THE APPLICATION OF STEM CELLS

Put simply, the unmet need is that clinical medicine needs to improve upon and expand current therapies that rely on donated tissue, as well as to be able to offer novel approaches to treat diseases that presently have relatively poor treatment options or are completely intractable. The urgency for such new ways to improve health through regenerative medicine is growing rapidly, especially because of the impact of lifestyle-related diseases and the fact that we are living longer and becoming more likely to succumb to age-associated loss of organ function.

### A growing aged population is contributing to an increase in age- and lifestyle-related disease

The global population is increasing, thanks in part to advances in medical treatments and access to care, but this is happening in the context of a steady fall in the overall birth rate, meaning that the proportion of older people is increasing. Based on current trends, the global population will become more evenly spread by age distribution as

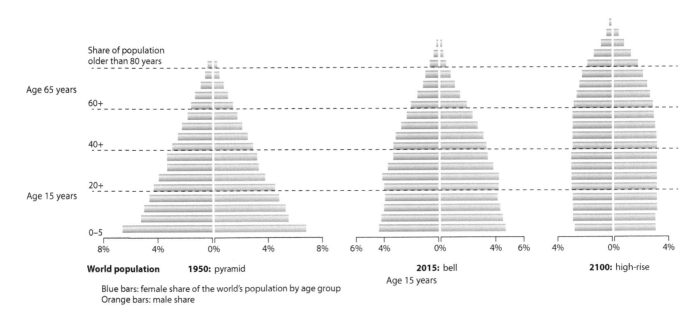

**Figure 17.1 The changing global balance of age.** The global distribution of age has typically resembled a population pyramid, with a greater population in the younger generations than the elderly. These population distribution charts show how the population pyramid distribution in 1970 has shifted to a dome-like profile in 2015 and the predicted columnar profile for 2060 as result of increased longevity and reduced global birth rates. (From Der Spiegel/Statista. (https://www.statista.com/chart/10366/age-structure-of-world-population/). Made available under the Creative Commons License CC BY-ND 3.0 (https://creativecommons.org/licenses/bynd/3.0/legalcode).)

longevity increases, with a prediction that by 2060 the current population pyramid will have evolved into a columnar profile (Figure 17.1). An increased aged population emerges hand in hand with an increase in age-related diseases, some being a natural consequence of the aging process, such as macular degeneration in the eye or muscle wastage, but many resulting from lifestyle burden, including smoking, alcohol, and diet (Figure 17.2). One of the biggest and most rapidly increasing lifestyle-related causes of death is that of liver disease (Box 17.1).

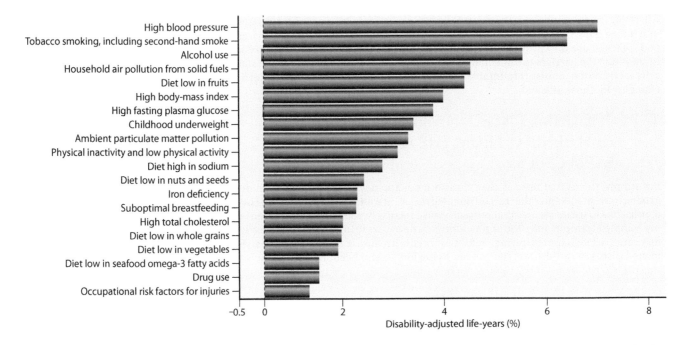

**Figure 17.2 Global lifestyle-related disease.** Lifestyle-related disease is a major burden on healthcare systems worldwide, and as the aged population increases, so does the burden of lifestyle-related disease. This bar chart shows the burden of disease attributable to 20 lifestyle factors in 2010 expressed as a percentage of global disability-adjusted life-years for both men and women. (Adapted from Lim S, Vos T, Flaxman A et al. (2012) *The Lancet* 380:2224–2260 [doi: 10.1016/S0140-6736(12) 61766-8].)

BOX 17.1 BACKGROUND: LIFESTYLE-RELATED LIVER DISEASE

The liver is the second largest organ of the body and shows a remarkable capacity to regenerate itself back to normal size even when up to 70% of it is removed (Chapter 11). The liver plays a crucial role in the removal of toxins from the blood, controlling cholesterol levels and releasing bile to aid the breakdown of fats and facilitate digestion. Failure of the liver to function normally results in a plethora of diseases, often with a high morbidity and mortality rate. Unfortunately, due to our modern way of life, the rate of liver disease is rapidly increasing (Figure 1). Several lifestyle choices contribute to the loss of liver function, among which are drug abuse, obesity, and chronic alcohol consumption. Alcohol is one of the major causes of liver disease, with approximately 15%–20% of chronic drinkers developing hepatitis or cirrhosis, both of which are serious diseases involving inflammation, apoptosis, and fibrosis, leading to severe scarring that inhibits regeneration by the liver. Once scarring occurs, the disease is irreversible and the only form of treatment is liver transplantation. However, as for most organs required for transplantation, there is a severe shortage of cadaveric donors, meaning that a large percentage of people remain on the organ donor waiting list, with ~20% never receiving a transplant and eventually succumbing to their disease.

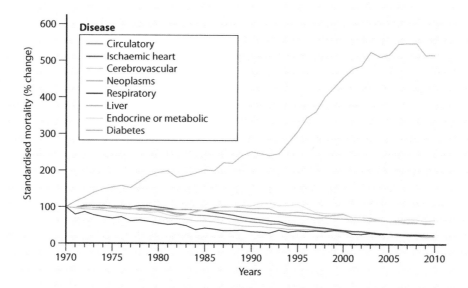

**Box 17.1 Figure 1 Increasing prevalence of liver disease in the United Kingdom.** As population dynamics have changed over the last few decades, so have the lifestyles that people have chosen to lead, in particular increased consumption of alcohol and drug use. This chart shows the incidence of various diseases in the United Kingdom between 1970 and 2010, highlighting a dramatic rise in the occurrence of liver disease in contrast to stable or falling incidences in other major disease categories. The data are standardized to 100% in 1970 and subsequent trends plotted from the WHO-HFA database. (From Williams R, Aspinall R, Bellis M et al. (2014) *The Lancet* 384: 1953–1997. [doi: 10.1016/S0140-6736(14) 61838-9]. Copyright 2014 with permission from Elsevier.)

Whether associated with aging processes or the consequence of self-induced damage, the standard approach to treatment for many failing organs has been donor tissue transplantation. However, the availability of donor organs is plainly not able to keep up with the demand for transplantations (Figure 17.3), a situation further compounded by the shifting age distribution of populations. The use of stem cells for the production of replacement cells and tissues offers an obvious solution but one that, as we shall see, requires significant hurdles to be overcome before becoming a realistic treatment, option let alone standard practice.

## 17.2 MODELING DISEASE USING STEM CELL DERIVED CELLS

One of the most immediate and effective applications of stem cells that has benefitted medical science has been their use in better understanding the origins of certain diseases and as novel tools to help in the screening for and testing of new drug candidates. The generation of **induced pluripotent stem (iPS) cells** (Chapter 5) from diseased tissues of an individual is the most obvious example of such benefit, although **embryonic stem (ES) cells** (Chapter 3) and iPS cells obtained from a diverse

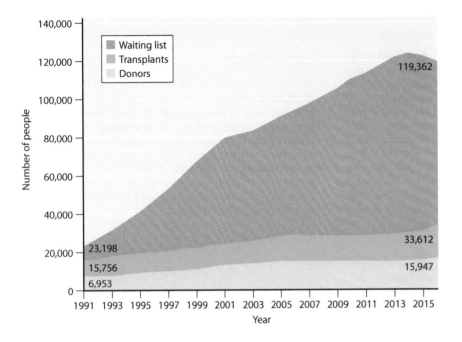

Figure 17.3 **The gap between transplantation need and donor availability.** With an increasing global population, there is a growing number of people waiting for organ transplantation without a consequent increase in the number of individuals willing to donate. This chart depicts the number of people in the United States awaiting organ transplantation (orange) compared to those who have received a transplant (blue) and those willing to donate (yellow) between 1991 and 2015. (Adapted from https://www.organdonor.gov/statisticsstories/statistics.html.)

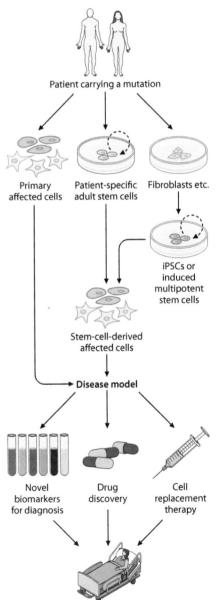

range of individuals are also potentially highly useful in the screening new drugs for toxicity or variations in responsiveness between different populations.

## Differentiation of induced pluripotent stem cells provides new avenues for the modeling of disease

An exciting use for stem cells in biomedicine is in the modeling of human diseases that are otherwise difficult or impossible to mimic *in vitro*. Historically, disease modeling utilizing a patient's own stem cells has been limited by the scarcity of available patient tissue for research and the inability to grow these cells in long-term *in vitro* culture.

The advent of iPS cell technology made it possible to derive individual-specific pluripotent stem cells and thereby a potentially endless source of differentiated cells with which to investigate how an inherited or acquired genetic disease affects the relevant tissue and ultimately to identify potential targets for drug therapy or to assess if **gene editing** could play a part in future treatment strategies (Figure 17.4). A major challenge of this strategy, as it is for any application of pluripotent cells requiring their **differentiation** into fully functional adult cells, is to direct differentiation toward the correct cell type and determine if the behavior of these cells *in vitro* corresponds with the pathological features observed in the patients. A large number of diseases have now been probed using iPS cells, most notably for neurological and cardiac diseases, probably reflecting the relative ease of deriving fully differentiated neurons and cardiomyocytes from pluripotent stem cells (Table 17.1).

## Using induced pluripotent stem cell derived cardiomyocytes has given detailed insight into the defects in long-QT syndromes

One of the first examples of the use of patient iPS cells as a means better to understand an inherited disease was for long-QT syndromes (LQTS), which are associated with prolongation of the QT interval on an electrocardiogram and a high risk of sudden

Figure 17.4 **Use of stem cells in disease modeling.** Cells derived from the tissue affected by a particular disease can be used to understand the mechanisms responsible for disease or can serve as a model for drug discovery or testing. Relevant cells (orange or green) can be obtained either directly from the tissue or by differentiation from specific adult stem cells (blue). Alternatively, if the disease has an inherited genetic origin, then any somatic cells (lime green) can be used to produce iPS cells, which can in turn be differentiated to the required tissue cell type. (From Sterneckert J, Reinhardt P & Schöler H. (2014) *Nat Rev Genet* 15:625–639 [doi:10.1038/nrg3764]. Reprinted by permission from Springer Nature.)

**TABLE 17.1 IPS CELLS USED TO MODEL NEURODEGENERATIVE AND CARDIAC DISEASES**

| Disease | Associated genes | Clinical features | Cell type differentiated from iPSC |
|---|---|---|---|
| *Neurodegenerative diseases* | | | |
| Spinal muscular atrophy | SMN1, SMN2 | Movement disorder | Motor neurons |
| Amyotrophic lateral sclerosis | SOD1, FUS, TDP-43, C9ORF72 | Movement disorder | Motor neurons |
| Parkinson's disease | GBA, LRRK2, PARK2, PARK7, PINK1, SNCA, UCHL1, MAPT, SNCAIP | Movement disorder | Midbrain dopaminergic neurons |
| Huntington's disease | CAG triplet repeats in huntingtin (HTT) | Movement dysfunction, cognitive abnormalities, psychiatric disturbances | Neural stem cells, astrocytes, striatal neurons |
| *Cardiovascular diseases* | | | |
| Long-QT syndrome LQT1 LQT2 LQT3 | KCNQ1 KCNH2 SCN5A | Ventricular tachyarrhythmias | Cardiomyocytes |
| Timothy syndrome | CACNA1C | Ventricular tachyarrhythmias, cutaneous syndactyly, intellectual disability and seizures, etc. | Cardiomyocytes |

*Source:* Adapted from Csöbönyeiová et al (2016) *Gen Physiol Biophys*, 35: 1–12.
*Note:* The table lists some key examples of neurological and cardiac diseases that are known or expected to originate through a single gene defect, for each highlighting the responsible gene(s) and the clinical features. The last column contains the relevant cell types generated from patient iPS cells that have been used to investigate the cell and molecular biology consequences of the individual gene defects.

cardiac death due to ventricular arrhythmia. Mutations in at least 13 genes are known to lead to LQTS, but the mechanisms underlying the disease were poorly understood until the advent of iPS cell technology. One such disease is Timothy syndrome (TS), which is caused by gain-of-function mutations in the L-type calcium channel *CACNA1C* gene. Live cell imaging and electrophysiological studies on TS ventricular-like cardiomyocytes derived from iPS cells revealed deficits in contraction, electrical signaling, and calcium handling (Figure 17.5).

## Differentiated cells obtained from pluripotent stem cells can be a useful resource on which to test candidate drugs prior to clinical trials

An iPS cell-based model of one particular inherited neurological disease nicely illustrates the benefits not only for mechanistic understanding but also the potential for novel drug identification. Familial dysautonomia (FD) is a progressive neurodegenerative disorder caused by a single point mutation in the I-κ-B kinase complex-associated protein gene (*IKBKAP*) that leads to mis-splicing and a marked reduction in IKAP protein. This reduction is particularly acute in FD iPS cell-derived neural crest, which may partly explain why the disease specifically affects the

**Figure 17.5 Disease phenotype of long-QT syndrome cardiac myocytes generated from iPS cells.** iPS cells from a patient with a specific genetic disorder can be differentiated into the cell type suspected to be affected in order to investigate the cell and molecular basis for the disease phenotype. In the example illustrated, iPS cells generated from a patient with Timothy syndrome (TS) were differentiated into cardiomyocytes (CM). (**A**) CM from control (left) and TS (right) iPS cells were stained using anti-α-actinin antibodies (red) and nuclei highlighted using Hoechst dye (blue). The insets show higher magnification images of the sarcomeres, revealing that the control and TS CM exhibit the same structure. Scale bar, 10 μm. (**B**) Measurement of electrical currents in control (black) and TS (red) CM reveal a defect in voltage-dependent channel inactivation following a voltage pulse. (From Yazawa M, Hsueh B, Jia X et al. (2011) *Nature* 471:230–234 [doi: 10.1038/nature09855]. Reprinted by permission from Springer Nature.)

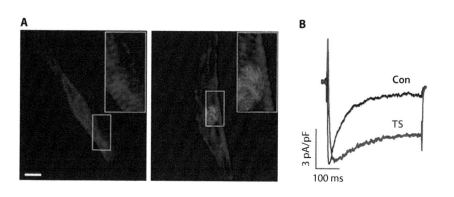

A

B

Con

TS

3 pA/pF

100 ms

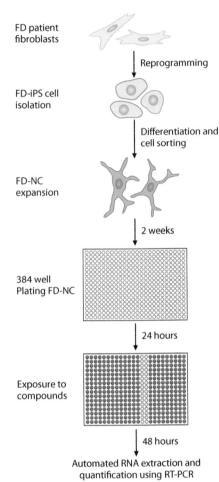

Figure 17.6 **Using iPS cells generated from familial dysautonomia patients as a way to screen for drugs that alleviate the disease phenotype.** Differentiated cells derived from patient iPS cells can be used screen for compounds able to reverse the aberrant cell phenotype. In the example illustrated, iPS cells generated from familial dysautonomia (FD) patient fibroblasts were differentiated into neural crest (NC) cells. Following purification and expansion, FD-NC were plated in 384 well plates and exposed to a compound library. Each plate contained control wells that were exposed only to the solvent in which the compounds were dissolved (yellow). Following treatment, RNA extracted from the cells in each well was used in quantitative reverse transcriptase (RT) PCR reactions to measure any increase in IKBKAP mRNA to near normal levels. (From Lee G, Ramirez C, Kim H et al. (2012) *Nat Biotechnol* 30:1244–1248 [doi: 10.1038/nbt.2435]. Reprinted by permission from Springer Nature.)

peripheral nervous system. Large scale production of neural crest cells from FD iPS cells made it possible to develop a high throughput screen for small molecules that are capable of elevating *IKBKAP* mRNA to near normal levels (Figure 17.6). One effective compound that was identified out of a total of nearly 7000 tested rescued IKAP protein expression by affecting IKBKAP transcription through modulation of intracellular cAMP levels and PKA dependent CREB phosphorylation.

## Differentiation of pluripotent stem cells into organoids can produce models that better reflect the situation in the native tissue

Within the chapters in this book in which we discuss adult stem cells, we have described how isolated stem cells can give rise to spheres or **organoids** *in vitro* that effectively mimic the features of the tissue from which they are derived. Organoids may also be produced from iPS cells that have been differentiated along a tissue-specific lineage that incorporates the stem cell component and therefore represent a potentially improved model in which to investigate the effects of inherited or acquired mutational differences and to screen for drugs to alleviate the defects, and there are a large number of examples of specific inherited conditions for which this has proved to be a very useful approach (Table 17.2).

## The ideal model scenario involves the engineering of specific disease mutations into induced pluripotent stem cells derived from a normal individual

Although obtaining iPS cells from the cells of a patient with a particular inherited disorder has shed light on many genetic diseases, one drawback can be that the

**TABLE 17.2  3D ORGANOID MODELS DIFFERENTIATED FROM PATIENT-DERIVED IPS CELLS**

| Tissue | Disease |
|---|---|
| Liver | Polycystic liver disease, Alagille syndrome, A1AT deficiency, cystic fibrosis |
| Lung | Idiopathic pulmonary fibrosis |
| Brain | Miller-Dieker syndrome, primary microencephaly, autism/macrocephaly, Alzheimer's disease, Parkinson's disease |
| Kidney | Polycystic kidney disease |
| Pancreas | Pancreatic facets of cystic fibrosis |
| Intestine | Hirschsprung's disease, cystic fibrosis |
| Retina | Retinal degeneration, Leber congenital amaurosis (LCA), retinitis pigmentosa |

*Source:*  Adapted from Ho et al (2018) *Int J Mol Sci*, 19: 936.
*Note:*  The table lists a number of major organs that are affected by inherited genetic disorders for which patient-derived iPS cells have been used to generate organoids to mimic the affected tissue and thereby facilitate study of the cell and molecular consequences of the disease.

"normal" iPS cells that are used to produce the control differentiated cells inevitably carry a large range of differences in their overall genetic background. Such genetic differences will always be the case when comparing two people who are not identical twins. What this means is that even though a phenotype may be observed that is seemingly attributable to the inherited mutation being studied, there remains the possibility that the background genome difference between the control cells and patient iPS cells could be playing a role. If only one, or even a few, mutations are thought to be responsible for the inherited defect, then the strategy of choice is to introduce these into the genome of the normal iPS cell control. Generation of such "isogenic" pairs of cell lines is now relatively easy through **"gene editing"** technology, especially using the CRISPR-Cas9 system.

## 17.3 MAKING USE OF AN INDIVIDUAL'S RESIDENT CELLS FOR REPAIR AND REGENERATION

Although regenerative medicine is often thought to involve the provision of cells from an external source, and we will go onto consider such strategies subsequently, it is best first to consider what might be achieved with stem cells through looking more to the body's own mechanisms or even into tricking resident cells to do things that evolution has hidden. As we describe in several chapters in this book, tissue homeostasis often relies on resident adult stem cell populations that, although not always able to repair extensive acute or chronic tissue loss, might be amenable to some form of manipulation that enhances their capability in such circumstances. In the category of repair using the body's own stem cells, we also include the very real prospect that some inherited disorders, in particular those involving a single gene defect, are close to being amenable to novel therapies combining genome editing and stem cell technologies. Even more ambitious is the prospect that our bodies have "forgotten" during evolution how to perform some types of repair that more "primitive" animals can achieve. Hence, as we describe in Chapter 7, certain amphibians can regenerate severed limbs, while fish can repair extensive damage to the structure of their heart musculature. If we understand the details of these evolutionarily earlier processes, then we might be able to re-ignite these for therapeutic purposes.

### Our resident stem cells might be made more effective at repairing tissue damage in our bodies

When considering stem cell therapy, we often think of the injection of donor or patient-derived stem cells or their derivatives rather than therapies that actually enhance the activation of stem cell populations residing within the tissue of interest. As our understanding of stem cell function is increasing for multiple stem cell types, it is now becoming feasible to target therapies for injury or disease at either the resident stem cells themselves or cells of the **niche** in which they reside in order to initiate or enhance repair. Such a therapeutic strategy is of course only an option for a particular tissue if the stem cells have not themselves been lost through disease or damage. Understanding the molecular pathways that control **tissue stem cell** maintenance and differentiation is expected to be the key to delivering this sort of clinical intervention. Much of what is known about the molecular regulation of stem cells and their interaction with their various environments is covered in Chapters 4 and 6.

Skeletal muscle is a good example of what may be possible, both because of the state of our knowledge and also because muscle wastage is a major compounding factor in the quality of life in the elderly. Normally, the stem cells in muscle respond to injury by the activation of Notch signaling promoting their proliferation; however, upon aging, the activation of Notch is delayed, but this might feasibly be overcome by pharmacological targeting. Much the same might be applicable to other tissue stem cell types if the specific activation cues from the different stem cell niches can be identified.

### Genome engineering of pluripotent or adult stem cells has the potential to provide life-long cures for many inherited diseases

A further goal of stem cell regenerative medicine beyond repair or regeneration of complex tissues affected by disease or damage is to be able to overcome the effects of inherited genetic defects that often have profound effects on the quality of life and lifespan but if rectified would mean that the affected individual would be cured for life. The feasibility of genetic correction using gene editing has been vastly improved by the application of nucleases that can be precisely targeted to specific DNA sequences and thereby facilitate the introduction of the desired nucleotide changes through recombination. These designer nucleases include zinc finger nucleases, transcription activator-like effector nucleases (TALENS), and clustered regularly interspaced short palindromic repeat (CRISPR)-associated nuclease Cas9, the latter being the most recent and by far the most versatile powerful addition to the armory. The use of such nucleases essentially eliminates problems associated with conventional gene therapy techniques, which have employed retroviral vectors that are associated with unacceptable levels of insertional mutation. At present, it is probably only realistic to consider gene editing as a way of tackling inherited defects in single genes.

Which cells should be targeted by gene editing depends mainly on the tissues that are impacted by the genetic defect. Adult stem cells might be a suitable cell type if the inherited defect affects a specific tissue that is maintained by them, with the proviso that it must be possible to transplant and engraft the stem cells. For example, inherited hematological diseases such as sickle cell anemia and thalassemia, which both affect hemoglobin in red cells, or certain immune deficiency diseases, such as severe combined immunodeficiency (SCID) caused by a mutation in the gene encoding the receptor for the cytokine Interleukin-2 (IL-2), should be curable if HSC can be gene edited and transplanted back into the patient. On the other hand, if an inherited gene mutation has a multi-organ influence, or if stem cells in the affected tissue are not accessible, then it may be necessary to consider gene editing in either the early embryo or in pluripotent stem cells. As might be expected, gene editing in embryo-derived cells raises considerable safety and ethical concerns. The main safety consideration is ensuring against the possibility of off-target effects of the editing process. Also, the gene editing process might be incomplete when performed on early stage embryos, leading to a situation of mosaicism, the effects of which are not known should the embryo progress to gestational term. From an ethical viewpoint, of course, there will be many who object to any intervention that leads to changes in the genetic potential of a human embryo (Chapter 20).

In spite of the various concerns surrounding gene editing, in 2015, a research group in China published that they had used CRISPR-Cas9 technology on non-viable human embryos to try to correct the human β-globin gene mutation responsible for β-thalassemia. Although these studies served as a major proof of principle, they also highlighted the flaws; less than half of the embryos tested positive for the genetic correction, and there were a large number of off-target mutations, showing therefore that this technology is far from entering the clinic.

## Reactivation of the capacity for regeneration that has been lost to us during evolution could be a therapeutic option

In the United States, it is estimated that there are around 2 million people who have had limb amputation. Each year, approximately 185,000 Americans undergo upper or lower limb amputation, and this figure is expected to triple by 2050 as the aging population increases. Traumatic injury is responsible for just under half of these amputation cases, vascular complications from diabetes making up the other half, and with the rising number of cases of diabetes, this proportion is set to increase. In Chapter 7, we describe in detail the concept of epimorphic regeneration, the formation of a **blastema** at the amputation plane in amphibians leading to the regeneration of complete and functional limbs and tails. Although mammals retain an element of regenerative capacity within the distal digit tip, particularly during development or the neonatal period, adults have lost this regenerative capability somewhere along the evolutionary pathway. As researchers are unraveling more about the mechanisms permitting complete limb regeneration in amphibians, there is a raised expectation that we could one day exploit these mechanisms to reactivate regeneration of the human limb. Numerous factors have been identified in the amphibian blastema that are central to the regeneration and patterning of the limb, many of which also govern the formation and patterning of the mammalian limb during embryogenesis. These factors include fibroblast growth factors (FGF), Wnt-β-catenin, retinoic acid, and Sonic Hedgehog (Shh). Researchers are now trying to engineer select steps of the limb regeneration process in order to promote the partial or complete restoration of a digit or limb in mammals, first exploring the possibility in rodents with the aim eventually to apply the findings to humans. This may involve introducing to the injury site factors such as soluble extracellular matrix (ECM), specific growth factors, or tissue-disrupting proteases. The use of external bioreactors might also be necessary in order to facilitate the appropriate conditions required for regeneration at the amputation plane. This could all, of course, be in vain, as mammals might quite simply be missing the crucial gene orthologues required for the full extent of limb regeneration seen in amphibians, one such gene potentially being *Prod-1*, which is known to be crucial for full limb regeneration in the salamander.

## 17.4   USING EXTERNALLY SOURCED STEM CELLS IN REGENERATIVE MEDICINE

Replacing a failing or damaged organ using stem cells either to produce cells and tissues outside of the patient's body or by introducing them into a tissue so that they might differentiate *in situ* to bring about functional improvement represents the principal approach to regenerative medicine. Like those potential approaches that rely on the patient's own stem cells or regenerative capacity as described previously, the provision of cells or tissues that are developed *ex vivo* and then introduced into the patient has its own particular challenges, ranging from the need for more fundamental understanding of specific cells and mechanisms to others that are very much more practical in nature (Table 17.3). In the context of cell replacement therapies, the disease itself might be an important factor for determining the likelihood of success. For instance, diseases resulting from small lesions in an organ are better candidates for cell therapy than organs that have undergone complete failure and would require a complex array of cell replacement. Similarly, the complexity of the affected tissue matters, such that if it is made up of multiple cell types arranged into a variety of structures, then it might be difficult selectively to repair the affected region.

In addition to the challenges associated with the practical derivation and application of cells for use in regenerative medicine, there are several other issues that receive attention elsewhere in this book. These include those ethical concerns related to the generation and use of pluripotent stem cells that have a provenance involving human embryos, which we discuss in Chapter 20. Another important consideration in the production of differentiated cells from stem cells is the safety aspect, especially the possibility of inadvertently introducing potentially tumorigenic cells or pathogenic agents such as viruses. The possibility of tumor formation from undifferentiated

### TABLE 17.3  CHALLENGES IN THE IMPLEMENTATION OF STEM CELLS IN REGENERATIVE MEDICINE

| Regenerative approach | Major challenges |
|---|---|
| Augmenting normal stem cell-based repair | 1. Understanding how stem cell activity is controlled by cell intrinsic and extrinsic (niche) mechanisms<br>2. Identifying drugs that can manipulate the molecular mechanisms in the stem cell or its niche to elicit an improved repair response |
| Reversal of genetic defect | 1. Accessing the appropriate stem cell for genome editing<br>2. Targeting the mutated gene without off target effects |
| Reactivating latent regenerative capacity | 1. Understanding how regeneration works in lower vertebrates<br>2. Determining if similar processes are feasible but "masked" in mammals<br>3. Identifying ways in which latent regenerative potential can be activated |
| Replacement of tissue | 1. Generating fully functional replacement cells<br>2. Producing sufficient cells<br>3. Avoiding rejection by the immune system<br>4. Incorporating cells into a complex tissue structure |

*Note:*  This table lists ways in which the manipulation of stem cells can potentially be used in regenerative medicine. For each of the four categories, some of the major challenges faced in being able to make progress toward feasible treatment options are provided in the second column.

pluripotent stem cells contaminating a differentiated population is a real one since these cells readily form **teratomas** *in vivo* (Chapter 3), so stringent processes have to be developed to ensure absolute purity. The risk of accidental infection with pathogens is also a concern, in particular with respect to any animal-derived components present in the culture media used in the isolation, culture, or differentiation of stem cells. Very strict regulations and testing regimes have been put in place by many countries to avoid such contaminations, although the development and adoption of totally synthetic media components considerably minimizes any dangers.

## Obtaining fully functional differentiated cells from stem cells is crucial if they are going to be effective in a regenerative therapy

It may sound trivial, but knowing what cell type is affected in a disease and that it is possible to generate precisely the same cell type *ex vivo* are crucial determinants of the likelihood of success for a stem cell-based regenerative therapy. Not only must the cell type or types lost to disease or injury be accurately identified and quantified, it is also important to know the reason behind the loss since this might persist and immediately destroy replacement cells, making the treatment effectively useless. This would be the case, for instance, if tissue damage has been caused by an autoimmune mechanism, such as that contributing to type I diabetes, in which pancreatic islet β cells are targeted through a reaction against pre-proinsulin, among other autoantigens.

Knowing what cell type needs to be produced then begs the question of the most appropriate stem cell source for these cells. A widely held expectation is that pluripotent stem cells (ES or iPS cells) represent the best starting point for the generation of most tissue cell types, although in specific instances, adult stem cells (Part IV, Chapters 10 to 16), or even **reprogramming** with no stem cell involvement (Chapter 5), might be the preferred option (Figure 17.7). In some special cases, directed differentiation *ex vivo* may not be necessary, as is most obviously illustrated by the transplantation of bone marrow-derived **hematopoietic stem cells** (HSC) (Chapter 9) in the treatment of a variety of blood disorders.

The choice of the most appropriate stem cells for regenerative medicine applications is governed by a number of considerations. Primarily, if there is more than one option, then the choice will very much depend upon the ability to differentiate the source cell toward a given cell type that has the same functional activity as the one that it is replacing. As we have discussed in Chapter 3, the differentiation of pluripotent stem cells to a multitude of cell types is now possible and being continuously refined, but there is still some way to go for many cell types potentially of therapeutic value. For example, cardiomyocytes can quite readily be generated from both ES cells and iPS cells; however, over 30 structural and functional properties need to be satisfied, and, importantly, how to specify between the three distinct subtypes needs to be resolved. Likewise, hepatocytes can readily be produced from pluripotent stem cells but are not phenotypically equivalent to cells freshly isolated from a liver.

## Differentiation of stem cells in three-dimensional culture conditions can help to improve the efficacy of the desired functional cell type

A major problem in achieving and maintaining desired differentiated cell characteristics lies in the fact that although some cells, typified by HSC, readily grow in single-cell suspension culture and others such as **mesenchymal stem cells** (MSC) grow best when they are firmly attached to a surface, this is by no means the norm. Most cells

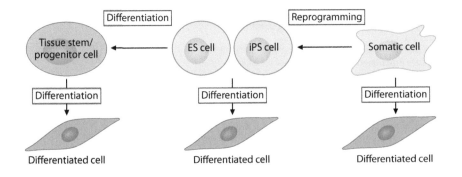

Figure 17.7 **Sources of human cells for regenerative medicine.** Schematic representation showing routes for the derivation of differentiated tissue cells either directly by induced transdifferentiation from somatic cells (green) or normal differentiation of isolated tissue stem cells/progenitors (pink) or from pluripotent stem cells (blue). The pluripotent stem cells can be obtained from embryos (ES cells) or by reprogramming of somatic cells (iPS cells).

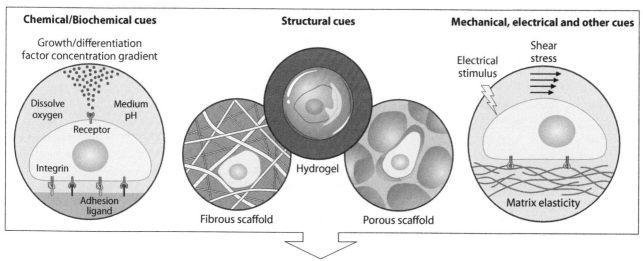

**Figure 17.8 Recreating an appropriate tissue environment using biomaterials.** Representative diagram showing the necessary requirements for construction of tissue environments using biomaterials, consisting of: (i) appropriate chemical/biochemical cues such as oxygen concentration, pH, growth or differentiation factor presentation; (ii) structural cues depending on the scaffold structure; and (iii) mechanical, electrical, and other cues, including matrix elasticity and shear stress. These three categories can all be combined when designing appropriate biomaterials for tissue regeneration. (From Mashayekhan S, Hajiabbas M & Fallah A. (2013) Stem cells in tissue engineering. In: Bhartiya D & Lenka N (eds) *Pluripotent Stem Cells*. IntechOpen [doi:10.5772/54371]. Published under the terms of the Creative Commons Attribution License (http://creativecommons.org/licenses/by/3.0).)

do not normally grow in suspension or on flat surfaces in contact with liquid media; rather they are embedded in a three-dimensional (3D) architecture involving other cell types and **extracellular matrix** components, which collectively provide biochemical signals, mechanical stress, and topographical cues. In order better to replicate the tissue environment, considerable effort has been expended to create suitable **3D** culture conditions so that stem cells can be correctly differentiated to cells that retain the desired function (Figure 17.8). The elaboration of such 3D cultures relies heavily on the use of artificial matrices or scaffolds, which are constituted from a variety of so-called **biomaterials** that form a porous substrate with a large surface area for cell attachment onto which can be added chemical or biochemical modifications to facilitate cell adhesion or to stimulate cell growth or differentiation. Biomaterials are important in most regenerative medicine contexts, and are relevant both in culture to prepare cells for use and in tissues undergoing repair. Biomaterials must satisfy a number of basic criteria, including biocompatibility, biodegradability, and appropriate mechanical properties, all of which must be balanced with the need for a highly porous structure to allow cell entry and, in the *in vivo* scenario, the prospect of vascularization and ECM formation. There are three broad types of biomaterials that are used in regenerative medicine, namely ceramics, synthetic polymers, and natural polymers (Table 17.4). Each type of biomaterial has particular advantages and disadvantages, and it is often the case that the ideal material is best generated as a composite; for example, the combination of a natural polymer such as collagen with a ceramic like hydroxyapatite provides biological signals for cell growth and differentiation and at the same time the sorts of mechanical properties that are required for tissues like bone (Figure 17.9).

## The large numbers of cells required to regenerate all or part of an organ creates a number of challenges for *ex vivo* differentiation of stem cells

Aside from being able to make cells that function perfectly, another big challenge for the regeneration of a tissue is to scale up the process of differentiation in order to obtain sufficient cells to produce all or part of the organ being regenerated. In the case of the liver, for example, this means generating 150 million hepatocytes for each gram of tissue required. In fact, many potential therapies would require cell numbers in the range of $10^{10}$–$10^{12}$ cells, numbers that are not feasible using conventional tissue culturing techniques. Standard tissue culture, as used in a research laboratory, for instance, utilizes static vessels in the form of flasks or plates that typically produce

**TABLE 17.4  BIOMATERIALS**

| Biomaterial type | Characteristics | Examples |
|---|---|---|
| Ceramics | High mechanical stiffness; hard, brittle surface; good for bone applications | Hydroxyapatite, tri-calcium phosphate |
| Synthetic polymers | Tailored architecture possible; degradation can be controlled | Polystyrene, Poly-L-lactic acid, Poly glycolic acid, Poly-DL-lactic coglycolic acid |
| Natural polymers | Promote cell adhesion and growth; compatible with ECM formation; biodegradable | Collage, proteoglycan, alginate-based compounds, chitosan |

*Note:* The table lists the various types of biomaterials in common use, their corresponding characteristics, and some examples of specific molecules.

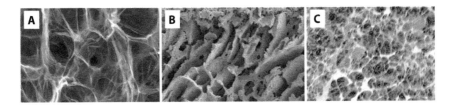

Figure 17.9 **Scanning electron microscopic images of biomaterials.** Scanning electron microscopic images showing the structure of various biomaterials including (**A**) a collagen-GAG (CG) scaffold, (**B**) hydroxyapatite (HA), and (**C**) a composite collagen-HA (CHA) scaffold. Comparison of images (**A**) and (**B**) illustrates the differing degrees of porosity and mechanical properties, while image (**C**) shows that a combination of the individual biomaterial properties can generate a unique and potentially optimal structure. (A: From O'Brien F, Harley B, Yannas I et al. (2004) *Biomaterials* 25: 1077–1086 [doi: 10.1016/s0142-9612(03)00630-6]. Copyright 2004 with permission from Elsevier. B: From Buckley C & O'Kelly K. (2010) *J Mater Sci Mater Med* 21: 1731–1738 [doi: 10.1007/s10856-010-4013-9]. Reprinted by permission from Springer Nature. C: From O'Brien F. (2011) *Materials Today* 14: 88–95 [doi:10.1016/S1369-7021(11)70058-X]. Copyright 2011 with permission from Elsevier.)

on the order of 10 million cells. To overcome this limitation, specialized cell culture vessels, usually referred to as bioreactors, have been developed to maintain an environment that optimizes cell proliferation and differentiation (Figure 17.10). Conditions generated within a given bioreactor are dictated by the characteristics of the source cells, the ultimate differentiated cell type required, and all cell stages between these two. Bioreactor culture conditions usually aim to incorporate a 3D scaffold aspect for the reasons outlined previously. Also, oxygen tension has to be precisely regulated, as this can have a profound effect on the balance between proliferation and differentiation of stem cells. In any cell culture system, there is a continuous nutrient consumption and production of toxic metabolic waste products; if this waste is not removed or nutrients replenished, then this has a detrimental effect on the cells in culture and can ultimately affect their proliferative and differentiation capacity. Bioreactors have therefore been designed that can allow the continuous perfusion of the stem cells during expansion or differentiation with culture medium containing essential nutrients while continuously removing the by-products of metabolism. In this way, a continually homogeneous culture environment can be established, but care also has to be taken that the mechanics of the bioreactor do not induce hydrodynamic shear stress, since this can influence both viability and differentiation.

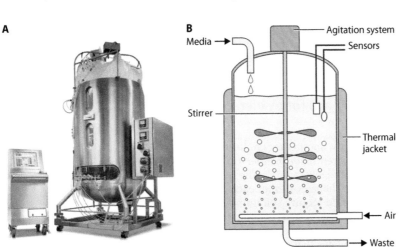

Figure 17.10 **Bioreactor.** Illustration of a typical bioreactor (**A**) and a cross section (**B**) to show the principal component parts. (A: From https://www.gelifesciences.com/en/us/shop/cellculture-and-fermentation/stirred-tank-bioreactors/stirredtank-bioreactor-systems/xcellerex-xdr-50-to-2000-singleuse-stirred-tank-bioreactors-p-06177. Used with permission from GE Healthcare Life Sciences. B: "Bioreactor principle" by YassineMrabet - Own work. Licensed under Creative Commons Attribution-ShareAlike 3.0 Unported via Wikimedia. (http://creativecommons.org/licenses/by-sa/3.0/deed.en).)

## Immunological incompatibility in regenerative medicine can be tackled as in conventional transplantation therapies but can be largely avoided through the use of personalized stem cells

Clinical transplantation of solid organs and bone marrow cells over the last half century has taught medical science about how best to deal with the immunological barriers that are present when the donor and the recipient are not perfectly matched in terms of their histocompatibility antigens, which without some form of intervention would cause the eventual rejection of the engrafted tissue. Most transplantation operations are performed using tissues that are immunologically non-identical (**allogeneic**), even if a degree of matching can be achieved. Only when the donor is an identical twin or is also the recipient, as is the case for some applications of HSC-based therapy, is the grafted tissue fully compatible (**autologous**) from an immunological perspective. The use of cells in regenerative medicine can be subject to the very same considerations, although there are potentially more opportunities to avoid immune rejection through the use of cells that are ultimately derived from the recipient (Figure 17.11).

Aside from trying to achieve the closest possible immunological match between the donor and recipient, the possibility of rejection is handled using one or more interventions that elicit a state of tolerance to "non-self" by suppressing the immune system (Table 17.5). Broadly speaking, immune suppression can be achieved using drugs that inhibit some aspect of immune cell function, with antibodies that block critical molecules such as receptors on immune cells, through the action of cells that naturally suppress the immune response, and using physical barriers that prevent the access of antibodies or effector immune cells to the tissue graft, for instance, by encapsulation in biomaterials like hydrogels.

The application of cells derived from pluripotent stem cells in regenerative medicine raises novel possibilities for the induction of tolerance or the almost complete avoidance of immune rejection. Human ES cell lines, and the cells derived from them, will individually be immunologically incompatible with most would-be recipients, although "banks" of lines representing a wide range of histocompatibility types are being generated, meaning that a degree of matching will be possible in an "off-the-shelf" manner, although immune suppression would almost inevitably still be required. One scenario that is being considered to avoid immune rejection of cells obtained by differentiation from ES cells is to derive dendritic cells that can be administered initially to induce a state of tolerance prior to transplantation of the required cell type. The greatest excitement, though, is without doubt the prospect of personalized transplantations using pluripotent stem cells, which are themselves obtained from cells of the patient, either by **somatic cell nuclear transfer** (SCNT

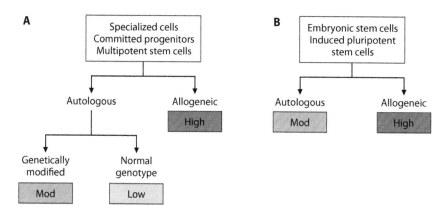

**Figure 17.11 Extent of immune reaction against different cells used in regenerative medicine strategies.** The likely reaction of the immune system in a patient receiving a regenerative therapy will depend on the type of cells and tissues used. (**A**) Any allogeneic cells are at highest risk of rejection. Autologous cells can be immunogenic if they have been altered through a genetic modification that leads to the expression of an exogenous protein. (**B**) Allogeneic cells derived from ES cells or autologous cells derived from iPS cells can acquire immunogenic features during prolonged culture. (From Zakrzewski J, van den Brink M & Hubbell J. (2014) *Nat Biotech* 32: 786–794 [doi: 10.1038/nbt.2960]. Reprinted by permission from Springer Nature.)

**TABLE 17.5  STRATEGIES TO OVERCOME IMMUNE-MEDIATED REJECTION IN STEM CELL-BASED REGENERATIVE THERAPIES**

| Approach | Mechanism | Examples of targets |
|---|---|---|
| Immune suppression using drugs | Inhibition of signaling pathways<br>Inhibition of immune cell metabolism<br>Inhibition of neovascularization | mTOR, NF-kB, JAK<br>HMG-CoA reductase<br>VEGF |
| Immune suppression using antibodies | Lymphocyte depletion<br>Cytokine inhibition<br>Blockade of co-stimulatory molecules | CD20<br>TNF-$\alpha$; IL-6<br>CD28, CD40 |
| Cell-mediated immune suppression | Mimicking natural tolerance mechanisms | Regulatory T-cells, dendritic cells, MSC |
| Physical blockade | Preventing access of immune mediators (cells or antibodies) to targets | Encapsulation in permeable biomaterial |

*Source:* Adapted from Zakrzewski et al (2014) *Nat Biotech*, 32: 786–794.
*Note:* This table lists ways in which rejection of cells by a patient's immune system can be overcome. For each of the four strategies, some of the principal immune mechanisms targeted are presented in the second column, and specific examples of molecules involved in each process that might be therapeutically influenced are indicated in the last column.
*Abbreviations:* IL-6, interleukin 6; JAK, Janus kinase; mTOR, mouse target of rapamycin; NF-$\kappa$B, nuclear factor $\kappa$B; TNF-$\alpha$, tumor necrosis factor $\alpha$; VEGF, vascular endothelial growth factor.

ES cells; Chapter 3) or by induction of pluripotency to yield iPS cells. However, even such personalized cells may not be the perfect solution since they have been seen to be immunogenic, perhaps because of retention of developmental antigens expressed during the differentiation process or aberrant antigen expression during the prolonged culture needed to establish the pluripotent lines.

## 17.5   TISSUE ENGINEERING

As we have just seen, although there are challenges to the generation of fully functional replacement cells in quantities comparable to those found in the relevant tissue, these are surmountable problems that researchers are well on the way to solving. Perhaps the most testing challenge for regenerative medicine is to be able to utilize these cells in the context of a complex whole tissue or organ in a patient. Most tissues and organs are highly complex, composed of several specialized cell types, each organized into specific structures and forming multiple interactions with neighboring cells and ECM components. Tissue engineering seeks to find a way to recapitulate what is normally created through the complex processes of development. How cells are combined with or without biomaterials to produce whole or partial replacements for tissues or organs is in itself is a huge discipline that encompasses a multitude of specialisms ranging from cell biology and materials science to machine engineering and mathematical modeling. As such, it is beyond the scope of this book to do more than give a flavor of the subject, and the reader is referred to specialized texts for greater insight.

### The complexity of most tissue structures can be a barrier to the incorporation of replacement cells

Although some tissues appear to have the ability to self-organize as their cell component is generated from their resident stem cells, as we describe, for instance, in Chapter 11 when considering the potential of **intestinal stem cells** to form organoids, this is not a feasible option for most organs, such as the heart and the liver. So, not only must cells such as cardiomyocytes or hepatocytes be fashioned into cardiac musculature or hepatic parenchyma, but the replacement tissue needs to be rebuilt to contain blood and lymph vessels, a network of nerve cells, and structural elements that help to shape and support the structure. The three components required for regeneration and repair of tissues are sometimes referred to as the "tissue engineering triad", namely the cells, the biomaterial scaffold, and the appropriate growth conditions.

## The development of biomaterials as scaffolds has fueled the advance of tissue engineering

As discussed previously, porous scaffolds for 3D culture of stem cells and their differentiated derivatives come in many forms constituted by one or more of the three types of biomaterials. If these are to be used in the reconstruction of organs, then the problem of how to get blood vessels to permeate the structure needs to be resolved together with provision of an environment that effectively mimics the natural ECM of the tissue. Such biomimetic biomaterials that instruct and support the correct placement and maturation of cells can provide interactive ligands, growth factor stimuli, and mechanical cues to influence cell behavior. Inclusion of blood vessels might be achieved by engineering microvasculature in scaffolds prior to implantation or alternatively could occur in the tissue itself if vascularization *in situ* could be stimulated.

The most significant advances in the use of biomaterial scaffolds for tissue engineering have without doubt been in the area of connective tissues, including tendon, ligament, bone, and cartilage. The latter has no natural ability to regenerate and so poses a particular problem, although it can be effectively produced by seeding chondrocytes or their precursors on a collagen-based matrix. One the other hand, tissues such as bone have some inherent capacity for repair, which might best be harnessed and enhanced *in vivo*. The capacity of MSC readily to yield the necessary differentiated cells (Chapter 10) for application in tissue engineering illustrates nicely how the nature of the scaffold can influence how cells can be promoted to differentiate into the desired tissue cell types. We have already seen that mechano-sensitive MSC differentiation can dictate whether these stem cells give rise to fat cells, nervous tissue, muscle cells, or bone forming cells, all of which can be of central importance in the recreation of structural organs and tissue support elements. Hence, when MSC are grown on firm gels that mimic the elasticity of muscle, differentiation down a myogenic (muscle-forming) lineage is observed, whereas when MSC are grown on rigid gels that mimic pre-calcified bone, the cells differentiate down an osteogenic pathway. Similarly, with **neural stem cells**, neuron differentiation is favored on soft scaffolds that mimic normal brain tissue, whereas differentiation into neuron-supporting glial cells is promoted on harder matrices.

## Additive manufacturing technology involving three-dimensional printing has created a novel opportunity for the creation of complex whole organ structures

The prospect of engineering complex tissues *in vitro* has received a considerable boost through a development stemming from additive manufacturing techniques involving 3D printing. The emergence of the concept of 3D-bioprinting exemplifies multidisciplinary approaches to medicine in the twenty-first century, involving as it does the coming together of medical science, cell biology, materials science, engineering, and computer science. In essence, 3D-bioprinting entails the spatial patterning of biocompatible materials, cells, and supporting components by computer-aided layer-by-layer deposition into functional living tissues. The central challenge of 3D-bioprinting is to reproduce the complex micro-architecture of ECM components and multiple cell types in sufficient resolution to generate tissue-like 3D structures and thereby recapitulate biological function. The essential requirements are to have sufficiently detailed structural information for the tissue that it can be recreated from successive two-dimensional layers using 3D printing technologies to dispense cells onto a biocompatible scaffold, which can then be matured post-printing using a bioreactor environment to provide mechanical and chemical stimulation necessary to control the remodeling and growth of tissues.

There are three major components to a 3D-bioprinter, namely the hardware, the "ink", and the substrate onto which materials are printed (Figure 17.12). Many 3D-bioprinters are under development, and some have even been made commercially available. The most commonly used printer is based on the ink jet principle and therefore is not so different from those that we use every day in our offices and homes. The ink can be one or more cell types in appropriate media, which may incorporate biomaterials such as hydrogels. The greatest successes to date have been in the creation of skin or tissues with a high dependence on cartilaginous structures, such as airways and heart valves (Figure 17.13). However, the synthesis of much more complicated tissues is being piloted, including efforts to print micro-channels that

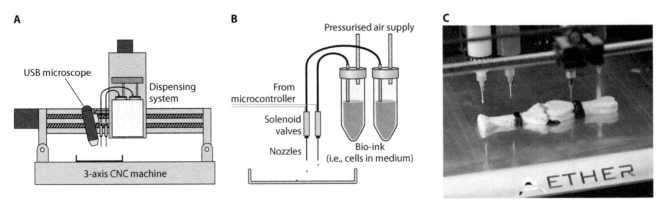

Figure 17.12 **3D-bioprinter. (A)** Schematic drawing of the cell printer system. The computer numerical controlled (CNC) platform is able to move in any direction along the three axes relative to the system that dispenses the cells and their supportive medium. **(B)** Detailed schematic of the arrangement of the dispensing systems. **(C)** A state-of-the-art commercially available bioprinter. (A and B: From Faulkner-Jones A, Greenhough S, King J et al. (2013) *Biofabrication* 5: 1–12 [doi: 10.1088/1758-5082/5/1/015013]. © IOP Publishing. Reproduced with permission. All rights reserved. C: Used with permission from Aether.)

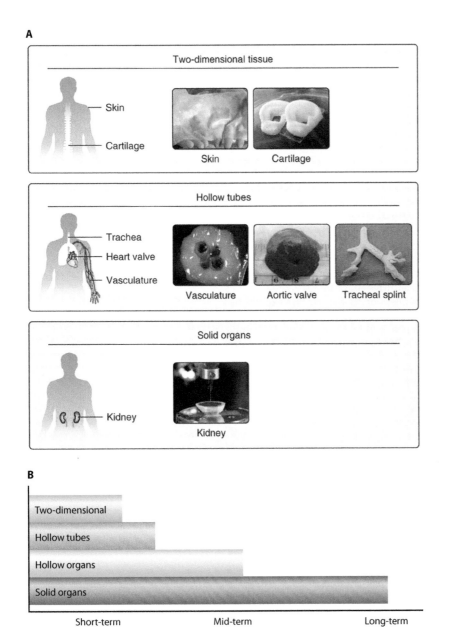

Figure 17.13 **Complexity of tissue structures amenable to 3D-bioprinting. (A)** Examples of attempts at human-scale bioprinted tissues include skin, cartilage, vasculature, aortic valves, tracheal splints, and an early stage kidney prototype, all of which required integration of multiple components for proper function. **(B)** The bars depict a predicted approximate timeframe for the development of various types of 3D bioprinted tissues depending on their complexity. Less complex two-dimensional tissues and hollow tubes are achievable in the near future, while the delivery of more complex hollow organs and solid organs is necessarily a more long-term prospect. (From Murphy S & Atala A. (2014) *Nat Biotech* 32:773–785 [doi: 10.1038/nbt.2958]. Reprinted by permission from Springer Nature. Skin and kidney photos: Used with permission from Wake Forest Institute for Regenerative Medicine. Vasculature: Vasculature photo: From Marga F, Jakab K, Khatiwala C et al. (2012) *Biofabrication* 4: 022001 [doi:10.1088/1758-5082/4/2/022001]. © IOP Publishing. Reproduced with permission. All rights reserved. Aortic valve photo: From Duan B, Hockaday L, Kang K et al. (2013) *J Biomed Mater Res A* 101:1255–1264 [doi:101002/jbm.a34420]. With permission from Wiley. Tracheal splint photo: Used with permission from University of Michigan Health System. Splint developers are Dr. Scott Hollister, University of Michigan and Dr. Glenn Green, CS Mott Children's Hospital.)

can serve as a form of vasculature to maximize the diffusion of nutrients and oxygen from neighboring tissues, as exemplified by the company Organovo's attempts to produce fully functional human liver and kidney (Figure 17.14).

## Decellularization of donor tissues can provide a framework on which to engineer reconstruction of an organ from stem cell-derived cells

Perhaps quite surprisingly, a large part of the constituent matter in an organ is not cellular but rather is made up of a variety of biological polymers that are secreted by the resident tissue cells and constitute the ECM. These polymers are largely protein in nature but often highly modified, for example, with sulfate groups or chains of sugar residues, and provide cues that affect cell migration, proliferation, and differentiation. The use of ECM derived from tissues that have been decellularized is increasing in regenerative medicine applications (Figure 17.15). The processes by which tissues can be decellularized are varied, depending on factors such as their size, density, and lipid content, but generally involve soaking or perfusing it with a combination of agents, which are aimed at minimizing damage to the ECM. These agents include acids and bases; hyper- and hypotonic solutions; ionic and non-ionic detergents; alcohols; and enzymes such as nucleases, lipases, and various proteases. Collectively, these treatments can take several weeks but ultimately produce structures that often still resemble the organ before decellularization, except that it will have become colorless and translucent. Originally explored using rodent organs, successful decellularizations have been performed on several human tissues, including liver, kidneys, and lungs. However, decellularization is perhaps most impressively illustrated by the current state of the art for the human heart. The team in the laboratory of Harald Ott at the Harvard Stem Cell Institute have successfully repopulated a human cardiac ECM scaffold with cardiomyocytes derived from human iPS cells (Figures 17.16 and 17.17). The myocardial tissue that Ott and his colleagues generated was maintained in biomimetic culture for 120 days and was shown to be force-generating, exhibiting electrical conductivity, left ventricular pressure development, and metabolic function that mimics a normal heart with a fair degree of fidelity (Figure 17.18).

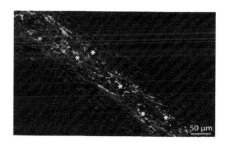

**Figure 17.14 Three-dimensional-bioprinter construction of a kidney tubule.** This image shows a successful attempt to 3D bioprint kidney proximal tubular tissues using multiple cell types. The tissue is completely cellular (no scaffolds) with clear cellular connections and tissue architecture. The tissue was stained with antibodies against the endothelial cell specific marker CD31 (green) and the fibroblast marker TE7 (red). The third cell type is the polarized renal proximal tubular epithelial cells. The endothelial cells associate with each other, forming networks, which over time "hollow out" to form microvessel lumens lined with epithelial cells (*). (From King S, Creasey O, Presnell S et al. (2015) *FASEB Journal* 29(1Suppl):LB426. Used with permission.)

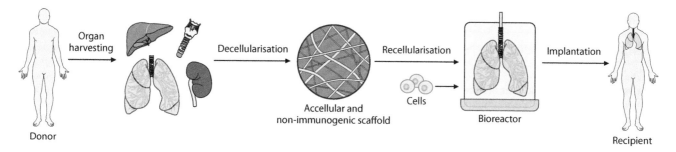

**Figure 17.15 Organ decellularization as a platform for tissue regeneration.** Schematic diagram depicting an idealized cycle for tissue regeneration involving donor organ decellularization and recellularization using a bioreactor prior to implantation into the recipient patient.

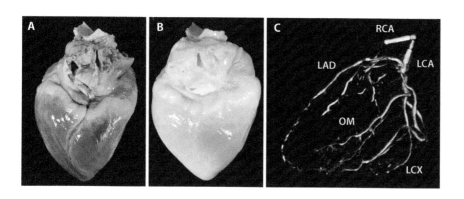

**Figure 17.16 Decellularized human heart.** A porcine heart prior to (**A**) and post (**B**) decellularization. The image in (**C**) shows computed tomography of a decellularized heart, revealing the right coronary artery (RCA), left coronary artery (LCA), left anterior descending (LAD) artery, left circumflex (LCX) coronary artery, and obtuse marginal (OM) coronary artery. (A and B: From Hodgson M, Knutson C, Momtahan N et al. (2018) *Methods Mol Biol* 1577: 95–102 [doi:10.1007/7651_2017_31]. Reprinted by permission from Springer Nature. C: From Guyette J, Charest J, Mills R et al. (2016) *Circ Res* 118: 56–72 [doi: 10.1161/CIRCRESAHA.115.306874]. With permission from Wolters Kluwer Health, Inc.)

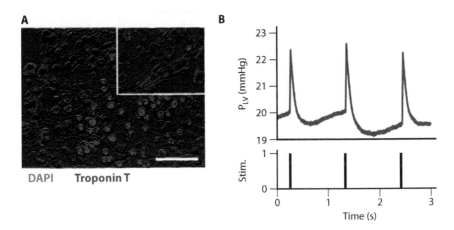

Figure 17.17 **Recellularization of a human heart scaffold in a bioreactor.** The image shows a decellularized human heart undergoing recellularization in a bioreactor organ chamber. The red arrows indicate the direction of media perfusion into the main coronary arteries, the blue arrows show fluid flow into the left ventricular balloon during mechanical stimulation, and the black arrows indicate the location of intramyocardial injections. (From Guyette J, Charest J, Mills R et al. (2016) *Circ Res* 118: 56–72 [doi: 10.1161/CIRCRESAHA.115.306874]. With permission from Wolters Kluwer Health, Inc.)

Figure 17.18 **Structure and function of a recellularized human heart.** Repopulation of a decellularized human heart with induced pluripotent stem cell-derived cardiomyocytes can produce a functional myocardium. (**A**) Immunofluorescent staining for the cardiac marker Troponin T (red) showing the expected structure associated with cardiomyocytes (nuclei stained blue with DAPI). Scale bars 50 μm. (**B**) Left ventricular pressure ($P_{LV}$) measurements reveal a proper cardiac output in the recellularized tissue under electrical stimulation at 0.8 Hz. (From Guyette J, Charest J, Mills R et al. (2016) *Circ Res* 118: 56–72 [doi: 10.1161/CIRCRESAHA.115.306874]. With permission from Wolters Kluwer Health, Inc).)

## 17.6  PROGRESS, FAILURES, AND PROSPECTS IN STEM CELL-BASED REGENERATIVE MEDICINE

After a decade and a half of intense research into the potential for the application of stem cells in regenerative medicine, where do we currently stand in terms of tangible success and patients that have benefitted? As might be expected for a developing field, especially in such a high-profile area where people's health stands to be improved, there has been much anticipation and lots of claims, some of them rather dubious or even blatantly false, and some failures and controversy, but also some very encouraging signs of progress.

### Many structural tissues derived from mesenchymal stem cells can be effectively regenerated

Perhaps the most significant advances have involved structural tissues, such as bone, tendons, and cartilage, which have been repaired or even created *de novo*. These tissues are particularly amenable to regenerative therapies, largely because of a combination of their relatively simple composition and structure (tendons and cartilage lack a nerve or blood supply) and the availability of the stem cell source (MSC) and its straightforward differentiation to the desired cell types. Three-dimensional bioprinting also lends itself well to reconstructions of these structural tissue elements, one exciting example of which employs a hand-held device referred to as the "BioPen" that has already found application in the clinic (Box 17.2).

Another attempted use of stem cells for the generation of cartilaginous tissue, which was hailed as a breakthrough but soon became mired in controversy, was in the replacement of damaged airways. This approach was first performed in 2008 on Colombian Claudia Castillo, who had a severely damaged bronchus as a result of bacterial infection, involving a multi-national team of European scientists and clinicians (Box 17.3).

### Regeneration of structures in the eye damaged by trauma or disease is proving highly feasible

The application of stem cell-based regenerative medicine to debilitating conditions of the eye is already having considerable impact and promises in the very near future to provide significant improvements in the quality of life of large numbers of people, especially among the older parts of the population. The first use of stem cells in the eye builds on the established procedure for corneal grafting using cadaveric donated

## BOX 17.2 CASE STUDY: BESPOKE IMPLANTS "HAND-CRAFTED" BY SURGEONS USING A HAND-HELD THREE-DIMENSIONAL BIOPRINTER

Every year in the United Kingdom, approximately 10 000 people develop problems with the cartilage in their knees, either through injury or osteoarthritis, among other causes. For the most severe cases, the only option is surgery to repair the damaged cartilage or replace the knee. The precise detail of the damage and shape of the patient's cartilage is often only realized once surgery is underway. So imagine the excitement among surgeons by the development of the BioPen, an adaption of the 3D-bioprinter that is now small enough to essentially draw stem cells and other biological materials directly into the patient's joint during surgery, creating a bespoke implant suited to the patient (Figure 1). This proof of concept has been developed by an Australian surgeon who used the BioPen to deposit a mix of hydrogel bioink and stem cells directly into the patient's body in order to draw and sculpt a customized cartilage implant. This is a novel concept that could reduce the requirement for invasive knee and hip replacement surgery in the future.

Box 17.2 Figure 1 **BioPen.** The picture shows how a 3D printing is used precisely to sculpt a bespoke cartilage replacement into a damaged joint area. Often, the exact geometry of a cartilage implant cannot be known prior to the surgery actually taking place. The BioPen has custom titanium nozzles that allow the 3D printing of multiple ink formulations side by side. The cartilage-repairing BioPen extrudes human stem cells from the patient's own body, surrounded by a protective hydrogel bioink. The hydrogel is made from a mixture of gelatin and hyaluronic acid that is hardened into a 3D scaffold with a low-powered UV light. The hydrogel protects the stem cells as they begin to multiply and grow new cartilage. Over time, the scaffold dissolves back into the body, leaving functional tissue behind. (Used with permission from Claudia Di Bella, MD, PhD, FRACS, St. Vincent's Hospital, Melbourne.)

## BOX 17.3 CONTROVERSY: ENGINEERING REPLACEMENT AIRWAYS USING STEM CELLS AND DECELLULARIZED SCAFFOLDS

In the first attempt to regenerate a damaged airway in Claudia Castillo, a team from the University of Padua decellularized a cadaveric trachea in a process lasting 6 weeks before another team at the University of Bristol used a specially designed bioreactor to seed it with cartilage cells derived from the patient's bone marrow MSC. The patient received the finished organ at the Hospital Clinic, Barcelona, in an operation performed by surgeon Paolo Macchiarini. A second transplantation of a regenerated trachea was subsequently performed on Ciaran Finn-Lynch at Great Ormond Street Children's Hospital in London. Although both Claudia and Ciaran initially showed considerable improvement in their health with no signs of graft rejection,

their health later deteriorated, requiring significant further interventions. In the meantime, Paolo Macchiarini pushed ahead with a modified strategy in which a synthetic tube was used in place of the decellularized tissue. In all, he treated nine patients in this way, none of whom showed lasting improvement, and most died between a few months and 2 years after the operation. The revelation of the way in which these treatments were conducted has caused a storm of controversy and, if nothing else, illustrates how important it is for proper pre-clinical animal experiments and clinical trials to be conducted in an environment governed by strict regulatory guidelines.

tissue as a way of treating blindness caused by damage to the surface of the eye as the result of injury, infection, or disease. As discussed in Chapter 15, the cornea is an epithelial tissue that is normally maintained by specialized stem cells that reside in the limbic regions on either side of the iris. These cells can now be extracted, preferably from the unaffected eye of the patient, and cultured to produce layers of cells that can be grafted, leading to full restoration of vision (Figure 17.19). The second stem cell-based therapy is presently in the clinical trials phase and relates to age-related macular degeneration (AMD), which is one of the most common causes of blindness (Box 17.4). AMD arises because of degradation of the retinal pigmented epithelium (RPE), which lies at the back of the eye and provides essential physical and metabolic support required for the functioning of the layers of photoreceptors

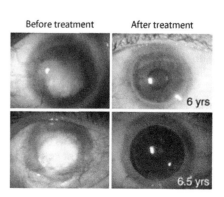

Before treatment    After treatment

6 yrs

6.5 yrs

Figure 17.19 **Treatment of corneal damage using limbal stem cell derived corneal epithelium.** Photographs of two patients with corneal damage before and after stem cell-based therapy. The eyes before treatment are shown on the left, and both can be seen to exhibit extensive damage to the cornea, which appear totally opaque. Following treatment with corneal epithelial cells differentiated from the patient's own limbal stem cells derived from their other eye (right), it can be seen that full transparency was restored and persisted for 6 years or more. (From Pellegrini G, Rama P & De Luca M. (2010) *Trends Mol Med* 17: 1–7 [doi:10.1016/j.molmed.2010.10.003]. Copyright 2010 with permission from Elsevier.)

BOX 17.4 BACKGROUND: AGE RELATED MACULAR DEGENERATION

One of the key contributors of vision loss in people aged 50 years or over is age-related macular degeneration (AMD) (Figure 1). There are two forms of the disease, namely neovascular (wet) AMD or non-neovascular (dry) AMD. The former is generally more severe form of the disease. Despite having some success at slowing the progression of wet AMD with the use of anti-vascular drugs, there is at present no cure for either form of the disease. The desire to establish a cure for AMD is driven by the escalating incidences of

the disease, with 3 million Americans predicted to be affected by 2020. AMD involves atrophy of the retina and retinal pigment epithelium (RPE) beneath the macula. Due to the central role of the RPE in the pathology of AMD, one of the ideal therapies is the replacement of RPE cells by transplantation. Autologous transplantation of RPE from donors has been performed but is technically very challenging and not routine practice.

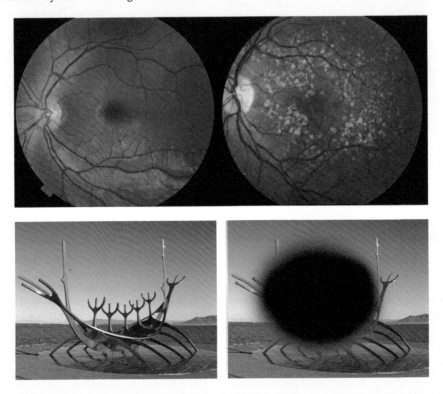

Box 17.4 Figure 1 **Age-related macular degeneration.** Photograph showing a normal macula (left)versus extensive macula degeneration (right), with corresponding images below of what the patient's visual field would be like. (Top: Source: National Eye Institute of the NIH. (http://www.nei. nih.gov/photo/eyedis/images/EDA22_72.jpg.)

(Chapter 14). RPE can very effectively be differentiated from pluripotent stem cells, both in terms of appropriate cell function and the numbers that would be required for regenerative therapy, and sheets of cells can be introduced at the back of the retina using a relatively straightforward surgical procedure. In 2018, a landmark clinical trial led by Pete Coffey and Prof Lyndon da Cruz in London was published describing the restoration of sight in two patients with advanced wet AMD. Their procedure involved depositing a human ES cell-derived monolayer of RPE onto a coated synthetic basement membrane to form a patch suitable for surgically grafting into the back of the patient's eye (Figure 17.20). The two patients both gained vision in the affected eye that was maintained at least up to a year post-treatment.

## Successful regeneration of a severed spinal cord using cells from the olfactory bulb

Traumatic damage to the spinal cord can result in permanent loss of motor and sensory functions below the injury, as well as other possible effects in the respiratory and renal systems. Such injuries to the central nervous system do not repair naturally, so that the synaptic connections from the damaged axons are lost forever. Animal studies have shown that there are cells in the sensory part of the nose (the olfactory bulb) that help

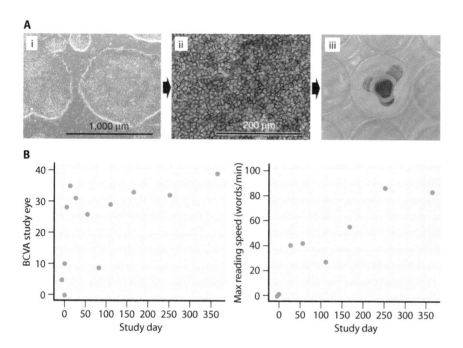

**Figure 17.20 Restored sight in age-related macular degeneration (AMD) clinical trial.** A disease for which stem cell-based therapy is showing strong promise is AMD. This figure illustrates the cell preparation and example treatment results in an early clinical trial for the treatment of wet AMD using ES cell-derived retinal pigmented epithelium (RPE). (**A**) Brightfield images illustrating stages in cell preparation: (i) ES cell colonies were expanded on recombinant human vitronectin; (ii) initial foci of differentiated cells were dissected and dissociated to achieve a pure RPE population that developed into the classical pigmented, cobblestone morphology when seeded onto dishes; and (iii) the advanced therapeutic medicinal product was supplied to the surgical team in a custom-manufactured, single-use sterile container. (**B**) The two graphs illustrate the improvement in visual parameters for Patient 1 over the 12 months of the trial. The graph on the left shows best-corrected visual acuity (BCVA), while the graph on the right represents reading speed. (A and B: From da Cruz L, Fynes K, Georgiadis O et al. (2018) *Nat Biotechnol* 36: 328–337 [doi: 10.1038/nbt.4114]. Reprinted by permission from Springer Nature.)

the growth of new nerves emanating from the brain. These cells, known as olfactory ensheathing cells (OEC), are specialized glia that can be isolated and cultured and have the potential to facilitate nerve repair. Initial experiments in models of spinal cord damage in the rat proved very encouraging, and then in 2014 were put to the test in a human patient. Darek Fidyka, a Polish fireman, was paralyzed in 2010 following a knife attack that severed his spinal cord. In a collaboration between a Polish surgical team and Professor Geoffrey Raisman from the Institute of Neurology in University College London, Darek was treated with a combination of OEC obtained from the back of his nose and nerve fibers obtained from his ankle. The nerves were positioned to span the breech in his spinal cord, and then the OEC were injected at either end, where it was hoped that nerve growth and new connections could be established. Although improvements were slow to appear, Darek was eventually able to walk with a frame and since then has regained additional function in his legs and lower body such that by 2016 he was able to ride a bicycle (Figure 17.21). Although not strictly speaking

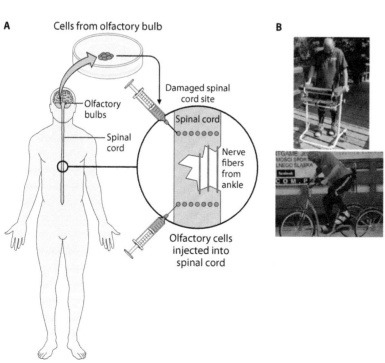

**Figure 17.21 Repair of spinal injury using patient's own cells.** Schematic (**A**) showing the process used for the regeneration of a spinal cord injury using cells from a patient's olfactory bulb that were injected into the site of damage together with nerve fibers from the patient's ankle. The result (**B**) was a regain of function in the patient's legs so that he was able to walk with the aid of a frame and eventually ride a bicycle. (**A**: If keeping: Adapted from The Guardian (21 October 2014. **B**: From the Wroclaw Walk Again Project, https://walk-again-project.org)

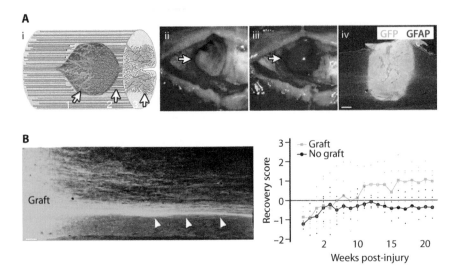

**Figure 17.22 Human NSC grafts improve spinal cord injury in rhesus monkeys.**
Repair of damaged spinal cord using stem cell derived neurons is a realistic prospect, as
illustrated in this trial using human cells in a rhesus monkey model. (**A**) (i) The schematic
on the left shows how an area of damage (brown) can be repaired using grafted neural
progenitor cells (green). Host axons (blue) grow into the graft (1), while graft axons
grow into the host spinal cord (2) and complete the relay circuit (3). The central two
images show the region of the lesion (ii) and the presence of transplanted human neural
progenitor cells in a fibrin matrix (iii, arrow pointing at ball of cells). The image on the
right (iv) is of a section through the lesion site following grafting of the human cells,
which can be visualized due to their expression of GFP. (**B**) Following transplantation
of the human neural progenitors into the lesion, axons labeled by GFP can be seen
emerging in linear arrays, traveling along the host white matter-gray matter interface
(arrowheads). (**C**) The graph shows a measure of recovery of forelimb behavior in control
(four animals) and treated animals (five animals) over the course of 20 weeks. The
large points (connected by lines) represent the mean scores, and the small points are
individual subjects. (From Rosenzweig E, Brock J, Lu P et al. (2018) *Nat Med* 24: 484–490
[doi: 10.1038/nm.4502]. Reprinted by permission from Springer Nature.)

a stem cell-based therapy, since the OEC and nerve fibers were both differentiated
cell types, this procedure illustrates how cells can be used to engineer new functional
nerve tissue.

The next logical step is to obtain the required cells from a stem cell source, for
example, deriving the nerve cells from neural stem cells, and in 2018, results were
published demonstrating the ability to successfully graft human neural stem cells
(NSC) into a spinal cord injury model in the rhesus monkey. The grafted human NSC
not only survived long term, but they gave rise to hundreds of thousands of axons
and synapses covering the injury site. Importantly, the grafted monkeys showed
significant improvement in their forelimb motor function in the weeks following
injury (Figure 17.22).

## Many more stem cell-based regenerative medicine therapies are on the horizon

It has been claimed throughout the recent history of the stem cell biology explosion
that the ability to produce pancreatic β-cells from ES cells will be a cure for type I
diabetes. However, the initial, perhaps overinflated, optimism gave way to the
realization that although differentiation resembling the path toward β-cells is
relatively straightforward, it would take far more effort to obtain cells that can
respond correctly to changes in blood glucose levels and release appropriate amounts
of insulin. The endeavor of many researchers, most notably in the laboratory of
Doug Melton (Box 17.5), has now achieved a functionality that in an experimental
context seems to be what would be required in a patient. Melton's team developed
a differentiation protocol capable of generating hundreds of millions of glucose-
responsive β-cells from human pluripotent stem cells *in vitro*. These β-cells package
and secrete insulin in response to glucose, both *in vitro* and into the serum of mice
shortly after transplantation, ameliorating hyperglycemia in diabetic animals
(Figure 17.23).

Douglas Melton (Figure 1) has had an illustrious career, culminating in his work at Harvard Stem Cell Institute, where has made some of the most significant advances in finding ways to encourage differentiation of near normal pancreatic beta cells from human ES cells.

Melton's path to this position started with an undergraduate degree in biology from the University of Illinois at Urbana-Champaign, followed by a period at the University of Cambridge, where he was awarded a Bachelor of Arts degree in the history and philosophy of science before embarking on PhD studies supervised by John Gurdon. In Gurdon's laboratory working on *Xenopus*, Melton picked up his interest and expertise in studying developmental biology, eventually leading him to focus on the pancreas, an area of intense personal interest given that both of his children were afflicted with type I diabetes.

When President George W. Bush restricted research on human ES cells, Melton used private donations to generate a large range of lines that he then made freely available to the research community. Melton worked on various strategies for the production of insulin-secreting islet beta cells, including by reprogramming of adult mouse exocrine pancreatic cells. Melton continues to work on ways of producing functional human beta cells, especially from pluripotent stem cells.

Box 17.5 Figure 1 **Douglas Melton.** (Photo credit: Justin Ide/ Harvard University. Used with permission.)

Similar advances in the differentiation of human ES cells to fully functional specialized cells can be expected in the next few years. We may see trials of replacement complex organs, such as the heart or liver. Although tissues built around artificial or decellularized scaffolds could become a reality, these are likely to be some way off. Perhaps more realistic in the medium term is the idea that partial, or "patched", tissue replacements might be possible. This type of approach could be particularly useful if the aim is to give the body's own mechanisms a chance to catch up, overcoming a life-threatening crisis period, such as that seen in acute liver failure after acetaminophen (paracetamol) overdose.

The major alternative approach to regenerative medicine through stem cell application, that is, making use of the body's own stem cells *in situ*, is becoming more likely as our knowledge of the biology of individual stem cell types and their

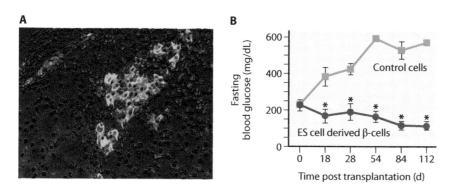

Figure 17.23 **Functional human pancreatic β cells obtained from pluripotent stem cells.** The generation of fully functional insulin-producing pancreatic β-cells from pluripotent stem cells *in vitro* has proved difficult. (**A**) This figure shows *in vivo* engraftment into mice of β-cells that were derived from human ES cells using a highly refined differentiation strategy. At 2 weeks post transplantation, the human graft was stained for cells expressing C-peptide (green) and glucagon (red). (**B**) Transplanted β-cells are rapidly able to overcome hyperglycemia in diabetic mice. The host NRG-Akita mice start to develop hyperglycemia around 4 weeks of age, becoming fully diabetic by about 30–40 weeks. The graph shows fasting blood glucose levels in transplanted with control cells (blue squares) or ES cell derived β-cells (red circles). Data are presented as mean ± SEM, *$p < 0.05$. (From Pagliuca F, Millman J, Gürtler M et al. (2014) *Cell* 159: 428–439 [doi:10.1016/j.cell.2014.09.040]. Copyright 2014 with permission from Elsevier.)

associated niches grows. Although presently only at the experimental stage, it would not be surprising if a novel therapeutic strategy involving manipulation of an adult stem cell or its niche, especially in relation to age-related deficits, were to reach the clinical trial phase within a few years.

## 17.7   SUMMARY

The scope of this chapter has been very broad, as we wanted to give a flavor of the ways in which stem cell technology can benefit both our understanding of disease and the prospects for novel therapies. Changing demographics and modern-day environmental and lifestyle impacts are combining to create a massive shortfall in delivery of new treatments, especially for diseases related to aging that require tissue repair. Broadly speaking, novel therapies that directly involve stem cells (rather than their use in drug discovery) aim to enhance endogenous stem cell capability or use external sources of cells to replace those lost through disease and trauma, the effects of aging, or to correct inherited genetic defects. Many challenges still stand in the way of full realization of the therapeutic potential of stem cells, notably their compatibility with the patient's immune system and, most significantly, the difficulties of incorporating cells into a generally complex tissue either as partial replacement or in generating a whole organ from scratch. These latter challenges rely heavily on the discipline of tissue engineering, for which we could only briefly touch on some of the key features. In spite of the complexities and the challenges, there are emerging examples of success, which promise only to increase in frequency as our knowledge and technological capabilities grow.

## KEY POINTS

- Stem cells provide several novel avenues for improvements in healthcare, particularly in the understanding of disease mechanisms, for the screening of new drugs, and as a source of cells for tissue repair and replacement.

- Longer life expectancy and our modern environment combine to create an increasing demand for repair or replacement of tissues, which is generally referred to as regenerative medicine.

- Stem cell technology offers the chance to overcome the current limitations of regenerative medicine either (1) by enhancing normal or "masked" repair processes or (2) by using cells derived from embryonic or adult stem cells.

- Interventions that enhance the body's own repair processes or trigger latent repair potential through the use of drugs or growth factors to influence stem cells are a real prospect in some situations.

- Gene editing of specific stem cells raises the possibility of treatment options for previously untreatable genetic diseases.

- Production of the desired cell type from stem cells is a major focus of current endeavor, both in terms of the precise functionality and the obtaining sufficient numbers of cells.

- If cells can be generated from personalized stem cells, then transplantation rejection, which is inherent to most transplantation strategies, can be avoided.

- Building cells into a viable, functioning organ structure is a major challenge that is being explored in a variety of ways, such as through the use of synthetic scaffolds or decellularized tissues or by 3D printing using a combination of cells and biomaterials.

- The need for properly regulated clinical trials before any new treatment is implemented in a clinical context is acutely illustrated by recent examples of tissue reconstruction using artificial scaffolds as the basis for replacement of diseased trachea.

- *Bona fide* clinical trials are presently being conducted for a number of regenerative therapies, and it can be expected that there will be an increasing number of ever more complex strategies reaching this stage over the next few years.

## FURTHER READING

### Disease modeling and drug discovery

Bellin, M., Marchetto, M.C., Gage, F.H., and Mummery, C.L. (2012) Induced pluripotent stem cells: the new patient? *Nat. Rev. Mol. Cell. Biol.*, 13: 713–726.

Freedman, B.S., Brooks, C.R., Lam, A.Q., Fu, H., Morizane, R., Agrawal, V., Saad, A.F. et al. (2015) Modelling kidney disease with CRISPR-mutant kidney organoids derived from human pluripotent epiblast spheroids. *Nat. Commun.*, 6: 8715.

Liu, G.H. (2014) Modelling Fanconi anemia pathogenesis and therapeutics using integration-free patient-derived iPSCs. *Nat. Commun.*, 5: 4330.

Liu, C., Oikonomopoulos, A., Sayed N., and Wu, J.C. (2018) Modeling human diseases with induced pluripotent stem cells: from 2D to 3D and beyond. *Development*, 145: dev156166.

Rowe, R.G. and Daley, G.Q. (2019) Induced pluripotent stem cells in disease modelling and drug discovery. *Nat. Rev. Genet.*, 20: 377–388.

Yazawa, M., Hsueh, B., Jia, X., Pasca, A.M., Bernstein, J.A., Hallmayer, J. and Dolmetsch, R.E. (2011) Using iPS cells to investigate cardiac phenotypes in patients with Timothy Syndrome. *Nature*, 471: 230–234.

## Replacement cells derived from pluripotent stem cells

Pagliuca, F.W., Millman, J.R., Gürtler, M., Segel, M., Van Dervort, A., Ryu, J.H., Peterson, Q.P. et al. (2014) Generation of functional human pancreatic β cells *in vitro. Cell*, 159: 428–439.
Szkolnicka, D. and Hay, D.C. (2016). Concise review: Advances in generating hepatocytes from pluripotent stem cells for translational medicine. *Stem Cells*, 34: 1421–1426.

## Immunological barriers to stem cell therapy

Zakrzewski, J.L., van den Brink, M.R., and Hubbell, J.A. (2014) Overcoming immunological barriers in regenerative medicine. *Nat. Biotech.*, 32: 786–794.

## Bioengineering

Guyette, J.P., Charest, J.M., Mills, R.W., Jank, B.J., Moser, P.T., Gilpin, S.E., Gershlak, J.R. (2016) Bioengineering human myocardium on native extracellular matrix. *Circ. Res.*, 118: 56–72.
Murphy, S.V. and Atala, A. (2014) 3D bioprinting of tissues and organs. *Nat. Biotech.*, 32: 773–785.

## Regenerative medicine pre-clinical and clinical trials

da Cruz, L., Fynes, K., Georgiadis, O., Kerby, J., Luo, Y.H., Ahmado, A., Vernon, A. et al. (2018) Phase 1 clinical study of an embryonic stem cell-derived retinal pigment epithelium patch in age-related macular degeneration. *Nat. Biotechnol.*, 36: 328–337.
Rosenzweig, E.S., Brock, J.H., Lu, P., Kumamaru, H., Salegio, E.A., Kadoya, K., Weber, J.L. et al. (2018) Restorative effects of human neural stem cell grafts on the primate spinal cord. *Nat. Med.*, 24: 484.

# Stem cells in biomedicine: Stem cells in aging and cancer

# 18

As organisms progress through their lives, every tissue, and for that matter every cell type, undergoes a gradual process of aging, resulting ultimately in the deterioration of body functions. Aging is accompanied by a decline in the homeostatic and regenerative capacity of all tissues and organs. With increasing age, wound healing is slower in the skin, hair is lost or turns gray, muscle mass and strength decrease, the ratio of different blood cell types is skewed, and neurogenesis declines. With the current demographic shift toward an ever more aged population, the diseases associated with human aging are becoming a global problem, so that there is an increasing need to understand the mechanisms behind cell and organ aging, eventually to improve quality of life well into the eighth decade of life and beyond.

Although cancer can afflict all age groups, it too can be considered a disease characterizing older age. Aging is typically a process involving loss of cell and organism fitness, whereas cancer is an apparent gain of cell fitness and function, and although these may seem to be at odds with one another, it is becoming ever more apparent that the processes underpinning aging and cancer share many common features. Understanding how stem cells are involved in the aging process and in cancer will have profound implications for the development of novel treatment strategies, perhaps even outstripping the potential uses of stem cell-based regenerative therapies (Chapter 17), at least in the near future.

In previous chapters, we have seen how stem cells have a unique combination of **self-renewal** capacity while retaining the ability to **differentiate** into one or more specific cell types and that this underlies their importance in both organ development and in the maintenance and repair of tissues throughout life. The special properties of stem cells and their crucial role in homeostasis is, however, something of a double-edged sword, and we will see in this chapter how the loss or gain of stem cell characteristics can be at the heart of both aging and cancer. Given the complex set of molecular controls required to enable stem cells to perform their unique functions, it is not surprising that any deterioration in these controls during aging can have disastrous consequences (Figure 18.1). If any balance of stem cell fate decision and differentiation is altered, then the result can be an inability to achieve homeostasis or repair. Stem cell exhaustion can also contribute to age-related tissue failure, either because self-renewal is compromised such that **asymmetrical** division does not yield a daughter stem cell, or the cells enter into senescence or undergo cell death. In contrast, if either cell division becomes **symmetrical** so that both daughter cells retain stem cell features in circumstances other than regenerative repair or the progression of the non-stem cell daughter following asymmetrical division is blocked, then uncontrolled expansion of the cell population can result, leading to cancer.

In this chapter, we will first illustrate how **adult stem cells** can become functionally compromised with age and discuss how such changes relate to current theories on the mechanisms of aging, including what prospects there might be to reduce or even reverse these in stem cells. We will then consider how stem cells or the acquisition of stem cell properties can explain the origins and development of cancer and how knowledge of this is changing the prospects for the discovery of novel therapies with the potential for complete eradication of tumors.

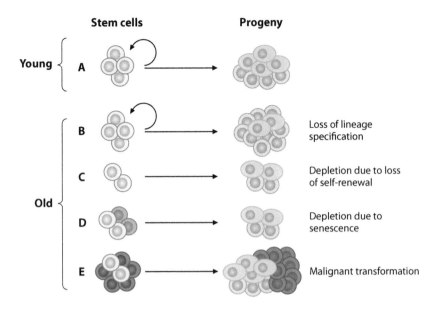

Figure 18.1 **Stem cell function changes with age.** Stem cell function is a contributory factor in aging, while inappropriate stem cell activity underpins the development of cancer. The diagram illustrates some of the possible changes in stem cells with age. (A) In young animals, tissue homeostasis or regeneration is achieved through tissue stem cells (blue cells) that divide asymmetrically to self-renew (curved arrow) and at the same time give rise (straight arrow) to a specific range of differentiated progeny (orange and green cells). (B) Lineage specification may be altered, even though the number of stem cells remains unaffected. (C) Some stem cells lose the capacity for self-renewal, resulting in consistently symmetric cell divisions that give rise only to two differentiated daughters and a gradual depletion of the stem cell pool. (D) Stem cell senescence (darker cells) can also contribute to a loss of functional stem cells. (E) The increase in malignancies (dark blue cells) seen with age, particularly in tissues with high turnover rates, such as blood and various epithelia, might arise from stem cells or early progenitors. (Adapted from Liu L & Rando T. (2011) *J Cell Biol* 193:257–266 [doi: 10.1083/jcb.201010131].)

## 18.1   THE ROLE OF STEM CELLS IN TISSUE AGING

### Different adult stem cell types exhibit distinct age-related changes

Studies of adult stem cells in a variety of tissues in both invertebrates and vertebrates have revealed a common trend of decreased functionality with age (Table 18.1), although the reasons behind this may differ between tissues, involving factors either in the stem cell itself (intrinsic aging) or within the stem cell's **niche** (Chapter 6) environment (extrinsic aging).

As we so often say throughout this book, the **hematopoietic stem cell** (HSC; Chapter 9) is one of the best understood stem cells, and this understanding includes the ways in which aging can influence its properties. Perhaps surprisingly,

### TABLE 18.1  CHARACTERISTICS OF ADULT STEM CELL AGING

| Stem cell population | Age-related changes | | Mechanism |
| --- | --- | --- | --- |
| | Frequency | Phenotypic characteristics | |
| Hematopoietic stem cells (m,h) | Increase | Increased myeloid commitment, decreased lymphoid commitment | Intrinsic |
| Intestinal stem cells (f,m) | Increase | Increased villus height, increased Paneth cell number | Intrinsic/extrinsic |
| Skeletal muscle stem cells (m) | Decrease | Decreased myogenesis, increased fibrogenesis and adipogenesis | Intrinsic/extrinsic |
| Neuronal stem cells (m,r) | Decrease | Decreased neurogenesis, increased gliogenesis | Extrinsic |
| Hair follicle stem cells (m,r) | No change | Lengthened rest phase in hair cycle | Intrinsic/extrinsic |
| Melanocyte stem cells (m) | Decrease | Terminal differentiation | Extrinsic |

*Note:* The characteristics of stem cell aging vary in terms of the numbers of cells present in a tissue, the way in which stem cell function changes, and the relative impact of intrinsic and extrinsic mechanisms.
*Abbreviations:*  m, mouse; h, human; f, fly; r, rat.

immunophenotypically defined HSC increase with age, at least in mice, but when cells are assayed individually, it is apparent that their ability to engraft and self-renew is severely compromised. Moreover, the differentiation potential of aged HSC is affected, skewing toward myeloid **commitment** at the expense of the lymphoid lineages, while examination of blood cells from old mice and elderly humans for mutations that must have been acquired in HSC has revealed that the complexity of the stem cells is very much reduced with age, possibly as the result of specific mutations driving **clonal** expansion. Such so-called clonal restriction increases the possibility that a particular mutation might have a damaging effect, especially in terms of the emergence of a cancerous phenotype (Figure 18.2).

Another well-studied adult stem cell from an aging perspective is the **satellite cell** in skeletal muscle (MuSC; Chapter 10). In contrast to HSC, both the number of MuSC and their individual engraftment potential decline with age, but like HSC, they exhibit some change in potential, becoming more prone to fibrogenic differentiation. Collectively, these changes limit the potential for muscle to repair following injury, but it is not clear that they are a contributory factor in the progressive decline in muscle mass (sarcopenia) that is usually associated with aging.

Comparison of the transplantation potential of HSC and MuSC nicely illustrates how the relative roles of intrinsic versus extrinsic aging can differ between adult stem cells. By performing combinations of transplantations between young or old donors into young or old recipients ("**heterochronic" transplantations**), it is possible to gain some insight into the effects of aging on the stem cell and its niche. In the case of HSC, such experiments indicate that the properties of HSC are largely compromised as the result of intrinsic changes, whereas in the case of MuSC, there seems to be a significant contribution from the niche (Figure 18.3).

Other adult stem cells for which there is information on the effect that aging has on their number or function include **neural stem cells** (NSC; Chapter 14), melanocyte stem cells in the skin (MCSC; Chapter 15), and **germ line stem cells** (GSC; Chapter 12),

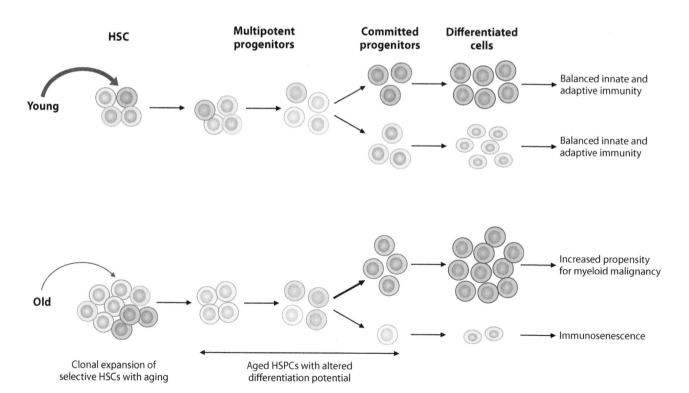

**Figure 18.2 Changes to hematopoietic stem cells during aging favor clonal expansion.** A drift toward clonality in hematopoiesis upon aging has been described as a major contribution to the expansion of the number of phenotypic hematopoietic stem cells (HSC) and to the age-associated myeloid bias in HSC differentiation. In the diagram, clonality of HSC and expanding multipotent progenitors is denoted by cells with a similar color. Old short-term repopulating HSC and multipotent progenitors with altered differentiation fates (denoted by a different color from young cells) lead to less differentiation toward lymphoid progenitors compared to toward myeloid progenitors. In this way, aged HSC drive, at least in part, the clinical outcomes of aging of the hematopoietic system, such as immune remodeling and increased propensity toward the initiation of myeloid malignancies. (From Akunuru S & Geiger H. (2016) *Trends Mol Med* 22: 701–712 [doi:10.1016/j.molmed.2016.06.003]. Copyright 2016 with permission from Elsevier.)

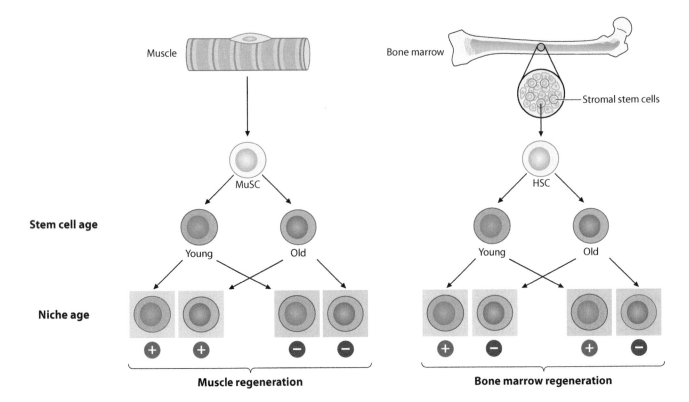

**Figure 18.3 Heterochronic transplantation as a means to compare the effects of aging on muscle satellite cells and hematopoietic stem cells.** Skeletal muscle satellite stem cells (MuSC) and hematopoietic stem cells (HSC) appear to have different aging mechanisms. The diagram outlines experiments in which young and old MuSC were transplanted into young and old hosts. Young and old stem cells are represented as red or blue circles, respectively, and young and old niches are indicated by pink or pale blue squares. Arrows indicate the transfer of stem cells into particular niches. The outcomes (successful versus ineffective regeneration) are shown as green "+ circles" or red "− circles," respectively. Young or old MuSC, when transferred into old niches, fail to generate productive tissue repair, whereas when introduced into a young microenvironment, they can both regenerate muscle tissue appropriately. These results imply that extrinsic mechanisms relating to the age of the niche most likely determine the regenerative potential of MuSC. In contrast, HSC from young animals, when transplanted into either young or old hosts, produce successful regenerative responses, while aged HSC fail to regenerate properly regardless of the age of the host. This suggests that HSC aging is regulated by intrinsic mechanisms. (Adapted from Silva H & Conboy I., Aging and stem cell renewal. (July 15, 2008) StemBook, ed. The Stem Cell Research Community, StemBook, doi/10.3824/stembook.1.11.1, https://www.stembook.org. Published under the terms of the Creative Commons Attribution License http://creativecommons.org/licenses/.)

each of which show a reduction in numbers with age that appears to be a consequence of environmental factors. **Mesenchymal stem cells** (MSC; Chapter 10) show an age-related decrease in number in the bone marrow and also shift their differentiation bias from the osteogenic lineage toward the formation of fat cells, in so doing contributing to the fatty marrow and osteoporosis that is commonly seen in the bones of the elderly. Like HSC, stem cells in the intestinal epithelium (ISC; Chapter 11) increase in number with age, although little else is known about the underlying causes or consequences, and hair follicle stem cells (HFSC; Chapter 15) seem unaffected in numbers terms but are effectively less active because they spend much longer in the resting phase during the hair cycle.

## The properties of adult stem cells make them vulnerable to aging

So what mechanisms underpin either the intrinsic or extrinsic influences that lead to loss of stem cell capability in different tissues? Stem cells are long lived and, like non-dividing cells such as heart muscle cells and nerves, are susceptible to chronological aging. This features the accumulation of environmentally driven changes, including genetic damage, **epigenetic** modifications, and chemical alterations to proteins, carbohydrates, and lipids, some of which can lead to the buildup of macromolecular aggregates that are toxic to the cell. Compounding the effect of cell longevity, most if not all **tissue stem cells** experience prolonged periods of **quiescence** under normal physiological conditions. Although quiescence represents a more stable state in

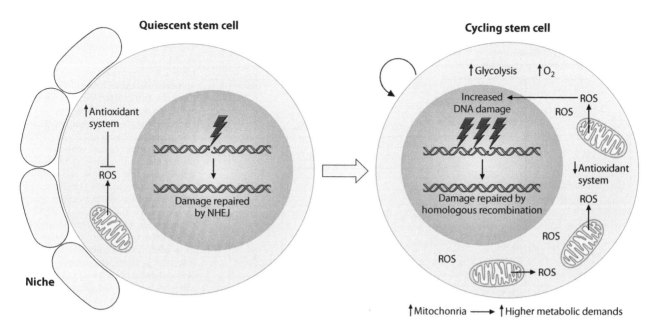

**Figure 18.4 Differences between quiescent and cycling stem cells.** The proliferation rate of a stem cell affects its exposure and responses to mechanisms of aging. Although quiescent stem cells are less susceptible to DNA damage, when it does occur, they are more likely to utilize error-prone non-homologous end-joining (NHEJ) as a repair mechanism, whereas cycling stem cells are more likely to use homologous recombination (HR). Cycling stem cells also have higher metabolic demands than quiescent stem cells, including both higher rates of glycolysis and increased oxidative metabolism, generating DNA damaging reactive oxygen species (ROS). (Adapted from Schultz M & Sinclair D. (2016) *Development* 143: 3–14 [doi: 10.1242/dev.130633].)

metabolic terms, for instance, because levels of oxygen-centered free radicals (reactive oxygen species or ROS) are much lower than in a metabolically active cell, the repair mechanisms operating are more prone to introduce errors and there are overall lower levels of stress-response and repair proteins (Figure 18.4). Furthermore, even though quiescence limits genome replications in a given period of time, avoiding replication-associated mutations, the long lifespan of stem cells means that they do still replicate many times and are thus subject to replicative aging processes, which collectively damage the genetic information. The fact that stem cells underpin tissue homeostasis means that the combined effects of chronological and replicative aging inevitably lead to defects in tissue repair and ultimately to tissue atrophy. In terms of the overall process of aging, the importance of stem cells is most relevant for tissues in which cell turnover is high, such as blood cells and the epithelia of the skin and gut, as compared to tissues such as the cerebral cortex and the heart, in which cellular turnover in adults is exceedingly low.

Aging involves a complex interplay of the effects of cumulative damage, altering stem cell function at several molecular levels (Figure 18.5). For instance, damage to the genome or epigenomic alterations may directly affect gene expression, or more indirectly could influence the transcriptome through changes to proteins involved in the transcription machinery, while protein abnormalities caused by gene mutations or by direct chemical modification might lead to faulty DNA-repair mechanisms or deficient cell maintenance. The importance of the stem cell niche is highlighted throughout this book, and just as alterations to intrinsic stem cell properties can lead to loss of function or aberrant behavior, so changes in the niche and the external environment beyond the niche may contribute to stem cell aging.

We will now look in a little more detail at some of the molecular aspects of stem cell aging, starting with gene expression and differentiation as readouts of the effect of cumulative changes, then focusing on the level of the genome/epigenome, before lastly considering the role of the niche from the perspective of the way it ages and contributes to changes in the stem cells.

## Adult stem cell transcriptomes and differentiation potential change with age

Changes in the transcriptome of a stem cell during aging will reflect both changes in their self-renewal and differentiation status and at least aspects of the mechanistic

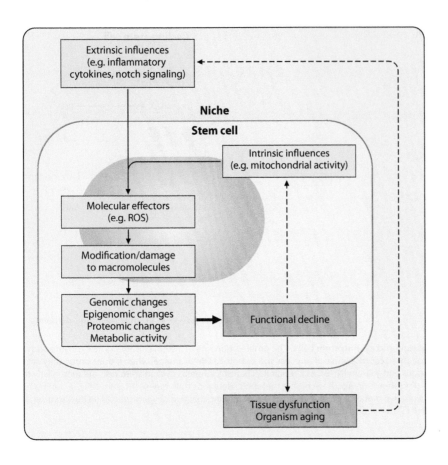

Figure 18.5 **Both extrinsic and intrinsic factors influence stem cell aging.** The systemic environment or immediate niche in which a stem cell sits can alter with age, leading to changes in signaling both within and between cells. Extrinsic influences, for example, an increased level of inflammatory cytokines or decreased Notch signaling, can affect the stem cells themselves or cells of the niche with which they interact. Intrinsic changes in stem cells, such as altered metabolic activity, can also occur during cell aging. Extrinsic and intrinsic influences converge at intracellular molecular effectors, such as reactive oxygen species (ROS), which in turn can cause reversible changes (e.g., epigenetic modifications) or irreversible changes (e.g., DNA mutations) to intracellular macromolecules. The combined effects of genomic, epigenomic, proteomic, and metabolic changes lead to a decline in stem cell function and ultimately contribute to tissue dysfunction and organism aging. (Adapted from Liu L & Rando T. (2011) *J Cell Biol* 193: 257–266 [doi: 10.1083/jcb.201010131].)

basis for any alterations. Yet again focusing on HSC, studies on these have led the way to understanding how overall stem cell gene expression can change with age. Using highly purified HSC from young and old mice, Margaret Goodell and colleagues at the Baylor College of Medicine identified 1500 genes that are age induced and 1600 that are age repressed (**Figure 18.6**). Genes associated with the stress response, inflammation, and protein aggregation dominated the up-regulated expression profile, while the down-regulated profile was marked by genes involved in the preservation of genomic integrity and chromatin remodeling.

Differentiation of tissue stem cells can be affected in several ways as stem cells age. Any form of block can of course be a precursor to the development of cancer, whereas premature or accelerated differentiation can be a cause of stem cell depletion. In addition, some stem cells that have more than one potential pathway of differentiation can exhibit a shift in the proportion of commitment to each of the alternative pathways. Continuing with the example of the HSC, this is skewed toward

Figure 18.6 **Microarray analysis of young and old hematopoietic stem cell transcriptomes.** Analysis of the genome-wide expression of genes can be used to assess the combined influence of intrinsic and extrinsic age-related changes on the phenotype of stem cells. The histogram shows one of the first such analyses performed on hematopoietic stem cells (HSC). The expression of more than 14,000 genes was examined by microarray analysis, using RNA from HSC purified from young (2-month old) and aged (21-month old) mice. 1600 genes were up regulated and 1500 were down regulated in the old compared to young HSC. As a way to determine what the significant changes are, genes are assigned to one or more gene ontology (GO) categories, which represent biological processes that they have been shown to be involved in. The occurrence of genes within the GO categories is then analyzed to determine how much more frequently they appear in one state versus another. The Up-with-Age list contains a large number of enriched GO categories that have been linked to aging in general, such as nitric oxide-mediated signal transduction, the stress response (protein folding), and the inflammatory response. Categories enriched for Down-with-Age genes include those involved in the preservation of genome integrity, such as chromatin remodeling and DNA repair. "Fold enrichment" indicates multiples of an expected baseline value; thus, 2 means a twofold increase over expected. (From Chambers S, Shaw C, Gatza C et al. (2007) *PLoS Biol* 5: e201 [doi: 10.1371/journal. pbio.0050201]. Published under the terms of the Creative Commons Attribution License http://creativecommons.org/licenses/.)

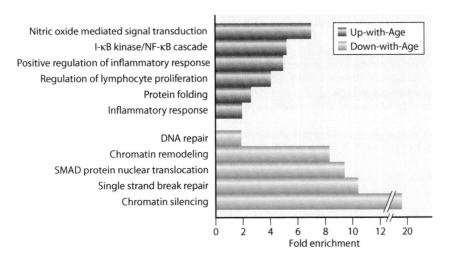

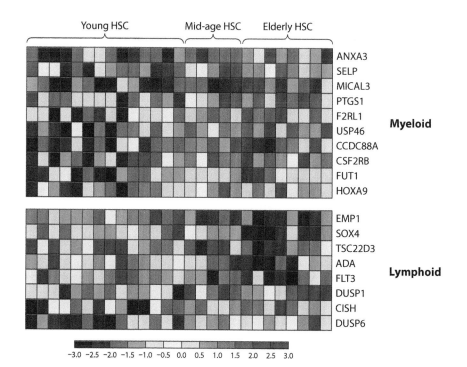

**Figure 18.7 Changes in hematopoietic stem cell gene expression with age.** Microarray analysis of gene expression in hematopoietic stem cells (HSC) in the bone marrow of young and old mice reveals a range of changes in gene expression, many of which correlate with the skewed differentiation potential of HSC with respect to the myeloid versus lymphoid lineages. Representing gene expression in the form of heat maps (blue and red respectively indicating decreased or increased expression), it can clearly be seen that aged HSC exhibit an increase in myeloid gene expression and a decrease in lymphoid gene expression. (From Pang W, Price E, Sahoo D et al. (2011) *Proc Natl Acad Sci USA*, 108: 20012–20017 [doi:10.1073/pnas.1116110108]. With permission from National Academy of Sciences, USA.)

myeloid cell commitment at the expense of lymphoid cells in later life (Figure 18.2), a change that is reflected in a general reduction in the expression of lymphoid related genes and an increase in genes associated with myeloid differentiation (Figure 18.7) resulting from changes at the level of the epigenetic profile of the stem cell.

The effects of aging on melanocyte stem cells is plain for all to see as hair becomes gray and is a typical sign of aging in mammals. Hair graying is an example of incomplete maintenance of MCSC caused by inappropriate differentiation (Figure 18.8). Premature differentiation of the stem cells *in situ* is seen as the appearance of ectopically pigmented melanocytes (EPM) in the hair follicle niche, a location that they would not normally occupy. Progeroid syndromes such as Werner's syndrome and Ataxia-telangiectasia and DNA damage caused by exposure to ionizing radiation lead to premature hair graying. Fate analysis of MCSC in the latter circumstance showing that they lose their stem cell immaturity and commit to differentiation to EPM in the niche, which results in stem cell depletion and subsequent hair graying.

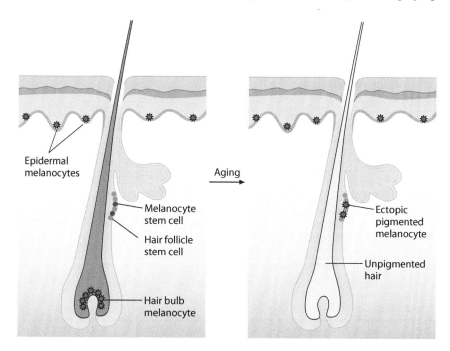

**Figure 18.8 Melanocyte stem cells are lost through differentiation during aging.** Melanocyte stem cells (MeSC; red circles) associated with the hair follicle provide pigmented cells to the growing hair. These follicular MeSC reside in the bulge region of the hair follicle and are supported in a niche by hair follicle stem cells (blue circles). Differentiated melanocytes (black) reside at the bulb to pigment the growing hair. Aging leads to the ectopic differentiation of MeSC in the follicular niche, resulting in a loss of the MeSC pool and hence loss of pigmentation of the hair. (From Mort R, Jackson I & Patton E. (2015) *Development* 142: 620–632 [doi: 10.1242/dev.106567]. Published under the terms of the Creative Commons Attribution License http://creativecommons.org/licenses/by/3.0/.)

Intrinsic loss of stem cell functionality across many different types of adult stem cell during aging could have at least some common grounding in changes in gene expression. However, one study sought to compare these changes between HSC, MuSC, ISC, and HFSC, using RNA transcriptome data sets derived from young (2–4 month) versus old (18–24 month) mice. Although some pairwise comparisons showed a degree of commonality (e.g., HSC versus HFSC), little to no overlap was seen across all four stem cell types. These results need to be viewed with caution, though, since pairwise comparisons between different data sets derived from a given adult stem cell exhibit often less than 10% overlap, suggesting that the methodology is inconsistent.

That there might be common features to changes in transcriptional regulation of gene expression comes from some gene-focused studies. A significant example of this is the FOXO (Forkhead O) group of transcription factors, which have been identified to have an association with longevity across species ranging from the nematode *Caenorhabditis elegans* to humans. FOXO protein activity, as a result of phosphorylation-mediated activation, is downstream of insulin and insulin-like growth factor signaling, and is importantly involved in the cellular response to oxidative stress. There is accumulating evidence that FOXO proteins have a central role in the aging of stem cells. Deletion of Foxo genes in HSC leads to reduced function or loss of these stem cells, which can be at least partially rescued by treatment of the mice with the antioxidant N-acetylcysteine (to lower ROS). In *Hydra*, the interstitial stem cell lineages (Chapter 7) seem also to require FoxO functions for its continuous self-renewal capacity.

## The epigenome of stem cells changes with age and influences gene expression

As in all cells, the stem cell epigenome encompasses DNA modifications such as methylation and post-translational modification of histones, for example, involving acetylation, methylation, and phosphorylation. Although it is difficult to dissect cause and effect in relation to such age-associated differences, and the direct measurement of epigenomic changes in stem cells is technically challenging, it is clear that the stem cell epigenome is a dynamic quantity. The stem cell epigenome is dependent on the cell environment and is susceptible to aging through changes in both the intrinsic machinery and extrinsic signals from the stem cell niche and beyond.

Circumstantial evidence for an age-related role for the epigenome in stem cells comes from analysis of the transcriptome in HSC and other adult stem cells. For example, comparative microarray screening of HSC RNA from young and old mice, as we have described previously, reveals an age-dependent reduction in the expression of various histone deacetylases and DNA methyltransferase. Several genes implicated in chromatin remodeling and transcriptional silencing, such as those coding for the Polycomb group (PcG) proteins and the Sirtuin (SIRT) family of lysine deacetylases, are expressed at much lower levels in aging HSC. PcG and SIRT act in young stem cells by silencing genes associated with aging. This silencing is achieved by trimethylation of histone 3 at lysine 27 (H3K27me3) and by de-acetylation of multiple residues. As the stem cells age, SIRTs diminish and the PcG proteins are replaced with trithorax (Trx) group proteins and the H3K27 demethylase JMJD3, with the result that active chromatin accumulates and the expression of pro-aging genes increases.

Analysis of epigenetic modifications in young and aged HSC to some extent supports the conclusions based on the expression of the epigenetic modifiers. Histone and DNA modifications in HSC show distinct age-related changes depending on whether a particular gene is associated with stemness or differentiation (Figure 18.9). DNA methylation at CpG residues increases, particularly at binding sites for transcription factors involved in HSC differentiation, and decreases at those associated with stemness. Increases in histone modifications that favor transcription (e.g., H3K27Ac, H3K4me1, and H3K4me3) are seen on genes associated with HSC identity. In contrast, some differentiation-associated gene promoters have increased H3K9me3 and H3K27me3, thereby limiting transcription and differentiation; this is more apparent on lymphoid compared to myeloid genes in HSC, perhaps explaining the skewing of differentiation toward the myeloid lineage.

As might be expected, a different adult stem cell type exhibits both similarities and differences in terms of the changes in epigenetic regulation during aging. Like HSC, MuSC exhibit an increase in the H3K27me3 modification, whereas the two stem cells types are distinguished by the change in the extent of the H3K4me3 modification, which is associated with active transcription of genes and becomes more extensive across the genome of HSC as they age but declines in MuSC.

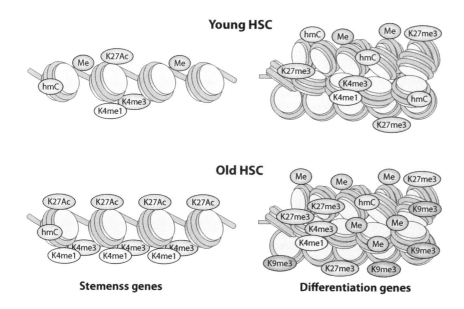

**Young HSC**

**Old HSC**

**Stemenss genes**

**Differentiation genes**

Figure 18.9 **Chromatin structure in HSC is affected differentially by aging according to the role of individual genes.** Histone and DNA modifications are altered in aging adult stem cells, leading to increases and decreases in gene expression, which go at least some way toward explaining altered stem cell self-renewal and differentiation potential. The diagram illustrates some of the changes that have been observed in chromatin modifications in hematopoietic stem cells (HSC). The schematic represents chromatin as DNA (brown lines) wrapped around nucleosomes (blue discs) in either an open or closed configuration on stemness or differentiation genes, respectively. Modifications to DNA and histones are marked by flags: Me – CpG methylation; hmC – hydroxymethyl C; and various methylations (me) and acetylations (Ac) of numbered lysine (K) residues on Histone 3. (From Cakouros D & Gronthos S. (2019) *Aging and Disease* 10:174–189 [doi:10.14336/AD.2017.1213]. Published under the terms of the Creative Commons Attribution License http://creativecommons.org/licenses/.)

## The maintenance of telomere length in adult stem cells declines with age

A possible factor in stem cell aging is the extent to which the ends of chromosomes, the **telomeres**, become shortened, thereby increasing the chances of large-scale chromosome losses, rearrangements or fusions, and ultimately cell senescence or death. Telomeres consist of characteristic repeat sequences of DNA that are copied by the enzyme telomerase during replication in a way that limits chromosome shortening. Telomerase is absent in the majority of adult somatic cells, with the exception of some cells of the immune system and the stem cells of most tissues. Nevertheless, stem cell telomeres gradually shorten with replicative age. It is fair to say, though, that a direct causal link has not been proven for an effect of telomere shortening on stem cell aging and that to date the evidence is somewhat circumstantial. Loss of telomerase function in mice lacking TERC (telomerase RNA component) is accompanied by reduced proliferative capacity in many organs, accelerated aging, reduced lifespan, and ultimately defects in stem cell function in the brain, skin, intestinal epithelium, and hematopoietic system. Naturally occurring mutations in humans in the telomerase core components TERT (telomerase reverse transcriptase) and TERC have been implicated in patients suffering from dyskeratosis congenita. This inherited disease is manifest early in life as tissues with a high proliferative demand on their resident stem cell populations exhibit accelerated shortening of their telomeres, leading to premature stem cell senescence and organ failure.

Intriguingly, though, some organisms seem to avoid aging through specialized mechanisms in their tissue stem cells, at least a component of which involves enhanced maintenance of telomeres. Most notably, some species of planarian flatworms have apparently limitless regenerative capacity supported by a population of highly proliferative, **pluripotent** adult stem cells, or **clonogenic neoblasts (cNeoblasts)** (Chapter 7). As a potential underpinning mechanism, somatic telomere maintenance in planarians appears to be adapted in species that do not require a sexual phase, telomere length being maintained through indefinite somatic telomerase activity in proliferating stem cells during regeneration or reproduction by fission. Interestingly, pluripotent cells derived from vertebrates, such as **embryonic stem (ES)** cells (Chapter 3), are like the planarian stem cells in that they maintain telomeres via active telomerase.

## Damage to macromolecules accumulates as stem cells age

Chronological and replicative stem cell aging are perhaps best understood in terms of DNA damage. Exposure to mutagens such as UV light, ionizing radiation (X-rays, $\gamma$-rays), chemicals like aromatic hydrocarbons, and ROS can all lead to DNA strand breaks and base pair defects. Replication-associated errors can give rise to point mutations but also gross chromosome rearrangements. A variety of mechanisms are in place both to prevent genome damage, for example, through tight regulation of ROS or maintenance of telomeres, and to repair DNA strand breaks and base pair defects. Alternatively, if the damage is too extensive and there is a risk that a cell with a mutated genome might survive and proliferate (e.g., as a cancer cell), then that cell may be eliminated through enforced suicide (programmed cell death, or apoptosis). In addition to effects on the genome, damaging agents like ROS can also impact macromolecules other than DNA, especially proteins, which can then be susceptible to misfolding or aggregation, leading to disrupted cell function. Such damage could itself impact the genome if it affects repair mechanisms and can certainly diminish the cell's ability to dispose of damaged proteins and organelles.

Oxidative damage by ROS is a natural consequence of an aerobic existence, highly reactive superoxide and hydroxyl radicals being generated as by-products of oxidative phosphorylation in the mitochondrial electron transport chain but also arising from sources as diverse as inflammation and through the action of environmental insults like UV light and cigarette smoke. It has been proposed that with age, ROS progressively accumulate within stem cells due to failure of mechanisms such as the enzymes superoxide dismutase and catalase that normally eliminate superoxide and hydrogen peroxide. The idea that ROS accumulation contributes to stem cell aging comes largely from studies on mouse HSC in which the levels of ROS are found to increase with age. Increased ROS levels activate the p38 mitogen-activated protein kinase (MAPK) pathway, which in turn promotes the conversion of quiescent HSC into a proliferative state. This enhanced proliferation likely explains the observed increase in the size of the stem cell pool for mammalian HSC but over time can lead to a gradual reduction or even exhaustion of the most immature stem cells in the hierarchy.

Autophagy, literally "self-eating," is a cellular process that can serve to eliminate damage, clearing away modified proteins and organelles like defective mitochondria, thereby decreasing ROS levels and reducing genome damage and cell senescence and playing a crucial role in enhancing cell longevity. Autophagy has been implicated in the maintenance and differentiation of various adult stem cells (Figure 18.10). For example, in human MSC, the level of constitutive autophagy is high but becomes undetectable upon differentiation into osteoblasts. Likewise, a high level of autophagy has been shown in mouse HSC, which if reduced by genetic ablation of two essential

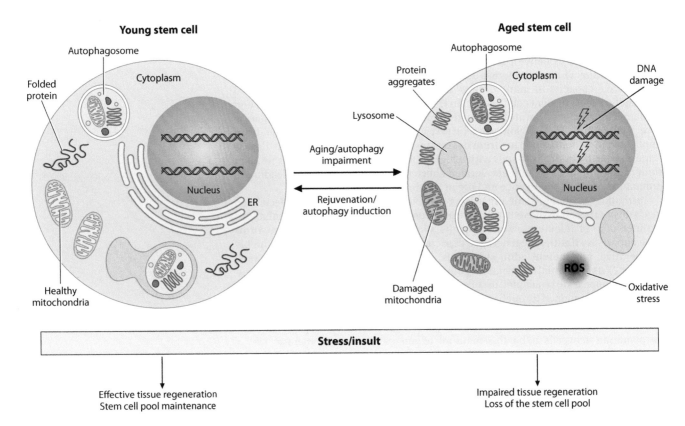

**Figure 18.10 Autophagy helps to keep stem cells young.** The aging of stem cells is very much influenced by alterations to cell metabolism. Aberrant mitochondrial activity leads to increased production of reactive oxygen species (ROS), which can lead to damage to macromolecules. Impairment to the process of autophagy causes accumulation of intracellular waste and proteotoxicity. (From Garcia-Prat L, Sousa-Victor P & Muñoz-Canoves P. (2017) *Cell Stem Cell* 20; 593–608 [doi: 10.1016/j. stem.2017.04.011]. Copyright 2017 with permission from Elsevier.)

autophagy genes, FIP200 or Atg7, leads to dysfunction of these stem cells. Various longevity-related signaling pathways are regulators of autophagy, the mTOR pathway being a key player. Evidence suggests that mTOR can slow down senescence and that loss of its function can impact on stem cells. Hence, conditional deletion of Tsc1 (tuberous sclerosis complex protein 1) leads to constitutively active mTOR and premature aging of HSC. The possible importance of FOXO proteins in stem cell aging that we mentioned previously is perhaps also reflected in the fact that FOXO3 is also an established regulator of autophagy in HSC.

## Metabolic activity is a key determinant of stem cell aging

We have already briefly covered how ROS, reflecting metabolic activity, can have a significant impact on aging stem cells, but this is only the tip of the iceberg in terms of the link between metabolism and stem cell aging. The availability of substrates, especially comparing nutrient poor and nutrient rich conditions, can influence the epigenome by affecting cofactors such as alpha ketoglutarate, NAD+, and S-adenosyl methionine (Figure 18.11). The consequence of these sorts of metabolite level differences will accumulate, leading to either slower or accelerated stem cell aging (Table 18.2), whereby caloric restriction has the effect of reducing stem cell aging, in contrast to nutrient rich conditions, which accelerate the decline in stem cell capacity.

A key component of metabolic change during stem cell aging is the mitochondrion. Mitochondrial DNA (mtDNA) mutations accumulate in stem cells during aging and cause mitochondrial dysfunction, causing respiratory chain deficiency and a consequent excess production of ROS. The development of a "mtDNA mutator" mouse strain nicely illustrates how accumulating mitochondrial mutations can affect stem cells with age. These animals, which lack mtDNA polymerase γ, have many features of premature aging and show functional defects in HSC and ISC.

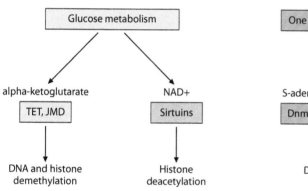
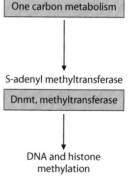

**Figure 18.11 Metabolites are co-factors of many chromatin regulators.** Metabolism is closely linked to epigenetic modification of chromatin, and thereby to gene expression, through the provision of co-factors of chromatin modifying enzymes. Abbreviations: TET – Ten-eleven translocation methylcytosine dioxygenase; JMD – Jumonji family demethylase; NAD – nicotinamide adenine dinucleotide; Dnmt – DNA methyltransferase). (From Brunet A & Rando T. (2017) *Curr Opin Cell Biol* 45:1–7 [doi:10.1016/j.ceb.2016.12.009]. Copyright 2017 with permission from Elsevier.)

## TABLE 18.2  DIETARY CONTROL OF REGULATORS OF STEM CELL FUNCTION

| Nutrient status/stem cell aging consequence | Metabolic effect | Affected pathway | Target process | Adult stem cells known to be influenced |
|---|---|---|---|---|
| Nutrient poor, aging slowed | Low insulin signaling | FOXO, Sirtuin | Autophagy | HSC, MuSC |
| | Caloric restriction | | Self renewal | HSC, NSC, MuSC |
| | | | DNA repair | HSC |
| | | | Quiescence | HSC, NSC, MuSC |
| Nutrient rich, aging accelerated | High insulin signaling | mTOR | Activation | MuSC |
| | High amino acids | | Proliferation | ISC |
| | High fat diet | | | |

*Source:* Adapted from Brunet A & Rando T. (2017) *Curr Opin Cell Biol* 45:1–7 [doi:10.1016/j.ceb.2016.12.009].
*Note:* Signaling pathways that influence key processes in stem cells can be affected by the availability of nutrients. When nutrients are limited, then aging is slowed, whereas in conditions of nutrient excess, stem cells are overstimulated and become depleted, and aging is accelerated.
*Abbreviations:* HSC, hematopoietic stem cell; MuSC, muscle stem cell; NSC, neural stem cell; ISC, intestinal stem cell.

## The effects of aging on stem cells is often mediated through changes in their niche

In Chapter 6, we described the sorts of interactions that take place between the stem cell and its microenvironment or niche. Since the niche plays a critical role in the maintenance of stem cell function, it is not surprising that it has a significant influence on the aging phenotype of several stem cell types. A notable example is the muscle satellite stem cell. As mentioned earlier in this chapter, studies of transplantation of satellite cells from either young or old mice into either young or old recipient animals reveal that the recipient muscle environment can guide the regenerative capability of the donor MuSC and that aging involves a component extrinsic to the stem cell.

Elegant evidence for the impact of niche age on stem cells has come from studies on the fruit fly *Drosophila*. We saw in Chapters 6 and 12 how studies on GSC in the male testes and the female ovaries of *Drosophila* have provided several paradigms for our wider understanding of stem cell biology, not least of which is the structure and function of the stem cell niche. The *Drosophila* GSC niches, constituted of the **hub** or **cap** cells in males and females, respectively (Chapter 6), lose their ability to maintain the vitality of the flies' stem cells over time, leading to a decline in the number of stem cells available. The hub cells of the male testes express a key self-renewal factor called Unpaired (Upd), which in young flies is protected by another protein, Imp. As the male GSC niche ages, the cells in the testis hub produce *let-7* micro RNA (Figure 18.12). The primary effect of increased let-7 is to initiate a chain of events that switch on aging by causing a decline in Imp, which in turn leads to the ultimate decline in Upd. The age-related loss of GSC can be reversed by experimentally increasing expression of Imp or Upd in the hub cells. What leads to accumulation of *let-7* in the niche of aged flies still remains an open question. In a similar way, the fall in GSC number and function within *Drosophila* ovaries correlates with a reduction in the expression of two key molecules: E-cadherin and bone morphogenetic protein (BMP). Just as in the case of restoration of Upd in the testes, restoration of high levels of E-cad and BMP enhance the function and lifetime of aged ovarian GSC. Mammalian male **spermatagonial stem cells** (SSC, Chapter 12) also decline dramatically in number with age, although when old mouse SSC are isolated and transplanted into a young male testis, they retain the same capacity as stem cells from young animals. The molecular explanation for these observations has not yet been described, but it is assumed that the capacity of the SSC niche declines as a result of age-related systemic hormonal factors.

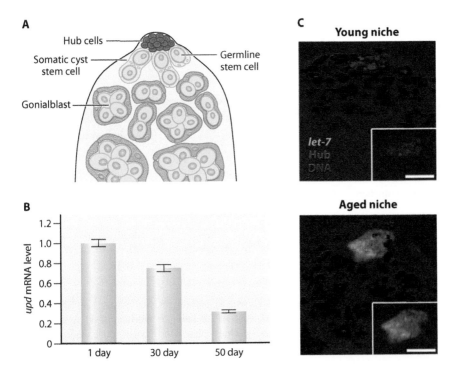

**Figure 18.12 Aging in the *Drosophila* male germ stem cell niche.** Aging of the stem cell niche can play a large part in the decline of stem cell function. The figure illustrates an example of such an age-related change in the *Drosophila* male germ stem cell niche. (**A**) The diagram depicts the apical tip of a *Drosophila* testis, showing the regulatory niche hub cells (red), a somatic cyst stem cell (CSC), a germline stem cell (GSC), and a committed progenitor, or gonialblast. The hub cluster secretes Unpaired (Upd), a self-renewal factor that is important for stem cell maintenance. (**B**) As measured by reverse transcriptase polymerase chain reaction, *upd* mRNA decreases with age. In young flies, *upd* mRNA is stabilized by the RNA binding protein Imp, but with age, Imp is increasingly targeted by the microRNA *let-7*. (**C**) These fluorescence microscope images of testes from young (upper) and old (lower) *Drosophila* show the effect of aging on the stem cell niche (top center in both photos). The hub cells (red), which function as part of the GSC supporting niche, express more *let-7* (green) in aged flies, which changes the signaling properties of hub cells, leading to fewer stem cells surrounding the hub that are available for tissue maintenance. (From Toledano H, D'Alterio C, Czech B et al. (2012) *Nature* 485:605–612 [doi: 1038/nature11061]. Reprinted by permission from Springer Nature.)

## Increasing knowledge about the effects of aging on adult stem cells will ultimately help to improve the quality of life in the elderly

The manifestation of stem cell aging does not have to be restricted to one of the previous mechanisms for a given stem cell type; indeed, it is more likely that multiple effects operate on most tissue stem cells. This is nicely illustrated by the changes that occur in HSC (see Figure 18.13). The most notable differences are the reduced regenerative potential upon transplantation of aged HSC and, as already mentioned, the skewed differentiation toward the myeloid lineage. The reduction in engraftment efficiency of older HSC is likely to be a result of an altered homing to and mobilization from the bone marrow, possibly due to changes in the niche selectivity of older HSC compared to young. *In vivo* imaging of the location of HSC within the bone marrow shows that older HSC are typically located further away from the endosteum (the lining of the bone marrow cavity), perhaps as a consequence of impaired adhesive properties of the stem cells.

The very fact that the **reprogramming** of adult somatic cells to give rise to **induced pluripotent stem cells** (iPSC; Chapter 5) can reverse features associated with aging demonstrates that it should be feasible to "rejuvenate" aged adult stem cells and, moreover, that reversible epigenetic modifications must play a large part in the aging process. Several experimental observations, mainly looking at HSC and involving manipulation of either metabolism or specific epigenetic modifiers, show that intrinsic stem cell age-related deficits can be overcome. Caloric restriction is well known for its ability to extend the lifespan of multiple organisms, including worms, flies, and mice, and at the stem cell level, prolonged fasting can recapture the young differentiation potential in aged HSC. As already mentioned, the expression of Sirtuins, which are mitochondrial histone deacetylases, decreases in old HSC, and when experimentally restored shifts the cell's reconstitution capacity and differentiation potential back to that of young cells. Another tractable candidate for adult stem cell rejuvenation is the nutrient-sensing

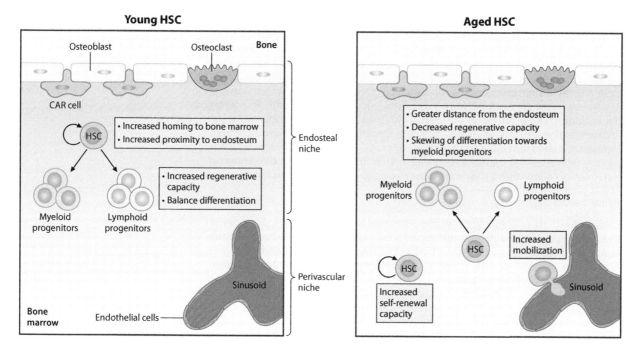

**Figure 18.13 Changes in hematopoietic stem cells with age.** Detailed characterization of hematopoietic stem cells (HSC) in the bone marrow of young and old mice reveals a range of differences that collectively reduce the fitness of the HSC with age. Young HSC home to the bone marrow and localize in close proximity to the endosteum, which forms the stromal lining of the marrow cavity and includes many osteoblasts, as well as other cells, such as bone-degrading osteoclasts and CXC-chemokine ligand 12-abundant reticular (CAR) cells. These HSC have high self-renewal and regenerative capacities and an equal propensity to differentiate toward lymphoid (blue) and myeloid (green) progenitor cells. The number of phenotypic HSC in the bone marrow increases with aging, probably as a consequence of an enhanced self-renewal activity, even though their individual regenerative capacity is reduced. Unlike young HSC, aged HSC localize away from the endosteal stem cell niche when they home to the bone marrow, which they do with roughly a twofold lower efficiency. This implies that aged HSC select for niches that are distinct from those that young HSC occupy. Aged HSC can be mobilized by cytokines in higher numbers than young HSC, and their differentiation is skewed toward the myeloid lineages. (From Geiger H, de Haan G & Florian M. (2013) *Nat Rev Immunol* 13:376–389 [doi: 10.1038/nri3433]. Reprinted by permission from Springer Nature.)

protein mTOR, which has been reported to be elevated in aged HSC, and when treated with the mTOR inhibitor rapamycin causes reversal of the age-associated increase in HSC numbers and restoration of self-renewal and reconstitution potential.

Perhaps not surprisingly, the other adult stem cell type that has received attention regarding the potential for reversal of aging is the MuSC. As we have described, aging of MuSC has both an intrinsic component and an influence of the niche. The former correlates with elevated activity of p38 MAPK, while at least some of the niche influence is the result of decreased supply of Notch ligands to the receptors on MuSC. Both of these influences have been alleviated to an extent by inhibition, either of p38, in combination with culture in a biomimetic matrix, or by forced activation of Notch.

More speculative possibilities will undoubtedly emerge given the importance of epigenetic modifications in stem cell aging combined with the fact that these depend upon the supply of the appropriate cofactors and substrates. It is not unreasonable to expect that the manipulation of specific metabolites, such as S-adenosyl methionine or alpha ketoglutarate, could be used, for instance, to influence the extent of certain histone lysine methylation or DNA methylation.

## 18.2   THE ROLE OF STEM CELLS IN CANCER

Cancer can be loosely characterized by aberrant proliferation, differentiation, and survival, with disease progression being accompanied by the acquisition of a more undifferentiated state. These general features of cancer are in many ways reminiscent of stem cell properties and, perhaps not unsurprisingly, aspects of stem cell signaling, fate determination, differentiation hierarchies, and niche dependency all appear to have some relevance to cancer biology.

In this section, we will see that stem cell properties can underpin the origin and behavior of tumors, potentially explaining their complexity and evolution, including progression to metastasis, and certainly may become a crucial factor in the design of new therapeutic anti-cancer strategies.

### Tumors are inherently heterogeneous

Heterogeneity of cell types within a tumor is believed by many to reflect a cell hierarchy analogous to that seen in a normal tissue. Just as tissue stem cells are at the top of a differentiation hierarchy, so tumors, at least during the earlier phases of their evolution, can be seen to behave as though derived from an immature cell component undergoing asymmetric divisions. A consequence of this model is that the majority of cells within a tumor are relatively harmless non-tumorigenic downstream derivatives of the immature cell at the top of the hierarchy. There is also strong evidence for the co-existence of genetically divergent cell clones within tumors that arise through secondary mutation and outgrowth of cells with some selective advantage, which is especially relevant when thinking about cancer therapy resistance. Such evolution of populations of cells within a tumor is often referred to as being "Darwinian."

### Tumor heterogeneity reflects a stem cell component

The realization that tumors are heterogeneous and contain immature cells, the presence of which often correlates with poor therapy response, spurred investigators to identify the specific cell types responsible. As far back as 1977, using a simple method to support human tumor stem cell growth in soft agar, Hamburger and Salmon demonstrated the presence of cells with high proliferation and low differentiation capacities. In the 1990s, seminal studies on leukemia revealed a small subpopulation of cells that have the properties of adult stem cells, including the ability to exist in a quiescent state, which because of their consequent slower metabolism are relatively unaffected by standard chemotherapeutic treatments. Probably the most significant observations were those made by John Dick (**Box 18.1**) and colleagues, looking for the tumorigenic component of acute myeloid leukemia (AML). They found that cells with the surface marker **cluster of differentiation** (CD) 34$^+$CD38$^-$ phenotype are capable of transferring leukemia to recipient animals. It was subsequently realized that the discovery of these quiescent leukemia cells, which were immediately termed **"cancer stem cells"** (CSC), served as a paradigm for many other cancers. An important consequence of the CSC hypothesis (**Figure 18.14**) that emerged from this paradigm was that it suggested a whole new approach to cancer treatment that would involve selective targeting of the stem cell component, thereby raising the prospect of total, life-long eradication of a tumor.

Following the ground-breaking studies on AML, similar studies involving transplantation of human breast cancer cells into immunocompromised mice showed that as few as 100 CD44$^+$CD24$^{-/low}$Lin$^-$ cells were able to form tumors that could be

John Dick (Figure 1) is an influential cancer research scientist who won international recognition for the development of the repopulation xenograft assay for the *in vivo* behavior of human hematopoietic cells in immunocompromised mice. The assay demonstrates that only a small minority of human acute myeloid leukemia (AML) cells have the potential for engraftment and indefinite proliferation. John Dick's findings ignited the field of cancer stem cell (CSC) investigation. He has shown that the AML cells responsible for initiating leukemia in xenografts are invariably CD34+/CD38− cells. His seminal paper, published in *Nature* in 1994, was largely dismissed as "interesting but not likely to apply to solid tumors" ("Meet the A-Team of Stem-Cell Science" by Carolyn Abraham, the *Globe and Mail*, 27 November, 2006). John Dick termed this dismissal "benign neglect," saying "the field needs controversy to move forward, controversy sparks better and better science, and eliminates sloppy thinking." His 1997 publication further developing the CSC hypothesis describes a hierarchy in which an abnormal stem cell is the key to forming and feeding cancers. In 2006, John Dick became a senior scientist in the division of Cellular and Molecular Biology at Toronto General Research Institute, and was the founding member of Canada's Stem Cell Network, one of the first such networks, which has persisted and significantly impacted stem cell research. In 2014, John Dick was elected a Fellow of the Royal Society.

Box 18.1 Figure 1 **John Dick.** (From Baker M. (2009) *Nat Rep Stem Cells* [doi:10.1038/stemcells.2009.47] Reprinted by permission from Springer Nature.)

serially passaged. At each passage, tumors generated additional CD44+CD24−/lowLin− tumorigenic cells as well as the phenotypically diverse mixed populations of non-tumorigenic cells seen in the original tumor, thereby providing strong support for the idea of a CSC-based hierarchy (Figure 18.15). Subsequently, many other solid tumors were also shown to contain cells with CSC-like characteristics, including those from colon, brain, and pancreas.

## Each cancer stem cell type is characterized by a set of surface markers

The first studies that defined CSC in AML relied on relatively simple identification on the basis of cell surface antigens. Surprisingly, with CSC being defined in more tumors, there has been little refinement in the process of their identification and little indication of there being a set of markers that consistently defines the CSC phenotype, even though there are some antigens that frequently associate with CSC, and claims have been made in relation to antigens such as CD44, CD133, and CD271 (Table 18.3). Further confounding the picture, investigation of melanoma cancer stem cells revealed almost 50 distinct

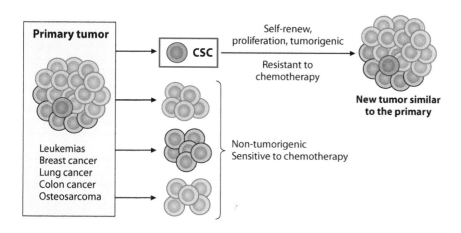

Figure 18.14 **The cancer stem cell hypothesis.** Tumors are heterogeneous, containing a hierarchy of immature cells, proliferating progenitors, and a mixture of differentiated cells. At the top of the hierarchy sits the cancer stem cell (CSC, red), which is relatively quiescent and, unlike the bulk of proliferating cells in the tumor, is resistant to traditional anti-cancer therapies, such as chemotherapy and irradiation. (From Romano M, De Francesco F, Pirozzi G et al. (2015) *Oncoscience* 2:443–456 [10.18632/oncoscience.163]. Published under the terms of the Creative Commons Attribution License http://creativecommons.org/licenses/.)

**A**

| Number of cells injected: | Tumors/injections | | | | | | | | | |
|---|---|---|---|---|---|---|---|---|---|---|
| | $5 \times 10^5$ | $10^5$ | $5 \times 10^4$ | $2 \times 10^4$ | $10^4$ | $5 \times 10^3$ | $10^3$ | 500 | 200 | 100 |
| Unsorted | 8/8 | 8/8 | 10/10 | — | 3/12 | — | 0/12 | — | — | — |
| CD44$^+$CD24$^+$ | — | — | — | 0/10 | 0/10 | 0/14 | 0/10 | — | — | — |
| CD44$^+$CD24$^{-/low}$ | — | — | — | 10/10 | 10/10 | 14/14 | 10/10 | — | — | — |
| CD44$^+$CD24$^{-/low}$ESA$^+$ | — | — | — | — | — | — | 10/10 | 4/4 | 4/4 | 1/6 |
| CD44$^+$CD24$^{-/low}$ESA$^-$ | — | — | — | — | — | — | 0/10 | 0/4 | 0/4 | 0/6 |

**B**

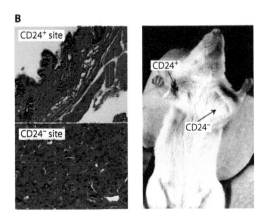

Figure 18.15 **Xenograft assay of breast cancer stem cells.** Following the identification of cancer stem cells (CSC) in leukemia, similar cells were sought in solid tumors. The first solid tumor CSC to be identified were derived from human breast cancer tissue. (A) The table shows the frequency of tumor formation following injection of different numbers of sorted human breast cancer cells into immune compromised mice. Cells were sorted using combinations of antibodies recognizing the surface antigens CD44, CD24, and epithelial surface antigen (ESA). The indicated number of cells of each phenotype were injected into the breast of NOD-SCID (non-obese diabetic/severe combined immunodeficiency) mice, and the tumor frequency after about 10 weeks is indicated as a fraction of the total number of mice tested. (B) This injected mouse shows a representative tumor at the CD44$^+$CD24$^{-/low}$ injection site but not at the CD44$^+$CD24$^+$ injection site. Examination of the histology at the site of injection of CD24$^+$ cells (upper stained section) revealed only normal mouse tissue, whereas the CD24$^{-/low}$ injection site (lower stained section) contained malignant cells. (From Al-Hajj M, Wicha M, Benito-Hernandez A et al. (2003) *Proc Natl Acad Sci USA*, 100: 3983–3988 [doi: 10.1073/pnas.0530291100]. With permission. Copyright (2003) National Academy of Sciences, USA.)

## TABLE 18.3  CANCER STEM CELL MARKERS FOR DISTINCT SOLID TUMOR TYPES

| | | | | Tumor | | | | |
|---|---|---|---|---|---|---|---|---|
| Breast | Colon | Glioma | Liver | Lung | Melanoma | Ovary | Pancreas | Prostate |
| | ABCB5 | | | ABCG2 | ABCB5 | | ABCG2 | |
| ALDH1 | ALDH1 | | | ALDH1 | ALDH1 | | ALDH1 | ALDH1 |
| | β-Catenin | | | | | | | |
| CD24 | CD24 | CD15 | CD13 | CD90 | CD20 | CD24 | CD24 | CD44 |
| CD44 | CD26 | CD90 | CD24 | CD117 | CD133 | CD44 | CD44 | CD133 |
| CD90 | CD29 | CD133 | CD44 | CD133 | CD271 | CD117 | CD133 | CD166 |
| CD133 | CD44 | Nestin | CD90 | | | CD133 | c-Met | Trop2 |
| | CD133 | | CD133 | | | | CXCR4 | |
| | CD166 | | | | | | Nestin | |
| | LGR5 | | | | | | | |

*Source:*  From Medema J (2013) *Nat Cell Biol* 15:338–344 [doi:10.1038/ncb2717]. Reprinted by permission from Springer Nature.
*Note:*  The table lists a number of biomarkers that have been used to isolate and characterize cancer stem cells. The upper row represents members of the ABC transporter family of proteins. Alcohol dehydrogenase (ALDH) is quite widely expressed in stem cells and is often used to enrich them. Many surface markers, indicated by their CD designation, are expressed on CSC. Although there is little tumor type specificity, the patterns of expression do allow a degree of discrimination.

subpopulations, as defined by different antigen combinations, which showed CSC activity in transplantation assays. Such findings and the variability seen when different immune-compromised hosts are employed in the transplantation assays has led some to speculate that we grossly overestimate the number of cells that have true CSC properties.

## The definition of cancer stem cell properties depends on the assay

Despite considerable focus on the concept of CSC and the examples that have been described in a variety of tumors, doubt remains about the conclusions that can be drawn from **xenotransplantation** assays, and thinking about the CSC model is being refined. Two linked principal issues confound the interpretation of xenotransplantation assays. The first is that the environment provided by the mouse host for the human tumor cells is artificial and may give a false indication of the stem cell properties of the transplanted populations. For example, mouse growth factors and adhesion molecules can fail to interact with the appropriate human receptors. Furthermore, given that transplantation requires the test cells to home to an appropriate niche, it has to be assumed that this is available, can be found, and, having reached there, that the behavior of the CSC has not been perturbed in the process. The second concern relates to the interaction between human cells and the mouse immune system, which is of necessity compromised to avoid rejection. Almost certainly, this scenario does not model what would be experienced by CSC in their natural environment. Therefore, it is not surprising that the apparent frequency of CSC varies depending on the degree to which the mouse host immune system is compromised. For instance, when Morrison and colleagues initially characterized malignant melanoma using a NOD/SCID host mouse, they found that CSC were present at a frequency of one in a million cells, but upon switching to a more profoundly immune compromised host (one that additionally lacked the IL2 receptor [NOD/SCID/IL2R$_\gamma$KO]), the frequency jumped to one in four!

With the previous reservations in mind, there has been a move toward using alternative experimental strategies to investigate CSC *in vivo*. In particular, **lineage tracing** offers the possibility to follow the behavior of CSC without extensive perturbation to the model system. This type of approach relies upon the identification of a gene whose expression is thought to be characteristic of the CSC and can be illustrated with the sorts of studies utilizing the Lgr5 gene that are being performed on colorectal cancer (CRC) and the relationship between CSC and the normal **intestinal stem cells** (Chapter 11). In 2017, Sato and colleagues working in Tokyo combined CRISPR/Cas9, organoid, and xenotransplantation technologies (Chapter 2) to show the dynamics between Lgr5+ CSC and differentiated cells in human CRC *in vivo*. Lineage tracing experiments with a tamoxifen-inducible Cre knock-in allele of Lgr5, introduced using CRISPR/Cas9 recombination into the tumor cells growing as organoids, revealed the self-renewal and differentiation capacity of the Lgr5+ tumor cells, the presumed CSC. Using a parallel second approach, which instead placed the apoptosis-inducing protein Caspase 9 under the control of the Lgr5 gene, they were able selectively to ablate the Lgr5+ CSC, leading to tumor regression. Interestingly, and adding significantly to our view of the nature of CSC, the tumor *in vivo* eventually regained Lgr5+ CSC, which derived from differentiated (Krt20+) cells that had not been ablated along with the original CSC. This result, and similar ones from other systems, suggest that there is flexibility within a tumor such that CSC can be derived from differentiated cells. This modified view of a hardwired hierarchy of CSC and downstream differentiated cells is also consistent with the ideas of Darwinian evolution and furthermore suggests ways in which environmental factors can affect the Darwinian evolution of the tumor.

## The origin of cancer is rooted in subversion or acquisition of stem cell properties

Cancer "driver" mutations, that is, genetic changes that are causal for the initiation of a tumor, are usually selective for transformation of undifferentiated cells, suggesting that stem or **progenitor** cells provide a permissive context. Furthermore, many of the frequently encountered driver mutations occur in or affect signaling pathways that are commonly associated with stem cell properties (Chapter 4). Some of the most frequently affected stem cell signaling pathways that are affected in cancer are Wnt, Notch, and Hedgehog (Figure 18.16). For example, Wnt signaling is important in several adult stem cells, including in HSC, the epidermis, and the intestinal epithelium, and correspondingly, it is frequently activated in tumors of the colon and skin. Individual

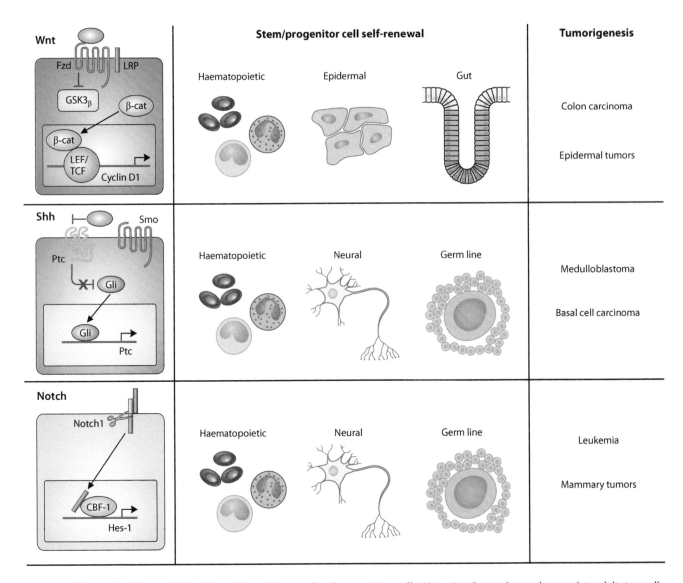

Figure 18.16 **Stem cell signaling pathways in normal and cancer stem cells.** Many signaling pathways that regulate adult stem cells also play an important part in the development of cancer stem cells (CSC). Three pathways are illustrated that have been particularly well studied and occur in a range of tumor scenarios. Wnt, Shh, and Notch pathways have been shown to contribute to the self-renewal of stem cells or progenitors in a variety of organs, including hematopoietic, epidermal (skin), epithelial (e.g., intestine), neural, and germ cells. When dysregulated, these pathways can contribute to tumorigenesis and the maintenance of CSC, for example, in colon carcinoma and epidermal tumors (Wnt), medulloblastoma and basal cell carcinoma (Shh), and T-cell leukemia (Notch). (Illustration from Reya T, Morrison S, Clarke M et al. (2001) *Nature* 414:105–111 [doi: 10.1038/35102167]. Reprinted by permission from Springer Nature.)

components in these pathways can be affected by mutations affecting the protein sequence, by upstream changes in the signaling pathway that impact them through modifications such as phosphorylation or degradation, or by epigenetic changes, such as the extent of gene promoter CpG methylation, which can alter their level of expression.

Another fundamental property of stem cells that is linked to the biology of tumors, in particular their progression, is that of asymmetric division. Normally, adult stem cells exhibit asymmetric division, whereby one daughter cell retains stem cell characteristics while the other assumes a more differentiated character. In a cancer, asymmetric cell division is generally seen in earlier stages of the disease, contributing to a tumor that is a mixture of immature and differentiated cell types. A switch toward symmetric division, such that neither daughter cell has progressed to a more differentiated phenotype, preludes a more aggressive state.

The first system in which a loss of asymmetric stem cell division was correlated with tumor evolution was in *Drosophila* neural stem cells. Neuroblast division in *Drosophila* is normally asymmetric, and elegant genetic experiments showed that this is under the control of a complex set of interacting proteins that become distributed

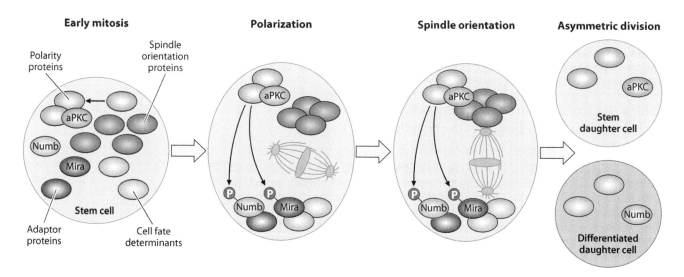

**Figure 18.17 The molecular mechanisms underpinning asymmetric stem cell division.** Asymmetric cell division is characteristic of adult stem cells and cancer stem cells through much of their development, although tumor progression often involves a switch to symmetric division. Mechanisms responsible for asymmetric division have been described in the *Drosophila* neuroblast. In a cell that is about to divide, atypical protein kinase C (aPKC) forms a complex with multiple other components in the apical part of the cell, a position inherited from a previous cell division. Polarization results when aPKC phosphorylates the cell fate determining protein Numb, which in complex with other factors accumulates in the basal part of the cell. The aPKC complex then becomes linked through another protein to the mitotic spindle, establishing its apical-basal orientation. Following cell division, the asymmetric inheritance of Numb acts to inhibit Notch signaling and promotes differentiation of the daughter cell in which it has been concentrated. (From Lytle N, Barber A & Reya T. (2018) *Nat Rev Cancer* 18:669–680 [doi: 10.1038/s41568-018-0056-x]. Reprinted by permission from Springer Nature.)

in two groups toward the apical and basal regions of the cell as it undergoes mitosis. As a result of phosphorylation mediated by one of the apical components (atypical protein kinase C), the protein Numb, which is a determinant of differentiation, becomes localized to the basal part of the cell. The end result is retention of stem cell characteristics by the apically derived daughter and differentiation of the daughter cell originating from the basal region (Figure 18.17). Importantly, genetic deletion of Numb and some of the other components of fate determination in the basal region of the dividing neuroblast resulted in the formation of brain tumors.

Since these findings with *Drosophila* neuroblasts, a number of factors influencing asymmetric versus symmetric stem cell division have now been found to be linked to cancer. As predicted, proteins favoring asymmetric division are found to be expressed at lower levels in some cancers compared to their normal cellular counterparts, while increased expression is seen for some factors linked with symmetric stem cell division. The asymmetry determinant Numb is down regulated in several tumors, including leukemia and breast and colon cancer, whereas the RNA-binding protein Musashi, which promotes stemness through symmetrical division, is seen to be up regulated, for example, in leukemia. Since Musashi inhibits Numb, it effectively mimics the loss of Numb expression.

Other stem cell associated characteristics also come into play in cancers as they evolve, especially under the highly selective pressures of chemotherapy or radiotherapy. These characteristics include drug efflux, heightened DNA damage repair capacity, and the supportive role of the niche microenvironment. Drug transporters of the ABC ("ATP-binding cassette") family are well known to be highly expressed in many stem cells, and, in fact, their presence is used as the basis of the **side population** isolation technique (Chapter 2). Similarly, some stem cells exhibit enhanced DNA repair capacity, for instance, through activation of checkpoint kinases (Chk1 and 2), which lessens the impact of ionizing radiation used in a therapeutic context. Lastly, the niche environment of a tumor, which can be partially elicited by the CSC itself, can provide ideal conditions for survival and self-renewal, whether this is because of proximity to cells that secrete supportive factors or that the relative hypoxia in some solid tumors provides a stimulus for the expression of molecules favoring stemness.

Whether there are features of CSC that specifically distinguish them is not a fully resolved issue. What we have described previously suggests that CSC simply adopt the features of adult stem cells, but in circumstances that are abnormal. A big problem in setting out to find evidence for a CSC gene expression "signature" is that, as we have already pointed out, the precise identification of a CSC depends on the assay being used and might be further compounded by **plasticity** of tumor cell phenotypes. Another difficulty is knowing what a CSC should be compared to, that is, what the equivalent normal cell type is. Even if a CSC signature from a particular type of tumor could be identified, it cannot be assumed that a given signature is useful for identifying CSC in a different tumor type.

## The nature of a given cancer stem cell can be determined by its cellular origin but can also be influenced by its environment and selective pressures

What cells give rise to CSC, and are there any underlying principles applicable to all tumors in terms of how the originating cell dictates the ultimate phenotype and the extent to which the tumor conforms to the CSC model? There are two straightforward possibilities for the origin of CSC; they may be either altered versions of normal stem cells, retaining the capacity for self-renewal but having lost some regulatory constraints, or they may be cells further down the differentiation path that have regained all or some of the self-renewal characteristics of stem cells. On top of this, the precise stem cell of origin within a tissue, the location of a committed cell within a differentiation hierarchy, or the developmental stage of a given tissue all could impact the phenotype of the CSC, the tumor it supports, and the driver mutations that underlie the stem cell properties (Figure 18.18).

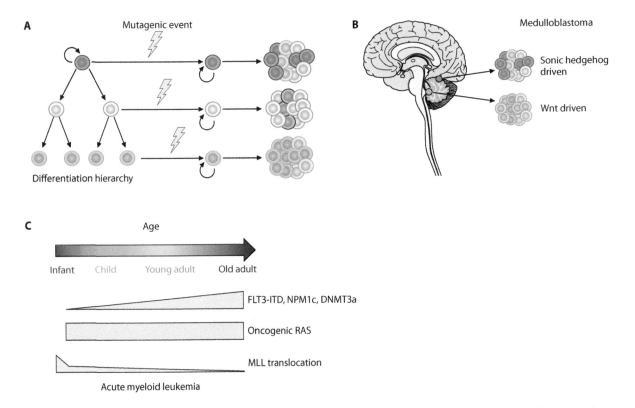

**Figure 18.18 Potential sources of heterogeneity in cancer stem cells.** Whether a tumor is heterogeneous and hierarchically organized depends at least partly on the originating cell. Potential variation in the cell of origin for a given tumor type could be a consequence of its position within a differentiation hierarchy, its precise location within a differentiated tissue, or the stage of development during which it arises. (**A**) Different cell types in a stem cell differentiation hierarchy within a normal tissue may be transformed into cancer cells. The specific properties of the cell of origin influence the types of mutations that are competent to transform and the properties of the resulting cancer. (**B**) The location of the cell of origin within tissues can influence the types of mutations that are competent to transform and the properties of the resulting cancer. In the example illustrated, medulloblastomas can arise either from Sonic Hedgehog pathway activation in granule neuron precursors of the cerebellum (pink) or from Wnt pathway activation in dorsal brain-stem progenitors (green). (**C**) Temporal differences in the cell of origin also influence the types of mutations that are competent to initiate tumor formation and the properties of the resulting cancer, consistent with the observation that the driver mutation spectrum changes with age in patients. This can be illustrated through the types of oncogenically activated proteins acting as drivers of acute myeloid leukemias (AML). Those AML that arise early in life often carry a translocation involving the *MLL* (myeloid/lymphoid leukemia) gene. Activated oncogenic Ras can be associated with AML from almost any stage of life, whereas mutations in *Flt3*, *NPM1*, and *DNMT3a* are characteristic of disease arising later in life. (From Magee J, Piskounova E & Morrison S. (2012) *Cancer Cell* 21: 283–296 [doi: 10.1016/j.ccr.2012.03.003]. Copyright 2012 with permission from Elsevier.)

## The niche plays an important role in cancer stem cell plasticity and tumor evolution

Recent data suggest that CSC rely on a niche and signaling pathways in much the same way as normal stem cells do to control their self-renewal and differentiation. However, the interaction of CSC with a niche is certainly not a simple matter of hijacking one occupied by the equivalent normal stem cell type. Overall, the emerging picture is one of mutual interactions, flexibility, and the possibility of remodeling during tumor evolution. CSC might even be generated by the microenvironment acting on more differentiated tumor cells and inducing them to acquire CSC features, and indeed we have already described such a situation in the case of CRC.

A number of interactions between CSC and their tumor microenvironment have been characterized, as shown in Figure 18.19. An interesting example of the type of interactions that are likely to be crucial and the fact that this can be a two-way process is seen in glioblastoma. Normally neural stem cells and endothelial cells (EC) localize together, and a similar interaction between the tumor equivalents is reflected by a higher incidence of xenograft tumor formation when CSC and EC derived from a primary glioblastoma are injected together. At least one reason behind this cooperation between the CSC and a niche cell is that the EC produces nitric oxide that leads to activation of Notch signaling and consequently self-renewal of the CSC. Intriguingly, glioblastoma stem cells may well support their own niche through differentiation into EC. CRC represents a different situation but also shows how normal cells play a role in the behavior of CSC. As in normal intestinal stem cells, Wnt signaling is involved in maintaining stemness in CSC. It has been suggested that CRC niche cells, possibly myofibroblasts, produce Type I collagen, which stimulates conversion of non-CSC cancer cells into CSC, and hepatocyte growth factor, which then enhances the Wnt signaling cascade in the CSC. As a final example of CSC cross-talk with normal cells in the tumor niche, in skin cancer, endothelial cells in a perivascular niche have been suggested to interact with CSC. VEGF produced by the CSC activates neuropilin-1 by an autocrine and paracrine loop, which then acts as a co-receptor with VEGF receptor, resulting in proliferation. This also increases the area of the vascular niche, thereby enhancing tumor growth and maintenance.

It is not just solid tumors evolving from a normal tissue involving cell–cell interactions that demonstrate CSC dependence on a niche. Malignant hematopoietic cells may modify the normal bone marrow HSC niche to provide optimal conditions for their maintenance and expansion. Investigation of a

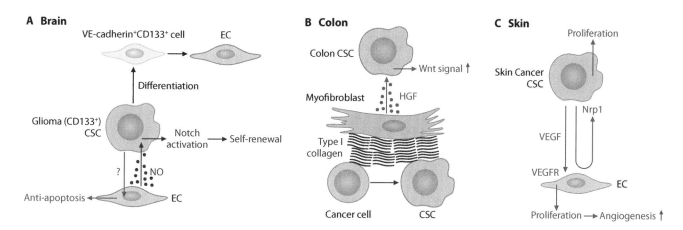

**Figure 18.19 Formation and regulation of the cancer stem cell niche.** Cancer cells, including cancer stem cells (CSC), and normal tissue cells interact with each other to create a supportive niche. The niche in turn modifies the CSC. Shown here are putative examples of such reciprocally dynamic cell-cell interactions in tumors of the brain, colon, and skin. (**A**) In generating a perivascular niche for glioma stem cells, the CD133+ CSC localize near endothelial cells (EC). These EC produce nitric oxide (NO), which induces self-renewal through activated Notch signaling. At the same time, the CSC prevents EC from undergoing apoptosis. Additionally, CSC in glioblastoma can differentiate into EC via CD133+VE-cadherin+ endothelial progenitors. (**B**) In colon cancer, myofibroblasts seem to be the source of the CSC niche. These myofibroblasts may promote the dedifferentiation of cancer cells into CSC, probably by means of the Type I collagen that they produce. Moreover, hepatocyte growth factor (HGF) produced by the myofibroblasts can enhance Wnt signaling in the CSC. (**C**) In skin cancer, the CSC produce vascular endothelial growth factor (VEGF), which stimulates both EC and the CSC themselves. VEGF induces angiogenesis, resulting in expansion of the vascular niche region and induction of self-renewal of the CSC. The cell-surface protein Nrp1 on the CSC is a possible receptor for VEGF. (From Takakura N. (2012) *Cancer Sci* 103: 1177–1181 [doi: 10.1111/j.1349-7006.2012.02270.x]. With permission from Wiley.)

mouse model of BCR-ABL dependent chronic myeloid leukemia (CML) has shown that factors produced by the myeloid CSC (including thrombopoietin and the chemokine CCL3) act on bone marrow mesenchymal stem cells to elicit expansion of osteoblastic lineage cells, which are normally a component of the endosteal HSC niche. However, the tumor cells effect more than just niche expansion, since the osteoblast cells generated preferentially support the CML stem cells over normal HSC. The current view, based on gene expression data, is that this shift in niche function is due to a change in growth factor expression in the osteoblast cells, including a reduction in HSC supportive factors (e.g., stem cell factor; SCF) and an increase in molecules that enhance myeloid cells (e.g., transforming growth factor-β2, TGFβ2).

The degree to which a given CSC type is dependent on a specific niche may relate to the cell type's aggressiveness. Less malignant tumors may depend more on their niche, whereas highly metastatic tumors such as melanoma have perhaps lost this dependency, possibly as the result of continued mutation. Such distinctions could also relate to the differences we have already mentioned that are seen when using the xenotransplantation assay to assess the proportion of cancer stem cells. A final complicating observation is that more differentiated cells within a tumor appear to inhibit tumor growth, suggesting the possibility of feedback signals. One such signal is bone morphogenetic protein, which can limit the cancer stem cell pool by driving differentiation.

## Understanding cancer stem cells and their niche interactions opens up new therapeutic possibilities for cancer

The treatment of cancer relies heavily on three basic approaches, that is, surgical removal, exposure to ionizing radiation, and chemotherapy, which are often used in sequence or combination depending on the type and stage of the tumor. Chemotherapy mainly targets and destroys proliferative and highly active cells, which inevitably also affects normal cells that have similar properties, and although tumors usually disappear as a result of such treatments, they often recur again several months or years later as the result of residual cells that are resistant to the effects of the cytostatic drugs. Tumor heterogeneity, whether between related tumors in different individuals or as a given tumor evolves, is likely to be reflected in the degree of response to a specific anti-cancer treatment. Such differences in responsiveness to cancer therapy are a major factor in the recent rise of the concept of personalized therapy through patient stratification, that is, matching a particular treatment regime to maximize the likelihood of a positive response.

The recent advances in our knowledge of CSC offer some exciting prospects for new approaches to the treatment of cancer. Two principal areas for therapeutic targeting seem to be emerging. First, genes (or gene products) that are associated with CSC might provide valuable targets for drug development. For instance, the CSC component of AML expresses higher levels of activated NF-kB family transcription factors and the receptor for the cytokine interleukin-3 (IL3-Rα, CD123). Much effort is going into targeting these two molecules; NF-kB through the action of plant-derived small molecules known as parthenolides and IL3-Rα using an immunotherapy approach that entails coupling a cytotoxic molecule (diphtheria toxin) to an antibody fragment capable of recognizing the receptor. The second big area of interest around cancer therapy design is focused on the CSC niche. The dominant role played by the tumor microenvironment in determining CSC characteristics is particularly important, especially since this can potentially vary over time, including the niche-induced **dedifferentiation** of non-tumorigenic tumor cells toward stem cells.

Strategies targeting the niche aim to interfere with cell signaling within the niche so that the cancer cells will either start to cycle and thus become more susceptible to conventional chemotherapy or move out of the niche, where, again, they might begin to cycle and be vulnerable to chemotherapy. An encouraging early result in this direction again came from studies on AML by John Dick using an activating monoclonal antibody directed to the adhesion molecule CD44. When administered to xenograft models of human AML, the antibody block of CD44 markedly reduced leukemic repopulation, and the absence of leukemia in serially transplanted mice demonstrated that the AML cancer stem cells had been directly targeted. Mechanisms underlying this eradication included interference with transport to stem cell–supportive microenvironmental niches and alteration of AML CSC fate.

## 18.3  SUMMARY

In this chapter, we have looked at two of the greatest challenges to health improvement, namely aging and cancer, taking the perspective of the involvement of adult stem cells. Although not restricted to aging, cancer is a significant component of age-related morbidity and mortality, and both aging and cancer have a significant dependence on stem cells. In the broadest sense, the effect of stem cells in aging is one of loss of capacity or capability, which can involve changes in the stem cells, their niche, or both. In contrast, cancer has a stem cell basis to its origin and evolution and generally involves a modification or misappropriation of stem cells or stem cell characteristics. As for the effects of aging on some adult stem cells operating through their niche, so CSC are much influenced by their environment, often evolving to a more aggressive state through reciprocal cell-cell interactions or even being induced from non-stem cells present within the heterogeneous tumor cell population. Our growing understanding of the stem cell features of aging and cancer is opening up considerable opportunities for novel therapies aimed either at the stem cells or their niche in order to alleviate the effects of aging or to eradicate cancer.

---

## KEY POINTS

- Aging and cancer both have a close link to the function of cells with stem cell properties.

- Stem cell depletion or loss of function can lead to tissue deterioration and aging, whereas excess proliferation of cells with stem cell properties can be the basis of malignant tumor formation.

- Stem cells are subject to both chronological and replicative aging as a consequence of a complex relationship between cell intrinsic and extrinsic influences.

- DNA damage plays a major role in both cell and tissue aging and in the initiation of cancer.

- Stem cells exhibit changes in their epigenome and transcriptome as they age, with a reduction in Sirtuin and Polycomb group proteins and consequent accumulation of active chromatin leading to expression of pro-aging genes. The conserved FOXO family of transcription factors are likely to be important determinants of stem cell capability during aging.

- Shortening of chromosome ends, or telomeres, may be associated with stem cell age. Mechanisms involving telomerase counteract chromosome shortening.

- Reactive oxygen species (ROS) play an important role in stem cell aging and cancer development, with increasing ROS levels as cells age leading to an increase in oxidative damage.

- The niche is a crucial regulator of stem cell function and can mediate changes in aging and cancer formation.

- Cancers are inherently heterogeneous and evolve as a result of both genetic and environmental influences.

- Defining what is a cancer stem cell can depend on the assay used.

- Cancer stem cells have no known uniquely defining phenotype.

- Tumor initiation can occur in the stem cell population or in a mature cell that regains stem cell characteristics.

- Cancer stem cells can depend on a niche like other stem cells, although this may be distinct from the normal stem cell counterpart and may be formed through a reciprocal interaction between the cancer stem cell and its environment.

- Discovery of cancer stem cell-specific features and the essential requirements of their niche is providing new avenues for anti-cancer therapy.

---

## FURTHER READING

### General

McDonald, R.B. (2014) *Biology of Aging.* Garland Science, New York & London.

Weinberg, R.A. (2014) *The Biology of Cancer,* 2nd ed. Garland Science, New York & London.

### The role of stem cells in aging

Akunuru, S. and Geiger, H. (2016) Aging, clonality, and rejuvenation of hematopoietic stem cells. *Trends Mol Med,* 22; 701–712.

Brunet, A. and Rando, T.A. (2017) Interaction between epigenetic and metabolism in aging stem cells. *Curr Opin Cell Biol,* 45: 1–7

Cakouros, D. and Gronthos, S. (2019) Epigenetic regulation of bone marrow stem cell aging: Revealing epigenetic signatures associated with hematopoietic and mesenchymal stem cell aging. *Aging and Disease,* 10: 174–189.

Geiger, H., de Haan, G. and Florian, M.C. (2013) The ageing haematopoietic stem cell compartment. *Nat Rev Immunol,* 13: 376–389.

Jung, Y. and Brack, A.S. (2014) Cellular mechanisms of somatic stem cell aging. *Curr Top Dev Biol*, 107: 405–438.

Liu, L. and Rando, T.A. (2011) Manifestations and mechanisms of stem cell aging. *J Cell Biol*, 193: 257–266.

Schultz, M.B. and Sinclair, D.A. (2016) When stem cells grow old: Phenotypes and mechanisms of stem cell aging. *Development* 143; 3–14.

Zhang, H., Menzies, K.J. and Auwerx, J. (2018) The role of mitochondria in stem cell fate and aging. *Development*, 145.

## The role of stem cells in cancer

Al-Hajj, M., Wicha, M.S., Benito-Hernandez, A., Morrison, S.J. and Clarke, M.F. (2003) Prospective identification of tumorigenic breast cancer cells. *Proc Natl Acad Sci USA*, 100: 3983–3988.

Batlle, E. and Clevers, H. (2017) Cancer stem cells revisited. *Nat Med.*, 23: 1124–1134.

Bonnet, D. and Dick, J. E. (1997) Human acute myeloid leukemia is organized as a hierarchy that originates from a primitive hematopoietic cell. *Nature Medicine*. 3: 730–737.

Huntly, B.J.P. and Gilliland, D.G. (2005) Leukaemia stem cells and the evolution of cancer-stem-cell research. *Nat Rev Cancer*, 5: 311–321.

Lytle, N.K., Barber, A.G. and Reya, T. (2018) Stem cell fate in cancer growth, progression and therapy resistance. *Nat Rev Cancer*, 18: 669–680.

Magee, J.A., Piskounova, E. and Morrison, S.J. (2012) Cancer stem cells: Impact, heterogeneity, and uncertainty. *Cancer Cell*, 21: 283–296.

Shimokawa, M., Ohta, Y., Nishikori, S., Matano, M., Takano, A., Fujii, M., Date, S. et al. (2017) Visualization and targeting of LGR5+ human colon cancer stem cells. *Nature*, 545: 187–192.

Zhao, Y., Dong, Q., Li, J., Zhang, K., Qin, J., Zhao, J., Sun, Q. et al. (2018) Targeting cancer stem cells and their niche: Perspectives for future therapeutic targets and strategies. *Semin Cancer Biol.*, 53: 139–155.

# 19

# Stem cell applications in domesticated and wild animals and in horticulture

Although a major driving force for research on stem cells has been the belief that they represent the possible basis for a variety of novel treatments for human disease, the potential for the application of stem cell science to domesticated and wild animals and economically important plant crops is also enormous and perhaps ultimately will have a bigger impact on people's lives, especially in underdeveloped parts of the world.

The rapidly increasing global population, from 6 billion in the year 2000 to a predicted 9 billion in 2050, is imposing greater demands on our existing resources and ability to generate sufficient food, while the consequent environmental influences impact mankind's fellow species, putting many in danger of extinction. There is considerable promise for the application of stem cell science in the fields of veterinary and human medicine, agriculture production, and species conservation, and in this chapter, we will illustrate the current status with a few of the examples where progress has been made and speculate on what the future may hold. Solutions to the challenge of feeding the world's population and improving its health will also rely heavily on our ability better to utilize plants. We will consider how plants might be altered through their stem cells to improve crop productivity and how they can be harnessed in the production of specific drugs for human therapy.

## 19.1 ANIMAL HEALTH, AGRICULTURAL PRODUCTIVITY, AND BIOMEDICAL ADVANCEMENT INVOLVING ANIMALS

Many of the potential applications of stem cells to domesticated companion animals (e.g., horses, dogs, and cats) are self evident, being very much the parallel of what is being adopted in humans. Less obvious perhaps are the ways in which stem cells are being utilized as a means to improve desirable characteristics of different livestock species and non-mammalian food sources, including milk and meat production. An extreme solution to the latter, which has been proposed and partially demonstrated, is to produce a meat substitute directly from stem cells. Stem cell technologies also offer the prospect of the creation of animal models tailored to particular human diseases, to be used, for example, in the screening of new drug therapies prior to first-in-man trials or in the production of drugs, or of producing replacement human organs in animals for transplantation purposes.

### Mesenchymal stem cells have been used successfully to treat tendon injuries in racehorses

Veterinary medicine has seen considerable interest in the prospect of utilizing stem cells in new therapeutic strategies. The most relevant species, our companion animals, suffer many of the same diseases as humans that are thought to be potentially tractable through stem cell-based approaches. Indeed, human and veterinary medicine are often closely integrated, and animals such as horses and dogs are effectively serving as pre-clinical models for human cell therapies. Certainly, animals such as rabbits, dogs, cats, and sheep have often played a key role in advancing our understanding of specific aspects of stem cell biology.

A somewhat relaxed regulatory environment around the application of stem cells in the veterinary context has encouraged a more rapid translation into clinical veterinary practice and widespread commercialization. A significant degree of caution is needed, though, in interpreting claims of benefit, since almost all such studies have failed to adhere to the principles of fully controlled clinical trials.

Although several **adult stem cell** types are being considered in the treatment of injuries and diseases in domesticated animals, the one that has received the greatest attention to date, and for which the results in various animal species have direct relevance to potential application in humans, is the **mesenchymal stem cell** (MSC; Chapters 10 and 16). As for current clinical trials on MSC in humans, the most common sources are bone marrow, fat tissue, and umbilical cords, but it remains unclear which source is most appropriate for individual therapeutics.

To date, the headline-grabbing application of MSC in veterinary medicine has been in the treatment of musculoskeletal injuries in horses. A particularly common injury is tendon damage (tendinopathy) that, although treatable, can often end a horse's career. This is because fibrous scar tissue with inferior biomechanical properties forms at the tendon-bone interface, lessening resilience to the normal level of stress (Box 19.1). Despite the lack of controlled clinical trials, MSC appear to have a beneficial effect on equine tendinopathy, the return to pre-injury function and reduction in re-injury rates being more pronounced than can be achieved with conventional therapy. It appears that in the presence of MSC, the regenerating tendon

## BOX 19.1 BACKGROUND: TENDONOPATHIES IN RACEHORSES

The equine industry is vast; in the United States alone, there are over 9 million horses that have a value within the economy of around $40 billion. Sports such as racing put tremendous strain on a horse's legs, musculoskeletal injuries accounting for 97% of horse fatalities linked to racing. Nearly half of all injuries in racehorses are to tendons and ligaments. Tendons are highly elastic, dynamic structures that connect muscle to bone and aid in movement, shock absorption, and support to the limb. The most commonly affected tendon is the superficial digital flexor tendon, which resides just under

the skin running down the back of the cannon bone and into the pastern (Figure 1). Musculoskeletal injuries such as those affecting tendons usually show poor response to conventional treatments that involve rest and controlled exercise and often leave the animal prone to re-injury. Normal tendons have great tensile strength due to the high proportion of Type I collagen (>90%); however, scar tissue formation following injury results in the thinner Type III collagen being formed in larger amounts, resulting in a weaker tendon that is prone to re-injury.

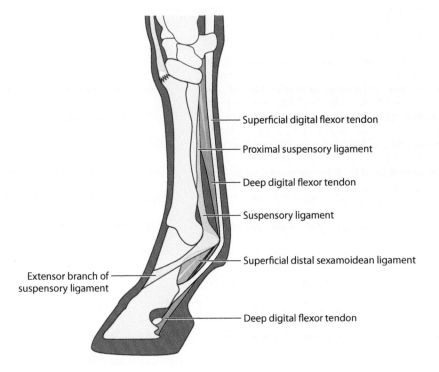

Box 19.1 Figure 1 **The lower limb tendons of the horse.** The most common injury in race horses involves the tendons of the legs, often resulting in an end to the horse's racing career. The schematic represents a longitudinal section through the lower limb of the horse showing the major tendons and their connections to the bones. The most frequently damaged tendon, the superficial digital flexor tendon (SDFT), can be seen in the upper part of the diagram.

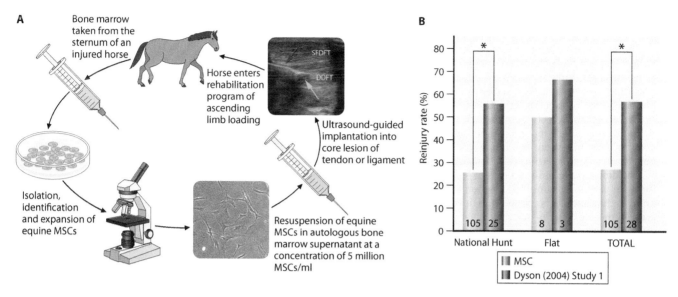

**Figure 19.1 Treatment of tendon injury with mesenchymal stem cells.** Conventional therapies for horses with tendon injuries often result in the formation of scar tissue that lessens the degree of further stress that the limb can undergo before re-injury occurs. An approach to the treatment of superficial digital flexor tendon injury using mesenchymal stem cells (MSC) is illustrated that has revolutionized the treatment. (**A**) Cells are harvested from the bone marrow in the sternum and the MSC are isolated, cultured, and then suspended in a protective medium ready for injection into the central core of the lesion under guidance by ultrasonography. SFDT: superficial digital flexor tendon; DDFT: Deep digital flexor tendon. (**B**) Re-injury rates (blue) in racehorses that have received MSC therapy are shown compared to data from other treatment strategies (red). * denotes a statistically significant difference at the $P < 0.05$ level. (A: From Smith R. (2008) *Disabil Rehabil* 30:1752–1758 [doi:10.1080/09638280701788241]. Reprinted by permission of Taylor & Francis Ltd, http://www.tandfonline.com. Photos: From https://sites.tufts.edu/hlanews/2013/01/topics-ofdiscussion-5/. Used with permission from Dr. Andy Hoffman and Dr. Jose Garcia-Lopez, and From Goodrich L, Werpy N, Barrett M et al. (2012) *AAEP Proceedings* 58:49–53. Used with permission from American Association of Equine Practitioners. B: From Godwin E, Young N, Dudhia J et al. (2012) *Equine Vet J* 44:25–32 [doi: 10.1111/j.2042-3306.2011.00363.x]. © 2011 EVJ Ltd. Adapted with permission from Wiley.)

heals via the formation of fibrocartilage rather than scar tissue. One month following the injection of the MSC into the site of the damaged tendon, apparent increases in the tendon density have been observed, and by 6 months, the tendon has been seen fully to regenerate, resulting in a reduced re-injury rate (Figure 19.1). The regenerated tendon is accompanied by a quick reduction in lesion size, improved collagen fiber orientation, and expression of higher levels of Type I collagen that is associated with uninjured tendons. How the MSC mediate their effect is unclear, although it is likely that key is the secretion of various trophic factors and modulation of inflammatory responses (Chapter 16).

The use of MSC in veterinary medicine extends beyond the repair of tendons in horses, having been used to treat related musculoskeletal injuries in other domesticated animals such as dogs, cats, and pigs. MSC may offer the prospect of novel stem cell-based therapies for the improvement of bone and wound healing. There are examples where MSC have been used in an attempt to treat non-union fractures and bone replacement in the legs and jaw of dogs, and MSC derived from the umbilical cord have been tested in goats on skin wounds.

Through their combined ability to **differentiate** into mesenchymal lineages, including chondrocytes (Chapter 10), and to moderate inflammatory responses (Chapter 16), MSC also have potential in the treatment of joint diseases, including ligament rupture and arthritis. These conditions are often encountered in some companion species, either as a genetic predisposition, as seen in dogs, or as a consequence of wear and tear.

There are even a few sporadic instances of MSC being trialed in more exotic circumstances beyond domesticated animals, ranging from treatment of soft tissue injuries in the bottlenose dolphin to the repair of a lumbar fracture in a Bengal tiger!

## Alterations to the genome to improve livestock characteristics and enhance agricultural productivity

Livestock make up a heterogeneous group of domesticated mammals essential for agricultural commodities such as food, fiber, and labor. The hoofed animals (ungulates) constitute the most important group, including those that are odd-toed, such as the horse, and the even-toed sheep, pigs, and cattle. Classically, since the dawn of agriculture in places like the Middle East, the human race has been enhancing desirable characteristics in animals through selective breeding. Without these interventions, which have effectively subverted or accelerated Darwinian evolution, we would not have dairy cattle that can produce almost 2000 gallons of milk in a year or racehorses that can sustain speeds of near 40 miles per hour for several minutes. Selecting for specific features or traits, of course, continues as the demands to produce more food and improve performance increase, but mankind is now far more impatient and is eager to find ways to speed up the process and to be able to introduce new traits that cannot be achieved through selective breeding alone.

The development in the 1980s of techniques for the introduction of foreign DNA into the genome of animals opened up new possibilities for the modification of animal characteristics without the need for selective breeding. The creation of such **transgenes** was first restricted to laboratory mammals, but very soon began to be applied to livestock species, with the initial demonstration in sheep and pigs by Ralph Brinster and colleagues at the University of Pennsylvania in 1985.

The essential method for introduction of transgenes involves the injection of the foreign DNA, which usually consists of a gene encoding a protein that will affect the desired trait driven by regulatory elements that ensure expression in the right tissues (Chapter 2), into the pronuclei of a fertilized oocyte that is then implanted into the uterus of a foster mother. The livestock animal traits that have been the focus of such modifications include growth rate and body composition and disease resistance (Table 19.1).

Although transgenic technology is tried and tested, it is inefficient and time consuming, and because transgene integration into the genome is random, can lead to variable expression and unwanted mutation of the genome. The discovery of **pluripotent stem cells**, first in mice and later other mammalian and non-mammalian species, and the use of these cells in bringing about very precise changes in gene structure through homologous recombination (Chapter 3), opened up the prospect of more sophisticated and efficient alterations to livestock and other commercially relevant species. Despite the best efforts to derive **embryonic stem (ES) cells** from livestock ungulates, all attempts over the last 20 years or so have failed. This failure could simply be a result in differences in the method of isolating the embryos or the inability to prevent spontaneous differentiation of the ES cells in culture. The logistics of retrieving embryos from such large animals further compound the problem by making it difficult accurately to capture embryo hatching from the zona pellucida. Fundamental differences in early embryonic development in ungulates compared to the mouse might also play a part. Hence, although the chronological sequence of embryogenesis is the same, the timing of fertilization, cleavage, compaction, and blastula formation differs, taking 1 week in ungulates as opposed to 4 days in

## TABLE 19.1  EXAMPLES OF TRANSGENIC MODIFICATIONS IN LIVESTOCK ANIMALS

| Feature improved | Specific change introduced | Gene introduced or modified | Species |
|---|---|---|---|
| Enhanced nutrition | Increased levels of omega-3 fatty acids | *C. elegans* gene encoding an n-3 fatty acid desaturase | Pig |
| Reduced environmental impact | Better utilization of dietary phosphate | Salivary phytase | Pig |
| Enhancing milk | Salivary phytase | Lactoferrin, insulin-like growth factor (IGF-1), anti-thrombin, thrombopoietin | Cattle, sheep, goat |
| Enhancing growth rate and carcass composition | Enhanced growth and feed efficiency | Growth hormone | Pig |
| Improved disease resistance | Secretion of anti-bacterial peptide in milk | Lysostaphin | Cattle |
| Enhanced reproductive performance | Litter size | Estrogen receptor or FecB gene | Pig, sheep |
| Improving hair and fiber | Increased wool yield | Insulin-like growth factor (IGF-1) | Sheep |

*Note:* Since the 1980s, the production of transgenic animals has enhanced the agricultural industry. The table lists some examples of the ways in which specific transgene expression in livestock animals can be used to modify their phenotype.

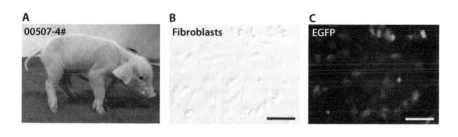

Figure 19.2 **Pig induced pluripotent stem cell chimeras.** Induced pluripotent stem (iPS) cells generated from pigs have been generated under conditions that make them suitable for SCNT, resulting in the birth of completely normal piglets. The figure shows one example of the use of such pig iPS cells. (**A**) Live piglet produced by SCNT cloning using EGFP-expressing pig iPS cells. (**B**) Ear fibroblasts derived from the piglet and established in culture. (**C**) The same fibroblasts viewed under fluorescent light to reveal expression of EGFP encoded by the donor-derived cells. Scale bars are 100 μm. (From Fan N, Chen J, Shang Z et al. (2013) *Cell Res* 23:162–166 [doi: 10.1038/cr.2012.176]. Reprinted by permission from Springer Nature.)

the mouse. Furthermore, some ungulate embryos do not immediately attach to the uterine lining like those in rodents, instead drawing their nutrition from uterine secretions, which in the case of the pig involves extension at day 5 of gestation of the 150 μm spherical **blastocyst** into a thin filamentous embryo, which can grow up to a meter in length by day 15!

The failure to obtain livestock ES cells may have been overcome to a large extent by the more successful application of the technique for the generation of **induced pluripotent stem (iPS) cells**, which again was first worked out in mouse cells (Chapters 3 and 5). In 2009, the first iPS cells were derived from pig fibroblasts using either the human or murine **reprogramming** genes. The resulting iPS cell colonies morphologically resembled human ES cells and iPS cells, rather than the murine counterparts, and like human iPS cells were dependent on fibroblast growth factor-2 and activin/nodal signaling rather than leukemia inhibitory factor (LIF) to maintain their pluripotent state (Chapter 5). Pig iPS cells therefore represent **epiblast stem cells** (EpiSC) rather than the naïve ES cell type. This has a potential practical consequence in that, similar to human ES cells and iPS cells, pig iPS cells might not be suitable for the generation of **chimeric** embryos. However, some researchers have employed a LIF-based medium containing protein kinase inhibitors to alter signaling pathways, making it possible to select for pig iPS cell lines that resemble naïve ES cells. Using such iPS cells, there has been demonstration that they can be incorporated into embryos and subsequently contribute to part of chimeric tissues in live-born piglets.

Despite the ability to generate iPS cells from livestock animals including pig, sheep, goat, and cattle, the promise for their use in biotechnology applications has yet to be realized. In order for these iPS cell lines to be of use, it needs to be shown that they have a significant advantage in transgenic technologies over and above that of the somatic cells already employed. Progress is being made toward this latter goal, for instance, by showing that pig iPS cells can be generated under conditions that make them ideal donors for **somatic cell nuclear transfer** (SCNT) **cloning** (Chapter 5) giving rise to perfectly normal piglets (Figure 19.2). The advantage of this scenario is that the iPS cell can be used as the vehicle in which to make genetic modifications to introduce favorable features (e.g., muscle quality) or delete undesirable characteristics (e.g., susceptibility to viral infection).

## Mammary stem cells might be used to enhance milk production in cows to meet increasing demand

Since the desirable features of livestock animals generally involve the quantity and quality of their tissues (e.g., muscle or hair) or production of milk, it is reasonable to assume that the roles played by adult stem cells at the top of the relevant differentiation hierarchies might provide an opportunity for improvements through increases in the production of cells. This is certainly being given serious consideration in regard to the production of milk by cows, although this is currently under investigation and for now remains a theoretical prospect.

Since there is worldwide increased demand for milk production from cattle (Box 19.2), enhancing the milk yield from lactating cattle is an exciting prospect. In Chapter 15, we describe **mammary stem cells** (MaSC) and how they are responsible for the net growth, renewal, and turnover of mammary epithelial cells (MEC). Since milk yield is a function of the number and secretory activity of MEC, approaches to regulate their number and maintenance through manipulation of MaSC represent a potentially good strategy to increase the efficiency of milk production.

MaSC are responsible for the huge increase in mammary gland mass in cattle at the onset of lactation from just a few 100 mg at birth to 4–6 kg at lactation. Manipulation of MaSC and downstream progenitors during first gestation may allow cows to enter their first lactation with a greater number of fully developed cells. The decline in

The dairy industry forms a mainstay of the global food industry, not just for the production of milk, butter, and cheese but also for use in infant formula milk as a substitute for breastfeeding. The demand for the latter has increased significantly in recent years, particularly in China. The dairy cow has a typical milk production cycle that produces approximately 3000 kg of milk per year, with milk increasing just after birth, maintained for variable times, and then decreasing throughout the milking period. When a lactating cow becomes pregnant, lactation terminates at around day

40 in preparation for the successive lactation, resulting in a "dry period" where no milk is produced. The dry period typically lasts between 50 and 60 days, a break that has huge financial implications for the farmer. Some lactating cows maintain peak milk production for longer periods of time, and these cows are classed as being "persistent." Increased persistence results in higher milk yield, with benefits to the health and welfare of the cow, as they are exposed to less stress caused by consecutive pregnancies.

milk production during lactation appears to be due to a steady loss of MEC through apoptosis, so, again, increasing the activity of MaSC may increase the duration of high-level milk production. During the dry period, there is extensive turnover of secretory cells, so that an ability to promote regeneration of MEC from MaSC should shorten or eliminate the time when cows are not lactating.

Clinical mastitis in dairy cattle occurs quite frequently by the invasion of bacteria through the teat and can have a severe impact on udder tissue, significantly reducing animal value and milk production. Mammary tissue becomes damaged by the release of bacterial toxins and the breakdown of the epithelial lining of the alveoli, resulting in atrophy and thus a reduction in the number and activity of MEC that consequently contributes to decreased milk production (**Figure 19.3**). It is possible that MaSC may be adopted therapeutically to correct the structural and cytological defects in the bovine udder caused by mastitis, leading to restoration of milk production.

So, what progress has been made toward this clear ambition to increase MaSC activity and MEC production in dairy cattle to enhance milk yields and as a regenerative approach following disease in the udder? Most MaSC research has been carried out in mice and humans, but bovine mammary tissue is similar. Akin to MaSC isolated from human and mouse mammary tissue (Chapter 15), surface markers such

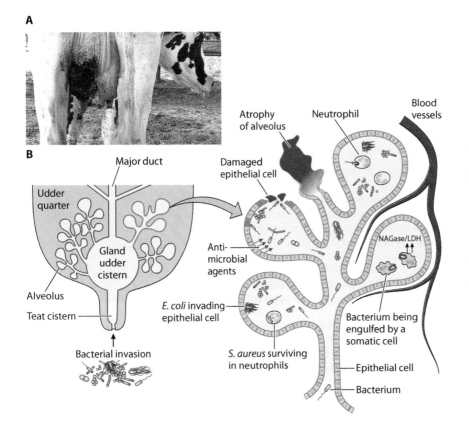

Figure 19.3 **Clinical mastitis in dairy cattle.** Mastitis is a major endemic disease of dairy cattle and can cause milk production to plummet, with serious financial implications. (**A**) Mastitis is caused by a bacterial infection of the mammary gland and can cause serious and lasting damage to a cow's udder. (**B**) Bacteria enter the mammary gland through the teat and colonize the alveoli, causing inflammation. After becoming infected, cellular defense mechanisms that eradicate the bacteria lead to the release of bacterial toxins N-acetyl-β-d-glucosaminidase (NAGase) and lactate dehydrogenase (LDH), with consequent damage to the epithelial cell lining, eventually resulting in atrophy of the damaged alveolus. (A: "Dairy cow with gangrenous mastitis, rear quarter" by Mohammad Golkar – Own work. Licensed under the Creative Commons Attribution Share Alike 4.0 International license via Wikimedia (https://creativecommons.org/licenses/by/4.0/). B, C: From Viguier C, Arora S, Gilmartin N et al. (2009) *Trends Biotechnol* 27:486–93 [doi:10.1016/j.tibtech.2009.05.004]. Copyright 2009 with permission from Elsevier.)

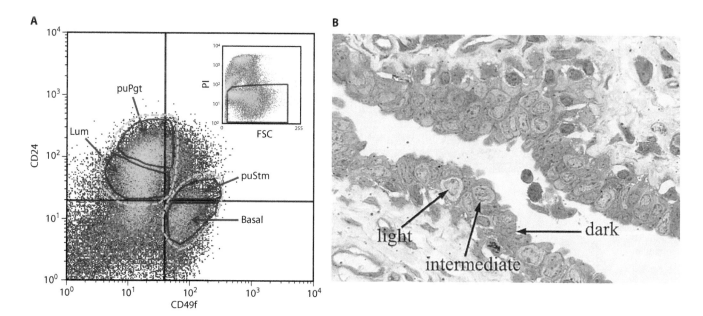

**Figure 19.4 Identification of bovine mammary stem cells.** Bovine mammary stem cells (MaSC) can be isolated from the mammary gland in a similar manner to other species. (**A**) Four populations of epithelial cells can be identified in the bovine mammary gland based on their expression of CD24 and CD49f. The diagram shows a flow cytometry plot of the intensity of immunofluorescent staining for CD24 (y-axis) and CD49f (x-axis). Populations enriched for putative stem cells (puStm) and their progenitors (puPgt), as well as their complementary basal and luminal (Lum), differentiated derivatives are encircled by solid red lines. (**B**) Additionally, MaSC can be visualized in sections of mammary gland tissue by histological staining. The degree of staining correlates with the stem cell potential of the MaSC. In this section of a bovine mammary gland, putative MaSC can be identified as "pale" or "light" cells in the epithelial cell population. (A: From Rauner G & Barash I. (2012) *PLOS ONE* 7:e30113 [doi: 10.1371/journal.pone.0030113]. Published under the terms of the Creative Commons Attribution License (http://creativecommons.org/licenses/). B: From Ellis S & Capuco A. (2002) *Tissue Cell* 34:155–63 [doi: 10.1016/s0040-8166(02)00025-3]. Copyright 2002 with permission from Elsevier.)

as **cluster of differentiation** 24 (CD24) and CD49f can be used to isolate the equivalent bovine MaSC. Similarly, bovine MaSC are also identified as "pale" or "light" cells that are present throughout all stages of mammary gland development and capable of giving rise to both luminal and basal myoepithelial cell types (Figure 19.4). The light cells make up 10% of the epithelial cells and are highly proliferative, whereas intermediate and darker stained cells represent a greater proportion of the total but have a lower proliferative capacity. How might MaSC function be altered? Well, there is no answer at the moment, but an indication of what might be possible comes from the use of the hormone bovine somatotrophin (bST). Administration of bST to dairy cattle can reduce the dry period from 60 days to 30 days without altering the volume of milk produced in the consecutive lactation cycle. The hormone is thought to act by increasing the numbers of MEC produced from MaSC by increasing their **self-renewal** potential, and in so doing may be leading toward a positive gain in milk production.

## Laboratory grown meat could offer a solution to increasing global meat demand and reduce the impact on the environment

Farming of livestock for meat production utilizes 70% of agricultural land, and it will not be possible to meet increasing demand for meat (predicted to increase by 73% by 2050) through current farming techniques, especially because of the limited availability of land. Enhancing meat production through genetic modification as described previously is a potential solution, and this can be via the route of either classical transgenesis or through genetic modification in pluripotent stem cells that are capable of becoming part of the germ line in an embryonic chimera. Presently, the availability of such pluripotent cells (iPS cells) for livestock species is restricted to pigs.

An alternative approach being considered to improve the supply of meat is to do away with the need for animals by directly culturing muscle from stem cells. The concept of cultured meat is not new, perhaps rather surprisingly having been first proposed in 1931 by the British Prime Minister Winston Churchill (Figure 19.5). In an essay entitled "Fifty Years Hence," Churchill predicted a future in which meat would

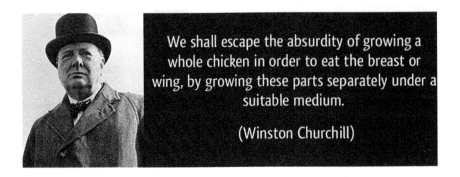

**Figure 19.5 Winston Churchill: An unexpected stem cell visionary.** In 1931, Winston Churchill proposed that one day we would be able to culture meat without the need for large scale animal farming. (Adapted from http://izquotes.com/quote/219031. Used with permission from izquotes.com.)

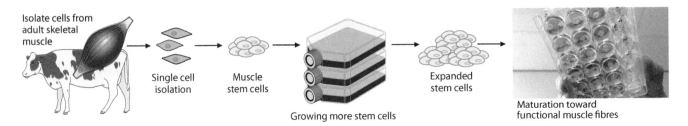

**Figure 19.6 Culture of bovine muscle stem cells.** The process required for the production of muscle in the laboratory is relatively straightforward. The schematic diagram summarizes the process required in order to produce muscle from bovine skeletal muscle stem cells. The stem cells are isolated from muscle tissue, expanded in culture, and then differentiated into muscle cells, which can be further matured by placing under mechanical strain. The photograph shows one way in which such strain can be applied through attachment of each end of a bundle of immature muscle cells. (Photo credit: Seth Bannon of Fifty Years. Used with permission.)

be cultured on an industrial scale and would ultimately be indistinguishable from the natural products. More than 80 years later, Churchill's prediction is being put into reality.

The technological concept behind the production of laboratory-grown meat is quite simple. A small biopsy taken from the muscle of a living animal such as a cow is enzymatically and mechanically disrupted to liberate the resident **muscle stem cells**, also known as **satellite cells** (Chapter 10). These satellite cells can be grown using standard cell culture techniques, the inclusion of high serum concentrations allowing the cells to divide up to 50 times over a 7- to 8-week period. Once sufficient numbers of cells are obtained, they are separated into batches and set up in culture under tension strain, which can be elicited through attachment to the culture dish, providing a stimulus for their maturation and the production of muscle proteins (Figure 19.6). After 3 weeks, the muscle is harvested. In this way, Dutch researcher Mark Post (Figure 19.7) made news headlines in August 2013 by producing the first laboratory-grown burger that was cooked and eaten on live television in London (Figure 19.8). In order to produce an 85 g burger, at least 10,000 muscle strips were required, at a cost of about $300,000! The food critics who tasted it commented: "The mouth feel is like meat. I miss the fat, there's a leanness to it, but the general bite feels like a hamburger" and "What was consistently different was flavor"! It will be important to work on the fat content in order to maximize flavor but also to improve the protein composition and the ability to create larger muscle fibers for thicker "cuts of meat." It is being explored how MSC might be utilized together with satellite cells to produce both muscle and fat in combination. The color of the stem cell-derived meat is also an issue. As produced by Mark Post, the meat was white due to the lack of myoglobin and had to be colored with beetroot juice to make it look like the real

**Figure 19.7 Mark Post.** The Dutch researcher made headline news by producing the first laboratory grown beef burger in 2013 ("Mark Post at the SingularityU The Netherlands Summit 2016" by Sebastian ter Burg. Licensed under the Creative Commons Attribution 2.0 Generic License via Wikimedia. (https://creativecommons.org/licenses/by/2.0/).)

Figure 19.8 **Television debut of the stem cell burger.** On 5 August 2013, the first burger produced from stem cell derived muscle was cooked on live television on the BBC. Presenter Nina Hossain (left) was joined by Mark Post of Maastricht University; Hanni Ruetzler, a food researcher from the Future Food Studio; and food author Josh Schonwald, who sampled the stem cell derived burger. (Top: "Hanni Rützler tastes world's first cultured hamburger" by World Economic Forum. (https://creativecommons.org/licenses/by/3.0/deed.en). Bottom: "First cultured hamburger unbaked" by World Economic Forum. Both images licensed under Creative Commons Attribution 3.0 Unported license. (https://creativecommons.org/licenses/by/3.0/deed.en).)

thing. So, for sure, this served as a proof of concept, but there is a long way to go before laboratory grown meat hits the supermarket shelves. Despite the existing limitations, numerous companies have jumped onto the idea of meat production from cells, with several attempting to produce laboratory-grown beef, chicken, duck, and fish. Although these efforts have brought the costs down, these are still prohibitively high from a commercial perspective, but the hope is to bring affordable, environmentally friendly, animal friendly meat to market in the next 3–4 years.

## Genetic manipulation of stem cells from non-mammalian animals that are important as food sources might enhance particular qualities

As the global demand for proteins increases, especially in developing countries experiencing population growth and rising consumer incomes, there will be an increasing demand for meat from sources other than the relatively expensive, high maintenance livestock species. Forecasts suggest that poultry meat uptake per person will increase by some 9% between 2013 and 2022, compared with gains of 3% to 4% for beef and pig meat.

The production of transgenic lines, as described previously with respect to livestock animals, has long been seen as a potential way to introduce desirable traits into chickens. Different methods have been employed to generate transgenic chickens, including microinjection, use of retroviruses, and transfection of pluripotent ES cells and primordial germ cells, which have been relatively easy to produce from chicken embryos. The use of ES cells in this way has already enabled the generation of novel chicken strains (Figure 19.9). An interesting example relates to the ovomucoid protein, which is present in eggs and can cause a devastating allergy because it is very heat stable and therefore resistant to denaturation during food production. Genetic ablation of the ovomucoid gene in chicken ES cells means that it will be possible to reduce the risk of this particular allergic response. Unlike some of the livestock species, advances in the creation of iPS cells from birds, including chickens, are somewhat behind, although chimeric embryos have been produced from chicken cells reprogrammed using mammalian pluripotency factors (Chapter 5).

Fish represent another readily available food source, particularly as species previously harvested only through fishing are adapted to production in fish farms. ES-like cells have been generated from a number of commercially relevant fish species such as the Atlantic cod and turbot, but at the time of writing, iPS cell technology had only been applied to the experimental model zebrafish and, like bird iPS cells, the resulting reprogrammed cells have so far only successfully contributed to chimeras in embryos (Chapter 5). Transgenic methods have already been used in some commercially important fish species as a way to enhance the muscle content, for example, in trout through the inactivation of myostatin expression or over expression of follistatin, and it is easy to envisage that these and more sophisticated genetic alterations will soon be made through the use of pluripotent stem cells.

Figure 19.9 **Use of embryonic stem cells to generate genetically modified chickens.** Chicken pluripotent embryonic stem (ES) cells can be incorporated into chimeric animals. (**A**) ES cells are injected through a fine needle into the embryonic blastodisc lying on top of the yolk. (**B**) The yolk and embryo are returned to the shell and incubated. (**C**) Chimeric chicks can be recognized by their mixed down colors. In the picture shown, the black down is derived from cells originating from the ES cells, while the white down is from the fertilized egg. (A and C: From Kagami H. (2016) *Anim Sci J* 87:1065–1075 [doi: 10.1111/asj.12620]. With permission from Wiley.)

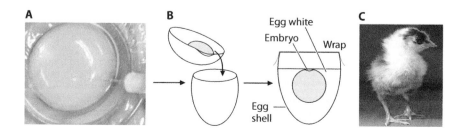

## The prospect of using animals to produce human biologics could both aid in the development of novel therapies and tackle the problem of organ shortage

The use of animals in biomedical research and the production of medically relevant proteins has a long history, but more recent advances in stem cell biology are opening up new and exciting possibilities both in the provision of biologics (tissues and proteins) for therapeutic use and for better models of disease and improved pre-clinical testing.

As for improvements in livestock and other domesticated animals important in agriculture, trangenesis has proved an effective way of producing large amounts of proteins for medical application. For example, by targeting expression of a protein of interest to the mammary gland, for example, by insertion into the gene encoding the milk protein β-casein, recombinant proteins such as human anti-thrombin III and thrombopoietin have been produced in large amounts in the milk of transgenic goats, sheep, or cattle. Again, the prospect of using pluripotent stem cells as a starting point for the genetic modification offers the potential of vastly improved precision and efficiency.

Chickens, and more specifically their eggs, are another exciting prospect for the pharmaceutical industry as a means to produce proteins for medicinal purposes, such as erythropoietin to be used to combat anemia or specific antibodies against therapeutic targets. Chicken pluripotent cell lines that are able to contribute to a chimeric adult and be passed through the germ line make this an even more attractive option. A more indirect link between stem cell-based genetic manipulation of chickens and human health involves prevention of the spread of avian influenza, a virus that poses a global threat both to poultry production and human health. An example of a transgenic solution entails engineering chickens to express a short RNA sequence that knocks down the virus polymerase, thereby interfering with virus propagation.

Rodents, especially mice, have been at the heart of biomedical research for over half a century and have proved invaluable in defining gene function in relation to disease and as a means to test innumerable drug candidates. However, mouse physiology differs from humans in many significant ways, and their lifespan, at a maximum of 2 years, is only a tiny fraction of what most people can expect to reach. Large animal models are therefore often more appropriate. The pig is particularly relevant due to its large size and the striking similarity with humans in terms of anatomy, physiology, and genetics. Pigs therefore often represent excellent models for human diseases and can be very useful for testing of new therapies initially developed in rodents. As we have already discussed, the pig is the only large domesticated animal for which pluripotent stem cell lines are available that might be useful in the context of genetic modification. These cells are in the form of iPS cells and show potential for both chimera generation and as a source of nuclei for SCNT cloning. Primates are arguably even better as models of human disease, and for several species, both ES cells and iPS cells have been successfully produced (Chapters 3 and 5). Although this would offer the prospect of the creation of disease models through genetic modification, this confronts cultural sensitivities and is unlikely to be widely adopted. The pig does not elicit such controversy and is much more likely to become the gold standard for human disease models.

Although raising a whole host of ethical issues, another step in the evolving field of biomedical application of stem cells, which would overcome a major hurdle in the global shortage of organ donors, is the generation of transplantable human organs grown inside of chimeric animals. In the United States alone, there are presently more than 75,000 patients on the organ waiting list, and about 20 of these will die every day before receiving the much-needed organ. As an example of current progress, a team of scientists at the Salk Institute in California using human pluripotent stem cells have succeeded in generating a human-pig chimeric embryo that was allowed to progress to 4 weeks of development. In these studies, the researchers were able to show that the human pluripotent stem cells gave rise to colonizing cells at a frequency of around 1 human cell in 100,000 pig cells (Figure 19.10). This established a proof of principle, but ethical restraints have thus far restricted further development of such chimeras.

Other inter-species chimeras have been generated that have enabled demonstration that organs produced in one species can go on to treat a disease in another species. For example, in 2017, a group of scientists successfully generated mouse islet cells in apancreatic rats that upon isolation and transplantation into a diabetic mouse

Figure 19.10 **Establishing human pluripotent stem cells in pig embryos.** As a proof of principle, human pluripotent stem cells were injected into the blastocyst of a pig, which was then allowed to develop for 4 weeks before assessment of the degree of chimerism. The confocal microscopy image of the pig embryo reveals the presence of the human cells (green), which are seen to have contributed to the development of the heart. (Used with permission from the Salk Institute.)

**A**

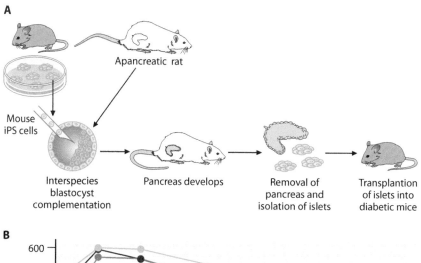

**B**

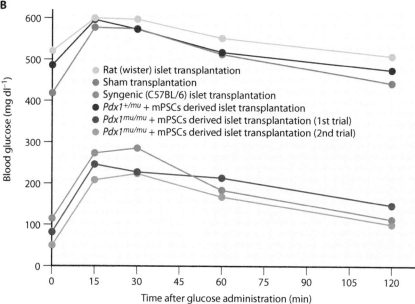

Figure 19.11 **Generation of mouse islet cells in apancreatic rats.** The generation of human organs in animals could overcome the shortage of tissues from registered organ donors. (**A**) As a proof of concept that interspecies organogenesis is feasible and yields cells or organs that could be used for therapy, researchers generated mouse induced pluripotent stem (iPS) cells expressing EGFP and injected them into a blastocyst from an apancreatic rat. A pancreas develops in the adult rat derived almost entirely from mouse cells. The islet cells can then be harvested and transplanted into a diabetic mouse model to reverse hyperglycemia. (**B**) Transplantation of mouse cell-derived islet cells into a diabetic mouse model showed a reduction in blood glucose when receiving mouse-derived EGFP+ islet cells from Pdx1$^{mu/mu}$ rats compared to control transplants. (From Yamaguchi T, Sato H, Kato-Itoh M et al. (2017) *Nature* 542:191–196 [doi: 10.1038/nature21070]. Reprinted by permission from Springer Nature.)

model achieved reversal of hyperglycemia in the recipients (Figure 19.11). Such a development points to the real possibility that treatment of human diabetes might be revolutionized if similar interspecies chimerism could be achieved.

## 19.2  SPECIES CONSERVATION

The growing human population not only sets challenges around food supply and animal wellbeing but also impacts the lives of many plant and animal species. There are many reasons for the decline in individual species, not least of which are loss of habitat, overexploitation, and competition with invasive species. There are currently over 16,000 species, including both animals and plants, on the endangered list around the world, a number that has increased significantly in the last decade. In order to protect endangered species from extinction, numerous plans have been put into practice such as captive breeding, habitat protection, increased protection from poachers, and increased stringency at borders to prevent the introduction of non-indigenous species. However, despite such efforts, many critically endangered species have since become extinct in the wild (in the last 500 years, 800 species have become extinct at least), and alternative ways are needed to prevent the continuing losses to the world's biodiversity.

Stem cell-based approaches are providing some new and exciting prospects for the preservation of endangered animals and the maintenance of their genetic diversity. Most notably, the generation of IPS cells, already discussed in this chapter in the context of agriculturally important animals, provides an opportunity to be able to produce germ cells for *in vitro* fertilization. Knowledge of the biology of **spermatogonial stem cells** (SSC; Chapter 12) may also be useful.

## Stem cell technology now offers the possibility of saving endangered animals or even resurrecting extinct species

The novel *Jurassic Park* by Michael Crichton was made into the blockbuster film of the same title by Steven Spielberg in 1993. The story revolved around the creation of a wildlife park of cloned dinosaurs on a Pacific island off Costa Rica. The cloning was supposedly achieved by extracting the DNA of dinosaurs from blood ingested by insects that had subsequently been preserved in amber. The DNA strands were incomplete, but by supposedly integrating them into frog chromosomes, the gaps were filled, enabling the development of fully grown dinosaurs—the details of this last part are a bit sketchy!

Although Jurassic Park may be a fanciful vision, it is certainly true that the advent of iPS cell technology suggests ways in which the individual genetic diversity of endangered or even recently extinct animals might be captured. Work beginning in the 1970s initiated the cryopreservation of cells from endangered species at the San Diego Zoo's Institute for Conservation Research (the "Frozen Zoo"). This collection holds skin fibroblast cells from approximately 8600 individual animals representing some 800 different vertebrate species. Using cells from two of these species, the drill (*Mandrillus leucophaeus*) and the northern white rhinoceros (*Ceratotherium simum cottoni*) (Figure 19.12), the Israeli scientist Inbar Friedrich Ben-Nun has given hope that stem cell technology can indeed aid in the conservation of endangered species (Figure 19.13). She created iPS cells from these two species with the intention of one

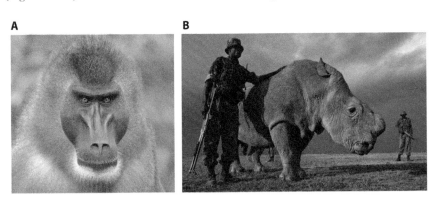

**A**   **B**

**Figure 19.12 The drill and northern white rhinoceros.** There are thousands of species of both plants and animals on the endangered species list. The quest now is to attempt a radical conservation strategy to prevent them from becoming extinct. Induced pluripotent stem cells have been generated from at least two of these critically endangered animals. (**A**) The drill (*Mandrillus leucophaeus*) is a large African monkey that is critically endangered, largely due to the illegal bushmeat trade and loss of habitat. (**B**) Pictured here is Sudan, the last remaining male northern white rhinoceros (*Ceratotherium simum*), that died in early 2018. (A: "Drill" by Clément Bardot—Own work. Licensed under the Creative Commons Attribution Share Alike 4.0 International license via Wikimedia (https://creativecommons.org/licenses/by/4.0/). B: From CB2/ZOB/BrentStirton/National Geographic/Newscom.)

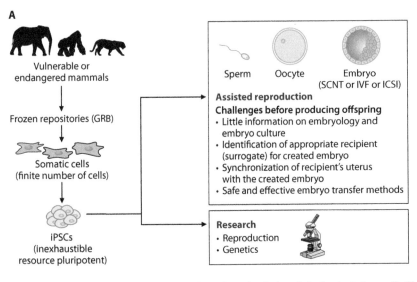

**Figure 19.13 Saving endangered species using induced pluripotent stem cells.** The advent of induced pluripotent stem (iPS) cell technology could hold the answer for conservationists struggling to save critically endangered species. (**A**) The process on the face of it is much like that for generating iPS cells from mice or humans. Cells collected from endangered species, usually fibroblasts, can be reprogrammed to iPS cells using pluripotency factors from mouse or human. These iPS cells can then be deposited in a genome resource bank (GRB) and subsequently utilized as a source of genetic material in order to maintain the species and its genetic diversity. The iPS cells could, for example, be differentiated into male or female gametes. Alternatively, iPS cell nuclei can be used directly to generate an embryo, for example by SCNT. There are, however, many challenges facing this technology when it comes to species where little is known about culture conditions and finding surrogates for the created embryo that need to be overcome before this can be successfully used to save an endangered species. (**B**) Inbar Friedrich Ben-Nun, a pioneer in the creation of large animal iPS cells. (A: From Selvaraj V, Wildt D & Pukazhenthi B. (2011) *Nat Methods* 8: 805–807 [doi:10.1038/nmeth.1715]. Reprinted by permission from Springer Nature. B: Used with permission of Dr. Inbar Friedrich Ben-Nun.)

day utilizing them to derive viable germ cells in order to be able to introduce new genetic material into existing breeding programs. Using retroviral vectors encoding the human reprogramming factors Oct4, Sox2, Klf4, and Myc to transduce skin fibroblasts from a 15-year-old male drill, Ben-Nun was able to generate four iPS cell lines that were karyotypically normal and exhibited the typical features of pluripotent stem cells, including **embryoid body** and **teratoma** formation following transplantation into immune deficient mice. Similarly, Ben-Nun created three iPS cell lines from a 10-year-old northern white rhinoceros. The race is on though, especially for the northern white rhinoceros, which is in a desperate situation. At the time of writing, there are now only two of them remaining, both of which are females (daughter and grand-daughter) that have been unable to breed naturally! So realistically, even with rapid technological advances, the northern white rhinoceros might soon follow the fates of the dodo, the passenger pigeon, and the plethora of other animals lost forever.

## Spermatogonial stem cells can be used for species conservation

In Chapter 12, we described the existence of the SSC and how transplantation has been used to demonstrate the stem cell function of these cells. SSC provide an alternative method to preserve the breeding potential of genetically valuable animals. If a male animal dies as an adult prior to breeding, mature sperm can be harvested and cryopreserved for later use in artificial insemination. However, should the animal die in the neonatal or juvenile stage before contributing their genes to the next generation, theoretically a small segment of testes could be removed and surgically grafted into immune deficient mice that would nurture the foreign tissue and support spermatogenesis. The efficiency of such **xenografts** differs between species, but the fertilizing ability of the sperm generated from the graft has been shown for some domesticated species through the production of viable offspring using assisted reproductive technology. However, this method has not yet produced offspring for highly endangered animals.

Direct transplantation of SSC could prove to be another valuable approach to preserve the male germ line. This method involves isolation of SSC from the donor testis and injection directly into the recipient testes that have SSC **niche** vacancy following focal testicular irradiation. Once spermatogenesis from the donor SSC is established, semen is collected and assessed for mature donor sperm. This method has proven successful despite involving many challenging steps; however, it requires the donor and recipient species to be closely related, hindering its application in the conservation of highly endangered species.

## 19.3  HORTICULTURAL DEVELOPMENT

The horticultural world also has exciting opportunities for the application of stem cell science and in fact is some way ahead in having introduced stem cell-based procedures into what are essentially standard practices in some areas. As we will outline, plant stem cells are being exploited in several areas of horticulture related to food production, to enhance plant health, and in the production of biologics for human health and other applications.

As we discuss in Chapter 8, plant stem cells are present in the **meristems**, which are regions of growth and development principally in shoots and roots and associated with the vasculature. In addition, tissue known as **callus** that forms following wounding is a heterogeneous mix of cells, some of which transiently exhibit a number of stem cell-like properties. Callus has often been utilized as an alternative source of plant stem cells in the context of plant improvement through genetic modification or the production of useful plant compounds. However, only plant stem cells embedded in meristems can divide, yielding cells that differentiate while at the same time giving rise to new stem cells.

## Genetically modified plants have been grown for many decades to enhance specific traits

Much the same as for animals used for food, the conventional breeding of plants is hardly able to keep pace with the increasing demands for fruit, vegetables, and cereal crops, especially in developing countries. Just as accelerated improvements in livestock and other food source animals has been sought through transgenic modification, so economically important plants are being altered by genetic engineering. In the case of plants, though, stem cells or stem cell-like behavior have been central to the

**TABLE 19.2  EXAMPLES OF TRANSGENIC MODIFICATIONS IN PLANTS**

| Feature improved | Specific change introduced | Gene introduced or modified | Species |
|---|---|---|---|
| Improved nutritional quality | Ability to produce β-carotene | Three enzymes in the β-carotene biosynthetic pathway | Rice |
| Insect resistance | Production of toxic bacterial proteins | *Bacillus thuringiensis* δ-endotoxins | Potato, cotton, maize |
| Disease resistance | Resistance to tobacco mosaic virus (TMV) | TMV coat protein | Tobacco, tomato, potato |
| Herbicide tolerance | Resistance to glyphosate through overproduction of enzyme involved in its synthesis | 5-enolpyruvylshikimate-3-phosphate synthase | Many |
| Salt resistance | Ability to sequester excess salt in leaf cell vacuoles | Sodium/H+ anti-port pump | Tomato |

*Note:*  Plants have been in the news a lot longer than for animals with respect to genetic modification, particularly for enhancing specific traits in crops. The table lists some examples of the ways in which specific transgene expression in plants can be used to modify their phenotype.

process for much longer. This has been because many of the means of introducing novel genetic modifications into the plant genome as transgenes have relied upon meristems or cells reprogrammed into callus. The scope of engineered changes to plant genomes is broad (Table 19.2), ranging from the creation of resistance to specific pathogens such as tobacco mosaic virus; combating resistance to stresses including water, temperature, and heavy metals; and tolerance to herbicides.

## Plant stem cells can be manipulated to eradicate infections

Plant stem cells located in the meristems also represent a direct means, not requiring the introduction of genetic changes, to rid plants of pathogens. Many plants that are grown commercially suffer systemic diseases caused by viral, bacterial, and fungal infections. These infections are traditionally treated with chemicals, although no such agents are effective against plant viruses. The effectiveness of such bactericides and fungicides commonly declines as the pathogens are selected for resistance. As already mentioned, genetic engineering provides some opportunities to combat viruses and resistance to bactericidal and fungicidal compounds. However, a particular feature of meristems has offered a relatively simple solution, in many cases curing plants of disease.

**Apical meristems**, present in the shoots and roots, are generally either free or carry a very low concentration of the pathogen. The reason for this includes the absence of the vascular system in the meristems, which hinders the free passage of the pathogen. It is also speculated that the high metabolic activity or specific hormone or metabolite content inherent to the meristem might inhibit pathogen replication. To rid a plant of an infection, horticulturalists have therefore resorted to meristem culture. Quite simply, this involves micro-dissection of the meristem (usually shoot tip) and aseptic culture in a specialized synthetic growth medium (Figure 19.14).

## Meristematic cells can be used to produce plant products on an industrial scale

Many plants produce a diverse range of secondary metabolites that can be harnessed and exploited in the field of medicine. Indeed, plant derived products have been used as medicines for millennia and remain the basis for many traditional remedies. The full potential of these compounds, and many others yet to be identified, remains to be realized. Often the plants are rare wild plants, and the active molecules are only present in very small quantities. They may also be difficult to purify, and the alternative option of chemical synthesis is not feasible because the chemistry is just not available to make this possible. Potentially, the *in vitro* culture of dedifferentiated cells from a plant that generate the compound of interest could circumvent some of these problems, but achieving this on an industrial scale can be problematic.

A solution has recently emerged based on the application of the specific plant stem cells, that is, the cambial meristematic cells (CMC). As described in Chapter 8, these cambial cells constitute the stem cells that give rise to the vascular tissues of a plant. The particular circumstance that led to the first demonstration of this potential use of CMC involved a secondary metabolite called paclitaxel, which is widely used as an anti-cancer therapeutic in the treatment of breast, lung, and other non-small cell cancers.

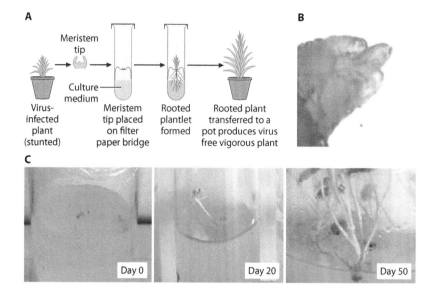

Figure 19.14 **Pathogen eradication through shoot apical meristem culture.** Viral infections of plants can be devastating, risking population and financial consequences if crops are affected and effectively destroyed. (**A**) Often the apical meristems of a plant, which houses the stem cells for the shoots, are pathogen free and can be used to propagate a virus-free plant when cultured in the right media for rooting. (**B**) A dissected meristem. The square is 5 mm in dimension. (**C**) Time course for the regrowth of a plant from a dissected apical meristem. (B and C: From Quiroz K, Berríos M, Carrasco B et al. (2017) *Biol Res* 50:20 [doi:10.1186/s40659-017-0125-8]. Licensed under the terms of the Creative Commons Attribution 4.0 International License (http://creativecommons.org/licenses/by/4.0).)

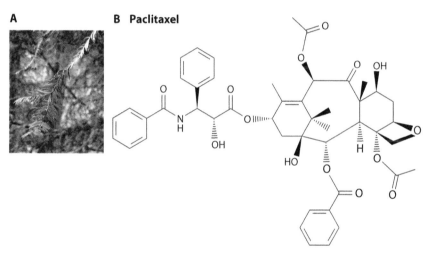

Figure 19.15 **The yew tree derived drug paclitaxel.** (**A**). Paclitaxel, found at very low concentrations in the bark of the yew tree (*Taxus brevifolia*), is an important anti-cancer drug in the treatment of breast, lung, and other non-small cell cancers. (**B**) The chemical structure of paclitaxel. (A: "Pacific Yew" by Jason Hollinger. Licensed under the Creative Commons Attribution 2.0 Generic License via Wikimedia. (https://creativecommons.org/licenses/by/2.0/).)

Paclitaxel is found at low concentrations (0.001%–0.05%) in the bark of the yew tree *Taxus brevifolia* (Figure 19.15). Extraction of the compound from the tree results in its death, which is a particular problem because the yew is notoriously slow growing. Since the discovery of paclitaxel in 1962, researchers have endeavored to find alternative sources of the chemical. The chemical synthesis proved to be commercially non-viable. The then alternative of plant cell culture was also not a solution, not least because long-term plant cell culture is inherently variable, suffering from cell heterogeneity, genetic instability, and low growth rates. A solution lies in the isolation of CMC, which can be cultured to produce numbers that are sufficient for the practical isolation of even very low abundance secondary metabolites. CMC culture will most likely provide the key platform technology in the utilization of plant natural products (Figure 19.16). In 2010,

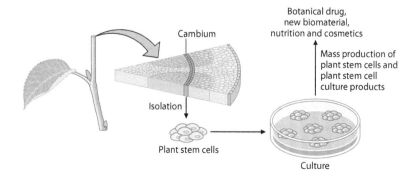

Figure 19.16 **Cambial meristematic cells as a platform for the production of plant compounds.** Many plant-derived drugs are obtained from notoriously slow growing plants, and their extraction results in the death of the plant, making the isolation of large quantities of drug both expensive and time consuming. A solution involves the culture of cambial meristematic cells (CMC). Shown is the general scheme for the isolation and culture of CMC for their potential applications for industry.

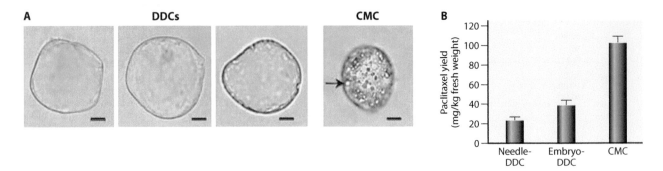

Figure 19.17 **Production of paclitaxel from cambial meristematic cells.** In 2010, researchers in Edinburgh developed a system for producing paclitaxel from cambial meristematic cells (CMC). (**A**) Comparison of dedifferentiated cells (DDC) and a cambial meristematic cell (CMC). CMC are smaller and possess characteristic small vacuole-like structures (black arrow). Scale bars, 20 μm. (**B**) Paclitaxel production by 3-month-old DDC and CMC following batch culture in a flask for 10 days. (From Lee E, Jin Y, Park J et al. (2010) *Nat Biotechnol* 28:1213–1217 [doi: 10.1038/nbt.1693]. Reprinted by permission from Springer Nature.)

researchers in Edinburgh developed a culture system for the yew utilizing CMC, based on the observation that the production of paclitaxel was at its highest within this region (Figure 19.17). Culturing of the CMC yielded approximately five times the quantity when compared to conventional plant cell cultures utilizing dedifferentiated cells.

## 19.4   SUMMARY

Domesticated animals, both as companions and for economic benefit, and horticulture are areas in which stem cell technology is already being applied and will play an increasingly important role. As in humans, animal health and wellbeing can benefit from stem cell related treatments, but in addition, the associated technologies create opportunities to improve upon conventional breeding strategies for strain improvement, including both enhancement of existing features as well as making possible the production of human biologics and even organs for transplantation. Likewise, stem cell-based approaches have the potential both to improve plant characteristics and to engineer the production of various biologics, such as drugs.

## KEY POINTS

- The application of stem cell technology to economically important plants and animals is likely to be as important to overall world health as the more direct uses in medicine.

- Many of the uses of stem cells in domesticated animals parallel actual or potential applications in humans, as exemplified by the use of MSC in the repair of tendon or cartilage injuries.

- Although difficulties in generating ES cells from many domesticated species has meant that easier genetic manipulation to enhance desirable traits cannot rely on these cells, the ability to generate iPS cells from most species offers a realistic solution.

- The production of human biologics mediated through a stem cell-based approach is being developed through genetic manipulation in stem cells that are then used to generate animals.

- Mixed species chimerism, elicited by introduction of stem cells of one species into the embryo of another species,

is a potential route for the generation of transplantable replacement tissues.

- A number of applications of adult stem cells in domesticated animals are being considered, such as the use of MaSC to enhance milk production in cattle or the use of muscle stem cells to produce meat *ex vivo*.

- The use of stem cells in genetic engineering is not restricted to mammals and is being actively pursued in animals such as chickens and various fish species.

- The conservation of endangered species can potentially be facilitated through iPS cell technology or the rescue and transfer of spermatogonial stem cells.

- Plant stem cells derived from meristems or reprogrammed cells in callus tissue potentially provide a convenient means for genetic engineering in plants and for the production of biologics.

## FURTHER READING

Campbell, K.H., McWhir, J., Ritchie, W.A. and Wilmut, I. (1996) Sheep cloned by nuclear transfer from a cultured cell line. *Nature*, 380: 64–66.

Ochoa-Villarreal, M. Howat, S., Jang, M.I., Kim, I.S., Jin, Y-W., Lee, E-K. and Loake, G.J. (2015) Cambial meristematic cells: A platform for the production of plant natural products. *N. Biotechnol.*, S1871–6784.

Roberts, R.M. Yuan, Y., Genovese, N. and Ezashi, T. (2015) Livestock models for exploiting the promise of pluripotent stem cells. *ILAR J.*, 56(1), 74–82.

Wheeler, M.B. (2013) Transgenic animals in agriculture. *Nat. Educ. Knowl.*, 4: 1.

Wu, J. Wu, J., Paltero-Luengo, A., Sakurai, M., Sugawara, A., Gil, M.A., Yamauchi, T., Suzuki, K. et al. (2017) Interspecies chimerism with mammalian pluripotent stem cells. *Cell*, 168(3), 473–486.

# Stem cells and society: A philosophical, ethical, and economical view

# 20

Like any area of research, the factors that drive the progress of stem cell science are many and varied, but the public has perhaps never been quite so engaged in science, surpassing even the excitement that was engendered by the first manned missions to the moon at the end of the 1960s. What stem cell science is perceived to offer to society has appealed to fundamental human feelings around health and wealth, although different cultures, both within individual countries and across international divides, have responded in distinct ways to the recent advances. The disproportionate emphasis on human embryos in the context of stem cell science has been driven by ethical concerns, although the way in which different religions have approached these varies widely. The prospect of wealth generation through stem cell science has inevitably led to a dark side to stem cell science, unscrupulous and frankly fraudulent people exploiting those desperate to find treatments for intractable illnesses, offering stem cell-based cures that have no scientific foundation.

Politicians want to do the right thing, of course, and their gauging of public opinion about stem cell science has had both positive and negative effects on the research community. Not surprisingly, the great promise that stem cell science holds for health, as well as the religious and political controversies, has given the media the opportunity to speculate, criticize, and condemn, and in so doing has further fueled public expectations, fears, and debate.

Any exploitation of new scientific discoveries is limited by business imperatives, and stem cell science is no exception. Even though the potential of the application of stem cells to human and animal health and biotechnology is vast, the route to market is generally far from clear. Stem cell science is complex and often far removed from a deliverable product, it is expensive, and what can be protected as intellectual property (IP) is often not straightforward, so it is hardly surprising that after a decade and a half of the stem cell science boom, the number of products on the market, especially those that directly involve cells, is very limited. The burden that new, highly costly therapies will undoubtedly place upon healthcare systems raises further questions for society, especially since the benefits of stem cell science are likely to be more available to the rich. In a world with an increasing population, the cost of medical interventions is being stretched to the limit.

In this chapter, we will explore these broad issues, illustrating with examples the way in which society perceives and is responding to the new challenges and opportunities that stem cell science raises.

## 20.1 PHILOSOPHICAL AND MORAL PERSPECTIVES

By this stage, after 19 chapters describing the role of stem cells throughout biology, you will appreciate that they are not only very exciting from the perspective of their unique properties, but they also take many forms in different tissues of animals and plants and at different stages in life. However, the focus of ethical, religious, and political debate, and consequently the public's view, has been on the use of embryos in the generation of stem cells and the implications of human **cloning** and the generation of human tissues from **pluripotent stem cells**. To many, the term "stem cell" equates with "**embryonic stem (ES) cell**," and any work involving stem cells, whatever their source or type, comes under the same scrutiny.

## Reports in the media influence the public perception of stem cells

It is impossible to move through life without being impacted by the media, whether through newspapers, television, or the Internet. The influence of the Internet on public thinking and opinion can be profound, especially if there are no other avenues for informed debate. The portrayal of stem cell science by the media has without doubt done a lot to color the view of the public at large, which of course in turn impacts political decision-making affecting regulatory legislation and research funding. Reporting can range from reasoned provision of information and debate of critical issues through to blatant sensationalism. The principal drivers for the media are all too often the desire to highlight a major "breakthrough" that will improve the treatment of a disease that is presently seen as incurable or to "expose" unscrupulous behavior or fraudulent claims by researchers or those offering supposed novel treatments. This situation as it has related to stem cell science is probably no different than any other where the media affect our lives, and the consequences can be both positive and negative. However, there can be no doubt that the reporting of stem cell research and a disproportionate emphasis on certain aspects, largely around the moral debate over the derivation of human ES cells, has significantly influenced its direction of travel, at least when taking a view over the short to medium term.

From a positive perspective, media-generated public interest has ensured in many countries that funding for stem cell research has been increased, or at least has been protected in times of economic crisis, although many in the broader scientific community would argue that research in less "trendy" areas has suffered through the diversion of funds. To a great extent, the media reflect what the public want to hear, and in that sense, stem cell science is bound to engender sensationalist reporting when the promise appears to be a longer, healthier life. It seemed during the early part of the twenty-first century that a week would not pass without a story of a breakthrough that was going to lead to a "cure" for diseases like diabetes, heart failure, multiple sclerosis, and Alzheimer's (Figure 20.1). There were seldom any reality checks about the actual significance of a particular scientific finding, the fact that what was being reported was probably preliminary and an incremental change, or that the timescales that apply, if indeed the discovery would lead to a new therapy, might extend over many years. As a consequence of this "hype," public expectations were raised to unprecedented levels. Unfortunately, many people were given false hope and in the worst case had been drawn into seeking treatments that had not gone through the usual rigor of clinical testing and had absolutely no scientifically proven efficacy.

There are not many of us actively engaged in stem cell research that have not had calls from desperate individuals with incurable diseases and no treatment options asking if they can be put into a stem cell clinical trial. Such situations had a positive benefit in that many researchers, funders, and governmental bodies set about finding better ways to inform through direct public engagement, involving two-way discussions between scientists and the general public. At least in the United Kingdom, the majority of scientists now participate in some form of outreach or public

Figure 20.1 **Front page headlines about stem cell therapy "breakthroughs."** Stem cell breakthroughs are reported all over the world, whether it is a potential stem cell therapy for normally incurable diseases or new and often controversial techniques being published. These are two representative British newspaper headlines reporting of such breakthroughs. (Adapted with permission from Daily Express.)

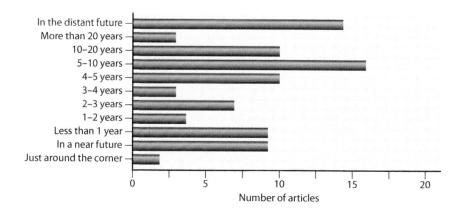

**Figure 20.2 Media portrayal of stem cell therapy timelines and perspectives.** The reports of stem cells in the media significantly affect the general public's opinion of the potential for future therapies using stem cells. Analysis of 87 newspaper articles for their assessment of stated time until stem cell therapy is common practice revealed that this could be a reality in just 5–10 years from now and highlights the use of ambiguous terminology. (From Kamenova K & Caulfield T (2015) *Sci Trans Med* 7:278 [doi: 10.1126/scitranslmed.3010496]. Reprinted with permission from AAAS.)

engagement, the overall climate being one in which researchers seek to gain trust with the public and share knowledge and correct any misconceptions.

Should the media take the majority of the blame for overhyping of the prospects emerging from stem cell science? Probably not, especially since they are generally only reporting on the information that they can get their hands on, even if hyperbole does come into play quite often to increase the interest value. Some of the blame has to be laid at the door of the researchers, often because they have been reticent about explaining their science or sometimes through making overinflated claims either unwittingly or for reasons of self-promotion and the next grant funding opportunity. As part of the realization that something had to be done through public engagement, so too the approach of researchers in their communication with the media needed to be examined. One clear pattern that emerged from an assessment of newspaper reporting about potential new stem cell therapies, in addition to the use of ambiguous terminology such as "just around the corner," was that these could be implemented and become common practice within 5–10 years or sooner (Figure 20.2). For someone diagnosed with a degenerative disorder, such a timeframe is within the realms of hope that their disease could one day be cured. As the role of the scientists themselves in reporting accuracy became clearer, many of them were encouraged to participate in media training so that we could provide accurate and accessible information, properly contextualized, and importantly avoiding any false impression of immediate treatment benefits. The advice on the latter point was quite clear: however tempting it might be to respond to a reporter with a definitive timescale for when a treatment might be available, always give a non-specific impression of "quite some time in the future."

## The moral status of the embryo is at the heart of debate surrounding embryonic stem cell research

Opinions on the acceptability of the use of ES cells vary widely and are often very controversial. Nevertheless, surveys conducted in several countries have indicated that the majority of the public consider the derivation of stem cells from human embryos morally acceptable, with roughly only a third of people opposing the idea (Figure 20.3).

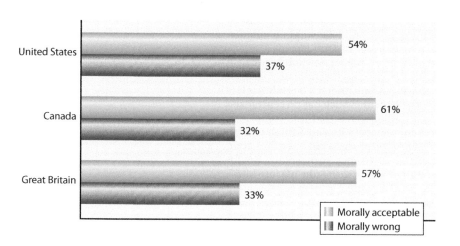

**Figure 20.3 Moral acceptability of stem cell research.** Opinions on the moral acceptability of the use of embryonic stem (ES) cells varies widely. Gauging the general public's opinion through the use of polling in the United States, Canada, and Great Britain revealed that the majority of people feel that it is morally acceptable to use ES cells, only a third of people being opposed to the prospect. (Adapted from Lyons L. (October 19, 2004) *Gallup News* Permission conveyed through Copyright Clearance Center, Inc.)

As you may know, the federal government currently provides very limited funding for medical researh that uses stem cells obtained from human embryos. Which would you prefer the governement do ...?

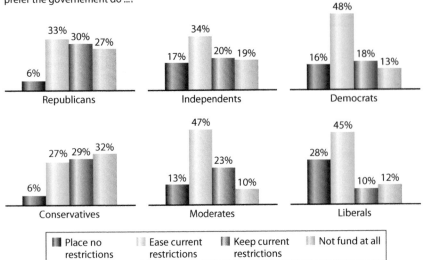

Figure 20.4 **Political party views on government funding of embryonic stem cell research.** The general public's opinion on the use of embryonic stem (ES) cells correlates with their political affiliation. Results of a poll in the United States from 2009 on the views of the use of government funds to fuel ES cell research reveals that the Democrats on the whole favor this, whereas the Republicans would prefer tighter restrictions placed on these funds. (Adapted from Morales L. (March 9, 2009) *The Gallup Poll Briefing* Permission conveyed through Copyright Clearance Center, Inc.)

The opinion of pro-life campaigners in the United States is that all human life should be valued until natural death and that deliberate destruction of human life through abortion is morally wrong, and therefore derivation of ES cells from the **inner cell mass** of a **blastocyst** at day 5–7 of development (Chapter 3) is considered tantamount to murder. As might be expected, then, when the first isolation of human ES cells was described in 1998, this initiated massive ethical debate about the use of embryos for this purpose, even if they were surplus to *in vitro* fertilization programs and would have been destroyed anyway. Alternative viewpoints include that before day 14 of human embryonic development, the point at which implantation into the uterine wall takes place, the embryo could not survive, or that human life begins once sensation is possible with the formation of the nervous system at week 4 of gestation. Within the United States, there is a clear parallel between political affiliation and the acceptance of ES cell research (Figure 20.4). On the surface, it appears that Democrats have been in favor of fewer restrictions on the policies governing ES cell research, whereas Republicans largely prefer that rigorous guidelines be maintained to restrict ES cell research.

## Religious doctrines can influence levels of acceptance for stem cell research

The religious dimension, often influencing national policies, plays a huge part in the public's overall opinion on the use of human embryos to derive stem cells. Among the Christian faiths (Figure 20.5), Roman Catholics, Orthodox Christians, and conservative Protestants consider fertilization to mark the beginning of human life and that destruction of the human embryo violates their right to life. Pope Paul VI's landmark encyclical *Humanae vitae* (1968) stated that "scientific research is important, but it can never come at the expense of the weakest among us." Less conservative Protestants consider the embryo to have potential for human status but only following implantation and therefore allow some embryonic research prior to day 14 of development.

The Abrahamic religions other than Christianity have different viewpoints. Judaism places the emphasis on the importance of saving life, which is the ultimate goal of all ES cell research and therefore makes it acceptable to perform research on embryos less than 40 days old. A similar view is held by the Islamic faith, which has no centralized authority on the matter. Some agree with the Catholic view that it is completely immoral, whereas others believe that the fetus does not gain its soul until the fourth month of development, as the Qur'an states, "each of you possesses his own formation within his mother's womb, first as a drop of matter for 40 days, then a blood clot for 40 days, then as a blob for 40 days, and then the angel is sent to breathe life into him."

Buddhist beliefs are to some extent conflicted with respect to stem cell research. On the one hand, they place great emphasis on the central virtues of knowledge (prajña) and compassion (karva) and have a long tradition of practicing medicine in monasteries, so that the prospect of any cure to alleviate human suffering should be

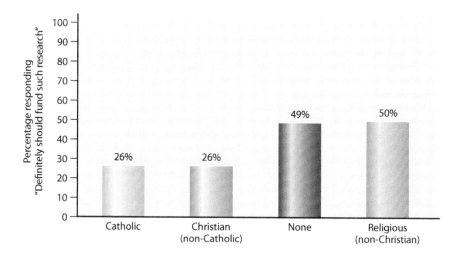

Figure 20.5 **Survey of religious groups and their opinions on government funding for embryonic stem cell research.** Religious belief plays a large role in determining the opinion of the public with regards to the use of embryonic stem (ES) cells for research. A poll on whether the government should fund ES cell research highlights the difference in opinion across some of the religions, emphasizing that the Christian faith in particular sways the public in their opinion that these funds should not be used for this purpose. The results show the percentage that responded "definitely should fund such research." (From Gray M. The New "Catholic Vote": The Quiet Rise of the None/Other. *Nineteen Sixty-four* March 29, 2012, http://nineteensixty-four.blogspot.com/2012/03/new-catholicvote-quiet-rise-of.htm/ Used with permission of Mark M. Gray.)

welcomed. However, Buddhists also value the importance of not harming (ahimsa) and therefore have grave reservations about ES cell research that results in the destruction of life, human or animal, but should have no ethical issues regarding potential life preserving uses of human **adult stem cells**.

Hinduism also has aspects to its belief structures that can lead to mixed views on stem cell research. The majority Hindu view is that any harm to a sentient being is prohibited, thereby not only ruling out any use of embryos but also severely limiting research that can be carried out on human, animal, or plant stem cells. The Vedas, which is one of the oldest of the sacred Hindu scripts, emphasizes the sanctity of life as the heart of the Hindu doctrine of non-violence and that all life is God's creation and should be respected. However, another concept adhered to by a minority of Hindus, namely the law of nature that one must "kill to survive," sacrificing a few for the greater good, creates a paradox for the acceptance of stem cell research.

Perhaps ironically, certain aspects of adult or **tissue stem cell** biology received extra support and impetus because of the very fact that there were no implications in terms of embryos. An interesting example has been the view taken by The Vatican, which, not surprisingly given the doctrine of the Catholic Church, made a strong stand against any research involving human embryos and cells derived from them. Many were surprised, though, that they vigorously promoted investigation of adult stem cell biology. In 2011, the Pontifical Council for Culture established an alliance with the United States biopharmaceutical company NeoStem, which includes a non-profit arm called the Stem for Life Foundation. This enabled The Vatican to support and award adult stem cell researchers and, most notably, to organize international conferences on the subject (Figure 20.6).

In retrospect, the imperative to demonstrate that stem cells derived from non-embryonic sources might have potential for use in **regenerative medicine** that could equal that of pluripotent stem cells probably encouraged the initial overinterpretation of findings such as those purporting to show stem cell **"plasticity"** (Chapter 5) or the pluripotent nature of cells such as **multipotent adult progenitor cells (MAPC)** isolated from adult tissues (Chapter 10). More insidiously, and something that we will come back to later, are the disingenuous or blatantly fraudulent claims that some sources of adult stem cells, for example, deciduous teeth or umbilical cords, can be sources of cells that can be frozen to provide a reserve set of stem cells for treatments later in life. Although for sure **hematopoietic stem cells** (HSC) can be utilized from umbilical cords (Chapter 9), and there may be some therapeutic efficacy in the derivation of **mesenchymal stem cells** (MSC) from either source (Chapters 10 and 16), the validity of most of the claims that banking cells from these tissues will serve as some sort of panacea for any future disease to which the donor might succumb is plainly unfounded based on our current level of understanding.

## International human embryonic stem cell laws influence the progress of stem cell research

There is considerable country-to-country variation in the policies that govern research on human embryos and the generation of ES cells from them. Within the European Union, human ES cell line derivation is permitted in countries such as Sweden and

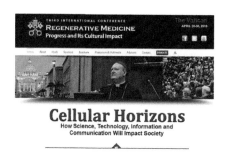

Figure 20.6 **The Vatican Third International Conference on Regenerative Medicine.** The conference was held in April 2016, organized by the Pontifical Council for Culture's Office for Science and Faith and the Stem Cell for Life Foundation (Used with permission from Stem for Life Foundation. (http://celltherapyconference2016.com.)

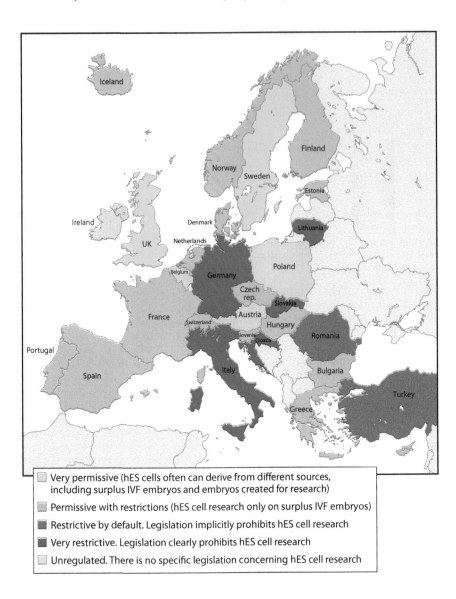

Figure 20.7 **European stem cell policies.** Across Europe, there is a great deal of variation in the policies governing the use of human embryos and the generation of embryonic stem cells from them. This map highlights this variation. Green – Allow the use of embryos from different sources as well as creating embryos solely for research purposes; Orange – Permit ES cell research but only on embryos that are surplus to IVF clinic requirements; Purple and Red – Various degrees of restrictions on ES cell research; and Blue – Countries with no specific legislation on ES cell research. (From Campos-Ruiz V. (2013) *Human Stem Cell Research and Regenerative Medicine.* Strasbourg, France: European Science Foundation. Used with permission.)

Very permissive (hES cells often can derive from different sources, including surplus IVF embryos and embryos created for research)

Permissive with restrictions (hES cell research only on surplus IVF embryos)

Restrictive by default. Legislation implicitly prohibits hES cell research

Very restrictive. Legislation clearly prohibits hES cell research

Unregulated. There is no specific legislation concerning hES cell research

the United Kingdom but is banned in many, including in Germany and Italy, where any research has to be carried out on imported ES cell lines (Figure 20.7). Although politicians and legislators try to reflect the view of the public at large, or at least they think they do, individual governmental stances on stem cell ethics tend to follow their country's religious and ethical foundations. Some governments have passed laws that totally prohibit research involving human embryos, while others take more of a facilitative approach built on clear regulatory guidance. The long running, and often research limiting debate in the United States focused on human ES cells, especially whether research on these cells should be funded by the federal government, has dominated the international stem cell arena, but many equally significant decisions have impacted greatly the trends and dynamics of stem cell research around the world. Highly restrictive research environments exist in countries such as Germany, where the law prohibits the importation and utilization of human ES cells derived before May 2007 and prevents the derivation of stem cells from embryos in Germany. Unlike most of the restrictive guidelines in other countries, the German legislation contains harsh penal provisions: the importation of stem cells without approval, or "deliberately giving false information" to gain approval, can incur fines or up to 3 years imprisonment, although the original provision that made it a criminal offense for German researchers to use human ES cells in other countries has been eliminated. The European Biopatent Directive specifies that use of human embryos for industrial or commercial purposes cannot be patented, and although this legislation does not necessarily affect the ability of researchers' individual member states to work on human ES cells, it has, as we will see later, had a profound effect on the ability of

European researchers to lodge patent applications for discoveries that are in some way related to human ES cells, thereby limiting many directions of investigation.

In contrast, many countries, including the United Kingdom and Australia, have relatively liberal regulations for research involving human ES cells. Hence, in the United Kingdom, it is possible under the conditions of the 2008 Human Fertilisation and Embryology Act to utilize unused *in vitro* fertilization (IVF) embryos, embryos created by IVF specifically for research purposes, embryos created by **somatic cell nuclear transfer** (SCNT), hybrid embryos created from animal and human gametes, "cybrids" (created by SCNT using human nuclei and animal oocytes), and **chimeric** human embryos involving injection of animal cells into human embryos. In all cases, embryo development should not be allowed to progress beyond 14 days of gestation. Implementation of the 2008 Act is tightly regulated and monitored by the Human Fertilisation and Embryology Authority, which will only grant licenses for embryo research that seeks to improve knowledge and therapeutic opportunities related to human reproduction and development.

China has long been known for its mostly unrestricted policies surrounding human ES cell research. China also had very relaxed guidelines governing the use of stem cells for therapeutic purposes and, as a result, numerous for-profit stem cell clinics were established selling unproven therapies to patients. However, in 2015, the Chinese government implemented new rules for stem cell therapies in order to restrict the illegitimate stem cell trials offered. These new rules require informed patient consent, the use of clinical-grade stem cells, treatment only at authorized hospitals, and the prohibition of advertising or charging for such therapies.

## Legislation imposed in the United States has profoundly affected the progression of stem cell research

We will take a slightly more detailed view of the way in which the issue of human ES cell research has been handled from the regulatory perspective in the United States because this has had a profound influence on the ability of researchers to perform experiments, even encouraging some like Roger Pedersen (Figure 20.8) to take the bold step of emigrating to the United Kingdom to take advantage of its more relaxed regulatory environment. Restrictive legislation applied in the United States from the mid 1990s throughout the first decade of the twenty-first century. Following the birth in 1978 of Louise Brown in the United Kingdom, the first to be born following IVF, the United States federal government imposed legislation barring the use of federal funds for research on human embryos. In 1995 this ban was reviewed, and the administration of President Bill Clinton proposed that federal funding could once again be used for research on human embryos surplus to IVF requirements, although no embryos should be specifically created for research. However, Congress intervened through the Dickey-Wicker Amendment, which prohibited any federally funded research that would result in the destruction of human embryos, regardless of their source. However, the limitations on how United States federal funding could be utilized did not extend to non-governmental funding sources, and for this reason it was still possible in 1998 for James Thomson to do research on human embryos and to isolate ES cells from them (Chapter 3). Following this major breakthrough, the Clinton Administration guidelines were re-examined and in 1999, the president's National Bioethics Advisory Committee recommended that federal funds could now be used for research on human ES cells derived from surplus embryos stored at IVF clinics.

This relaxation of the guidelines was short-lived, though, since when President George W. Bush (Figure 20.9) was elected, he soon made his own pro-life convictions felt. Under the Bush administration, new legislation was implemented in 2001 that only allowed the release of federal funds for research on human ES cell lines created prior to 2001 and banned the creation of new lines, even from surplus embryos obtained for IVF treatment. This more restrictive legislation remained in place throughout Bush's term in office, flying in the face of the majority of public opinion (Figure 20.10). A reprieve came for human ES cell research in the United States

**Figure 20.8 Roger Pedersen: A stem cell science migrant.** Restrictive legislation in the United States on the use of human ES cells led to the emigration of some scientists to pursue their careers in other countries. Stem cell researcher Roger Pedersen left the United States in 2001 to set up a laboratory in Cambridge, UK. (Photo credit: Christopher Vaughan / Stanford Medicine. Used with permission.)

"Embryonic stem-cell research requires the destruction of life to create a stem cell. That's why I think we've got to be very careful in balancing the ethics and the science."

— Georg W. Bush —

**Figure 20.9 President George W. Bush was strongly against human embryonic stem cell research.** George Bush's election as president of the United States in 2001 saw rigorous restrictions placed on the release of federal funds for human embryonic stem cell research, largely founded by his own pro-life views.

**Figure 20.10 Public approval of President Bush's veto of the bill for embryonic stem cell funding.** The strict legislation imposed by President Bush on the release of federal funds for embryonic stem cell research was put to the public through an opinion poll in 2006. President Bush's decision to veto the bill to expand federal funding for stem cell research was disapproved of by a majority of the public. (Adapted from Carroll J. (July 27, 2006) *The Gallup Poll Briefing* Permission conveyed through Copyright Clearance Center, Inc.)

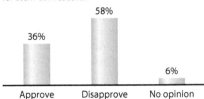

Public approval for President Bush's decision to veto bill to expand federal funding for stem cell research

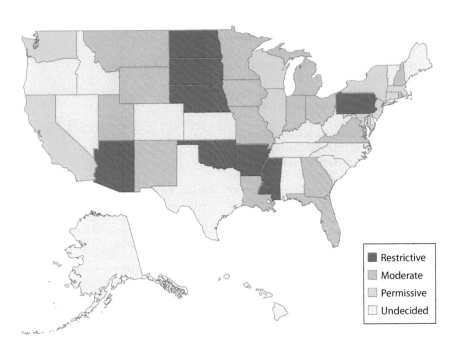

Restrictive
Moderate
Permissive
Undecided

**Figure 20.11 Differences in individual state legislation governing stem cell research involving human embryos in the United States.** This map of the United States highlights the states with restrictive (red), moderate (orange), and permissive (green) policies on stem cell research involving human embryos. States colored gray have no specific stem cell policies in place. (From Matthews K & Rowland M. (2011) *Stem Cells and Biomedical Research in Texas.* Houston, Texas: James A. Baker III Institute for Public Policy of Rice University. Used with permission from the Baker Institute.)

when President Barack Obama was elected in 2009. President Obama removed the restriction for federal funding to be used on the creation of new human ES cell lines from surplus embryos from IVF clinics.

The balance of opinion in the United States is very much state dependent (Figure 20.11). This is most notably highlighted by the reaction of the State of California to President Bush's restrictive legislation, when on November 2, 2004, 59.1% of voters supported California Proposition 71 to initiate a constitutional amendment affecting stem cell research. The measure, also known as the California Stem Cell Research and Cures Initiative, was added as Article XXXV of the California Constitution and made conducting stem cell research a state constitutional right. The initiative authorized the sale of general obligation bonds to allocate $3 billion dollars to stem cell research over a period of 10 years, with a priority to promote human ES cell research, leading to the establishment of the California Institute for Regenerative Medicine (CIRM).

## High-profile figures have helped promote and progress stem cell research for the treatment of disease

Largely as a result of the position taken by the administration of President George W. Bush, a number of well-known figures from worlds as diverse as politics and entertainment put themselves into the limelight as strong advocates of stem cell research, especially where that research required the use of human embryo-derived cells. These included Christopher Reeve, best known for his role in the *Superman* films, Michael J. Fox of *Back to the Future* fame, and Nancy Reagan, wife of President Ronald Reagan (Figure 20.12). In 1995, Christopher Reeve suffered paralysis from the neck down following a riding accident. He and his wife set up the Christopher & Dana Reeve

**A**

**B**

**Figure 20.12 High profile advocates of human embryonic stem cell research.** Christopher and Dana Reeve set up a foundation advocating stem cell research (**A**). Michael J. Fox and Nancy Reagan (**B**) together were advocates of stem cell research for the treatment of neural degenerative disease (A: "Reeve Foundation logo" by Rob Gerth. Licensed under the Creative Commons Attribution Share Alike 4.0 International license via Wikimedia (https://creativecommons.org/licenses/by/4.0/ B: From s_bukley/Newscom.)

On April 11, 2001, Nancy Reagan wrote a heartfelt letter to President Bush, detailing her reasons for supporting stem cell research, and asking him to see the issue from the point of view of those whose lives might be saved as a result of the research. She wrote:

*Dear Mr. President,*

*As you know, Ronnie recently celebrated his ninetieth birthday. In earlier times, we would have been able to share our mutual pride in a life filled with wonderful memories. Now, while I can draw strength from these memories, I do it alone as Ronnie struggles in a world unknown to me or the scientists who devote their lives to Alzheimer's research. Because of this, I am determined to do what I can to save others from this pain and anguish. I'm writing, therefore, to ask your help in supporting what appears to be the most promising path to a cure—stem cell research.*

*I also know that this is not the first you have heard of this issue. And I know there are others who feel just as*

*strongly in opposition to this. But I ask your help to ensure that this embryonic stem cell research, under appropriate guidelines, be protected as scientists pursue medical miracle possibilities.*

*Ronnie was very brave in writing to the public about his condition. It was his way of sharing with the thousands of families who are already afflicted. He always believed in man's ability to make this a better world, and I know he would be gratified to know that his own suffering might spare others the same wrenching family journey.*

*Mr. President, I have some personal experience regarding the many decisions you face each day. I do not want to add to that burden, but I'd be very grateful if you would take my thoughts and prayers into your consideration on this critical issue.*

*Most sincerely,*
*Nancy Reagan*

Foundation, with the expressed aim to improve the prospects for people with serious spinal injury and to accelerate research, including that involving stem cell-based approaches. Likewise, Michael J. Fox, who in 1991 at the age of only 30 was diagnosed with early onset Parkinson's disease, upon disclosing his condition in 1998 committed himself to the campaign for increased research into the disease. He established the Michael J. Fox Foundation for Parkinson's Research, which played an early role in supporting work in stem cell research for Parkinson's disease, including funding the original proof-of-principle demonstration that ES cells could provide a robust source of dopamine neurons. Both actors openly criticized the Bush administration for blocking research that they believed could improve their conditions.

Going against traditional Republican party views, former United States First Lady Nancy Reagan became an advocate for research using human ES cells, believing strongly that the research could lead to a cure for Alzheimer's disease, which afflicted her husband Ronald Reagan for the last ten years of his life. In 2001, in a personal letter, she urged President Bush to reverse his ban on the use of federal funding for human ES cell research (Box 20.1). Although the ban remained in place, Nancy Reagan continued to help promote stem cell research, and in 2004, she went as far as to make a public speech on the issue at a fundraising dinner for the Juvenile Diabetes Research Foundation in Hollywood. She said she believed stem cell research "may provide our scientists with many answers that for so long have been beyond our grasp."

## Academic stem cell societies and organizations bring scientists together and help to engage the public's understanding of stem cell research

National and international societies and institutes focusing on stem cell research have played an ever-increasing role in providing better information to the public and specific patient groups and in influencing politics and decision-making processes. The most prominent and influential is probably the International Society for Stem Cell Research (ISSCR), which is an independent non-profit organization at the forefront of stem cell science. Founded in 2002, the ISSCR actively promotes the exchange of information between scientists in particular through its annual stem cell conference that has over 3000 researchers from over 55 countries attending. The ISSCR also has a strong public facing aspect, including online resources to inform patients about stem cell therapies (Figure 20.13) and guides in over ten different languages aimed at highlighting issues such as stem cell tourism. Likewise, in Europe, the EuroStemCell project, funded through the European Union's Horizon 2020 research and innovation program, provides independent, expert-reviewed information and tested educational resources on stem cells and their impact on society.

Figure 20.13 **Public information provided by International Society for Stem Cell Research.** The International Society for Stem Cell Research (ISSCR) is an organization at the forefront of stem cell science. It serves as a hub of information for the scientific community and publishes essential patient information regarding the use of stem cells to treat disease (public information provided by International Society for Stem Cell Research). Information on stem cell research and its application to human health can be found at www.closerlookatstemcells.org.

## 20.2 EXPLOITING STEM CELLS

Since the first successful bone marrow transplantation treatment for leukemia in 1957, there has been an ever-increasing desire to utilize stem cells from other organs to improve human health and tackle disease. Bone marrow transplantations are now common practice for the treatment of various blood disorders, but the same cannot be said for other therapies involving stem cells. However, a quick search of the Internet for stem cell news yields hits on a daily if not hourly basis. It is therefore impossible not to raise the expectations of those patients who are desperate for a cure for their particular condition. In this section, we will discuss how the general public perceive and act upon the information surrounding stem cell research, as is predominantly conveyed to them through newspapers, television, or the Internet. Importantly, we want to highlight the ways in which stem cells are being exploited for medical, economical, ecological, and simply personal gain and the impacts that these applications, whether genuine or misguided, are having on the future prospects for the stem cell field.

### As the promise of stem cell therapies increases, so does stem cell tourism

Traveling to other countries for medical treatment is not a new concept. "Medical tourism," as it became known, is a growing industry globally, becoming more commonplace for many reasons, including the relative ease of international travel, increasing healthcare costs in many countries, lengthy waiting times for treatment in countries with public healthcare systems, and the increase in accessibility to information through the Internet and social media. When stem cell science appeared to offer fresh hope for incurable diseases, people inevitably wanted to seek out those claiming to provide novel treatments. As we have already discussed, most if not all of such expectations were based on misconceptions and unscrupulous promotion of unproven or even untested therapies, the only possible exception being HSC transplantation for the treatment of a range of diseases for which efficacy has a long-established track record.

The majority of stem cell "treatments" available through specialized stem cell clinics employ cells taken from the patient's own body. Some clinics also claim to use donated cord blood or embryonic or fetal cells, but it is often hard to know what is being administered. Such clinics are deemed illegal in the United States, Canada, and Europe, and the majority of stem cell tourism is centered in countries such as Thailand, Costa Rica, Mexico, Argentina, China, India, and Russia. These clinics, of which there are many hundreds, offer treatments for conditions such as Alzheimer's, multiple sclerosis, Parkinson's disease, cardiovascular disease, and spinal cord injury, among a whole host of other conditions.

In these clinics, which can charge anything between $5000 and $40,000 for treatments that are unfounded, why are patients so willing to travel, pay such sums of money, and risk potentially harmful effects of stem cell procedures? The answers lie in the fact that many taking this route are desperate and view the treatment as a last resort (Figure 20.14). Compounding desperation, appealing websites highlight the supposed benefits of the treatments offered by a clinic, supported by positive patient testimonials but failing to point out the lack of effect on most patients and the significant associated risks. Due to the illegality of these clinics, many of the problems associated with such treatments go largely undocumented, and patients are reluctant to come forward with their experiences. However, the very same ease of global

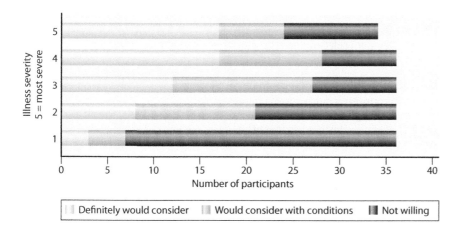

Figure 20.14 **Public opinion on stem cell treatment relative to the severity of the illness.** Incurable diseases often lead both patients and their loved ones to desperation and to a willingness to try anything. This public survey gauges whether people would consider stem cell therapies and how their response correlates with the severity of the disease or injury, the results unsurprisingly highlighting the fact that the more severe the illness is, the more willing people would be to try stem cell therapy. (From Einsiedel E & Adamson H. (2012). *Dev World Bioeth* 12:35–44 [doi: 10.1111/j.1471-8847.2012.00319.x]. With permission from Wiley.)

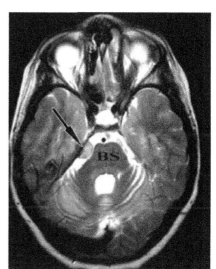

Figure 20.15 **Tumor deriving from transplanted fetal neural stem cells.** Some unregulated procedures using stem cells have been performed, but often with dire consequences. This image is an MRI scan of the brain of a young Israeli boy suffering from a hereditary neurodegenerative disorder, who 4 years previously had been treated in Russia with fetal neural stem cells. The scan reveals the presence of a tumor (arrow) next to the brain stem (BS). The tumor was subsequently shown to have derived from the transplanted stem cells. (From Amariglio N, Hirshberg A, Scheithauer B et al. (2009) *PloS Med* 6:e1000029 [doi: 10.1371/journal.pmed.1000029]. Published under the terms of the Creative Commons Attribution License http://creativecommons.org/licenses/.)

communication means that some stories do inevitably emerge. For example, in 2005, doctors in Israel discovered a tumor next to the brain stem (Figure 20.15) in a young boy who 4 years earlier had received fetal **neural stem cells** in Russia as treatment for a hereditary neurodegenerative disorder (ataxia telangiectasia). The tumor was confirmed by histological and genetic analyses as deriving from the injected fetal neural stem cells, in fact originating from two donors.

One of the early destinations for stem cell tourism was actually in Europe, namely the now infamous X-Cell Center based in Cologne and Düsseldorf in Germany (Box 20.2). The story of this clinic illustrates not only how desperate people seek out treatments, whatever the cost, and how there are always disreputable individuals who take every advantage they can of such desperation but also shows how national patterns of regulation can influence and be influenced by such operations. The Dutch founder, Cornelius Kleinbloesem, set up his clinic in Germany following imposition in 2007 of a moratorium in Holland banning unproven stem cell treatments, exploiting a loophole in German law that allowed him to operate. The center continued to take in patients until 2011, but following an adverse event involving bleeding in the brain of one young patient receiving stem cells by injection into his brain, the German authorities finally had a reason to be able to enforce a requirement for special application in order to continue therapies, a requirement that the clinic was unable to satisfy. Although this forced the closure of the X-Cell Centre, Cornelius

## BOX 20.2 CONTROVERSY: THE X-CELL CENTER

During the early years of stem cell tourism, one of the foremost destinations was the X-Cell Center, established in 2007 in Germany by Dutch entrepreneur Cornelius Kleinbloesem. The X-Cell Center offered stem cell treatments for numerous disorders, charging up to $12,000. Over the 4 years of its operation, the clinic treated thousands of people that were almost exclusively from outside of Germany. These people were attracted by reports of "miracles" from former patients published on the X-Cell Center's website, which made no mention of the risks associated with the treatments or statements of patients who experienced no benefit. Playing upon the tourist theme, paying patients were treated to greetings at airports and stations and chauffeured to the center,

making the whole experience "pleasurable." This pleasure was sometimes short lived when patients realized that this was a business driven solely by money rather than patient welfare, payment being demanded before therapy could start. Eventually though, the X-Cell Center was forced to close by German authorities following two reported incidences. The first was in 2010, when a 10-year-old boy from Azerbaijan suffered severe internal bleeding in the brain following a stem cell injection. Then, just a few months later, an 18-month-old boy died from complications relating to a similar procedure. German authorities imposed strict guidelines for any future stem cell treatments that the X-Cell Center could not uphold, and it was therefore forced to cease trading.

Kleinbloesem was not deterred, moving to Beirut in Lebanon, where he set up a clinic called Cells4health offering the very same range of treatments. This time, he avoided some legal issues by collaborating with a London-based stem cell company (Precious Cells International) that extracted stem cells from patients in London that were then flown to Beirut for injection back into the patient.

Until stem cell therapies enter mainstream medical practice, it is inevitable that clinics such as the X-Cell Center and Cells4health will continue to find ways to operate somewhere in the world and that there will be plenty of people who will be convinced that they have some chance of cure, however much this may cost. In the meantime, numerous stem cell researchers, bioethicists, policy makers, and professional societies such as the ISSCR are trying to develop strategies to combat the demand for stem cell tourism. Together, such groups are encouraging governments and regulatory bodies to implement guidelines that filter out the fraudulent clinics and are also producing accessible and understandable information to discourage would-be patients from being coerced into seeking out untested and risky therapies.

## Tissue stem cell banking for potential future use as a pay-for service

Another area of exploitation evolving out of the public interest in the potential of stem cell-based therapies centers on the banking of stem cells, with the expectation that these can be held as an insurance policy for the treatment of a variety of diseases at some point in later life. The contention here is not that cells can be banked and recovered for use, even decades later, since this has been common practice in the transplantation of HSC for over half a century. The problem is again one of public misconception of the benefits that can be expected from stem cell banking and the way in which commercial interests have played upon the proven success of HSC transplantations.

Tissues that have proved to be attractive in this context are usually regarded as "waste," including umbilical cords, deciduous teeth, and fat. You are by now familiar with the stem cells that have been characterized in these tissues, umbilical cord being a good source of HSC in the blood component (Chapter 9) and MSC (Chapter 10) contained in Wharton's jelly; shed infant teeth containing MSC that are in some respects distinct from those found in the bone marrow or umbilical cord; and fat, often collected from cosmetic liposuction procedures, containing MSC. However, the claims made by commercial stem cell banks for the potential application of stem cells from these tissues far exceed the known capacity of the stem cells present. The implication is often that any diseased or damaged tissue can be treated, but there is no evidence to support this broad assertion. Stem cell banking services come at a high premium, with businesses charging large sums of money to store stem cells frozen for several decades but also because they can limit donation of tissues such as umbilical cord into public stem cell banks for conventional use in **allogeneic** HSC transplantation. In the United Kingdom alone, there are approximately a dozen businesses that store umbilical cord derived cells, and at least 100,000 parents have opted to do so. Many companies are now promoting the storage of children's deciduous ("milk") teeth (Figure 20.16), a company with a base in the United Kingdom being one of the first in the world to collect these teeth within 24 hours of them falling out (following the purchase of the $60 kit), then isolating (for a fee of $320) and storing the stem cells housed within them (for an annual fee of $200) and absolutely no guarantee that the stem cells can ever actually be used!

## Stem cell cosmetics

There is a rapidly increasing interest in the use of stem cells in cosmetic products, claims being made that their regenerative potential makes them ideal either to prevent or to reverse the effects of aging (Figure 20.17), but most of the claims for benefit are frankly absurd. The cosmetic industry has definitely jumped on the bandwagon where stem cell technology is concerned and is benefitting from the media hype surrounding stem cells and their use in the treatment of human disease.

Aging is an inevitable part of life, bringing with it numerous consequences such as cognitive function decline, muscle mass reduction, hearing and eyesight loss, and osteoarthritis. Although many of these age-related changes might one day be alleviated by stem cell based regenerative treatments, more immediate concerns about appearance and the occurrence of wrinkles, gray hair, and balding have led to a burgeoning cosmetics industry using, or claiming to use, stem cells as the basis for its

**Figure 20.16 Tooth stem cell banking.** BioEden was the world's first company to offer the service of banking stem cells isolated from children's decidual "milk" teeth, playing on the children's tooth fairy belief. (Adapted from https://uk.bioeden.com.)

**Figure 20.17 Stem cells and cosmetics.** To many, maintaining their youth is as important as maintaining their health, and claims that stem cells can aid in this are fueling the growth of products that claim to be rejuvenating through the inclusion of stem cells, extracts from them, or agents that in some way enhance the body's own stem cells.

## BOX 20.3 CASE STUDY: WRINKLE TO BONE

In 2009, a Los Angeles woman in her 60s underwent cosmetic surgery involving the removal of abdominal fat by liposuction for the purpose of isolation of autologous MSC for injection into the skin around her eyes, the intention being that these cells would "rejuvenate" her skin. Some 3 months later she complained of immense pain in her right eye and reported a clicking noise when she attempted to open it. Upon examination, surgeons discovered small fragments of bone embedded around her eye resulting from

the uncontrolled **differentiation** of the injected MSC. In addition to the injection of unapproved MSC as the cosmetic agent, the surgeons also injected the dermal filler calcium hydroxyapatite that has been used for more than 20 years in cosmetic surgery. However, they failed to realize that this mineral is used to promote the differentiation of MSC toward bone! This skin "rejuvenation" not only cost this woman over $20,000, but also meant that she had to endure both further surgery and financial loss to correct the problems.

products. The cosmetic industry, which was predicted to be worth $265 billion in 2017, sees an opportunity of using the stem cell brand and is now frequently advertising cosmetics and cosmetic procedures that claim to reverse or prevent aging effects through the use of stem cells. A simple search on the Internet for stem cell creams reveals numerous products purportedly containing stem cells or extracts from them; however, in many instances the customer is being deceived about what they actually contain. Similarly, most stem cell cosmetic surgery treatments are actively promoted by private clinics without any substantiating clinical evidence that these therapies work, even though they usually cost tens of thousands of dollars and carry with them no explanation of potential risks.

Both the United States and the European Union authorities have to date not approved stem cells or extracts that promote their survival for use by the cosmetic industry; however, in spite of this there are still companies pursuing this, sometimes with dire consequences (Box 20.3). There are numerous cosmetic industry websites that advertise and advocate the use of stem cells, many of which neglect to inform the consumer about the risks associated with such treatments. To circumvent the restrictions placed on the use of human stem cells in cosmetics, many companies claim to incorporate plant stem cells (Chapter 8) and their extracts in their products. Numerous plants have been suggested to have potentially beneficial extracts such as apple, grape, alpine rhododendron, and jasmine gardenia, among others; however, the clinical evidence that they impact the aging process is at best speculative, and what any of this actually has to do with stem cells is little more than fantasy.

## The stem cell promise is also being exploited in animals

Not all controversy and misrepresentation relating to stem cells is restricted to their application in humans. Our animal companions, whether domesticated pets, farm animals, or wild species, provide both genuine and spurious opportunities for the application of stem cell-based products or knowledge and equally can raise ethical issues.

As we have described in Chapter 19, there are already examples of stem cell-based applications in domesticated animals ranging from treatments for conditions such as damaged tendons in horses to enhanced methods for the introduction of desirable genetic traits and **reproductive cloning**, as exemplified by Dolly the sheep. The introduction of novel therapies is nowhere near as tightly regulated for animals as it is for humans, which raises questions of its moral correctness, while any form of genetic modification is always bound to stir up controversy. Human reproductive cloning, which is universally banned irrespective of a particular country's view on ES cell research, has been widely successful for domesticated animals, including both companion species and farmed species. The first commercially cloned pet was a cat from North Texas named "Little Nicky," which in 2004 cost the owner $50,000. The first cloned dog, an Afghan hound called "Snuppy" (Figure 20.18), was announced in 2005 by Hwang Woo-Suk, the notorious stem cell scientist who fraudulently claimed the year before to have produced an ES cell line from a cloned human embryo (Chapter 3). Snuppy was created using a cell from the ear of an adult Afghan hound and was the sole survivor out of 1095 transplanted embryos implanted into 123 surrogate bitches, clearly a very high number of both newly created cloned embryos and surrogate mothers for the production of just one healthy surviving pup. There are now established companies that offer to prolong canine companionship through cloning using a small tissue sample obtained while the dog is living or shortly after

**Figure 20.18 The world's first cloned dog family.** Reproductive cloning of domestic animals is now available as a commercial service. The first cloned pet dog, an Afghan hound, was produced from a cell from the ear of the donor animal. The pup produced through this procedure, named Snuppy, was the only survivor from 1095 transplanted embryos but in turn was cloned to generate the second generation of pups that are shown in this picture. (From Kim M, Oh H, Kim G et al. (2017) *Sci Rep* 7:15235 [doi: 10.1038/s41598-017-15328-2]. Licensed under a Creative Commons Attribution 4.0 International License (http://creativecommons.org/licenses/by4.0).)

its death, all for the princely sum of $100,000. This business, like stem cell tourism, feeds on the vulnerability of people, who likewise are often misled by an intentional shortage of facts, and do not understand what lies behind the process or the fact that the cloning process cannot result in an exact replica of their original dog, particularly from the perspective of its behavior and character.

Another moral dilemma for humans raised by what stem cell technology has the potential to make possible in the very near future is the prospect that **induced pluripotent stem (iPS) cells** can be generated from tissues of any animal, alive or dead. As we discussed in Chapter 19, this is being seriously considered as a way to help save endangered species and in itself can be considered part of the broad ethical debate for those who object to the principle of creating new life by such a route. Specifically, there could be seen to be issues, with either a moral or philosophical basis, around the idea of "rescuing" a species from imminent extinction. In particular, if the decline in a species appeared to be due purely to Darwinian principles, rather than mankind's influence, then is it right to intervene in what is essentially a natural process? A more challenging question concerns the possibility that the same approach could soon be applied to species that have become extinct. Although the Hollywood view of bringing back dinosaurs through cloning and stem cell technology as portrayed in the 1993 hit film *Jurassic Park* can make people's imagination run wild, there is a realistic opportunity to resurrect extinct species such as the woolly mammoth, for which intact cells can be recovered from bodies preserved in permafrost since the Ice Age. In 2014, an almost perfectly preserved 43,000-year-old mammoth was discovered in Siberia containing intact blood cells (Figure 20.19). The cloning of the mammoth would require conversion of the isolated tissue cells into stem cells either by iPS cell technology or by SCNT, both of which are feasible. In order to produce a living animal by this route, it would be necessary to use a closely related living species both as a donor of embryos or eggs for the cloning procedures and as a surrogate in which the cloned embryo could develop to birth. In the case of the mammoth, this would be one of the extant species of elephant. Aside from the fact that the procedures would do harm, probably to a large number of animals, elephants are of course themselves endangered, making it unlikely that this would be viewed by humanity as an acceptable price to pay. Moreover, there are broader philosophical questions around the right of humans effectively to reverse what in all likelihood was a consequence of natural evolution, that is, the mammoth became extinct because it was bound to as its environment changed or other species outcompeted it. Also, would it be right to bring an extinct species back to be placed into either a zoo or an ecosystem in which it most certainly no longer has a place and which it may itself disturb? Perhaps the arguments will vary depending on the nature of a particular species' demise. The mammoth disappeared because that's what evolution effectively dictated, but what about a species such as the dodo, which disappeared rapidly as a result of man's activities? One could pose the question "For what purpose should we bring back extinct species?" For sure, there will be much debate on this over the coming years.

Figure 20.19 **Blood from a preserved mammoth could be a source of DNA for cloning to generate a live animal.** The woolly mammoth (**A**) became extinct around 10,000 years ago. In 2014, an almost perfectly preserved 43,000-year-old mammoth was discovered (**B**) containing intact blood cells. These blood cells (**C**) could theoretically be used to derive induced pluripotent stem cells that might then be used to bring the mammoth back to life. (A: "Woolly mammoth" by Flying Puffin. Licensed under the Creative Commons Attribution 2.0 Generic License via Wikimedia. (https://creativecommons.org/licenses/by/2.0/) B and C: Photo credit: Semyon Grigoriev.)

## 20.3   THE ECONOMIC PERSPECTIVE

Stem cells and the technologies associated with them are no different than any other product or process when it comes to their practical and commercial exploitation. From initial discovery or invention through to delivery and application can be a long path, with a huge amount of drop off along the way. Just as for drugs and medical devices being developed in the pharmaceutical and biotechnology industries, the commercialization of stem cell related products for use in humans is bound not only by IP rights and economic benefit but also has to be seen in the context of a massively complex regulatory environment that is in place to ensure that the benefits of a new product far outweigh the risks. As we have already seen, the chance of immediate economic benefit on a large scale has led many unscrupulous individuals to completely avoid the regulatory processes in delivering untested stem cell "therapies" to unwitting patients.

### Strict legislation across the globe determines the translational reality of stem cell research

The term "translation" has been widely adopted as a way of describing the process of taking a research discovery or invention through to a point that it can be practically applied. In the biomedical arena, which is the principal although not exclusive home for the application of stem cell science, this has recently given birth to the discipline of "translational medicine." The so-called "bench to bedside" approach that is represented by translational medicine is founded on more traditional drug discovery involving the interplay between clinics, the pharmaceutical industry, and government regulations and the bodies that implement these. On top of this structure, specific legislation and processes have been imposed because of the cellular nature of many of these new products. It is beyond this scope of this book to cover the details of the processes and regulations behind the clinical trials that have to be undertaken for any new drug, cell material, device, or procedure that is being proposed for human application, but further information on this and specific aspects relating to cell based therapies can be found in the reading list at the end of the chapter.

As for legislation covering research on embryo-derived cells, the regulations governing clinical application of stem cell therapies vary widely between different countries. In Europe, the European Medicines Agency (EMA) Committee for Advanced Therapies has created a strong regulatory framework for dealing with unproven stem cell therapies, placing the protection of patients at the center of all regulation. The EMA has developed legislation covering safety and efficacy for all medicinal products and maintains that the quality and manufacturing of these products should meet good manufacturing practice (GMP) requirements. Medicinal products derived from stem cell manipulation are tightly controlled by the advanced therapy medicinal products (ATMP) legislation, which states: "Cell therapies are defined as medicinal products when there is more than minimal manipulation of any cell type destined for clinical application or where the intended use of the cells is different to their normal function in the body. Any use of such cell-based medicines is subject to authorisation and controls, including their manufacture. Permitting manufacturers to avoid compliance with quality standards, for example by inappropriate reclassification of the treatment beyond the mandate of competent authorities for control of medicines, could risk exposing patients to cross-contamination and inadequate characterisation of the cell preparations, resulting in short and long-term risks for individual patients." Two exemptions have made it possible, though, for commercial interests to exploit unproven stem cell treatments. The first is through "compassionate" use programs, which allow doctors to obtain treatment for a patient while the medicine is still being developed or undergoing certification, asserting the patient's "right to try." The second concerns a hospital or medical practice exemption, which allows for medicinal products containing stem cells to be made available to an individual patient under the professional responsibility of a doctor, but only under the regulatory authority of the member state where the product was made.

### In the United States, the Food and Drug Administration oversees the safety and efficacy of drugs and stem cell treatments

In very much the same way, the United States Food and Drug Administration (FDA) uses regulations to ensure the safety and effectiveness of drugs, biological products,

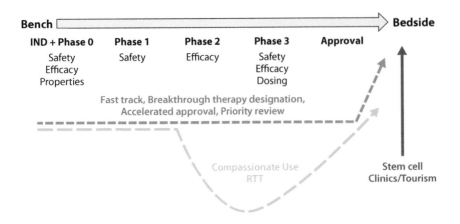

**Figure 20.20 The United States Food and Drug Administration approval process for a new stem cell therapy.** The introduction of new stem cell therapies into clinical practice is tightly regulated by the United States Food and Drug Administration (FDA). The diagram highlights the phases involved in moving stem cell therapies from the research phase (bench) to the treatment of patients (bedside). Mechanisms (orange) have been put in place by the FDA to fast track the process by accelerating the clinical trial phases (yellow arrow). The green route shows the use of a drug for compassionate use, or right-to-try (RTT), which bypasses clinical trial Phases 2 and 3. The red route shows how stem cell tourism can effectively bypasses all of the safety and efficacy testing. (From Knoepfler P. (2015) *Adv Drug Deliv Rev* 82–83:192–196 [doi: 10.1016/j.addr.2014.12.001]. Copyright 2015 with permission from Elsevier.)

and medical devices under the Biologics Control Act of 1902 and the Food, Drug, and Cosmetic Act of 1938. The review process requires evidence of the safety and effectiveness of the product in adequate and well-controlled human clinical trials, which are underpinned by sufficient data generated from animal studies to aid in evaluating any potential risks (Figure 20.20). It must also be shown how the safety, purity, and potency of the product will be ensured during its manufacture. In 1997, the FDA amended its regulations to encompass cell and tissue-based products, and, as a result, human cells, with the exception of those deemed "minimally manipulated" and used for **autologous** treatments (e.g., bone marrow or umbilical cord blood used for HSC transplantations) would be subject to FDA premarket approval as biologics. As for interpretation of the European regulations, determining the degree of manipulation and homology of use has been somewhat exploited by many profit-making stem cell clinics as a means to avoid the necessity of FDA approval and by classifying their treatments as minimally manipulated. In some instances, such as terminal illness, the FDA can approve a therapy for trial on a case-by-case basis provided it has met Phase 1 safety testing in what is termed as compassionate use. Some states in America, such as Colorado, also have a "right to try law" (RTT), which likewise allows gravely ill patients to enter into clinical testing of a stem cell therapy without the need for the full process of FDA approval.

The FDA generally sends warning letters to clinics that claim to offer treatments, or even cures, using stem cell therapies that are not approved. The FDA will also initiate legal proceedings, as it did, for example, in 2009, when it acted to prevent Regenerative Sciences, LLC, from manufacturing a product called Regenexx, involving manipulation of autologous MSC outside of the body and subsequent re-injection in order to treat orthopedic injuries. Similarly, in 2011, three men were arrested and charged with 15 counts of criminal activity related to the illegal use of stem cells. In this latter case, a licensed midwife running a maternity care clinic in Texas obtained umbilical cord blood from birth mothers, telling them it was for research purposes, but sold it on to a laboratory in Arizona that passed it to a paid consultant at a university in South Carolina, who used university facilities to manufacture stem cell products. These products were then sold to a man representing himself as a physician licensed in the United States who traveled to Mexico to perform unapproved stem cell procedures on people suffering from cancer, multiple sclerosis, and other autoimmune diseases.

The FDA also works with the Bureau of Consumer Protection of the Federal Trade Commission (FTC), whose role is to protect United States consumers from false, misleading, or deceptive advertising. Under the Food, Drug, and Cosmetic Act, advertising for treatments based on stem cells that are not approved by the FDA is a serious infraction of the law. The FTC itself can only enforce its own statute, not the Food, Drug, and Cosmetic Act, and then only if it can show that a claim is false or deceptive. With emergent scientific issues like stem cell therapies, the FTC looks to the FDA for guidance to establish the absence of scientific validity. To date, the FTC has not brought up charges in an enforcement case regarding the advertising of unapproved or unsubstantiated stem cell therapies. A typical website has a litany of disclaimers tailored to escape United States regulation, an example of which is: "We do not claim that our treatment protocols are approved by the United States FDA or proven to be effective in the United States for any condition that appears on this site or for any other condition."

## The patenting of stem cell discoveries is controversial

A major determinant of the progression of a discovery or invention towards translation and commercial application is the ability to secure IP ownership through patenting. Some would argue that patenting a discovery can have an inhibitory effect on research progress, and for that reason the authorities controlling the patenting laws and their interpretation have expended considerable effort to create the right balance so as to encourage both fundamental discovery and commercial application. The particular problem in the area of stem cells and associated research tools is the nature and source of the cells, raising questions about whether they should be patentable on philosophical or ethical grounds and to what extent manipulations to these cells might render them patentable.

Both United States and European patent laws have been tested in some very high-profile cases involving stem cell related materials. The United States patent law clearly states that the laws of nature, natural phenomena, and abstract ideas are not eligible for patent, but even with that backdrop, the United States Patent and Trademark Office has recognized inventions involving stem cells. In deliberating on some contested cases, the United States courts have noted that there is a "delicate balance between creating incentives that lead to creation, invention, and discovery" versus "impeding the flow of information that might permit, indeed spur invention," and making it clear that much depends on the extent of manipulation of the stem cell in the laboratory. In this context, three patents held by the Wisconsin Alumni Research Foundation (WARF) have been center stage. These patents cover the isolation of the first non-human primate and human ES cell lines and have been the subject of legal challenge for a decade. They were originally challenged in 2006 by the Foundation for Taxpayer and Consumer Rights and the Public Patent Foundation, the challenges asserting that the patents are overly broad and inhibit research access. The challenge finally ended in February 2015, when the Supreme Court declined to hear the case, upholding the decision of the United States Patent and Trade Office to maintain the patents. The decision means the WARF will get to keep its patent rights for the cells, created in 1998 by University of Wisconsin-Madison scientist James Thompson, even though their claim that the methods used to isolate them from embryos were novel has been vigorously disputed as being obvious and therefore not patentable. Nevertheless, the challengers succeeded in preventing WARF from gaining rights over iPS cells.

In the European Union, the 1998 Biopatent Directive is meant to harmonize patent protection for biotechnology inventions across Europe. An important exemption detailed in the Directive has had, and continues to have, a particularly profound influence on the ability of researchers in Europe to patent inventions involving human ES cells. In Article 6(2), it states that "uses of human embryos for industrial or commercial purposes" shall be excluded from patentability. This has famously led to another long running challenge, this time to a patent granted to Oliver Brüstle in 1999 by the German Patent and Trademark Office covering neural precursor cells derived from human ES cells. Greenpeace filed a case against the patent in 2004, and in 2006, the German Federal Patent Court ruled that the patent was mainly invalid based on interpretation of Article 6(2)(c). Brüstle appealed in the German Supreme Court, which in turn requested a ruling by the Court of Justice of the European Union (CJEU) on the interpretation of Article 6(2)(c). The CJEU held that the concept of "human embryo" must be understood to encompass any ovum once fertilized, and from this, the German Federal Supreme Court ruled that the Brüstle patent would be invalid with respect to claims encompassing neural precursor cells that had been obtained by destroying embryos. The ban on patenting human ES cells in the European Union is currently being challenged by groups in the United Kingdom.

## The high cost of stem cell therapies needs to be balanced against the overall economic benefit

A major current question that hangs over the application of stem cell science in healthcare is one of health economics. Essentially this poses the question of whether the benefits that a new therapy are likely to provide in the foreseeable future are justifiable given the current high costs. This has both a purely economic dimension in times when healthcare budgets are stretched and also a moral dimension when the costs of stem cell therapies are compared to relatively inexpensive treatments that could be delivered to far greater numbers of people, especially in developing countries.

BOX 20.4  BACKGROUND: COST OF DIABETES TREATMENT IN A LIFETIME

Diabetes mellitus is predicted to become one of the world's biggest disabling diseases in the next 25 years. In 1985, there were approximately 30 million people worldwide with the disease, and this figure is predicted to increase to 300 million by 2025. The number of deaths worldwide due to diabetes is estimated at around 800,000 per year, and this is thought likely to be a considerable underestimate, with actual numbers somewhere in the region of 4 million, representing 9% of the global total of deaths. Diabetes is an extremely costly disease with no current cure. The costs incurred by the affected individual, the healthcare system, and wider aspects of society encompass the costs of drugs, health and life insurance, hospital residences, and regular laboratory tests, not to mention the economic component through loss of workforce. In the United Kingdom, treatment for diabetes constitutes 10% of the overall National Health Service (NHS) budget, representing approximately $14 billion per year (Figure 1), and this figure is thought to project to an estimated 17% in the not so distant future, the majority of this cost resulting from complications associated with the disease such as cardiovascular disease, kidney disease, and neural conditions.

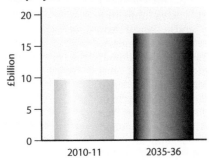

**The costs of diabetes in 2010-11 versus the projected costs for 2035-36**

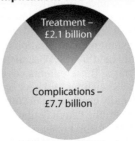

**The proportion of diabetes spending on complications and treatment**

Treatment – £2.1 billion

Complications – £7.7 billion

Total – £9.8 billion

Box 20.4 Figure 1 **Cost of diabetes in the United Kingdom.** The incidence of diabetes is increasing worldwide, with a corresponding increase in healthcare costs. Analysis of the medical cost of diabetes in the United Kingdom between 2010–2011 was around £10 billion and is predicted to nearly double by 2035 (left). The majority of the medical costs associated with diabetes are as a result of complications of the disease (right). (Adapted from www.diabetes.org.uk.)

With an ever-increasing global population and an increasing life expectancy, it is inevitable that the cost of healthcare is set to escalate dramatically. In the United States, the total healthcare costs per capita almost doubled between 2000 and 2011, and the costs were expected to reach $10,000 per capita by the end of 2015. A key factor in this rise in costs is that although life expectancy is increasing, this is not being matched by improvements in the health status of people in their old age, and the number of people with age-associated chronic disorders, often involving multiple co-morbidities, is severely stretching healthcare systems. Countering the argument that stem cell therapies are inherently costly compared to many other treatments, there is also a view that they could alleviate some of the pressure on medical costs associated with the long-term care of the chronically sick elderly by providing long-lasting cures for major and economically costly diseases such as diabetes (Box 20.4). At present it is impossible to predict exactly how stem cell therapies could benefit us financially. However, based upon costs of current treatments for heart disease, diabetes, and neurological disease and estimates of the costs of stem cell-based therapies, some predict that there could be significant savings overall (Figure 20.21). For the treatment of diabetes with stem cell therapy, the estimated savings, even if only 20% of diabetes patients undergo the treatment, would already be over $100 billion per year. This savings figure is likely to increase as stem cell therapies become more common practice, effectively releasing the financial burden associated with lengthy medical treatments.

## 20.4  SUMMARY

This chapter marks the end of our book in which we have described the massively diverse nature of stem cells, a topic that is already vast but which will undoubtedly expand into areas that are presently mere science fiction. Taking everything that we currently know about stem cells, it seemed fitting to end with this chapter by

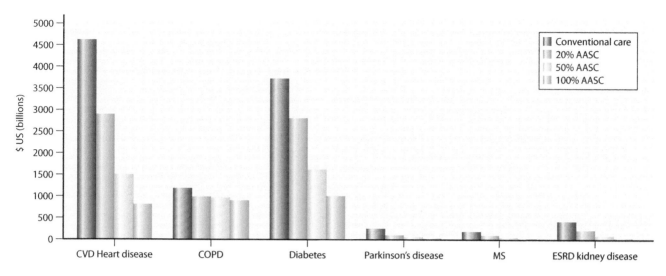

**Figure 20.21 Estimated reduction in medical costs through application of stem cell therapies.** The overall cost of using stem cells as a therapy has to be balanced against the medical costs for current treatments. Analysis of the predicted cost of therapies over the course of 8 years for various diseases using conventional therapy (red), 20% treatment with an autologous adult stem cell (AASC) therapy (blue), 50% AASC (yellow), or 100% AASC (green) shows that stem cell therapies should eventually reduce medical costs. Abbreviations: CVD – cardiovascular disease; COPD – chronic obstructive pulmonary disease; MS – multiple sclerosis; ESRD – end stage renal disease. (From www.repairstemcells.org. With permission from Don Margolis, Repair Stem Cells Institute.)

reflecting on some of the philosophical, ethical, and financial boundaries that might be considered before taking aspects of stem cell science into society. We have highlighted how important the media is in swaying the public's opinion on stem cell research and the medical tragedies that have resulted in the downfall of stem cell trial programs. The general opinion on the use of ES cells is heavily influenced by both religious and political beliefs, but despite the various doubts, numerous private organizations and centers have been founded that use stem cells in patients. Few if any of these uses have undergone rigorous scrutiny for their efficacy and safety, resulting in unnecessary suffering or disillusionment in the patients, the demise of the unscrupulous purveyors, and a loss of trust among the general public. Finally, we ended on the financial implications of stem cell applications, since all of the scientific research and clinical trials might be in vain if the cost of novel therapies and the actual improvement in life quality cannot be squared with the cost and benefit of conventional treatments.

## KEY POINTS

- The public perception of stem cells has been very much skewed toward ES cells and issues relating to human embryos.

- Factors influencing individual and population views on stem cells, especially when human embryos are involved, are many, including religion, political inclination, and the media.

- When reporting stem cell science, the media are often driven by the need to sensationalize, but scientists themselves are sometimes responsible for the accuracy of the information provided.

- National reactions to stem cell science involving human embryos, including the legislative environment, vary widely between countries, often guided by the predominant religious views.

- There is no national regulatory framework that permits human reproductive cloning.

- Restrictive legislation during the presidency of George W. Bush limited how federal funding could be used for human ES cell research but stimulated a culture of advocacy, state-specific support, and international migration of researchers to more permissive environments.

- Public overexpectation about what stem cell science can presently achieve in terms of patient benefit, combined with people in desperate health circumstances, created a stem cell tourism market in countries with little or no regulatory framework able to control unscrupulous promotion of untested therapies.

- In countries with a good regulatory framework, there is not always complete clarity about the real benefits of commercial stem cell products, ranging from stem cell banking to cosmetics.

- The application of stem cell science to animals also raises ethical and philosophical concerns.

- As for conventional development of new therapies, the path of stem cell products or technologies from the research laboratory to use in patients in a clinic is costly, heavily regulated, and lengthy.

- A major hurdle for commercialization of stem cell related products is their patentability, which is complex and often hangs upon subtle interpretation of legislation.

- Ultimate uptake of stem cell products and technologies by healthcare systems will depend on health economic considerations.

## FURTHER READING

Appendix E: Overview of International Human Embryonic Stem Cell Laws. (2012) The Stem Cell Debates: Lessons for Science and Politics. A Report of the Witherspoon Council on Ethics and the Integrity of Science. *The New Atlantis*: 129–146.

Board on Health Sciences Policy; Board on Life Sciences; Division on Earth and Life Studies; Institute of Medicine; National Academy of Sciences. (2014 Jun 18) *Stem Cell Therapies. Opportunities for Ensuring the Quality and Safety of Clinical Offerings: Summary of a Joint Workshop*. National Academies Press (US), Washington (DC).

Friedman, L.M., Furberg, C.D. and DeMets, D. (Eds) (2010) *Fundamentals of Clinical Trials*, 4th ed. Springer.

Huss, R., Guenther, C. and Hauser, A. (Eds). *Advances in Pharmaceutical Cell Therapy: Principles of Cell-Based Biopharmaceuticals*, 1st ed. World Scientific.

Zachariades, N.A. (2013) Stem cells: Intellectual property issues in regenerative medicine. *Stem Cells & Development*, 22: 59–62.

# Glossary

| Chap | Term | Definition |
|------|------|------------|
| 1 | Adult stem cell | An undifferentiated cell found in a differentiated tissue that can both self-renew and differentiate to yield all or some of the specialized cell types of the tissue from which it originated. |
| 8 | Apical meristem | The growth points of plants that contain stem cells, positioned at the tips of shoots (shoot apical meristem, SAM) or roots (root apical meristem, RAM). |
| 1 | Asymmetric | A term applied to the division of a stem cell when the two daughter cells are non-equivalent, most often being one stem cell and one cell committed to differentiation. |
| 17 | Biomaterial | A chemical material that mimics the structure and properties of naturally occurring biological molecules, usually to serve as a support for the growth and differentiation of cells in a way that recapitulates a body tissue. |
| 7 | Blastema | A region of regeneration, usually at a site of truncation of a limb or digit, which involves de-differentiation and trans-differentiation of cells to replace those lost. |
| 3 | Blastocyst/blastula | A preimplantation embryo of 30A–150 cells. The blastocyst consists of a sphere made up of an outer layer of cells (the trophectoderm), a fluid filled cavity (the blastocoel), and a cluster of cells on the interior (the inner cell mass). A variant term is "blastodisc," in which the developing embryo is formed as a flat disc with no cavity. |
| 15 | Bulge | A region surrounding a hair follicle that contains stem cells for hair follicle, skin epidermis, and melanocyte generation. |
| 8 | Callus | A disorganized growth of plant cells, containing de-differentiated cells that are able to generate new meristems capable of producing fully differentiated plant tissues. |
| 18 | Cancer stem cell (CSC) | A cell found in a tumor that is capable of recapitulating the hierarchy of more differentiated cells that constitute the tumor. Also known as a "cancer initiating cell." |
|  | Cap cell | This cell forms a crucial part of the stem cell niche in *Drosophila* gonads, providing essential signals in the form of both direct molecular contacts and through the production of growth factors. |
| 13 | Cardiosphere | A 3D spherical structure of cells produced *in vitro* through the self-renewal and differentiation of cardiac progenitor/stem cells. |
| 3 | Chimera | An organism composed of cells derived from at least two genetically different zygotes. Theoretically, the zygote could be from separate species. |
| 3 | Clone/cloning | A clone can refer to an organism or an expansion of cells; in both cases, the term indicates the genetic equivalence of the component cells. Cloning is the process in which an organism produces one or more genetically alike copies of itself by asexual means. Cloning may occur by propagation of cuttings, as in the case of plants; continual budding, as in the case of hydra; fission, as in the case of bacteria and protozoa; parthenogenic asexual reproduction, as in the case of aphids; or somatic cell nuclear transfer, as in the case of higher-order animals such as mammals. The term cloning can also be applied to a group of cells undergoing replication by repetitive mitoses (cell divisions). Also see entries for Reproductive Cloning and Therapeutic Cloning. |

*(Continued)*

| Chap | Term | Definition |
|------|------|------------|
| 7 | Clonogenic neoblast | A pluripotent stem cell found in invertebrates that is able to generate all of the tissues of the organism. |
| 2 | Cluster of differentiation (CD) | Cell membrane molecules that can be bound by specific antibodies, enabling their use in the identification and isolation of cells. |
| 2 | Colony-forming unit (CFU) | Groups of cells growing on a culture plate surface or in within a semi-solid nutrient with each group being created from the multiplication of an individual stem/progenitor cell. |
| 1 | Commitment | Refers to the step in a cell's existence when its fate is determined along a particular pathway of differentiation. |
| 2 | Cre Recombinase/ Cre-loxP | This is a system for the targeted recombination of DNA in an experimental organism allowing deletion or insertion of portions of the genome. The Cre recombinase protein is normally expressed in bacteria and recognizes sequences (loxP sites) that are present in the DNA of viruses that infect bacteria, thereby limiting the ability of the virus to replicate. Through genetic engineering, loxP sites can be introduced into a desired gene location. Expression of Cre recombinase can be elicited in specific cells at an appropriate time or stage in their differentiation using a variety of approaches. As a consequence, the genome of the genetically modified organism is cut at the loxP site. |
| 7 | Dedifferentiation | Changes in the differentiation status of a cell so that it resumes the phenotype of a stage earlier in the lineage of which it is a part. |
| 10 | Dental pulp stem cell (DPSC) | A mesenchymal stem cell-like cell that resides in the inner pulp core of adult teeth and is able to differentiate into odontoblasts, the precursors of dentin-producing cells. |
| 1 | Differentiation | The process whereby a stem cell, progenitor, or precursor cell assumes a more mature phenotype. |
| 3 | Ectoderm | The upper, outermost of the three primitive germ layers of the embryo; it gives rise to skin, nerves, and brain. |
| 3 | Embryoid body (EB) | Clumps of cellular structures that arise when embryonic stem cells are cultured – they do not have an *in vivo* equivalent, although they are analogous to the pre-implantation blastocyst. Embryoid bodies contain tissue from all three of the germ layers: endoderm, mesoderm, and ectoderm. |
| 3 | Embryonal carcinoma cell (EC cell) | A type of pluripotent stem cell derived from a teratocarcinoma (usually a testis tumor). |
| 3 | Embryonic germ cell (EG cell) | Cells found in a specific part of the embryo/fetus called the gonadal ridge that normally develop into mature gametes. |
| 1 | Embryonic stem cell (ES cell) | Pluripotent cells obtained by culture of cells derived from the blastocyst inner cell mass (in rodents) or the epiblast of the post implantation embryo (primates). |
| 3 | Endoderm | Lower layer of a group of cells derived from the inner cell mass of the blastocyst; it later becomes the lungs and digestive organs. |
| 13 | Endothelial progenitor cell (EPC) | A cell that serves to regenerate the endothelial lining of blood vessels. |
| 3 | Epiblast | Gives rise to the ectoderm and mesoderm. The mesoderm then displaces the hypoblast cells and forms the entodermal cell layer on its inner surface. |
| 3 | Epiblast derived stem cell (EpiSC) | Pluripotent cells obtained by culture of cells derived from the epiblast, which arises from the inner cell mass of the blastocyst in mammals or the blastodisc in birds/reptiles. |
| 1 | Epigenetic(s) | Non-genetic effects, that is, not involving the genomic DNA sequence, which can influence gene expression. |
| 13 | Epithelial to mesenchymal transition (EMT) | The process by which cells with epithelial characteristics lose their cell contacts and polarity to become migratory and adopt a mesenchymal, fibroblast-like phenotype. |
| 3 | Extracellular matrix (ECM) | The macromolecular microenvironment next to a cell that allows for structural support, orientation, and connections for cell-to-cell interactions and formation of connective tissues. |
| 2 | Fluorescence activated cell sorting | A means of purifying cells based on the attachment of fluorescent-labeled antibodies to cell surface proteins that are characteristic of a specific cell type. Such marking of cells is combined with passage of the cells through a laser beam as a single cell stream to activate the specific fluorescent labels, followed by steering of cells with the desired fluorescence into a collection vessel. |

*(Continued)*

| Chap | Term | Definition |
|---|---|---|
| 17 | Gene editing | This term encompasses any form of genetic manipulation that is subsequently inherited by cells in either culture or *in vivo*. More recently, the term has become synonymous with the precise modifications that can be introduced using CRISPR-cas9 gene modification. |
| 12 | Germ (line) stem cell (GSC) | Adult stem cells residing in the gonads that are the source of gametes (sperm and oocytes). Although monopotent, these stem cells might also be regarded as totipotent since their progeny can contribute to the genetic make-up of the zygote. |
| 9 | Hematopoietic stem cell (HSC) | The precursor of mature blood cells that are defined by their ability to replace the bone marrow system, following its obliteration (for example, by irradiation), and re-establish mature blood cell production. |
| 5 | Heterokaryon | A multinucleated cell formed by the fusion of genetically distinct cells. |
| 6 | Hub cell | Also known as cap cells, these are cells that contribute to certain stem cell niches, forming a point of attachment and orientation for stem cells. Specifically applied to describe cells in the germ stem cell niche in *Drosophila* gonads. |
| 5 | Induced pluripotent stem cell (iPS cell) | Pluripotent ES cell-like cells that have been derived from somatic cells (adult or embryonic) that have been reprogrammed by various means, including ectopic expression of pluripotency transcription factors or by chemical and growth factor treatment. |
| 3 | Inner cell mass (ICM) | A small group of cells attached to the wall of the blastocyst (the embryo at a very early stage of development that looks like a hollow ball). Embryonic stem cells are made by isolating and culturing the cells that make up the inner cell mass. During development, it is the inner cell mass that will eventually give rise to all the organs and tissues of the future embryo and fetus but does not give rise to the extra-embryonic tissues, such as the placenta. |
| 11 | Intestinal stem cell (ISC) | Stem cells located at the base of the intestinal villi crypts that serve continuously to replace the differentiated cells of the intestinal epithelium. |
| 2 | Knockout | A mouse that has had one or both copies of a specific gene deleted or inactivated by homologous recombination or targeting by gene editing techniques. |
| 1 | Label-retaining cell (LRC) | Labels, such as radioactive thymidine, incorporated as a pulse into newly synthesized DNA, are passed to daughter cells and subsequent cell generations, leading to overall dilution of the amount of label per cell. If a daughter cell enters a quiescent state immediately after labeling, as can be the case for a stem cell following asymmetric division, then this cell will retain the label for a longer time than the daughter cell that commits to further proliferation and differentiation. |
| 11 | +4 Label-retaining cell | A label-retaining cell in the intestinal crypt that has stem cell properties but is downstream of the stem cells at the top of the hierarchy. Probably serves a reserve or backup role. |
| 15 | Limbic stem cell | Stem cell located in the cornea in the limbic region that generates the corneal epithelium. |
| 2 | Lineage tracing | The identification of all progeny of a single cell. Usually employs lineage-restricted or inducible genetic labeling involving expression of an enzyme that acts on a substrate, giving a colored product, or a DNA recombinase that activates a colored reporter gene product. |
| 15 | Mammary stem cell | Multipotential adult stem cell residing in the mammary gland that both self-renews and gives rise to the specialized mammary epithelial cells. |
| 15 | Mammosphere | A 3D spherical structure of cells produced *in vitro* through the self-renewal and differentiation of mammary gland progenitor/stem cells. |
| 8 | Meristem | Regions of a plant containing undifferentiated cells from which growth can occur. Differentiation of meristem cells gives rise to the organs of a plant, including shoots, roots, leaves, and flowers. |
| 5 | Mesenchymal to epithelial transition (MET) | Process by which migratory mesenchymal cells (often spindle-shaped) adopt an epithelial morphology, that is, planar and polarized into an apical face and basal surface. |
| 10 | Mesenchymal stem cell (MSC) | Also known as mesenchymal stromal cells, they can give rise to a large number of tissue types such as bone, cartilage (the lining of joints), fat tissue, and connective tissue (tissue that is in between organs and structures in the body). |
| 3 | Mesoderm | The middle layer of the embryonic disc, which consists of a group of cells derived from the inner cell mass of the blastocyst. This middle germ layer is known as gastrulation and is the precursor to bone, muscle, and connective tissue. |

(Continued)

| Chap | Term | Definition |
|---|---|---|
| 1 | Monopoten(t/cy) | Describes a stem cell that is able to self-renew and commit to differentiation along a single lineage to yield one type of functional mature cell (e.g., skeletal muscle). |
| 4 | Morphogen | A signaling molecule that can influence the fate and differentiation of cells. Often produced from a discrete group of cells to create a gradient that can lead to distinct fates depending on concentration at particular points within the tissue. |
| 1 | Multipoten(t/cy) | Describes a stem cell that is able to self-renew and commit to differentiation along multiple lineages, but all within a particular tissue, organ, or physiological system (e.g., hematopoietic stem cells giving rise to the cells of the blood). |
| 10 | Muscle stem cell (MuSC) | See *Satellite Cell* |
| 14 | Neural stem cell (NSC) | A type of stem cell that resides in the brain, which can make new nerve cells (neurons) and other cells that support nerve cells (glia). In the adult, neural stem cells can be found in very specific and very small areas of the brain where replacement of nerve cells is seen. |
| 14 | Neurosphere | An organoid growing in culture that originates from neural stem cells and contains a mixture of stem cells and their differentiated progeny, including neurons, astrocytes, and oligodendrocytes. |
| 11 | Neutral drift | A term from evolutionary biology that, when applied to stem cell populations, describes how they change with time. If a stem cell is lost from an equipotent population, this will be replaced by the division of a neighbor, leading to a dynamic behavior in which clones expand and contract at random so that they eventually take over or are eliminated. |
| 6 | Niche | In the context of stem cells, the niche is an area within a tissue that provides a specific microenvironment that maintains a stem cell's potential for self-renewal and differentiation. The microenvironment may have one or more cell components, including neural inputs, as well as extracellular matrix and assorted soluble growth factors. |
| 3 | Nuclear cloning | See *Cloning* |
| 6 | Organizer cell | Cells in plant stem cell niches in the meristems that are in close contact with the stem cells and serve to control self-renewal and differentiation. They are analogous to hub cells in some animal stem cell niches. |
| 2 | Organoid | 3D masses of cells growing in culture that are derived from stem cells through a process of expansion, commitment, differentiation, and self-organization. These structures mimic the tissues from which the stem cells are derived, including the presence of stem cells in their niche. |
| 11 | Oval cell | A cell type described from the liver that has been proposed as a bipotent progenitor for hepatocytes and biliary epithelium, but there is controversy about their generality in species other than rodents and their true role in liver regeneration. |
| 11 | Paneth cell | Cell present in the intestinal epithelium at the base of the villi crypts. They are secretory, producing antimicrobial proteins. Their location adjacent to the intestinal stem cell supports that they contribute to the stem cell niche. |
| 6 | Parabiosis | In the context of investigative stem cell biology, the term usually refers to experiments on rodents involving the joining of the circulatory systems of two animals for the purpose of determining rates of stem cell mobilization, migration, and the influence of the systemic environment on stem cell maintenance and aging. |
| 3 | Parthenogenesis | A form of reproduction where an egg develops without the fusion of sperm with the egg cell. Parthenogenesis occurs commonly among insects and other arthropods. Artificially inducing parthenogenesis using human eggs may be a means to isolate stem cells from an embryo without fertilization. |
| 10 | Pericyte | Fibroblast-like cell, related to mesenchymal stem cells, which is located on the outer face of arterioles, capillaries, and venules in direct contact with the endothelial cells of the vessel wall. |
| 5 | Plasticity | A proposed, yet controversial process by which a given tissue stem cell can also become a specialized cell type characteristic of a different tissue. For example, when a hematopoietic stem cell is seen to generate both blood cells and muscle cells. |
| 3 | Pluripoten(t/cy) | A characteristic of stem cells that can become all the cell types that are found in an implanted embryo, fetus, or developed organism but not embryonic components of the trophoblast and placenta (extra-embryonic tissues). |

(Continued)

| Chap | Term | Definition |
|------|------|------------|
| 1 | Potency | In relation to stem cells, this term refers to their differentiation potential, in particular the number of distinct lineages or cell types that they are capable of generating. |
| 3 | Primordial germ cell (PGC) | These originate early in embryogenesis of invertebrates and vertebrates and populate the developing gonads, where they serve as the source of the stem cells that give rise to the gametes. |
| 1 | Progenitor (precursor) cell | A progenitor cell is an early descendant of a stem cell that can differentiate but is no longer able to self-renew. A progenitor cell is often more limited in the kinds of cells it can become than a stem cell. |
| 1 | Quiescence | The state in which a cell is not dividing but retains the ability to re-enter the cell cycle. Many adult stem cells exist in their niche in a quiescent state, being triggered to self-renew and differentiate when they receive a signal input that indicates a demand for replacement cells. |
| 7 | Regenerat(e/ion) | In biology, the process of regeneration leads to the regrowth of a damaged or lost tissue from remnants of that tissue or from adjacent structures. The term is also used to mean ways of medical intervention to artificially achieve the same. |
| 17 | Regenerative medicine | Medical interventions that aim to repair damaged organs, most often by using stem cells to replace cells and tissues damaged by aging and by disease. |
| 3 | Reproductive cloning | Somatic cell nuclear transfer used for the production of a fetus and delivery of a live offspring that is genetically identical the donor of the somatic cell DNA. Also see entries for Cloning and Therapeutic Cloning. |
| 5 | Reprogramming | Used in the context of cell reprogramming, this refers to the process of changing the phenotype of a differentiated cell back to a developmentally more immature state, which usually means to a pluripotent state. |
| 14 | Retinal stem/progenitor cell | Refers to cells from the retina that are able to differentiate into photoreceptor (rod or cone) cells. In most vertebrates, a small group of stem cells are maintained at the margin of the retina. |
| 10 | Satellite cell | Also known as skeletal muscle stem cells, these monopotent stem cells, which are located as single cells between the muscle fibers and the basal lamina, give rise to replacement muscle cells throughout life. |
| 1 | Self-renewal | This describes a defining feature of a stem cell, that is, its ability upon cell division to generate daughter cells, one of which is in all ways measurable exactly the same as itself in terms of phenotype as determined by gene expression and protein activity. |
| 10 | Skeletal stem cell | See *Mesenchymal Stem Cell* |
| 2 | Side population | A definition of stem cell populations based on a particular protocol for FACS. Applies to a wide variety of stem cell types. The property is related to the presence of a protein that expels small molecules from the cell. |
| 3 | Somatic cell nuclear transfer (SCNT) | A technique in which the nucleus of a somatic cell is injected into an egg that has had its nucleus removed. If the new egg is then implanted into the womb of an animal, an individual will be born that is a clone. The clone has the identical genetic material as the somatic cell, which supplied the nucleus that carries the genetic material. |
| 12 | Spermatogonial stem cell | Monopotent germ stem cells of male gonads that differentiate into sperm. |
| 3 | Stromal cell | A non-blood cell that is derived from blood organs, such as bone marrow or fetal liver, which is capable of supporting growth of blood cells *in vitro*. Stromal cells that make this matrix within the bone marrow are also derived from mesenchymal stem cells. |
| 1 | Symmetrical division | A term applied to the division of a stem cell when both daughter cells are equivalent, usually retaining the stem cell features of the mother cell, although, in some circumstances, division can lead to commitment of both daughters. |
| 18 | Telomere | This is the terminal region at the ends of a chromosome, which is constituted of multiple repeats (thousands) of short DNA sequences. They serve to protect the chromosome ends from enzymatic damage or fusion with other chromosomes. |
| 3 | Teratoma/ teratocarcinoma | A tumor composed of tissues from the three embryonic germ layers. Usually found in ovary and testis. Produced experimentally in animals by injecting pluripotent stem cells in order to determine the stem cells' abilities to differentiate into various types of tissues. |

*(Continued)*

| Chap | Term | Definition |
|---|---|---|
| 3 | Therapeutic cloning | Somatic cell nuclear transfer for the isolation of embryonic stem cells. The embryonic stem cells are derived from the blastocyst and can be instructed to form particular cell types to be implanted into damaged tissue to restore its function. If the stem cells are placed back into the individual who gave the DNA for the somatic cell nuclear transfer, the embryonic stem cells and their derivatives are genetically identical and thus will not be rejected. Also see entries for Cloning and Reproductive Cloning. |
| 17 | Tissue engineering | This combines multiple disciplines/approaches to recreate replacement tissues for medical application. This can encompass the amalgamation of cell biology, materials science, and engineering to produce scaffolds on which cells can adopt an organization analogous to the tissue to be repaired/replaced. |
| 1 | Tissue stem cell | See *Adult Stem Cell* |
| 1 | Totipoten(t/cy) | Stem cells that can give rise to all cell types that are found in an embryo, fetus, or developed organism, including the embryonic components of the trophoblast and placenta required to support development and birth. The zygote and the cells at the very early stages following fertilization are considered totipotent. |
| 7 | Transdifferentiation | The ability of a particular cell of one tissue, organ, or system, including stem or progenitor cells, to differentiate into a cell type characteristic of another tissue, organ, or system; for example, blood stem cells changing to liver cells. |
| 2 | Transgene/Transgenesis | A gene that has been incorporated from one cell or organism and passed on to successive generations. |
| 1 | Transit amplifying cell | These cells lie on the pathway of differentiation from a stem cell to its differentiated progeny and represent a stage at which cell number can be expanded through proliferation. |
| 3 | Trophectoderm | The outer layer of the developing blastocyst that will ultimately form the embryonic side of the placenta. |
| 4 | Trophoblast | The extraembryonic cell type responsible for formation of the trophectoderm. |
| 2 | Xeno-transplantation/ xenograft | Cell, tissue, or organ transplanted from one species into another. |

# Index